STEIN'S
# ALGEBRA
# IN EASY STEPS

Edwin I. Stein

**ALLYN AND BACON, INC.**

Boston    Rockleigh, N.J.    Atlanta    Dallas    San Jose
London    Sydney    Toronto

*Editor:* Andrew P. Mastronardi
*Designer:* L. Christopher Valente
*Photo Researchers:* Mary Ruzila
                    Carmen Johnson
*Cover Design:* L. Christopher Valente
*Buyers:* Martha Ballentine
          Roger Powers


*Photographs are acknowledged below:*

*Cover photograph:* Charles Kemper

pp. 11—Bohdan Hrynewych. 41—Stock, Boston/George Bellerose.
95—Peter Menzel. 119—Andrew Brilliant. 139—Talbot D. Lovering (Allyn
& Bacon Staff Photographer). 201—E. P. Jones/H. Armstrong Roberts.
245—The Picture Cube/Ken Robert Buck. 301—Allen Ruid. 355—Frank
Siteman. 391—Mike Porcellino. 439—Stock, Boston/Mike Mazzachi.
463—Boeing Co. 499—Talbot D. Lovering (Allyn & Bacon Staff
Photographer). 533—The Picture Cube/Nancy Wasserman. 565—Stock,
Boston/Peter Menzel. 589—Frank Siteman. 613—Frank Siteman.

ISBN   0-205-06956-8

Library of Congress Catalog Card Number   80-70121

Printed in the United States of America

     4   5   6   7   8   9     90  89  88  87  86  85  84  83

# PREFACE

ALGEBRA IN EASY STEPS is an individualized basal text in elementary algebra. It features a wealth of carefully graded practice material, individualized assignments, a step-by-step approach to problem solving, a comprehensive treatment of the formula, a complete testing program, and an abundance of interesting applications from informal geometry, arithmetic, commerce, science, industry, and aviation showing the practical value of algebraic principles and skills. All these materials are so organized that diagnostic and prescriptive techniques may easily be used.

ALGEBRA IN EASY STEPS is specially designed so that each pupil by a simple testing procedure is directed quickly to specific assignments based on individual needs. Students do not waste time drilling on examples that they can solve. This system of individualized assignments accelerates individual achievement.

Each exercise contains its aim, the procedure to be followed in simple non-technical language, and completely worked-out sample solutions. Following this developmental material come the diagnostic tests, keyed to the immediately succeeding practice examples. The examples in the diagnostic tests and the related practice materials are closely calibrated in difficulty. Since one example is not perceptibly more difficult than the preceding one, the pupil is led to think clearly, to develop a full understanding, and to experience the pride of success. The final examples in each exercise provide an automatic end test.

The carefully planned, flexible organization of ALGEBRA IN EASY STEPS allows for individual differences and at the same time provides maximum and minimum materials for modern group methods of instruction including practice material for class, home, and optional assignments.

The testing program includes the automatic testing in each exercise, chapter reviews, cumulative reviews, competency check tests, keyed achievement tests, and maintenance practice in arithmetic. The final reviews include not only separate tests on the equation, the formula, problem solving, and algebraic operations, but also keyed inventory and achievement tests which can be used in remedial algebra classes.

Each chapter begins with a short motivating introduction explaining the purpose and use of the material in the chapter.

The author acknowledges the assistance of his wife, Elaine, and of Charlotte Jaffe and Marilyn Lieberman.

*Edwin I. Stein*

# CONTENTS

# 7 Problem Solving 244

# 8 Inequalities in One Variable 300

# 9 Special Products and Factoring 354

# CONTENTS

# 17 Trigonometry 610

# Review Tests

# INTRODUCTION

The language of algebra is symbolic. Both in arithmetic and informal geometry we have used mathematical symbols including the *addition* or plus symbol $+$, the *subtraction* or minus symbol $-$, the *multiplication* or times symbol $\times$, and the raised dot $\cdot$, the *division* symbols $\div$ or $\overline{)}$, the *equality* symbol $=$, the *inequality* symbols $<$ or $>$, the *parentheses* ( ) or *brackets* [ ], the *square root* symbol $\sqrt{\phantom{x}}$, and the raised numerals indicating *exponents*. Algebra continues to use symbols and arithmetic numbers, plus variables that may represent a specific number or many numbers.

*Arithmetic numbers* are numbers represented by numerals like 6 and 43. Each arithmetic number is definite in value and is sometimes called a *constant*.

A *variable* holds a place open for a number. See p. 23.

A *numerical expression,* sometimes called a *numerical phrase,* consists of a single numeral with or without operational symbols, like 41 or 57, or of two or more numerals with operational symbols like $6 + 4$, $82 - 19$, etc.

An *algebraic expression,* sometimes called an *algebraic phrase,* may be a numerical expression or an expression containing one or more variables joined by operational symbols like $r$, $9y$, $3c - 7d$, $2x^2 - 4x + 5$, etc.

Numerical expressions and algebraic expressions are *mathematical expressions*. The equality and the inequality symbols are verb symbols used in *mathematical sentences*.

A sentence that uses the equality symbol as its verb connecting two mathematical expressions like $5b = 30$ or $8y - 7 = y + 14$ is called an *equation*.

A sentence that uses an inequality symbol as its verb connecting two mathematical expressions like $6n > 42$ or $x + 9 < 11$, is called an *inequality*.

A formula is a special kind of equation. See page 33.

# Language of Algebra

# Number Symbols— Arithmetic Numbers

**EXERCISE 1-1**

**I. Aim**    To review the numbers of arithmetic and their numerals (number symbols) and to learn to read and write lists of numerals.

## II. Procedure

1. Use the numbers, beginning with zero, named by the numerals 0, 1, 2, 3, 4, 5, 6, 7, 8, 9, 10, 11, 12, . . . *as whole numbers.* See sample solution 1(a).
2. Use the numbers that include the common fractions, like $\frac{3}{4}$; the mixed numbers, like $4\frac{5}{8}$; the decimal fractions, like 0.07; the mixed decimals, like 6.239; and percents, like 8% as rational numbers of arithmetic. See sample solution 1(b).
3. Consider the whole numbers also as rational numbers since they can be written in fraction form. See sample solutions 1(b) and 2.
4. Use square roots (see page 644) of numbers other than perfect squares (numbers having an exact square root) as irrational numbers of arithmetic. See sample solution 1(c).
5. To indicate that an unlimited number of numbers follow a pattern and continue in this pattern without ending, use three dots after the numeral naming the last listed number. See sample solution 3.
6. To indicate the missing numbers when there are a limited number of numbers that follow a pattern but too many to list, use three dots before the numeral naming the last number. See sample solution 4.

## III. Sample Solutions

1. **a.** Which of the following numerals name whole numbers?
   **b.** Which name rational numbers?
   **c.** Which name irrational numbers?

   $2\frac{1}{2}$        .168        $\sqrt{29}$        11        $\frac{7}{16}$        43%        56.7

*Answers:* **1. a.** whole numbers: 11

   **b.** rational numbers: $2\frac{1}{2}$, .168, 11, $\frac{7}{16}$, 43%, 56.7

   **c.** irrational numbers: $\sqrt{29}$

2. Is 5 a rational number?
   Since $5 = \frac{5}{1}$, it may also be considered a rational number.
   *Answer:* Yes.

12

**3.** Read: 0, 5, 10, . . .

*Answer:* Zero, five, ten, and so on without end.

**4.** Write the list of numerals indicating: One, three, five, and so on up to and including thirty-one.

*Answer:* 1, 3, 5, . . ., 31

## DIAGNOSTIC TEST

**1.** Select numerals naming whole numbers: $\frac{9}{10}$; 45; $\sqrt{37}$

**2.** Select numerals naming rational numbers: $\sqrt{18}$; 2.3; $\frac{14}{2}$

**3.** Select numerals naming irrational numbers: .007; $\sqrt{49}$; 6

**4.** Read or write in words: 1, 3, 5, . . .

**5.** Read or write in words: 0, 2, 4, . . . , 28

**6.** Write the list of numerals indicating: Zero, six, twelve, and so on up to and including sixty.

**7.** Write the list of numerals indicating: One, two, three, and so on without end.

**8.** Is 42 included in the following list: 0, 4, 8, . . . ?

**9.** Write the numerals naming all the numbers in the list: 15, 17, 19, . . . , 35.

## RELATED PRACTICE EXAMPLES

Select numerals naming:

**1.** Whole numbers
  **a.** 152; $\sqrt{23}$; 4
  **b.** $8\frac{1}{2}$; $\frac{56}{8}$; .003
  **c.** 29%; 70; $\frac{4}{5}$
  **d.** $\sqrt{57}$; 8.9; 100%
  **e.** 12.5; $\frac{8}{3}$; $\sqrt{81}$

**2.** Rational numbers
  **a.** $\frac{11}{12}$; 6; .39
  **b.** $\sqrt{8}$; 9.25; 40%
  **c.** 71; $8\frac{1}{6}$; $\sqrt{51}$
  **d.** 85.7; $\sqrt{25}$; $\frac{13}{4}$
  **e.** 49; .001; $\sqrt{40}$

**3.** Irrational numbers
  **a.** $\sqrt{31}$; .8; $6\frac{3}{4}$
  **b.** 5.291; $\sqrt{16}$; $\frac{9}{8}$
  **c.** $66\frac{2}{3}\%$; $\frac{2}{4}$; $\sqrt{2}$
  **d.** $3\frac{5}{6}$; $\sqrt{45}$; .0009
  **e.** 78.06; 5%; $\sqrt{11}$

Read, or write in words, each of the following:

**4. a.** 0, 1, 2, . . .
  **b.** 1, 4, 9, . . .
  **c.** 0, 6, 12, . . .
  **d.** 9, 10, 11, . . .
  **e.** 0, 10, 100, . . .

**5. a.** 1, 3, 5, . . . , 49
  **b.** 0, 1, 2, . . . , 25
  **c.** 0, 7, 14, . . . , 98
  **d.** 2, 4, 6, . . . , 100
  **e.** 2, 4, 8, . . . , 256

Write the list of numerals indicating:

6. **a.** One, two, three, and so on up to and including eighteen.
   **b.** Zero, five, ten, and so on up to and including ninety.
   **c.** Two, three, five, and so on up to and including seventy-nine.
   **d.** Zero, three, six, and so on up to and including forty-eight.
   **e.** Zero, twenty, forty, and so on up to and including three hundred.

7. **a.** Zero, one, two, and so on without end.
   **b.** Three, nine, twenty-seven and so on without end.
   **c.** Zero, ten, twenty, and so on without end.
   **d.** Four, sixteen, sixty-four, and so on without end.
   **e.** Zero, six, twelve, and so on without end.

8. **a.** Is 27 included in the list: 0, 3, 6, . . . ?
   **b.** Is 56 included in the list: 0, 7, 14, . . . , 70?
   **c.** Is 90 included in the list: 0, 20, 40, . . . ?
   **d.** Is 12 included in the list: 0, 4, 8, . . . ?
   **e.** Is 81 included in the list: 1, 4, 9, . . . , 144?

9. Write the numerals naming all the numbers in each of the following lists:
   **a.** 1, 2, 3, . . . , 12
   **b.** 2, 3, 5, . . . , 29
   **c.** 0, 5, 10, . . . , 45
   **d.** 10, 12, 14, . . . , 24
   **e.** 0, 9, 18, . . . , 108
   **f.** 3, 6, 9, . . . , 42

## MISCELLANEOUS EXAMPLES

1. **a.** Is .33 a rational number?
   **b.** Is $\sqrt{25}$ an irrational number?
   **c.** Is 40% a whole number?
   **d.** Is 65 a rational number?

2. **a.** Does $\frac{15}{3}$ name a whole number?
   **b.** Does 1.00 name a rational number?
   **c.** Does $\sqrt{49}$ name a whole number?
   **d.** Does 50% name a rational number?
   **e.** Does $3\frac{4}{4}$ name a whole number?

3. **a.** Is 16 included in the list: 0, 3, 6, . . . , 30?
   **b.** Is 75 included in the list: 0, 5, 10, . . . ?
   **c.** Is 169 included in the list: 1, 4, 9, . . . , 400?
   **d.** Is 64 included in the list: 2, 4, 8, . . . , 128?
   **e.** Is 500 included in the list: 10, 100, 1000, . . . ?

# Symbols of Operation— Numerical Expressions

**I. Aim**   To translate a mathematical verbal phrase into a numerical expression.

## II. Procedure

Write the numerals and operational symbols as required in the proper order to form the numerical expression.

**1.** Use the *addition* or plus ($+$) symbol to mean: sum, add, more than, increased by or exceeded by.

**2.** Use the *subtraction* or minus ($-$) symbol to mean: difference, subtract, take away, less than, decreased by or diminished by.

**3.** Use the *multiplication* or times ($\times$) symbol or the raised dot ($\cdot$) to mean: product, times, or multiply.

**4.** Use the *division* ($\div$ or $\overline{)}$ ) symbol or the fraction bar to mean: quotient, or divide.

In algebra the fraction form is generally used to indicate division. The word "over" is sometimes used to express the relationship of the numerator to the denominator.

## III. Sample Solutions

Write as a numerical expression:

**1.** The sum of four and seven

*Answer:* $4 + 7$

**2.** The difference between ten and three

*Answer:* $10 - 3$

**3.** The product of six and nine

*Answer:* $6 \times 9$ or $6 \cdot 9$

**4.** The quotient of forty divided by eight

*Answer:* $40 \div 8$, $8\overline{)40}$, or $\frac{40}{8}$

---

### DIAGNOSTIC TEST

Write each of the following as a numerical expression:

**1.** Seven increased by two

**2.** The difference between twelve and nine

**3.** The product of five and eleven

**4.** The quotient of eighteen divided by six

## RELATED PRACTICE EXAMPLES

Write each of the following as a numerical expression:

1. **a.** Five plus four
   **c.** Add three and eight
   **e.** Eleven more than nine
   **g.** Twenty increased by two
   **i.** The sum of twelve and seven

   **b.** The sum of eight and six
   **d.** Six added to ten
   **f.** Nine exceeded by one
   **h.** Seven more than three
   **j.** Add six and fifteen

2. **a.** Four minus one
   **c.** Twelve less nine
   **e.** Five subtracted from sixteen
   **g.** Take ten from fourteen
   **i.** Four less than thirty

   **b.** The difference between seven and two
   **d.** From eight subtract two
   **f.** From twenty take six
   **h.** Nine diminished by eight
   **j.** Eighteen decreased by twelve

3. **a.** Eight times nine
   **b.** Multiply seven by five
   **c.** The product of six and two
   **d.** Six multiplied by ten
   **e.** Twice seven

4. **a.** Divide ten by five
   **b.** Twelve divided by three
   **c.** The quotient of sixty divided by fifteen
   **d.** Three over five
   **e.** The quotient of twenty divided by four

## CUMULATIVE REVIEW

1. Select the numerals naming:
   **a.** Whole numbers: $6\frac{3}{4}$, 42%, 103, 9.8, $\sqrt{58}$
   **b.** Rational numbers: 8%, $\sqrt{36}$, .07, $\frac{6}{5}$, $2\frac{3}{8}$
2. Read or write in words:
   **a.** 0, 5, 10, . . .          **b.** 9, 18, . . . , 108
3. Write the list of numerals indicating:
   **a.** Two, four, six, and so on up to and including thirty.
   **b.** Zero, eight, sixteen, and so on without end.
4. **a.** Write the numerals naming all the numbers in the list: 6, 12, 18, . . . , 72.
   **b.** Is 56 included in the list: 0, 7, 14, . . . , 91?
5. Write each of the following as a numerical expression:
   **a.** The sum of nineteen and thirty-nine
   **b.** Eighteen decreased by seven
   **c.** Fifteen multiplied by ten
   **d.** One hundred divided by twenty-five
   **e.** The difference between forty and fourteen

# Binary Operations; Inverse Operations–Arithmetic Review

### I. Aim

To learn what binary operations and inverse operations are and how they operate.

### II. Procedure

The operations of addition, subtraction, multiplication, and division are *binary operations*. In these the operation is performed on only two numbers at a time.

1. To find the sum, add the addends. See sample solution 1.
2. To find the difference, subtract the subtrahend from the minuend. See sample solution 2.
3. To find the product, multiply the multiplicand by the multiplier. See sample solution 3. Numbers that you multiply are called factors.
4. To find the quotient, divide the dividend by the divisor. See sample solution 4.
5. Use addition and subtraction as *inverse* operations. They undo each other. See sample solutions 5 and 6.
6. Use multiplication and division as inverse operations. They undo each other. See sample solutions 7 and 8.

### III. Sample Solutions

**1.** Add: $185 + 253$

$$\begin{array}{r} 185 \\ +253 \end{array} \Big\rangle \text{addends}$$

$$\overline{438} \text{ sum}$$

*Answer: 438*

**2.** Subtract: $928 - 219$

928 minuend
$-219$ subtrahend

709 difference or remainder

*Answer: 709*

**3.** Multiply: $39 \times 15$

$$\begin{array}{r} 39 \text{ multiplicand} \\ \times 15 \text{ multiplier} \end{array} \Big\rangle \text{factors}$$

$$\begin{array}{r} 195 \\ 39 \\ \hline 585 \text{ product} \end{array}$$

*Answer: 585*

**4.** Divide: $675 \div 25$

$$\begin{array}{r} 27 \text{ quotient} \\ 25\overline{)675} \text{ dividend} \\ 50 \\ \hline 175 \\ 175 \end{array}$$

divisor

*Answer: 27*

Find the missing operation and number in each of the following:

**5.** $(8 + 3)? \Box = 8$

*Answer:* $(8 + 3) - 3 = 8$ Subtracting 3 undoes adding 3.

**6.** $(7 - 4)? \Box = 7$

*Answer:* $(7 - 4) + 4 = 7$ Adding 4 undoes subtracting 4.

**7.** $(9 \times 6)? \Box = 9$

*Answer:* $(9 \times 6) \div 6 = 9$ Dividing by 6 undoes multiplying by 6.

**8.** $(12 \div 2)? \Box = 12$

*Answer:* $(12 \div 2) \times 2 = 12$ Multiplying by 2 undoes dividing by 2.

## DIAGNOSTIC TEST

Complete each of the following as indicated:

**1.** Add:
$687 + 4598 + 79$

**2.** Subtract:
$430,070$
$-89,169$

**3.** Multiply:
$806 \times 950$

**4.** Divide:
$692)\overline{544,604}$

**5.** Add:
$4\frac{5}{6} + 3\frac{2}{3}$

**6.** Subtract:
$7\frac{1}{4} - \frac{4}{5}$

**7.** Multiply:
$3\frac{3}{4} \times 5\frac{1}{3}$

**8.** Divide:
$16 \div 1\frac{1}{8}$

**9.** Add:
$4.5 + .62 + 24$

**10.** Subtract:
$1 - .001$

**11.** Multiply:
$.018 \times .5$

**12.** Divide:
$24 \div .03$

**13.** Find 8% of $15.75.

**14.** What percent of 80 is 52?

**15.** 12% of what number is 30?

Find the missing operation and number in each of the following:

**16.** $(4 + 8)? \Box = 4$

**17.** $(11 - 3)? \Box = 11$

**18.** $(5 \times 7)? \Box = 5$

**19.** $(24 \div 6)? \Box = 24$

## RELATED PRACTICE EXAMPLES

**1.** Add:
  **a.** $63 + 125$
  **b.** $74 + 9 + 287$
  **c.** $835 + 924 + 378$
  **d.** $429 + 89 + 5863$
  **e.** $914 + 6985 + 67 + 595$

**2.** Subtract:
  **a.** $7305 - 856$
  **b.** $10,000 - 9083$
  **c.** $43,528 - 14,519$
  **d.** $109,847 - 39,927$
  **e.** $800,041 - 790,562$

3. Multiply:
   a. $76 \times 25$   b. $307 \times 94$   c. $89 \times 643$   d. $491 \times 528$
4. Divide:
   a. $95\overline{)3420}$   b. $731{,}042 \div 806$   c. $493\overline{)387{,}498}$   d. $1{,}995{,}840 \div 5280$
5. Add:
   a. $\frac{3}{4} + \frac{7}{8}$   b. $2\frac{1}{2} + 5\frac{2}{3}$   c. $9 + 1\frac{4}{5}$   d. $3\frac{1}{3} + 7\frac{5}{6} + 4\frac{1}{4}$
6. Subtract:
   a. $\frac{3}{5} - \frac{1}{10}$   b. $1\frac{2}{3} - \frac{5}{6}$   c. $7 - 5\frac{3}{8}$   d. $6\frac{5}{16} - 1\frac{11}{12}$
7. Multiply:
   a. $\frac{2}{5} \times \frac{25}{36}$   b. $\frac{11}{12} \times \frac{7}{8}$   c. $3\frac{5}{6} \times 24$   d. $5\frac{1}{4} \times 9\frac{2}{3}$
8. Divide:
   a. $\frac{13}{16} \div \frac{3}{4}$   b. $18 \div 2\frac{2}{3}$   c. $4\frac{1}{2} \div 6$   d. $7\frac{1}{3} \div 3\frac{1}{7}$
9. Add:
   a. $3.2 + 6.8$   b. $.95 + 8.2$   c. $.1 + .11$   d. $2 + .003 + 1.4$
10. Subtract:
    a. $5.8 - 3.9$   b. $6.3 - .46$   c. $.032 - .03$   d. $5 - .0005$
11. Multiply:
    a. $1.4 \times .23$   b. $.65 \times .04$   c. $.807 \times 1.61$   d. $500 \times .002$
12. Divide:
    a. $2.5\overline{)8.75}$   b. $.6\overline{).3}$   c. $.02\overline{)10}$   d. $.36 \div 1.8$

Find the following:
13. a. 15% of $9800            b. 3% of 24.9
    c. 108% of 67             d. 6.19% of $25,000
14. a. What percent of 8 is 5?        b. What percent of $.96 is $.72?
    c. 18 is what percent of 48?      d. $45 is what percent of $75?
15. Find the missing numbers:
    a. 50% of what number is 28?
    b. 4 is 16% of what number?
    c. 3% of what amount is $45?
    d. $83\frac{1}{3}$% of what number is 63?

Find the missing operation and number in each of the following:
16. a. $(7 + 2) - \square = 7$           17. a. $(10 - 4) + 4 = \square$
    b. $(15 + 5) - 5 = \square$              b. $(9 - 7) + \square = 9$
    c. $(4 + 9) ? \square = 9$               c. $(5 - 1) ? \square = 5$
    d. $(11 + 3) ? \square = 11$             d. $(14 - 6) ? \square = 14$
    e. $(x + 8) ? \square = x$               e. $(n - 2) ? \square = n$
18. a. $(8 \times 2) \div 2 = \square$    19. a. $(6 \div 3) \times 3 = \square$
    b. $(3 \times 10) \div \square = 3$      b. $(20 \div 5) \times \square = 20$
    c. $(12 \times 4) ? 4 = 12$              c. $(32 \div 8) ? 8 = 32$
    d. $(5 \times 9) ? \square = 5$          d. $(63 \div 9) ? \square = 63$
    e. $(n \times 7) ? \square = n$          e. $(x \div 10) ? \square = x$

# Symbols; Exponents; Square Root; Parentheses

### I. Aim

To study how to express repeated factors in exponential form and to use the square root and parentheses symbols in numerical expressions.

### II. Procedure

1. To write repeated factors briefly, use the exponential form where a small numeral (called the *exponent*) is written to the upper right (superscript) of the repeated factor (called the *base*). This exponent indicates how many times the factor is being used in multiplication. See sample solution 1.

2. To find the *power* of a factor, multiply the equal factors to determine their product. Sometimes the power may be expressed as an indicated product using exponents. See sample solutions 1 and 2.

3. Observe that the second power of a number is usually called the *square* of the number and the third power of a number is usually called the *cube* of the number. See sample solution 3.

4. Write the symbol over the numeral to indicate the *square root* of a number named by the numeral. The *square root* of a number is that number which when multiplied by itself produces the given number. It is one of two equal factors. See sample solution 4.

5. Use parentheses ( ) or brackets [ ] to group together two or more numerals so that the number they name may be treated as a single quantity. See sample solution 5.

6. Parentheses are sometimes used to separate one numeral from another. See page 23.

7. Numerals written within parentheses next to each other without any operational symbol indicates multiplication. See sample solution 6.

### III. Sample Solutions

1. Write in exponential form: $7 \times 7 \times 7 \times 7 \times 7$  *Answer:* $7^5$

   In $7^5$, the base is 7 and the exponent is 5.

2. Express as a numeral using exponents: Three to the sixth power.
   *Answer:* $3^6$

3. **a.** Read $8^2$. *Answer:* "Eight squared" or "the square of eight" or "eight to the second power" or "the second power of eight."

**b.** Read $5^3$.  *Answer:* "Five cubed" or "the cube of five" or "five to the third power" or "the third power of five."

4. Write symbolically: The square root of thirty.  *Answer:* $\sqrt{30}$

5. Write as a numerical expression: Four times the quantity seven plus nine.  *Answer:* $4(7 + 9)$

6. Find the value of: **a.** (8)(5) **b.** 8(5) **c.** (8)5

    **a.** $(8)(5) = 8 \times 5 = 40$ **b.** $8(5) = 8 \times 5 = 40$ **c.** $(8)5 = 8 \times 5 = 40$

*Answers:* 40

## DIAGNOSTIC TEST

1. Read or write in words: **a.** $11^6$ **b.** $7^2$ **c.** $6^3$
2. Express as a numeral: Ten to the ninth power.
3. What is the exponent in $15^7$?
4. What is the base in $17^4$?
5. Write in exponential form: $6 \times 6 \times 6 \times 6 \times 6 \times 6 \times 6$
6. Which power of 3 is 81?
7. Write symbolically: The square root of ninety-one.
8. Write as a numerical expression: Eleven times the sum of eight and two.
9. Find the value of: (6)(9)

## *RELATED PRACTICE EXAMPLES*

1. Read or write in words, each of the following:
    **a.** $5^3$     **b.** $9^2$     **c.** $7^5$     **d.** $3^8$     **e.** $2^1$     **f.** $8^9$
    **g.** $10^4$    **h.** $15^6$    **i.** $24^{10}$    **j.** $6^{12}$    **k.** $4^{17}$    **l.** $1^{20}$
2. Express each of the following as a numeral, using exponents:
    **a.** Five to the sixth power        **b.** Ten to the ninth power
    **c.** Eight to the eleventh power    **d.** Four cubed
    **e.** Twelve squared              **f.** Nine to the fifth power
    **g.** Seventeen to the tenth power   **h.** Twenty to the fourth power
3. What is the exponent in each of the following?
    **a.** $8^3$     **b.** $15^{11}$     **c.** $7^8$     **d.** $25^{16}$     **e.** $40^{23}$     **f.** $19^{100}$
4. What is the base in each of the following?
    **a.** $5^7$                   **b.** $11^3$
    **c.** $18^9$                  **d.** $54^4$
    **e.** $10^{11}$                **f.** $72^{30}$
5. Write in exponential form each of the following:
    **a.** $2 \times 2 \times 2$          **b.** $8 \times 8 \times 8 \times 8$
    **c.** $17 \times 17$            **d.** $6 \times 6 \times 6 \times 6 \times 6$

   **e.** $3 \times 3 \times 3 \times 3 \times 3 \times 3 \times 3$
   **f.** $5 \times 5 \times 5 \times 5 \times 5 \times 5 \times 5 \times 5$
   **g.** $4 \times 4 \times 4 \times 4 \times 4 \times 4 \times 4 \times 4 \times 4 \times 4$
   **h.** $10 \times 10 \times 10 \times 10 \times 10 \times 10 \times 10 \times 10 \times 10$
   **i.** $7 \times 7 \times 7 \times 7 \times 7 \times 7 \times 7 \times 7 \times 7 \times 7 \times 7 \times 7$
   **j.** $1 \times 1 \times 1 \times 1 \times 1 \times 1 \times 1 \times 1 \times 1 \times 1 \times 1 \times 1 \times 1 \times 1 \times 1$

**6.** Which power:
   **a.** Of 2 is 32?           **b.** Of 5 is 125?
   **c.** Of 4 is 64?           **d.** Of 10 is 10,000?
   **e.** Of 8 is 64?           **f.** Of 2 is 64?

**7.** Write symbolically:
   **a.** The square root of 7
   **b.** The square root of forty-five
   **c.** The square root of one hundred
   **d.** The square root of twenty
   **e.** The square root of eighty-nine

**8.** Write each of the following as a numerical expression:
   **a.** Nine times the sum of four and six
   **b.** Four times the difference between eleven and three
   **c.** Twelve times the quantity fifteen plus twenty
   **d.** Five times the quantity seven minus eight
   **e.** Ten plus the quantity nine minus one
   **f.** Seventeen minus the quantity eight plus sixty

**9.** Find the value of each of the following:
   **a.** (6)(7)     **b.** 4(12)     **c.** (9)3
   **d.** 10(15)    **e.** (8)(20)    **f.** (45)2

## COMPETENCY CHECK TEST

In each of the following, select the letter corresponding to your answer.
Write as a numerical expression:

**1.** Product of eight and five: **a.** $\frac{8}{5}$ **b.** $8 \times 5$ **c.** $8 + 5$ **d.** $8 - 5$

**2.** Sum of twelve and seven: **a.** $12 \cdot 7$ **b.** $12 - 7$ **c.** $12 \div 7$ **d.** $12 + 7$

**3.** Six to the tenth power: **a.** $10^6$ **b.** $6^{\frac{1}{10}}$ **c.** $6^{10}$ **d.** $60$

**4.** Three divided by five: **a.** $3\overline{)5}$ **b.** $\frac{3}{5}$ **c.** $5 \div 3$ **d.** $\frac{5}{3}$

**5.** Seven less than nine: **a.** $7 - 9$ **b.** $7 \div 9$ **c.** $9 - 7$ **d.** answer not given

**6.** Square root of eight: **a.** $2^8$ **b.** $8^2$ **c.** $8^4$ **d.** $\sqrt{8}$

**7.** Two times the quantity ten minus five:
   **a.** $2 \times 10 - 5$    **b.** $10 - 2 \times 5$    **c.** $2(10 - 5)$    **d.** answer not given

# Algebraic Expressions

I. Aim    To translate a mathematical verbal phrase containing a variable into an algebraic expression.

II. Procedure

1. Write the numerals, variables, and operational symbols as required in the proper order to form the algebraic expression. Note that a *variable* holds a place open for a number, and is expressed by one of the following:

   a. A small letter
   b. A capital letter
   c. A letter with subscript, such as $w_1$ or $w_2$, read "$w$ sub one or $w$ sub two."
   d. A letter with a prime mark, such as $P'$ or $P''$, read as "$P$ prime or $P$ double prime."
   e. A frame, such as □, △, ○, and ☐.
   f. A blank space
   g. A question mark
   h. Sometimes the first letter of a key word, especially in formulas.
   A variable may represent any number, but under certain conditions it may represent a specific number or numbers.

2. No multiplication symbol is necesssary when the factors are two variables named by letters or are a number and a variable named by a letter. In the latter case the numeral always precedes the variable. This numeral is usually called the *numerical coefficient* of the variable. In $5y^2$, the 5 is the numerical coefficient of $y^2$.

3. Note that $c$ times $d$ may be expressed as:
   $c \times d$ or $c \cdot d$ or $cd$(preferred) or $(c)(d)$ or $c(d)$ or $(c)d$

III. Sample Solutions

Write as an algebraic expression:

1. The sum of $n$ and six
   *Answer:* $n + 6$

2. The difference between the perimeter
   ($p$) and the circumference ($c$)
   *Answer:* $p - c$

**3.** The product of the base ($b$) and the height ($h$)

*Answer: bh*

**5.** The distance ($d$) divided by the rate of speed ($r$)

*Answer: $d \div r$*

**7.** The square of the radius ($r$)

*Answer: $r^2$*

**9.** The square root of the area ($A$)

*Answer: $\sqrt{A}$*

**4.** $x$ multiplied by nine

*Answer: $9x$*

**6.** One-half of $ab$

*Answer: $\frac{1}{2}ab$ or $\dfrac{ab}{2}$*

**8.** The cube of the side ($s$)

*Answer: $s^3$*

**10.** Seven times the quantity $b$ minus three

*Answer: $7(b - 3)$*

## DIAGNOSTIC TEST

Write each of the following phrases as an algebraic expression, using where necessary the first letter of each key word to represent a variable:

1. The quotient of $x$ divided by $y$.
2. The product of $m$ and six.
3. The discount ($d$) subtracted from the list price ($l$).
4. The cost increased by the margin.
5. The difference between the product of the length and width and the product of the length and height.
6. Write one-third $Bh$ in two ways.

Write each of the following as an algebraic expression:

7. The sum of the squares of the altitude ($a$) and the base ($b$).
8. One-sixth the product of Pi and the cube of the diameter ($d$).
9. The square root of the difference between the square of the hypotenuse ($h$) and the square of the altitude ($a$).
10. Pi times the radius ($r$) times the sum of the radius ($r$) and the slant height ($s$).

## *RELATED PRACTICE EXAMPLES*

Write each of the following as an algebraic expression:

1. **a.** $m$ divided by $b$     **b.** $x$ multiplied by $n$

    **c.** $v$ less $g$

    **e.** $r$ times $t$

    **g.** The difference between $m$ and $s$

    **i.** The product of $b$ and $h$

    **d.** $y$ added to $a$

    **f.** The sum of $f$ and $g$

    **h.** The quotient of $c$ divided by $d$

    **j.** $a$ increased by $w$

2. **a.** The sum of $x$ and nine

    **c.** The product of $y$ and eight

    **e.** Ten diminished by $b$

    **g.** Seven more than $n$

    **i.** Five times $x$

    **b.** $a$ subtracted from five

    **d.** $n$ divided by two

    **f.** The quotient of sixteen divided by $r$

    **h.** $b$ decreased by four

    **j.** $m$ added to fifty

3. **a.** The sum of the principal ($p$) and the interest ($i$).

    **b.** The difference between the selling price ($s$) and the gain ($g$).

    **c.** The product of the force ($F$) and the distance ($d$).

    **d.** The quotient of the distance ($d$) divided by the time ($t$).

    **e.** 180° decreased by angle $C$.

4. Write each of the following as an algebraic expression, using the first letter of each key word to represent the variable:

    **a.** The profit divided by the cost.

    **b.** 273° more than the Celsius temperature reading.

    **c.** The amount less the interest.

    **d.** Two times Pi times the radius.

    **e.** The number of articles multiplied by the price of one article.

5. Write each of the following as an algebraic expression. Be careful, there is more than one operation to be represented.

    **a.** The sum of the bases $b_1$ and $b_2$ divided by two.

    **b.** Twice the length of the equal side ($e$) subtracted from the perimeter ($p$).

    **c.** The product of the acceleration of gravity ($g$) and time ($t$) added to the velocity ($V$).

    **d.** The difference between the product of the altitude and base and the product of the length and width.

    **e.** The sum of the product of twice Pi and the radius and the product of Pi, the radius, and the height.

6. **a.** Write $\frac{1}{4}v$ in another way.

    **b.** Write $\frac{n}{6}$ in another way.

    **c.** Write one half of $bh$ in two ways.

    **d.** Write four thirds $r$ in two ways.

    **e.** Write one fifth of an amount ($a$) in two ways.

Write each of the following as an algebraic expression:
7. **a.** The square of side (*s*).
   **b.** *V* raised to the second power.
   **c.** *R* times the square of *I*.
   **d.** The difference between the square of the hypotenuse (*h*) and the square of the altitude (*a*).
   **e.** The product of Pi, the square of the radius (*r*), and the height (*h*).
8. **a.** The cube of the edge (*e*).
   **b.** The third power of *m*.
   **c.** The constant *k* times the cube of the velocity (*v*).
   **d.** Seven times the cube of *x* increased by nine times the cube of *y*.
   **e.** The cube of the diameter (*d*) decreased by four-thirds the product of Pi and the cube of the radius (*r*).
9. **a.** The square root of 47.
   **b.** The square root of the density (*d*).
   **c.** The square root of the quotient of the power (*P*) divided by the resistance (*R*).
   **d.** The square root of the product of *b* and *c*.
   **e.** The square root of the sum of the squares of the altitude (*a*) and the base (*b*).
10. **a.** Four times the sum of *y* and eight.
    **b.** Twice the quantity *l* plus *w*.
    **c.** *d* times the quantity *n* minus *l*.
    **d.** The square of the sum of *x* and *y*.
    **e.** The product of the sum of *a* and *x* and the difference of *a* and *x*.

## CUMULATIVE REVIEW

1. Write each of the following as a numerical expression:
   **a.** Five raised to the ninth power.
   **b.** The square root of ninety-four.
   **c.** Seven times the difference between forty and sixteen.
2. **a.** In $7^{11}$, what is the exponent?
   **b.** In $15^8$, what is the base?
3. Write each of the following as an algebraic expression:
   **a.** The difference between 180° and *B*.
   **b.** The discount (*d*) divided by the list price (*l*).
   **c.** The Celsius temperature reading (*C*) increased by 273°.
   **d.** The product of Pi ($\pi$) and the square of the radius (*r*).
   **e.** The square root of the weight (*w*).
   **f.** One half of the height (*h*) times the sum of the bases (*B,b*).

# Number Sentences—Symbols of Equality and Inequality

**I. Aim**   To learn to read number sentences and to write them symbolically.

## II. Procedure

Observe that each of these symbols, $=$, $>$, and $<$ represents a verb phrase in a number sentence (a sentence dealing with numbers).

1. Use the equality symbol $=$ to mean "is equal to." The expressions on both sides of the equality sign designate the same number.
2. Use the inequality symbol $>$ to mean "is greater than."
3. Use the inequality symbol $<$ to mean "is less than."
4. Use a line drawn through the symbols $=$, $>$, and $<$ to reverse their meaning.

  The symbol $\neq$ means "is not equal to."
  The symbol $\not>$ means "is not greater than."
  The symbol $\not<$ means "is not less than."

5. Number sentences may be true or false. See sample solution 8.

## III. Sample Solutions

Read each of the following:

1. $8 + 4 = 12$   *Answer:* Eight plus four is equal to twelve.

2. $15 > 9$   *Answer:* Fifteen is greater than nine.

3. $24 < 50 - 20$   *Answer:* Twenty-four is less than fifty minus twenty.

4. $60 \div 2 \neq 10 \times 4$   *Answer:* Sixty divided by two is not equal to ten times four.

5. $35 \not> 29 + 7$   *Answer:* Thirty-five is not greater than twenty-nine plus seven.

6. $11 - 3 \not< 6$   *Answer:* Eleven minus three is not less than six.

7. Write symbolically: Five times nine is less than fifty minus four.
*Answer:* $5 \times 9 < 50 - 4$

8. **a.** Is the sentence $16 > 9 + 8$ true?   *Answer:* No. 16 is not greater than 17.

   **b.** Is the sentence $5 + 7 \not< 21 - 3$ false?   *Answer:* Yes. 12 is less than 18.

27

## DIAGNOSTIC TEST

1. Select the verb phrase (is greater than, is equal to, or is less than) that will make the sentence true:

   Number 439 (is greater than, is equal to, is less than) the number 943.

Read, or write in words, each of the following:

2. $14-5=3\times3$

3. $.6+.7>1$

4. $\frac{7}{3}<\frac{9}{10}$

5. $4+11 \not> 6\times3$

6. $4.2 \neq .042$

7. $12\div3 \not< 9-2$

Write symbolically each of the following:

8. Seventeen plus four is greater than twenty.

9. Forty-five is equal to nine times five.

10. Sixty divided by three is less than forty minus ten.

11. Seventy-two is not equal to four times fifteen.

12. One hundred minus eight is not less than seven times nine.

13. Twelve plus eleven is not greater than ninety divided by three.

14. Which of the following sentences are true?

    **a.** $23+5>29$
    **b.** $.08<.008$
    **c.** $2\frac{1}{2}+\frac{3}{4} \not> 3\frac{1}{2}-\frac{1}{4}$

15. Which of the following sentences are false?

    **a.** $9+11 \neq 7\times3$
    **b.** $32-19>22\div2$
    **c.** $8\times.4 \not< .3+.02$

## *RELATED PRACTICE EXAMPLES*

1. In each of the following, select the one verb phrase (is greater than, is equal to, or is less than) that will make a true statement:

   **a.** Number 89 (is greater than, is equal to, is less than) number 98.

   **b.** Number 2.7 (is greater than, is equal to, is less than) number 2.45.

   **c.** Number $5\frac{2}{3}$ (is greater than, is equal to, is less than) number $\frac{17}{3}$.

   **d.** Number .06 (is greater than, is equal to, is less than) number .6.

   **e.** Number $\frac{7}{8}$ (is greater than, is equal to, is less than) number $\frac{15}{16}$.

   **f.** Number 1.6 (is greater than, is equal to, is less than) number .18.

   **g.** Number 5 (is greater than, is equal to, is less than) number 4.9.

   **h.** Number .31 (is greater than, is equal to, is less than) number .31.

   **i.** Number $\frac{3}{5}$ (is greater than, is equal to, is less than) number $\frac{16}{30}$.

   **j.** Number 100 (is greater than, is equal to, is less than) number 100.

Read, or write in words, each of the following:

2. **a.** $12=12$
   **b.** $37=37$
   **c.** $6+9=15$

   **d.** $18-5=13$
   **e.** $7\times8=56$
   **f.** $72\div9=8$

   **g.** $39+16=11\times5$
   **h.** $100\div25=51-47$
   **i.** $4\times7\times2=2\times4\times7$

3. **a.** $11 > 7$          **b.** $32 > 28$          **c.** $4 > 0$
   **d.** $6 \times 2 > 10$          **e.** $.2 > .36 - .17$          **f.** $65 + 19 > 52$
   **g.** $\frac{1}{2} + \frac{1}{2} > \frac{1}{2} \times \frac{1}{2}$          **h.** $125 \div 25 > 23 - 14$          **i.** $\frac{1}{4} \times 8 > 6 \div 6$

4. **a.** $17 < 24$          **b.** $.09 < .1$          **c.** $\frac{3}{5} < \frac{2}{3}$
   **d.** $58 + 26 < 100$          **e.** $39 < 6 \times 7$          **f.** $54 \div 6 < 13$
   **g.** $19 \times 16 < 500 - 89$          **h.** $107 + 59 < 924 \div 4$          **i.** $642 - 171 < 325 + 210$

5. **a.** $4 \not> 7$          **b.** $6 + 11 \not> 50$          **c.** $25 \times 8 \not> 47 - 5$

6. **a.** $16 \neq 10$          **b.** $24 \div 3 \neq 15$          **c.** $12 - 4 \neq 20 \times 1$

7. **a.** $32 \not< 8$          **b.** $29 \not< 47 - 19$          **c.** $6.3 \div 9 \not< .1 + .05$

Write symbolically each of the following:

8. **a.** Fourteen is greater than nine.
   **b.** Six minus five is greater than zero.
   **c.** Ten times four is greater than ten plus four.

9. **a.** Seventy is equal to seventy.
   **b.** Seven times eight is equal to fifty-six.
   **c.** Eighteen divided by three is equal to eight minus two.

10. **a.** Nine is less than twenty-one.
    **b.** Sixteen plus two is less than nineteen.
    **c.** Ninety-four divided by two is less than eighty minus fifteen.

11. **a.** Sixteen is not equal to forty-one.
    **b.** Fifty is not equal to nine times eleven.
    **c.** Eight divided by three is not equal to six times zero.

12. **a.** Twenty-three is not less than twenty.
    **b.** Thirteen times seven is not less than seventy plus forty.
    **c.** Four sevenths is not less than five elevenths.

13. **a.** Three tenths is not greater than thirty hundredths.
    **b.** Nine plus four is not greater than thirty minus twelve.
    **c.** Ten minus ten is not greater than ten divided by ten.

14. Which of the following sentences are true?
    **a.** $205 > 186$          **b.** $.65 < .7$          **c.** $\frac{42}{63} \neq \frac{36}{54}$
    **d.** $.95 \not< 8$          **e.** $3 - 7 \not> 13$          **f.** $87 \neq 103 - 17$
    **g.** $6 \times 9 > 54$          **h.** $\frac{2}{3} \not> \frac{9}{16} \div 1\frac{1}{2}$          **i.** $19 + 14 < 30$

15. Which of the following sentences are false?
    **a.** $63 < 36$          **b.** $.3 \neq .03$          **c.** $\frac{7}{12} > \frac{4}{7}$
    **d.** $4 + .4 \not> .44$          **e.** $14 \not< 21 - 9$          **f.** $51 \div 17 \not> 5 - 2$

# Open Sentences–
# Equations and Inequalities

## I. Aim

To recognize equations and inequalities and to learn to read and write open sentences.

## II. Procedure

1. To recognize whether the *open sentence* (one which contains a variable) is an equation or an inequality, observe whether an equality symbol ($=$) or an inequality symbol ($<, >, \neq, \not<, \not>$) is used.

    When the equality symbol is used, the open sentence is an equation. When the inequality symbol is used, the open sentence is an inequality. See sample solution 1.

2. To read, or write an open sentence, read or write each variable, numerical symbols of operation, and equality or inequality symbol in the required order.

    Note in sample solution 2 how a variable is read.

## III. Sample Solutions

1. Which of the following are equations? Which of the following are inequalities?

    $$d + 5 > 9 \qquad a - 9 = 10 \qquad 7n < 16 \qquad 6x \neq 12$$

    *Answer:* Equation, $a - 9 = 10$
    Inequalities, $d + 5 > 9$, $7n < 16$, $6x \neq 12$

2. Read each of the following:

    $$2b = 8 \qquad y < 7 \qquad r + 4 \not> 6 \qquad n - 9 \neq 15$$

    *Answers:* $2b = 8$ is read as follows: "Two times *some* number $b$ is equal to eight."

    $y < 7$ is read as follows: "*Each* number $y$ is less than seven."

    $r + 4 \not> 6$ is read as follows: "*Each* number $r$ plus four is not greater than six."

    $n - 9 \neq 15$ is read as follows: "*Each* number $n$ minus nine is not equal to fifteen."

## DIAGNOSTIC TEST

1. Which of the following are open sentences?

   $9 - 3 < 12$ $\qquad$ $25 > 16 + n$ $\qquad$ $20 \times 6 = 85 + 35$

2. Which of the following are equations? Which are inequalities?

   $c + 11 > 1$ $\qquad$ $6d - 4 = 9$ $\qquad$ $27 < 5y$

Read, or write in words, each of the following:

3. $6x - 1 = 23$ $\qquad$ 4. $t + 9 < 7$ $\qquad$ 5. $25 > n + 2$

6. $150 \not< 60$ $\qquad$ 7. $\dfrac{3b}{2} \not> 12$ $\qquad$ 8. $9a - 5 \neq 4a + 13$

Write symbolically each of the following:

9. Some number $c$ decreased by one is equal to twelve.
10. Each number $y$ increased by six is greater than four.
11. Each number $t$ divided by four is less than thirteen.
12. Ten times each number $r$ is not equal to thirty-two.
13. Four times each number $x$ is not less than fifty-one.
14. Eight times each number $b$ is not greater than twenty-five.

## RELATED PRACTICE EXAMPLES

1. Which of the following are open sentences?

   **a.** $6 + 8 = 14$ $\qquad$ **b.** $b + 15 = 19$ $\qquad$ **c.** $24 < 3x$

   **d.** $16 - 7 > 2 + 5$ $\qquad$ **e.** $4 \times 1 < 5$ $\qquad$ **f.** $y - 14 > 46$

   **g.** $35 < 28 + m$ $\qquad$ **h.** $\frac{20}{4} > 0$ $\qquad$ **i.** $7 = \dfrac{x}{5}$

2. Which of the following are equations? Which are inequalities?

   **a.** $4c < 36$ $\qquad$ **b.** $r + 3 = 7$ $\qquad$ **c.** $\dfrac{h}{12} > 6$

   **d.** $10 = 30a$ $\qquad$ **e.** $g - 2 > 15$ $\qquad$ **f.** $21 < d + 11$

   **g.** $58 < 9 + 2t$ $\qquad$ **h.** $.04c > 90$ $\qquad$ **i.** $5b - 6 = 2$

Read, or write in words, each of the following:

3. **a.** $12x - 5 = 20$ $\qquad$ **b.** $0 = 16c - 8$ $\qquad$ **c.** $26 = 3m + 4$
4. **a.** $3y < 21$ $\qquad$ **b.** $m - 19 < 7$ $\qquad$ **c.** $6n + 7 < 31$
5. **a.** $x > 9$ $\qquad$ **b.** $37 > b + 18$ $\qquad$ **c.** $9x - 18 > 0$
6. **a.** $n + 4 \not< 15$ $\qquad$ **b.** $14 \not< 5d - 6$ $\qquad$ **c.** $18y \not< 54$
7. **a.** $8b \not> 48$ $\qquad$ **b.** $\dfrac{n}{5} \not> 10$ $\qquad$ **c.** $40 \not> 3r - 8$
8. **a.** $c - 6 \neq 13$ $\qquad$ **b.** $4n + 3 \neq 39$ $\qquad$ **c.** $10x - 8 \neq 2x + 27$

Write symbolically each of the following:

9.  **a.** Some number $x$ increased by four is equal to twenty.

     **b.** Five times some number $s$ is equal to fifteen.

     **c.** Twelve times some number $n$ minus two is equal to ten.

10.  **a.** Each number $h$ decreased by nine is greater than eleven.

     **b.** Seven times each number $a$ plus six is greater than forty-eight.

     **c.** Six times each number $d$ increased by thirteen is greater than sixty.

11.  **a.** Each number $n$ decreased by nine is less than two.

     **b.** Twenty-five is less than each number $x$ plus eight.

     **c.** Three times each number $y$ divided by four is less than thirty.

12.  **a.** Four times each number $c$ is not equal to zero.

     **b.** Fifty-four is not equal to two times each number $n$ minus five.

     **c.** each number $g$ increased by twelve is not equal to nineteen.

13.  **a.** Sixteen times each number $m$ is not less than seventy.

     **b.** Each number $b$ plus nine is not less than ten.

     **c.** Eighteen is not less than five times number $k$.

14.  **a.** Each number $d$ divided by five is not greater than four.

     **b.** Eleven times each number $y$ decreased by one is not greater than eight.

     **c.** Seven times each number $n$ is not greater than fifty-six.

## *CUMULATIVE REVIEW*

1. Which of the following may be represented by an algebraic sentence?

     **a.** 5 more than the number $n$

     **b.** 20 is more than the number $x$

     **c.** 8 is less than the number $y$

     **d.** 13 less than the number $d$

2. Which is an open sentence: $30 + 7 = 37$ or $6n + 7 = 37$?

3. Which is an equation: $5x + x = 18$ or $8x + 5 < 18$?

4. Which is an inequality: $-3y > -21$ or $9y - 3 = -21$?

Read, or write in words, each of the following:

5.  **a.** $10 \div 10 > 10 - 10$    **b.** $.8 + .7 \not< .8 \times .7$    **c.** $\frac{3}{4} \times 5 \neq \frac{4}{5} \times 3$

6.  **a.** $2n + 6 < -8$        **b.** $3y - 5 = -2y$     **c.** $4x - 3 > -1$

Write symbolically each of the following:

7.  **a.** Sixty-four less twelve is not equal to fifty-one.

     **b.** Twice eighteen is less than ninety divided by two.

     **c.** The sum of six and two is not greater than the product of six and two.

8.  **a.** Some number $m$ divided by four is equal to five.

     **b.** Each number $x$ decreased by two is less than twelve.

     **c.** Six times each number $n$ plus nine is greater than eighty.

     **d.** Eight times each number $y$ minus one is not less than sixty.

# Formulas

I. **Aim**    To express a verbal statement of some mathematical or scientific principle as a formula and to translate a formula into a word statement.

II. **Procedure**

1. To express a verbal statement as a formula, write numerals, operating symbols, and letters (variables) representing the given quantities in the required order to show the relationship between quantities. A quantity may be represented by the first letter of a key word. See sample solution 1.

   Observe that a *formula* is a special kind of equation. It is a mathematical rule expressing the relationship of two or more quantities by means of numerals, variables, and operating symbols.

2. To translate a formula into a word statement, write the statement in words that describe the relationship between the quantities given in the formula. See sample solution 2.

III. **Sample Solutions**

1. Express as a formula:
   The discount (*d*) is equal to the difference between the list price (*l*) and the net price (*n*).   *Answer:* $d = l - n$

2. Translate the formula $A = lw$ when $A =$ area of a rectangle, $l =$ length of a rectangle, and $w =$ width of a rectangle.

   *Answer:* The area of a rectangle is equal to the length of the rectangle times the width of the rectangle.

---

## DIAGNOSTIC TEST

1. Express each of the following as a formula:
   a. The horsepower (*H.P.*) required for the wing of an airplane is equal to the product of the drag of the wing (*D*) and the velocity (*V*) divided by 550.
   b. The perimeter of a rectangle (*p*) is twice the sum of the length (*l*) and the width (*w*).
   c. The base of a right triangle (*b*) is equal to the square root of the difference between the square of the hypotenuse (*h*) and the square of the altitude (*a*).

2. Translate each of the following formulas as a word statement:
   a. $A = \pi r^2$, where $A$ = area of circle, $\pi$ = Pi or 3.14, and $r$ = radius of circle

   b. $d = \dfrac{m}{v}$, where $d$ = density, $m$ = mass, and $v$ = volume

   c. $A = 2\pi r(r + h)$, where $A$ = total area of a cylinder, $\pi$ = Pi or 3.14, $r$ = radius, and $h$ = height

## RELATED PRACTICE EXAMPLES

1. Express each of the following mathematical, scientific, industrial, or commercial principles as a formula:
   a. The volume of a rectangular solid ($V$) equals the product of the length ($l$), width ($w$), and height ($h$).
   b. The average ($A$) of two numbers $c$ and $d$ equals the sum of the two numbers divided by two.
   c. The distance ($d$) traveled at a uniform rate of speed equals the rate ($r$) times the time ($t$).
   d. The cost of goods sold ($c$) is equal to the inventory at beginning ($I_1$) added to the purchases ($p$) less the inventory at the end ($I_2$).
   e. The central angle of a regular polygon ($a$) equals 360° divided by the number of sides ($n$).
   f. The number of gallons ($g$) in a given number of pints ($p$) is equal to the number of pints divided by 8.
   g. The circumference of a circle ($c$) is equal to twice the product of Pi and the radius ($r$).
   h. The capital ($C$) of a business is the difference between the assets ($A$) and the liabilities ($L$).
   i. The rate of commission ($r$) is equal to the commission ($c$) divided by the sales ($s$).
   j. The temperature reading on the Fahrenheit scale ($F$) is equal to nine fifths of the reading on the Celsius scale ($C$) increased by 32°.
   k. The surface speed ($S$) of a revolving pulley in feet per minute is equal to Pi times the diameter ($d$) in feet times the number of revolutions per minute ($R$).
   l. In the study of levers, the product of one weight ($W$) by the length of its arm ($L$) equals the product of the other weight ($w$) by the length of its arm ($l$).

m.   The entire surface area of a cylinder ($A$) is equal to twice Pi times the radius ($r$) times the sum of the radius ($r$) and the height ($h$).

n.   The tip speed of a propeller ($T.S.$) is equal to Pi times the diameter ($d$) times the number of revolutions per second ($N$).

o.   Centripetal force ($F$) equals the product of the weight of the body ($w$) and the square of the velocity ($v$) divided by the product of the acceleration of gravity ($g$) and the radius of the circle ($r$).

p.   The valence number ($N$) is the quotient of the atomic weight ($A.W.$) divided by the equivalent weight ($E.W.$).

q.   The horsepower of a steam engine ($H.P.$) equals the product of the steam pressure in lb. ($P$), the length of the stroke in ft. ($L$), the area of the piston in sq. in. ($A$), and the number of strokes per minute ($N$) divided by 33,000.

r.   The distance ($d$) a freely falling body drops is one half the product of the acceleration due to gravity ($g$) and the square of the time of falling ($t$).

s.   The hypotenuse ($h$) of a right triangle equals the square root of the sum of the squares of the altitude ($a$) and the base ($b$).

t.   The radius of action of an airplane ($R$) is equal to the total flying time ($T$) multiplied by the quotient obtained by dividing the product of the ground speed out ($GSO$) and the ground speed in ($GSI$) by their sum.

2. Translate each of the following formulas as a word statement:

a.   $W = IE$, where $W$ = power in watts, $I$ = current in amperes, and $E$ = electromotive force in volts

b.   $s = c + p$, where $s$ = selling price, $c$ = cost, and $p$ = profit

c.   $i = A - p$, where $i$ = interest, $A$ = amount, and $p$ = principal

d.   $r = \dfrac{d}{t}$, where $r$ = average rate of speed, $d$ = distance traveled, and $t$ = time of travel

e.   $p = b + 2e$, where $p$ = perimeter of isosceles triangle. $b$ = base, and $e$ = length of each equal side

f.   $A = p + prt$, where $a$ = amount, $p$ = principal, $r$ = rate of interest per year, and $t$ = time in years

g.   $A = \frac{1}{4}\pi d^2$, where $A$ = area of circle. $\pi$ = Pi or 3.14, and $d$ = diameter of circle

h.   $C = \frac{5}{9}(F - 32)$, where $C$ = Celsius temperature reading and $F$ = Fahrenheit temperature reading

i.   $a = \sqrt{h^2 - b^2}$, where $a$ = altitude of right triangle, $h$ = hypotenuse, and $b$ = base

j.   $A = \pi r(r + s)$, where $A$ = entire area of cone, $\pi$ = Pi, $r$ = radius, and $s$ = slant height

# REVIEW

1. Write each of the following as a numerical expression:
   a. The sum of ten and four
   b. The product of three and nine
   c. Twenty decreased by twelve
   d. Sixty divided by five
   e. Seven to the sixth power
   f. The square root of fifty-seven
   g. Six times the difference between eleven and two
   h. $2 \times 2 \times 2 \times 2 \times 2 \times 2 \times 2 \times 2 \times 2$ in exponential form
2. Write each of the following as an algebraic expression:
   a. The annual depreciation ($d$) divided by the original cost ($c$)
   b. Eight less than twice a certain number ($n$)
   c. The sum of angles $R$, $S$, and $T$
   d. The square of $t$ multiplied by 16
   e. The square root of the product of $x$ and $y$
   f. The product of the sum of $c$ and $d$ and the difference of $a$ and $b$
3. Read, or write in words, each of the following:
   a. $6 - 2 < 17$
   b. $1.5 \div 3 \neq 5 \div 2.5$
   c. $8\frac{3}{4} > 11 - 3\frac{1}{4}$
4. Multiply symbolically each of the following:
   a. Forty divided by five is not greater than six plus four.
   b. Seven times twelve is equal to one hundred minus sixteen.
   c. Sixty-three is not less than nine times seven.
5. Which of the following sentences are true?
   a. $8 = 6$
   b. $7 > 5$
   c. $4 < 10$
   d. $1 \not> 0$
   e. $3 \neq 2$
   f. $9 \not< 15$
   g. $\frac{3}{4} \not< \frac{5}{12}$
   h. $14 = 14$
   i. $\frac{5}{6} < \frac{2}{3}$
   j. $13 > 16$
   k. $5 \times .6 \neq 5.6$
   l. $\frac{7}{8} \not> \frac{8}{7}$
6. Which are open sentences? Which are inequalities?
   a. $8 + 11 = 19$
   b. $5x > 21$
   c. $4n - 5 = 45$
   d. $33 < 18 + 16$
   e. $y - 6 > 10$
   f. $\frac{2r}{7} = 5$
7. Read, or write in words, each of the following:
   a. $8x + 6 > 80$
   b. $10c - 9 \not< 19$
   c. $3n - 7 \neq n + 4$
8. Write symbolically each of the following:
   a. Each number $n$ increased by four is less than twelve.
   b. Twice each number $x$ minus ten is not greater than three.
   c. Twenty-one is equal to three times some number $y$ plus nine.
9. Express each of the following as a formula:
   a. The displacement of the piston ($D$) equals the area of the piston ($A$) times the strokes ($s$).
   b. The sum of the measures of the two complementary angles $C$ and $D$ equals 90°.
   c. The pitch of a roof ($p$) is equal to the rise ($r$) divided by the span ($s$).

**10.** Translate each of the following formulas as a word statement:
    **a.** $p = 3s$, where $p$ = perimeter of a rectangular octagon and $s$ = length of side
    **b.** $W = I^2R$ where $W$ = power in watts, $I$ = current in amperes, and $R$ = resistence in ohms
    **c.** $A = \frac{1}{2}h(b + b')$ where $A$ = area of trapezoid, $h$ = height, $b$ = length of lower base, and $b'$ = length of upper base

The numeral at the end of each problem indicates the Exercise where help may be found.

**1.** Write each of the following as a numerical expression: 1-2
    **a.** The product of eleven and nine     **b.** Eight increased by four

**2.** Write each of the following as an algebraic expression: 1-5
    **a.** The difference of $d$ and $r$     **b.** The quotient of ten divided by $t$

**3.** Read, or write in words, each of the following: 1-6 1-7
    **a.** $75 \div 5 < 8 \times 4$     **b.** $6x + 7 > 43$     **c.** $r - 3 \neq 4r$

**4.** Write symbolically each of the following:
    **a.** Twenty-five less eleven is less than fifteen. 1-6
    **b.** Some number $n$ increased by five is equal to thirty. 1-7

**5.** Write as a formula: The number of screw threads per inch ($n$) is equal to one (1) divided by the pitch (p). 1-8

## MAINTENANCE PRACTICE IN ARITHMETIC

**1.** Add:
85,967
49,586
39,679
97,859

**2.** Subtract:
630,051
269,755

**3.** Multiply:
9,008
804

**4.** Divide:
509$\overline{)3,077,923}$

**5.** Add: $6\frac{2}{3} + 1\frac{7}{8} + 3\frac{1}{2}$

**6.** Subtract: $16 - 9\frac{5}{6}$

**7.** Multiply: $12\frac{1}{2} \times 1\frac{4}{5}$

**8.** Divide: $10 \div 1\frac{3}{4}$

**9.** Add: .582 + 96.5 + 8.73

**10.** Subtract: 9.6 − .64

**11.** Multiply: .02 × .003

**12.** Divide: 1.4$\overline{)7.28}$

**13.** Find $87\frac{1}{2}\%$ of $720.

**14.** What percent of 60 is 36?

**15.** 5% of what number is 19?

# INTRODUCTION

## *Extension of the Number System*

The weather stations in the larger cities report daily both the high and low temperature readings. This information and local hourly temperature readings are published in many newspapers. In the winter, some of the numerals indicating reported temperature readings contain "−" signs. Stock quotation changes, generally found in the financial section of a newspaper, also consist of numerals containing "+" or "−" signs.

**TEMPERATURES IN CHICAGO**

MAXIMUM, 1 P.M. . . . . . . 8
MINIMUM, 7 A.M. . . . . . . −7

| | | | | | |
|---|---|---|---|---|---|
| 3 A.M. | −5 | Noon | 2 | Unofficial— | |
| 4 A.M. | −5 | 1 P.M. | 5 | 8 P.M. | 2 |
| 5 A.M. | −5 | 2 P.M. | 5 | 9 P.M. | 2 |
| 6 A.M. | −6 | 3 P.M. | 4 | 10 P.M. | 0 |
| 7 A.M. | −7 | 4 P.M. | 4 | 11 P.M. | −2 |
| 8 A.M. | −5 | 5 P.M. | 3 | Midnight | −2 |
| 9 A.M. | −4 | 6 P.M. | 2 | 1 A.M. | −2 |
| 10 A.M. | −2 | 7 P.M. | 2 | 2 A.M. | −4 |
| 11 A.M. | 0 | | | | |

**For 24 hours ended 7 p.m., Jan. 27:**

Mean temperature, −1; normal, 24; deficiency for January, 66 degrees.
Precipitation, none; deficiency for January, .12 of an inch; total since Jan. 1, 1.59 inches.
Highest wind velocity, 22 miles an hour, from the northwest at 12:47 A.M.
Barometer, 7 A.M.; 30-44; 7 P.M., 30-49.

| and Div. Mars. | Sts. 100c, | First. | High. | Low. | Last. | Net Chg. |
|---|---|---|---|---|---|---|
| A–B–C–D | | | | | | |
| 4cf | | | | | | |
| 1.10 | 22 | 34¾ | 34¾ | 34¼ | 34½ | |
| 2.40 | 105 | 71 | 71½ | 71 | 71¼ | + ½ |
| 1.40 | 531 | 47¾ | 50 | 47¼ | 49½ | |
| 1.50 | 9 | 26½ | 26¾ | 26½ | 26¾ | + ½ |
| 1.41 | x10 | 49 | 49 | 48½ | 18½ | + ¼ |
| .20 | 9 | 18½ | 18½ | 18¾ | 18½ | + ½ |
| 1.40 | 18 | 18½ | 18¼ | 18¾ | 19 | |
| | 76 | 73½ | 75 | 73 | 75 | +1¾ |
| 1.40 | 86 | 20½ | 20¾ | 19¼ | 20½ | + ¼ |
| Cn | 241 | 48½ | 41¾ | 47¼ | 48 | − 2¼ |
| 275 | 10 | 19¼ | 19¾ | 19 | 19 | − ½ |
| 150 | 50 | 43¼ | 43¾ | 42¼ | 42¼ | −1¼ |
| 1 | 243 | 13½ | 14¼ | 13¼ | 14½ | +1½ |
| 32 | 8 | 18½ | 18¼ | 18¼ | 18¾ | −¼ |
| 110 | 18 | 46 | 46 | 45 | 45 | −1 |
| 420 | 241 | 31 | 31 | ... | | |
| | ..... | | | | | |

These signs preceding the digits in the numerals do not mean addition or subtraction. They are used to represent quantities which are opposites. The "−" sign in the temperature reading indicates that the temperature is below zero, whereas the numerals which have no signs represent readings above zero. The "+" sign in the stock quotation indicates a gain or rise in price, whereas the "−" sign indicates a loss or fall in price.

We found in arithmetic that the numerals naming the whole numbers, rational numbers, and irrational numbers contained no signs. We are now extending our number system to include new numbers which are named by numerals that contain either the symbol "+" or the symbol "−". They are called *positive numbers* and *negative numbers* respectively.

A number named by a numeral containing a prefixed "+" sign is a *positive number*. All numbers greater than 0 are positive numbers and their numerals may be written without a sign as we have seen in arithmetic.

A number named by a numeral containing a prefixed "−" sign is a *negative number*.

Zero (0) is neither a positive nor a negative number.

Sometimes the "+" and "−" signs in writing numerals for positive and negative numbers are placed to the upper left of the digits instead of being centered immediately preceding the digits.

Both +3 and ⁺3 name the same number.

Both −3 and ⁻3 name the same number.

Positive and negative numbers are sometimes called *signed numbers* or *directed numbers*.

Positive and negative numbers have many practical uses. They are used in science and related fields, statistics, weather reports, stock reports, sports, and many other fields to express opposite meanings or directions.

## *Extension of the Number Line*

In arithmetic the number line has been restricted to points which are associated with the whole numbers and fractions of arithmetic and are located to the right of the point labeled.

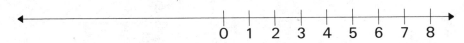

Observe that the numerals are arranged in a definite order on the number line so that they correspond one-to-one with points on the line.

Now let us extend the number line to the left of the point labeled 0. Using the interval between 0 and 1 as the unit measure, we locate equally spaced points to the left of

0. The first new point is labeled −1, read "negative one"; the second point −2, read "negative two"; the third −3, read "negative three"; etc.

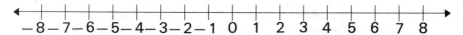

The numbers corresponding to the points to the left of 0 are *negative numbers*.

The numbers corresponding to the points to the right of 0 are *positive numbers*.

When the numerals naming both the positive and negative numbers contain signs, the number line would look like this:

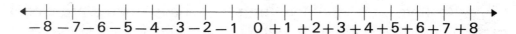

## Opposites

Examination of the number line below will show that points corresponding to +2 and −2 fall on the opposite sides of the point marked 0 but are the same distance from 0.

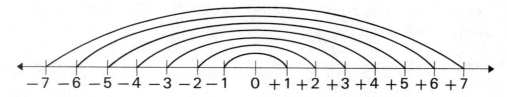

A pair of numbers, one positive and the other negative, such as +2 (or 2) and −2, which have the same absolute value (see section 2-3) are called *opposites*. The opposite of zero is zero. Also see section 2-4.

# Positive and Negative Numbers

# Reading and Writing Numerals
# Naming Positive and Negative
# Numbers

### I. Aim
To learn to recognize, read and write numerals naming positive and negative numbers.

### II. Procedure

1. Read the $+$ or $^+$ sign when part of a numeral as "positive" and not as "plus."
2. Read the $-$ or $^-$ sign when part of a numeral as "negative" and not as "minus."
3. To write a numeral naming a positive number, prefix a positive sign ($+$ or $^+$) to the numeral naming the number.
4. To write a numeral naming a negative number, prefix a negative sign ($-$ or $^-$) to the numeral naming the number.
5. Zero (0) is neither a positive nor a negative number.

*Note:* While both sign locations (centered and raised) are correct in naming positive and negative numbers, the numerals having centered signs are more generally used. Consequently, they have been chosen for use in this book starting with Exercise 2-2.

### III. Sample Solutions

1. Read: $+.5$   *Answer:* Positive five tenths
2. Read: $-78$   *Answer:* Negative seventy-eight
3. Write symbolically: Positive four   *Answer:* $+4$ or $^+4$
4. Write symbolically: Negative two thirds   *Answer:* $-\frac{2}{3}$ or $^-\left(\frac{2}{3}\right)$ or $\frac{^-2}{3}$

---

## DIAGNOSTIC TEST

Read, or write in words, each of the following:

1. $-9$  2. $+16$  3. $^+48$  4. $^-31$  5. $-.96$  6. $\dfrac{^-7}{12}$  7. $+2\frac{3}{4}$  8. $+5.3$

Which of the following name:

9. Positive numbers? $-35$; $\dfrac{^+9}{10}$; $+87$; $^-5.8$; $0$; $+6\frac{2}{3}$

10. Negative numbers? $^+60$; $-.029$; $0$; $^-8\frac{3}{5}$; $+39.1$; $-\frac{11}{12}$

42

Write symbolically:
11. Positive twenty-one          12. Negative fourteen
13. Negative five ninths         14. Positive four and six tenths

## *RELATED PRACTICE EXAMPLES*

Read, or write in words, each of the following:

1. **a.** $-13$      **b.** $-2$       **c.** $-47$      **d.** $-150$      **e.** $-86$
2. **a.** $+8$       **b.** $+34$      **c.** $+210$     **d.** $+72$       **e.** $+95$
3. **a.** $^+20$     **b.** $^+69$     **c.** $^+86$     **d.** $^+103$     **e.** $^+52$
4. **a.** $^-11$     **b.** $^-3$      **c.** $^-400$    **d.** $^-97$      **e.** $^-74$
5. **a.** $+.06$     **b.** $-5.2$     **c.** $+10.9$    **d.** $-.018$     **e.** $+20.45$
6. **a.** $\dfrac{-3}{4}$   **b.** $\dfrac{+5}{8}$   **c.** $\dfrac{-1}{6}$   **d.** $\dfrac{+3}{10}$   **e.** $\dfrac{-11}{16}$
7. **a.** $+9\frac{1}{2}$   **b.** $-14\frac{9}{10}$   **c.** $-5\frac{2}{5}$   **d.** $6\frac{5}{12}$   **e.** $-11\frac{8}{15}$
8. **a.** $^+.8$    **b.** $^+15.2$    **c.** $^-.0012$    **d.** $^-4.09$    **e.** $^+23.75$
9. Which of the following numerals name positive numbers:
   **a.** $-7$      **b.** $+51$      **c.** $-\frac{9}{20}$      **d.** $+3.8$      **e.** $-4\frac{3}{5}$
   **f.** $\dfrac{-11}{20}$   **g.** $+46$   **h.** $0$   **i.** $-2\frac{5}{16}$   **j.** $\dfrac{+19}{5}$
   **k.** $-16.3$   **l.** $+7\frac{3}{4}$   **m.** $-125$   **n.** $+\frac{17}{20}$   **o.** $^+.003$
10. Which of the following numerals name negative numbers:
    **a.** $-5$      **b.** $+3.4$      **c.** $-1.002$      **d.** $^-6\frac{3}{8}$      **e.** $^+.004$
    **f.** $\dfrac{-5}{6}$   **g.** $0$   **h.** $^-12$   **i.** $+4\frac{1}{3}$   **j.** $+1,000$
    **k.** $-.25$   **l.** $\dfrac{-10}{3}$   **m.** $^-64$   **n.** $+14.9$   **o.** $\dfrac{-9}{10}$

Write symbolically:
11. **a.** Positive six      **b.** Positive fifteen      **c.** Positive forty-nine
    **d.** Positive one hundred      **e.** Positive eighty-six
12. **a.** Negative one      **b.** Negative twenty      **c.** Negative fifty-three
    **d.** Negative two hundred sixty      **e.** Negative ninety-four
13. **a.** Negative one third      **b.** Positive three fourths
    **c.** Negative nine eighths      **d.** Positive seven and two fifths
    **e.** Negative seventeen and one half
14. **a.** Positive nine tenths
    **b.** Negative eighteen hundredths
    **c.** Positive seven and three tenths
    **d.** Positive eleven and five hundredths
    **e.** Negative twenty and eight tenths

# Use of Positive and Negative Numbers— Opposite Meaning

I. **Aim**   To develop the use of positive and negative numbers to represent opposite meanings and directions.

## II. Procedure

1. Represent positive and negative numbers by words or statements opposite in meaning or direction.
2. Represent words or statements opposite in meaning or direction by positive and negative numbers.

## III. Sample Solutions

1. If $+6$ kilometers means 6 kilometers north, what does $-4$ kilometers mean?   *Answer:* 4 kilometers south
2. If 9 degrees below zero is represented by $-9°$, how can 37 degrees above zero be represented?   *Answer:* $+37°$

---

### DIAGNOSTIC TEST

1. If $+50$ pounds represents an upward force of 50 pounds, what does $-70$ pounds represent?
2. If $-3°$ means 3 degrees south latitude, what does $+7°$ mean?
3. If $5 deposited is represented by $+$5$, how can $3 withdrawn from the bank be represented?
4. If 4 yards lost in a football game is indicated by $-4$ yards, how can 6 yards gained be indicated?

---

### RELATED PRACTICE EXAMPLES

1. **a.** If $+4$ pounds means 4 pounds overweight, what does $-3$ pounds mean?
   **b.** If $+10$ floors indicate 10 floors up, what does $-1$ floor indicate?
   **c.** If $+8$ miles represents 8 miles east, what does $-2$ miles represent?
   **d.** If $+3$ points means a 3-point rise in the market price of a stock, what does $-5$ points mean?
   **e.** If $+$60$ represents a gain of $60 in a business transaction, what does $-$75$ represent?

2. **a.** If −100 feet represents 100 feet below sea level, what does +200 feet represent?

   **b.** If −5 years indicates 5 years ago, what does +3 years indicate?

   **c.** If −23 items means an inventory shortage of 23 items, what does +49 items mean?

   **d.** If −2 amperes represents a discharge of 2 amperes of electricity, what does +8 amperes represent?

   **e.** If −12° means 12 degrees below zero, what does +67° mean?

3. **a.** If 24 degrees north latitude is indicated by +24°, how can 43 degrees south latitude be indicated?

   **b.** If $17 earned is represented by +$17, how can $6 spent be represented?

   **c.** If +2% indicates an increase of 2% in the cost of living, how can a 3% decrease in the cost of living be indicated?

   **d.** If 9 degrees above normal temperature is represented by +9°, how can 7 degrees below normal be represented?

   **e.** If a tail wind of 25 miles per hour is indicated by +25 m.p.h., how can a head wind of 39 miles per hour be indicated?

4. **a.** If 20 degrees west longitude is indicated by −20°, how can 32 degrees east longitude be indicated?

   **b.** If 16 feet to the left is indicated by −16 feet, how can 18 feet to the right be indicated?

   **c.** If the business liabilities of $1,375 is represented by −$1,375, how can assets of $5,200 be represented?

   **d.** If 8 miles south is represented by −8 miles, how can 7 miles north be indicated?

   **e.** If the deficiency of 1.25 inches of rainfall is indicated by −1.25 inches, how can an excess of .75 inch rainfall be indicated?

## CUMULATIVE REVIEW

1. Which of the following name:

   **a.** Positive numbers?     −57     +21     $+\frac{5}{8}$     −.38     0     $+4\frac{1}{2}$

   **b.** Negative numbers?     $+\frac{2}{3}$     −.95     0     $+1\frac{5}{6}$     −4.07     −84

2. Write symbolically:

   **a.** Negative eighteen                 **b.** Positive fifty

   **c.** Positive seven eighths           **d.** Negative five and eight tenths

3. If +8° means 8 degrees above zero, what does −7° mean?

4. If −3 miles indicates 3 miles west, what does +6 miles indicate?

5. If $+2\frac{1}{2}$ points represents a $2\frac{1}{2}$-point rise in stock, how can a drop of $1\frac{3}{4}$ points in a stock be represented?

# Absolute Value

I. **Aim**   To determine the absolute value of a number when it is positive, negative, or zero.

II. **Procedure**

1. **a.** Use the value of the arithmetic number with no sign that corresponds to the given number as its absolute value.

   The *absolute value* of any number is the corresponding arithmetic number which has no sign.

   **b.** The absolute value of 0 is 0.

2. Use a pair of vertical bars ‖ as the symbol to designate absolute value.

3. When operating with indicated absolute values first find the values and then operate as indicated. See sample solution 5.

III. **Sample Solutions**

1. Read: $|-8|$   *Answer:* The absolute value of negative eight
2. What is the absolute value of $+7$?   *Answer:* 7
3. What is the absolute value of $-7$?   *Answer:* 7
4. Find the indicated absolute value: $|-14| = ?$   *Answer:* 14
5. Multiply as indicated: $|-3| \times |+6|$

   Since $|-3| = 3$ and $|+6| = 6$, then $|-3| \times |+6| = 3 \times 6 = 18$

   *Answer:* 18

---

## DIAGNOSTIC TEST

1. Read or write in words: $|-29|$

2. Write symbolically, using the absolute value symbol: The absolute value of positive sixty

3. What is the absolute value of $+15$?

4. Find the indicated absolute value: $|-3.62|$

In each of the following, first determine the absolute value of each number, then perform the required operation:

5. $|-9| + |+7|$          6. $|+.8| - |-.5|$

7. $|-\frac{2}{3}| \times |-\frac{9}{10}|$          8. $|+24| \div |-4|$

---

**46**

## *RELATED PRACTICE EXAMPLES*

1. Read, or write in words, each of the following:
   **a.** $|-40|$   **b.** $|+71|$   **c.** $|-\frac{5}{6}|$   **d.** $|+9\frac{2}{5}|$   **e.** $|+.03|$   **f.** $|-8.6|$

2. Write symbolically, using the absolute value symbol:
   **a.** The absolute value of negative ten
   **b.** The absolute value of positive twelve
   **c.** The absolute value of positive two thirds
   **d.** The absolute value of negative eight tenths
   **e.** The absolute value of negative five and three fourths

3. What is the absolute value of each of the following?
   **a.** $+13$   **b.** $-\frac{5}{8}$   **c.** $+17.3$   **d.** $-2\frac{5}{6}$   **e.** $-110$   **f.** $+.009$

4. Find the indicated absolute value of each of the following:
   **a.** $|+4| = ?$   **b.** $|-1| = ?$   **c.** $|-\frac{3}{8}| = ?$   **d.** $|-1.26| = ?$
   **e.** $|-3\frac{1}{4}| = ?$   **f.** $|0| = ?$   **g.** $|+\frac{17}{6}| = ?$   **h.** $|-200| = ?$

In each of the following, first determine the absolute value of each number, then perform the required operation:

5. **a.** $|-6| + |+8|$   **b.** $|-\frac{2}{3}| + |-\frac{2}{3}|$   **c.** $|+4\frac{1}{2}| + |+1\frac{1}{4}|$
   **d.** $|+.4| + |-.7|$   **e.** $|-9| + |-12|$   **f.** $|-8.5| + |+2.8|$

6. **a.** $|-9| - |-5|$   **b.** $|+\frac{3}{4}| - |-\frac{5}{8}|$   **c.** $|+4\frac{7}{10}| - |-3\frac{5}{6}|$
   **d.** $|+2.3| - |+.89|$   **e.** $|-10.6| - |+3.4|$   **f.** $|-6\frac{1}{3}| - |-2\frac{1}{2}|$

7. **a.** $|-4| \times |+7|$   **b.** $|+\frac{5}{8}| \times |+\frac{2}{5}|$   **c.** $|-1\frac{1}{8}| \times |-\frac{2}{3}|$
   **d.** $|+.7| \times |-.006|$   **e.** $|-10| \times |-15|$   **f.** $|-3\frac{5}{6}| \times |+12|$

8. **a.** $|-48| \div |-6|$   **b.** $|-\frac{15}{16}| \div |+\frac{2}{3}|$   **c.** $|+7\frac{1}{2}| \div |+1\frac{1}{4}|$
   **d.** $|+.8| \div |-.02|$   **e.** $|-40| \div |+5\frac{1}{3}|$   **f.** $|-5.6| \div |-.14|$

## *MISCELLANEOUS EXAMPLES*

In each of the following, first determine the absolute value of each number, then perform the required operation:

1. **a.** $|-8| + |-5|$   **b.** $|+18| \div |-3|$   **c.** $|-9| - |+7|$
   **d.** $|-\frac{5}{6}| \times |+\frac{2}{5}|$   **e.** $|+.7| + |-1.4|$   **f.** $|-1\frac{1}{3}| \div |-\frac{1}{2}|$
   **g.** $|+6.3| - |-2.9|$   **h.** $|-10| \times |-7|$   **i.** $|-4\frac{3}{4}| + |+3\frac{2}{5}|$

2. **a.** $\dfrac{|-45|}{|-9|}$   **b.** $\dfrac{|+6| \times |-4|}{|-8|}$   **c.** $\dfrac{|-60|}{|-8| - |-7|}$

   **d.** $\dfrac{|-6| \times |-3|}{|-2| + |-7|}$   **e.** $\dfrac{|-14| - |+2|}{|+3| \div |-1|}$   **f.** $\dfrac{|-9| + |+6|}{|-7| - |-2|}$

# Opposites;
# Opposite of a Number;
# Additive Inverse

I. Aim    To determine the opposite or additive inverse of a number.

II. Procedure

1. On page 40 we found that a pair of numbers, one positive and one negative, such as $+2$ and $-2$, which have the same absolute value are called *opposites*.

   a. For the opposite of a given positive number, use the negative number having the same absolute value.
   b. For the opposite of a given negative number, use the positive number having the same absolute value.
   c. The opposite of zero is zero.

2. Use a dash sign prefix (symbol $-$) to indicate "the opposite of."

3. When the sum of two addends is zero, each addend is said to be the *additive inverse* of the other addend. On page 69 we shall see that the sum of a number and its opposite is zero. Therefore, the additive inverse of a given number is the opposite of this number.

III. Sample Solutions

1. Find the opposite of $+9$   *Answer:* $-9$
2. Find the additive inverse of $-2.5$   *Answer:* $+2.5$
3. Read: $-(-36)$   *Answer:* The opposite of negative thirty-six
4. Write symbolically: The opposite of positive two is negative two
   *Answer:* $-(+2) = -2$
5. Find the value: $-(-6\frac{3}{4}) = ?$   *Answer:* $+6\frac{3}{4}$

## DIAGNOSTIC TEST

1. What is the opposite of: **a.** $+23$? **b.** $-(\frac{3}{8})$?
2. Read, or write in words, each of the following:
   **a.** $-(+.9)$      **b.** $-(-37)$      **c.** $-20$      **d.** $-(16) = -16$
3. Write symbolically:
   **a.** The opposite of negative seven

48

  **b.** The opposite of positive eighteen
  **c.** The opposite of negative twelve is positive twelve
 **4.** What is the additive inverse of:
  **a.** $-21$?        **b.** $+6\frac{4}{5}$?
 **5.** What is the opposite of the opposite of:
  **a.** $+58$?        **b.** $-8.3$?
 **6.** Determine each of the following:
  **a.** $-(-11) = ?$  **b.** $-[+(\frac{15}{16})] = ?$  **c.** $-(-74) = ?$

## RELATED PRACTICE EXAMPLES

1. What is the opposite of each of the following numbers?
 **a.** 3    **b.** $+14$   **c.** $-19$   **d.** $-30$   **e.** $-(\frac{7}{8})$
 **f.** $+2\frac{1}{2}$   **g.** $+3.5$  **h.** $-.06$  **i.** $-5\frac{3}{4}$  **j.** $+(\frac{17}{9})$

2. Read, or write in words, each of the following:
 **a.** $-17$   **b.** $-(5)$    **c.** $-(+28)$ **d.** $-(-9)$
 **e.** $-[-(\frac{3}{5})]$  **f.** $-(+.7)$   **g.** $-(-12)$ **h.** $-(-31)$
 **i.** $-(+5) = -5$ **j.** $-(-20) = +20$ **k.** $-(8) = -8$ **l.** $-(-100) = 100$

3. Write symbolically:
 **a.** The opposite of negative forty
 **b.** The opposite of positive fifteen
 **c.** The opposite of sixty-three
 **d.** The opposite of the opposite of seventeen
 **e.** The opposite of negative fourteen is positive fourteen
 **f.** The opposite of the opposite of sixty is sixty
 **g.** The opposite of negative three eighths is positive three eighths
 **h.** The opposite of positive ninety is negative ninety

4. What is the additive inverse of each of the following numbers?
 **a.** $+1$   **b.** $-24$   **c.** $-(\frac{3}{11})$  **d.** $+5\frac{7}{8}$  **e.** $+.9$
 **f.** $-(\frac{15}{4})$  **g.** $-6.81$  **h.** 0   **i.** $7\frac{1}{3}$   **j.** $-63$

5. What is the opposite of the opposite of:
 **a.** $-11$?    **b.** $+17$?   **c.** 8?    **d.** $-.04$?
 **e.** $+5.2$?   **f.** $-\frac{1}{3}$?   **g.** 60?   **h.** $-93$?

6. Determine each of the following:
 **a.** $-(-9) = ?$ **b.** $-(+7) = ?$ **c.** $-[+(\frac{4}{11})] = ?$ **d.** $-(-.6) = ?$
 **e.** $-(15) = ?$ **f.** $-(-3\frac{1}{2}) = ?$ **g.** $-[-(\frac{19}{4})] = ?$ **h.** $-(100) = ?$

# Integers and Real Numbers

**I. Aim**  To learn to recognize integers, rational numbers, irrational numbers, and real numbers.

## II. Procedure

Use the following descriptions of integers, rational numbers, irrational numbers, and real numbers:

1. The group of numbers which consists of all the whole numbers and their opposites is called the *integers*.

   ..., $-2$, $-1$, 0, 1, 2, ... and ..., $-2$, $-1$, 0, $+1$, $+2$, ... describe the integers. Observe that the three dots are used at both ends since the integers continue in both directions without ending.

   1, 2, 3, ... describes the positive integers.

   0, 1, 2, ... describes the non-negative integers (or whole numbers).

   ..., $-3$, $-2$, $-1$, 0 describes the non-positive integers.

2. All the fractional numbers and their opposites are called the *rational numbers*.

   These include the integers since each integer may be named in fraction form. For example: $8 = \frac{8}{1}$.

3. An irrational number is a number that is both a non-terminating and non-repeating decimal like the square root of any positive number other than perfect squares. For example: $\sqrt{15} = 3.873 \ldots$. An irrational number cannot be expressed as a quotient of two integers (with division by zero excluded). All the positive and negative irrational numbers are called the *irrational numbers*.

4. All the rational numbers and all the irrational numbers are called the *real numbers*.

   The real numbers include all the integers, all the positive and negative fractional numbers, and all the positive and negative irrational numbers. There are an infinite number of real numbers in our number system.

## III. Sample Solutions

1. Which of the following name: **a.** integers? **b.** rational numbers? **c.** irrational numbers? **d.** real numbers?

$$+\tfrac{5}{8}; \; -.45; \; +\sqrt{29}; \; -5; \; +\tfrac{12}{3}; \; -3\tfrac{1}{2}; \; +6.72; \; -36$$

*Answers:*

**a.** Integers: $-5$; $\frac{12}{3}$; $-36$

**b.** Rational numbers: $+\frac{5}{8}$; $-.45$; $-5$; $+\frac{12}{3}$; $-3\frac{1}{2}$; $+6.72$; $-36$

**c.** Irrational numbers: $+\sqrt{29}$

**d.** Real numbers: all the given numbers

**2.** Write the missing integers of the list: $-16, -15, -14, \ldots, -9$

*Answers:* $-13, -12, -11, -10$

---

### DIAGNOSTIC TEST

Which of the following numerals name:

**1.** Integers? $+8.3$ $\quad -4$ $\quad -\frac{30}{5}$ $\quad +5\frac{2}{3}$ $\quad -.09$ $\quad +\sqrt{25}$

**2.** Rational numbers? $-\frac{5}{6}$ $\quad -2.57$ $\quad -4\frac{3}{4}$ $\quad +14$ $\quad +\sqrt{70}$ $\quad +.375$

**3.** Irrational numbers? $-19$ $\quad +\sqrt{81}$ $\quad -2\frac{1}{2}$ $\quad -\sqrt{95}$ $\quad +.111$ $\quad -\frac{31}{4}$

**4.** Real numbers? $+\sqrt{24}$ $\quad +49$ $\quad -\frac{3}{8}$ $\quad +.276$ $\quad -9\frac{7}{10}$ $\quad -63.8$

**5.** Which integers are missing in the list: $-5, -4, -3, \ldots, 5$?

---

### RELATED PRACTICE EXAMPLES

1. Which of the following numerals name integers?

   **a.** $-\frac{7}{12}$ $\quad +\sqrt{19}$ $\quad -8$ $\quad +5.7$ $\qquad$ **b.** $+\frac{15}{3}$ $\quad +47$ $\quad -.03$ $\quad -\sqrt{54}$

   **c.** $-\sqrt{81}$ $\quad -.734$ $\quad +2\frac{5}{6}$ $\quad +21$ $\qquad$ **d.** $-63$ $\quad +10.6$ $\quad +\sqrt{27}$ $\quad +\frac{11}{16}$

2. Which of the following numerals name rational numbers?

   **a.** $-\sqrt{7}$ $\quad +46$ $\quad -.75$ $\quad +\frac{8}{9}$ $\qquad$ **b.** $-19$ $\quad +\sqrt{9}$ $\quad -\frac{32}{4}$ $\quad +4.067$

   **c.** $+\frac{11}{2}$ $\quad -1.8$ $\quad +\sqrt{51}$ $\quad -47$ $\qquad$ **d.** $+.3075$ $\quad -8\frac{1}{2}$ $\quad +100$ $\quad -\sqrt{99}$

3. Which of the following numerals name irrational numbers?

   **a.** $-5$ $\quad +\sqrt{89}$ $\quad -\frac{7}{10}$ $\quad +.96$ $\qquad$ **b.** $-\sqrt{67}$ $\quad +93$ $\quad -53.009$ $\quad +\frac{54}{9}$

   **c.** $+18.4$ $\quad -8\frac{5}{8}$ $\quad +11$ $\quad -\sqrt{49}$ $\qquad$ **d.** $+.684$ $\quad -\frac{5}{6}$ $\quad +\sqrt{28}$ $\quad -45$

4. Which of the following numerals name real numbers?

   **a.** $+\sqrt{17}$ $\quad -32$ $\quad -8.525$ $\quad +\frac{3}{5}$ $\qquad$ **b.** $-6\frac{1}{4}$ $\quad +.44$ $\quad -\sqrt{9}$ $\quad +14$

   **c.** $-9$ $\quad -.53$ $\quad +\frac{11}{3}$ $\quad -\sqrt{105}$ $\qquad$ **d.** $+21.06$ $\quad +\sqrt{65}$ $\quad -86$ $\quad -5\frac{3}{10}$

5. Which integers are missing in each of the following lists?

   **a.** $+1, +2, +3, \ldots, +8$ $\qquad$ **b.** $-9, -8, -7, \ldots, -1$

   **c.** $-6, -5, -4, \ldots, 0$ $\qquad$ **d.** $-3, -2. -1, \ldots, +3$

   **e.** $-10, -9, -8, \ldots, +10$ $\qquad$ **f.** $-9, -7, -5, \ldots, 5$

   **g.** $-50, -40, -30, \ldots, 40$ $\qquad$ **h.** $-24, -20, -16, \ldots, 12$

# Even and Odd Integers; Consecutive Integers

**I. Aim** To learn to recognize even integers, odd integers, consecutive integers, and even and odd consecutive integers.

## II. Procedure

Use the following descriptions of even and odd integers, consecutive integers, and even and odd consecutive integers:

1. An *even integer* is an integer that can be divided exactly by two (2). Numbers whose numerals end with a 0, 2, 4, 6 or 8 are even integers. The group of all even integers is described by:

$$\ldots, -4, -2, 0, 2, 4, \ldots$$

2. An *odd integer* is an integer that cannot be divided exactly by two (2). Numbers whose numerals end with a 1, 3, 5, 7, or 9 are odd integers. The group of all odd integers is described by:

$$\ldots, -3, -1, 1, 3, \ldots$$

3. *Consecutive integers* are integers that differ by one (1). Usually the greater integer is written second as: 6 and 7; −1 and 0; or −10 and −9.

4. *Consecutive even integers* are even integers that differ by two (2). Usually the greater integer is written second as: 4 and 6; −2 and 0; or −10 and −8.

5. *Consecutive odd integers* are odd integers that differ by two (2). Again the greater integer is written second as: 5 and 7; −1 and 1; or −13 and −11.

## III. Sample Solutions

1. Which of the following name even integers? +9, −4, +6, −1, 0, +12
*Answer:* −4, +6, 0, +12

2. Which of the following name odd integers? −6, +5, −29, +70, −136
*Answer:* +5, −29

3. Name the four consecutive integers just following −7.
*Answer:* −6, −5, −4, −3

4. Name the three consecutive odd integers just preceding −9.
*Answer:* −15, −13, −11

5. Name the consecutive even integers between −8 and +2.
*Answer:* −6, −4, −2, 0

## *RELATED PRACTICE EXAMPLES*

1. Which of the following name even integers?
   a. $-18$  $+3$  $-36$  $+75$  $-110$
   b. $+4$  $-27$  $0$  $-46$  $+112$
   c. $+51$  $-14$  $-9$  $-278$  $+82$
   d. $-33$  $+6$  $-94$  $-50$  $+367$
2. Which of the following name odd integers?
   a. $+7$  $-16$  $+52$  $+103$  $-81$
   b. $-15$  $-30$  $+49$  $-127$  $+475$
   c. $-10$  $-1$  $+25$  $-79$  $+196$
   d. $+8$  $-11$  $+35$  $-42$  $-229$
3. a. Name the four consecutive integers just following:
      (1) $+3$              (2) $-6$              (3) $-1$
   b. Name the consecutive integers between:
      (1) $+3$ and $+11$    (2) $-13$ and $-3$    (3) $-5$ and $+5$
   c. Name the five consecutive integers just preceding:
      (1) $+9$              (2) $-3$              (3) $+2$
4. a. Name the five consecutive odd integers just following:
      (1) $+7$              (2) $-17$             (3) $-5$
   b. Name the consecutive odd integers between:
      (1) $+3$ and $+19$    (2) $-21$ and $-5$    (3) $-11$ and $+9$
   c. Name the two consecutive odd integers just preceding:
      (1) $+15$             (2) $-9$              (3) $-1$
5. a. Name the three consecutive even integers just following:
      (1) $+8$              (2) $-10$             (3) $-4$
   b. Name the consecutive even integers between:
      (1) $0$ and $+12$     (2) $-16$ and $-4$    (3) $-6$ and $+6$
   c. Name the four consecutive even integers just preceding:
      (1) $+10$             (2) $-10$             (3) $0$

# THE REAL NUMBER LINE

The *real number line* is the complete collection of points which correspond to all the real numbers.

The real number line is endless in both directions and only a part of it is shown at any time.

There are an infinite number of points on the real number line. However, there is one and only one point that corresponds to each real number and one and only one real number that corresponds to each point on the real number line.

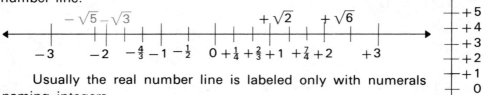

Usually the real number line is labeled only with numerals naming integers.

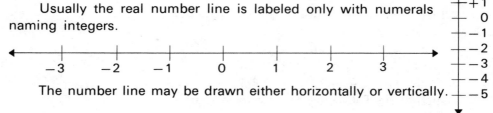

The number line may be drawn either horizontally or vertically.

When the horizontal and vertical number lines are combined, then a rectangular grid (Cartesian Coordinate System) is formed, which is useful in graphic interpolation and construction studied later in algebra.

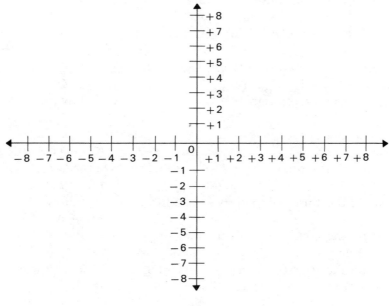

# Graph of a Group of Numbers on the Number Line

**I. Aim**   (a) To identify particular points on a number line.
(b) To draw the graph of a number or group of numbers on the number line.

**II. Procedure**

**1.** To identify particular points on the number line, use capital letters. See sample solutions 1 and 2.

**2.** To draw the graph of a number or a group of numbers on the number line:

 **a.** Draw an appropriate number line.

 **b.** Locate the point or points whose coordinate or coordinates are listed in the given numbers.

 **c.** Use heavy solid or colored dots to indicate these points.

 **d.** Observe that:

 (1) Each point on the number line is called the _graph_ of the number to which it corresponds, and

 (2) Each number is called the _coordinate_ of the related point on the line.

 (3) The _graph of a number_ is a point on the number line whose coordinate is the number.

 (4) The _graph of a group of numbers_ consists of all the points on the number line whose coordinates are the number.

**III. Sample Solutions**

**1.** Name the letter that corresponds to the point whose coordinate is −2.

```
    R   N   B   T   G   A   E   H   C   S   P
◄───┼───┼───┼───┼───┼───┼───┼───┼───┼───┼───┼───►
   −5  −4  −3  −2  −1   0  +1  +2  +3  +4  +5
```

_Answer:_ T

**2.** Name the coordinate of point _C_.

_Answer:_ +3

**3.** Draw the graph of: −3, −1, +2, +5

_Answer:_
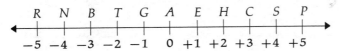

**55**

## DIAGNOSTIC TEST

1. Name the number that corresponds to the point on the number line below labeled:
   a. *D*　　　　　b. *C*　　　　　c. *M*
   d. *L*　　　　　e. *B*

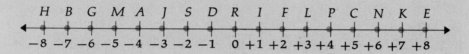

2. What letter labels the point on the above number line corresponding to each of the following numbers:
   a. −2?　　　　b. +4?　　　　c. −6?
   d. 0?　　　　　e. +7?

3. On the number line, draw the graph of −3, −1, 0, +2, +4.

4. Write the coordinates of which the following is the graph:

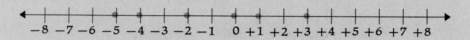

## RELATED PRACTICE EXAMPLES

1. Name the number that corresponds to the point on the following number line labeled:

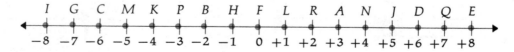

|       |       |       |
|-------|-------|-------|
| a. *B* | b. *D* | c. *G* |
| d. *F* | e. *N* | f. *C* |
| g. *H* | h. *I* | i. *E* |
| j. *L* | k. *M* | l. *A* |
| m. *K* | n. *R* | o. *P* |

2. What letter labels the point on the following number line corresponding to each of the following numbers?

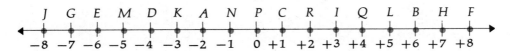

| | | | | |
|---|---|---|---|---|
| a. +3 | b. −7 | c. −4 | d. +6 | e. 0 |
| f. −8 | g. +5 | h. −3 | i. +2 | j. −1 |
| k. +7 | l. +1 | m. −2 | n. +8 | o. −5 |

3. On a number line, draw the graph of:

a. +4, −6
b. −3, 0, +6
c. −4
d. −2, −1, 1, 4
e. −4, −1, 0, 3, 5
f. −6, −5, −3, −2, −1
g. 4, −3, 7, −6, 2, −4
h. −1, 0, +1, +3, +4, +7
i. −5, −2, 0, 1, 2, 4, 8
j. +2, −3, +4, −5, 0, −2, −1
k. 6, −2, 0, −4, 3, −5, 1, −3, 4
l. −4, −3, −2, . . ., 5

4. Write the coordinates of which the following is the graph:

a.

b.

c.

d.

e.

f.

g.

h.

# Comparing Positive and Negative Numbers

### I. Aim    To compare positive and negative numbers.

### II. Procedure
1. Observe that:
   a. Of two numbers on a horizontal number line, the number corresponding to the point on the line farther to the right is the greater number.
   b. Of two numbers on a vertical number line, the number corresponding to a particular point is greater than the number corresponding to a point located below it.
   c. Zero is regarded sometimes as a quantity and sometimes as a point of reference.
2. When comparing two positive numbers, take the number whose absolute value is greater as the greater number.
3. When comparing two negative numbers, take the number whose absolute value is smaller as the greater number.
4. Any positive number is greater than zero or any negative number.
5. Zero is greater than any negative number.

### III. Sample Solutions
1. Which is greater: $+6$ or $+4$?
   $|+6| = 6 \quad |+4| = 4 \quad 6 > 4$
   *Answer:* $+6$

2. Which is greater: $-6$ or $-4$?
   $|-6| = 6 \quad |-4| = 4 \quad 4 < 6$
   *Answer:* $-4$

3. Which is greater: $+3$ or $-5$?
   *Answer:* $+3$

4. Which is greater: $-2$ or $0$?
   *Answer:* $0$

5. Which is greater: the absolute value of $+7$ or the absolute value of $-8$?
   *Solution:*
   $|+7| = 7 \qquad |-8| = 8 \qquad 8 > 7$
   *Answer:* The absolute value of $-8$ is greater.

6. Which has a greater opposite number: $-2$ or $+9$?
   *Solution:*
   The opposite of $-2$ is $+2$. The opposite of $+9$ is $-9$. $+2 > -9$
   *Answer:* The opposite of $-2$ is greater.

## DIAGNOSTIC TEST

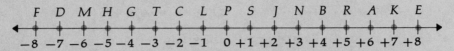

1. On the above horizontal number line, which point corresponds to the:
   a. Greater number: *R* or *M*?
   b. Smaller number: *P* or *A*?
2. Which number on the horizontal number line is associated with the point farther to the right:
   a. +3 or −7?        b. −4 or 0?
3. On the vertical number line at the right, which point corresponds to the:
   a. Smaller number: *J* or *B*? b. Greater number: *F* or *H*?
4. Which number on the vertical number line is associated with the point that is lower: a. −1 or −3? b. 0 or +4?
5. Select the greater of the two given numbers:
   a. +8 or +3        b. −8 or −3        c. 0 or +9
   d. −5 or 0        e. −7 or +1
6. Select the smaller of the two given numbers:
   a. +7 or +4        b. −11 or −6        c. +12 or 0
   d. 0 or −10        e. +6 or −8
7. Which of the following sentences are true:
   a. −14 < −4? b. 0 > −1? c. +8 ≮ −5? d. −9 ⋡ +1?
8. Rewrite each of the following and replace each question mark with whichever of these symbols, =, <, or > that will make a true sentence:
   a. −10 ? +10        b. −3 ? 0        c. −20 ? −23        d. +11 ? +10
Name the following numbers in order of size:
9. Greatest first: −9  +6  −11  +2  −4  +1        0  +17  −2  +8
10. Smallest first: +7  −3  −8  +5  −6        0  −16  +21  −21  +2
11. Which is greater: the absolute value of +7 or the absolute value of −8?
12. Which has the greater opposite number: −5 or +10?

The vertical number line:

E +5
I +4
F +3
B +2
K +1
D 0
A −1
J −2
C −3
H −4
G −5

## *RELATED PRACTICE EXAMPLES*

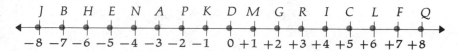

1. **a.** On the above horizontal number line, which point corresponds to the greater number:
    (1) Point *A* or point *B*?
    (2) Point *K* or point *M*?
    (3) Point *C* or point *H*?
    (4) Point *P* or point *D*?
   **b.** Which point corresponds to the smaller number:
    (1) Point *G* or point *L*?
    (2) Point *I* or point *E*?
    (3) Point *D* or point *N*?
    (4) Point *F* or point *R*?

2. Which number is associated with the point farther to the right on a horizontal number line:
    **a.** −5 or −4?     **b.** −9 or +9?     **c.** 0 or −3?
    **d.** +2 or +5?     **e.** +3 or −2?     **f.** 0 or +1?

3. **a.** On the vertical number line at the right, which point corresponds to the smaller number:
    (1) Point *F* or point *M*?
    (2) Point *B* or point *D*?
    (3) Point *G* or point *A*?
    (4) Point *S* or point *M*?
   **b.** Which point corresponds to the greater number:
    (1) Point *E* or point *R*?
    (2) Point *D* or point *H*?
    (3) Point *M* or point *B*?
    (4) Point *G* or point *P*?

| | |
|---|---|
| *P* | +6 |
| *F* | +5 |
| *C* | +4 |
| *M* | +3 |
| *S* | +2 |
| *E* | +1 |
| *B* | 0 |
| *W* | −1 |
| *H* | −2 |
| *A* | −3 |
| *G* | −4 |
| *D* | −5 |
| *R* | −6 |

4. Which number is associated with the point that is lower on the vertical number line:
    **a.** +3 or +1?     **b.** 0 or −4?     **c.** −8 or +7?
    **d.** −2 or −1?     **e.** +6 or 0?     **f.** +2 or −10?

5. Select the greater of the two given numbers:
    **a.** +4 or +7     +5 or +3     +9 or +6     +8 or +10     +4 or +1
    **b.** −2 or −1     −8 or −5     −3 or −4     −7 or −10     −6 or −5
    **c.** 0 or +6     +8 or 0     0 or +1     +3 or 0     +5 or 0
    **d.** 0 or −3     −2 or 0     0 or −7     0 or −6     −9 or 0
    **e.** −2 or +3     +4 or −6     −5 or +5     −3 or +2     +6 or −8

6. Select the smaller of the two given numbers:
   - **a.** $+3$ or $+5$    $+6$ or $+10$    $+12$ or $+1$    $+7$ or $+8$    $+2$ or $+4$
   - **b.** $-7$ or $-4$    $-3$ or $-1$    $-9$ or $-5$    $-2$ or $-8$    $-10$ or $-12$
   - **c.** $\;0$ or $+5$    $+4$ or $0$    $+10$ or $0$    $0$ or $+15$    $0$ or $+7$
   - **d.** $\;0$ or $-8$    $-3$ or $0$    $0$ or $-5$    $0$ or $-17$    $-14$ or $0$
   - **e.** $-3$ or $+4$    $+1$ or $-7$    $-8$ or $+5$    $+6$ or $-3$    $+10$ or $-12$

7. Which of the following sentences are true?
   - **a.** $-4 < 0$
   - **b.** $-2 > -6$
   - **c.** $+1 < -3$
   - **d.** $+3 > +4$
   - **e.** $0 < +9$
   - **f.** $-1 < -7$
   - **g.** $-6 > +2$
   - **h.** $0 > -1$
   - **i.** $-3 \not> -9$
   - **j.** $-10 \not< -12$
   - **k.** $-8 \not> +2$
   - **l.** $-5 \not< -5$
   - **m.** $-\frac{3}{4} > -\frac{5}{7}$
   - **n.** $+.06 \not> -.99$
   - **o.** $-1\frac{1}{4} < -2\frac{1}{2}$
   - **p.** $0 \not< -10$
   - **q.** $-42 > -15$
   - **r.** $-1 < +1$

8. Rewrite each of the following and replace each question mark with whichever of these symbols, $=$, $<$, or $>$ that will make a true sentence:
   - **a.** $+1\,?\,+3$
   - **b.** $0\,?\,-5$
   - **c.** $-8\,?\,-8$
   - **d.** $-6\,?\,-7$
   - **e.** $-9\,?\,0$
   - **f.** $+4\,?\,+2$
   - **g.** $-5\,?\,-4$
   - **h.** $0\,?\,+8$
   - **i.** $-2\,?\,-11$
   - **j.** $+17\,?\,-17$
   - **k.** $-20\,?\,+19$
   - **l.** $+35\,?\,+35$

9. Name the following numbers in order of size (greatest first):
   - **a.** $-2$   $+6$   $+3$   $-9$   $-1$   $0$   $+7$   $+1$   $+16$   $-8$
   - **b.** $+38$   $+20$   $-16$   $+3$   $-7$   $-5$   $+23$   $-35$   $+8$   $-2$
   - **c.** $+15$   $-12$   $-4$   $+5$   $+2$   $-8$   $+10$   $0$   $-6$   $+21$
   - **d.** $-5$   $-3$   $+7$   $+1$   $+15$   $-9$   $-16$   $-11$   $+20$   $-14$

10. Name the following numbers in order of size (smallest first):
    - **a.** $+5$   $0$   $-3$   $-1$   $-10$   $+9$   $+15$   $-12$   $+25$   $-17$
    - **b.** $-14$   $+9$   $+19$   $-15$   $-22$   $+7$   $+10$   $-6$   $+2$   $-30$
    - **c.** $-2$   $+5$   $-7$   $+18$   $-26$   $-4$   $-13$   $+10$   $+1$   $-50$
    - **d.** $+3$   $-9$   $+7$   $0$   $-1$   $+11$   $-8$   $+6$   $-12$   $+14$

11. Which is greater:

    **a.** The absolute value of $-12$ or the absolute value of $+6$?

    **b.** The absolute value of $-3$ or the absolute value of $-8$?

    **c.** The absolute value of $+9$ or the absolute value of $-10$?

    **d.** The absolute value of $+10$ or the absolute value of $-11$?

    **e.** The absolute value of $-4$ or the absolute value of $+5$?

12. Which has the greater opposite number:
    - **a.** $+7$ or $+9$?
    - **b.** $-4$ or $-6$?
    - **c.** $+8$ or $+5$?
    - **d.** $-8$ or $+5$?
    - **e.** $+10$ or $-12$?
    - **f.** $0$ or $-3$?
    - **g.** $+4$ or $0$?
    - **h.** $-2$ or $+2$?
    - **i.** $-14$ or $-11$?

# Opposite Directions; Vectors

**I. Aim**  To use a vector to picture the size and direction of the movement indicated by a positive or a negative number.

**II. Procedure**

**1.** Observe that:

  **a.** Positive and negative numbers, used as directed numbers, indicate movements in opposite directions. Generally a movement to the right of a particular point is considered as moving in a positive direction from the point, and a movement to the left of a point as moving in a negative direction from the point.

  **b.** The sign in the numeral naming a directed number indicates the direction, and the absolute value of the directed number represents the magnitude (distance in units) of the movement.

  **c.** An arrow representing this directed line segment is generally called a *vector*.

**2.** Using the number line as a scale, draw a vector to picture the direction and size of the movement indicated by the signed number. Use the sign in the numeral to indicate the direction and the absolute value to indicate the length of the arrow.

**3.** To find the numeral that describes the movement from one point to another on a number line:

  **a.** Draw the vector between the points with the arrowhead pointing in the direction of the movement.

  **b.** Use the number of units of length in the vector as the absolute value of the number named by the numeral.

  **c.** Select the sign according to the direction of the vector. If it points to the right, the sign is positive (+), if it points to the left, the sign is negative (−).

III. Sample Solutions

1. A movement of how many units of distance and in what direction is represented by: **a.** +5? **b.** −3?

   *Answers:* **a.** +5 means moving 5 units to the right.
             **b.** −3 means moving 3 units to the left.

2. Using the number line as a scale, draw a vector representing: **a.** +3 **b.** −4

   *Answers:* **a.**

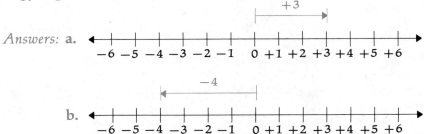

   **b.**

3. Write the numeral that is represented by the vector that illustrates:
   **a.** A movement from −3 to +5.

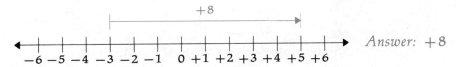

   *Answer:* +8

   **b.** A movement from +2 to −4.

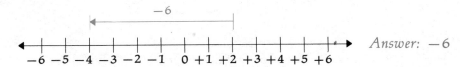

   *Answer:* −6

## DIAGNOSTIC TEST

1. A movement of how many units of distance and in what direction is represented by:
   **a.** −14?    **b.** +23?
2. Represent each of the following movements by a numeral naming a signed number:
   **a.** Moving 6 units to the right
   **b.** Moving $2\frac{1}{2}$ units to the left

3. Write the numeral that is represented by each of the following vectors:

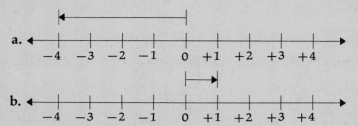

4. Using a number line as a scale, draw a vector representing:
   a. +8    b. −2

5. Write the numeral that is represented by each of the following vectors:

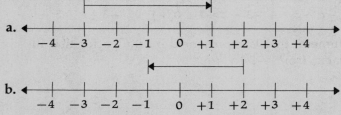

6. Using a number line as a scale, draw a vector and write the numeral represented by this vector that illustrates a movement from:
   a. +1 to +8    b. −7 to −2    c. +3 to −5    d. −1 to −6

## RELATED PRACTICE EXAMPLES

1. Each of the following numbers represents a movement of how many units of distance and in what direction?
   a. −6          b. +11         c. +1          d. −20         e. −38
   f. 17          g. $+\frac{13}{16}$       h. $-3\frac{1}{2}$       i. −4.5         j. +92

2. Represent each of the following movements by a numeral naming a signed number:
   a. Moving 7 units to the right          b. Moving 14 units to the left
   c. Moving 9 units to the left           d. Moving 23 units to the right
   e. Moving $6\frac{3}{4}$ units to the left         f. Moving $10\frac{1}{3}$ units to the right

3. Write the numeral that is represented by each of the following vectors:

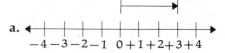 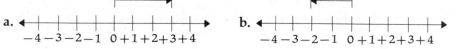

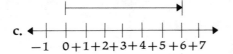
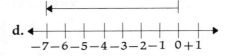

c.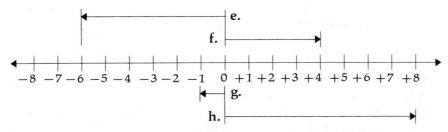
d.

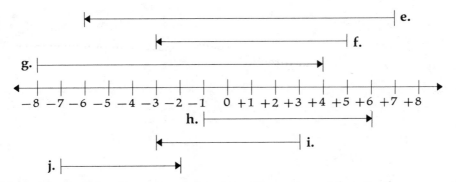

4. Using the number line as a scale, draw a vector representing:
   a. −4            b. +7            c. +2            d. −6            e. −5
   f. +1            g. +9            h. −8            i. −1½           j. +5¾

5. Write the numeral that is represented by each of the following vectors:

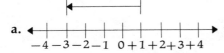
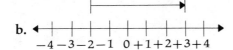

a.
c.

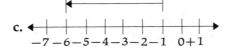
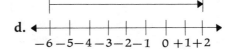

b.
d.

e.
f.
g.
h.
i.
j.

6. Using the number line as a scale, draw the vector and write the numeral
   represented by this vector that illustrates a movement from:
   a. −2 to +6        b. +3 to −1       c. −5 to 0        d. +4 to −7
   e. +2 to +8        f. −1 to −8       g. −6 to +7       h. +6 to 0
   i. −9 to +2        j. +5 to +1       k. −10 to −3      l. +4 to −4

# Addition of Positive and Negative Numbers on the Number Line

**I. Aim**  To add positive and negative numbers on the number line by using vectors.

**II. Procedure**

1. Draw a vector to represent each addend, the first beginning at the point 0, sometimes called the *origin*, and each of the other vectors from the point reached by the vector representng the previous addend.
2. Then measure the resultant vector from 0 to the point reached by the vector representing the last addend.
3. The signed number corresponding to this resultant vector is the sum. The coordinate of the final point reached indicates this number.
4. To avoid confusion, draw the vectors representing the addends above the number line and the vector representing the sum below the number line.

**III. Sample Solutions**

**1.** $(+2) + (+4) = ?$
*Answer:* $+6$

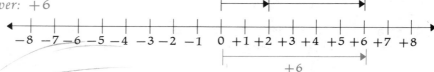

**2.** $(-3) + (-5) = ?$
*Answer:* $-8$

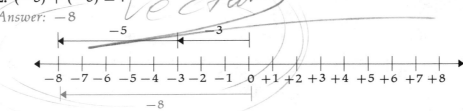

**3.** $(+6) + (-2) = ?$
*Answer:* $+4$

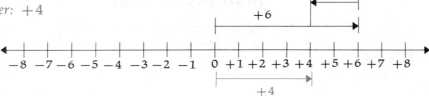

**4.** $(+6) + (-8) = ?$

*Answer:* $-2$

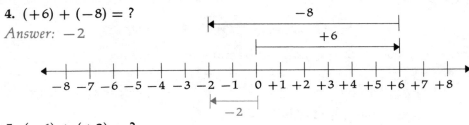

**5.** $(-6) + (+2) = ?$

*Answer:* $-4$

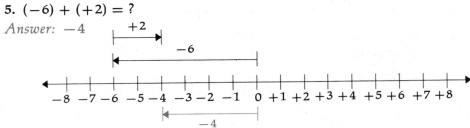

**6.** $(-6) + (+8) = ?$

*Answer:* $+2$

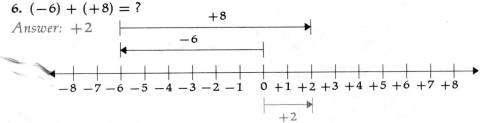

**7.** $(-6) + (+6) = ?$

*Answer:* $0$

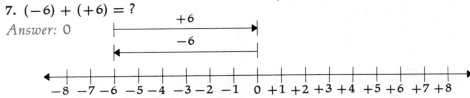

**8.** $(-5\frac{1}{2}) + (+3\frac{1}{2}) = ?$

*Answer:* $-2$

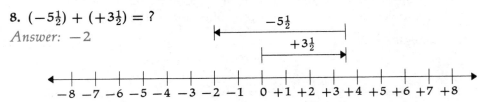

## DIAGNOSTIC TEST

In each of the following, add on the number line by using vectors:

**1.** $(+2)+(+5)$   **2.** $(-1)+(-6)$   **3.** $(+4)+(-2)$   **4.** $(+3)+(-7)$
**5.** $(-8)+(+3)$   **6.** $(-4)+(+9)$   **7.** $(-5)+(+5)$   **8.** $(+3\frac{2}{3})+(-5\frac{1}{3})$

## RELATED PRACTICE EXAMPLES

In each of the following, add on the number line by using vectors:

1. **a.** $(+6) + (+3)$     **b.** $(+4) + (+1)$     **c.** $(+5) + (+8)$
    **d.** $(+3) + (+2)$     **e.** $(+2) + (+9)$     **f.** $(+1) + (+7)$

2. **a.** $(-5) + (-4)$     **b.** $(-2) + (-8)$     **c.** $(-6) + (-3)$
    **d.** $(-2) + (-2)$     **e.** $(-7) + (-5)$     **f.** $(-3) + (-10)$

3. **a.** $(+8) + (-3)$     **b.** $(+7) + (-5)$     **c.** $(+9) + (-1)$
    **d.** $(+6) + (-4)$     **e.** $(+11) + (-2)$     **f.** $(+5) + (-3)$

4. **a.** $(+5) + (-9)$     **b.** $(+2) + (-11)$     **c.** $(+3) + (-7)$
    **d.** $(+1) + (-8)$     **e.** $(+6) + (-10)$     **f.** $(+4) + (-5)$

5. **a.** $(-4) + (+1)$     **b.** $(-9) + (+6)$     **c.** $(-11) + (+5)$
    **d.** $(-7) + (+2)$     **e.** $(-6) + (+3)$     **f.** $(-8) + (+4)$

6. **a.** $(-3) + (+7)$     **b.** $(-1) + (+8)$     **c.** $(-2) + (+12)$
    **d.** $(-5) + (+10)$     **e.** $(-7) + (+9)$     **f.** $(-8) + (+11)$

7. **a.** $(-3) + (+3)$     **b.** $(+7) + (-7)$     **c.** $(+8) + (-8)$
    **d.** $(-2) + (+2)$     **e.** $(-10) + (+10)$     **f.** $(+9) + (-9)$

8. **a.** $(-2\frac{1}{2}) + (-4\frac{1}{2})$     **b.** $(+6) + (-3\frac{1}{4})$     **c.** $(-7\frac{2}{3}) + (+9\frac{1}{3})$
    **d.** $(+4\frac{3}{4}) + (+5\frac{1}{2})$     **e.** $(-8\frac{1}{4}) + (+5)$     **f.** $(+6\frac{1}{2}) + (-6\frac{1}{2})$

## CUMULATIVE REVIEW

1. What number is neither positive nor negative?
2. On a number line draw the graph of: $-6, -3, -2, 0, +1, +4$.
3. Write the coordinates of which the following is the graph:

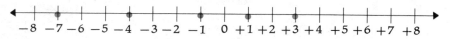

4. If $+7$ yards means 7 yards gained, what does $-12$ yards mean?
5. Which sentences are true? **a.** $-8 < -9$ **b.** $+10 > -15$ **c.** $0 \not> -9$
6. Which has the greater absolute value: $-9$ or $+7$?
7. Which has the greater opposite value: $-18$ or $-20$?
8. What signed number is represented by each of the following vectors?

**a.**         **b.**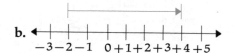

# Addition of Integers and Rational Numbers

### I. Aim     To add integers and rational numbers.

### II. Procedure

Analysis of Exercise 2-10 indicates that:

1. To add two positive numbers, find the sum of their absolute values and prefix the numeral with a positive sign.

   Note: Addition may be indicated horizontally as: $(+3) + (+9)$     or vertically as:     add: $+3$
   $\underline{+9}$

2. To add two negative numbers, find the sum of their absolute values and prefix the numeral with a negative sign.

3. To add **a.** a positive number and a negative number
   or **b.** a negative number and a positive number,
   subtract the smaller absolute value from the greater absolute value and prefix the difference with a sign of the number having the greater absolute value.

4. The numeral naming a positive number does not require the positive sign.

5. To add rational numbers, follow the same procedures as used with integers.

6. The sum of any number and its opposite is 0.

7. The sum of any non-zero number and zero is the non-zero number.

8. To add three or more numbers, either add successively or add first the positive numbers, then the negative numbers, and finally the two sums.

### III. Sample Solutions

Add as indicated in each of the following:

1. $(+7)+(+4)=?$ *Solution:* $|+7|=7$; $|+4|=4$; and $7+4=11$. Use $+$ sign.
   *Answer:* $+11$

2. $(-8)+(-2)=?$ *Solution:* $|-8|=8$; $|-2|=2$; and $8+2=10$. Use $-$ sign.
   *Answer:* $-10$

3. **a.** $(+5)+(-4)=?$ *Solution:* $|+5|=5$; $|-4|=4$; and $5-4=1$. Use $+$ sign.
   *Answer:* $+1$

   **b.** $(+1)+(-6)=?$ *Solution:* $|+1|=1$; $|-6|=6$; and $6-1=5$. Use $-$ sign.
   *Answer:* $-5$

   **c.** $(-3)+(+9)=?$ *Solution:* $|-3|=3$; $|+9|=9$; and $9-3=6$. Use $+$ sign.
   *Answer:* $+6$

   **d.** $(-7)+(+3)=?$ *Solution:* $|-7|=7$; $|+3|=3$; and $7-3=4$. Use $-$ sign.
   *Answer:* $-4$

## DIAGNOSTIC TEST

**1.** Add as indicated:
$(+23) + (+18)$

**2.** Add: $\begin{array}{r} +16 \\ +\ 7 \\ \hline \end{array}$

**3.** Add as indicated:
$(-39) + (-15)$

**4.** Add: $\begin{array}{r} -\ 5 \\ -12 \\ \hline \end{array}$

Add:

**5.** $\begin{array}{r} +8 \\ -5 \\ \hline \end{array}$      **6.** $\begin{array}{r} +\ 2 \\ -10 \\ \hline \end{array}$      **7.** $\begin{array}{r} -16 \\ +11 \\ \hline \end{array}$      **8.** $\begin{array}{r} -19 \\ +23 \\ \hline \end{array}$

**9.** Add as indicated:   $(-13) + (+7)$

Add:

**10. a.** $\begin{array}{r} -10 \\ +10 \\ \hline \end{array}$      **11. a.** $\begin{array}{r} 0 \\ -6 \\ \hline \end{array}$      **12. a.** $\begin{array}{r} -17 \\ 8 \\ \hline \end{array}$

     **b.** $(+1) + (-1)$      **b.** $(+2) + (0)$      **b.** $(4) + (-3)$

**13. a.** $\begin{array}{r} -2\frac{3}{8} \\ +3\frac{1}{4} \\ \hline \end{array}$      **14. a.** $\begin{array}{r} -9.35 \\ +8.72 \\ \hline \end{array}$

     **b.** $(-\frac{5}{6}) + (-\frac{4}{5})$      **b.** $(-.34) + (-1.6)$

**15.** Add as indicated:   $(-5) + (+8) + (-6) + (+2)$

**16.** Find the sum of the following:   $(-7) + [(+12) + (-8)]$

**17.** Add: $\begin{array}{r} -10 \\ +\ 8 \\ -15 \\ +19 \\ \hline \end{array}$      **18.** First find the absolute values, then add as indicated: $|-5| + |-2|$

**19.** First add, then find the absolute value of the sum: $|(-7) + (+3)|$

**20.** Does $(-2) + (+7) = |-2| + |+7|$?

## RELATED PRACTICE EXAMPLES

**1.** Add as indicated:

     **a.** $(+6) + (+5)$      **b.** $(+2) + (+11)$      **c.** $(+7) + (+9)$
     **d.** $(+10) + (+3)$      **e.** $(+18) + (+5)$      **f.** $(+24) + (+37)$

**2.** Add:

$\begin{array}{r} +5 \\ +3 \\ \hline \end{array}$   $\begin{array}{r} +7 \\ +8 \\ \hline \end{array}$   $\begin{array}{r} +9 \\ +2 \\ \hline \end{array}$   $\begin{array}{r} +4 \\ +4 \\ \hline \end{array}$   $\begin{array}{r} +2 \\ +8 \\ \hline \end{array}$   $\begin{array}{r} +9 \\ +3 \\ \hline \end{array}$   $\begin{array}{r} +7 \\ +12 \\ \hline \end{array}$   $\begin{array}{r} +11 \\ +13 \\ \hline \end{array}$

**3.** Add as indicated:

     **a.** $(-6) + (-2)$      **b.** $(-4) + (-9)$      **c.** $(-1) + (-1)$
     **d.** $(-7) + (-10)$      **e.** $(-23) + (-8)$      **f.** $(-54) + (-37)$

Add:

| | | | | | | | |
|---|---|---|---|---|---|---|---|
| 4. $-3$ | $-7$ | $-8$ | $-2$ | $-6$ | $-9$ | $-12$ | $-36$ |
| $\underline{-1}$ | $\underline{-4}$ | $\underline{-5}$ | $\underline{-7}$ | $\underline{-10}$ | $\underline{-9}$ | $\underline{-8}$ | $\underline{-29}$ |
| 5. $+4$ | $+8$ | $+10$ | $+7$ | $+9$ | $+2$ | $+19$ | $+25$ |
| $\underline{-2}$ | $\underline{-3}$ | $\underline{-5}$ | $\underline{-4}$ | $\underline{-3}$ | $\underline{-1}$ | $\underline{-9}$ | $\underline{-18}$ |
| 6. $+1$ | $+6$ | $+3$ | $+7$ | $+4$ | $+15$ | $+9$ | $+23$ |
| $\underline{-5}$ | $\underline{-7}$ | $\underline{-6}$ | $\underline{-9}$ | $\underline{-11}$ | $\underline{-18}$ | $\underline{-16}$ | $\underline{-35}$ |
| 7. $-5$ | $-7$ | $-8$ | $-12$ | $-15$ | $-22$ | $-18$ | $-60$ |
| $\underline{+3}$ | $\underline{+1}$ | $\underline{+2}$ | $\underline{+8}$ | $\underline{+9}$ | $\underline{+16}$ | $\underline{+12}$ | $\underline{+27}$ |
| 8. $-2$ | $-3$ | $-5$ | $-1$ | $-7$ | $-15$ | $-23$ | $-37$ |
| $\underline{+9}$ | $\underline{+4}$ | $\underline{+11}$ | $\underline{+10}$ | $\underline{+14}$ | $\underline{+21}$ | $\underline{+32}$ | $\underline{+49}$ |

9. Add as indicated:

a. $(+8) + (-7)$     b. $(-3) + (+5)$     c. $(-6) + (+4)$
d. $(+9) + (-17)$     e. $(-2) + (+14)$     f. $(+10) + (-21)$
g. $(-18) + (+6)$     h. $(+25) + (-19)$     i. $(+6) + (-7)$
j. $(-11) + (+2)$     k. $(+34) + (-23)$     l. $(-42) + (+81)$

Add:

10. a.

| | | | | | | | |
|---|---|---|---|---|---|---|---|
| $+5$ | $-8$ | $+12$ | $-27$ | $-36$ | $+51$ | $+60$ | $-75$ |
| $\underline{-5}$ | $\underline{+8}$ | $\underline{-12}$ | $\underline{+27}$ | $\underline{+36}$ | $\underline{-51}$ | $\underline{-60}$ | $\underline{+75}$ |

b. $(-9) + (+9)$    $(+43) + (-43)$    $(-18) + (+18)$    $(+80) + (-80)$

11. a.

| | | | | | | | |
|---|---|---|---|---|---|---|---|
| $+4$ | $0$ | $-2$ | $0$ | $+15$ | $0$ | $-47$ | $0$ |
| $\underline{0}$ | $\underline{+3}$ | $\underline{0}$ | $\underline{-8}$ | $\underline{0}$ | $\underline{+26}$ | $\underline{0}$ | $\underline{-100}$ |

b. $(-11) + (0)$    $(0) + (+10)$    $(+23) + (0)$    $(0) + (17)$

12. a.

| | | | | | | | |
|---|---|---|---|---|---|---|---|
| $6$ | $4$ | $15$ | $-12$ | $2$ | $-29$ | $34$ | $96$ |
| $\underline{2}$ | $\underline{-7}$ | $\underline{-4}$ | $\underline{9}$ | $\underline{-11}$ | $\underline{17}$ | $\underline{59}$ | $\underline{-42}$ |

b. $(5) + (-9)$    $(-16) + (20)$    $(33) + (-18)$    $(-51) + (41)$

13. a.

| | | | | | | | |
|---|---|---|---|---|---|---|---|
| $-\frac{3}{16}$ | $+\frac{1}{4}$ | $+1\frac{1}{8}$ | $-2\frac{2}{3}$ | $-4\frac{1}{2}$ | $+\frac{5}{9}$ | $-5\frac{1}{6}$ | $-2\frac{11}{12}$ |
| $\underline{-\frac{9}{16}}$ | $\underline{-\frac{3}{4}}$ | $\underline{+\frac{7}{8}}$ | $\underline{+3\frac{1}{3}}$ | $\underline{-2\frac{7}{8}}$ | $\underline{-\frac{5}{9}}$ | $\underline{-4\frac{3}{5}}$ | $\underline{+6\frac{2}{3}}$ |

b. $(+\frac{5}{6}) + (-\frac{2}{3})$    $(-3\frac{4}{5}) + (-1\frac{1}{2})$    $(-7) + (+5\frac{3}{4})$    $(+2\frac{3}{8}) + (-1\frac{7}{12})$

14. a.

| | | | | | | | |
|---|---|---|---|---|---|---|---|
| $+1.3$ | $-.96$ | $+.2$ | $-3.01$ | $+1.625$ | $+83.5$ | $-2.8$ | $-.008$ |
| $\underline{+2.8}$ | $\underline{-.07}$ | $\underline{-.9}$ | $\underline{+4.16}$ | $\underline{-\ .875}$ | $\underline{+\ 7.3}$ | $\underline{+\ .97}$ | $\underline{-.192}$ |

b. $(-.5) + (-.6)$    $(+18.3) + (+8.4)$    $(+.12) + (-.7)$    $(-.94) + (+.9)$

15. Add as indicated:
    **a.** $(+6) + (-2) + (+7)$            **b.** $(-3) + (-5) + (+4)$
    **c.** $(-4) + (+8) + (-6)$            **d.** $(+2) + (+7) + (-5) + (-4)$
    **e.** $(+1) + (-9) + (+4) + (-3)$     **f.** $(-8) + (-6) + (+15) + (-2)$

16. Find the sum of each of the following. First determine the sum of the numbers named by the numerals within the brackets, then the final sum.
    **a.** $[(+2) + (-12)] + (+8)$            **b.** $[(-5) + (+7)] + (-6)$
    **c.** $[(-1) + (+1)] + (+2)$            **d.** $(-25) + [(+11) + (+16)]$
    **e.** $(+10) + [(-3) + (-7)]$           **f.** $(-4) + [(+6) + (-9)]$

17. Add:

**a.**

| +8 | +4 | +10 | +14 | + 3 | +14 | +20 | +26 |
|----|----|-----|-----|-----|-----|-----|-----|
| +3 | +1 | + 8 | + 9 | +12 | + 8 | +15 | +14 |
| +7 | +9 | +15 | + 6 | +23 | +29 | +16 | +11 |

**b.**

| −2 | −6 | − 5 | −15 | −12 | −16 | −25 | −14 |
|----|----|-----|-----|-----|-----|-----|-----|
| −5 | −8 | −10 | − 6 | −19 | −17 | −30 | −12 |
| −4 | −3 | −23 | −32 | −24 | −18 | −40 | −13 |

**c.**

| +6 | + 8 | + 9 | −8 | −12 | −15 | +8 | + 9 |
|----|-----|-----|----|-----|-----|----|-----|
| +4 | +15 | + 2 | +4 | + 6 | + 9 | −3 | −25 |
| −2 | − 7 | −11 | +5 | + 5 | +13 | +5 | + 1 |

**d.**

| −3 | − 9 | −15 | +6 | +18 | +20 | − 8 | −25 |
|----|-----|-----|----|-----|-----|-----|-----|
| −6 | − 1 | − 3 | −9 | −15 | − 3 | +15 | +29 |
| +8 | +10 | +25 | −7 | − 5 | −19 | − 7 | − 8 |

**e.**

| +8 | −9 | +6 | − 8 | +14 | +7 | −12 | + 5 |
|----|----|----|-----|-----|----|-----|-----|
| +5 | −6 | +8 | − 9 | −13 | −2 | + 9 | −19 |
| +6 | −5 | −3 | + 2 | − 5 | +6 | − 2 | + 6 |
| +3 | −9 | −7 | +15 | + 1 | −9 | + 4 | +25 |

18. First find the absolute values, then add as indicated:
    **a.** $|-6| + |-9|$            **b.** $|+7| + |-7|$
    **c.** $|-3| + |+4|$            **d.** $|-8| + |0|$

19. First add, then find the absolute value of the sum:
    **a.** $|(-2) + (+1)|$            **b.** $|(-4) + (-5)|$
    **c.** $|(+7) + (-1)|$            **d.** $|(-9) + (+9)|$

20. **a.** Does $(+9) + (-13) = |+9| + |-13|$?
    **b.** Does $(-8) + (-3) = |-8| + |-3|$?
    **c.** Does $(+6) + (+10) = |+6| + |+10|$?
    **d.** Does $(-25) + (+25) = |-25| + |+25|$?
    **e.** Does $(-4) + (-5) = |(-4) + (-5)|$?

## *APPLICATIONS*

### PART 1—GROUND SPEEDS

Just as strong winds force a person to walk either more slowly or quickly depending upon the direction of the wind, so winds decrease or increase the speed developed by a plane in flight through the air. The speed of the plane in still air is called *air speed*. A *"tail"* wind increases the air speed of a plane because it blows in the direction the plane is flying. A *"head"* wind, blowing in the opposite direction, decreases the air speed. The actual speed of the plane measured by land markings is called *ground speed*.

Find the answers to the following by means of signed numbers and illustrate graphically:

1. What is the ground speed of a plane if the air speed is 160 miles per hour (m.p.h.) and the tail wind has a velocity (speed) of 20 m.p.h.?

2. A plane develops an air speed of 340 m.p.h. Find the ground speed if it flies against a head wind of 35 m.p.h.

3. Find the actual speed of a plane headed due east at an air speed of 190 m.p.h. when a wind of 28 m.p.h. is blowing from the west.

4. What is the actual speed of a plane flying north at an air speed of 415 m.p.h. when a wind of 30 m.p.h. is coming from the north?

5. Find the ground speed of a plane meeting a head wind of 40 m.p.h. when the air speed of the plane is 225 m.p.h.

### PART 2—MISCELLANEOUS

Find the answers to the following by means of signed numbers.

1. A football team gained 3 yards on the first play, lost 4 yards on the second play, gained 6 yards on the third play, and gained 5 yards on the fourth play. Did the team make a first down?

2. How far below the surface of the water is the top of a submarine mountain called San Juan Seamount off the coast of California if the ocean floor depth is 12,000 feet and the mountain rises 10,188 feet?

3. At 8 A.M. the temperature was $-2°$. If the temperature rose 5 degrees in the next hour, what did the thermometer register at 9 A.M.?

4. A certain stock closed on Monday at the selling price of $37\frac{3}{4}$. Find the closing price of the stock on Friday if it gained $\frac{7}{8}$ point on Tuesday, lost $1\frac{1}{4}$ points on Wednesday, lost $2\frac{1}{2}$ points on Thursday, and gained $\frac{3}{8}$ point on Friday.

# Subtraction of Integers and Rational Numbers

**I. Aim**    To subtract integers and rational numbers.

**II. Procedure**

1. Subtraction is the inverse operation of addition. In subtraction we find the number (missing addend) which when added to the subtrahend (given addend) will equal the minuend (sum).

$$(+6) - (-2) = ? \text{ is thought of as } (-2) + ? = (+6)$$

2. To subtract on the number line by using vectors, see sample solution 1.
   **a.** Draw from the point marked 0 a vector above the number line for the given subtrahend and a vector below the number line for the given minuend.
   **b.** Above the number line draw a vector *from* the point reached by the vector representing the subtrahend *to* the point reached by the vector representing the minuend.
   **c.** The length of this vector is the absolute value and the direction of the vector indicates the sign of the missing number.

3. A comparison of:
   **a.** $(+6) - (-2) = +8$      **b.** $(+3) - (+9) = -6$
       $(+6) + (+2) = +8$          $(+3) + (-9) = -6$
   **c.** $(-2) - (+7) = -9$      **d.** $(-1) - (-8) = +7$
       $(-2) + (-7) = -9$          $(-1) + (+8) = +7$
   shows that *subtracting a number gives the same answer as adding its opposite.* Thus,

4. To subtract any rational number (this includes integers) from another rational number, add to the minued the opposite (or additive inverse) of the number which is to be subtracted. See sample solutions 2 through 6 indicating both the vertical and horizontal arrangements.

5. Or just change the sign of the given subtrahend mentally and add this number to the minuend.
   This method generally used in traditional mathematics, is a brief way of finding the opposite and adding it to the minued.

6. To find the value of a numerical expression that contains the subtraction of a larger arithmetic number from a smaller arithmetic number, consider the given expression as a listing of signed numbers with the addition operation understood. See sample solutions 7 through 9.

7. To find the value of a numerical expression having its first numeral prefixed by a negative sign, regard this number as a negative number. See sample solution 10. However, in sample solution 11 this negative indicates the opposite.
8. When there is successive subtraction, work in order from left to right. See sample solution 12.

## III. Sample Solutions

1. Subtract on the number line by using vectors:   $(+6) - (-2) = ?$

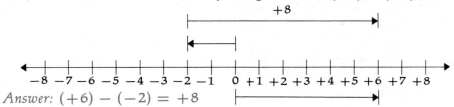

*Answer:* $(+6) - (-2) = +8$

Vertically:

2. **a.** Subtract:     Add:
   $\begin{array}{r} +5 \\ +7 \end{array} \longrightarrow \begin{array}{r} +5 \\ -7 \\ \hline -2 \end{array}$

   *Answer:* $-2$

3. **a.** Subtract:     Add:
   $\begin{array}{r} +5 \\ -7 \end{array} \longrightarrow \begin{array}{r} +5 \\ +7 \\ \hline +12 \end{array}$

   *Answer:* $+12$

4. **a.** Subtract:     Add:
   $\begin{array}{r} -5 \\ +7 \end{array} \longrightarrow \begin{array}{r} -5 \\ -7 \\ \hline -12 \end{array}$

   *Answer:* $-12$

5. **a.** Subtract:     Add:
   $\begin{array}{r} -5 \\ -7 \end{array} \longrightarrow \begin{array}{r} -5 \\ +7 \\ \hline +2 \end{array}$

   *Answer:* $+2$

6. **a.** Subtract:     Add:
   $\begin{array}{r} 0 \\ +7 \end{array} \longrightarrow \begin{array}{r} 0 \\ -7 \\ \hline -7 \end{array}$

   *Answer:* $-7$

Horizontally:

2. **b.** $(+5) - (+7) = (+5) + (-7) = -2$   *Answer:* $-2$
3. **b.** $(+5) - (-7) = (+5) + (+7) = +12$   *Answer:* $+12$
4. **b.** $(-5) - (+7) = (-5) + (-7) = -12$   *Answer:* $-12$
5. **b.** $(-5) - (-7) = (-5) + (+7) = +2$   *Answer:* $+2$
6. **b.** $0 - (+7) = 0 + (-7) = -7$   *Answer:* $-7$
7. Subtract as indicated:   $5 - 8$
   $5 - 8 = (5) - (8) = (5) + (-8) = -3$   *Answer:* $-3$

**8.** Find the value of $12 - 5 + 6 - 3$. This value may be found by arithmetic as follows:

$$12 - 5 + 6 - 3$$
$$= (12) - (5) + (6) - (3)$$
$$= \quad 7 \quad + (6) - (3)$$
$$= \quad\quad 13 \quad\quad - (3)$$
$$= \quad\quad\quad 10$$

*Answer:* 10

**9.** Find the value of $5 - 12 + 6 - 3$

$$5 - 12 + 6 - 3$$
$$= (5) - (12) + (6) - (3)$$
$$= (5) + (-12) + (6) + (-3)$$
$$= \quad (-7) \quad + (6) + (-3)$$
$$= \quad\quad (-1) \quad\quad + (-3)$$
$$= \quad\quad\quad\quad -4$$

*Answer:* $-4$

**10.** Find the value of $-5 - 7$.

$$-5 - 7$$
$$= (-5) - (7)$$
$$= (-5) + (-7)$$
$$= \quad\quad -12$$

*Answer:* $-12$

**11.** Find the value of $-(-2) - (8)$.

$$-(-2) - (8)$$
$$= (+2) - (8)$$
$$= (+2) + (-8)$$
$$= \quad\quad -6$$

*Answer:* $-6$

**12.** Subtract as indicated:

$$(+6) - (-7) - (+8)$$
$$= (+6) + (+7) - (+8)$$
$$= \quad (+13) \quad\quad - (+8)$$
$$= \quad (+13) \quad\quad + (-8)$$
$$= \quad\quad\quad\quad +5$$

*Answer:* $+5$

## DIAGNOSTIC TEST

**1.** Subtract on the number line by using vectors:

    **a.** $(+3) - (-6)$      **b.** $(+1) - (+9)$      **c.** $(-2) - (+2)$

    **d.** $(-8) - (-3)$      **e.** $(0) - (-4)$      **f.** $(-7) - (-7)$

Subtract:

| **2.** $+10$ | **3.** $+6$ | **4.** $-14$ | **5.** $-8$ | **6.** $+12$ | **7.** $+16$ |
|---|---|---|---|---|---|
| $+\ 1$ | $+9$ | $-11$ | $-21$ | $-\ 4$ | $-35$ |

| **8.** $-19$ | **9.** $-7$ | **10.** $-17$ | **11.** $0$ | **12.** $+18$ | **13.** $-32$ |
|---|---|---|---|---|---|
| $+13$ | $+30$ | $0$ | $-6$ | $+18$ | $+32$ |

Subtract as indicated:

**14.** $(+1) - (-7)$      **15.** $(-10) - (+6)$      **16.** $(+19) - (+14)$

        **17.** $(-11) - (-12)$          **18.** $(0) - (-9)$

Subtract:

**19. a.** $-1\frac{7}{8}$          **20. a.** $+\ 6.82$          **21. a.** $-26$
      $+2\frac{1}{4}$                        $+11.07$                      $\underline{\phantom{+}17}$
      $\underline{\phantom{+2\frac{1}{4}}}$          $\underline{\phantom{+11.07}}$

  **b.** $(-\frac{4}{5}) - (-\frac{5}{6})$     **b.** $(-.91) - (+1.3)$     **b.** $(50) - (-8)$

**22.** Subtract $-9$ from $-3$.

**23.** Subtract as indicated:  $7 - 16$

**24.** From 5 subtract 11.

**25.** Find the value of:  **a.** $4 - 6 + 1$   **b.** $8 - 3 - 4 - 2 + 5 - 7$

**26.** Find the value of:  $-(-9) - (11)$

**27.** Find the value of:  **a.** $-5 - 12$
  **b.** $-1 + 7 - 2 + 3 - 6 - 4 + 5$

**28.** Subtract as indicated:  $(-8) - (+9) - (-16)$

**29.** Find the value of each of the following numerical expressions. First combine the numbers in each set of parentheses or brackets:
  **a.** $(8 - 14) - (5 + 1)$          **b.** $(-7) - [(+5) - (-9)]$

**30.** Simplify by performing the indicated operations and by finding the absolute values:

$$|(-2) + (-10)| - |(-3) - (-8)|$$

**31.** Does $(+6) - (-8) = |+6| - |-8|$?

## RELATED PRACTICE EXAMPLES

**1.** Subtract on the number line by using vectors:
  **a.**  $(+7) - (+3) = ?$     **b.** $(-2) - (-8) = ?$     **c.** $(+5) - (+9) = ?$
  **d.**  $(-4) - (-1) = ?$     **e.** $(+6) - (-5) = ?$     **f.** $(-3) - (+2) = ?$
  **g.**  $(0) - (-3) = ?$      **h.** $(+7) - (+7) = ?$     **i.** $(+4) - (-4) = ?$
  **j.**  $(+6) - (+10) = ?$    **k.** $(-1) - (+6) = ?$     **l.** $(+5) - (0) = ?$
  **m.** $(+1) - (-6) = ?$     **n.** $(-9) - (-2) = ?$     **o.** $(-5) - (+8) = ?$
  **p.** $(+3) - (-1) = ?$     **q.** $(-4) - (-2) = ?$     **r.** $(+2) - (+9) = ?$

Subtract:

**2.**  $+6$      $+8$      $+7$      $+9$      $+3$      $+11$      $+15$      $+23$
      $\underline{+4}$      $\underline{+3}$      $\underline{+1}$      $\underline{+6}$      $+2$      $\underline{+\ 5}$      $\underline{+\ 8}$      $\underline{+17}$

**3.**  $+2$      $+1$      $+\ 8$      $+4$      $+\ 9$      $+\ 3$      $+16$      $+25$
      $\underline{+4}$      $\underline{+9}$      $\underline{+12}$      $\underline{+5}$      $\underline{+14}$      $\underline{+10}$      $\underline{+21}$      $\underline{+34}$

**4.**  $-8$      $-9$      $-15$      $-14$      $-7$      $-18$      $-23$      $-46$
      $\underline{-1}$      $\underline{-2}$      $\underline{-\ 8}$      $\underline{-\ 6}$      $\underline{-3}$      $\underline{-17}$      $\underline{-14}$      $\underline{-39}$

**5.**  $-3$      $-1$      $-2$      $-\ 4$      $-\ 9$      $-\ 7$      $-13$      $-24$
      $\underline{-5}$      $\underline{-6}$      $\underline{-3}$      $\underline{-14}$      $\underline{-13}$      $\underline{-18}$      $\underline{-19}$      $\underline{-31}$

| 6. | $+4$ | $+9$ | $+7$ | $+11$ | $+14$ | $+39$ | $+28$ | $+51$ |
|---|---|---|---|---|---|---|---|---|
| | $-1$ | $-3$ | $-2$ | $-10$ | $-11$ | $-27$ | $-19$ | $-43$ |

| 7. | $+6$ | $+4$ | $+9$ | $+1$ | $+7$ | $+15$ | $+46$ | $+84$ |
|---|---|---|---|---|---|---|---|---|
| | $-7$ | $-8$ | $-14$ | $-2$ | $-18$ | $-23$ | $-58$ | $-93$ |

| 8. | $-7$ | $-6$ | $-8$ | $-10$ | $-12$ | $-17$ | $-59$ | $-62$ |
|---|---|---|---|---|---|---|---|---|
| | $+2$ | $+5$ | $+6$ | $+9$ | $+4$ | $+12$ | $+41$ | $+25$ |

| 9. | $-3$ | $-4$ | $-1$ | $-8$ | $-10$ | $-16$ | $-36$ | $-47$ |
|---|---|---|---|---|---|---|---|---|
| | $+4$ | $+9$ | $+7$ | $+19$ | $+15$ | $+28$ | $+42$ | $+61$ |

| 10. | $+2$ | $+8$ | $-1$ | $-9$ | $+5$ | $-14$ | $+23$ | $-31$ |
|---|---|---|---|---|---|---|---|---|
| | $0$ | $0$ | $0$ | $0$ | $0$ | $0$ | $0$ | $0$ |

| 11. | $0$ | $0$ | $0$ | $0$ | $0$ | $0$ | $0$ | $0$ |
|---|---|---|---|---|---|---|---|---|
| | $+3$ | $+5$ | $-7$ | $-10$ | $+2$ | $-11$ | $-16$ | $+35$ |

| 12. | $+9$ | $+8$ | $-4$ | $-1$ | $+15$ | $-13$ | $+49$ | $-20$ |
|---|---|---|---|---|---|---|---|---|
| | $+9$ | $+8$ | $-4$ | $-1$ | $+15$ | $-13$ | $+49$ | $-20$ |

| 13. | $+7$ | $+12$ | $-18$ | $-8$ | $+26$ | $-1$ | $-30$ | $+47$ |
|---|---|---|---|---|---|---|---|---|
| | $-7$ | $-12$ | $+18$ | $+8$ | $-26$ | $+1$ | $+30$ | $-47$ |

Subtract as indicated:

14. **a.** $(+12) - (-8)$  15. **a.** $(-6) - (+7)$  **a.** $(+11) - (+3)$
**b.** $(+10) - (-13)$  **b.** $(-16) - (+21)$  **b.** $(+10) - (+21)$
**c.** $(+32) - (-32)$  **c.** $(-47) - (+31)$  **c.** $(+9) - (+5)$
**d.** $(+6) - (-1)$  **d.** $(-18) - (+18)$  **d.** $(+1) - (+13)$

17. **a.** $(-9) - (-7)$  **a.** $(+17) - (0)$
**b.** $(-5) - (-18)$  **b.** $(0) - (-9)$
**c.** $(-24) - (-24)$  **c.** $(-8) - (0)$
**d.** $(-15) - (-19)$  **d.** $(0) - (+25)$

Subtract:

| 19. **a.** | $+\frac{3}{4}$ | $-\frac{5}{6}$ | $-\frac{9}{16}$ | $+2\frac{1}{6}$ | $+1\frac{7}{8}$ | $-\frac{5}{12}$ | $+6\frac{3}{5}$ | $-2\frac{5}{6}$ |
|---|---|---|---|---|---|---|---|---|
| | $+\frac{1}{4}$ | $+1\frac{1}{6}$ | $-\frac{7}{8}$ | $-4\frac{1}{3}$ | $+2\frac{1}{2}$ | $+1\frac{2}{3}$ | $-4$ | $-3\frac{3}{8}$ |

**b.** $(-\frac{1}{2}) - (+\frac{1}{2})$   $(-3\frac{2}{3}) - (-7\frac{3}{4})$   $(+\frac{5}{16}) - (-\frac{11}{12})$   $(+1\frac{3}{7}) - (+2\frac{1}{9})$

| 20. **a.** | $-.5$ | $-1.6$ | $+.4$ | $+4.13$ | $-7.86$ | $+.312$ | $-62.1$ | $+81.24$ |
|---|---|---|---|---|---|---|---|---|
| | $-.8$ | $+.9$ | $+8.3$ | $-3.69$ | $-2.57$ | $+.059$ | $+59.9$ | $-74.58$ |

**b.** $(+.03) - (-.07)$ $(+1.9) - (+.21)$ $(-.84) - (+.16)$ $(-5.9) - (-5.9)$

| 21. **a.** | $4$ | $8$ | $-3$ | $9$ | $-17$ | $21$ | $35$ | $28$ |
|---|---|---|---|---|---|---|---|---|
| | $7$ | $11$ | $6$ | $-5$ | $13$ | $9$ | $-40$ | $52$ |

**b.** $(2) - (14)$    $(-4) - (5)$    $(16) - (-25)$    $(-31) - (27)$

22. **a.** From $-2$ subtract $+10$    **b.** Subtract $-5$ from $+9$
**c.** From $-16$ take $-26$    **d.** Take $-11$ from $0$

   e. From $+18$ subtract $+51$  
   g. Subtract $+144$ from $-56$  
   i. From $-19$ subtract $+105$  
   f. From $+40$ take $-40$  
   h. Take $-67$ from $-67$  
   j. Take $+287$ from $+198$  

Subtract as indicated:

23. **a.** $6 - 9$    **b.** $2 - 11$    **c.** $1 - 8$    **d.** $12 - 21$  
    **e.** $5 - 17$    **f.** $18 - 40$    **g.** $37 - 53$    **h.** $14 - 32$  
    **i.** $56 - 85$    **j.** $79 - 97$    **k.** $128 - 200$    **l.** $67 - 105$  

24. **a.** From 4 subtract 7         **b.** From 9 take 13  
    **c.** Subtract 16 from 5       **d.** Take 20 from 14  
    **e.** From 36 subtract 95     **f.** Take 100 from 46  
    **g.** Subtract 53 from 18     **h.** From 328 take 509  
    **i.** Subtract 162 from 26    **j.** From 291 subtract 912  

25. Find the value of each of the following numerical expressions:  
    **a.** $6-5-7$               **b.** $4-9+2$  
    **c.** $3+2-10$             **d.** $11-4+3$  
    **e.** $1-6-3$               **f.** $3-5-1+6$  
    **g.** $7-8+4-8$           **h.** $1-7+2-4$  
    **i.** $2-5-9+3$          **j.** $5+8-12+1$  
    **k.** $8-2-5+3-7$      **l.** $2-4-3+9-1$  
    **m.** $6+1-8-3-5$      **n.** $4-2-6+5+9$  
    **o.** $3-8-4+7+6-2$    **p.** $5-1+3-10+2-4$  
    **q.** $7+11-9-8+3-5+1$    **r.** $2-3-5+6-12+3-7+2$  
    **s.** $4-6-3+15-2-8+7-6$    **t.** $9+2-7+5-4-6-2+3-8$  

26. Find the value of each of the following numerical expressions:  
    **a.** $-(-7) - (4)$    **b.** $-(-9) - (-5)$    **c.** $-(2) - (-10)$  
    **d.** $-(16) - (9)$    **e.** $-(-6) + (-8)$    **f.** $-(0) - (-12)$  
    **g.** $-(-10) - (15)$    **h.** $-(-5) + (-11)$    **i.** $-(-1) - (-1)$  

27. Find the value of each of the following numerical expressions:  
    **a.** $-2-3$                  **b.** $-9+6$  
    **c.** $-12-8$               **d.** $-7-2-4$  
    **e.** $-8+3-10+5$       **f.** $-5+9-3-6-12$  
    **g.** $-1-4-2+5-15+11$   **h.** $-3+7-3-4-8+6-4$  
    **i.** $-2-4+8-9-1+13-5$   **j.** $-5+6-1-4-5+7-10$  

28. Subtract as indicated:  
    **a.** $(-7) - (+3) - (-9)$      **b.** $(-4) - (-12) - (-8)$  
    **c.** $(2) - (-10) - (6)$        **d.** $-(+1) - (+1) - (-2)$  
    **e.** $(-11) - (+6) - (-5)$     **f.** $-(-17) - (-9) - (+8)$  

29. Find the value of each of the following numerical expressions. First combine the numbers in each set of parentheses or brackets.  
    **a.** $(3 + 2) - (4 + 6)$      **b.** $(6 - 7) - (3 - 1)$  
    **c.** $(4 - 4) - (8 + 5)$      **d.** $(2 + 10) - (1 - 6)$

    **e.** $(15 - 21) + (3 - 12)$         **f.** $(5 - 7) + (1 - 4) - (8 - 2)$
    **g.** $(6 - 1) - (-3 + 5) + (4 - 7)$   **h.** $(9 - 5) + (4 + 1) - (2 - 8)$
    **i.** $8 + (4 - 9) - 3 - (7 - 2)$     **j.** $7 - (5 - 1) - (2 - 6) - 4$
    **k.** $[(-8) - (-7)] - (-10)$        **l.** $(-15) - [(-2) - (+9)]$
    **m.** $[(-4) + (-12)] - (-15)$     **n.** $(+2) - [(0) - (-4)]$

30. Simplify:

    **a.** $|-6| - |-10|$             **b.** $|+9| - |-3|$
    **c.** $|2 - 12|$               **d.** $|-4 + 7|$
    **e.** $-|17 - 33|$           **f.** $|8 - 11| - |5 + 4|$
    **g.** $|-9 - 8| - 16$        **h.** $7 - |6 - 14|$
    **i.** $|(0 - 3) - (9 - 10)|$     **j.** $|(-1) - (-8)| - |(-2) + (-5)|$

31. **a.** Does $(-8) - (+9) = |-8| - |+9|$?
    **b.** Does $(-3) - (-7) = |-3| - |-7|$?
    **c.** Does $(+10) - (-1) = |+10| - |-1|$?
    **d.** Does $(+13) - (+20) = |+13| - |+20|$?
    **e.** Does $(+7) - (-9) = |(+7) - (-9)|$?

## APPLICATIONS

PART 1—INTERESTING FACTS

Find the answers by means of signed numbers:

1. An ocean depth of 35,640 feet in the Marianas Trench near the island of Guam is the deepest place in the Pacific Ocean. The greatest depth in the Atlantic Ocean is 30,246 feet located near Puerto Rico. Find the difference in their depths.

2. In Asia, Mount Everest is the highest point with an elevation of 29,002 feet and the Dead Sea is the lowest point at 1,286 feet below sea level. What is the difference in their altitudes?

3. Mount McKinley in Alaska with an elevation of 20,300 feet is the highest point in North America. Death Valley in California at 282 feet below sea level is the lowest point. Find the difference.

4. If 1 degree of latitude is equivalent to 60 nautical miles, how far does a ship sail due north from Perth, Australia, 32° south latitude to Hong Kong, China, 22° north latitude?

5. What is the difference between the longitude of Stockholm, Sweden, 18° east longitude and Denver, Colorado, 105° west longitude. Find the difference in sun time if 1 degree of longitude is equivalent to 4 minutes of time.

## PART 2—CHANGES AND DEVIATIONS

**1.** Find the net changes in the prices of the following stocks and indicate them by means of signed numbers:

| Stock | Close Tues. | Close Wed. | Net Change | Stock | Close Tues. | Close Wed. | Net Change |
|---|---|---|---|---|---|---|---|
| Allied Chemical | $50\frac{3}{4}$ | $50\frac{7}{8}$ | | Gulf Oil | $41\frac{1}{4}$ | $41\frac{1}{2}$ | |
| Atlantic Richfield | 94 | $94\frac{7}{8}$ | | RCA | $22\frac{5}{8}$ | $22\frac{5}{8}$ | |
| General Electric | $50\frac{1}{4}$ | $50\frac{1}{2}$ | | Sears | $16\frac{7}{8}$ | 17 | |
| General Motors | $47\frac{1}{2}$ | $47\frac{7}{8}$ | | Sperry Rand | $47\frac{1}{8}$ | $47\frac{3}{4}$ | |
| Grace | $36\frac{3}{4}$ | 37 | | U.S. Steel | 19 | $19\frac{3}{8}$ | |

**2.** Using the hourly temperature report, find:

**a.** How many degrees the temperature dropped during the hour from 3 A.M. to 4 A.M.

| Time | Temp. | Time | Temp. |
|---|---|---|---|
| 1 A.M. | 8 | 7 A.M. | −6 |
| 2 A.M. | 6 | 8 A.M. | −5 |
| 3 A.M. | 2 | 9 A.M. | −2 |
| 4 A.M. | −1 | 10 A.M. | 2 |
| 5 A.M. | −3 | 11 A.M. | 5 |
| 6 A.M. | −7 | Noon | 9 |

**b.** What temperature change occurred from 6 A.M. to 7 A.M.

**c.** How many degrees the temperature rose during the hour from 9 A.M. to 10 A.M.

**d.** What temperature change occurred from 2 A.M. to 3 A.M.

**3. a.** The record high temperature in Cleveland is 103° and the record low is −17°. Find the difference in temperatures.

**b.** The lowest temperature recorded in Omaha is −32°. The record low in Boston is −18°. What is the difference in the temperature lows?

**c.** The normal temperature in an eastern city for the month of February is 31.2°. The temperature for the same month this year is 29.6°. Represent this deviation from normal by means of a signed number.

**d.** The normal precipitation for the month of April in a western city is 1.6 inches. Represent the deviation from normal by a signed number if the precipitation for April of last year was 2.3 inches.

# Multiplication of Integers and Rational Numbers

I. **Aim**    To multiply integers and rational numbers.

II. **Procedure**

1. Observe that a $\times$ symbol or a raised dot ($\cdot$) may be used to indicate the operation of multiplication or the factors may be written next to each other within parentheses without any multiplication symbol.
   For example:
   Eight times seven may be written as $8 \times 7$, $8 \cdot 7$, $(8) \times (7)$, $(8)(7)$, $8(7)$ or $(8)7$.

2. Multiplication may be thought of as repeated addition.
   $3 \times 2$ is the sum of three 2's or $(2) + (2) + (2) = 6$

3. Since a numeral without a sign represents a positive number, $3 \times 2$ may be represented as $(+3)(+2)$ or $(3)(+2)$
   Thus, $(+3)(+2)$ or $(3)(+2) = (+2) + (+2) + (+2) = +6$
   That is: $(+3)(+2) = +6$
   This is illustrated on the number line as:

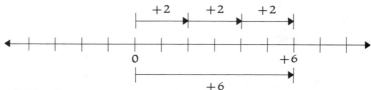

The product of two positive numbers is a positive number.

4. **a.** Also $(+3)(-2)$ or $(3)(-2) = (-2) + (-2) + (-2) = -6$
   That is: $(+3)(-2) = -6$
   This is illustrated on the number line as:

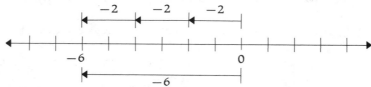

   **b.** Since $(+2)(-3) = (-3)(+2)$ and $(+2)(-3) = -6$, it follows that $(+2)(-3) = -6$
   The product of a positive number and a negative number is a negative number.

5. Observe in the following sequence that the product increases 2 units from left to right as the multiplier decreases 1 unit.

$$
\begin{array}{cccc}
-2 & -2 & -2 & -2 \\
\underline{+3} & \underline{+2} & \underline{+1} & \underline{\phantom{+}0} \\
-6 & -4 & -2 & 0
\end{array}
$$

It is logical to expect that the product should increase 2 units each time the multiplier decreases 1 unit.

$$
\begin{array}{cccc}
-2 & -2 & -2 & -2 \\
\underline{\phantom{-}0} & \underline{-1} & \underline{-2} & \underline{-3} \\
0 & +2 & +4 & +6
\end{array}
$$

Thus, $(-3)(-2) = +6$

The product of two negative numbers is a positive number.

6. _To multiply two positive numbers or two negative numbers,_ find the product of their absolute values and prefix the numeral for this product with a positive sign.

7. _To multiply a positive number by a negative number_ or _a negative number by a positive number,_ find the product of their absolute values and prefix the numeral for this product with a negative sign.

8. The product of zero and any positive or negative number is zero.

9. To multiply three or more factors, multiply successively. Observe that provided zero is not a factor.

    **a.** An odd number of negative factors produces a negative number as the product. Thus, an odd power of a negative number is a negative number.

    **b.** An even number of negative factors produces a positive number as the product. Thus, an even power of a negative number is a positive number.

        In each case the absolute value of the product is equal to the product of the absolute values of the given factors.

10. If the product of two numbers is one (1), then each factor is said to be the _multiplicative inverse_ or the _reciprocal_ of the other like 6 and $\frac{1}{6}$ or $\frac{2}{5}$ and $\frac{5}{2}$.

## III. Sample Solutions

Multiply in each of the following:

$$
\begin{array}{llll}
\textbf{1.} \ +4 & \textbf{2.} \ -8 & \textbf{3.} \ +6 & \textbf{4.} \ -3 \\
\underline{\phantom{1}+3} & \underline{\phantom{2}-2} & \underline{\phantom{3}-5} & \underline{\phantom{4}-7} \\
+12 & +16 & -30 & +21 \\
\end{array}
$$

| | | | | | |
|---|---|---|---|---|---|
| **1.** $+4$ | **2.** $-8$ | **3.** $+6$ | **4.** $-3$ | **5.** $\phantom{+}0$ | **6.** $+7$ |
| $\underline{+3}$ | $\underline{-2}$ | $\underline{-5}$ | $\underline{-7}$ | $\underline{-8}$ | $\underline{\phantom{+}0}$ |
| $+12$ | $+16$ | $-30$ | $+21$ | $0$ | $0$ |
| _Answers:_ $+12$ | $+16$ | $-30$ | $+21$ | $0$ | $0$ |

**7.** $(-9)(+2) = -18$
_Answer:_ $-18$

**8.** $(-3)(-1)(-4)(-2) = +24$
_Answer:_ $+24$

**9.** $(-4)(+3)(-2)(-5) = -120$
_Answer:_ $-120$

**10.** $(-2)(-4)(-1)(0)(-6) = 0$
_Answer:_ $0$

**11.** $(-5)^3 = (-5)(-5)(-5) = -125$
_Answer:_ $-125$

## DIAGNOSTIC TEST

1. Read or write in words:
   **a.** $9 \cdot 6$      **b.** $(-11)(+15)$      **c.** $8(-5)$
2. Find the answers by means of signed numbers:
   **a.** How many more amperes of electricity will a battery have in 5 hours if it is being charged at the rate of 8 amperes per hour?
   **b.** How many more amperes of electricity did a battery have 7 hours ago if it is being discharged at the rate of 6 amperes per hour?

   Multiply:
3. **a.** $+9$
   $\underline{+7}$

   **b.** $(+18)(+3)$
4. **a.** $-2$
   $\underline{+8}$

   **b.** $(+1)(-7)$
5. **a.** $+10$
   $\underline{-\ 6}$

   **b.** $(-12)(+15)$
6. **a.** $-11$
   $\underline{-11}$

   **b.** $(-5)(-9)$
7. **a.** $\ \ 0$
   $\underline{-2}$

   **b.** $(0)(+14)$
8. **a.** $\ \ 16$
   $\underline{-\ 6}$

   **b.** $(20)(-10)$
9. **a.** $(-\frac{5}{8})(+72)$

   **b.** $(-2\frac{2}{5})(-3\frac{3}{4})$
10. **a.** $+4.9$
    $\underline{-7.5}$

    **b.** $(+.04)(+.02)$
11. **a.** $(-1)(+2)(-1)(-2)$      **b.** $(-4)(-3)(-3)(-4)$
12. Find the value of:
    **a.** $(-2)^5$      **b.** $(-3)^4$
13. Find the value of each of the following numerical expressions:
    **a.** $[(-2) \times (+1)] \times (-4)$
    **b.** $(+10) \times [(-9) + (+3)]$
    **c.** $[(-5)(-6)] + [(-5)(+8)]$
14. Simplify by finding the absolute values and performing the indicated operations:
    $$|(+8)(-1)| \times |(-2)(-4)|$$
15. Find the value of each of the following numerical expressions:
    **a.** $6(-12)$      **b.** $-7(4-5)$
16. Does $(-8) \times (+9) = |-8| \times |+9| = ?$
17. What is the multiplicative inverse of: $-\frac{3}{4}$?   $+10$?   $-1\frac{2}{3}$?

## *RELATED PRACTICE EXAMPLES*

**1.** Read, or write in words, each of the following:

**a.** $4 \cdot 5$             **b.** $(+6) \times (-2)$

**c.** $(-20)(+34)$       **d.** $9(-12)$

**e.** $15 \times 23$         **f.** $(-72)(-17)$

**g.** $(46)7$            **h.** $30 \cdot 59$

**i.** $(81) \times (29)$      **j.** $(-58)(+63)$

**2. a.** Representing a rise of 2 degrees in temperature by $+2°$, a fall of 2 degrees in temperature by $-2°$, 5 hours from now by $+5$, and 5 hours ago by $-5$, find the answers to the following situations using signed numbers.

(1) The temperature is rising at the constant rate of 2 degrees per hour $(+2°)$. How will the temperature 5 hours from now $(+5)$ compare with the present temperature?

(2) The temperature is falling at the constant rate of 2 degrees per hour $(-2°)$. How will the temperature 5 hours from now $(+5)$ compare with the present temperature?

(3) The temperature rose at the constant rate of 2 degrees per hour $(+2°)$. How did the temperature 5 hours ago $(-5)$ compare with the present temperature?

(4) The temperature fell at the constant rate of 2 degrees per hour $(-2°)$. How did the temperature 5 hours ago $(-5)$ compare with the present temperature?

**b.** Representing a bank deposit of $6 by $+6$, a withdrawal of $6 by $-6$, 3 weeks from now by $+3$, and 3 weeks ago by $-3$, find the answers to the following situations using signed numbers:

(1) If you deposit $6 each week in your bank account, how will your bank account 3 weeks from now compare with your present bank account?

(2) If you deposited $6 each week in your bank account, how did your bank account 3 weeks ago compare with your present bank account?

(3) If you withdraw $6 each week from your bank account, how will your bank account 3 weeks from now compare with your present bank account?

(4) If you withdrew $6 each week from your bank account, how did your bank account 3 weeks ago compare with your present bank account?

Multiply:

**3. a.**

| $+6$ | $+9$ | $+7$ | $+8$ | $+5$ | $+3$ | $+8$ | $+12$ |
|------|------|------|------|------|------|------|-------|
| $+5$ | $+4$ | $+3$ | $+8$ | $+9$ | $+6$ | $+1$ | $+\ 4$ |

**b.** $(+8)(+7)$     $(+5)(+15)$     $(+10)(+9)$     $(+36)(+40)$

4. **a.**

| $-9$ | $-8$ | $-12$ | $-7$ | $-3$ | $-24$ | $-17$ | $-19$ |
|------|------|-------|------|------|-------|-------|-------|
| $+6$ | $+7$ | $+5$ | $+11$ | $+12$ | $+8$ | $+4$ | $+10$ |

   **b.** $(+8)(-9)$   $(+2)(-18)$   $(+16)(-5)$   $(+30)(-60)$

5. **a.**

| $+8$ | $+3$ | $+7$ | $+9$ | $+5$ | $+2$ | $+10$ | $+15$ |
|------|------|------|------|------|------|-------|-------|
| $-2$ | $-9$ | $-6$ | $-4$ | $-5$ | $-14$ | $-8$ | $-3$ |

   **b.** $(-5)(+6)$   $(-9)(+14)$   $(-22)(+12)$   $(-17)(+20)$

6. **a.**

| $-5$ | $-7$ | $-6$ | $-2$ | $-8$ | $-12$ | $-13$ | $-20$ |
|------|------|------|------|------|-------|-------|-------|
| $-4$ | $-7$ | $-8$ | $-6$ | $-9$ | $-3$ | $-5$ | $-8$ |

   **b.** $(-3)(-3)$   $(-11)(-10)$   $(-17)(-5)$   $(-48)(-31)$

7. **a.**

| $+4$ | $0$ | $0$ | $-5$ | $0$ | $-16$ | $0$ | $+32$ |
|------|-----|-----|------|-----|-------|-----|-------|
| $0$ | $+8$ | $-1$ | $0$ | $+9$ | $0$ | $-20$ | $0$ |

   **b.** $(0)(-11)$   $(+6) \times 0$   $(0)(+25)$   $0 \times (-19)$

8. **a.**

| $8$ | $6$ | $4$ | $-3$ | $12$ | $-27$ | $35$ | $81$ |
|-----|-----|-----|------|------|-------|------|------|
| $4$ | $-2$ | $-9$ | $7$ | $-12$ | $8$ | $14$ | $-50$ |

   **b.** $(-5)(8)$   $(11)(-9)$   $(-13)(30)$   $(48)(-25)$

9. $(-\frac{1}{3})(-12)$   $(-\frac{5}{8})(+\frac{1}{2})$   $(+1\frac{4}{5})(-4\frac{5}{6})$   $\frac{1}{2}(-\frac{11}{16})$

   $(-1\frac{1}{2})(+2\frac{2}{3})$   $(-8\frac{3}{4})(-3\frac{1}{7})$   $(-2\frac{3}{16})(+1\frac{1}{4})$   $\frac{3}{4}(-4\frac{4}{9})$

10. **a.**

| $-.3$ | $+.8$ | $-.2$ | $-5.8$ | $+1.06$ | $-96.5$ | $-3.79$ | $-.018$ |
|-------|-------|-------|--------|---------|---------|---------|---------|
| $-.3$ | $-.9$ | $+7$ | $-2.4$ | $-.32$ | $+85.6$ | $-.07$ | $+.049$ |

   **b.** $(10)(-.4)$   $(-.8)(-.01)$   $(+.03)(-1.9)$   $(-.25)(100)$

11. **a.** $(+3)(-2)(-3)$   **b.** $(-1)(-1)(-1)$

   **c.** $(-2)(+1)(+8)$   **d.** $(-4)(-5)(+6)(-3)$

   **e.** $(-6)(-4)(-4)(-2)$   **f.** $(+5)(-4)(0)(-2)$

   **g.** $(-3)(-3)(-3)(-3)(-3)$   **h.** $(+2)(-1)(-2)(+5)(-1)$

   **i.** $(-2)(+1)(-3)(-1)(+2)(-4)$   **j.** $(-1)(-1)(-1)(-1)(-1)(-1)$

12. Find the value of each of the following:

   **a.** $(-3)^3$   **b.** $(-5)^2$   **c.** $(-1)^2$   **d.** $(-4)^3$   **e.** $(-2)^4$

   **f.** $(-1)^3$   **g.** $(-6)^3$   **h.** $(-2)^6$   **i.** $(-3)^5$   **j.** $(-1)^{10}$

13. Find the value of each of the following numerical expressions:

   **a.** $[(+3) \times (-2)] \times (-4)$   **b.** $(-1) \times [(+7) \times (-3)]$

   **c.** $[(-7) \times (-5)] \times (+2)$   **d.** $(+4) \times [(+2) \times (-6)]$

   **e.** $(-3) \times [(-5) + (-8)]$   **f.** $(+5) \times [(-9) + (-2)]$

   **g.** $[(-6) \times (-1)] + [(-6) \times (-3)]$   **h.** $(-4) \times [(-3) \times (+1)]$

   **i.** $(-4) \times [(-3) + (+1)]$   **j.** $[(+7)(-2)] + [(+7)(-4)]$

14. Simplify by finding the absolute values and performing the indicated operations:

   **a.** $|(-6)(+2)|$   **b.** $|-5| \times |-8|$

   **c.** $(-7) \times |-9|$   **d.** $|1 - 5| \times |3 + 4|$

   **e.** $|(-4)(+3)| \times |(-9)(-1)|$   **f.** $(-16) \times |(-2)(-4)(-3)|$

15. Find the value of each of the following numerical expressions:
    a. $4(-6)$                   b. $8(-9)$                    c. $-2(-8)$
    d. $-5(+7)$                  e. $\frac{2}{7}(-14)$         f. $-\frac{1}{6}(-9)$
    g. $6(2 + 1)$                h. $7(4 - 4)$                 i. $-5(8 + 2)$
    j. $-12(8 - 11)$             k. $\frac{2}{3}(3 - 7)$       l. $-\frac{3}{5}(15 + 5)$
    m. $(9 + 7)(6 - 8)$          n. $(4 - 9)(2 - 11)$          o. $(-1 - 2)(3 - 5)$

16. a. Does $(-4) \times (-2) = |-4| \times |-2|$?
    b. Does $(-3) \times (+15) = |-3| \times |+15|$?
    c. Does $(+6) \times (+7) = |+6| \times |+7|$?
    d. Does $(+1) \times (-4) = |+1| \times |-4|$?
    e. Does $(-9) \times (+3) = |(-9) \times (+3)|$?

17. a. What is the product of each of the following?
    $(-7)(-\frac{1}{7})$            $(-\frac{3}{2})(-\frac{2}{3})$            $(+\frac{5}{8})(+\frac{8}{5})$            $(-2\frac{3}{4})(-\frac{4}{11})$
    b. In each case is the product of the two non-zero rational numbers one (1)?
    c. What is a multiplicative inverse?
    d. Does each non-zero rational number have a multiplicative inverse?
    e. What is the multiplicative inverse of:
    $-8$?        $+13$?        $-\frac{3}{5}$?        $+\frac{7}{12}$?        $-\frac{19}{6}$?        $-4\frac{2}{9}$?        $+3\frac{1}{4}$?

## CUMULATIVE REVIEW

1. On a number line, draw the graph of: $-3,\ -1,\ 0,\ +3$.

2. Which sentences are true?   a. $-6 > -4$   b. $+1 < -5$   c. $+7 \not< +7$

3. Find the additive inverse of:   a. $-17$   b. $+\frac{3}{10}$   c. $-.06$

4. Find the multiplicative inverse of:   a. $+20$   b. $-\frac{4}{5}$   c. $+2\frac{2}{3}$

5. Find the value of:   a. $|-9| = ?$   b. $-(-5) = ?$

6. Compute as indicated:
    a. $(-12) + (-8)$              b. $(-4) - (-6)$              c. $(-6)(-9)$
    d. $(+7) - (-10)$             e. $-3(+11)$                  f. $(+5) + (-7)$
    g. $(+6) - (+9)$             h. $(+4)(-5)$                  i. $(-2)^3$
    j. $(-3)(-3)(-3)(-3)$         k. $(-1)(+5)(-2)(-1)(+3)$

7. Subtract the product of $-6$ and $+8$ from the product of $-5$ and $-9$.

8. Multiply the sum of $-7$ and $+3$ by the difference between $-11$ and $-1$.

# Division of Integers and Rational Numbers

**I. Aim**    To divide integers and rational numbers.

**II. Procedure**

1. Division is the inverse operation of multiplication. In division we find the factor (quotient) which multiplied by the given factor (divisor) will equal the given product (dividend).

2. **a.** Since $(+2) \times (+3) = +6$
    then $(+6) \div (+2) = +3$
    or $\dfrac{+6}{+2} = +3$

   **b.** Since $(-2) \times (+3) = -6$
    then $(-6) \div (-2) = +3$
    or $\dfrac{-6}{-2} = +3$

   Observe that the quotient of two positive numbers is a positive number and the quotient of two negative numbers is a positive number.

3. **a.** Since $(+2) \times (-3) = -6$
    then $(-6) \div (+2) = -3$
    or $\dfrac{-6}{+2} = -3$

   **b.** Since $(-2) \times (-3) = +6$
    then $(+6) \div (-2) = -3$
    or $\dfrac{+6}{-2} = -3$

   Observe that the quotient of a negative number divided by a positive number or the quotient of a positive number divided by a negative number is a negative number.

4. *To divide a positive number by a positive number* or *a negative number by a negative number,* divide their absolute values and prefix the numeral for this quotient with a positive sign.

5. *To divide a positive number by a negative number* or *a negative number by a positive number,* divide their absolute values and prefix the numeral for this quotient with a negative sign.

6. Any non-zero number divided by itself is one (1).

7. Zero divided by either a positive number or by a negative number is zero. We cannot divide a positive number or a negative number by zero.

**III. Sample Solutions**

Divide each of the following:

1. $\dfrac{+10}{+5} = ?$    *Answer:* $+2$

2. $\dfrac{-21}{-7} = ?$    *Answer:* $+3$

3. $\dfrac{-20}{+4} = ?$    *Answer:* $-5$

4. $\dfrac{+24}{-12} = ?$    *Answer:* $-2$

5. $\dfrac{-9}{+9} = ?$    *Answer:* $-1$

6. $\dfrac{0}{-7} = ?$    *Answer:* $0$

7. $(-15) \div (-3) = ?$    *Answer:* $+5$

## DIAGNOSTIC TEST

Divide:

**1. a.** $\dfrac{+16}{+2}$  **2. a.** $\dfrac{-60}{-5}$  **3. a.** $\dfrac{-63}{+9}$

  **b.** $(+42) \div (+7)$  **b.** $(-28) \div (-4)$  **b.** $(-85) \div (+17)$

**4. a.** $\dfrac{+15}{-1}$  **5. a.** $\dfrac{-9}{-9}$  **6. a.** $\dfrac{0}{-12}$

  **b.** $(+52) \div (-3)$  **b.** $(+15) \div (-15)$  **b.** $(0) \div (+1)$

**7. a.** $\dfrac{-24}{6}$  **8. a.** $\left(-\frac{7}{16}\right) \div \left(+\frac{1}{2}\right)$  **9. a.** $\dfrac{-5.4}{-.6}$

  **b.** $(56) \div (-8)$  **b.** $\left(-3\frac{3}{5}\right) \div \left(-1\frac{1}{8}\right)$  **b.** $(-.09) \div (+.3)$

**10. a.** $-5\overline{)-45}$
  **b.** $+2\overline{)-100}$

Find the value of each of the following numerical expressions:

**11. a.** $[(-125) \div (-25)] \div (+5)$  **b.** $(-60) \div [(+20) + (-10)]$

**12. a.** $\dfrac{14 - 28}{-7}$  **b.** $\dfrac{(-15) + (-27)}{(-2)(-3)}$

**13.** Simplify by finding the absolute values and performing the indicated operations:

$$\dfrac{|6 - 4| - |3 - 10|}{|(-5)(-2)|}$$

**14.** Does $(+56) \div (-8) = |+56| \div |-8|$?

## RELATED PRACTICE EXAMPLES

Divide:

**1. a.** $\dfrac{+8}{+4}$  $\dfrac{+6}{+1}$  $\dfrac{+25}{+5}$  $\dfrac{+24}{+6}$  $\dfrac{+42}{+14}$  $\dfrac{+44}{+11}$  $\dfrac{+100}{+20}$  $\dfrac{+80}{+16}$

  **b.** $(+50) \div (+5)$  $(+36) \div (+4)$  $(+72) \div (+12)$  $(+350) \div (+25)$

**2. a.** $\dfrac{-21}{-7}$  $\dfrac{-56}{-8}$  $\dfrac{-27}{-3}$  $\dfrac{-60}{-10}$  $\dfrac{-48}{-2}$  $\dfrac{-54}{-9}$  $\dfrac{-85}{-5}$  $\dfrac{-132}{-12}$

  **b.** $(-20) \div (-4)$  $(-84) \div (-6)$  $(-37) \div (-1)$  $(-105) \div (-15)$

3. **a.** $\dfrac{-30}{+6}$ $\quad$ $\dfrac{-14}{+7}$ $\quad$ $\dfrac{-8}{+1}$ $\quad$ $\dfrac{-45}{+9}$ $\quad$ $\dfrac{-84}{+12}$ $\quad$ $\dfrac{-72}{+24}$ $\quad$ $\dfrac{-90}{+30}$ $\quad$ $\dfrac{-98}{+49}$

$\quad$ **b.** $(-75) \div (+5)$ $\quad$ $(-81) \div (+3)$ $\quad$ $(-54) \div (+18)$ $\quad$ $(-135) \div (+45)$

4. **a.** $\dfrac{+35}{-5}$ $\quad$ $\dfrac{+72}{-9}$ $\quad$ $\dfrac{+28}{-4}$ $\quad$ $\dfrac{+46}{-2}$ $\quad$ $\dfrac{+45}{-15}$ $\quad$ $\dfrac{+64}{-8}$ $\quad$ $\dfrac{+81}{-27}$ $\quad$ $\dfrac{+100}{-25}$

$\quad$ **b.** $(+9) \div (-3)$ $\quad$ $(+63) \div (-7)$ $\quad$ $(+112) \div (-16)$ $\quad$ $(+108) \div (-12)$

5. **a.** $\dfrac{+4}{+4}$ $\quad$ $\dfrac{-4}{-4}$ $\quad$ $\dfrac{-4}{+4}$ $\quad$ $\dfrac{+4}{-4}$ $\quad$ $\dfrac{-1}{-1}$ $\quad$ $\dfrac{+10}{-10}$ $\quad$ $\dfrac{-8}{+8}$ $\quad$ $\dfrac{+17}{+17}$

$\quad$ **b.** $(-7) \div (+7)$ $\quad$ $(-12) \div (-12)$ $\quad$ $(+23) \div (+23)$ $\quad$ $(+36) \div (-36)$

6. **a.** $\dfrac{0}{+5}$ $\quad$ $\dfrac{0}{-9}$ $\quad$ $\dfrac{0}{+2}$ $\quad$ $\dfrac{0}{-1}$ $\quad$ $\dfrac{0}{-25}$ $\quad$ $\dfrac{0}{+11}$ $\quad$ $\dfrac{0}{-73}$ $\quad$ $\dfrac{0}{+100}$

$\quad$ **b.** $(0) \div (-6)$ $\quad$ $0 \div (+3)$ $\quad$ $0 \div (-15)$ $\quad$ $(0) \div (+42)$

7. **a.** $\dfrac{-10}{2}$ $\quad$ $\dfrac{81}{-9}$ $\quad$ $\dfrac{23}{-1}$ $\quad$ $\dfrac{-45}{15}$ $\quad$ $\dfrac{63}{-7}$ $\quad$ $\dfrac{84}{-12}$ $\quad$ $\dfrac{-200}{25}$ $\quad$ $\dfrac{96}{-16}$

$\quad$ **b.** $(18) \div (-3)$ $\quad$ $(-50) \div (10)$ $\quad$ $(91) \div (-13)$ $\quad$ $(-144) \div (24)$

8. $(+\tfrac{3}{4}) \div (-\tfrac{1}{8})$ $\quad$ $(-\tfrac{2}{3}) \div (+4)$ $\quad$ $(+1\tfrac{1}{2}) \div (+\tfrac{1}{6})$ $\quad$ $(-1\tfrac{7}{8}) \div (-1\tfrac{1}{4})$

$\quad$ $(-\tfrac{5}{8}) \div (-\tfrac{5}{6})$ $\quad$ $(-20) \div (-3\tfrac{1}{3})$ $\quad$ $(+6\tfrac{1}{4}) \div (-2\tfrac{1}{2})$ $\quad$ $(-9\tfrac{1}{3}) \div (+2\tfrac{4}{5})$

9. $\dfrac{-7.5}{+5}$ $\quad$ $\dfrac{+8.1}{-.3}$ $\quad$ $\dfrac{+.48}{-4}$ $\quad$ $\dfrac{-8}{-1.6}$ $\quad$ $\dfrac{-1.9}{100}$ $\quad$ $\dfrac{+9.5}{+.05}$ $\quad$ $\dfrac{-36}{+.04}$ $\quad$ $\dfrac{-.45}{-.15}$

$\quad$ $(+3.8) \div (-.2)$ $\quad$ $(-.56) \div (-1.4)$ $\quad$ $(-.06) \div (+.6)$ $\quad$ $(-1.7) \div (-.85)$

10. $+7\overline{)+42}$ $\quad$ $-9\overline{)+72}$ $\quad$ $-6\overline{)-54}$ $\quad$ $+12\overline{)-96}$ $\quad$ $-.2\overline{)+2}$

$\quad$ $-10\overline{)+90}$ $\quad$ $-8\overline{)-56}$ $\quad$ $5\overline{)-65}$ $\quad$ $+.4\overline{)-2.8}$ $\quad$ $-3\overline{)-51}$

11. Find the value of each of the following numerical expressions:

$\quad$ **a.** $[(-16) \div (-8)] \div (-2)$ $\qquad$ **b.** $(-27) \div [(-9) \div (-3)]$

$\quad$ **c.** $(+64) \div [(-8) + (+4)]$ $\qquad$ **d.** $[(-10) + (-2)] \div (+6)$

$\quad$ **e.** $[(-36) \div (+9)] \div (-4)$ $\qquad$ **f.** $(-72) \div [(-24) \div (+3)]$

$\quad$ **g.** $[(+100) \div (-20)] \div (-5)$ $\qquad$ **h.** $[(-80) + (+30)] \div (-10)$

$\quad$ **i.** $(-90) \div [(+45) + (+15)]$ $\qquad$ **j.** $(+144) \div [(-16) \div (-4)]$

12. Find the value of each of the following numerical expressions:

$\quad$ **a.** $\dfrac{8 + 6}{7}$ $\qquad$ **b.** $\dfrac{3 - 7}{2}$ $\qquad$ **c.** $\dfrac{5 - 1}{-4}$

$\quad$ **d.** $\dfrac{3 - 15}{-6}$ $\qquad$ **e.** $\dfrac{-8 - 2}{-2}$ $\qquad$ **f.** $\dfrac{-11 + 5}{-3}$

$\quad$ **g.** $\dfrac{(-6)(-12)}{(-4) + (-4)}$ $\qquad$ **h.** $\dfrac{(-9) - (-5)}{(+3)(-2)}$ $\qquad$ **i.** $\dfrac{(11 - 3) + (-6 - 9)}{(4 - 5) - (8 - 2)}$

13. Simplify by finding the absolute values and performing the indicated operations:

a. $\dfrac{|-12|}{6}$  b. $\dfrac{|-27|}{|-3|}$  c. $\left|\dfrac{-56}{-7}\right|$

d. $\dfrac{|-10| \times |-5|}{|-2| + |+3|}$  e. $\dfrac{|(-6)(+9)|}{|3 - 8| - |-7|}$  f. $\dfrac{|9 - 1| + |4 - 5|}{|(-1)(-1)|}$

14. a. Does $(-20) \div (+4) = |-20| \div |+4|$?
 b. Does $(+45) \div (+9) = |+45| \div |+9|$?
 c. Does $(-28) \div (-2) = |-28| \div |-2|$?
 d. Does $(+63) \div (-7) = |+63| \div |-7|$?
 e. Does $(-100) \div (-20) = |(-100) \div (-20)|$?

## CUMULATIVE REVIEW

1. Compute as indicated:
 a. $(-3)(-4)$  b. $(+6) + (-11)$  c. $(+56) \div (-7)$
 d. $(0) - (-9)$  e. $(-72) \div (-8)$  f. $-9(+3)$
 g. $(-12) + (-6)$  h. $(-1)^4$  i. $(-2) - (+5)$
2. Add the quotient of $-63$ divided by $+7$ to the product of $-3$ and $+8$.
3. Subtract the product of $-4$ and $-2$ from the sum of $-6$ and $-7$.
4. Find the value of $-8 - 5 + 3 - 4 + 6 + 9$.
5. The hourly temperatures for a 12-hour period were 12°, 7°, 5°, 2°, 0°, $-1°$, $-3°$, $-4°$, $-2°$, 0°, 3°, and 5°. What was the average temperature?

## COMPETENCY CHECK TEST

In each of the following, select the letter corresponding to your answer:
1. The sum of $+6$ and $-9$ is:
 a. $+15$  b. $-15$  c. $+3$  d. $-3$
2. When $-8$ is subtracted from $-4$, the answer is:
 a. $-12$  b. $-4$  c. $+4$  d. $+12$
3. The product of $-2$ and $-6$ is:
 a. $-12$  b. $-8$  c. $+12$  d. $+8$
4. The quotient of $-42$ divided by $+7$ is:
 a. $-35$  b. $-6$  c. $-49$  d. $+5$
5. The fullback in five attempts gained 6 yards, gained 3 yards, lost 2 yards, gained 7 yards, and lost 4 yards. His average gain per try is:
 a. 10 yd.  b. 2 yd.  c. 3 yd.  d. answer not given

# 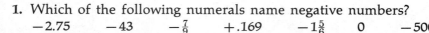REVIEW

1. Which of the following numerals name negative numbers?

   $-2.75$    $-43$    $-\frac{7}{9}$    $+.169$    $-1\frac{5}{6}$    $0$    $-500$

2. Which of the following numerals name integers? Rational numbers? Irrational numbers? Real numbers?

   $-.1$    $+2\frac{1}{8}$    $-13$    $+\frac{25}{5}$    $-\sqrt{13}$    $-\frac{11}{16}$

3. On a number line, draw the graph of: $-8, -4, -1, 0, +3, +5$.

4. Write the coordinates of which the following is the graph:

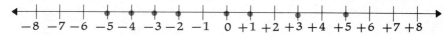

$$\begin{array}{ccccccccccccccccc} -8 & -7 & -6 & -5 & -4 & -3 & -2 & -1 & 0 & +1 & +2 & +3 & +4 & +5 & +6 & +7 & +8 \end{array}$$

5. If $+500$ feet represents 500 feet above sea level, what does $-225$ feet represent?

6. Which of the following sentences are true?

   **a.** $-9 < -7$    **b.** $-15 > 0$    **c.** $-24 \not< -24$    **d.** $-6 \not> +1$

7. Name the following numbers in order of size (greatest first):

   $-1$    $+6$    $-10$    $+12$    $0$    $-8$    $+14$    $-11$    $+9$    $-5$

8. **a.** What is the opposite of $-70$?

   **b.** What is the additive inverse of $+.2$?

   **c.** Find the absolute value:   (1) $|+57| = ?$    (2) $|-\frac{7}{12}| = ?$

   **d.** Find the value of:   (1) $-(+1\frac{2}{3}) = ?$    (2) $-(-64) = ?$

   **e.** Find the multiplicative inverse of:   (1) $-\frac{1}{5}$   (2) $-20$   (3) $+\frac{8}{9}$

9. Using the number line as a scale, draw the vector and write the numeral represented by it that indicates a movement from:

   **a.** $-2$ to $-4$    **b.** $+8$ to $0$    **c.** $+1$ to $+9$    **d.** $-3$ to $+2$

10. Add:   **a.** $\begin{array}{r} -3 \\ -3 \\ \hline \end{array}$    **b.** $\begin{array}{r} +14 \\ -21 \\ \hline \end{array}$    **c.** $(-27) + (+27)$    **d.** $(-47) + (+54)$

11. Subtract:   **a.** $\begin{array}{r} -5 \\ -18 \\ \hline \end{array}$    **b.** $\begin{array}{r} 0 \\ +20 \\ \hline \end{array}$    **c.** $(-37) - (+37)$    **d.** $(16) - (-23)$

12. Multiply:   **a.** $\begin{array}{r} -15 \\ +4 \\ \hline \end{array}$    **b.** $\begin{array}{r} -9 \\ -9 \\ \hline \end{array}$    **c.** $(-1)(+40)$    **d.** $(10)(-35)$

13. Divide:   **a.** $\dfrac{-100}{-4}$    **b.** $\dfrac{-24}{+6}$    **c.** $(18) \div (-1)$    **d.** $(-60) \div (12)$

Find the value of each of the following numerical expressions:

14. **a.** $-9 + 4 - 7 + 10 - 8$    **b.** $(3 - 4) - (12 - 2) - (1 - 7)$

15. $\dfrac{(-40) - (+16)}{(+2)(-4)}$

## Keyed Achievement Test

The numeral at the end of each problem indicates the section where help may be found.

1. Write symbolically: Five times each number $x$ decreased by seven is greater than forty-eight. 1-7
2. Write as a formula: The cutting speed ($s$) of a bandsaw in feet per minute is equal to Pi ($\pi$) times the diameter ($d$) times the number of revolutions per minute ($R$). 1-8
3. On a number line, draw the graph of: $-7, -3, -2, 0, 1, 2, 4.$ 2-7
4. a. Find the absolute value of: (1) $|-36| = ?$ (2) $|+\frac{9}{25}| = ?$ 2-3
   b. Find the opposite of: (1) $+57$ (2) $-8$ (3) $-\frac{4}{7}$ 2-4
   c. Find the value of: (1) $-(+2\frac{1}{2}) = ?$ (2) $-(-61) = ?$ 2-4
   d. Find the additive inverse of: (1) $-5\frac{3}{4}$ (2) $+.04$ (3) $-101$ 2-4
   e. Find the multiplicative inverse of: (1) $-15$ (2) $+\frac{11}{13}$ (3) $-1\frac{1}{2}$ 2-13
5. Compute as indicated:
   a. Add:
   $(+7) + (-15)$     $(-9) + (-11)$     $(-21) + (+21)$ 2-11
   b. Subtract:
   $(-12) - (+9)$     $(-10) - (-16)$     $14 - 35$ 2-12
   c. Multiply:
   $(+8)(-6)$     $(-5)(-7)$     $(-4)(-3)(-2)$ 2-13
   d. Divide:
   $(-20) \div (+5)$     $(+36) \div (-36)$     $(-54) \div (-6)$ 2-14

## MAINTENANCE PRACTICE IN ARITHMETIC

1. Add:
93,874
28,659
57,969

2. Subtract:
405,216
387,156

3. Multiply:
836
967

4. Divide:
$498\overline{)400,890}$

5. Add:
$\frac{7}{16} + 3\frac{3}{4} + 1\frac{5}{8}$

6. Subtract:
$12 - 5\frac{2}{3}$

7. Multiply:
$8\frac{1}{4} \times 2\frac{5}{6}$

8. Divide:
$56 \div \frac{7}{8}$

9. Add:
$3.9 + .26 + 17$

10. Subtract:
$.04 - .004$

11. Multiply:
$.003 \times 500$

12. Divide:
$12 \div .8$

13. Find 9% of $8,530.

14. 22 is what per cent of 40?

15. 90% of what number is 63?

# INTRODUCTION

Many pupils in their seventh and eighth year mathematics have substituted numerical values for variables in a formula to determine a required value. This process is called *evaluating a formula*. In the formula $A = lw$, $A = 10$ when $l = 5$ and $w = 2$. However, when $l = 4$ and $w = 3$, $A$ is equal to 12. The value of the expression $lw$ depends upon the values of the variables $l$ and $w$. In general, the value of any algebraic expression depends upon the values of the variables it contains. If these values change, the value of the expression usually changes.

Since this process of substituting given values in a given expression is used throughout the study of algebra, it is of utmost importance that algebra students acquire skill in its use. Evaluation will be used in conjunction with formulas, dependence, graphs of equations and formulas, checking equations, inequalities, and checking answers of examples done by algebraic processes.

In this unit we study the evaluation of elementary expressions, expressions with parentheses, expressions with exponents, and formulas. It should be noted that parentheses are used to group two or more numbers so that they may be treated as a single quantity. Exponents are used to indicate the power of a given number.

# Evaluation

**3**

# Simple Expressions–Using Numbers of Arithmetic

**I. Aim**  To find the value of simple algebraic expressions when numbers of arithmetic are assigned to the variables.

**II. Procedure**

1. Rewrite the given expression.

2. Substitute in order the given numerical value for each variable.

3. Perform the necessary operations as indicated to get the answer. See sample solutions 1, 2, 3 and 4.

4. When there is a numerical coefficient, it is rewritten and used as a factor to find the numerical value of the expression. See sample solution 5.

5. When there is more than one term (part of expression separated) by the + or − sign, first find the value of each part. Then combine as indicated. See sample solutions 6 and 7.

6. If the expression is a fraction, simplify both the numerator and denominator separately, and then express the fraction in simplest terms. See sample solution 8.

**III. Sample Solutions**

Find the value of each of the following expressions when $a = 4$ and $b = 2$.

| **1.** $a + b$ | **2.** $a - b$ | **3.** $ab$ | **4.** $\dfrac{a}{b}$ |
|---|---|---|---|
| $= 4 + 2$ | $= 4 - 2$ | $= 4 \cdot 2$ | $= \frac{4}{2}$ |
| $= 6$ | $= 2$ | $= 8$ | $= 2$ |
| *Answer: 6* | *Answer: 2* | *Answer: 8* | *Answer 2* |

**5.** $5ab$
$= 5 \cdot 4 \cdot 2$
$= 40$

*Answer: 40*

**6.** $a + 6b$
$= 4 + 6 \cdot 2$
$\quad\downarrow \qquad \downarrow$
$= 4 + \;\; 12$
$= 16$

*Answer: 16*

**7.** $2ab - 3b + 5$
$= 2 \cdot 4 \cdot 2 - 3 \cdot 2 + 5$
$\qquad\downarrow \qquad\quad \downarrow \qquad \downarrow$
$= \quad 16 \quad - \quad 6 \; + 5$
$= 15$

*Answer: 15*

**8.** $\dfrac{4a - 3b}{3ab - 4} = \dfrac{4 \cdot 4 - 3 \cdot 2}{3 \cdot 4 \cdot 2 - 4} = \dfrac{16 - 6}{24 - 4} = \dfrac{10}{20} = \dfrac{1}{2}$  *Answer: $\frac{1}{2}$*

## DIAGNOSTIC TEST

Find the value of:
1. $a + x$, when $a = 11$ and $x = 17$
2. $n - y$, when $n = 12$ and $y = 3$
3. $cd$, when $c = 4$ and $d = 19$    4. $\dfrac{v}{a}$, when $v = 54$ and $a = 18$

5. $12bc$, when $b = 7$ and $c = 2$    6. $n + 11$, when $n = 9$
7. $18 - x$, when $x = 15$    8. $b + 6h$, when $b = 3$
9. $11x + 5y$, when $x = 6$ and $y = 9$    and $h = 8$
10. $9t - 3z$, when $t = 7$ and $z = 12$
11. $4cm + 7m$, when $c = 10$ and $m = 3$
12. $8bn - 4cn$, when $b = 7$, $c = 2$, and $n = 5$
13. $15a - 6b + 6$, when $a = 12$ and $b = 9$
14. $3cx + 11c - 2x$, when $c = 8$ and $x = 1$

15. $\dfrac{18a}{y}$, when $a = 15$ and $y = 6$    16. $\dfrac{5c - d}{c + 2d}$, when $c = 9$ and $d = 1$

17. $\dfrac{8m - 2n + 4}{m + 4n - 7}$, when $m = 3$ and $n = 2$

18. Using each replacement for the variable in the expression $2x - 5$, find the corresponding list of values of the algebraic expression when the replacements for $x$ are: 3, 4, 5, . . . , 10.

## *RELATED PRACTICE EXAMPLES*

Find the value of:
1. **a.** $x + y$, when $x = 8$ and $y = 5$    **b.** $c + d$, when $c = 1$ and $d = 9$
  **c.** $m + n$, when $m = \frac{3}{4}$ and $n = \frac{5}{8}$    **d.** $b + h$, when $b = 1.6$ and $h = .82$
2. **a.** $x - y$, when $x = 8$ and $y = 5$    **b.** $r - s$, when $r = 3$ and $s = 10$
  **c.** $m - n$, when $m = 2\frac{1}{2}$    **d.** $b - a$, when $b = .46$
  and $n = 1\frac{2}{3}$    and $a = -.7$
3. **a.** $xy$, when $x = 8$ and $y = 5$    **b.** $cp$, when $c = 12$ and $p = 9$
  **c.** $rt$, when $r = 30$ and $t = 4\frac{1}{2}$    **d.** $bh$, when $b = 2.5$ and $h = 1.5$
4. **a.** $\dfrac{x}{y}$, when $x = 8$ and $y = 5$    **b.** $\dfrac{c}{d}$, when $c = 6$ and $d = 24$

  **c.** $\dfrac{a}{m}$, when $a = 90$ and $m = 18$    **d.** $\dfrac{v}{t}$, when $v = 3.6$ and $t = .15$

5. **a.** $8x$, when $x = 23$    **b.** $\frac{1}{2}bh$, when $b = 14$ and $h = 3\frac{1}{2}$
  **c.** $19ac$, when $a = 9$ and $c = 12$    **d.** $1.3mny$, when $m = .08$, $n = 5.4$, and $y = 20$

6. **a.** $b+7$, when $b=25$     **b.** $n+9$, when $n=15$
  **c.** $2\frac{1}{2}+a$, when $a=3\frac{5}{6}$     **d.** $.05+c$, when $c=1.75$

7. **a.** $d-19$, when $d=31$     **b.** $2-x$, when $x=0$
  **c.** $n-4$, when $n=5\frac{1}{2}$     **d.** $7-y$, when $y=1.8$

8. **a.** $12+9g$, when $g=32$     **b.** $h+5n$, when $h=10$ and $n=6$
  **c.** $m-\frac{1}{3}x$, when $m=2\frac{1}{4}$ and $x=\frac{3}{8}$     **d.** $a+6b$, when $a=4.9$ and $b=.3$

9. **a.** $5b+3x$, when $b=9$ and $x=2$     **b.** $2n+7r$, when $n=3$ and $r=8$
  **c.** $9a+4t$, when $a=\frac{1}{2}$ and $t=\frac{2}{3}$     **d.** $12v+1.6y$, when $v=2.1$ and $y=.06$

10. **a.** $3c-5n$, when $c=8$ and $n=2$     **b.** $5g-4t$, when $g=11$ and $t=6$
  **c.** $7h-2a$, when $h=3$ and $a=1\frac{1}{2}$     **d.** $.06s-1.1b$, when $s=20$ and $b=.7$

11. **a.** $5bx+6x$, when $b=7$ and $x=3$
  **b.** $3ax+4anxy$, when $a=6$, $n=3$, $x=4$, and $y=1$
  **c.** $7dw+\frac{1}{2}tw$, when $d=\frac{1}{3}$, $t=1\frac{1}{2}$, and $w=\frac{3}{4}$
  **d.** $10cnr+.7cr$, when $c=.8$, $n=1.4$, and $r=.5$

12. **a.** $6ny-4d$, when $d=5$, $n=7$, and $y=4$
  **b.** $15cx-7dxy$, when $c=14$, $d=2$, $x=6$, and $y=5$
  **c.** $\frac{3}{4}bt-12$, when $b=1\frac{1}{3}$ and $t=15$
  **d.** $8.3mn-.7abn$, when $a=.2$, $b=.1$, $m=.9$, and $n=1.3$

13. **a.** $6a+5x-4$, when $a=11$ and $x=3$
  **b.** $12c-17d-20$, when $c=5$ and $d=2$
  **c.** $3h-10m+\frac{5}{16}$, when $h=\frac{7}{8}$ and $m=\frac{1}{5}$
  **d.** $.9b-.4c-3.8$, when $b=5$ and $c=.7$

14. **a.** $6by-7b+8y$, when $b=12$ and $y=2$
  **b.** $18a-5ac-2acd$, when $a=4$, $c=2$, and $d=1$
  **c.** $\frac{1}{2}mn-\frac{5}{8}nr+\frac{3}{4}mnr$, when $m=\frac{1}{2}$, $n=\frac{1}{4}$, and $r=\frac{1}{3}$
  **d.** $9p+.7st-4pt$, when $p=4.6$, $s=.08$, and $t=.5$

15. **a.** $\dfrac{b}{2}$, when $b=56$     **b.** $\dfrac{15x}{y}$, when $x=7$ and $y=5$

  **c.** $\dfrac{m}{10}$, when $m=2\frac{1}{2}$     **d.** $\dfrac{6.3a}{b}$, when $a=.8$ and $b=1.4$

16. **a.** $\dfrac{m+n}{m-n}$, when $m=18$ and $n=6$     **b.** $\dfrac{5cd-7d}{15}$, when $c=20$ and $d=10$

  **c.** $\dfrac{4a+9c}{a-c}$, when $a=1\frac{1}{2}$ and $c=\frac{1}{2}$     **d.** $\dfrac{5xy-4y}{3x-2xy}$, when $x=8$ and $y=.6$

17. **a.** $\dfrac{3a-7x+6}{2a+6x-5}$, when $a=8$ and $x=2$

**b.** $\dfrac{6b - 3y + 5}{4b + 9y - 12}$, when $b = \frac{1}{2}$ and $y = 2$

**c.** $\dfrac{7c - 2cd + 11d}{9c - 4}$, when $c = 9$ and $d = 1$        **d.** $\dfrac{8m - 2mx}{4m - 5mx + 2x}$, when $m = .5$ and $x = .4$

18. Using each given replacement for the variable, find the corresponding list of values of each of the following algebraic expressions:

   **a.** $x + 8$, when replacements for $x$ are: 3, 6, 9, 12

   **b.** $4n$, when replacements for $n$ are: 0, 1, 2, . . . , 6

   **c.** $6b + 11$, when replacements for $b$ are: 2, 4, 6, 8, 10

   **d.** $y - 5$, when replacements for $y$ are: 5, 6, 7, . . . , 12

   **e.** $10 - a$, when replacements for $a$ are: 0, 1, 2, . . . , 10

   **f.** $3m - 2$, when replacements for $m$ are: 1, 5, 11, 18

   **g.** $21 - 9c$, when replacements for $c$ are: 0, 1, 2

   **h.** $\frac{1}{4}d$, when replacements for $d$ are: 2, 4, 6, . . . , 16

   **i.** $\dfrac{5t}{7}$, when replacements for $t$ are: 0, 7, 21, 35, 63

   **j.** $\dfrac{3a - 7}{2}$, when replacements for $a$ are: 3, 4, 5, . . . , 9

## *CUMULATIVE REVIEW*

1. Compute as indicated:

   **a.** $(-5)(-8)$         **b.** $(+9) + (-10)$         **c.** $(-3) - (-3)$

   **d.** $(-21) \div (+7)$         **e.** $(-6)(+9)$         **f.** $(-15) - (+8)$

   **g.** $(-2) + (-4)$         **h.** $(-60) \div (-12)$         **i.** $(-5)(+5)(-5)$

2. Find the value of each of the following numerical expressions:

   **a.** $8 - 5 + 3$         **b.** $-4 + 3 + 7 - 9 - 5 + 8$

   **c.** $(6 + 9) + (8 - 8) + (3 - 10)$         **d.** $\frac{2}{3}(2 - 8)$

   **e.** $(7 - 4) + (4 - 7) - (6 - 14)$

   **f.** $(2 - 5) - (11 - 3) - (1 - 8)$         **g.** $\dfrac{(-35) - (-21)}{(-2)(-7)}$

   **h.** $10 - (5 - 7) - 6 + (2 - 9)$

   **i.** $(-4 - 3)(-4 - 2)$         **j.** $\dfrac{6(-4) - 8(-9)}{(-5) + (-2)}$

   **k.** $-3(1 - 8) - (3 - 4)$

   **l.** $9 - 4(7 - 5) + 2(6 - 1)$         **m.** $\dfrac{7(8 - 2) + 2(1 - 7)}{-3(-8 - 2)}$

3. Find the value of:

   **a.** $8cd$ when $c = 7$ and $d = 5$         **b.** $6s - 5t$ when $s = 8$ and $t = 3$

4. Using each replacement for the variable in the expression $4n + 1$, find the corresponding list of values of the algebraic expression when the replacements for $n$ are: 0, 1, 2, . . . , 8.

# Simple Expressions–Using Positive and Negative Numbers

**EXERCISE 3-2**

**I. Aim**    To evaluate an algebraic expression when positive and negative numbers are assigned to the variables.

**II. Procedure**

1. Follow the same procedure as with numbers of arithmetic Exercise 3-1.
2. Use parentheses to separate the signed numbers.
3. When necessary, use operation rules for signed numbers.

**III. Sample Solutions**

Find the value of the following expressions if $a = 4$, $b = -2$, and $c = -5$:

**1.**  $b + c$
$= (-2) + (-5)$
$= -7$
*Answer: $-7$*

**2.**  $ac$
$= (4)(-5)$
$= -20$
*Answer: $-20$*

**3.**  $-4ab$
$= -4(4)(-2)$
$= 32$
*Answer: $32$*

**4.**  $b - c$
$= (-2) - (-5)$
$= +3$
*Answer: $+3$*

**5.**  $\dfrac{a}{b}$
$= \dfrac{4}{-2}$
$= -2$
*Answer: $-2$*

**6.**  $2a - 3bc + 5ac$
$= 2(4) - 3(-2)(-5) + 5(4)(-5)$
$= 8 - 30 - 100$
$= -122$
*Answer: $-122$*

---

## DIAGNOSTIC TEST

Find the value of the following algebraic expressions if $a = 8$, $b = -2$, and $c = -1$:

**1.** $a + b$        **2.** $a - b$        **3.** $ab$        **4.** $\dfrac{a}{b}$

**5.** $b + c$        **6.** $b - c$        **7.** $bc$        **8.** $\dfrac{b}{c}$

**9.** $3b$        **10.** $-5c$        **11.** $3ac$        **12.** $2a - 5b + 3c$

**13.** $5ac - 2ab - 3bc$    **14.** $4bc - ac + 5$    **15.** $\dfrac{2ab - 3bc}{2c}$    **16.** $\dfrac{3ac - 4bc + 8}{3a - ab}$

**100**

## *RELATED PRACTICE EXAMPLES*

Find the value of each of the following algebraic expressions:

**A.** When $m=6$, $n=-4$, and $x=-2$

1. $m+n$
2. $m-n$
3. $mn$
4. $\dfrac{m}{n}$
5. $n+x$
6. $n-x$

7. $nx$
8. $\dfrac{n}{x}$
9. $5n$
10. $-4x$
11. $2mx$
12. $4m-6n$

13. $2mn-nx+3mx$
14. $5mx-4x-2$
15. $\dfrac{2m-3x}{3}$
16. $\dfrac{4n-5x+16}{m-n}$

**B.** When $a=-2$, $x=-10$, and $y=5$

1. $x+y$
2. $y-x$
3. $xy$
4. $\dfrac{x}{y}$
5. $a+x$
6. $a-x$

7. $ax$
8. $\dfrac{a}{x}$
9. $4x$
10. $-5a$
11. $-6ay$
12. $5a-4x-3y$

13. $3ax-2xy-6ay$
14. $4ay-ax-1$
15. $\dfrac{3ax-2xy}{2ax}$
16. $\dfrac{2xy+ax-5ay}{4ax-3xy}$

## *MISCELLANEOUS EXAMPLES*

1. Find the value of:
   a. $b-r$, when $b=-\frac{3}{8}$ and $r=-\frac{5}{6}$
   b. $-5am$, when $a=-2\frac{1}{2}$ and $m=-8$
   c. $3c-4x+6y$, when $c=-1\frac{2}{3}$, $x=-3\frac{1}{2}$, and $y=-\frac{3}{4}$
   d. $4dn+5dx-2nx$, when $d=2\frac{1}{2}$, $n=-1\frac{1}{4}$, and $x=-\frac{2}{5}$
   e. $2g-3n$, when $g=-.8$ and $n=-.4$
   f. $bc-bd$, when $b=-1.4$, $c=.9$, and $d=-2.3$
   g. $2r-s+5t$, when $r=2.5$, $s=-3$, and $t=-1.8$
   h. $3cg+8ch-7gh$, when $c=-.2$, $g=-.7$, and $h=.3$
2. Using each given replacement for the variable, find the corresponding list of values of each of the following algebraic expressions:
   a. $n+9$, when replacements for $n$ are: $-4$, $-2$, $0$, $1$, $3$
   b. $-7y$, when replacements for $y$ are: $-9$, $-4$, $-1$, $0$, $2$, $6$
   c. $12+10b$, when replacements for $b$ are: $-6$, $-5$, $\ldots$, $-1$
   d. $5z-8$, when replacements for $z$ are: $-7$, $-3$, $-1$, $0$, $1$, $2$, $6$
   e. $16-2y$, when replacements for $y$ are: $-8$, $-5$, $-2$, $0$, $1$, $5$, $8$, $10$
   f. $-\frac{2}{3}t$, when replacements for $t$ are: $-6$, $-3$, $0$, $1$, $6$, $9$
   g. $\dfrac{6d+3}{7-8d}$, when replacements for $d$ are: $-3$, $-2$, $-1$, $0$, $1$, $2$

# Expressions with Parentheses— Using Numbers of Arithmetic

**I. Aim**   To evaluate an algebraic expression containing parentheses when numbers of arithmetic are assigned to the variables.

## II. Procedure

1. Follow the same procedure as explained in Exercise 3-1.
2. However, first evaluate each set of parentheses separately and use the result as a single term.
3. Then complete as directed. A numeral written immediately next to the parentheses implies multiplication.

## III. Sample Solutions

Find the value of each of the following expressions when $b=5$, $c=3$, and $x=2$:

**1.** $b(3c+x)$
$=5(3\cdot3+2)$
$=5(9+2)$
$=5(11)$
$=55$
*Answer: 55*

**2.** $4x-(2b-7)$
$=4\cdot2-(2\cdot5-7)$
$=8-(10-7)$
$=8-(3)$
$=5$
*Answer: 5*

**3.** $b+c(c+x)$
$=5+3(3+2)$
$=5+3(5)$
$=5+15$
$=20$
*Answer: 20*

**4.** $(b+c)(c+x)$
$=(5+3)(3+2)$
$=(8)(5)$
$=40$
*Answer: 40*

---

### DIAGNOSTIC TEST

Find the value of each of the following algebraic expressions when $a=6$, $c=4$, and $x=2$:

1. $3(a+c)$
2. $6(c-x)$
3. $(2a+3x)$
4. $2(4a+3c)$
5. $c(a+x)$
6. $x(4c-5x)$
7. $5c(2a+7)$
8. $a(ac+cx)$
9. $ax(6c-7)$
10. $3acx(2ac-6cx+5)$
11. $2a+(c+x)$
12. $2a-(c+x)$
13. $a+c+(a+x)$
14. $a+c(a+x)$
15. $(a+c)(a+x)$
16. $7a-c(a+x)$
17. $4(a+c)-x(2a-5x)$
18. $\dfrac{3(2a+3cx)}{4(7c-ac)}$

---

## RELATED PRACTICE EXAMPLES

Find the value of each of the following algebraic expressions:

**A.** When $b=4$, $c=3$, and $d=1$

1. $4(b+d)$
2. $6(c-d)$
3. $(3b+5d)$
4. $2(3b-4c)$
5. $c(b-d)$
6. $b(3c+2d)$
7. $3d(2c-b)$
8. $c(bd-cd)$
9. $bd(2b-3)$
10. $4bcd(3cd-b)$
11. $5b+(c-d)$
12. $5b-(c-d)$

**102**

13. $b+d+(b+c)$     14. $b+d(b-c)$     15. $(b-d)(b-c)$

16. $b-d(b-c)$     17. $3(5bd-4d)-c(cd+d)$     18. $\dfrac{4(bc-2d)}{9(2bc-4)}$

**B.** When $a=8$, $x=4$, and $y=2$     1. $3(x+y)$     2. $2(a-1)$

3. $(3a-2y)$     4. $4(a-3y)$     5. $x(a+6)$     6. $a(3x-2y)$

7. $5x(2a-5y)$     8. $x(ax+ay+a)$     9. $xy(2a+3x-2)$

10. $6axy(2ay-x+ax)$     11. $3a+(ax-y)$     12. $4x-(xy+2)$

13. $2a-x+(2x-6)$     14. $2a+x(2a-6)$     15. $(2a-x)(2x-6)$

16. $2a-x(2x-6)$     17. $2a(xy+2)-3x(a-2y)$     18. $\dfrac{5(x+3y)}{3x(a+3x-2xy)}$

## MISCELLANEOUS EXAMPLES

1. Find the value of:

  **a.** $\frac{1}{2}(c+h)$, when $c=5\frac{2}{3}$ and $h=2\frac{1}{6}$

  **b.** $a(d-n)$, when $a=\frac{3}{8}$, $d=1\frac{1}{5}$, and $n=\frac{1}{4}$

  **c.** $(b-\frac{1}{2})(c-\frac{3}{4})$, when $b=1\frac{1}{2}$ and $c=2\frac{1}{4}$

  **d.** $2n+x(n+x)$, when $n=3\frac{1}{3}$ and $x=\frac{1}{2}$

  **e.** $\dfrac{(a-n)(a+n)}{an}$, when $a=4\frac{1}{8}$ and $n=\frac{5}{8}$

  **f.** $.9(m+y)$, when $m=.8$ and $y=.4$

  **g.** $.02b(b-x)$, when $b=1.2$ and $x=1.1$

  **h.** $a+b(a-b)$, when $a=4.3$ and $b=3.7$

  **i.** $(m+n)(m-n)$, when $m=1.01$ and $n=.62$

  **j.** $\dfrac{2(a+3c)}{3(3a-c)}$, when $a=.4$ and $c=.2$

2. Using each given replacement for the variable, find the corresponding list of values of each of the following algebraic expressions:

  **a.** $3(a+2)$, when replacements for $a$ are: 0, 1, 2, . . . , 8

  **b.** $4b(b+7)$, when replacements for $b$ are: 0, 1, 2, . . . , 6

  **c.** $2x(3x-2)$, when replacements for $x$ are: 2, 5, 8, 12, 20

  **d.** $(y+4)(y+8)$, when replacements for $y$ are: 0, 1, 2, . . . , 8

  **e.** $n+6(n-1)$, when replacements for $n$ are: 1, 2, 3, . . . , 9

  **f.** $3m+(2m+9)$, when replacements for $m$ are: 0, 2, 4, . . . , 14

  **g.** $8s-(3s-2)$, when replacements for $s$ are: 1, 2, 3, 6, 7, 10

# Expressions with Parentheses—Using Positive and Negative Numbers

## I. Aim

To evaluate an algebraic expression containing parentheses when positive and negative numbers are assigned to the variables.

## II. Procedure

1. Follow the same procedure as explained in Exercise 3-3.
2. A numeral written immediately next to the parentheses implies multiplication.

## III. Sample Solutions

Find the value of each of the following expressions when $m = -6$, $n = -2$, and $x = 3$.

1. $2m(3n - 5x)$
   $= (2[-6])(3[-2] - 5[3])$
   $= (-12)(-6 - 15)$
   $= (-12)(-21)$
   $= +252$

   *Answer: 252*

2. $m + x + (m - n)$
   $= (-6) + (3) + ([-6] - [-2])$
   $= (-6) + (3) + (-4)$
   $= -7$

   *Answer: −7*

---

### DIAGNOSTIC TEST

Find the value of each of the following algebraic expressions when $a = 4$, $m = -5$, and $x = -2$.

1. $2(a + m)$
2. $-5(a - m)$
3. $-4(m + x)$
4. $(2a + 5x)$
5. $a(m - 2x)$
6. $x(am + 5m)$
7. $2mx(3m - 4ax)$
8. $5x + (4m - 6mx)$
9. $a + m + (m - x)$
10. $a + m(m - x)$
11. $(a + m)(m - x)$
12. $\dfrac{2(3a - 6x)}{a(m - x)}$

---

### RELATED PRACTICE EXAMPLES

Find the value of each of the following algebraic expressions:

A. When $a = 2$, $b = -4$, and $c = -1$

1. $3(a + b)$
2. $-4(a - c)$
3. $-2(b + c)$

4. $(2a - 5b)$

5. $a(b - 6c)$

6. $b(3a + b)$

7. $3ac(2a - 5bc)$

8. $2b + (a - c)$

9. $a + b + (a - c)$

10. $a + b(a - c)$

11. $(a + b)(a - c)$

12. $\dfrac{4(a + b)}{a(b - 2c)}$

**B.** When $x = 8$, $y = -4$, and $z = -2$

1. $4(y + z)$

2. $-5(x - z)$

3. $-6(y + z)$

4. $(3x + 7y)$

5. $x(y - 2z)$

6. $z(xz - 2y)$

7. $-4yz(xy + 2xz)$

8. $3x + (y + 5z)$

9. $x - z + (2y + z)$

10. $x - z(2y + z)$

11. $(x - z)(2y + z)$

12. $\dfrac{x(y + 2z)}{4(x + xz)}$

### *MISCELLANEOUS EXAMPLES*

Find the value of:

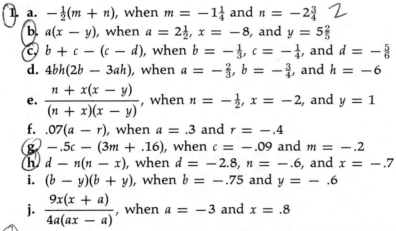

**1.**  **a.** $-\frac{1}{2}(m + n)$, when $m = -1\frac{1}{4}$ and $n = -2\frac{3}{4}$

**b.** $a(x - y)$, when $a = 2\frac{1}{2}$, $x = -8$, and $y = 5\frac{2}{5}$

**c.** $b + c - (c - d)$, when $b = -\frac{1}{3}$, $c = -\frac{1}{4}$, and $d = -\frac{5}{6}$

**d.** $4bh(2b - 3ah)$, when $a = -\frac{2}{3}$, $b = -\frac{3}{4}$, and $h = -6$

**e.** $\dfrac{n + x(x - y)}{(n + x)(x - y)}$, when $n = -\frac{1}{2}$, $x = -2$, and $y = 1$

**f.** $.07(a - r)$, when $a = .3$ and $r = -.4$

**g.** $-.5c - (3m + .16)$, when $c = -.09$ and $m = -.2$

**h.** $d - n(n - x)$, when $d = -2.8$, $n = -.6$, and $x = -.7$

**i.** $(b - y)(b + y)$, when $b = -.75$ and $y = -.6$

**j.** $\dfrac{9x(x + a)}{4a(ax - a)}$, when $a = -3$ and $x = .8$

**2.** Using each given replacement for the variable, find the corresponding list of values of each of the following algebraic expressions:

**a.** $2(x + 3)$, when replacements for $x$ are: $-4, -3, \ldots, 0$

**b.** $-5(n + 1)$, when replacements for $n$ are: $-3, -2, \ldots, 3$

**c.** $y(y - 2)$, when replacements for $y$ are: $-3, -2, -1, 0$

**d.** $7x(8 - 2x)$, when replacements for $x$ are: $-6, -3, -1, 0, 5, 7$

**e.** $-3r(5r + 6)$, when replacements for $r$ are: $-2, -1, 0, 1, 2$

**f.** $-8c(12 - 2c)$, when replacements for $c$ are: $-10, -7, -3, 0, 1, 8$

**g.** $(b - 5)(3b + 4)$, when replacements for $b$ are: $-8, -5, -2, 0, 3, 5, 6$

**h.** $2d - (5d - 7)$, when replacements for $d$ are: $-2, -1, 0, 1, 3$

**i.** $8g - 4(3g + 2)$, when replacements for $g$ are: $-4, -3, 0, 2, 5$

**j.** $\dfrac{5(a - 1)}{a(8 - a)}$, when replacements for $a$ are: $-10, -8, -5, -1, 3, 9$

# Expressions with Exponents— Using Numbers of Arithmetic

## I. Aim
To evaluate an algebraic expression containing exponents when numbers of arithmetic are assigned to the variables.

## II. Procedure
1. Follow the same procedure as explained in Exercise 3-1.
2. However, raise each number as required to the proper power.
3. Compute as directed.

## III. Sample Solutions
Find the value of each of the following expressions when $b = 2$, $x = 4$, and $y = 3$:

**1.** $x^2$
$= (4)^2$
$= 4 \cdot 4$
$= 16$
*Answer:* 16

**2.** $b^3$
$= (2)^3$
$= 2 \cdot 2 \cdot 2$
$= 8$
*Answer:* 8

**3.** $y^4$
$= (3)^4$
$= 3 \cdot 3 \cdot 3 \cdot 3$
$= 81$
*Answer:* 81

**4.** $3by^2$
$= 3 \cdot 2 \cdot (3)^2$
$= 3 \cdot 2 \cdot 9$
$= 54$
*Answer:* 54

**5.** $(bx)^2$
$= (2 \cdot 4)^2$
$= (8)^2$
$= 64$
*Answer:* 64

**6.** $x^2 - 3x + 6$
$= (4)^2 - 3 \cdot 4 + 6$
$= 16 - 12 + 6$
$= 10$
*Answer:* 10

---

### DIAGNOSTIC TEST

Find the value of each of the following algebraic expressions when $a = 4$, $b = 3$, and $c = 2$:

1. $b^2$
2. $2b^2$
3. $bc^2$
4. $a^2b$
5. $a^2b^2$
6. $5a^2bc^2$
7. $a^3$
8. $2a^2bc^3$
9. $(ab)^2$
10. $c^4$
11. $b^5$
12. $a^2 + b^2$
13. $3a^2 - 2b^2$
14. $b^2c + bc^2$
15. $a^2 + 6$
16. $a^2 + 2a - 3$
17. $6b^2 - 2bc + c^2$
18. $5a^3 - 7a^2 + 9a - 2$
19. $\dfrac{b^3}{c^2}$
20. $(a + b)^2$
21. $\dfrac{a^2}{4} + \dfrac{b^2}{9}$
22. $4(3a - c)^2$
23. $\dfrac{b^2 - 4b + 4}{b^2 - 4}$

**106**

## RELATED PRACTICE EXAMPLES

Find the value of each of the following algebraic expressions:

**A.** When $c = 5$, $m = 3$, and $x = 2$

1. $c^2$
2. $2m^2$
3. $cx^2$
4. $m^2x$
5. $c^2x^2$
6. $4c^2mx$
7. $c^3$
8. $3c^3m^2x$
9. $(mx)^2$
10. $x^4$
11. $m^5$
12. $c^2 + m^2$
13. $2m^2 - 3x^2$
14. $c^2x + m^2x$
15. $m^2 + 7$
16. $c^2 + 6c - 5$
17. $7m^2 + 2mx + 5x^2$
18. $9x^3 - 3x^2 - 4x + 8$
19. $\dfrac{c^2}{x^3}$
20. $\dfrac{c^2 + 6c + 9}{c^2 - 9}$
21. $\dfrac{c^2}{9} + \dfrac{x^2}{4}$
22. $(m + x)^2$
23. $6(7c - 4m)^2$

**B.** When $b = 6$, $d = 2$, and $y = 1$

1. $d^2$
2. $3y^2$
3. $bd^2$
4. $b^2y$
5. $d^2y^2$
6. $3bd^2y^2$
7. $d^3$
8. $4b^2dy^2$
9. $(by)^2$
10. $b^4$
11. $y^5$
12. $b^2 - d^2$
13. $3d^2 - 2y^2$
14. $b^2d - dy^2$
15. $b^3 - 5$
16. $b^2 - 4b + 4$
17. $d^2 + 5dy - 6y^2$
18. $7b^4 - 3b^3 + 5b^2 - 2b$
19. $\dfrac{d^3}{y^2}$
20. $\dfrac{d^2 - 2d + 1}{d - 1}$
21. $\dfrac{2d^2}{3} - y^2$
22. $(b - y)^2$
23. $2b(d + 5y)^2$

## MISCELLANEOUS EXAMPLES

1. Find the value of:

   **a.** $y^2$, when $y = \frac{3}{4}$

   **b.** $6s^2$, when $s = 7\frac{1}{2}$

   **c.** $3m^2n - 2mn^2$, when $m = \frac{2}{3}$ and $n = \frac{1}{2}$

   **d.** $\frac{4}{3}\pi r^3$, when $\pi = \frac{22}{7}$ and $r = 21$

   **e.** $9d^2 - 4dx + x^2$, when $d = 1\frac{1}{3}$ and $x = \frac{1}{4}$

   **f.** $V^2$, when $V = 140.5$

   **g.** $.7854d^2$, when $d = 12.8$

   **h.** $\pi r^2h$, when $\pi = 3.14$, $r = 6.2$, and $h = 3.9$

   **i.** $m^4 - 2m^3 + 9m^2 - 3m + 8$, when $m = .4$

   **j.** $\dfrac{E^2}{R}$, when $E = 110$ and $R = 2.5$

2. Using each given replacement for the variable, find the corresponding list of values of each of the following algebraic expressions:

a. $x^2$ when replacements for $x$ are: 0, 1, ... , 6

b. $y^3$ when replacements for $y$ are: 2, 4, 6, 8, 10

c. $a^4$ when replacements for $a$ are: 1, 3, 5, 7, 9

d. $16t^2$ when replacements for $t$ are: 1, 2, 5, 6, 9, 10, 12

e. $\frac{44}{7}r^2$ when replacements for $r$ are: 7, 10, 21, 35

f. $2c^2 - 1$ when replacements for $c$ are: 1, 2, 3, ... , 8

g. $m^3 - m^2$ when replacements for $m$ are: 1, 2, 3, 4, 5

h. $5b^2 + 7b - 2$ when replacements for $b$ are: 1, 2, 3, ... , 7

i. $\dfrac{h^2 - 9}{h - 3}$ when replacements for $h$ are: 4, 6, 8, 9, 12

j. $\dfrac{x^2 + 6x + 5}{x + 1}$ when replacements for $x$ are: 0, 1, ... , 7

### CUMULATIVE REVIEW

1. Multiply the sum of $-6$ and $+7$ by the difference between $+8$ and $-3$.

2. Subtract the product of $-4$ and $-5$ from the quotient of $+9$ divided by $-1$.

3. Find the value of:

a. $6 - 14 + 9 - 2$

b. $(8 - 4) - (3 - 7) + (-5 + 1)$

c. $\dfrac{-12 - 4}{-4}$

d. $-10(8 - 15)$

e. $6 + 5(2 - 9) - 3(-8 - 4)$

f. $\dfrac{2(5 - 1) - (3 - 6)}{5(-11 + 9)}$

4. Find the value of:

a. $-7mn$ when $m = 8$ and $n = -3$

b. $4x - 2y$ when $x = -1$ and $y = -5$

c. $(3r - 2)(r - 3s)$ when $r = -8$ and $s = 6$

d. $2n^2 - 5n - 12$ when $n = 4$

e. $\dfrac{c^2 - d^2}{(c - d)^2}$ when $c = 6$ and $d = 3$

5. Using each given replacement for the variable, find the corresponding list of values of each of the following algebraic expressions:

a. $3n - 4$ when replacements for $n$ are: 0, 1, 2, ... , 10

b. $9 - 5y$ when replacements for $y$ are: $-5, -2, -1, 0, 3, 6$

c. $x + 2(x - 2)$ when replacements for $x$ are: $-4, -3, -2, ... , 4$

d. $16t^2$ when replacements for $t$ are: 1, 2, 3, ... , 9

e. $\dfrac{5c - 6}{2}$ when replacements for $c$ are: 2, 4, 6, 8, 10

# Expressions with Exponents—Using Positive and Negative Numbers

**I. Aim**   To evaluate an algebraic expression containing exponents when positive and negative numbers are assigned to the variables.

**II. Procedure**

1. Follow the same procedure as explained in Exercise 3-5.
2. Use parentheses to separate the signed numbers.
3. When necessary, use operation rules for signed numbers. Observe that: An even power of any number, positive or negative, is a positive number. See sample solution 3. An odd power of a negative number is a negative number. See sample solution 2.

**III. Sample Solutions**

Find the value of the following expressions if $a = -4$ and $x = -3$.

**1.** $a^2$
$= (-4)^2$
$= 16$
*Answer: 16*

**2.** $x^3$
$= (-3)^3$
$= -27$
*Answer: -27*

**3.** $x^4$
$= (-3)^4$
$= 81$
*Answer: 81*

**4.** $6ax^2$
$= 6(-4)(-3)^2$
$= 6(-4)(9)$
$= -216$
*Answer: -216*

**5.** $a^2 - 2ax + x^2$
$= (-4)^2 - 2(-4)(-3) + (-3)^2$
$= 16 - 24 + 9$
$= 1$
*Answer: 1*

---

## DIAGNOSTIC TEST

Find the value of each of the following algebraic expressions, when $c = 4$, $d = -6$, and $x = -3$:

**1.** $d^2$

**2.** $-2x^2$

**3.** $c^2 d$

**4.** $x^3$

**5.** $d^3 \times x^5$

**6.** $2c^2 - 3d^2$

**7.** $d^2 x$

**8.** $d^2 - 3d - 4$

**9.** $d^2 - 2dx + x^2$

**10.** $(c + x)^2$

**11.** $-5(3d - 4x)^2$

**12.** $\dfrac{c^2}{4} - \dfrac{d^2}{9}$

**13.** $\dfrac{3d^2}{2x^3}$

**14.** $\dfrac{x^2 - 4x - 5}{x^2 - 1}$

## RELATED PRACTICE EXAMPLES

Find the value of each of the following algebraic expressions:

**A.** When $d = 8$, $m = -6$, and $n = -2$

1. $m^2$     2. $-2n^2$     3. $mn^2$
4. $-n^3$    5. $m^4n^2$    6. $2d^2 - 3mn$
7. $m^2n - mn^2$     8. $m^2 - 5m + 6$
9. $m^2 - 3mn - 4n^2$

10. $\dfrac{d^2}{2mn}$     11. $\dfrac{m^2 + mn - 6n^2}{m^2 - 2n^2}$

12. $\dfrac{m^2}{3} - \dfrac{5n^2}{2}$

13. $(d + n)^2$     14. $-7(4m - 13n)^2$

**B.** When $b = 4$, $x = -2$, and $y = -1$

1. $x^2$     2. $-3y^2$     3. $b^2x$
4. $y^3$     5. $x^4y^5$     6. $2b^2 - 5y^2$
7. $x^2y - xy^2$     8. $x^2 - 8x - 2$
9. $x^2 - 5xy - 2y^2$

10. $\dfrac{2b^2}{3}$     11. $\dfrac{x^2 - 4xy - 3y^2}{x - y}$

12. $\dfrac{x^2}{2} - \dfrac{b^2}{4}$

13. $(b - y)^2$     14. $-3y(5x + 2b)^2$

## MISCELLANEOUS EXAMPLES

1. Find the value of:
   **a.** $c^2$, when $c = -\frac{1}{2}$     **b.** $-2b^3$, when $b = -\frac{3}{4}$
   **c.** $a^4x$, when $a = -\frac{1}{3}$ and $x = -\frac{9}{10}$     **d.** $10n^2 - 5n - 18$, when $n = -1\frac{1}{5}$
   **e.** $\frac{1}{2}x^2 - \frac{2}{3}y^2$, when $x = -\frac{5}{6}$ and $y = -\frac{1}{2}$     **f.** $-m^2$, when $m = -.05$
   **g.** $b^2 + c^2$, when $b = 3.4$ and $c = -4.1$     **h.** $3t^2 - 12.6$, when $t = -.7$
   **i.** $-.06a^2b^3$, when $a = -.3$ and $b = -5.4$
   **j.** $2x^2 - 3xy - 5y^2$, when $x = -1.8$ and $y = -2.6$

2. Using each given replacement for the variable find the corresponding list of values of each of the following algebraic expressions:
   **a.** $b^2$, when replacements for $b$ are: $-5$, $-4$, $-3$, $-2$, $-1$
   **b.** $r^3$, when replacements for $r$ are: $-3$, $-2$, $-1$, $0$, $1$
   **c.** $d^5$, when replacements for $d$ are: $-4$, $-2$, $0$, $2$, $4$
   **d.** $-7n^4$, when replacements for $n$ are: $-2$, $-1$, $0$, $1$, $2$
   **e.** $(6b^3)^2$, when replacements for $b$ are: $-3$, $-2$, $-1$, $3$
   **f.** $y^2 - 2y$, when replacements for $y$ are: $-4$, $-3$, $-2$, ..., $4$
   **g.** $4x^2 - 8x + 1$, when replacements for $x$ are: $-9$, $-7$, $-4$, $-1$, $0$, $1$, $3$
   **h.** $9a^3 - 5a^2 - 3a$, when replacements for $a$ are: $-3$, $-2$, $-1$
   **i.** $\dfrac{b^2 - 64}{b + 8}$, when replacements for $b$ are:

   $-10$, $-7$, $-6$, $-4$, $-3$, $-2$, $-1$
   **j.** $\dfrac{n^2 + 3n - 28}{n - 4}$, when replacements for $n$ are:

   $-6$, $-5$, $-2$, $-1$, $0$, $3$, $5$, $6$

# Evaluation of Formulas

**I. Aim**   To evaluate a formula. *define – give values*

**II. Procedure**

1. Rewrite the formula.
2. Substitute the known numerical values for the variable.
3. Perform the required operations to find the numerical value. Observe that to evaluate a formula means to find the numerical value of an indicated variable when the numerical values of all the other variables of a formula are known. Also see Exercise 6-10.

**III. Sample Solutions**

1. Find the value of $A$ when $l = 7$ and $w = 4$.

$$A = lw$$
$$A = 7 \cdot 4$$
$$A = 28$$

*Answer: 28*

2. Find the value of $A$ when $p = 120$, $r = .04$, and $t = 5$.

$$A = p + prt$$
$$A = 120 + 120 \times .04 \times 5$$
$$A = 120 + 24$$
$$A = 144$$

*Answer: 144*

3. Find the value of $I$ when $W = 100$ and $E = 20$.

$$I = \frac{W}{E}$$
$$I = \frac{100}{20}$$
$$I = 5$$

*Answer: 5*

4. Find the value of $V$ when $\pi = 3.14$, $r = 4$, and $h = 6$.

$$V = \pi r^2 h$$
$$V = 3.14 \times (4)^2 \times 6$$
$$V = 3.14 \times 16 \times 6$$
$$V = 301.44$$

*Answer: 301.44*

---

## DIAGNOSTIC TEST

Find the value of:

1. $d$ when $r = 52$ and $t = 4$, using the formula $d = rt$
2. $p$ when $l = 29$ and $w = 16$, using the formula $p = 2l + 2w$
3. $i$ when $A = 125$ and $p = 98$, using the formula $i = A - p$
4. $I$ when $E = 110$ and $R = 5$, using the formula $I = \frac{E}{R}$

5. $A$ when $\pi = 3.14$ and $r = 6$, using the formula $A = \pi r^2$

6. $S$ when $n = 6$, $a = 3$, and $l = 13$, using the formula $S = \dfrac{n}{2}(a + l)$

7. $C$ when $F = -13$, using the formula $C = \frac{5}{9}(F - 32)$

8. Copy the following table:

| $x$ | $-2$ | $-1$ | $0$ | $1$ | $2$ |
|---|---|---|---|---|---|
| $y$ | | | | | |

Determine the missing corresponding values of $y$ by using the sentence $y = 3x - 4$, then complete your table of values.

## RELATED PRACTICE EXAMPLES

Find the value of:

1. **a.** $p$ when $s = 6$, using the formula $p = 4s$
   **b.** $A$ when $l = 8$ and $w = 5$, using the formula $A = lw$
   **c.** $V$ when $B = 50$ and $h = 7$, using the formula $V = Bh$
   **d.** $A$ when $a = 16$ and $b = 9$, using the formula $A = ab$
   **e.** $c$ when $\pi = 3.14$ and $d = 12$, using the formula $c = \pi d$
   **f.** $E$ when $I = 11$ and $R = 20$, using the formula $E = IR$
   **g.** $A$ when $a = 18$ and $b = 26$, using the formula $A = ab$
   **h.** $i$ when $p = 200$, $r = .06$, and $t = 4$, using the formula $i = prt$
   **i.** $V$ when $l = 12$, $w = 8$, and $h = 7$, using the formula $V = lwh$
   **j.** $c$ when $\pi = \frac{22}{7}$ and $r = 21$, using the formula $c = 2\pi r$
   **k.** $d$ when $r = 25$ and $t = 7$, using the formula $d = rt$
   **l.** $P$ when $H = 100$ and $D = 64$, using the formula $P = HD$
   **m.** $T.S.$ when $\pi = 3.14$, $d = 7$, and $N = 30$, using the formula $T.S. = \pi dN$
   **n.** $W$ when $I = 6$ and $E = 220$, using the formula $W = IE$
   **o.** $W$ when $F = 125$ and $d = 8$, using the formula $W = Fd$

2. **a.** $A$ when $p = 450$ and $i = 83$, using the formula $A = p + i$
   **b.** $s$ when $c = 25$ and $g = 7$, using the formula $s = c + g$
   **c.** $A$ when $c = 43$, using the formula $A = C + 273$
   **d.** $D_t$ when $D_w = 97.2$ and $D_p = 51.9$, using the formula $D_t = D_w + D_p$
   **e.** $A$ when $L = 4{,}250$ and $C = 6{,}500$, using the formula $A = L + C$

**f.** $p$ when $a = 19$, $b = 23$, and $c = 18$, using the formula $p = a + b + c$

**g.** $p$ when $l = 37$ and $w = 26$, using the formula $p = 2l + 2w$

**h.** $A$ when $p = 130$, $r = .03$, and $t = 8$, using the formula $A = p + prt$

**i.** $v$ when $V = 29$, $g = 32$, and $t = 4$, using the formula $v = V + gt$

**j.** $F$ when $C = 82$, using the formula $F = 1.8C + 32$

3. **a.** $B$ when $A = 58$, using the formula $B = 90 - A$

**b.** $C$ when $A = 352$, using the formula $C = A - 273$

**c.** $l$ when $c = 164$ and $s = 139$, using the formula $l = c - s$

**d.** $C$ when $A = 8{,}925$ and $L = 3{,}672$, using the formula $C = A - L$

**e.** $p$ when $A = 475$ and $i = 86$, using the formula $p = A - i$

**f.** $g$ when $s = 134$ and $c = 117$, using the formula $g = s - c$

**g.** $i$ when $A = 266$ and $p = 210$, using the formula $i = A - p$

**h.** $b$ when $p = 58$ and $e = 13$, using the formula $b = p - 2e$

**i.** $c$ when $s = 81$ and $g = 29$, using the formula $c = s - g$

**j.** $C$ when $A = 47$ and $B = 105$, using the formula $C = 180 - A - B$

4. **a.** $a$ when $n = 5$, using the formula $a = \dfrac{360}{n}$

**b.** $r$ when $d = 42$, using the formula $r = \dfrac{d}{2}$

**c.** $A$ when $a = 18$ and $b = 6$, using the formula $A = \dfrac{ab}{2}$

**d.** $r$ when $d = 135$ and $t = 3$, using the formula $r = \dfrac{d}{t}$

**e.** $V$ when $B = 27$ and $h = 16$, using the formula $V = \dfrac{Bh}{3}$

**f.** $I$ when $E = 110$ and $R = 22$, using the formula $I = \dfrac{E}{R}$

**g.** $d$ when $m = 117$ and $v = 13$, using the formula $d = \dfrac{m}{v}$

**h.** $W$ when $w = 84$, $l = 16$, and $L = 12$, using the formula $W = \dfrac{wl}{L}$

**i.** $H.P.$ when $D_t = 225$ and $V = 130$, using the formula $H.P. = \dfrac{D_t V}{375}$

**j.** $I$ when $E = 55$, $n = 6$, $R = 10$, and $r = 2$, using the formula $I = \dfrac{nE}{R + nr}$

5. **a.** $A$ when $s = 27$, using the formula $A = s^2$

**b.** $A$ when $d = 13$, using the formula $A = .7854d^2$

**c.** $V$ when $e = 8$, using the formula $V = e^3$

**d.** $A$ when $\pi = 3.14$ and $r = 16$, using the formula $A = \pi r^2$

**e.** $V$ when $\pi = 22/7$, $r = 42$, and $h = 15$, using the formula $V = \pi r^2 h$

**f.** $s$ when $t = 6$, using the formula $s = 16t^2$

**g.** $W$ when $I = 7$ and $R = 4$, using the formula $W = I^2 R$

**h.** $F$ when $W = 48$, $v = 60$, $g = 32$, and $r = 12$, using the formula $F = \dfrac{Wv^2}{gr}$

**i.** $K$ when $m = 125$ and $v = 50$, using the formula $K = \frac{1}{2}mv^2$

**j.** $L$ when $C_L = .64$, $d = .002$, $S = 180$, and $V = 100$, using the formula

$$L = C_L \frac{d}{2} S V^2$$

6. **a.** $p$ when $l = 38$ and $w = 45$, using the formula $p = 2(l + w)$

**b.** $A$ when $p = 50$, $r = .06$, and $t = 5$, using the formula $A = p(1 + rt)$

**c.** $C$ when $A = 62$ and $B = 97$, using the formula $C = 180 - (A + B)$

**d.** $A$ when $\pi = 3.14$, $r = 6$, and $h = 14$, using the formula $A = 2\pi r(r + h)$

**e.** $S$ when $n = 8$, $a = 2$, and $l = 16$, using the formula $S = \dfrac{n}{2}(a + l)$

**f.** $C$ when $F = 68$, using the formula $C = \frac{5}{9}(F - 32)$

**g.** $A$ when $h = 8$, $b = 10$, and $b' = 4$, using the formula $A = \dfrac{h}{2}(b + b')$

**h.** $l$ when $a = 2$, $n = 5$, and $d = 3$, using the formula $l = a + (n - 1)d$

**i.** $A$ when $\pi = 22/7$, $r = 35$, and $s = 72$, using the formula $A = \pi r(r + s)$

**j.** $R$ when $T = 3$, $GSO = 236$, and $GSI = 214$, using the formula

$$R = T \left( \frac{GSO \times GSI}{GSO + GSI} \right)$$

7. **a.** $A$ when $C = -8$, using the formula $A = C + 273$

**b.** $F$ when $C = -40$, using the formula $F = 1.8C + 32$

**c.** $C$ when $A = 725$ and $L = 950$, using the formula $C = A - L$

**d.** $v$ when $a = -30$ and $t = 2$, using the formula $v = at$

**e.** $M$ when $a = -8$, $b = 3$, and $c = -10$, using the formula $M = \dfrac{a+b+c}{3}$

**f.** $C$ when $A = 100$, using the formula $C = A - 273$

**g.** $C$ when $A = 3{,}980$ and $L = 5{,}245$, using the formula $C = A - L$

**h.** $M$ when $a = 8$, $b = -4$, $c = 10$, and $d = -2$, using the formula

$$M = \frac{a + b + c + d}{4}$$

**i.** $C$ when $F = -4$, using the formula $C = \dfrac{5F - 160}{9}$

**j.** $F$ when $C = -35$, using the formula $F = \frac{9}{5}C + 32$

8. In each of the following, copy the given table, determine the missing corresponding values of $y$ by using the given sentence, then complete your table of values:

**a.**
$y = x$

| x | 0 | 1 | 2 | 4 | 9 |
|---|---|---|---|---|---|
| y |   |   |   |   |   |

**b.**
$y = 6x$

| x | −2 | −1 | 0 | 1 | 3 |
|---|---|---|---|---|---|
| y |   |   |   |   |   |

**c.**
$y = -3x$

| x | −4 | −3 | 0 | 1 | 2 |
|---|---|---|---|---|---|
| y |   |   |   |   |   |

**d.**
$y = x + 1$

| x | −3 | −1 | 0 | 2 | 5 |
|---|---|---|---|---|---|
| y |   |   |   |   |   |

**e.**
$y = x - 5$

| x | −2 | 0 | 1 | 5 | 7 |
|---|---|---|---|---|---|
| y |   |   |   |   |   |

**f.**
$y = 2x + 3$

| x | −3 | −1 | 0 | 1 | 2 |
|---|---|---|---|---|---|
| y |   |   |   |   |   |

**g.**
$y = 4x - 9$

| x | −4 | −1 | 0 | 2 | 3 |
|---|---|---|---|---|---|
| y |   |   |   |   |   |

**h.**
$y = 7x - 2$

| x | −2 | −1 | 0 | 1 | 2 |
|---|---|---|---|---|---|
| y |   |   |   |   |   |

**i.**
$y = 8 - x$

| x | −10 | −8 | 5 | 8 | 10 |
|---|---|---|---|---|---|
| y |   |   |   |   |   |

**j.**
$y = 12 - 5x$

| x | −3 | 0 | 1 | 3 | 5 |
|---|---|---|---|---|---|
| y |   |   |   |   |   |

**k.**
$y = \dfrac{x}{4}$

| x | −12 | −4 | 0 | 2 | 8 |
|---|---|---|---|---|---|
| y |   |   |   |   |   |

**l.**
$y = \dfrac{2x + 1}{3}$

| x | −4 | −2 | 0 | 1 | 4 |
|---|---|---|---|---|---|
| y |   |   |   |   |   |

# REVIEW

1. Find the value of:
   a. $c + 9d$, when $c = 21$ and $d = 4$
   b. $(m + n)(m - n)$, when $m = -2$ and $n = 6$
   c. $4x^2 - 7xy + 3y^2$, when $x = -3$ and $y = 2$
   d. $5a - b(a - b)$, when $a = -4$ and $b = -6$
2. Using each given replacement for the variable, find the corresponding list of values of each of the following algebraic expressions:
   a. $3x + 7$, when replacements for $x$ are: $-4, -3, -2, \ldots, 4$
   b. $5w - 4(7w - 8)$ when replacements for $w$ are: $-6, -4, -2, \ldots, 6$
   c. $n^2 - 4n - 5$ when replacements for $n$ are: $-3, -1, 0, 2, 3, 5, 6$
3. a. When $r = 3$, what is the value of $2r^4$? Of $(2r)^4$?
   b. When $x = -6$, what is the value of $8x + 3$? Of $8(x + 3)$?
   c. When $b = 2$, what is the value of $(b + 7)(b - 4)$? Of $b + 7(b - 4)$?
4. a. Find the value of $E$ when $I = 10$ and $R = 18$, using the formula $E = IR$.
   b. Find the value of $A$ when $B = 67$, using the formula $A = 90 - B$.
   c. Find the value of $s$ when $t = 3$, using the formula $s = 16t^2$.
   d. Find the value of $A$ when $p = 100$, $r = .08$, and $t = 4$, using the formula $A = p(1 + rt)$
   e. Find the value of $F$ when $C = -20$, using the formula $F = 1.8C + 32$.

## COMPETENCY CHECK TEST

In each of the following select the letter corresponding to your answer:
1. When $x = -4$, the value of $8 - 2x$ is:  a. $-16$  b. 0  c. 6  d. 16
2. The value of $(c + d)(c - d)$ when $c = 7$ and $d = 2$ is:
   a. 14  b. 45  c. 5  d. answer not given
3. When $a = 6$ and $c = 10$, the value of $c^2 - a^2$ is:
   a. 16  b. 4  c. 64  d. 256
4. Using formula $c = 2\pi r$, when $\pi = \frac{22}{7}$ and $r = 21$, the value of $c$ is:
   a. 132  b. 66  c. 264  d. 33
5. Using formula $E = I(R + r)$, when $I = 8$, $R = 9$, and $r = 3$, the value of $E$ is:  a. 20  b. 96  c. 75  d. answer not given

116

## Keyed Achievement Test

The numeral at the end of each problem indicates the exercise where help may be found.

1. Write as an algebraic expression: The square of the difference between the product of $b$ and $h$ and the product of $l$ and $w$. $\boxed{\text{1-5}}$

2. Express as a formula: The wavelength of sound waves ($l$) equals the velocity of the waves ($v$) divided by the number of waves per second ($n$). $\boxed{\text{1-8}}$

3. On a number line, draw the graph of: $-7, -5, -3, \ldots, 5$. $\boxed{\text{2-7}}$

4. Compute as directed:
   a. $(-5)+(-17)$ $(-11)+(+11)$ $(+7)+(-4)$ $(+8)+(+6)$ $\boxed{\text{2-11}}$
   b. $(+7)-(-3)$ $(-6)-(-4)$ $1-20$ $(-8)-(+10)$ $\boxed{\text{2-12}}$
   c. $(-3)(-9)$ $(+6)(-10)$ $(-4)(+2)$ $(-1)(-7)(-5)$ $\boxed{\text{2-13}}$
   d. $(-30)(+2)$ $(-45)(-5)$ $(+27)(-3)$ $(-16)(-16)$ $\boxed{\text{2-14}}$

5. a. Find the absolute value: $|-3.2| = ?$ $\boxed{\text{2-3}}$
   b. Find the opposite of $+5\frac{3}{4}$. Of $-93$. $\boxed{\text{2-4}}$
   c. Find the additive inverse of $-49$. Of $+\frac{5}{12}$. $\boxed{\text{2-4}}$
   d. Find the multiplicative inverse of $+20$. Of $-8\frac{1}{3}$. $\boxed{\text{2-13}}$

6. Find the value of: a. $m - 6n$, when $m = 10$ and $n = -8$ $\boxed{\text{3-2}}$
   b. $a^2 + 3a - 15$, when $a = -9$ $\boxed{\text{3-6}}$
   c. $\dfrac{x + y(x - y)}{(x + y)(x - y)}$, when $x = 4$ and $y = -2$ $\boxed{\text{3-4}}$
   d. $v$ when $V = 75$, $g = 32$, and $t = 7$, using formula $v = V + gt$ $\boxed{\text{3-7}}$
   e. $A$ when $\pi = \frac{22}{7}$, $r = 14$, and $h = 10$, using formula $A = 2\pi r(r + h)$ $\boxed{\text{3-7}}$

## MAINTENANCE PRACTICE IN ARITHMETIC

1. Add:
87,469
35,874
56,629

2. Subtract:
580,107
95,188

3. Multiply:
6,409
8,379

4. Divide:
$659\overline{)3,820,882}$

5. Add:
$1\frac{7}{12} + 3\frac{1}{3} + 5\frac{3}{4}$

6. Subtract:
$20\frac{5}{16} - 19$

7. Multiply:
$5\frac{1}{3} \times 4\frac{3}{8}$

8. Divide:
$6\frac{1}{4} \div 1\frac{7}{8}$

9. Add:
$7.94 + 25.8 + .832$

10. Subtract:
$4.6 - .541$

11. Multiply:
$.045 \times .008$

12. Divide:
$.05\overline{).1}$

13. Find 36% of $287.50.

14. What percent of 160 is 60?

15. $33\frac{1}{3}\%$ of what number is 849?

# INTRODUCTION

Algebra students are generally familiar with formulas that are studied in arithmetic and informal geometry. A *mathematical formula* is a rule that expresses the relationship between quantities by means of numerals, variables, and symbols of operation and equality. For example, the formula $p = 4s$ shows the relationship between the perimeter of a square and the length of its side.

Tables of related values are sometimes used to show the relationship between quantities. Very often data given in tabular form may be analyzed and their relationship determined. It is then possible to express this relationship by a formula.

Where the same mathematical situation occurs frequently, it is sometimes economical and practical to develop a formula covering the relationship involved.

In this unit formulas are derived from tables of related numerical facts, from similar mathematical situations, and from diagrams.

# Patterns Developing Formulas by Using Tables of Related Numerical Facts

I. **Aim**    To develop a formula by using a table containing two rows of related numbers.

II. **Procedure**

1. Compare each pair of corresponding numbers given in the two rows.
2. Check whether one number is:
    a. Some number times the other.
    b. Some number more or less than the other.
    c. Some number times the other, plus or minus some other number.
    d. Or whether the sum of the two numbers is a fixed number.
3. If an identical relationship can be found between each pair of corresponding numbers, express this relationship as a word statement or rule.
4. Then express this word rule as a formula.

III. **Sample Solutions**

1. In the following table, a. find the missing numbers, b. write a word statement describing the relationship between the rows of numbers and then, c. write the formula for this relationship.

| Row $r$ | 1 | 2 | 3 | 7 | 10 | 15 |
|---------|---|----|----|---|----|----|
| Row $s$ | 8 | 16 | 24 |   |    |    |

Comparing 8 to 1, 16 to 2, and 24 to 3, we find each number in the bottom row is 8 times its corresponding number in the top row.

(a) Therefore the missing numbers are $8 \times 7 = 56$, $8 \times 10 = 80$, and $8 \times 15 = 120$.

(b) The relationship expressed as a word statement is: Each number in row $s$ is 8 times its corresponding number in row $r$.

(c) Expressed as a formula this relationship is: $s = 8r$

2. How many months are in 2 years? 5 years? Write a formula expressing the number of months ($m$) in $y$ years.

    In 2 years there are $12 \times 2$ or 24 months or $m = 24$.
    In 5 years there are $12 \times 5$ or 60 months or $m = 60$.
    In $y$ years there are $12 \times y$ or $12y$ months or $m = 12y$.
    Therefore the formula is: $m = 12y$

*Answers:* 24 months, 60 months, $m = 12y$

**120**

## DIAGNOSTIC TEST

In each of the following tables of related numerical facts find the missing numbers. Write a word statement describing the relationship between the rows of numbers. Then write the formula for this relationship.

(Copy the tables below and fill in the missing numbers. Do not write in this book.)

1. Relationship between air pressure at the earth's surface and area:

| Area in square inches (A) | 1 | 2 | 3 | 8 | 24 |
|---|---|---|---|---|---|
| Air pressure in pounds (P) | 15 | 30 | 45 | | |

2. Comparative ages of two boys:

| Tom s age in years (T) | 2 | 3 | 6 | 9 | 16 |
|---|---|---|---|---|---|
| John's age in years (J) | 9 | 10 | 13 | | |

3. Interest for 60 days at 6%:

| Principal (p) | $100 | $250 | $395 | $418 | $940 |
|---|---|---|---|---|---|
| Interest (i) | $1.00 | $2.50 | $3.95 | | |

4. Number of triangles in a polygon:

| Number of sides (s) | 3 | 4 | 7 | 12 | 17 |
|---|---|---|---|---|---|
| Number of triangles (t) | 1 | 2 | 5 | | |

5. Cost of developing and printing a roll of color film:

| Number of prints (n) | 1 | 2 | 3 | 8 | 12 |
|---|---|---|---|---|---|
| Cost (c) | | $1.19 | $1.48 | $1.77 | |

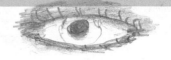

**6.** Annual statements of inventory:

| Inventory at start (*I.S.*) | $5,000 | $7,500 | $8,200 | $6,900 | $16,900 |
|---|---|---|---|---|---|
| Purchases (*P*) | $8,000 | $10,000 | $9,600 | $20,000 | $25,420 |
| Sales (*S*) | $9,000 | $15,500 | $12,000 | $17,400 | $23,780 |
| Inventory at end (*I.E.*) | $4,000 | $2,000 | | | |

**7.** Cost of second class mail: $c = .10 + .06(w - 2)$

| Weight in ounces (*w*) | 2 or less | 3 | 4 | 9 | 11 |
|---|---|---|---|---|---|
| Cost (*c*) | $.10 | $.16 | $.22 | | |

**8. a.** How many days are in 72 hours? In 216 hours? Write a formula expressing the number of days (*d*) in *h* hours.
  **b.** If 1 mile equals 1.61 kilometers, how many kilometers are in 5 miles? In 17 miles? Write a formula expressing the number of kilometers (*k*) in *m* miles.

### RELATED PRACTICE EXAMPLES

In each of the following tables of related numerical facts find the missing numbers. Write a word statement describing the relationship between the rows of numbers. Then write the formula for this relationship. (Copy the tables on the next page and fill in the missing numbers. Do not write in this book.)

**1. a.** Changing days to hours:

| Number of days (*d*) | 1 | 2 | 3 | 7 | 13 |
|---|---|---|---|---|---|
| Number of hours (*h*) | 24 | 48 | 72 | | |

**b.** Wages earned:

| Number of hours (*h*) | 5 | 6 | 7 | 14 | 25 |
|---|---|---|---|---|---|
| Wages (w) | $15 | $18 | $21 | | |

**c.** Changing meters to yards:

| Number of meters (*m*) | 2 | 5 | 8 | 15 | 22 |
|---|---|---|---|---|---|
| Number of yards (*y*) | 2.2 | 5.5 | 8.8 | | |

**d.** Fire insurance rates:

| Rate per $100 for 1 yr. (*r*) | $.08 | $.12 | $.14 | $.20 | $.34 |
|---|---|---|---|---|---|
| Rate per $100 for 3 yr. (*R*) | $.20 | $.30 | $.35 | | |

**e.** Cost of gasoline:

| Number of gallons (*g*) | 2 | 5 | 6 | 8 | 19 |
|---|---|---|---|---|---|
| Cost (*c*) | $2.40 | $6.00 | $7.20 | | |

2. **a.** Total salary, fixed salary, and commission:

| Commission (*c*) | $7 | $19 | $28 | $47 | $55 |
|---|---|---|---|---|---|
| Total salary (*s*) | $63 | $75 | $84 | | |

**b.** Interest and total amount paid back:

| Interest (*i*) | $4 | $5 | $8 | $14 | $20 |
|---|---|---|---|---|---|
| Total amount (*A*) | $59 | $60 | $63 | | |

**c.** Comparison of two rows of numbers:

| Row *a* | 12 | 13 | 15 | 21 | 36 |
|---|---|---|---|---|---|
| Row *b* | 18 | 19 | 21 | | |

**d.** Total salary, fixed salary, and bonus:

| Bonus (*b*) | $5 | $18 | $31 | $45 | $50 |
|---|---|---|---|---|---|
| Total salary (*s*) | $54 | $67 | $80 | | |

**e.** Making coffee:

| Number of cups of coffee required ($C$) | 4 | 5 | 7 | 10 | 14 |
|---|---|---|---|---|---|
| Number of tablespoons of coffee ($T$) | 5 | 6 | 8 | | |

**3. a.** Changing millimeters to centimeters:

| Number of millimeters (mm) | 20 | 30 | 38 | 40 | 96 |
|---|---|---|---|---|---|
| Number of centimeters (cm) | 2 | 3 | 3.8 | | |

**b.** Changing nautical miles to degrees of latitude:

| Number of nautical miles ($m$) | 60 | 180 | 420 | 540 | 1,200 |
|---|---|---|---|---|---|
| Latitude in degrees ($L$) | 1 | 3 | 7 | | |

**c.** Annual and monthly rent:

| Rent per year ($R$) | $4,200 | $4,320 | $4,500 | $6,000 | $9,000 |
|---|---|---|---|---|---|
| Rent per month ($r$) | $350 | $360 | $375 | | |

**d.** Weight and capacity of gasoline:

| Weight in pounds ($w$) | 30 | 42 | 48 | 90 | 162 |
|---|---|---|---|---|---|
| Number of gallons ($g$) | 5 | 7 | 8 | | |

**e.** Perimeter and length of a side of a regular hexagon:

| Perimeter in feet ($p$) | 24 | 42 | 48 | 96 | 168 |
|---|---|---|---|---|---|
| Length of side in feet ($s$) | 4 | 7 | 8 | | |

**4. a.** Number of diagonals in a polygon:

| Number of sides ($s$) | 4 | 5 | 6 | 7 | 9 | 12 | 13 | 16 |
|---|---|---|---|---|---|---|---|---|
| Number of diagonals ($d$) | 1 | 2 | 3 | | | | | |

**b.** Comparison of two rows of numbers:

| Row $x$ | 17 | 21 | 29 | 34 | 43 | 52 | 75 | 86 |
|---|---|---|---|---|---|---|---|---|
| Row $y$ | 9 | 13 | 21 | | | | | |

**c.** Down payment and amount of mortgage on a house costing $72,000:

| Down payment ($p$) | $12,000 | $16,000 | $9,000 | $21,200 | $26,000 |
|---|---|---|---|---|---|
| Mortgage ($m$) | $60,000 | $56,000 | $63,000 | | |

**d.** Ages of two girls:

| Sally's age in years ($S$) | 12 | 13 | 14 | 17 | 22 |
|---|---|---|---|---|---|
| Dorothy's age in years ($D$) | 5 | 6 | 7 | | |

**e.** Complementary angles:

| Angle $A$ in degrees | 10 | 18 | 30 | 36 | 45 | 57 | 60 | 75 |
|---|---|---|---|---|---|---|---|---|
| Angle $B$ in degrees | 80 | 72 | 60 | | | | | |

5. **a.** Bank discount:

| Face of note ($F$) | $100 | $250 | $980 | $340 | $850 |
|---|---|---|---|---|---|
| Discount ($D$) | $6 | $15 | $49 | $17 | $51 |
| Proceeds of note ($P$) | $94 | $235 | | | |

**b.** Net profit:

| Sales ($s$) | $79 | $580 | $450 | $900 | $1,200 |
|---|---|---|---|---|---|
| Cost of goods ($c$) | $60 | $400 | $275 | $750 | $925 |
| Expenses ($e$) | $8 | $60 | $75 | $80 | $150 |
| Net profit ($p$) | $11 | $120 | $100 | | |

**c.** Real estate taxes:

| Assessed value ($v$) | $1,500 | $2,000 | $6,700 | $8,000 | $11,500 |
|---|---|---|---|---|---|
| Tax rate per $100 ($r$) | $2 | $3 | $2.50 | $3.60 | $5.20 |
| Amount of tax ($T$) | $30 | $60 | | | |

**d.** Area of trapezoid:

| Lower base in feet ($b_1$) | 8 | 13 | 15 | 12 | 20 |
|---|---|---|---|---|---|
| Upper base in feet ($b_2$) | 10 | 7 | 13 | 18 | 37 |
| Height in feet ($h$) | 5 | 8 | 10 | 7 | 6 |
| Area in square feet ($A$) | 45 | 80 | 140 | | |

6. **a.** Cost of developing and printing a roll of film:

| Number of prints ($n$) | 1 | 2 | 3 | 8 | 12 |
|---|---|---|---|---|---|
| Cost ($c$) | $.52 | $.64 | $.76 | | |

**b.** Comparison of two rows of numbers:

| Row $x$ | 1 | 2 | 3 | 6 | 11 |
|---|---|---|---|---|---|
| Row $y$ | 8 | 13 | 18 | | |

**c.** Comparison of two rows of numbers:

| Row $A$ | 1 | 2 | 3 | 7 | 16 |
|---|---|---|---|---|---|
| Row $B$ | 3 | 10 | 17 | | |

**d.** Air Parcel Post—cost of sending a package between two particular zones:

| Weight in pounds ($w$) | 1 | 2 | 3 | 6 | 15 |
|---|---|---|---|---|---|
| Cost ($c$) | $4.06 | $6.10 | $9.34 | | |

7. **a.** Parcel Post—cost of sending a package between two particular zones:

| Weight in pounds ($w$) | 2 or less | 3 | 4 | 6 | 11 |
|---|---|---|---|---|---|
| Cost ($c$) | $2.22 | $2.62 | $3.02 | | |

**b.** Long distance telephone charges between two particular zones:

| Number of minutes ($m$) | 3 or less | 4 | 5 | 7 | 12 |
|---|---|---|---|---|---|
| Cost (c) | $.50 | $.65 | $.80 | | |

$$c = .50 + .15(m - 3)$$

**c.** Laundry charges:

| Number of pieces ($n$) | 12 or less | 13 | 14 | 18 | 23 |
|---|---|---|---|---|---|
| Cost ($c$) | $2.50 | $2.70 | $2.90 | | |

**d.** Cost of sending a telegram between two particular zones:

| Number of words ($n$) | 15 or less | 16 | 17 | 20 | 25 |
|---|---|---|---|---|---|
| Cost ($c$) | $1.50 | $1.58 | $1.66 | | |

**e.** Laundry charges:

| Number of pounds ($w$) | 8 or less | 9 | 10 | 15 | 18 |
|---|---|---|---|---|---|
| Cost ($c$) | $2.25 | $2.44 | $2.63 | | |

8. **a.** How many feet are in 24 inches? 60 inches? Write a formula expressing the number of feet ($f$) in $i$ inches.
   **b.** How many milliliters are in 20 centiliters? 3.5 centiliters? Write a formula expressing the number of milliliters ($mL$) in $cL$ centiliters.
   **c.** How many seconds are in 120 minutes? 80 minutes? Write a formula expressing the number of seconds ($s$) in $m$ minutes.
   **d.** How many kilowatts are in 4,000 watts? 7,000 watts? Write a formula expressing the number of kilowatts ($kW$) in $W$ watts.
   **e.** One cubic foot of seawater weighs 64 pounds. Write a formula expressing the weight in pounds ($w$) of $n$ cubic feet of seawater.

# Business Formulas and Geometric Formulas from Diagrams

**EXERCISE 4-2**

I. Aim    To develop a formula from a diagram or geometric figure.

II. Procedure
1. Study the diagram or figure to determine the relationship between the part of the diagram representing the quantity to be determined and the parts representing the known quantities.
2. Write the word statement or rule expressing this relationship.
3. Then express this word rule as a formula.

III. Sample Solutions

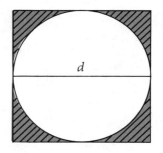

The diagram at the right shows that a circle with diameter $d$ is inscribed in a square whose side also measures $d$. Write a formula for the shaded area.
1. Study shows that the shaded area can be found by subtracting the area of the circle from the area of the square. The formula for the area of the circle is $A = \frac{1}{4}\pi d^2$. The formula for the area of a square whose side measures $d$ is $A = d^2$.
2. The shaded area is equal to the area of the square minus the area of the circle.
3. Expressing this as a formula: $A = d^2 - \frac{1}{4}\pi d^2$
*Answer: $A = d^2 - \frac{1}{4}\pi d^2$*

## DIAGNOSTIC TEST

1. Study the diagram on the left side of the next page, then write the word statement that describes and the formula that expresses:
   a. Face of note in terms of proceeds and bank discount.
   b. Bank discount in terms of face of note and proceeds.
   c. Proceeds in terms of face of note and bank discount.

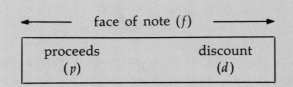

2. An inside horse on a merry-go-round makes path $A$ and an outside horse makes path $B$ in one turn. Study the diagram at the above right, then write the word statement and the formula expressing the difference in the length of ride ($L$) if the diameter of the small circle is $d$ and of the large circle is $D$.

3. Facts: When one straight line meets another, the adjacent angles, which have the same vertex and a common side, are supplementary (the sum of their measures is 180°). Also, when two straight lines intersect, the opposite angles have the same measure.

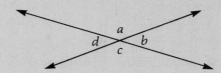

Using the above facts and the diagram, write the word statement that describes and the formula that expresses:
a. $m\angle a$ in terms of $m\angle b$.
b. $m\angle a$ in terms of $m\angle c$.
c. $m\angle b$ in terms of $m\angle d$.
d. $m\angle c$ in terms of $m\angle d$.
e. The sum of $m\angle b$ and $m\angle c$.

4. Write the formula for the entire area of the following figure:

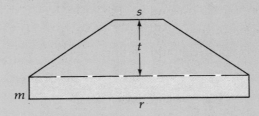

### *RELATED PRACTICE EXAMPLES*

1. **BUSINESS**

In each of the following, first study the diagram and write a word statement describing each required relationship as indicated in the diagram. Then write the formula for this relationship.

**a.** (1) List price in terms of net price and discount.

(2) Net price in terms of list price and discount.

(3) Discount in terms of list price and net price.

| list price (*l*) | |
|---|---|
| net price (*n*) | discount (*d*) |

**b.** (1) Net proceeds in terms of sales and commission.

(2) Commission in terms of net proceeds and sales.

(3) Sales in terms of commission and net proceeds.

| sales (*s*) | |
|---|---|
| net proceeds (*p*) | commission (*c*) |

**c.** (1) Interest in terms of principal and amount.

(2) Principal in terms of amount and interest.

(3) Amount in terms of principal and interest.

| amount (*A*) | |
|---|---|
| principal (*p*) | interest (*i*) |

**d.** (1) Loss in terms of cost and selling price.

(2) Selling price in terms of cost and loss.

(3) Cost in terms of loss and selling price.

| cost (*c*) | |
|---|---|
| selling price (*s*) | loss (*l*) |

**e.** (1) Selling price in terms of cost, operating expenses, and profit.

(2) Margin in terms of operating expenses and profit.

(3) Profit in terms of selling price, cost, and operating expenses.

(4) Cost in terms of selling price and margin.

(5) Operating expenses in terms of margin and profit.

(6) Selling price in terms of cost and margin.

(7) Profit in terms of operating expenses and margin.

(8) Operating expenses in terms of cost, selling price, and profit.

(9) Margin in terms of selling price and cost.

(10) Cost in terms of selling price, operating expenses, and profit.

2. **MEASUREMENT OF LINES**

In each of the following, study the diagram, then write:

a. the word statement that describes;

and b. the formula that expresses:

**a.** (1) Dimension $c$ in terms of $a$ and $b$.

(2) Dimension $b$ in terms of $a$ and $c$.

(3) Dimension $a$ in terms of $b$ and $c$.

**b.** (1) Line segment $AD$ in terms of the 3 smaller segments. (Generally expressed as $\overline{AD}$; m$\overline{AD}$ means the "measure of the line segment $AD$.")

(2) Line segment $AB$ in terms of $\overline{AD}$, $\overline{BC}$, and $\overline{CD}$.

(3) Line segment $BC$ in terms of $\overline{AD}$, $\overline{AB}$, and $\overline{CD}$.

(4) Line segment $BD$ in terms of $\overline{AD}$ and one other segment.

(5) Line segment $BC$ in terms of $\overline{AC}$ and $\overline{AB}$.

**c.** (1) Perimeter (*p*) of the isosceles triangle in terms of the three sides.

(2) Side *b* in terms of the perimeter and the other two sides.

(3) Side *e* in terms of the perimeter and side *b*.

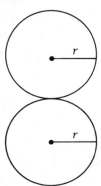

**d.** (1) Dimension *y* in terms of two other dimensions.

(2) Dimension *x* in terms of two other dimensions.

(3) Dimension *c* in terms of two other dimensions.

(4) Dimension *a* in terms of two other dimensions.

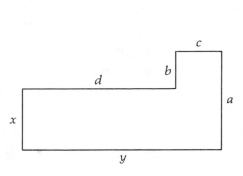

**e.** Distance one must skate in making a figure 8 in terms of the radius of the two equal circles.

3. **MEASUREMENT OF ANGLES**

(Symbol ∠ means "angle"; symbol m∠ means "measure of an angle").

**a.** *Fact:* The sum of the measure of the angles of a triangle is equal to 180°. In each of the following, use the given fact and diagram to write the word statement that describes and the formula that expresses, in triangle *ABC*:

(1) The sum of the measures of the 3 angles of the triangle.

(2) m∠*A* in terms of the measures of the other angles of the triangle.

(3) m∠*B* in terms of the measures of the other angles of the triangle.

(4) m∠*C* in terms of the measures of the other angles of the triangle.

(5) The sum of m∠*A* and m∠*C* in terms of the third angle measure.

**b.** *Fact:* Complementary angles are two angles whose sum of measures is 90°. In each of the following, use the given fact and diagram to write the word statement that describes and the formula that expresses, in right angle *DEF*:

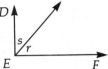

(1) the sum of m∠r and m∠s.
(2) m∠r in terms of m∠s.
(3) m∠s in terms of m∠r.

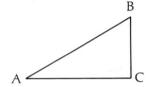

In right triangle *ABC*:
    (4) The sum of m∠A and m∠B.
    (5) m∠B in terms of m∠A.
    (6) m∠A in terms of m∠B.

**c.** *Fact:* Supplementary angles are two angles whose sum measures 180°.

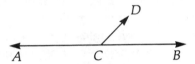

In each of the following, use the given fact and diagram to write the word statement that describes and the formula that expresses, in straight angle *ACB*:
    (1) The sum of the measures of angles *BCD* and *DCA*.
    (2) m∠BCD in terms of m∠DCA.
    (3) m∠DCA in terms of m∠BCD.

In the figure to the right, m∠a = m∠c, m∠b = m∠d, and ∠a and ∠b are supplementary angles. Express the following:
    (4) The sum of m∠c and m∠d.
    (5) m∠c in terms of m∠b.
    (6) m∠a in terms of m∠d.
    (7) m∠c in terms of m∠d.

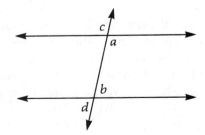

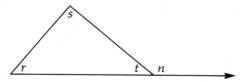

(8) In the figure above, ∠t and ∠n are supplementary angles, and the sum of m∠r, m∠s, and m∠t is 180°. Express the relationship of m∠n in terms of m∠r and m∠s.

**d.** *Fact:* The sum of the measure of all the angles about a point on one side of a line passing through this point is 180°.

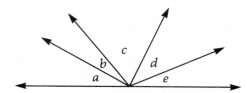

In each of the following, use the given fact and diagram to write the word statement that describes and the formula that expresses:
(1) The sum of measures of angles a, b, c, d, and e.
(2) m∠c in terms of the measures of angles a, b, d, and e.
(3) The sum of the measures of angles b, c, d, and e, when m∠a = 30°.
(4) The sum of the measures of angles a, b, and e, when the sum of m∠c and m∠d is 105°.
(5) The sum of the measures of angles b, d, and e in terms of m∠a and m∠c.

**e.** *Fact:* The measures of the opposite angles of a parallelogram are equal, and the sum of the measures of the angles of any quadrilateral is 360°.

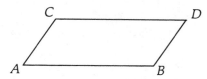

In each of the following, use the given fact and diagram to write the word statement that describes and the formula that expresses:
(1) The sum of the measures of angles A, B, C, and D.
(2) m∠A in terms of the measures of angles B, C, and D.
(3) The sum of measures of the angles B and D.
(4) m∠C in terms of m∠A and m∠D.
(5) m∠D in terms of m∠C and m∠B.

## 4. MEASUREMENT OF AREAS

**a.** A square whose side measures $s$ is cut out of a square whose side measures $S$. Write the word statement and then the formula expressing the area that remains.

**b.** Write the word statement and then the formula expressing the area of a flat washer if its radius measures $R$ and the radius of the hole measures $r$.

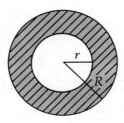

Write the formula for the entire area of each of the following figures:

**c.**

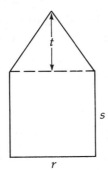

**d.**

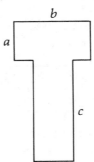

**e.**

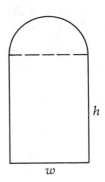

# REVIEW

1. In each of the following tables of numerical facts find the missing numbers. Then write the word statement and the formula that expresses the relationship between the rows of numbers.

a.

| Row $x$ | 1 | 2 | 3 | 5 | 9 |
|---------|---|---|---|---|---|
| Row $y$ | 6 | 12 | 18 | | |

b.

| Row $m$ | 8 | 9 | 10 | 13 | 21 |
|---------|---|---|----|----|----|
| Row $n$ | 13 | 14 | 15 | | |

c.

| Row $R$ | 2 | 3 | 4 | 6 | 11 |
|---------|---|---|---|---|----|
| Row $S$ | 11 | 15 | 19 | | |

2. How many meters are in 3 kilometers? 15 kilometers? Write a formula expressing the number of meters ($m$) in $k$ kilometers.
3. Write a formula for the area (A) of the following figure:

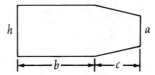

## COMPETENCY CHECK TEST

In each of the following select the letter corresponding to your answer:
1. The missing numbers in the table below are:
   a. 24 and 36      b. 20 and 29      c. 17 and 20      d. 21 and 33
2. The formula showing the relationship between the rows of numbers is:
   a. $y = 4x$      b. $y = 2x + 4$      c. $y = 3x + 2$      d. $y = x + 6$

| Row $x$ | 2 | 3 | 4 | 6 | 9 |
|---------|---|---|---|---|---|
| Row $y$ | 8 | 11 | 14 | | |

3. The formula that expresses the number of items ($n$) in $d$ dozen is:
   a. $d = \dfrac{12}{n}$      b. $n = \dfrac{d}{12}$      c. $d = 12n$      d. $n = 12d$

## Keyed Achievement Test

The colored numeral indicates the Exercise where help may be found.

1. Which of the following numerals name rational numbers? Which name integers? [2-5]
   $$-\tfrac{5}{6} \qquad \tfrac{27}{9} \qquad -\sqrt{47} \qquad 16\% \qquad -53 \qquad +7.2 \qquad +8\tfrac{2}{3} \qquad -.009$$

2. On a number line draw the graph of: $-4, -1, 0, 3, 5$ [2-7]

3. Compute as indicated: [2-11, 2-12, 2-13, 2-14]
   a. $(-4) + (-13)$    b. $(-12)(+2)$    c. $(-7) - (+5)$
   d. $(56) \div (-7)$    e. $6 - 11$    f. $(-18) \div (-3)$
   g. $(+14) + (-9)$    h. $(-2)(-4)(-1)$

4. Find the value of: a. $m + 7n$ when $m = 10$ and $n = -2$. [3-2]
   b. $3x^2 - 4x + 8$ when $x = -5$. [3-6]

5. Find the value of $A$ when $p = 600$, $r = .08$, and $t = 3$, using the formula $A = p(1 + rt)$. [3-7]

6. Determine the missing numbers. Then write the word statement and the formula that expresses the relationship between the following rows of numbers. [4-1]

   | Row $A$ | 7 | 8 | 9 | 10 | 15 |
   |---------|---|---|---|----|----|
   | Row $B$ | 2 | 3 | 4 |    |    |

7. Write the formula for the perimeter $(p)$ of the figure at the right. [4-2]

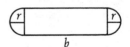

8. How many minutes are in 5 hours? In 12 hours? Write the formula expressing the number of minutes $(m)$ in $h$ hours. [4-1]

## MAINTENANCE PRACTICE IN ARITHMETIC

1. Add:
   863,978
   394,896
   506,949

2. Subtract:
   5,063,245
   992,866

3. Multiply:
   98,675
   76,489

4. Divide:
   $897\overline{)884,442}$

5. Add:
   $1\tfrac{7}{16} + 4\tfrac{5}{12}$

6. Subtract:
   $9\tfrac{1}{6} - 8\tfrac{3}{4}$

7. Multiply:
   $2\tfrac{2}{3} \times 6\tfrac{1}{8}$

8. Divide:
   $12 \div \tfrac{5}{6}$

9. Add:
   $24.6 + 8.63 + .597$

10. Subtract:
    $.793 - .41$

11. Multiply:
    $.005 \times .016$

12. Divide:
    $.45\overline{).9}$

13. Find $66\tfrac{2}{3}\%$ of $1,200.

14. What percent of 56 is 35?

15. 8% of what number is 108?

# INTRODUCTION

Before a student may proceed further in the study of algebra, the addition, subtraction, multiplication, and division of algebraic expressions should be learned. This knowledge is essential in the solutions of the equation and inequality in both one and two variables. The solution of practical problems also requires the use of algebraic operational techniques and skills since the solution of verbal problems and the evaluation of formulas depend upon the equation.

In this unit many technical algebraic words are used. Although some of them have been used in the preceding units, their meanings are reviewed together with the meanings of the new words in Exercise 5-1.

**Operations with Polynomials**

5

**I. Aim**    To understand the vocabulary used in the operations with algebraic expressions.

## II. Procedure

Read each of the following definitions carefully and study the illustrated corresponding examples.

1. An *algebraic expression* is an expression composed of arithmetic numbers, variables, and signs of operation.

    $7$, $d$, $a^2xy$, $-7m^2n$, $x - 5y$, $2a^2 - 7a + 8$, and $4b^3 - 2b^2c + 5bc^2 - 10c^3$ are algebraic expressions.

2. A *term*, also called a *monomial*, is a number, or a variable, or a product of variables, or a product of a number and a variable or variables.

    $9$, $s$, $cd$, and $-8xy^2$ are terms.

3. When a term consists only of a numeral naming a number, it is called a *constant term*.

    $12$ and $-31$ are constant terms.

4. *A term of an expression* is that part of it which is connected to the other parts of the expression by either a plus or minus sign.

    In $2x + 3$, $2x$ is one term and $3$ is another term.

5. A *monomial* is an algebraic expression of one term.

    $6y^2$ and $-mx$ are monomials.

6. A *binomial* is an algebraic expression of two terms.

    $7b + 5$ and $3c - 2d$ are binomials.

7. A *trinomial* is an algebraic expression of three terms.

    $3c^2 - 5cd + 6d^2$ is a trinomial.

8. A *polynomial* is an algebraic expression of one term or the sum of two or more terms. Usually the word polynomial is used to describe an expression of two or more terms and the word monomial for an expression of only one term.

    $5x^3 + 7x^2 + 3x + 8$ is a polynomial.

    Since $n^2 - 3n - 5$ may be written as $n^2 + (-3n) + (-5)$ it is also a polynomial in which $n^2$, $-3n$, and $-5$ are the terms.

9. A *factor* is any one of two or more numbers that are multiplied.
   Since $3 \times 7 = 21$, the 3 and 7 are factors of 21.

10. When a product consists of a number factor and a variable, the number factor is called the *numerical coefficient* of the variable.
    In $10b^5$, 10 is the numerical coefficient of $b^5$.
    The numerical coefficient 1 is usually not expressed. x means $1x$ or $+1x$ and $-x$ means $-1x$.

11. The small numeral written to the upper right (superscript) of a repeated factor is called an *exponent*. It indicates how many times the factor is being used in multiplication.
    In $n^7$, the exponent is 7. $n^7 = n \cdot n \cdot n \cdot n \cdot n \cdot n \cdot n$

12. The product of equal factors is called a *power* of the factor. It may be expressed as an indicated product using exponents.
    Since $2 \cdot 2 \cdot 2 = 8$, then 8 is said to be the third power of 2.
    This could be written also as $2^3$, read "two to the third power."

13. Terms which contain exactly the same variables and the same exponents for the corresponding variables are called *like terms*, or *similar terms*. Only the numerical coefficients may be different.
    $3a^2b^3c$ and $-5a^2b^3c$ are like terms.

14. Terms which contain different variables or the same variables but with different exponents are *unlike terms*.
    $7a$ and $-5b$ are unlike terms because the variables are different.
    $3a$ and $3a^2$ are unlike terms because the powers of the variables are different.

15. The expression $-x$ means the opposite of the variable $x$ and not negative $x$. It is read as "the opposite of $x$" or the additive inverse of $x$ or as "negative one times $x$" since $-x = -1x$.
    When $x$ is a positive number, then $-x$ is a negative number, but when $x$ is a negative number, then $-x$ is a positive number.

16. A polynomial is expressed in *descending powers* of one of the variables if the terms are arranged so that the exponents of this variable decrease from term to term, starting from the left.
    The terms of the polynomial $x^3 - 5x^2 + 3x - 4$ are arranged in descending powers of $x$.

17. When the exponents of the variable increase from term to term, starting from the left, the polynomial is said to be expressed in *ascending powers* of that variable.

    The terms of the polynomial $6 + 4y - 7y^2 + 2y^3$ are arranged in ascending powers of $y$.

18. When a polynomial contains two variables, it may be possible to arrange the terms so that the powers of one variable will decrease from term to term and the powers of the other variable will increase.

    $a^3 + 2a^2b - 5ab^2 + b^3$ is arranged with the powers of $a$ decreasing and the powers of $b$ increasing from term to term.

19. The degree of a monomial is the sum of the exponents of the variables in the monomial.

    The degree of $x^3$ is 3. The degree of $4m^5n^2$ is $5 + 2$ or 7.

20. The degree of a polynomial is the degree of the monomial with the highest degree.

    The degree of the polynomial $5x^4 - 3x^3 + 7x^2 - 10x + 12$ is 4.
    The degree of the polynomial $2a^4b + 5ab - 9ab^2$ is the degree of the monomial $2a^4b$ which is 5.

## III. Sample Solutions

1. Which of the following is a binomial?
   $6x^2 - 3x + 7$      $-9y$     $9 - y$   *Answer:* $9 - y$
2. Which of the following is a monomial?
   $4a^2 + 9b^2$    $49a^2b^2$     $4a^2 - 12ab + 9b^2$   *Answer:* $49a^2b^2$
3. Write the terms of the polynomial $8m^3 - 5m^2 + 7m - 3$
   *Answer:* $8m^3, \ -5m^2, \ 7m, \ -3$
4. Name the constant term in $3b^2 - 7b + 5$   *Answer:* 5
5. What is the numerical coefficient of $c^2d$ in $-6c^2d$?   *Answer:* $-6$
6. Which of the following pairs of terms are like terms?
   **a.** $m^2$ and $2n$     **b.** $6ab$ and $6ac$     **c.** $-x^3y$ and $4x^3y$
   *Answer:* $-x^3y$ and $4x^3y$
7. Rearrange the polynomial $4c^2 - 15 + 3c^3 - c$ so that it is expressed in descending powers of the variable.   *Answer:* $3c^3 + 4c^2 - c - 15$
8. Rearrange the polynomial $6n^4 - 2n + 7 - 5n^2 + 8n^3$ so that it is expressed in ascending powers of the variable.
   *Answer:* $7 - 2n - 5n^2 + 8n^3 + 6n^4$
9. Write the degree of the monomial $3r^6t^2$.   *Answer:* 8
10. Write the degree of the polynomial $5b^4d - 2b^3d^3 + 9b^2d^5$   *Answer:* 7

## DIAGNOSTIC TEST

1. How many terms does each of the following expressions contain?
   **a.** $18 + y$          **b.** $4b^3 - 3b^2c + bc^2 - 5c^3$          **c.** $8a(3a - 4)$

2. Which of the following expressions are monomials? Which are binomials?
   **a.** $-25mn$                              **b.** $-25m + n$
   **c.** $10d^3 - 6d^2 + 3d + 24$             **d.** $7a^2 - 9ax + 2x$

3. Name the constant term in each of the following:
   **a.** $20c - 8d$          **b.** $25 - 10a$          **c.** $8x^2 - 3x + 1$

4. Write the polynomial $2y^2 + (-7y) + (-12)$ without the plus signs indicating addition and without the parentheses.

5. Write the terms of the following polynomial:
   $3a^4 - 5a^3b + 2a^2b^2 - ab^3 + 10b^4$

6. Write the degree of the:
   **a.** Monomial: $6b^4x^7$          **b.** Polynomial: $9x^5y^2 - 5x^4y^4 + xy^6$

7. What is the numerical coefficient of $b^3n$ in $-11b^3n$?

8. What is the numerical coefficient in: **a.** $c^3$? **b.** $-a^2t$?

9. Read, or write in words, the following: $-by^2$

10. Which of the following pairs of terms are like terms?
    **a.** $x^4$ and $4x$          **b.** $5b^6d^2$ and $4b^2d^6$          **c.** $8a^5c^3$ and $-3a^5c^3$

Rearrange each of the following polynomials so that it is expressed:

11. In descending powers of the variable: $3c^2 - 5c^4 - c + 7 + 4c^3$

12. In ascending powers of the variable:
    $6d^5 - 4d^7 + 9d - 11 + 2d^9 - 10d^3$

13. In descending powers of one variable and ascending powers of the second variable: $10m^3x - 3mx^3 + 5m^4 - 10m^2x^2 + 12x^4$

## RELATED PRACTICE EXAMPLES

1. How many terms does each of the following expressions contain?
   **a.** $6 + x$          **b.** $9y$          **c.** $3n^2 - 4n + 8$
   **d.** $12c^4 + 7c^3d - 5c^2d^2 + 2cd^3 - d^4$          **e.** $-64a^5y^8$          **f.** $7d(8 - 3d^2)$

2. Which of the following expressions are monomials? Which are binomials? Which are trinomials?
   **a.** $8x - y$                          **b.** $-8xy$
   **c.** $3m^2 - 4m + 1$                   **d.** $20 - c$
   **e.** $12n(n^2 - 5)$                    **f.** $25$
   **g.** $a - b$                           **h.** $10c^2 + 3cd + 9d^2$
   **i.** $2b^3 - b^2 + 5b - 7$             **j.** $-a^2$

3. Name the constant term in each of the following:
   a. $5b - 2$       b. $7x - 3y$       c. $17 - 9c$
   d. $8n^2 - 4n + 6$   e. $42 - 11r + r^2$   f. $5a^2 - 4ax - x^2$

4. Write each of the following polynomials without the plus signs indicating addition and without parentheses:
   a. $n^2 + (-4n)$       b. $3c^2 + (-10)$
   c. $a^2 + (-10a) + (-1)$   d. $x^2 + (-5x) + (-6)$
   e. $2m^2 + (-m) + (-3)$   f. $y^3 + (-y^2) + (-2y) + (-7)$

5. Write the terms of each of the following polynomials:
   a. $6m^2 + 3mn + 2n^2$       b. $7a^4 - 5a^3 + 6a^2 - a + 1$
   c. $9 - r^2$               d. $25a^4b^4 - 16x^2y^2$
   e. $3x^3 - 2x^2y - xy^2 + 4y^3$   f. $5 + n - 4n^2 + 8n^3 - 7n^4 + n^5$

6. Write the degree of each of the following:
   a. Monomials:
      (1) $m^5$   (2) $a^2x^4$   (3) $2b^7$   (4) $8c^6d^3$   (5) $24$   (6) $3b^2x^3y^5$
   b. Polynomials:
      (1) $6b^2 - 2bc + c^2$              (2) $3n^6 - 2n^4 + 5n^2 - 7$
      (3) $4a^2b^4 - 25c^3y^5$            (4) $15c^5 - c^4x + 3c^3x^2 - 2c^2x^3 +$
      (5) $8x^7y^2 - 5x^6y^5 + 7x^3y^{10} -$     $4x^5$
          $2xy^{11}$                      (6) $4a^2b^2x - 9a^2bx + 5ab^3x^2$

7. What is the numerical coefficient of:
   a. $n$ in $6n$?              b. $r$ in $-12r$?
   c. $a^4$ in $25a^4$?          d. $x^3y^2$ in $-1.7x^3y^2$?
   e. $c^5d^4$ in $\frac{3}{4}c^5d^4x$?   f. $m^2t^6$ in $-100m^2t^6$?

8. What is the numerical coefficient in each of the following?
   a. $d$     b. $r^2$     c. $-m$     d. $-t^3$     e. $xy$     f. $-a^4b^2x$

9. Read, or write in words, each of the following:
   a. $-y$   b. $-h$   c. $-ax$   d. $-b^2$   e. $-c^3d^4$   f. $-nx^6y^5$

10. Which of the following pairs of terms are like terms?
    a. $-x$ and $5x$              b. $b^3$ and $3b$
    c. $4a^2y$ and $4ay^2$        d. $2c^4$ and $-c^4$
    e. $2c^6d^3$ and $5c^3d^6$    f. $-8x^4y^2z^3$ and $-x^4y^2z^3$
    g. $-11a^5b^3c^2$ and $-7a^5b^2c^3$   h. $9m^2x$ and $4m^2xy$

11. Rearrange each of the following polynomials so that it is expressed in descending powers of the variable:
    a. $5x + 7x^2 + 6$
    b. $8a - 11a^3 + 9$
    c. $2 - 9y + 5y^2$
    d. $16m^2 - 9m^4 - 8m - 10 + 4m^3$
    e. $3t^5 - t^9 + 2t^3 + 7t^{12} - 14t^6 + 6t^{10}$

12. Rearrange each of the following polynomials so that it is expressed in ascending powers of the variable:
    a. $12b - 14b^2 + 9$
    b. $2x^3 - 9x^2 + 3x - 11$
    c. $6y^8 - 3y^3 + 8 - y^{10} + 2y^2 - 7y^4$
    d. $9a^2 + 7 + 6a^4 - 2a^3 + 15a + a^5$
    e. $16d - 8d^5 - 15 - 4d^2 + 4d^6 - d^3$

13. Rearrange each of the following polynomials so that it is expressed in descending powers of one variable and ascending powers of the second variable:
    a. $5xy - y^2 + 3x^2$
    b. $6cd^2 + d^2 - 7c^2 - 3c^2d$
    c. $2am^3 + 6a^2m^2 - a^3m + a^4 - 9m^4$
    d. $11b^2x^3 - b^4x + 3b^5 - 9bx^4 + 7x^5 - 4b^3x^2$
    e. $12r^3s^3 + 8s^6 - r^2s^4 + 2r^5s - 10r^6 + 3r^4s^2 - 7rs^5$

## CUMULATIVE REVIEW

1. Add:

| | | | | | | | |
|---|---|---|---|---|---|---|---|
| $-7$ | $-8$ | $+4$ | $0$ | $-13$ | $+36$ | $+5.8$ | $-\frac{3}{8}$ |
| $-9$ | $+11$ | $-5$ | $-28$ | $+13$ | $-29$ | $+3.3$ | $+\frac{7}{8}$ |

2. Subtract:

| | | | | | | | |
|---|---|---|---|---|---|---|---|
| $-2$ | $+18$ | $-3$ | $-59$ | $0$ | $-61$ | $+.29$ | $-1\frac{1}{2}$ |
| $+9$ | $+5$ | $-16$ | $+59$ | $-10$ | $0$ | $-.07$ | $-2\frac{2}{3}$ |

3. Multiply:

| | | | | | | | |
|---|---|---|---|---|---|---|---|
| $+5$ | $-9$ | $+12$ | $-21$ | $0$ | $+14$ | $-.03$ | $-2.5$ |
| $-8$ | $+6$ | $+12$ | $-7$ | $-17$ | $-4$ | $-.2$ | $+1.6$ |

4. Divide:

| | | | | | | | |
|---|---|---|---|---|---|---|---|
| $\dfrac{-27}{+3}$ | $\dfrac{-40}{-5}$ | $\dfrac{+19}{-1}$ | $\dfrac{+36}{-9}$ | $\dfrac{-80}{-16}$ | $\dfrac{0}{-4}$ | $\dfrac{-.9}{3}$ | $\dfrac{12}{-.2}$ |

5. Compute as indicated:
    a. $(+13) + (-8)$      b. $(+17) - (+25)$      c. $(-6)(-11)$
    d. $(-21) \div (-7)$      e. $\frac{3}{4}(-20)$      f. $(-9) + (+5.4)$
    g. $(-.18) \div (-.6)$      h. $(+1\frac{1}{2}) - (-\frac{5}{8})$

6. What is the numerical coefficient: a. $15b^2x^5$? b. $y^3$? c. $-m^7n$?
7. Name the constant term in the expression: $12x^2 - 11x + 6$.
8. Write in descending powers of the variable: $4c^2 - 17 + 3c^3 - c + 2c^4$
9. Write in ascending powers of the variable:
    $8x^4 - 5x + 9x^5 - 11 + x^3 - 6x^2$
10. Which are like terms: a. $5c^2$ and $5c$? b. $7x^3y^5$ and $7x^5y^3$? c. $4a^4b^6$ and $-a^4b^6$?

# Addition of Monomials

**I. Aim**    To add monomials.

**II. Procedure**

1. *To add like terms,* find the algebraic sum of the numerical coefficients of the addends and prefix it to their common factor of variables. See sample solutions 1, 2, and 3.

2. The numerical coefficient 1 is usually not expressed.
   $y$ means $1y$ or $+1y$ and $-y$ means $-1y$. See sample solutions 4 and 5.

3. When the algebraic sum is zero, do not write the common factor of variables. $0y$ is $0$. See sample solution 6.

4. The numerical coefficients of the common quantity are added algebraically and not the coefficients of the terms inside the parentheses. See sample solution 7.

5. *To add unlike terms,* indicate the addition. The sum is generally not written as a single term. See sample solutions 8 and 9.

6. In shortened form the terms of the expression are listed with the addition operation understood. See sample solution 10.

7. Check by adding again.

**III. Sample Solutions**

1. The sum of $8x$ and $3x$ is:
   Horizontally:                OR    Vertically:
   $$8x + 3x = (8 + 3)x = 11x$$

   *Answer: $11x$*

   $$\begin{array}{r} 8x \\ 3x \\ \hline (8+3)x = 11x \end{array}$$

2. $\begin{array}{r} -3a^2 \\ +5a^2 \\ \hline +2a^2 \end{array}$
   *Answer: $+2a^2$*

3. $\begin{array}{r} -4bc^3 \\ -2bc^3 \\ \hline -6bc^3 \end{array}$
   *Answer: $-6bc^3$*

4. $\begin{array}{r} +6ab \\ -ab \\ \hline +5ab \end{array}$
   *Answer: $+5ab$*

5. $\begin{array}{r} -5x^2y \\ +4x^2y \\ \hline -x^2y \end{array}$
   *Answer: $-x^2y$*

6. $\begin{array}{r} -3ax \\ +3ax \\ \hline 0 \end{array}$
   *Answer: $0$*

7. $\begin{array}{r} 5(a + 3b) \\ -8(a + 3b) \\ \hline -3(a + 3b) \end{array}$
   *Answer: $= -3(a + 3b)$*

**8.** The sum of $x^2$ and $3x$ is: $x^2 + (3x)$ or $x^2 + 3x$.   *Answer: $x^2 + 3x$*

**9.** The sum of $2a$ and $-4b$ is: $2a + (-4b)$ or $2a - 4b$.   *Answer: $2a - 4b$*

**10.** Combine: $5x - 9x + 2x - x$

$$5x - 9x + 2x - x$$
$$= 5x + (-9x) + 2x + (-x)$$
$$= -3x$$

*Answer: $-3x$*

---

## DIAGNOSTIC TEST

Add.

| | | | |
|---|---|---|---|
| **1.** $\begin{array}{r} +6m \\ +4m \\ \hline \end{array}$ | **2.** $\begin{array}{r} -5x \\ -3x \\ \hline \end{array}$ | **3.** $\begin{array}{r} +3b \\ -7b \\ \hline \end{array}$ | **4.** $\begin{array}{r} -8a \\ +2a \\ \hline \end{array}$ |
| **5.** $\begin{array}{r} -3b^2 \\ 5b^2 \\ \hline \end{array}$ | **6.** $\begin{array}{r} -6ax \\ -3ax \\ \hline \end{array}$ | **7.** $\begin{array}{r} -x \\ -x \\ \hline \end{array}$ | **8.** $\begin{array}{r} 3c \\ -c \\ \hline \end{array}$ |
| **9.** $\begin{array}{r} -5s^2 \\ +4s^2 \\ \hline \end{array}$ | **10.** $\begin{array}{r} +2dx \\ -2dx \\ \hline \end{array}$ | **11.** $\begin{array}{r} 0 \\ +y \\ \hline \end{array}$ | **12.** $\begin{array}{r} -4a^3b^2x \\ +7a^3b^2x \\ \hline \end{array}$ |
| **13.** $\begin{array}{r} -4a^2b \\ +6a^2b \\ -5a^2b \\ \hline \end{array}$ | **14.** $\begin{array}{r} -7x \\ 2x \\ 4x \\ -x \\ \hline \end{array}$ | **15.** $\begin{array}{r} 6(a+b) \\ -3(a+b) \\ 4(a+b) \\ \hline \end{array}$ | **16.** $\begin{array}{r} +\frac{1}{2}ax \\ +\frac{1}{2}ax \\ \hline \end{array}$ |

**17.** $\begin{array}{r} -.4b \\ +.2b \\ \hline \end{array}$

**18.** Write $5a + (-7a)$ without parentheses.

**19.** Find the sum of $b^3$, $b^2$, and $-2b$.

**20.** Add:
$(-6x) + (11x) + (-9x)$

**21.** Combine:
$2y - 5y - 6y - 4y$

**22.** Combine:
$p + .02p$

---

## *RELATED PRACTICE EXAMPLES*

Add:

| | | | | | | |
|---|---|---|---|---|---|---|
| **1.** $\begin{array}{r} +8a \\ +2a \\ \hline \end{array}$ | $\begin{array}{r} +5x \\ +7x \\ \hline \end{array}$ | $\begin{array}{r} +15c \\ +8c \\ \hline \end{array}$ | $\begin{array}{r} +6b \\ +6b \\ \hline \end{array}$ | $\begin{array}{r} 5m \\ 4m \\ \hline \end{array}$ | $\begin{array}{r} 21y \\ 14y \\ \hline \end{array}$ | $\begin{array}{r} 27a \\ 34a \\ \hline \end{array}$ |
| **2.** $\begin{array}{r} -2d \\ -5d \\ \hline \end{array}$ | $\begin{array}{r} -3b \\ -2b \\ \hline \end{array}$ | $\begin{array}{r} -7x \\ -4x \\ \hline \end{array}$ | $\begin{array}{r} -8m \\ -8m \\ \hline \end{array}$ | $\begin{array}{r} -12r \\ -7r \\ \hline \end{array}$ | $\begin{array}{r} -6t \\ -14t \\ \hline \end{array}$ | $\begin{array}{r} -50y \\ -20y \\ \hline \end{array}$ |

3.
$$+6x \qquad +8c \qquad +7m \qquad +8b \qquad +15y \qquad +23z \qquad +13t$$
$$\underline{-4x} \qquad \underline{-5c} \qquad \underline{-9m} \qquad \underline{-11b} \qquad \underline{-5y} \qquad \underline{-10z} \qquad \underline{-25t}$$

4.
$$-8y \qquad -9a \qquad -3k \qquad -5d \qquad -16t \qquad -17r \qquad -33x$$
$$\underline{+3y} \qquad \underline{+6a} \qquad \underline{+7k} \qquad \underline{+10d} \qquad \underline{+24t} \qquad \underline{+12r} \qquad \underline{+41x}$$

5.
$$+3a^2 \qquad -6x^2 \qquad -8c^2 \qquad -7x^3 \qquad 5y^4 \qquad -15b^3 \qquad 12m^4$$
$$\underline{+2a^2} \qquad \underline{-4x^2} \qquad \underline{-6c^2} \qquad \underline{+9x^3} \qquad 5y^4 \qquad 7b^3 \qquad \underline{-16m^4}$$

6.
$$+4xy \qquad -6ab \qquad -4cd \qquad +5bc \qquad 6mn \qquad 9x^2y \qquad -5ab^2$$
$$\underline{+5xy} \qquad \underline{+8ab} \qquad \underline{-4cd} \qquad \underline{-8bc} \qquad 7mn \qquad \underline{-13x^2y} \qquad 11ab^2$$

7.
$$+a \qquad -d \qquad c \qquad -x^2 \qquad -bx \qquad +abc \qquad -x^2y^2$$
$$\underline{+a} \qquad \underline{-d} \qquad \underline{c} \qquad \underline{-x^2} \qquad \underline{-bx} \qquad \underline{+abc} \qquad \underline{-x^2y^2}$$

8.
$$+4b \qquad +3ab \qquad -7x^2 \qquad +x \qquad -a \qquad -9xy \qquad -a^2x$$
$$\underline{-b} \qquad \underline{+ab} \qquad \underline{-x^2} \qquad \underline{-5x} \qquad \underline{+9a} \qquad \underline{-xy} \qquad 5a^2x$$

9.
$$+3x \qquad -5b \qquad -4y \qquad +6m \qquad -2d \qquad 11a^2b \qquad -15m^2n^2$$
$$\underline{-2x} \qquad \underline{+6b} \qquad \underline{+3y} \qquad \underline{-7m} \qquad \underline{+d} \qquad \underline{-10a^2b} \qquad 14m^2n^2$$

10.
$$+b \qquad x^2 \qquad -a^2b^2 \qquad +5x \qquad 9bx \qquad -6m^2 \qquad -3x^2y^2$$
$$\underline{-b} \qquad \underline{-x^2} \qquad \underline{a^2b^2} \qquad \underline{-5x} \qquad \underline{-9bx} \qquad 6m^2 \qquad 3x^2y^2$$

11.
$$+x \qquad -b \qquad -2b \qquad -5a^2x^2 \qquad 0 \qquad 0 \qquad 0$$
$$\underline{0} \qquad \underline{0} \qquad \underline{0} \qquad \underline{0} \qquad \underline{+c} \qquad \underline{-4x} \qquad 2y^2$$

12.
$$+6a^2bx \qquad -9ax^4y^2 \qquad -5b^2c^5d^6 \qquad -5m^2x^4y \qquad -6a^4b^2x^2 \qquad +3m^3x^2y$$
$$\underline{+4a^2bx} \qquad \underline{-2ax^4y^2} \qquad \underline{+3b^2c^5d^6} \qquad \underline{-m^2x^4y} \qquad \underline{+6a^4b^2x^2} \qquad \underline{-4m^3x^2y}$$

13.
$$+5x \qquad -3a \qquad -7a^2 \qquad -3x^2y^2 \qquad -8a^2b^2c^2 \qquad +ab^3c$$
$$+2x \qquad -4a \qquad -5a^2 \qquad +5x^2y^2 \qquad +4a^2b^2c^2 \qquad -2ab^3c$$
$$\underline{+4x} \qquad \underline{-a} \qquad \underline{+3a^2} \qquad \underline{-8x^2y^2} \qquad \underline{+5a^2b^2c^2} \qquad \underline{-2ab^3c}$$

14.
$$-2x \qquad +4b^2 \qquad -5a^2x$$
$$+8x \qquad +2b^2 \qquad +2a^2x$$
$$-5x \qquad -5b^2 \qquad -4a^2x$$
$$\underline{+4x} \qquad \underline{-b^2} \qquad \underline{+9a^2x}$$

15.
$$5(c+d) \qquad -9(x-7y) \qquad -12(a+2b-5c)$$
$$\underline{3(c+d)} \qquad \underline{6(x-7y)} \qquad -4(a+2b-5c)$$
$$\underline{9(a+2b-5c)}$$

16.
$$+\tfrac{1}{2}x \qquad -\tfrac{3}{4}b^2 \qquad -4\tfrac{1}{8}ab$$
$$\underline{+\tfrac{1}{2}x} \qquad \underline{-1\tfrac{1}{4}b^2} \qquad \underline{+3\tfrac{5}{8}ab}$$

17.
$$+.3a \qquad -1.6x \qquad +.8m^2n$$
$$\underline{+.9a} \qquad \underline{-.4x} \qquad \underline{-4.2m^2n}$$

18. Write each of the following expressions without parentheses:
    **a.** $3x + (-2y)$     **b.** $12a + (-9)$     **c.** $5m + (-8)$     **d.** $7b + (6)$
    **e.** $11y + (3z)$     **f.** $14h + (-t)$     **g.** $10n + (8x)$     **h.** $3x + (-15)$

19. Find the sum of:
    **a.** $+5c$ and $+3d$          **b.** $-8r$ and $+5$          **c.** $-a$ and $+b$
    **d.** $3x$ and $2y$          **e.** $6x^2$ and $4x$          **f.** $+4a$ and $-2b$
    **g.** $x^3, 2x^2,$ and $-3x$          **h.** $-7b^3, -2b^2,$ and $-5b$
    **i.** $8w, -2x, -5y,$ and $6z$

20. Add horizontally:
    **a.** $5x + 3x$          **b.** $(-8n) + (7n)$          **c.** $(-4x^2) + (-2x^2)$
    **d.** $9a + (-6a)$          **e.** $2x + (-x)$          **f.** $(3c^2y) + (-c^2y)$
    **g.** $12x + (8)$          **h.** $6y + (-3)$          **i.** $11n + (-5n)$
    **j.** $23b + (-14)$          **k.** $7y + (-8y)$          **l.** $(9abx) + (-9abx)$
    **m.** $(4y) + (6y) + (-3y)$   **n.** $(-bc) + (-bc) + (-4bc)$
    **o.** $(5m^3r^2) + (-2m^3r^2) + (-9m^3r^2) + (6m^3r^2)$
    **p.** $(8c) + (-3d) + (2d) + (-5c)$
    **q.** $(3ax^2) + (4a^2x^2) + (-5ax^2) + (-ax^2)$
    **r.** $(6x) + (-7) + (8x) + (4) + (-20x) + (3)$

21. Combine:
    **a.** (1) $8b + 3b$          (2) $9xy + 15xy$          (3) $\frac{1}{2}c + \frac{3}{4}c$
         (4) $1.5a + 2.3a$          (5) $6y^2 - 4y^2$          (6) $3d - 9d$
         (7) $x - 7x$          (8) $4.1n - 2.6n$          (9) $4b^2 - 4b^2$
         (10) $8ax - 9ax$
    **b.** (1) $5x + 6x + 2x$          (2) $8m - 7m + m$   (3) $2a^2 - 5a^2 - 3a^2$
         (4) $cd + 6cd - 9cd$          (5) $x + 3x + 7x$   (6) $-xy^2 + 2xy^2 - 4xy^2$
         (7) $6b^2 + 2b^2 - 8b^2$          (8) $-l - 3l - 2l$   (9) $4m^2n^2 - m^2n^2 + 3m^2n^2$
         (10) $8r^3 - 10r^3 + 6r^3$
    **c.** (1) $2x - 3x - x + 4x$
         (2) $-4y - 8y + 6y - 4y$
         (3) $5a + 6a - 3a - a$
         (4) $-x^2 + 2x^2 - x^2 - 3x^2 + 6x^2$
         (5) $ab + 7ab - 2ab + ab - 4ab$
         (6) $2x^2y - 3x^2y - x^2y + x^2y - 3x^2y$
         (7) $4t - 2t - 6t + 3t + t - 5t$
         (8) $6R + 8R - 4R - 7R + 3R - 9R$
         (9) $7m^4x^2 - m^4x^2 - 2m^4x^2 + 3m^4x^2 - m^4x^2$
         (10) $2cr - 5cr + 7cr + cr - 6cr + 2cr$

22. Combine:
    **a.** $c + .45c$     **b.** $a + .05a$     **c.** $p + .3p$     **d.** $s + .4s$
    **e.** $a + .06a$     **f.** $c + (-.25c)$     **g.** $s + (-.18s)$     **h.** $p + .15p$
         **i.** $a + (-.2a)$     **j.** $s + .375s$

# Addition of Polynomials

**I. Aim**   To add two or more polynomials.

**II. Procedure**

1. When the polynomials are already arranged in columns so that like terms are under each other, add each column separately.
2. Otherwise, write one polynomial under the other so that like terms are in vertical columns, then add each column.
3. In some cases it may be necessary to rewrite the polynomials in descending powers or ascending powers of the variables.
4. If the sum of a column is zero, do not write this zero in the sum if the sum of any other column is not zero.
5. If the sums of all the columns are zero, write only one zero in the result.
6. To simplify a shortened form of an algebraic expression, add like terms mentally and then write the sum. See sample solution 5.
7. Polynomials may be added vertically or horizontally. See sample solution 6.
8. Check by adding again.

**III. Sample Solutions**

1.
$$\begin{array}{r} 2x^2 - 3xy + 5y^2 \\ x^2 + 2xy - 3y^2 \\ -5x^2 + 5xy - y^2 \\ \hline -2x^2 + 4xy + y^2 \end{array}$$

*Answer:* $-2x^2 + 4xy + y^2$

2.
$$\begin{array}{r} 3a + 5b - 4c \\ 2a - 5b - 4c \\ \hline 5a \quad\;\; - 8c \end{array}$$

*Answer:* $5a - 8c$

3.
$$\begin{array}{r} 3x - 2y + 7 \\ x - 3y - 4 \\ -4x + 5y - 3 \\ \hline 0 \end{array}$$

*Answer:* 0

4. Find the sum of $a^2 - b^2$, $2a^2 - 3ab + b^2$, and $5ab - a^2 + b^2$.

$$\begin{array}{r} a^2 \qquad\quad - b^2 \\ 2a^2 - 3ab + b^2 \\ -a^2 + 5ab + b^2 \\ \hline 2a^2 + 2ab + b^2 \end{array}$$

*Answer:* $2a^2 + 2ab + b^2$

150

5. Add like terms mentally, then write the sum:

$2c^2 - 3c + 6 - 2c + 7c^2 - 3 + c - 4c^2 - 4 + c^2$

Think:

$= 2c^2 + 7c^2 - 4c^2 + c^2 - 3c - 2c + c + 6 - 3 - 4$

$= 2c^2 + 7c^2 + (-4c^2) + c^2 + (-3c) + (-2c) + c + 6 + (-3) + (-4)$

$= 6c^2 + (-4c) + (-1)$

$= 6c^2 - 4c - 1$

*Answer:* $6c^2 - 4c - 1$

6. Find the sum of $(3x + 4y) + (5x - 7y)$.

Vertically:    Horizontally:

$$(3x + 4y) + (5x - 7y)$$

$$3x + 4y$$
$$5x - 7y$$
$$\overline{8x - 3y}$$

$= 3x + 4y + 5x - 7y$

$= 3x + 5x + 4y - 7y$

$= 8x - 3y$

*Answer:* $8x - 3y$

## DIAGNOSTIC TEST

Add:

1. $2a + 3$
   $3a - 5$

2. $4x - 2y$
   $2x - 5y$

3. $5b - 1$
   $-4b + 2$

4. $2a + b$
   $a - b$

5. $8m^2 - 7n^2$
   $-8m^2 + 7n^2$

6. $5a^2 - 4ab - 3b^2$
   $2a^2 + 2ab - 5b^2$

7. $7r - 3s - 4t$
   $9r + 3s - 2t$

8. $2b - 2c - d$
   $-6b - 3c$

9. $9m + 8r^2$
   $3m - 5r^2$
   $-15m + 2r^2$

10. $6a^2b^2 + 3ab - 5$
    $a^2b^2 - 5ab - 2$
    $-5a^2b^2 - 4ab + 9$

11. Find the sum of $4x^2 + 6x - 3$ and $2x^2 - 2x + 5$.
12. Find the sum of $x^2 - y^2$, $2x^2 - 4xy + y^2$, and $2y^2 - 3xy - x^2$.
13. Simplify $2x^2 - 5 - 2x + x^2 - 3x + 6 - x^2 - 4x - x^2 + 2$.
14. Add $(c^2 - 2cd) + (2cd + d^2) + (d^2 - c^2)$.

## RELATED PRACTICE EXAMPLES

Add:

1.

| | | | |
|---|---|---|---|
| $5x+2$ | $3b-2$ | $6m-3$ | $y+3$ |
| $3x+4$ | $4b-5$ | $7m+9$ | $y-7$ |

| | | | |
|---|---|---|---|
| $2x^2-5$ | $-8ab+5$ | $x^2+2$ | $-6x-2$ |
| $2x^2+3$ | $6ab-2$ | $-6x^2+3$ | $-x-5$ |

2.

| | | | |
|---|---|---|---|
| $3a+2b$ | $x+3y$ | $5c-2d$ | $2x^2-5x$ |
| $2a+4b$ | $2x+y$ | $c-3d$ | $9x^2+2x$ |

| | | | |
|---|---|---|---|
| $3ab-9xy$ | $9a^2-4b^2$ | $-c^2+5c$ | $-2a^2b-7ab^2$ |
| $7ab+4xy$ | $-2a^2+7b^2$ | $-4c^2+8c$ | $4a^2b-3ab^2$ |

3.

| | | | |
|---|---|---|---|
| $4a+6b$ | $-2x-9$ | $-b^2+5$ | $9-3cd$ |
| $3a-5b$ | $x-2$ | $2b^2-4$ | $6+2cd$ |

4.

| | | | |
|---|---|---|---|
| $5x+3$ | $x+y-z$ | $a^2-3b^2$ | $-2x^2-5xy+2y^2$ |
| $2x-3$ | $x-y-z$ | $-a^2-3b^2$ | $x^2+5xy-2y^2$ |

5.

| | | | |
|---|---|---|---|
| $x-4$ | $4x^2+9y^2$ | $-x^2-2xy-y^2$ | $2ab-3bc+6cd$ |
| $-x+4$ | $-4x^2-9y^2$ | $x^2+2xy+y^2$ | $ab-bc-5cd$ |
| | | | $-3ab+4bc-cd$ |

6.

| | | | |
|---|---|---|---|
| $2a+3b+1$ | $3b-4m+2x$ | $x^2+2x-7$ | $x^2+2xy+y^2$ |
| $4a+2b+7$ | $5b-6m-4x$ | $x^2-5x+3$ | $2x^2-xy-2y^2$ |

| | | | |
|---|---|---|---|
| $4a^2-2ab-9b^2$ | $9m^2-3m-5$ | $3m^3+2m^2-4m$ | $7a^3b-3a^2b^2+4ab^3$ |
| $-a^2+4ab-b^2$ | $2m^2-9m+6$ | $2m^3+8m^2-5m$ | $2a^3b-2a^2b^2-9ab^3$ |

7.

| | | | |
|---|---|---|---|
| $3a-4b+5c$ | $9x^2-4x+6$ | $-4x^2+2xy-9$ | $2y^2-y+6$ |
| $7a-5b-5c$ | $5x^2+4x-3$ | $4x^2+3xy+8$ | $-2y^2+y-6$ |

8.

| | | | |
|---|---|---|---|
| $4x^2-8$ | $5b^2-3b+2$ | $4x^2-5x$ | $5x^2-y^2$ |
| $2x^2$ | $4b^2-7b$ | $2x^2+3x-4$ | $2x^2+2xy+5y^2$ |

| | | | |
|---|---|---|---|
| $-3a-2c$ | $x^2+xy$ | $4c^2-6cd$ | $b^3+3b^2-4ab$ |
| $5a+4b$ | $+xy+y^2$ | $+6cd-9d^2$ | $-3b^2+4b-27$ |

9.

| | | | |
|---|---|---|---|
| $4a-3b$ | $8x^2-3xy$ | $-3a^2-4ab$ | $10b^2-c^2$ |
| $5a-4b$ | $6x^2-4xy$ | $-5a^2$ | $-4b^2-c^2$ |
| $-2a-b$ | $-9x^2+7xy$ | $+10a^2-4ab$ | $-6b^2+2c^2$ |

10.
$$2c^2 + 4c + 5$$
$$4c^2 + 5c + 2$$
$$8c^2 + 3c + 1$$

$$4a^2 - 3ab + b^2$$
$$2a^2 - 5ab - 2b^2$$
$$3a^2 - ab + b^2$$

$$a^3 \qquad\qquad + 4$$
$$a^2 + 5a - 1$$
$$3a^3 + a^2 - 3a + 6$$

$$2x^2 \qquad - y^2$$
$$x^2 + 2xy + 3y^2$$
$$- xy - 6y^2$$

$$5c^2 - 5cd - 2d^2$$
$$3c^2 - cd$$
$$+ 9cd - d^2$$

$$2d^2 + 3d - 9$$
$$- d^2 + d + 5$$
$$- d^2 - 5d + 3$$

$$5x^2 - 2x + 3$$
$$2x^2 - 4x - 5$$
$$- x^2 + 3x - 7$$

$$x^2 - 2xy + y^2$$
$$2x^2 - 2xy - y^2$$
$$- x^2 + 5xy - y^2$$

$$4x^4 + 3x^2y^2 - 2y^4$$
$$-5x^4 - x^2y^2 + y^4$$
$$- x^4 - 2x^2y^2 + y^4$$

$$2b^3 - 4b^2c$$
$$4b^2c - 3bc^2$$
$$3bc^2 - 6c^3$$

$$a^4 + 5a^3b$$
$$- 3a^3b - 2a^2b^2$$
$$- a^2b^2 + 2ab^3$$
$$- 2ab^3 - 6b^4$$

Find the sum of:

11. **a.** $6x - 2y$ and $3x + 5y$
    **b.** $2a - 3b + 4c$ and $7a - 5b - 3c$
    **c.** $3a - 4x$, $5a - 3x$, and $a - 2x$
    **d.** $2x^2 - 3x + 6$, $x^2 - 5x - 1$, and $2x^2 - 2x - 4$
    **e.** $8a^2 - 2ab - b^2$, $4a^2 - ab - b^2$, and $a^2 - ab - b^2$

12. **a.** $x^2 + x$ and $x + 3$        **b.** $a^2 + 2ab$ and $-3ab - b^2$
    **c.** $5x + y$ and $8y + 7$        **d.** $x^2 + 3x$ and $-2x + 9$
    **e.** $a^2 - 6ax$ and $6ax - x^2$        **f.** $6a^2 - 2a + 7$, $2a^2 - 3$, and $5a + 1$
    **g.** $8c^2 - 4d^2$, $3c^2 - 5cd + d^2$, and $4c^2 - 2cd + d^2$
    **h.** $2x^4 + 3x^3 + x^2$, $2x^3 - 3x^2 + 6$, and $x^2 - 5x + 7$
    **i.** $5ab - 3ab^2 + 5ab^3$, $6ab^2 - 2ab - 3ab^3$, and $6ab^3 - 2ab^2$
    **j.** $8x^3y - 2x^2y^2$, $5x^2y^2 - 2xy^3$, and $6x^3y - 5xy^3$

13. Simplify:
    **a.** $3b + 5 + 4b$        **b.** $a + 3 + 2a + 6$
    **c.** $9 - 4x - 3$        **d.** $c^2 - 7c + 6 + 2c^2 - 12$
    **e.** $m^2 - 2mn + n^2 - m^2 + n^2$        **f.** $2x^2y - 3xy^2 + x^2y - 4x^2y - 2xy^2$
    **g.** $5s - 3 + 7s + 8 - s - 9s - 5$
    **h.** $4n^2 - 8n + 7 - 6 + 2n - n^2 - 3n^2 + n - 8 - n^2$
    **i.** $9x^2 - 3xy - y^2 - 3x^2 - 5xy - x^2 - 4xy + y^2 - 7y^2$

14. Add each of the following as indicated:
    **a.** $(3m - 9) + (4m + 7)$        **b.** $(8r - 2s) + (-5r - 4s)$
    **c.** $(5a^2x - by^2) + (11a^2x - by^2)$        **d.** $(6b^2 - 3b - 5) + (-b^2 + 3b - 8)$
    **e.** $(4c^2 - cn - 2n^2) + (n^2 - cn - 5c^2)$        **f.** $(x - 5) + (2x - 3) + (x - 4)$
    **g.** $(c + d) + (2c - 3d) + (c - d)$        **h.** $(a - b) + (b - a) + (2b - a)$
    **i.** $(x^2 - 3x + 7) + (2x + x^2 - 3) + (8 - x - 2x^2)$

# Subtraction of Monomials

## I. Aim  To subtract monomials.

## II. Procedure

1. To subtract like terms, add to the minuend the additive inverse (or opposite) of the subtrahend.
2. Or simply change the sign of the given subtrahend mentally and proceed as in addition since the change in sign determines the additive inverse.
3. Note in Exercise 2-12 that both *subtracting a number* and *adding the opposite of this number* give the same answer.
4. Do not write variables after zero. See sample solutions 2, 3, and 4.
5. To subtract unlike terms, indicate their difference (sample solution 6).
6. Check by adding the answer to the given subtrahend. The sum should be the given minuend.

## III. Sample Solutions

1. From $-6xy$ subtract $9xy$.

   Horizontally:

   $$(-6xy) - (9xy) = (-6xy) + (-9xy)$$
   $$= -15xy$$

   Vertically:

   Subtract:
   $$-6xy$$
   $$\underline{\phantom{-}9xy}$$

   Add:
   $$-6xy$$
   $$\underline{-9xy}$$
   $$-15xy$$

   *Answer:* $-15xy$

2. Subtract:
   $$-5ax$$
   $$\underline{\phantom{-}0}$$
   $$-5ax$$

   *Answer:* $-5ax$

3. Subtract:   Add:
   $$0 \qquad\quad 0$$
   $$\underline{-2b} \quad \underline{+2b}$$
   $$\qquad\quad +2b$$

   *Answer:* $+2b$

4. Subtract:   Add:
   $$-3b^2c \qquad -3b^2c$$
   $$\underline{-3b^2c} \quad \underline{+3b^2c}$$
   $$\qquad\qquad 0$$

   *Answer:* 0

   Note: 0 in problem 2 represents $0ax$, 0 in problem 3 represents $0b$, and 0 in problem 4 represents $0b^2c$.

154

**5.** Subtract $5x^2$ from $4x^2$

Subtract:   Add:

$$
\begin{array}{c@{\qquad}c}
4x^2 & 4x^2 \\
\underline{5x^2} & \underline{-5x^2} \\
& -x^2
\end{array}
$$

*Answer:* $-x^2$

**6.** Subtract $-3n$ from $-7m$.

$(-7m) - (-3n)$
$= (-7m) + (+3m)$
$= -7m + 3n$

Vertically:

$$
\begin{array}{r}
-7m \\
-\,3n \\
\hline
-7m + 3n
\end{array}
$$

*Answer:* $-7m + 3n$

## *PRELIMINARY EXAMPLES*

Find the additive inverse of:

**1.** $8y$      **2.** $+11m^2$      **3.** $-6b$      **4.** $-x$      **5.** $+7r^2t^6$
**6.** $-3a^3x^2$      **7.** $-9c^4$      **8.** $2nxy$      **9.** $5ad^7$      **10.** $-8b^5x^3z$

## DIAGNOSTIC TEST

Subtract the lower term from the upper term:

**1.** $\begin{array}{r} +6x \\ \underline{+4x} \end{array}$      **2.** $\begin{array}{r} -3y \\ \underline{-8y} \end{array}$      **3.** $\begin{array}{r} +5abc \\ \underline{-7abc} \end{array}$

**4.** $\begin{array}{r} -6b^2 \\ \underline{+8b^2} \end{array}$      **5.** $\begin{array}{r} -2d \\ \underline{-\,d} \end{array}$      **6.** $\begin{array}{r} 6xy \\ \underline{0} \end{array}$

**7.** $\begin{array}{r} 0 \\ \underline{-5a} \end{array}$      **8.** $\begin{array}{r} 9xy^2 \\ \underline{9xy^2} \end{array}$      **9.** $\begin{array}{r} -4a^2 \\ \underline{+4a^2} \end{array}$

**10.** $\begin{array}{r} \frac{1}{4}a^3 \\ \underline{-\frac{3}{4}a^3} \end{array}$      **11.** $\begin{array}{r} -3.6x^2 \\ \underline{1.5x^2} \end{array}$      **12.** $\begin{array}{r} -7(x - y - z) \\ \underline{4(x - y - z)} \end{array}$

**13.** Find the missing number: $6x - 8x = 6x + (\quad)$

Subtract horizontally as indicated:

**14.** $(c^2y^3) - (-2c^2y^3)$      **15.** $p - .06p$      **16.** From $4r$ take $-6r$.

**17.** Subtract $-x$ from $-8x$.      **18.** From $0$ subtract $2y$.

**19.** Find the missing number: $5b - (-7c) = 5b + (\quad)$

**20.** Subtract the lower term from the upper term:

$$
\begin{array}{r}
+x^2 \\
\underline{+\,x}
\end{array}
$$

**21.** Subtract as indicated:

$10x - (-9y)$

**22.** Subtract $3b$ from $6a$.

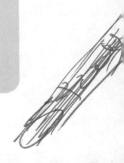

## RELATED PRACTICE EXAMPLES

Subtract the lower term from the upper term:

1. $+8x$ $\quad$ $+6a$ $\quad$ $+9c$ $\quad$ $+4b^2$ $\quad$ $+2xy$ $\quad$ $7m$ $\quad$ $6ab^2$
   $+3x$ $\quad$ $+2a$ $\quad$ $+3c$ $\quad$ $+8b^2$ $\quad$ $+9xy$ $\quad$ $4m$ $\quad$ $8ab^2$

2. $-5b$ $\quad$ $-9a^2$ $\quad$ $-8m^2$ $\quad$ $-2a$ $\quad$ $-3b^3$ $\quad$ $-7cd$ $\quad$ $-8x^2y^2$
   $-2b$ $\quad$ $-4a^2$ $\quad$ $-6m^2$ $\quad$ $-8a$ $\quad$ $-6b^3$ $\quad$ $-12cd$ $\quad$ $-10x^2y^2$

3. $+6a$ $\quad$ $+7m$ $\quad$ $+2c^2$ $\quad$ $+4bx$ $\quad$ $+3d^3y$ $\quad$ $5cy$ $\quad$ $6xyz$
   $-2a$ $\quad$ $-3m$ $\quad$ $-8c^2$ $\quad$ $-6bx$ $\quad$ $-4d^3y$ $\quad$ $-3cy$ $\quad$ $-9xyz$

4. $-5x$ $\quad$ $-9d^3$ $\quad$ $-3mx$ $\quad$ $-5r^2t$ $\quad$ $-6b$ $\quad$ $-2ry^2$ $\quad$ $-3b^2c^2$
   $-3x$ $\quad$ $+5d^3$ $\quad$ $+7mx$ $\quad$ $+8r^2t$ $\quad$ $3b$ $\quad$ $5ry^2$ $\quad$ $10b^2c^2$

5. $+6m$ $\quad$ $+3x^2$ $\quad$ $2a^2c$ $\quad$ $2ay^2$ $\quad$ $-9x$ $\quad$ $-4y$ $\quad$ $-axy$
   $+5m$ $\quad$ $+4x^2$ $\quad$ $a^2c$ $\quad$ $3ay^2$ $\quad$ $-8x$ $\quad$ $-3y$ $\quad$ $-2axy$

6. $3x$ $\quad$ $2b^4$ $\quad$ $a$ $\quad$ $5m^2x$ $\quad$ $-2b$ $\quad$ $-x$ $\quad$ $-5c^2x^2$
   $0$ $\quad$ $0$ $\quad$ $0$ $\quad$ $0$ $\quad$ $0$ $\quad$ $0$ $\quad$ $0$

7. $0$ $\quad$ $0$ $\quad$ $0$ $\quad$ $0$ $\quad$ $0$ $\quad$ $0$ $\quad$ $0$
   $+4b$ $\quad$ $+6x$ $\quad$ $+3a^2$ $\quad$ $5b^2c$ $\quad$ $-5y$ $\quad$ $-8rt$ $\quad$ $-2x^2y^2$

8. $+a$ $\quad$ $+5x$ $\quad$ $b^2c^2$ $\quad$ $-x$ $\quad$ $-4b^3$ $\quad$ $-9abc$ $\quad$ $-6a^3x$
   $+a$ $\quad$ $+5x$ $\quad$ $b^2c^2$ $\quad$ $-x$ $\quad$ $-4b^3$ $\quad$ $-9abc$ $\quad$ $-6a^3x$

9. $+x$ $\quad$ $+3a$ $\quad$ $7x^4y$ $\quad$ $-b^2$ $\quad$ $-2y$ $\quad$ $-8c^2$ $\quad$ $-5x^2y$
   $-x$ $\quad$ $-3a$ $\quad$ $-7x^4y$ $\quad$ $+b^2$ $\quad$ $+2y$ $\quad$ $8c^2$ $\quad$ $5x^2y$

10. $+\frac{3}{4}x$ $\quad$ $-\frac{3}{2}m^4$ $\quad$ $\frac{1}{3}cd^2$ $\quad$ $\frac{7}{8}xyz$ $\quad$ $-\frac{3}{5}a^2xy$
    $+\frac{1}{4}x$ $\quad$ $+\frac{1}{2}m^4$ $\quad$ $-\frac{2}{3}cd^2$ $\quad$ $\frac{5}{8}xyz$ $\quad$ $-\frac{1}{10}a^2xy$

11. $+.8y$ $\quad$ $+4.5b^2$ $\quad$ $-7.3\ ac$ $\quad$ $2.6b^2m^3$ $\quad$ $-7.05b^2c$
    $+.3y$ $\quad$ $-3.7b^2$ $\quad$ $-.49ac$ $\quad$ $3.8b^2m^3$ $\quad$ $1.6\ b^2c$

Subtract the lower expression from the upper expression:

12. $3(2a+7b)$ $\quad$ $-2(x-5y)$ $\quad$ $8(a+b-c)$ $\quad$ $-7(a-6x-9y)$
    $5(2a+7b)$ $\quad$ $-7(x-5y)$ $\quad$ $-2(a+b-c)$ $\quad$ $9(a-6x-9y)$

Find the missing numbers:

13. **a.** $6x-8x=6x+(\ \ )$ $\qquad$ **b.** $15t-20t=15t+(\ \ )$
    **c.** $7y-(-5y)=7y+(\ \ )$ $\qquad$ **d.** $2b-(-3b)=2b+(\ \ )$
    **e.** $31n-49n=31n+(\ \ )$ $\qquad$ **f.** $x^2-(-x^2)=x^2+(\ \ )$

Subtract horizontally as indicated:

14. **a.** $(4d) - (3d)$
    **c.** $(+6y) - (-3y)$
    **e.** $(-5c) - (4c)$
    **g.** $(-4abx) - (-6abx)$
    **i.** $(-9x^2y) - (-9x^2y)$

    **b.** $(+5x) - (+2x)$
    **d.** $(-8a) - (5a)$
    **f.** $(x^2) - (-x^2)$
    **h.** $(a^2b^3) - (a^2b^3)$
    **j.** $(-5mnx) - (-6mnx)$

Subtract as indicated:

15. **a.** $p - .05p$
    **c.** $s - .4s$
    **e.** $p - .08p$
    **g.** $c - .47c$
    **i.** $p - .045p$

    **b.** $a - .25a$
    **d.** $c - .32c$
    **f.** $s - .1s$
    **h.** $a - .56a$
    **j.** $c - .375c$

16. **a.** From $4b$ subtract $3b$.
    **b.** From $-5m$ take $7m$.
    **c.** Find the difference between $2x$ and $-4x$.
    **d.** From $-8abc$ take $-9abc$.
    **e.** Find the difference between $-2x^3y$ and $x^3y$.
    **f.** From $9b^2c$ take $-5b^2c$.
    **g.** Find the difference between $-8x^2$ and $3x^2$.
    **h.** From $-7x^2y^3$ subtract $-8x^2y^3$.
    **i.** Find the difference between $-6m$ and $+6m$.
    **j.** From $-mx^2y$ subtract $4mx^2y$.

17. **a.** Take $2x$ from $7x$.
    **b.** Subtract $-x^2$ from $3x^2$.
    **c.** Take $-4ab$ from $-5ab$.
    **d.** Take $3c^2d$ from $-2c^2d$.
    **e.** Subtract $-8xy$ from $-8xy$.

18. **a.** From 0 take $-3xy$.
    **b.** From $2a^2c$ subtract 0.
    **c.** Subtract $-8ab^2$ from 0.
    **d.** Take $xy$ from 0.
    **e.** Find the difference between 0 and $-2x$.

19. Find the missing numbers:
    **a.** $4x - (11y) = 4x + (\quad)$
    **b.** $3m - (-n) = 3m + (\quad)$
    **c.** $11d - (-6) = 11d + (\quad)$
    **d.** $9b - (-2c) = 9b + (\quad)$
    **e.** $18x - (23) = 18x + (\quad)$
    **f.** $42a - (-17y) = 42a + (\quad)$

Subtract the lower term from the upper term:

20.

| $a$ | $x$ | $b$ | $5$ | $-b^3$ |
|---|---|---|---|---|
| $+\ b$ | $-\ y$ | $+\ 4$ | $-\ c$ | $-\ c^2$ |

| $5a^2b$ | $-8mx$ | $6a^2y$ | $a^4$ |
|---|---|---|---|
| $-\ 2ab^2$ | $-\ 2nx$ | $3ay$ | $+\ a^3$ |

21. Subtract as indicated:

    **a.** $8a - (4c)$       **b.** $12m - (-3n)$       **c.** $13y^2 - (-6)$

    **d.** $x^3 - (5x^2)$     **e.** $4a^2x - (5ax^2)$     **f.** $15 - (-n)$

22. **a.** From $x$ take $y$.               **b.** From $-2b$ subtract $3c$.

    **c.** Find the difference between $3y^2$ and $-y$.   **d.** From $-a$ subtract $-6$.

    **e.** From $3b^2$ take $-9c^2$.           **f.** Subtract $5y$ from $7x$.

    **g.** Subtract $-2$ from $6m$.          **h.** Take $4ab$ from $-3xy$.

    **i.** Take $-bc$ from $-2ab$.         **j.** Subtract $-9$ from $b^2$.

## *CUMULATIVE REVIEW*

1. Add:

    **a.**

| $-5ab$ | $-16c^3$ | $3x^2y^2$ | $-a^3c$ | $-5m^6y^3$ | $12r^9t^2$ |
|---|---|---|---|---|---|
| $7ab$ | $-21c^3$ | $-2x^2y^2$ | $-a^3c$ | $5m^6y^3$ | $23r^9t^2$ |

    **b.**

| $8m - 11n$ | $4x^2 - 6xy - 7y^2$ | $8b^2 - 3bc$ | $-9a^2 + 7ab +\ \ b^2$ |
|---|---|---|---|
| $9m + 10n$ | $-\ x^2 - 6xy + 7y^2$ | $5bc - 4c^2$ | $-\ a^2 - 3ab - 2b^2$ |

2. Find the sum of:

    **a.** $-9r$ and $6r$

    **b.** $x^2 - y^2$, $6x^2 - xy - 2y^2$, and $5y^2 - 2xy$

3. Add as indicated:

    **a.** $(-5y^2) + (-y^2) + (-8y^2)$

    **b.** $(a^2 - 2ab) + (-3ab - b^2) + (b^2 - 2a^2)$

4. Combine:

    **a.** $n + .25n$

    **b.** $5x - 8 - 9x + 6 - 3x + 7$

5. Subtract the lower term from the upper term:

| $12x$ | $-15bc$ | $-5m^3$ | $4x^4y^2$ | $0$ | $-2a^5x^2$ |
|---|---|---|---|---|---|
| $21x$ | $9bc$ | $-8m^3$ | $-4x^4y^2$ | $-21a$ | $-\ a^5x^2$ |

6. Subtract as indicated:

    **a.** $(-20c^4y^2) - (25c^4y^2)$     **b.** $10x^2 - (-5xy)$     **c.** $(-abc) - (-7abc)$

7. **a.** From $x$ subtract $-8x$.

    **b.** Subtract $-3ab^2$ from $-9ab^2$.

# Subtraction of Polynomials

**I. Aim**     To subtract polynomials.

**II. Procedure**

1. When the polynomials are already arranged so that like terms are under each other, subtract each column separately.
   a. By adding the additive inverse of each term in the subtrahend to the corresponding term in the minuend. See sample solution 1.
   b. Or without rewriting the expressions, by changing the sign of each term of the subtrahend mentally and proceeding as in addition.
2. If the difference in a column is zero, do not write this zero in the answer if the difference of any other column is zero.
3. If the differences in all columns are zero, write only one zero in the answer.
4. When arranging polynomials, place the subtrahend under the minuend so that like terms are under each other. In some cases it may be necessary to rewrite the polynomials in descending or ascending powers of the variable.
5. Check by adding the answer to the given subtrahend. The sum should be the given minuend.

**III. Sample Solutions**

1. Subtract

$$2x^2 - 5xy - y^2$$
$$\underline{x^2 + 2xy - 3y^2}$$

Add:

$$2x^2 - 5xy - y^2$$
$$\underline{-\ x^2 - 2xy + 3y^2}$$
$$x^2 - 7xy + 2y^2$$

*Answer:* $x^2 - 7xy + 2y^2$

2. Subtract:

$$4a^2 - 3a + 6$$
$$\underline{-2a^2 - 3a + 7}$$
$$6a^2 \qquad\ - 1$$

*Answer:* $6a^2 - 1$

3. Subtract:

$$c^2 + 2cd$$
$$\underline{\qquad 4cd - d^2}$$
$$c^2 - 2cd + d^2$$

*Answer:* $c^2 - 2cd + d^2$

**4.** Subtract $8 - 4b^2$ from $3b^2 - 4b + 3$.

$$3b^2 - 4b + 3$$
$$-4b^2 \qquad + 8$$
$$\overline{7b^2 - 4b - 5}$$

*Answer:* $7b^2 - 4b - 5$

**5.** Subtract as indicated:

$(b^2 - 3b) - (2b - 5)$
$$b^2 - 3b$$
$$\underline{\quad\quad 2b - 5}$$
$$b^2 - 5b + 5$$

*Answer:* $b^2 - 5b + 5$

## PRELIMINARY EXAMPLES

What is the additive inverse of each of the following polynomials?

**1.** $6n + 3$ **2.** $5a^2 - 4b^2$ **3.** $-7y - 8z$

**4.** $9c^2 - cd + 2d^2$ **5.** $10x^2 + 3xy - 4y^2$ **6.** $12a^3 - 5a^2 + 6a - 19$

## DIAGNOSTIC TEST

Subtract the lower expression from the upper expression:

**1.** $7a - 5$
$\quad 2a + 8$

**2.** $a + 2b$
$\quad -2a \pm 3b$

**3.** $6x^2 - 5x + 2$
$\quad 3x^2 - 4x - 6$

**4.** $2a^2 - 3ab - b^2$
$\quad a^2 - 3ab + b^2$

**5.** $6a^2 - 5a + 7$
$\quad 6a^2 - 5a + 7$

**6.** $4m^2 - 5mn + n^2$
$\quad 3m^2 + 3mn$

**7.** $x^3$
$\quad 2x^3 - x^2 + 4x$

**8.** $\qquad 0$
$\quad 2x - 3y + z$

**9.** Subtract as indicated: $(x^2 - x) - (x - 1)$.

**10.** Find the difference between $2a - 3b + 6c$ and $3a - 4b - 2c$.

**11.** From $x^2 - y^2$ subtract $2x^2 - 3xy - y^2$.

**12.** Take $3x - y$ from $x - 2xy$.

## RELATED PRACTICE EXAMPLES

Subtract the lower expression from the upper expression:

**1.**

| | | | | |
|---|---|---|---|---|
| $4a + 5$ | $6x + 3$ | $5y^2 + 1$ | $4r + 6$ | $2m + 4$ |
| $3a + 2$ | $4x + 7$ | $9y^2 + 8$ | $r - 3$ | $6m - 5$ |
| | | | | |
| $x^2 - 6$ | $2y^2 - 8$ | $6ab - 9$ | $6x^2y - 3$ | $-2b - 7$ |
| $3x^2 + 5$ | $y^2 + 1$ | $8ab - 2$ | $5x^2y - 7$ | $5b + 2$ |

**2.**

$$\begin{array}{ccccc}
4x + 7y & 5a + 2b & a - 3c & 2x^2 - y^2 & 4ab - 3c \\
2x + 5y & 3a + 6b & -4a + 5c & x^2 + 3y^2 & 5ab - 2c
\end{array}$$

$$\begin{array}{ccccc}
8m + 3n & 2c^2 - 3d^2 & x^2 + y^2 & 4cd - d^2 & 5m^3 - n^2 \\
6m - 5n & 5c^2 - 4d^2 & 2x^2 - 3y^2 & cd + 2d^2 & -2m^3 + 4n^2
\end{array}$$

**3.**

$$\begin{array}{cccc}
4a + 3b + 6c & 2x^2 - 3x + 7 & 7a^2 - 5ab - 6b^2 & 3b^2 - 4b - 7 \\
2a + b + 2c & 3x^2 + x - 9 & 3a^2 + 4ab + 2b^2 & b^2 - b - 6
\end{array}$$

$$\begin{array}{cccc}
8m^2 + mn - 3n^2 & 2x^3 - 4x^2 + 5x & -5cd - 2cd^2 + 3cd^3 & 8r^4 - 2r^2 + 6 \\
2m^2 - 3mn + 2n^2 & 4x^3 - 2x^2 + 7x & cd + cd^2 - 5cd^3 & 6r^4 - 8r^2 - 3
\end{array}$$

$$\begin{array}{cc}
5x^3 - 4x^2 + 3x - 1 & 4a^3 - 5a^2b - 6ab^2 + b^3 \\
3x^3 - 5x^2 - x + 6 & 6a^3 - 4a^2b + 8ab^2 - 2b^3
\end{array}$$

**4.**

$$\begin{array}{cccc}
a + 6 & 5x^2 - 2x + 7 & 2x^2 - 3xy + 2y^2 & 4a^4 - 5a^2b^2 - 9b^4 \\
a - 3 & 2x^2 - 2x + 8 & x^2 + 2xy + 2y^2 & 4a^4 - 6a^2b^2 - 9b^4
\end{array}$$

**5.**

$$\begin{array}{cccc}
x - 2 & -4x^2 + 9y^2 & 2x^2 - 3xy - 5y^2 & 3r^2 + 9r + 6 \\
x - 2 & -4x^2 + 9y^2 & 2x^2 - 3xy - 5y^2 & -3r^2 + 9r + 6
\end{array}$$

**6.**

$$\begin{array}{cccc}
7a + 6b & 2x^2 - 2x - 8 & 5b^2 - 2bc + 3c^2 & 4m^2 - mx - 3x^2 \\
4a & 7x^2 + 5x & -8c^2 & -mx
\end{array}$$

**7.**

$$\begin{array}{cccc}
a & a^2 \quad - b^2 & 6c^2 - 3cd & 3r^2 \quad\quad - 1 \\
a - 3 & a^2 - 2ab + b^2 & 4c^2 + 7cd - d^2 & 2r^3 - 4r^2 - 5r
\end{array}$$

**8.**

$$\begin{array}{cccc}
0 & 0 & 0 & 0 \\
4x^2 - 2y^2 & 3ab - 2ab^2 + 5ab^3 & -x^2 - xy + 2y^2 & 3m^3 - m^2 + 2m - 6
\end{array}$$

**9.** Subtract as indicated:
  **a.** $(3x + 5) - (x - 2)$
  **b.** $(a - b) - (a + b)$
  **c.** $(a + b) - (a + c)$
  **d.** $(2s - a) - (s + b)$
  **e.** $(x^2 - x + 2) - (2x^2 + x - 2)$
  **f.** $(2c^2 - cd + 3d^2) - (c^2 - 2d^2)$
  **g.** $(x^2 - y^2) - (2x^2 - 2xy)$
  **h.** $(m^3 - 3m^2 - 2m) - (m^3 - 2m^2 - 2)$
  **i.** $(5 - a + a^2) - (a^2 + 7 - 2a)$
  **j.** $(3b - c) - (x - y)$

**10.** **a.** From $8x - 3$ subtract $2x + 7$.
  **b.** From $2a - 7c + 4x$ take $-5a + 6c - 8x$.

   **c.** Find the difference between $4x^2 - 2x + 6$ and $x^2 - 5x - 9$.

   **d.** Subtract $6b^3 - 5b^2 - b$ from $3b^3 + 5b^2 - b$.

   **e.** Take $4c^2 - 3cd + 5d^2$ from $7c^2 - 3cd + 6d^2$.

**11. a.** From $a^2 - 2ab + 2b^2$ take $a^2 - 3b^2$.

   **b.** From zero subtract $5x - 3y + 7z$.

   **c.** From $c^2 - 5d^2$ take $3c^2 - 7cd - d^2$.

   **d.** From 8 take $4x - 2$.

   **e.** Find the difference between $a^2 + b^2$ and $a^2 - c^2$.

   **f.** From $2m^2 - 3mn + n^2$ subtract zero.

   **g.** Find the difference between $8x - 7$ and $2y - 9$.

   **h.** From $m^2$ take $2mn - n^2$.

   **i.** From $b^3 - 2b$ subtract $4b^2 - 5b + 2$.

   **j.** Find the difference between $3a^3 - 4a^2b + 3ab^2$ and $5ab^2 - 2a^2b + b^3$.

**12. a.** Subtract $2c - 4$ from $3c$.

   **b.** Subtract $8y - 3$ from $2x + 7$.

   **c.** Take $4x^3 - 3x + 8$ from zero.

   **d.** Take $x^2 - 5x + 27$ from $x^3 - 81$.

   **e.** Subtract $2a^2 - a - 1$ from $a^3 + 2a^2 - a$.

   **f.** Take $5x^2$ from $2x^2 - 3x$.

   **g.** Subtract $3m - 5$ from $-m$.

   **h.** Subtract $2c^2 - d^2$ from $c^2 - 2cd - d^2$.

   **i.** Take $2a - 7c$ from $4a + 3b$.

   **j.** Take $-3mn + n^2$ from $m^2 - 2mn$.

## MISCELLANEOUS EXAMPLES

**1.** What must be added to $7x^2 - 3x + 8$ to get $3x^2 + 5x - 9$?

**2.** What must be added to $b^2 - 4d^2$ to get $b^2 + 4bd + 4d^2$?

**3.** What must be added to $-x$ to get $x^2$?

**4.** From the sum of $m^2 - 2mn + n^2$ and $m^2 + 2mn + n^2$ subtract $m^2 - 4mn + 4n^2$.

**5.** From $5x^2 - 3x + 7$ subtract the sum of $2x^2 - 2x + 4$ and $x^2 - x - 2$.

**6.** Subtract $-4xy + y^2$ from the sum of $3x^2 - 2y^2$ and $2x^2 - 4xy$.

**7.** From the sum of $8n^2 - 5n - 6$ and $3n^2 + 3n - 1$ subtract the sum of $-2n^2 + 3$ and $6n - 5$.

**8.** Find the difference between the sum of $8a - 3b$ and $2b - a$ and the sum of $5a + b$ and $2b - 7a$.

**9.** Subtract the sum of $9b^2 - 3c^2$ and $2b^2 + bc - 2c^2$ from the sum of $2b^2 - 2bc - c^2$ and $c^2 + bc - b^2$.

## *GEOMETRIC APPLICATIONS*

1. Find the dimensions of the required lengths in the following figures in terms of the variables:

   **a.** Lines

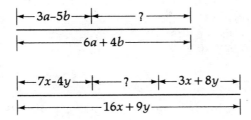

   **b.** Trapezoid

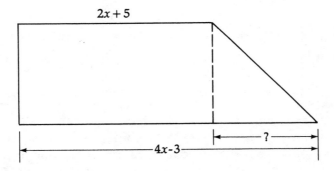

2. Find the angular measures of the required angles in the following figures in terms of the variables:

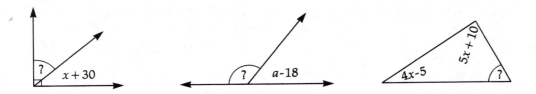

   **a.** Find the complement of an angle measuring $x + 30$ degrees.

   **b.** Find the supplement of an angle measuring $a - 18$ degrees.

   **c.** Find the measure of the third angle of a triangle in which one angle measures $5x + 10$ degrees, the second angle measures $4x - 5$ degrees, and the sum of the measures of the angles of the triangle is 180 degrees.

# Law of Exponents for Multiplication

### I. Aim

To develop an understanding and use of the law of exponents in multiplication.

Since $x^4$ is the shortened form of $x \cdot x \cdot x \cdot x$ and $x^5$ is the shortened form of $x \cdot x \cdot x \cdot x \cdot x$, then

$$x^4 \cdot x^5 = x \cdot x \cdot x \cdot x \cdot x \cdot x \cdot x \cdot x \cdot x = x^9$$

Thus $x^4 \cdot x^5 = x^9$

Observe that the exponent of the variable in the product (9) is obtained by adding the exponents of that variable in the factors (4 and 5). Thus, the law of exponents for multiplication may be expressed as:

$$x^m x^n = x^{m+n}$$

where $x$ represents any number and $m$ and $n$ are positive integers.

### II. Procedure

1. To multiply two factors when both consist of the same variable, apply the law of exponents for multiplication.
2. To multiply two factors when both consist of two or more identical variables, apply the law of exponents for multiplication to each variable separately. If the variables are represented by letters, arrange the letters in alphabetical order.
3. To multiply two factors which consist of two different variables, write the two variables, when represented by letters, alphabetically, next to each other without any multiplication symbol between them.
4. When one factor contains a variable not contained in the other factor, the variable is written unchanged in the product.

### III. Sample Solutions

**1.** Multiply:
$n^5 \cdot n^6 = n^{5+6} = n^{11}$
*Answer:* $n^{11}$

**3.** $b^5 \cdot a^4$
$= a^4 b^5$
*Answer:* $a^4 b^5$

**2.** $b^3 x^4 \cdot b^5 x$
$= (b^3 \cdot b^5)(x^4 \cdot x)$
$= b^8 x^5$
*Answer:* $b^8 x^5$

**4.** $a^2 b^4 \cdot a^7 c^3$
$= a^2 \cdot a^7 \cdot b^4 \cdot c^3$
$= a^9 b^4 c^3$
*Answer:* $a^9 b^4 c^3$

## DIAGNOSTIC TEST

**1.** Write in shortened form, using exponents:

  **a.** $c \cdot c \cdot c \cdot c \cdot c \cdot c \cdot c \cdot c$    **b.** $x \cdot x \cdot x \cdot x \cdot y \cdot y \cdot y \cdot y \cdot y \cdot z$

Multiply:

**2.** $b^{12} \cdot b^7$    **3.** $a^5$    **4.** $x$    **5.** $m^8n^2 \cdot m^3n$    **6.** $c^9dx^6$
  $\qquad\qquad\quad \underline{a^4}$   $\quad\underline{x^6}$   $\qquad\qquad\qquad\quad \underline{cd^8x^5}$

**7.** $m^2 \cdot x^4$    **8.** $r$    **9.** $bx \cdot cx$    **10.** $d^7n^4x^2$
  $\qquad\qquad\quad \underline{s^2}$   $\qquad\qquad\qquad\qquad\quad \underline{c^3d^5x^9}$

## *RELATED PRACTICE EXAMPLES*

**1.** Write in shortened form, using exponents:

  **a.** $r \cdot r \cdot r \cdot r \cdot r \cdot r$    **b.** $m \cdot m \cdot m \cdot m \cdot m \cdot m \cdot m \cdot m \cdot m$
  **c.** $10 \cdot 10 \cdot 10 \cdot 10 \cdot 10 \cdot 10 \cdot 10 \cdot 10$    **d.** $b \cdot b \cdot b \cdot b \cdot b \cdot c \cdot c \cdot c$
  **e.** $a \cdot a \cdot b \cdot b \cdot b \cdot n \cdot n \cdot n \cdot n \cdot n$    **f.** $d \cdot m \cdot m \cdot y \cdot y \cdot y \cdot y \cdot y \cdot y$
  **g.** $c \cdot c \cdot c \cdot d \cdot d \cdot d \cdot d \cdot h \cdot h \cdot h \cdot h$    **h.** $5 \cdot 5 \cdot 5 \cdot 5 \cdot 7 \cdot 7 \cdot 7 \cdot 11 \cdot 11$

Multiply:

**2.** **a.** $m^6 \cdot m^2$    **b.** $x^3 \cdot x^3$    **c.** $b^7 \cdot b^3$    **d.** $g \cdot g^5$
  **e.** $y^{12} \cdot y$    **f.** $10^6 \cdot 10^9$    **g.** $d^8 \cdot d^3 \cdot d^6$    **h.** $n^7 \cdot n^2 \cdot n^3 \cdot n^9$

**3.** $x^3$    $c^8$    $r^3$    $y^{10}$    $a^7$    $2^4$    $5^8$
  $\underline{x^2}$   $\underline{c^4}$   $\underline{r^3}$   $\underline{y^5}$   $\underline{a^{21}}$   $\underline{2^3}$   $\underline{5^6}$

**4.** $c$    $b^2$    $b$    $m$    $d^6$    $n$    $3^4$
  $\underline{c}$   $\underline{b}$   $\underline{b^5}$   $\underline{m^9}$   $\underline{d}$   $\underline{n^4}$   $\underline{3}$

**5.** **a.** $xy \cdot xy$    **b.** $a^3c^2 \cdot a^2c^4$    **c.** $m^5y^3 \cdot m^4y$    **d.** $c^6z^7 \cdot c^5z^{10}$
  **e.** $ad^2n \cdot a^2dn$    **f.** $b^2cx^3 \cdot bc^2x^3$    **g.** $abc^8 \cdot ab^2c$    **h.** $c^5d^3xy^7 \cdot c^5d^3xy^7$

**6.** $a^2c$    $m^4x^9$    $b^5cd$    $r^8s^4t^2$    $b^6d^4x^3$    $x^3yz^5$    $a^2c^3y^4$
  $\underline{ac^2}$   $\underline{m^3x^7}$   $\underline{bc^3d}$   $\underline{r^8s^4t^2}$   $\underline{b^7d^5x^9}$   $\underline{x^2yz^8}$   $\underline{a^4c^2y^3}$

**7.** **a.** $d^7 \cdot r^9$    **b.** $c \cdot x$    **c.** $s \cdot y^4$    **d.** $n^3 \cdot g^5$
  **e.** $x^2 \cdot a^2$    **f.** $t \cdot m$    **g.** $h^4 \cdot z^8$    **h.** $r \cdot c^7$

**8.** $m$    $d$    $a^3$    $t^2$    $x^6$    $c$    $n^5$    $p^2$    $y^9$    $w^2$
  $\underline{x}$   $\underline{b}$   $\underline{c^5}$   $\underline{r^2}$   $\underline{y^7}$   $\underline{d^4}$   $\underline{e^3}$   $\underline{a^8}$   $\underline{z^5}$   $\underline{g}$

**9.** **a.** $b^2c \cdot b^3d^2$    **b.** $m^3 \cdot m^4x^5$    **c.** $ax \cdot cy$    **d.** $c^2d \cdot b^2c$
  **e.** $n^6r^5 \cdot mr^4$    **f.** $t^2 \cdot gr^3$    **g.** $abc \cdot bcx$    **h.** $x^7y^9z^5 \cdot w^2z^9$

**10.** $x^2y$    $a^3b^9$    $c^2dr^4$    $b^4cx^5$    $m^5n^6r^3$    $d^2st^4$    $a^7d^5x^2$
  $\underline{dx}$   $\underline{b^2x^8}$   $\underline{a^3c^7}$   $\underline{b^2dy^2}$   $\underline{g^2mr^4}$   $\underline{c^5d^8s}$   $\underline{bx^2z^3}$

# Multiplication of a Monomial by a Monomial

**I. Aim**   To multiply a monomial by a monomial.

**II. Procedure**

1. Multiply their numerical coefficients.
2. Find the product of the variables using the law of exponents.
3. Prefix the numerical product to the product of the variables arranged in alphabetical order.
4. When the factor of variables does not have an expressed numerical coefficient, it is 1 understood. The numerical coefficient always precedes the factor of variables in the product. See sample solution 4 and 5. Observe: $x = +1x^1$ and $-x = -1x^1$
5. Write products of variables in alphabetical order. See sample solution.
6. When there are three or more factors, prefix the product of the numerical coefficients to the product of the variables. See sample solutions 7 and 8.

**III. Sample Solutions**

1. Multiply $7m^4n^3$ by $9m^2n^5$:
$$(7m^4n^3)(9m^2n^5)$$
$$= (7 \cdot 9)(m^4 \cdot m^2)(n^3 \cdot n^5)$$
$$= \quad 63 \qquad m^6 \qquad n^8$$
*Answer:* $63m^6n^8$

2. Multiply:
$$-5a^3b^4$$
$$\underline{+2ab^3}$$
$$-10a^4b^7$$
*Answer:* $-10a^4b^7$

3. Multiply:
$$-3c^3$$
$$\underline{-8c^8}$$
$$+24c^{11}$$
*Answer:* $+24c^{11}$

4. Multiply:
$$-bx$$
$$\underline{-3}$$
$$+3bx$$
*Answer:* $+3bx$

5. Multiply:
$$-4a^2$$
$$\underline{-c}$$
$$+4a^2c \quad$$ *Answer:* $+4a^2c$

6. Multiply:
$$-5bx^2$$
$$\underline{3xy^2}$$
$$-15bx^3y^2 \quad$$ *Answer:* $-15bx^3y^2$

7. Multiply:
$$(4a^2x)(-2ax^2)(-3ax^3)$$
$$= [4(-2)(-3)](a^2 \cdot a \cdot a)(x \cdot x^2 \cdot x^3)$$
$$= \qquad +24 \qquad a^4 \qquad x^6$$
*Answer:* $+24a^4x^6$

8. Multiply:
$$(-3x^3)^3$$
$$= (-3x^3)(-3x^3)(-3x^3)$$
$$= (-3)(-3)(-3)(x^{3+3+3})$$
$$= -27x^9$$
*Answer:* $-27x^9$

## DIAGNOSTIC TEST

Multiply:

1. $4c^2$
   $2c^4$

2. $-3a^2$
   $-4a^2$

3. $+5m^3$
   $-4m^7$

4. $-2r^4$
   $8r^3$

5. $-2ab^2c^2$
   $3a^2b^3c^3$

6. $-3x^5y$
   $x^6y$

7. $-ax$
   $5$

8. $-x^2$
   $-1$

9. $3c^2$
   $-a$

10. $2ab^2$
    $3b^4$

11. $-4a^3x$
    $7x^2y$

12. $-.8rt^2$
    $-1.2rt$

13. $\frac{1}{4}c^4d^2$
    $2c^3d^3$

14. $(-7ac)(8a^2c)$

15. $-3a(-5b)$

16. $(2ab)(-bc^2)(-3a^2c)$

17. $(-2b^2)^3$

## RELATED PRACTICE EXAMPLES

Multiply:

1. $3b^2$    $5a^3$    $2x^5$    $8y$
   $2b^2$    $7a$    $6x^3$    $4y$

   $6m^6$    $8d^4$    $6x$
   $8m^8$    $9d^8$    $12x^9$

2. $-5a^2$    $-6x$    $-4b^3$    $-9m^4$
   $-4a^3$    $-7x$    $-2b$    $-6m^6$

   $-3d^8$    $-6x^6$    $-3w^4$
   $-9d^3$    $-6x^6$    $-10w^5$

3. $+6s$    $+7r^2$    $+4x$    $+5b^2$
   $-2s$    $-3r^3$    $-8x^4$    $-5b^8$

   $2x^9$    $4t^4$    $12z^{12}$
   $-8x^2$    $-3t^5$    $-5z^5$

**4.** $\begin{array}{r} -9x^5 \\ +4x^3 \\ \hline \end{array}$ $\qquad$ $\begin{array}{r} -8a^2 \\ +7a^2 \\ \hline \end{array}$ $\qquad$ $\begin{array}{r} -6w^4 \\ +3w^3 \\ \hline \end{array}$ $\qquad$ $\begin{array}{r} -4r^5 \\ +4r^5 \\ \hline \end{array}$ $\qquad$ $\begin{array}{r} -2b \\ 10b \\ \hline \end{array}$ $\qquad$ $\begin{array}{r} -8m^3 \\ 12m^8 \\ \hline \end{array}$ $\qquad$ $\begin{array}{r} -4t^{10} \\ 20t^{10} \\ \hline \end{array}$

**5.** $\begin{array}{r} 2ab \\ 5ab \\ \hline \end{array}$ $\qquad$ $\begin{array}{r} -4x^2y \\ -8xy^2 \\ \hline \end{array}$ $\qquad$ $\begin{array}{r} -9a^2b^4 \\ 6a^3b^2 \\ \hline \end{array}$ $\qquad$ $\begin{array}{r} -8x^4y^3z \\ -8xyz^2 \\ \hline \end{array}$ $\qquad$ $\begin{array}{r} 4c^5d^6x^2 \\ 9c^8d^4x^7 \\ \hline \end{array}$ $\qquad$ $\begin{array}{r} 3x^3y^2z \\ -7x^4y^3z \\ \hline \end{array}$ $\qquad$ $\begin{array}{r} -9b^3d^2x^3 \\ -7b^5d^4x^6 \\ \hline \end{array}$

**6.** $\begin{array}{r} 4x^3 \\ x^2 \\ \hline \end{array}$ $\qquad$ $\begin{array}{r} -5x \\ -x \\ \hline \end{array}$ $\qquad$ $\begin{array}{r} -8b^4d^2 \\ b^2d \\ \hline \end{array}$ $\qquad$ $\begin{array}{r} -6axy \\ -a^2xy \\ \hline \end{array}$ $\qquad$ $\begin{array}{r} -a^2b^4 \\ -7a^3b \\ \hline \end{array}$ $\qquad$ $\begin{array}{r} -x \\ 3x^6 \\ \hline \end{array}$ $\qquad$ $\begin{array}{r} mn^2 \\ -9m^2n \\ \hline \end{array}$

**7.** $\begin{array}{r} a \\ 3 \\ \hline \end{array}$ $\qquad$ $\begin{array}{r} 8 \\ b^3c^2 \\ \hline \end{array}$ $\qquad$ $\begin{array}{r} x^2 \\ -2 \\ \hline \end{array}$ $\qquad$ $\begin{array}{r} -y \\ -3 \\ \hline \end{array}$ $\qquad$ $\begin{array}{r} -m^2n \\ 5 \\ \hline \end{array}$ $\qquad$ $\begin{array}{r} 5a \\ 2 \\ \hline \end{array}$ $\qquad$ $\begin{array}{r} -7abc \\ -5 \\ \hline \end{array}$

**8.** $\begin{array}{r} ax \\ 1 \\ \hline \end{array}$ $\qquad$ $\begin{array}{r} -5ab \\ 1 \\ \hline \end{array}$ $\qquad$ $\begin{array}{r} 1 \\ -b^2cd^3 \\ \hline \end{array}$ $\qquad$ $\begin{array}{r} 7x^4 \\ -1 \\ \hline \end{array}$ $\qquad$ $\begin{array}{r} -9x^2z \\ -1 \\ \hline \end{array}$ $\qquad$ $\begin{array}{r} -1 \\ 3by \\ \hline \end{array}$ $\qquad$ $\begin{array}{r} -1 \\ -2dx^4 \\ \hline \end{array}$

**9.** $\begin{array}{r} 4b \\ a \\ \hline \end{array}$ $\qquad$ $\begin{array}{r} -x \\ 3y \\ \hline \end{array}$ $\qquad$ $\begin{array}{r} -m^2 \\ -7n^3 \\ \hline \end{array}$ $\qquad$ $\begin{array}{r} 5a \\ -c \\ \hline \end{array}$ $\qquad$ $\begin{array}{r} -2b \\ ax \\ \hline \end{array}$ $\qquad$ $\begin{array}{r} by \\ 3cx \\ \hline \end{array}$ $\qquad$ $\begin{array}{r} -4a^2y \\ -bx^2 \\ \hline \end{array}$

**10.** $\begin{array}{r} 4x^2y \\ 2x^3 \\ \hline \end{array}$ $\qquad$ $\begin{array}{r} 5a \\ -ab \\ \hline \end{array}$ $\qquad$ $\begin{array}{r} -2x^2 \\ -3x^4y \\ \hline \end{array}$ $\qquad$ $\begin{array}{r} -9c^2d \\ 6d^2 \\ \hline \end{array}$ $\qquad$ $\begin{array}{r} -cm^2 \\ -8m^3 \\ \hline \end{array}$ $\qquad$ $\begin{array}{r} -4acxy \\ -ac^2y \\ \hline \end{array}$ $\qquad$ $\begin{array}{r} 6ab^2d^3 \\ -8b^3d^2 \\ \hline \end{array}$

**11.** $\begin{array}{r} 5bx \\ 3cx \\ \hline \end{array}$ $\qquad$ $\begin{array}{r} -6a^2m \\ 2an \\ \hline \end{array}$ $\qquad$ $\begin{array}{r} -9cy^3 \\ -3by^2 \\ \hline \end{array}$ $\qquad$ $\begin{array}{r} 8acy \\ -9abx \\ \hline \end{array}$ $\qquad$ $\begin{array}{r} 4a^2bc^2 \\ -b^3x \\ \hline \end{array}$ $\qquad$ $\begin{array}{r} -5am^2y \\ -6bx^2y^4 \\ \hline \end{array}$ $\qquad$ $\begin{array}{r} 4a^2bcx^2 \\ 2adxy^2 \\ \hline \end{array}$

**12.** $\begin{array}{r} .6x^8y \\ .3x^2y^3 \\ \hline \end{array}$ $\qquad$ $\begin{array}{r} -1.4a^2b \\ -.05abc^2 \\ \hline \end{array}$ $\qquad$ $\begin{array}{r} .07m^2x \\ -.01n^2y \\ \hline \end{array}$ $\qquad$ $\begin{array}{r} -3.7ab^2c \\ .15a^2bc^2 \\ \hline \end{array}$ $\qquad$ $\begin{array}{r} -1.04d^2m^3x \\ -7d^3x^2y \\ \hline \end{array}$

**13.** $\begin{array}{r} \frac{1}{2}ac^2 \\ 6a^3c \\ \hline \end{array}$ $\qquad$ $\begin{array}{r} -\frac{1}{4}\pi d^2 \\ 8\pi d \\ \hline \end{array}$ $\qquad$ $\begin{array}{r} -9m^2x^4 \\ -\frac{1}{3}m^2x^5 \\ \hline \end{array}$ $\qquad$ $\begin{array}{r} 20a^4c^2d^8 \\ -\frac{1}{10}a^3d^2x^5 \\ \hline \end{array}$ $\qquad$ $\begin{array}{r} -\frac{3}{8}m^5xy \\ -16n^2x^4z^2 \\ \hline \end{array}$

**Multiply as indicated:**

**14. a.** $6x \cdot x$ $\qquad$ **b.** $7cm^2 \times 6c^2m$ $\qquad$ **c.** $(3a)(4a^2)$ $\qquad$ **d.** $(-2b)(-5b)$
  **e.** $(5a^2x)(9x^5y^2)$ $\qquad$ **f.** $(-6bn^2)(3b^5n^2)$ $\qquad$ **g.** $(-x)(-5x^3)$
  **h.** $(4b^2c^3d)(-3b^3c^4d^5)$ **i.** $(-8x^2yz^3)(-2xy^2z^4)$ $\qquad$ **j.** $(-4m^4)(4m^4)$

**15. a.** $4(3b)$ $\qquad$ **b.** $2a(4a^2)$ $\qquad$ **c.** $-5b^2(b^3)$ $\qquad$ **d.** $-x(-y)$
  **e.** $-a(-a)$ $\qquad$ **f.** $3x^2(-3x^2)$ $\qquad$ **g.** $-4a^2b(5ab^2)$
  **h.** $3x(5xy)$ $\qquad$ **i.** $-1(-4x^2y)$ $\qquad$ **j.** $6m^3n^2(-am^2y)$

**16. a.** $3a^2 \cdot 2a^2 \cdot a^4$ $\qquad$ **b.** $4by^3 \cdot 5xy \cdot 2bx^2$ $\qquad$ **c.** $8c \cdot 3b \cdot 5c$ $\qquad$ **d.** $3a^2 \cdot 3a^2 \cdot 3a^2$
  **e.** $(-x)(-x)(-x)$ $\qquad$ **f.** $(2bx^2)(3b^2x^3)(-4bx)$ $\qquad$ **g.** $(-5ac)(-6bc)(-cd)$
  **h.** $(-7ab^2)(a^5b)(-3ab^3)$ $\qquad$ **i.** $(-2x)(-2x)(-2x)(-2x)$
  **j.** $(ab^2)(-2a)(+3ab)(-4a^2b)$

**17. a.** $(b^2x)^2$ $\qquad$ **b.** $(4a^6)^2$ $\qquad$ **c.** $(-3a^2b)^2$ $\qquad$ **d.** $(2mn^3x)^2$
  **e.** $(-5b^5xy^2)^2$ $\qquad$ **f.** $(a^5x^2z)^3$ $\qquad$ **g.** $(-4b^2x^3)^3$
  **h.** $(2b)^4$ $\qquad$ **i.** $(-4c^2d^3)^4$ $\qquad$ **j.** $(3x^2y^3)^5$

# Multiplication of a Polynomial by a Monomial

**I. Aim**   To multiply a polynomial by a monomial.

**II. Procedure**

1. Multiply each term of the polynomial by the monomial.
2. When the polynomial and the monomial are arranged vertically, multiply, beginning from the left.

**III. Sample Solutions**

1. Multiply:   $4b^2c^3(2b^4 + 5b^3c + 3b^3)$
   $$= (4b^2c^3 \cdot 2b^4) + (4b^2c^3 \cdot 5b^3c) + (4b^2c^3 \cdot 3bc^3)$$
   $$= 8b^6c^3 + 20b^5c^4 + 12b^3c^6$$

   *Answer:* $8b^6c^3 + 20b^5c^4 + 12b^3c^6$

2. Multiply:   $4a^2 - 5ad + d^2$
   $$\underline{-3ad^2\phantom{XXXXXXXXXXXXXXX}}$$
   $$-12a^3d^2 + 15a^2d^3 - 3ad^4$$

   *Answer:* $-12a^3d^2 + 15a^2d^3 - 3ad^4$

## DIAGNOSTIC TEST

Multiply:

1. $3(a + 4)$
2. $5(3x - 7)$
3. $-2(4m + 3)$
4. $-9(-d - 5)$
5. $6(3x - 4y)$
6. $-4(2c^2 + 8cd - 3d^2)$
7. $x(x + 1)$
8. $-x^4(2x^2 - 3)$
9. $b^2(3b^4 - b^3 - b^2 + 2b)$
10. $-m(m^3 - 5m)$
11. $ab(2a - 4b)$
12. $-c^3(-b^2 + 3x)$
13. $x^2y^5(4x^2y^2 - 3xy + 5)$
14. $2ab(3a^2b + 5ab^2)$
15. $-8c^3d^7(2c^2 - 4cd - 3d^2)$
16. $\frac{1}{4}r^2x(8r^2 + 12rx - 4x^2)$
17. $.5a^4b^2(2a^2b^2 + .6ab - .1)$
18. $4b^3 - 3b^2x + 6bx^2 - 5$
    $$\underline{-10b^2x^6\phantom{XXXXXXXXXXXXXX}}$$

## *RELATED PRACTICE EXAMPLES*

Multiply:

1. **a.** $2(x+5)$
   **b.** $9(a+3)$
   **c.** $5(4b+1)$
   **d.** $8(3m+9)$
   **e.** $4(5x^2+10)$

2. **a.** $3(c-5)$
   **b.** $8(m-6)$
   **c.** $7(3x-4)$
   **d.** $2(3ab-1)$
   **e.** $5(-x^2-9)$

3. **a.** $-4(w+8)$
   **b.** $-5(3x+2)$
   **c.** $-2(5y+6)$
   **d.** $-8(4y^2+1)$
   **e.** $-6(2xy+4)$

4. **a.** $-3(a-5)$
   **b.** $-9(2x-9)$
   **c.** $-4(b^2-3)$
   **d.** $-7(-5c-1)$
   **e.** $-2(-3ab-6)$

5. **a.** $4(a+b)$
   **b.** $7(2a+5x)$
   **c.** $10(3x-7y)$
   **d.** $9(-2x^2-4x)$
   **e.** $12(2b^2-3b+1)$

6. **a.** $-5(c+d)$
   **b.** $-3(2b-5x)$
   **c.** $-9(m^2-2m)$
   **d.** $-12(x^3-2x^2-5)$
   **e.** $-20(2b^2-bc+3c^2)$

7. **a.** $b(b+3)$
   **b.** $m(2m-4)$
   **c.** $x^2(3x^3+5)$
   **d.** $y^3(-2y^2-9)$
   **e.** $abc(3a^2bc+3)$

8. **a.** $-x(x-5)$
   **b.** $-a(a^2+3)$
   **c.** $-d^2(4d^4-1)$
   **d.** $-m^2n(-m^2n-6)$
   **e.** $-bx^3(-4bx+2)$

9. **a.** $c(c^2+c)$
   **b.** $x(x^3+3x^2-x)$
   **c.** $m^2(2m^6-5m^4-m^2)$
   **d.** $cd(c^3d-c^2d^2+2cd^3)$
   **e.** $x^2y^3(-x^2y+2xy-3xy^2)$

10. **a.** $-s(s-10s^3)$
    **b.** $-n(-n^2-n)$
    **c.** $-h^2(2h+5h^2)$
    **d.** $-xy(3x^2y-4xy^2)$
    **e.** $-b^3c^2(2b^3c-b^2c^2-4bc^3)$

11. **a.** $x(x+y)$
    **b.** $-b(3a-5b)$
    **c.** $ay(2a^2y-3ax)$
    **d.** $b^2x(-2bx^2+4cx)$
    **e.** $-c^2d^3(-5c^2+3cd-d^2)$

12. **a.** $a(x+3z)$
    **b.** $x(c^2-d^2)$
    **c.** $-b(5m-2x)$
    **d.** $m^2(-a-4d)$
    **e.** $a^2c(3x-5y)$

13. **a.** $abc(a^2b^2c^2-2abc+1)$
    **b.** $c^2d(c^3d-4c^2d^2-5cd^3)$
    **c.** $-x^4y^2(2x^3-3xy-4y^3)$
    **d.** $-b^3(4b^4-3b^3-2b^2+b-5)$
    **e.** $m^2n^3(m^3-m^2n+2mn^2-7n^2)$

14. **a.** $6a(4a^2 + 3a)$      **b.** $5x^2(3x^5 - 2x^2)$
     **c.** $4b^6(b^2 - 5b + 6)$      **d.** $3x^3(x^2 - 3x + 7)$
     **e.** $9mn(m^2 - 5mn + 1)$      **f.** $3x(4x^3 - 5x^2 + 6x)$
     **g.** $5my(2m^2 - 4my + y^2)$      **h.** $4a^2x^7(-5a^2 + ax - 5x^2)$
     **i.** $7b^2x^2(2b^2x^2 - 3bx + 7)$      **j.** $6bc^3y^5(2b^2 - 5c^2 - y^2 + 3)$

15. **a.** $-3b(b^2 - 5b)$      **b.** $-9a(2a^2 - a + 6)$
     **c.** $-4x^2(x^2 - 5x - 1)$      **d.** $-2s^3(-3r + 7s - 5t)$
     **e.** $-7xy^2(4x^2 - 3xy + 8)$      **f.** $-4y^4(2y^3 - 5y^2 + y)$
     **g.** $-2ab(3a^2 + 4ab - 5b^2)$      **h.** $-6m^5n^2(m^3 - m^2n^2 - 3n^4)$
     **i.** $-9a^2d^2(-2a^2d - 5a^2d^2 + 7a^2d^3)$      **j.** $-3x^4yz^3(x^3y - 6y^2z + xz^4 - 4)$

16. **a.** $\frac{1}{2}(4b^2 - 8b + 6)$      17. **a.** $3(.4x + .9y)$
     **b.** $\frac{1}{4}b(12b^2 - 4ab + 16)$      **b.** $.4(.5b - 2c)$
     **c.** $-\frac{1}{3}a^3x(6a^2 - 9ab - 21b^2)$      **c.** $.2a(a^2 - .3a - 7)$
     **d.** $\frac{2}{5}c^3(5c^3 - 10c^2 - 20c + 5)$      **d.** $-1.5b^2x(.2b^3 - 4bx + .1x^2)$
     **e.** $\frac{3}{4}x^2y^2(8x^3 + 4x^2y - 16xy^2 - 20y^3)$      **e.** $.6xy^3(.12x^2 + .7xy - 1.1y^2)$

18. Multiply:

     **a.** $4b + 5$
         $\underline{\phantom{4b+5}3\phantom{xxxx}}$

     **b.** $a^2 - 4ab$
         $\underline{\phantom{a^2-}-b\phantom{xx}}$

     **c.** $5x - 3y$
         $\underline{\phantom{5x-}7xy\phantom{x}}$

     **d.** $9c^2 - 4d^2$
         $\underline{\phantom{9c^2-}6cd\phantom{x}}$

     **e.** $2a^3b - 5ab^4$
         $\underline{\phantom{2a^3b-}-4a^2b^2\phantom{x}}$

     **f.** $a^2 + 2ab + b^2$
         $\underline{\phantom{a^2+}a\phantom{xxxxx}}$

     **g.** $7cd + d^2 + 6c^2$
         $\underline{\phantom{7cd+}-3cd\phantom{xx}}$

     **h.** $x^4 - 2x^3 - x^2 + 3$
         $\underline{\phantom{x^4-}4x^5\phantom{xxxx}}$

     **i.** $a^3 - 3a^2 + 3ax^2 - x^3$
         $\underline{\phantom{a^3-}-2a^3x^4\phantom{xx}}$

     **j.** $a^5 + 2a^4 - 3a^3 + a^2 - a + 1$
         $\underline{\phantom{a^5+}3a^2\phantom{xxxxxxxx}}$

     **k.** $2c^6 - 7c^5 + 4c^3 - 2c + 8$
         $\underline{\phantom{2c^6-}-8c^4\phantom{xxxxxx}}$

     **l.** $6x^4 - 3x^3y + 7x^2y^2 - 2xy^3 - 4y^4$
         $\underline{\phantom{6x^4-}-9x^6y^3\phantom{xxxxxxxx}}$

## CUMULATIVE REVIEW

1. Find the sum of: $2x^2 - 5y^2$, $3xy - y^2$, and $-5x^2 - 3xy$
2. Combine: $8 - 9 + 6n - 5 - 3n + 6 - 4n - n + 3$
3. Subtract $4c^2 - 3cd + 8d^2$ from $2c^2 - 5d^2$

     Multiply:

4. $4x^7y^4$      $-11a^6b^2c^5$      $-7m^4x^2z^3$      $8a^3 - 5a^2b + ab^2 - 3b^3$
    $\underline{-9x^8y^{15}}$      $\underline{-5a^3c}$      $\underline{6bn^6y^9}$      $\underline{-12a^3b^4}$

5. **a.** $-c(-3d)$      **b.** $-3x^2(2x^5 - x^4 + 5x^3)$      **c.** $4x^3y^2(3x^2 - xy^2 + 2y^3)$
6. Add the product of $(4n)(5ny^2)(-2n^2y)$ to the product of $-4n^2y(3n^2y^2 - ny^5)$.

# Multiplication of a Polynomial by a Polynomial

**I. Aim**    To multiply a polynomial by a polynomial.

**II. Procedure**

1. Arrange the terms of both multiplicand and multiplier so that the variables are in the same ascending or descending order of powers.
2. Going from left to right, multiply all the terms of the multiplicand by each term of the multiplier.
3. When writing the partial products, place only like terms under each other. Observe the alphabetical order (when the variables are represented by letters) and the ascending or descending order of the powers.
4. Add the partial products.
5. Do not write the numeral for zero for any column whose sum is zero if the sum of any other column is not zero.

**III. Sample Solutions**

1.
$$\begin{array}{r} x + 6 \\ \underline{x - 2} \\ x^2 + 6x \\ \underline{- 2x - 12} \\ x^2 + 4x - 12 \end{array}$$
*Answer:* $x^2 + 4x - 12$

2.
$$\begin{array}{r} 3x - 4y \\ \underline{5x - 2y} \\ 15x^2 - 20xy \\ \underline{- 6xy + 8y^2} \\ 15x^2 - 26xy + 8y^2 \end{array}$$
*Answer:* $15x^2 - 26xy + 8y^2$

3.
$$\begin{array}{r} 4a - d \\ \underline{4a + d} \\ 16a^2 - 4ad \\ \underline{+ 4ad - d^2} \\ 16a^2 \qquad - d^2 \end{array}$$
*Answer:* $16a^2 - d^2$

4.
$$\begin{array}{r} 2x^2 + 5 \\ \underline{x + 1} \\ 2x^3 \qquad + 5x \\ \underline{2x^2 \qquad + 5} \\ 2x^3 + 2x^2 + 5x + 5 \end{array}$$
*Answer:* $2x^3 + 2x^2 + 5x + 5$

5.
$$\begin{array}{r} 3x^2 - 5x + 4 \\ \underline{2x - 3} \\ 6x^3 - 10x^2 + 8x \\ \underline{- 9x^2 + 15x - 12} \\ 6x^3 - 19x^2 + 23x - 12 \end{array}$$
*Answer:* $6x^3 - 19x^2 + 23x - 12$

6.
$$\begin{array}{r} 2b^2 - b - 3 \\ \underline{4b^2 + 2b - 5} \\ 8b^4 - 4b^3 - 12b^2 \\ 4b^3 - 2b^2 - 6b \\ \underline{- 10b^2 + 5b + 15} \\ 8b^4 \qquad - 24b^2 - b + 15 \end{array}$$
*Answer:* $8b^4 - 24b^2 - b + 15$

## DIAGNOSTIC TEST

Multiply:   *Wright horizontally.*

**1.** $x + 4$
$x + 3$

**2.** $d + 8$
$d - 5$

**3.** $y - 7$
$y + 3$

**4.** $a - 5$
$a - 4$

**5.** $x + y$
$x + y$

**6.** $c - d$
$c - d$

**7.** $c - 2$
$c + 2$

**8.** $a + x$
$a - x$

**9.** $m^2 + 3$
$m^2 - 9$

**10.** $2x - 7$
$4x + 3$

**11.** $6 + b$
$5 - 2b$

**12.** $5ax - 3$
$3ax - 8$

**13.** $2s - 3$
$4 + 5s$

**14.** $2c - 3d$
$3c + 4d$

**15.** $3a^2 - 4b^2$
$2a^2 - 3b^2$

**16.** $8bc + x$
$3bc - 2x$

**17.** $x^2 + x$
$x + 1$

**18.** $5x^2 + 2$
$x + 7$

**19.** $a + b$
$a + c$

**20.** $y^2 - 2y + 1$
$y - 1$

**21.** $7x^2 - 6xy + 9y^2$
$4x + 3y$

**22.** $5s^2 - 2s + 6$
$2s^2 + 3s - 4$

**23.** $b^3 - 3b^2 + 9b - 27$
$b + 3$

**24.** $(5b + 2c)(3b - 7c)$

## *RELATED PRACTICE EXAMPLES*

Multiply:

**1.** $a + 7$
$a + 5$

$x + 1$
$x + 1$

$b + 6$
$b + 3$

$c + 2$
$c + 8$

$m + 5$
$m + 5$

**2.** $s + 4$
$s - 2$

$n + 7$
$n - 6$

$d + 5$
$d - 9$

$y + 3$
$y - 10$

$w + 1$
$w - 2$

**3.** $x - 8$
$x + 2$

$b - 5$
$b + 4$

$h - 9$
$h + 7$

$t - 2$
$t + 6$

$y - 6$
$y + 15$

**4.** $a - 3$
$a - 2$

$d - 6$
$d - 6$

$x - 1$
$x - 9$

$s - 3$
$s - 8$

$n - 9$
$n - 10$

**5.** $a + b$
$a + b$

$c + d$
$c + d$

$m + n$
$m + n$

$b + c$
$b + c$

$r + s$
$r + s$

6. $x - y$     $b - c$     $a - x$     $m - n$     $s - t$
$\underline{x - y}$     $\underline{b - c}$     $\underline{a - x}$     $\underline{m - n}$     $\underline{s - t}$

7. $x + 8$     $b + 4$     $r - 3$     $d - 1$     $y - 9$
$\underline{x - 8}$     $\underline{b - 4}$     $\underline{r + 3}$     $\underline{d + 1}$     $\underline{y + 9}$

8. $c + d$     $x + y$     $w - x$     $a - b$     $y - z$
$\underline{c - d}$     $\underline{x - y}$     $\underline{w + x}$     $\underline{a + b}$     $\underline{y + z}$

9. $b^2 + 4$     $a^2 - 3$     $x^2 - 4$     $x^3 + 6$     $c^4 - 3$
$\underline{b^2 + 2}$     $\underline{a^2 + 5}$     $\underline{x^2 - 7}$     $\underline{x^3 + 6}$     $\underline{c^4 + 3}$

10. $4c + 5$     $6b + 3$     $2a + 5$     $5y + 7$     $4x - 6$
$\underline{3c + 2}$     $\underline{6b + 3}$     $\underline{3a + 4}$     $\underline{2y - 3}$     $\underline{4x - 2}$

$5d + 4$     $3m - 4$     $6x - 9$     $-a - 1$     $2d + 7$
$\underline{6d + 7}$     $\underline{m + 5}$     $\underline{-2x - 5}$     $\underline{8a - 1}$     $\underline{2d - 7}$

11. $2 + x$     $8 + 2x$     $7 - b$     $1 - 4y$     $5 - 3b^2$
$\underline{3 - x}$     $\underline{5 + 7x}$     $\underline{7 + b}$     $\underline{6 + 7y}$     $\underline{5 - b^2}$

12. $3ab + 6$     $6xy - 3$     $2m^2n - 8$     $9abc + 3$     $1 - 2c^2y$
$\underline{2ab + 5}$     $\underline{8xy - 4}$     $\underline{-m^2n + 7}$     $\underline{5abc - 8}$     $\underline{4 - 3c^2y}$

13. $x + 4$     $b - 5$     $2c - 4$     $3b^2 + 8$     $4xy + 9$
$\underline{3 + x}$     $\underline{8 - b}$     $\underline{9 + 5c}$     $\underline{10 + 2b^2}$     $\underline{5 + 3xy}$

14. $2x + 3y$     $5m + 9n$     $6a + 2c$     $2a - b$     $8n - 3r$
$\underline{3x + 2y}$     $\underline{3m + 6n}$     $\underline{4a - 3c}$     $\underline{a + 3b}$     $\underline{9n + 6r}$

$6b - 5m$     $7x - 2y$     $5b + 3d$     $8c - 4x$     $4a - y$
$\underline{2b - 4m}$     $\underline{3x - 4y}$     $\underline{4b - 3d}$     $\underline{-c - 5x}$     $\underline{2a - 5y}$

15. $x^2 + y^2$     $a^4 - b^4$     $6m^3 - 4s^3$     $7x^2 - 2y$     $6a + 3d^2$
$\underline{x^2 + y^2}$     $\underline{a^4 + b^4}$     $\underline{2m^3 - 9s^3}$     $\underline{3x^2 - 4y}$     $\underline{-4a - 3d^2}$

16. $8ab + c$     $2x + 3xy$     $5mn - s$     $7cd^2 - 2d$     $4x^2y + 2xy^2$
$\underline{3ab + c}$     $\underline{3x + 4xy}$     $\underline{2mn + 3s}$     $\underline{6cd^2 - 5d}$     $\underline{3x^2y - 5xy^2}$

17. $a^2 + a$     $2x^2 - 3x$     $b^3 - b^2$     $c^3 - 2c^2$     $d^4 - d^3$
$\underline{a + 4}$     $\underline{x - 2}$     $\underline{2b - 5}$     $\underline{3c + 7}$     $\underline{d^2 - 2d}$

18. $4m^2 + 1$     $3y^2 - 9$     $2a^3 - 1$     $4x + 7$     $7 - 3a^2$
$\underline{m + 3}$     $\underline{5y + 2}$     $\underline{a - 1}$     $\underline{3x^2 - 9}$     $\underline{8 - a}$

19. $a + b$     $r - 2s$     $6x^2 + 3y$     $5ab - 2c$     $12b^2x - 4y$
$\underline{c + d}$     $\underline{c + 3d}$     $\underline{4x - 3y}$     $\underline{4ac - 7b}$     $\underline{3a^2x - 2z}$

**20.**
$$\begin{array}{r} x^2 + 2x + 1 \\ x + 1 \\ \hline \end{array} \qquad \begin{array}{r} 6a^2 + 4a + 4 \\ 5a - 2 \\ \hline \end{array} \qquad \begin{array}{r} 9 + 8y - y^2 \\ 2 + y \\ \hline \end{array} \qquad \begin{array}{r} 3a - 5 + 2a^2 \\ 4 - 7a \\ \hline \end{array}$$

**21.**
$$\begin{array}{r} a^2 + 2ab + b^2 \\ a + b \\ \hline \end{array} \qquad \begin{array}{r} 3c^2 + 2cd - 4d^2 \\ 5c - 3d \\ \hline \end{array} \qquad \begin{array}{r} 7xy - 4y^2 + 6x^2 \\ 5y + 2x \\ \hline \end{array} \qquad \begin{array}{r} 9m^2 - 4mn - 5n^2 \\ - 3m - n \\ \hline \end{array}$$

**22.**
$$\begin{array}{l} a + b + c \\ a - b + c \\ \hline \end{array} \qquad \begin{array}{l} 2x^2 + x + 3 \\ 7x^2 - x - 2 \\ \hline \end{array} \qquad \begin{array}{l} 4m^2 - 3m - 2 \\ 6m^2 + 5m + 4 \\ \hline \end{array}$$

$$\begin{array}{l} x^2 - 2xy + y^2 \\ x^2 + 2xy + y^2 \\ \hline \end{array} \qquad \begin{array}{l} 5a^3 - 6a^2 - 2a \\ 4a^2 + 2a - 3 \\ \hline \end{array} \qquad \begin{array}{l} 6c^4 - 3c^2d^2 + 8d^4 \\ 2c^3 - 4cd - 5d^3 \\ \hline \end{array}$$

**23.**
$$\begin{array}{r} x^3 - 2x^2 - x + 3 \\ 4x - 5 \\ \hline \end{array} \qquad \begin{array}{r} 3c^3 - 2c^2d + 5cd^2 - 4d^3 \\ 2c - 3d \\ \hline \end{array}$$

$$\begin{array}{r} x^3 - x^2 + x - 1 \\ x^2 - x + 1 \\ \hline \end{array} \qquad \begin{array}{r} 5a^3 - 3a^2x + 3ax^2 - x^3 \\ 6a^2 - 2ax + x^2 \\ \hline \end{array}$$

$$\begin{array}{r} 8a^5 + 2a^4 - 3a^3 + a^2 - a + 4 \\ 7a - 4 \\ \hline \end{array} \qquad \begin{array}{r} 3n^4 - 5n^3 + 6n^2 - 2n + 8 \\ 7n^3 - 5n^2 + n - 4 \\ \hline \end{array}$$

**24.** Multiply:

**a.** $(c + 3d)(2c + 4d)$      **b.** $(1 - x)(1 - x)$

**c.** $(2x - 4)(-3x + 9)$      **d.** $(5b^2 - 4x)(2b^2 + 9x)$

**e.** $(3a^2b + 5ab^2)(2a^2b - 7ab^2)$      **f.** $(3a^2 + 2a - 7)(-a - 8)$

**g.** $(2c^2 - 5c + 3)(3c - 4)$      **h.** $(b + c - 4x)(b + c - 3y)$

**i.** $(2b^2 - 4bc - c^2)(7b - 3c)$      **j.** $(x^4 + 2x^3 - 3x^2 + 7x + 5)(3x - 4)$

## CUMULATIVE REVIEW

**1.** Add $7g$ and $-8g$.

**2.** Combine: $9b^2 - 4bc + c^2 - 8b^2 + c^2 - 5bc - 3c^2 - bc$

**3.** Subtract $6x - 1$ from $10x - 5$.

Multiply:

**4.**
$$\begin{array}{r} 2x - 8 \\ 4x - 7 \\ \hline \end{array} \qquad \begin{array}{r} 7c - 9d \\ 3c - 2d \\ \hline \end{array} \qquad \begin{array}{r} 5y^4 - 11y^2 - 3 \\ 8y - 15 \\ \hline \end{array} \qquad \begin{array}{r} 2m^2 - 6mn + 4n^2 \\ 9m^2 + mn - 10n^2 \\ \hline \end{array}$$

**5. a.** $-k(-y)$      **b.** $6a^4bc^3(-4ab^4c^2)$      **c.** $-8x^2(x^3 - 7x^2 - 5x)$

     **d.** $-c^5d^3(4c^2d - 3c^7d^4 + 2c^3d^{10})$      **e.** $(4a - 3)(9a + 5)$      **f.** $(10m^2 - 7nx)(12n^2 - nx)$

**6.** Multiply the sum of $10x^2 - 5x$ and $x^2 - 3$ by the difference between $7x^2 - 11x - 5$ and $3x^2 + 4x - 2$.

Read

**I. Aim**  To develop and understand the use of the law of exponents in division.

Since $x^7 = x \cdot x \cdot x \cdot x \cdot x \cdot x \cdot x$ and $x^3 = x \cdot x \cdot x$, when $x \neq 0$, then

$$\frac{x^7}{x^3} = \frac{x \cdot x \cdot x \cdot x \cdot x \cdot x \cdot x}{x \cdot x \cdot x} = \frac{x}{x} \cdot \frac{x}{x} \cdot \frac{x}{x} \cdot x \cdot x \cdot x \cdot x$$

$$= 1 \cdot 1 \cdot 1 \cdot x^4 = x^4.$$

Thus, $x^7 \div x^3$ or $\dfrac{x^7}{x^3} = x^{7-3} = x^4$.

Observe that the exponent of the variable in the quotient (4) is obtained by subtracting the exponent of this variable in the divisor or denominator (3) from the exponent of this variable in the dividend or numerator (7). Thus, the law of exponents for division may be expressed as:

$$x^m \div x^n = \frac{x^m}{x^n} = x^{m-n}$$

where $x$ represents any nonzero number and $m$ and $n$ are positive integers such that $m > n$.

Since $\dfrac{x^m}{x^m} = x^{m-m} = x^0$ and $\dfrac{x^m}{x^m} = 1$, then $x^0 = 1$ when $x$ is any nonzero number. The expression $0^0$ is meaningless.

**II. Procedure**

1. To divide when the dividend and divisor both consist of the same variable, apply the law of exponents for division.
2. To divide when the dividend and divisor both consist of two or more identical variables, apply the law of exponents for division to each variable separately. If the variables are represented by letters, the letters are generally arranged in alphabetical order.
3. To divide when the dividend contains a variable not contained in the divisor, the variable is rewritten in the quotient, since the divisor of this variable is one.
4. Any power of a variable divided by the same power of the same variable, when it does not represent zero, is equal to one.

## III. Sample Solutions
Divide:

**1.** $\dfrac{m^{12}}{m^4} = m^{12-4} = m^8$

Answer: $m^8$

**2.** $\dfrac{a^4b^5}{a^3b^2} = \dfrac{a^4}{a^3} \cdot \dfrac{b^5}{b^2} = ab^3$

Answer: $ab^3$

**3.** $\dfrac{c^5d^2x^3}{c^3x^2} = \dfrac{c^5}{c^3} \cdot \dfrac{d^2}{1} \cdot \dfrac{x^3}{x^2} = c^2d^2x$

Answer: $c^2d^2x$

**4.** $\dfrac{c^6}{c^6} = 1$

Answer: $1$

## DIAGNOSTIC TEST

Divide:

**1.** $\dfrac{a^6}{a^2}$  **2.** $\dfrac{x^3}{x^3}$  **3.** $\dfrac{b^7d^4n^9}{b^4dn^5}$  **4.** $\dfrac{a^5c^2x^8}{a^4x^3}$  **5.** $\dfrac{m^2r^3s^4}{mrs^4}$

Divide as indicated:     **6.** $(n^9x^2y^8) \div (n^6xy^8)$

## *RELATED PRACTICE EXAMPLES*

**1.** Divide:

$\dfrac{a^5}{a^2}$   $\dfrac{y^9}{y^3}$   $\dfrac{m^{10}}{m^5}$   $\dfrac{s^3}{s^2}$   $\dfrac{x^8}{x}$   $\dfrac{v^2}{v}$   $\dfrac{2^{10}}{2^4}$   $\dfrac{5^8}{5^6}$

**2.** $\dfrac{b^2}{b^2}$   $\dfrac{x^5}{x^5}$   $\dfrac{a^3}{a^3}$   $\dfrac{d}{d}$   $\dfrac{m^2}{m^2}$   $\dfrac{w^{10}}{w^{10}}$   $\dfrac{3^7}{3^7}$   $\dfrac{8^{10}}{8^{10}}$

**3.** $\dfrac{m^4n^3}{m^2n^2}$   $\dfrac{a^8x^5}{a^4x^2}$   $\dfrac{a^2b^3c^4}{ab^2c^3}$   $\dfrac{x^4y^2z^7}{xyz^6}$   $\dfrac{m^4x^5y^6}{m^2xy^3}$   $\dfrac{a^4m^5x^9}{am^2x^5}$   $\dfrac{c^6x^4z^8}{c^4x^3z^5}$

**4.** $\dfrac{a^5x}{a^3}$   $\dfrac{c^3d^3}{c^2}$   $\dfrac{m^2n^5}{n^4}$   $\dfrac{a^2c^5d^3}{ac^3}$   $\dfrac{m^2n^3y^5}{ny}$   $\dfrac{bn^3t^7}{n^2t^4}$   $\dfrac{a^3x^3y}{x^2}$

**5.** $\dfrac{bc^2}{bc}$   $\dfrac{a^5b^4c^3}{ab^2c^3}$   $\dfrac{c^7d^2x^5}{c^2d^2x^3}$   $\dfrac{r^3s^8t^4}{r^3st^3}$   $\dfrac{m^7n^2x^6}{m^5x^6}$   $\dfrac{a^4c^2m}{a^3c^2m}$   $\dfrac{b^5d^4y^2}{b^5d^4y}$

**6.** Divide as indicated:

**a.** $(b^{12}) \div (b^3)$     **b.** $(a^5c^2) \div (a^5c^2)$     **c.** $(xy^3) \div (xy)$

**d.** $(c^3d^7) \div (d^2)$     **e.** $(m^6n^8y^5) \div (m^4n^2y^4)$     **f.** $(r^8s^2t) \div (r^8t)$

**g.** $(b^2h^7y^6) \div (bh^7y^6)$     **h.** $(a^8cx^3) \div (a^5c)$     **i.** $(x^{10}y^6z^5) \div (x^2y^3z^4)$

# Division of a Monomial by a Monomial

**I. Aim**    To divide a monomial by a monomial.

**II. Procedure**
1. Divide their numerical coefficients.
2. Find the quotient of the variables using the law of exponents for division.
3. Prefix the numerical quotient to the product of the quotient of the variables arranged in alphabetical order.
4. Since division by zero is excluded, the variables that appear in the divisor cannot represent zero.
5. Observe that when the divisor contains only a numerical factor and no variable, the quotient of the variables remains unchanged. See sample solution 3.
6. Check by multiplication.

**III. Sample Solutions**
Divide:

**1.** $\dfrac{-27a^6b^7}{3a^2b^4}$

$= \dfrac{-27}{3} \cdot \dfrac{a^6}{a^2} \cdot \dfrac{b^7}{b^4}$

$= -9a^4b^3$        *Answer:* $-9a^4b^3$

**2.** $\dfrac{32m^8y^7}{-4m^5y} = -8m^3y^6$    **3.** $\dfrac{-8x^2y^4}{-2} = +4x^2y^4$

*Answer:* $-8m^3y^6$    *Answer:* $4x^2y^4$

**4.** $\dfrac{-6b^4}{-2b^4} = +3$        **5.** $\dfrac{+5c^2mx^3}{-5cx^2} = -cmx$

*Answer:* $+3$        *Answer:* $-cmx$

**6.** $\dfrac{8x^2y}{-x} = -8xy$        **7.** $\dfrac{2a^2c}{-2a^2c} = -1$

*Answer:* $-8xy$        *Answer:* $-1$

**178**

## DIAGNOSTIC TEST

Divide:

1. $\dfrac{8b}{2}$

2. $\dfrac{-6m^4}{-3}$

3. $\dfrac{21m^3n^2}{-7}$

4. $\dfrac{-36h^6x^3y}{9}$

5. $\dfrac{16c^8}{4c^2}$

6. $\dfrac{-56a^9y^4}{-7a^3y^2}$

7. $\dfrac{-63b^8x^2}{9b^6x}$

8. $\dfrac{6x^4y^5z^2}{-2xyz}$

9. $\dfrac{7s^3t^2}{-1}$

10. $\dfrac{-12xy}{-12}$

11. $\dfrac{-21b^2}{3b^2}$

12. $\dfrac{6x^2y}{-6x^2y}$

13. $\dfrac{4a^6b^2}{-b}$

14. $\dfrac{-1.2h^2k^5}{.4hk^2}$

15. $(5x^2y^3) \div (-xy^2)$

## RELATED PRACTICE EXAMPLES

Divide:

1. $\dfrac{6b}{3}$ $\qquad$ $\dfrac{12ab}{4}$ $\qquad$ $\dfrac{20c^3}{5}$ $\qquad$ $\dfrac{18xyz}{3}$ $\qquad$ $\dfrac{32x^2y^3}{8}$ $\qquad$ $\dfrac{27a^2xy}{9}$ $\qquad$ $\dfrac{45b^4xy^3}{5}$

2. $\dfrac{-4a}{-2}$ $\qquad$ $\dfrac{-9x^2}{-3}$ $\qquad$ $\dfrac{-21bc}{-7}$ $\qquad$ $\dfrac{-12dy^2}{-6}$ $\qquad$ $\dfrac{-18m^2x^2}{-9}$ $\qquad$ $\dfrac{-48a^2my}{-12}$ $\qquad$ $\dfrac{-60b^4cx^2}{-5}$

3. $\dfrac{8x}{-4}$ $\qquad$ $\dfrac{10b^8}{-5}$ $\qquad$ $\dfrac{27ay}{-9}$ $\qquad$ $\dfrac{36m^4x^4}{-4}$ $\qquad$ $\dfrac{42cy^3}{-7}$ $\qquad$ $\dfrac{54bcd}{-6}$ $\qquad$ $\dfrac{80m^3n^2y^2}{-16}$

4. $\dfrac{-8x}{2}$ $\qquad$ $\dfrac{-10m}{5}$ $\qquad$ $\dfrac{-14x^3}{7}$ $\qquad$ $\dfrac{-15bd}{3}$ $\qquad$ $\dfrac{-40b^3c^2}{8}$ $\qquad$ $\dfrac{-32c^4d^4}{16}$ $\qquad$ $\dfrac{-56dm^2x^5}{8}$

5. $\dfrac{12x^8}{2x^2}$ $\qquad$ $\dfrac{15m^6}{3m^4}$ $\qquad$ $\dfrac{27a^3b^2}{9a}$ $\qquad$ $\dfrac{36c^5d^4}{18c^3d^2}$ $\qquad$ $\dfrac{24b^2x^7}{6bx^4}$

$\dfrac{49m^6x^3}{7m^3x}$ $\qquad$ $\dfrac{50a^3b^2c}{5a^2b}$ $\qquad$ $\dfrac{64m^4x^2}{8m^2x^2}$ $\qquad$ $\dfrac{72a^7b^5}{9a^4b^3}$ $\qquad$ $\dfrac{75m^2n^2y}{25mn^2y}$

6. $\dfrac{-4d^7}{-2d^3}$ $\qquad$ $\dfrac{-6b^8}{-3b^5}$ $\qquad$ $\dfrac{-15b^5c^3}{-5c^2}$ $\qquad$ $\dfrac{-18r^4t^2}{-3rt}$ $\qquad$ $\dfrac{-30s^3x}{-6s^2}$

$\dfrac{-42at^2}{-7at}$ $\qquad$ $\dfrac{-21m^5n^6}{-3m^2n^2}$ $\qquad$ $\dfrac{-54x^2yz}{-9xyz}$ $\qquad$ $\dfrac{-48a^4c^3d^9}{-8a^2c^3d^5}$ $\qquad$ $\dfrac{-64d^2r^2t^3}{-16dt^2}$

7. $\dfrac{-6a^9}{2a^3}$  $\dfrac{-12b^5}{3b^2}$  $\dfrac{-16m^4n}{4mn}$  $\dfrac{-32d^4r}{8d^3}$  $\dfrac{-35r^4t^2}{7rt^2}$

$\dfrac{-52gt^3}{4gt}$  $\dfrac{-72m^4n^2}{12mn^2}$  $\dfrac{-80a^3b^2x^4}{5abx}$  $\dfrac{-44c^5d^4x}{11c^2d^4}$  $\dfrac{-96m^5nx^8}{16m^2x^6}$

8. $\dfrac{10b^2}{-5b}$  $\dfrac{16c^5}{-8c^3}$  $\dfrac{21a^2c^4}{-7ac}$  $\dfrac{34b^2c}{-17b}$  $\dfrac{45x^4y}{-15x}$

$\dfrac{96a^5x^9}{-12a^2x^7}$  $\dfrac{63d^4y^7}{-7dy^6}$  $\dfrac{81c^4d^3x}{-9c^3d^3x}$  $\dfrac{54d^5x^2y^3}{-6dxy}$  $\dfrac{98a^4b^2c^5}{-49a^4b^2}$

9. $\dfrac{5s}{1}$  $\dfrac{-x}{1}$  $\dfrac{-5c^3x}{1}$  $\dfrac{2x}{-1}$  $\dfrac{7s^3}{-1}$  $\dfrac{-9b^2y^2}{-1}$  $\dfrac{-a^5}{-1}$

10. $\dfrac{3d}{3}$  $\dfrac{-6bd}{6}$  $\dfrac{-10y^2z^3}{10}$  $\dfrac{4x^6}{-4}$  $\dfrac{-9n^3r^4}{-9}$  $\dfrac{-25x^5y^4}{-25}$  $\dfrac{-32a^3b^2c}{-32}$

11. $\dfrac{16a^4}{8a^4}$  $\dfrac{-25c^3d}{5c^3d}$  $\dfrac{-30xy^2}{-3xy^2}$  $\dfrac{28abc}{-7abc}$  $\dfrac{48b^2x^3y}{16b^2x^3y}$

$\dfrac{6a^4}{6a^2}$  $\dfrac{-15bc^2}{-15bc}$  $\dfrac{-28c^3d}{28c}$  $\dfrac{9m^4x^3}{-9mx^2}$  $\dfrac{-20r^2s^3t}{-20rst}$

12. $\dfrac{2x}{2x}$  $\dfrac{-9b^2}{9b^2}$  $\dfrac{-4m^2n}{-4m^2n}$  $\dfrac{16bx^2}{-16bx^2}$  $\dfrac{-14c^8y}{14c^8y}$  $\dfrac{19x^3yz^4}{-19x^3yz^4}$

13. $\dfrac{3mn}{m}$  $\dfrac{4x^2}{-x}$  $\dfrac{-9x^3y}{-y}$  $\dfrac{-2b^8c^7}{b^5c^3}$  $\dfrac{-2xy^2}{-x}$  $\dfrac{15d^2rt^3}{-dt^3}$

14. $\dfrac{.6a^5x^2}{.2a^2x}$  $\dfrac{-1.8r^2t}{.3rt}$  $\dfrac{-.08m^3x^2y^4}{-.2mxy^2}$  $\dfrac{3.9a^4b^4c}{-1.3a^3b^4}$  $\dfrac{-8c^3xy^5}{-.5xy^2}$

15. Divide as indicated:
   a. $x^9 \div x^3$
   b. $(-3b^3x) \div (-3bx)$
   c. $(15m^5) \div (3m^2)$
   d. $(-32c^4y) \div (8c^2)$
   e. $(-42m^6n^6) \div (7m^3n^2)$
   f. $(14a^5b) \div (-ab)$
   g. $(-12b^7cx^2) \div (-4b^2c)$
   h. $(-9d^4x^3y^5) \div (9d^2xy^3)$
   i. $(60b^5m^4t^2) \div (15b^5m^4t^2)$
   j. $(-2.1c^6r) \div (-.3c^4)$

# Division of a Polynomial by a Monomial

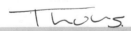

**I. Aim**   To divide a polynomial by a monomial.

**II. Procedure**

1. Divide each term of the polynomial by the monomial. The resulting polynomial consisting of the sum of these quotients is the required quotient.
2. When a term of the polynomial and the divisor are the same, write the numeral 1 for the quotient. See sample solutions.
3. Since division by zero is excluded, the variables that appear in the divisor cannot represent zero.
4. Check by multiplication.

**III. Sample Solutions**

Divide:

1.
$$\frac{42b^9x^3 - 30b^7x^4 + 36b^5x^8}{-6b^3x^2} = \frac{42b^9x^3}{-6b^3x^2} + \frac{-30b^7x^4}{-6b^3x^2} + \frac{36b^5x^8}{-6b^3x^2}$$
$$= (-7b^6x) + (+5b^4x^2) + (-6b^2x^6)$$
$$= -7b^6x + 5b^4x^2 - 6b^2x^6$$

*Answer: $-7b^6x + 5b^4x^2 - 6b^2x^6$*

2. $\dfrac{3x - 12}{3} = x - 4$

*Answer: $x - 4$*

3. $\dfrac{5m^2 - m}{-m} = -5m + 1$

*Answer: $-5m + 1$*

4. $\dfrac{4a^2b - 8ab + 12ab^2}{4ab} = a - 2 + 3b$   *Answer: $a + 3b - 2$*

## DIAGNOSTIC TEST

Divide:

1. $\dfrac{6x + 4}{2}$

2. $\dfrac{5a - 10}{-5}$

3. $\dfrac{3a + 6b}{3}$

4. $\dfrac{-16xy + 24yz}{-8}$

5. $\dfrac{12b^2 - 18bc + 24c^2}{6}$

6. $\dfrac{4b^2 + 3c^2 - d^2}{-1}$

7. $\dfrac{m^3 - m^2}{m}$

_Thurs._

8. $\dfrac{2x - 5x^2}{-x}$     9. $\dfrac{5d^4 + 3d^2}{d^2}$     10. $\dfrac{8b - 4}{4}$     11. $\dfrac{4c^2 - c}{-c}$

12. $\dfrac{x^5y - 2x^4y^2 + x^3y^3}{x^2y}$     13. $\dfrac{8a^3 - 6ab}{2a}$     14. $\dfrac{9c^5d^3 - 27c^2d^2 + 6c^3d^5}{-3cd^2}$

15. $\dfrac{7m^6n^3x^2 - 14m^5n^2x^4 + 21m^4n^2x^4 - 28m^3n^2x}{7m^3n^2x}$

16. $(6r^2t - 4rt + 8rt^2) \div (-2rt)$

## RELATED PRACTICE EXAMPLES

Divide:

1. $\dfrac{9a + 6}{3}$     $\dfrac{16x^2 + 24}{8}$     $\dfrac{5ay - 15}{5}$     $\dfrac{12r - 8}{4}$     $\dfrac{-50b - 30}{10}$

2. $\dfrac{8x + 10}{-2}$     $\dfrac{7m + 21}{-7}$     $\dfrac{18a - 81}{-9}$     $\dfrac{15c^2 - 24}{-3}$     $\dfrac{-24d + 36}{-12}$

3. $\dfrac{4c + 8d}{4}$     $\dfrac{21b^2 - 28c^2}{7}$     $\dfrac{36ab + 18cd}{9}$     $\dfrac{42m - 18m^2}{6}$     $\dfrac{-3b^2 + 9bc}{3}$

4. $\dfrac{10b + 6x}{-2}$     $\dfrac{15x + 25y}{-5}$     $\dfrac{64r^2 - 8d}{-8}$     $\dfrac{-20rs + 12st}{-4}$     $\dfrac{-56m^3 - 35m^2}{-7}$

5. $\dfrac{8a + 12b - 16}{4}$     $\dfrac{9a^2 - 18x^2 + 27y^2}{9}$

$\dfrac{6x^2 - 9xy + 36y^2}{-3}$     $\dfrac{5x^3 - 10x^2 - 25x + 15}{-5}$

$\dfrac{16b^4 - 8b^3 + 40b^2 - 24b}{8}$     $\dfrac{30c^4 - 12c^3d + 6c^2d^2 - 42cd^3 + 54}{-6}$

6. $\dfrac{2a + 5d}{1}$     $\dfrac{3x^2 - 7y^2}{1}$     $\dfrac{4c - 6}{-1}$     $\dfrac{x^2 - 4xy + 2y^2}{-1}$     $\dfrac{a^3 + 4a^2 - 6a - 1}{-1}$

7. $\dfrac{b^4 + b^3}{b}$     $\dfrac{cx - cy}{c}$     $\dfrac{5m^6 - m^4}{m}$     $\dfrac{3a^2 - a^3 + 2a^4}{a}$

8. $\dfrac{ad + bd}{-d}$     $\dfrac{2h^2 - h^4}{-h}$     $\dfrac{5c^8 + c^6 - 3c^2}{-c}$     $\dfrac{-xy + x^2y - x^2y^3}{-x}$

9. $\dfrac{a^4 - a^6}{a^2}$     $\dfrac{5xz^4 - y^2z^3}{-z^2}$     $\dfrac{n^8x - 2n^4y}{-n^4}$     $\dfrac{-4c^8 + 3c^7 - c^6}{-c^5}$

10. $\dfrac{5y+5}{5}$ $\qquad$ $\dfrac{24cd+8}{8}$ $\qquad$ $\dfrac{30m-6}{-6}$ $\qquad$ $\dfrac{3s-3}{-3}$ $\qquad$ $\dfrac{-6a^2-2}{2}$

11. $\dfrac{m^2+m}{m}$ $\qquad$ $\dfrac{4r^2-r}{r}$ $\qquad$ $\dfrac{2x^4-x}{-x}$ $\qquad$ $\dfrac{s^8+s^2}{s^2}$ $\qquad$ $\dfrac{-3y^6+y^3}{-y^3}$

12. $\dfrac{a^3b+a^2b^2-5ab^3}{ab}$ $\qquad$ $\dfrac{5c^4d^2-c^3d^3+2c^2d^4}{-c^2d}$ $\qquad$ $\dfrac{-3xy^4+4xy^3-xy^2}{xy^2}$

$\dfrac{6m^6n^4-2m^3n^2-m^4n^6}{m^3n^2}$ $\qquad$ $\dfrac{b^4c^2d^3+2b^2c^3d^4-b^3c^4d^4}{-b^2cd^3}$

13. $\dfrac{6b^2-12b^6}{3b}$ $\quad$ $\dfrac{15m^4-25m^3}{-5m}$ $\quad$ $\dfrac{-4a^2b+12ab^2}{4a}$ $\quad$ $\dfrac{8x^5y-2xy}{-2x}$ $\quad$ $\dfrac{16y^3-24xy}{-8y}$

$\dfrac{4a^2b-6ab^2}{2ab}$ $\qquad\qquad$ $\dfrac{12m^4n^3-6mn}{-6mn}$ $\qquad\qquad$ $\dfrac{54x^6+81x^4}{9x^2}$

$\dfrac{7c^5-21c^3d^2}{-7c^3}$ $\qquad\qquad$ $\dfrac{-8b^3d^4-12b^2d^3}{4bd^2}$

14. $\dfrac{16b^4-8b^3+24b^2}{4b}$ $\qquad$ $\dfrac{3a^5+6a^4-9a^2}{3a^2}$ $\qquad$ $\dfrac{15b^2c^2+30b^2c-15b^2}{-15b^2}$

$\dfrac{14ax^2-21a^2x+28a^3x}{7ax}$ $\qquad$ $\dfrac{-18x^2y^3z+9x^3yz^2-27xy^2z^3}{9xyz}$

$\dfrac{12g^2t^4-36g^2t^3-18g^2t^2}{6gt^2}$ $\qquad$ $\dfrac{36x^3y^3+72x^2y^2-54x^2y}{-18x^2y}$

$\dfrac{13c^2d^2x-26cd^3x+39cd^2x}{13cd^2}$ $\qquad$ $\dfrac{10r^8s^4-15r^6s^5-20r^3s^3}{-5r^2s^3}$

$\dfrac{64d^6r^3t^5-24d^3r^5t^4+72d^2r^4t^6}{8d^2r^3t^4}$

15. $\dfrac{2x^4-4x^3+6x^2-4x}{-2x}$ $\qquad$ $\dfrac{3a^4y-9a^3y^2-6a^2y^3+12ay^4}{3ay}$

$\dfrac{4a^4b^3c^2+8a^3b^4c-8a^2b^4c^3-4a^2b^3c}{-4a^2b^3c}$ $\qquad$ $\dfrac{6mn+12m^2n^2-18m^3n^3+24m^4n^4}{6mn}$

$\dfrac{5ab^3-10a^2b^2+5ab^2-15ab^4}{-5ab^2}$ $\qquad$ $\dfrac{-72x^{10}y^6-81x^8y^8+45x^7y^9-36x^4y^{11}}{-9x^4y^5}$

16. Divide as indicated:
    **a.** $(-c^2+c)\div(-c)$ $\qquad\qquad$ **b.** $(6b^2-5b+3)\div(-1)$
    **c.** $(ax+bx)\div(x)$ $\qquad\qquad$ **d.** $(2x-4x^2)\div(2x)$
    **e.** $(\pi r^2-2\pi r)\div(\pi r)$ $\qquad\qquad$ **f.** $(9cd^2-12c^2d)\div(-3cd)$
    **g.** $(7ab-14ac)\div(7a)$ $\qquad\qquad$ **h.** $(-9x^2-18x^3y^2)\div(-9x^2)$
    **i.** $(4a^2b^2c-20a^2bc^2+8ab^2c^2)\div(4abc)$ $\qquad$ **j.** $(25c^2x-15c^3y+10c^4z)\div(-5c^2)$

# Division of a Polynomial by a Polynomial

**I. Aim**    To divide a polynomial by a polynomial.

**II. Procedure**

1. Arrange the terms of both the dividend and the divisor in either descending or ascending powers of some common variable.

2. Divide the first term of the dividend by the first term of the divisor. Write the answer as the first term of the quotient.

3. Multiply the entire divisor by the first term of the quotient. Write this product under the dividend, keeping like terms under each other.

4. Subtract this product from the dividend.

5. Considering the remainder and any additional terms of the original dividend as a new dividend, repeat steps, 2, 3, and 4 until the remainder is no longer divisible.

6. The missing terms of an expression arranged in ascending or descending order of powers have zero as their numerical coefficients. See sample solution.

7. When the division is not exact, the remainder is written generally over the divisor to form an algebraic fraction. See sample solution.

8. Check by multiplication.

**III. Sample Solutions**

Divide:

1. $x^2 + 7x + 12$ by $x + 3$

STEP 1

$$x + 3 \overline{)x^2 + 7x + 12}$$

STEP 2

$$\begin{array}{r} x \phantom{+ 7x + 12} \\ x + 3 \overline{)x^2 + 7x + 12} \end{array}$$

STEP 3

$$\begin{array}{r} x \phantom{+ 7x + 12} \\ x + 3 \overline{)x^2 + 7x + 12} \\ x^2 + 3x \phantom{+ 12} \end{array}$$

STEP 4

$$\begin{array}{r} x \phantom{+ 7x + 12} \\ x + 3 \overline{)x^2 + 7x + 12} \\ \underline{x^2 + 3x} \phantom{+ 12} \\ + 4x \phantom{+ 12} \end{array}$$

**184**

## STEP 5

a) Repeat step 2

$$x+3 \overline{)\begin{array}{l} x+4 \\ x^2+7x+12 \end{array}}$$
$$\underline{x^2+3x}$$
$$+4x$$

b) Repeat step 3

$$x+3 \overline{)\begin{array}{l} x+4 \\ x^2+7x+12 \end{array}}$$
$$\underline{x^2+3x}$$
$$+4x+12$$
$$+4x+12$$

c) Repeat step 4

$$x+3 \overline{)\begin{array}{l} x+4 \\ x^2+7x+12 \end{array}}$$
$$\underline{x^2+3x}$$
$$4x+12$$
$$\underline{4x+12}$$

Answer: $x+4$

**2.** Divide $x^2-9x+20$ by $x-4$

$$x-4 \overline{)\begin{array}{l} x-5 \\ x^2-9x+20 \end{array}}$$
$$\underline{x^2-4x}$$
$$-5x+20$$
$$-5x+20$$

Answer: $x-5$

**3.** Divide $x^2-3xy-10y^2$ by $x-5y$

$$x-5y \overline{)\begin{array}{l} x+2y \\ x^2-3xy-10y^2 \end{array}}$$
$$\underline{x^2-5xy}$$
$$+2xy-10y^2$$
$$+2xy-10y^2$$

Answer: $x+2y$

**4.** Divide $15b^3+14b^2+2b-28$ by $3b+7$

$$3b+7 \overline{)\begin{array}{l} 5b^2+2b-4 \\ 15b^3+41b^2+2b-28 \end{array}}$$
$$\underline{15b^3+35b^2}$$
$$+6b^2+2b$$
$$\underline{+6b^2+14b}$$
$$-12b-28$$
$$-12b-28$$

Answer: $5b^2+2b-4$

**5.** Divide $a^4-b^4$ by $a-b$

$$a-b \overline{)\begin{array}{l} a^3+a^2b+ab^2+b^2 \\ a^4 \qquad\qquad\qquad -b^4 \end{array}}$$
$$\underline{a^4-a^3b}$$
$$+a^3b$$
$$\underline{+a^3b-a^2b^2}$$
$$+a^2b^2$$
$$\underline{+a^2b^2-ab^3}$$
$$+ab^3-b^4$$
$$+ab^3-b^4$$

Answer: $a^3+a^2b+ab^2+b^3$

**6.** Divide $3a^2-11a-18$ by $a-5$

$$a-5 \overline{)\begin{array}{l} 3a+4 \\ 3a^2-11a-18 \end{array}}$$
$$\underline{3a^2-15a}$$
$$+4a-18$$
$$\underline{+4a-20}$$
$$+2$$

Answer: $3a+4+\dfrac{2}{a-5}$

## DIAGNOSTIC TEST

Divide:

1. $c^2 + 10c + 21$ by $c + 3$
2. $x^2 + 3x - 10$ by $x - 2$
3. $m^2 - 4m - 12$ by $m - 6$
4. $b^2 - 2b - 63$ by $b + 7$
5. $s^2 + 2s - 15$ by $s + 5$
6. $a^2 - 10a + 24$ by $a - 4$
7. $x^2 + x - 6$ by $x + 3$
8. $18c^2 - 34c - 4$ by $9c + 1$
9. $8b^6 + 22b^3 + 15$ by $2b^3 + 3$
10. $a^2 - 2ax + x^2$ by $a - x$
11. $42m^2 + 16mn - 8n^2$ by $6m + 4n$
12. $8 + 2b - b^2$ by $2 + b$
13. $8a^4b^2 + 16a^2bx - 10x^2$ by $4a^2b - 2x$
14. $a^2 - 3b^2 + 2ab$ by $3b + a$
15. $n^3 + 4n^2 - 27n + 18$ by $n - 3$
16. $9c^3 - 21c^2d + 16cd^2 - 4d^3$ by $3c - 2d$
17. $a^2 + ab - ac + bc$ by $a + b$
18. $x^4 - x^3 - 8x^2 + 16x - 8$ by $x^2 + 2x - 4$
19. $b^4 - 16$ by $b + 2$
20. $m^2 + 11m + 32$ by $m + 5$

## *RELATED PRACTICE EXAMPLES*

Divide:

1. **a.** $x^2 + 4x + 4$ by $x + 2$
   **b.** $y^2 + 8y + 15$ by $y + 5$
   **c.** $a^2 + 4a + 3$ by $a + 1$
   **d.** $m^2 + 20m + 96$ by $m + 8$
   **e.** $r^2 + 21r + 80$ by $r + 16$

2. **a.** $b^2 + 2b - 15$ by $b - 3$
   **b.** $x^2 + 5x - 84$ by $x - 7$
   **c.** $d^2 + 3d - 28$ by $d - 4$
   **d.** $t^2 + 7t - 18$ by $t - 2$
   **e.** $z^2 + 8z - 9$ by $z - 1$

3. **a.** $c^2 - 5c - 24$ by $c - 8$
   **b.** $s^2 - 6s - 27$ by $s - 9$
   **c.** $v^2 - 4v - 5$ by $v - 5$
   **d.** $x^2 - 3x - 108$ by $x - 12$
   **e.** $a^2 - 20a - 96$ by $a - 24$

4. **a.** $x^2 - 5x - 36$ by $x + 4$
   **b.** $d^2 - 2d - 48$ by $d + 6$
   **c.** $s^2 - 15s - 54$ by $s + 3$
   **d.** $r^2 - 10r - 75$ by $r + 5$
   **e.** $y^2 - 2y - 120$ by $y + 10$

5. **a.** $m^2 + 5m - 14$ by $m + 7$
   **b.** $y^2 + 7y - 8$ by $y + 8$
   **c.** $c^2 + 13c - 48$ by $c + 16$
   **d.** $h^2 + 7h - 60$ by $h + 12$
   **e.** $t^2 + 6t - 135$ by $t + 15$

6. **a.** $x^2 - 11x + 24$ by $x - 8$
   **b.** $h^2 - 8h + 16$ by $h - 4$
   **c.** $n^2 - 14n + 45$ by $n - 9$
   **d.** $b^2 - 24b + 143$ by $b - 11$
   **e.** $z^2 - 20z + 100$ by $z - 10$

7. **a.** $a^2 - a - 2$ by $a + 1$
   **b.** $r^2 + r - 30$ by $r - 5$
   **c.** $m^2 - m - 56$ by $m - 8$
   **d.** $g^2 - g - 90$ by $g + 9$
   **e.** $v^2 - v - 132$ by $v - 12$

8. **a.** $2b^2 + 13b + 6$ by $b + 6$
   **b.** $25x^2 - 5x - 6$ by $5x - 3$
   **c.** $6y^2 - 13y - 28$ by $3y + 4$
   **d.** $28a^2 + 22a - 30$ by $7a - 5$
   **e.** $12x^2 - 22x + 8$ by $4x - 2$

9.  **a.** $x^4 + 9x^2 + 20$ by $x^2 + 4$
    **b.** $a^4 - 2a^2 - 8$ by $a^2 + 2$
    **c.** $c^6 - 2c^3 + 1$ by $c^3 - 1$
    **d.** $6y^4 + 9y^2 - 42$ by $3y^2 - 6$
    **e.** $10b^6 + b^3 - 24$ by $5b^3 + 8$

10. **a.** $x^2 + 2xy + y^2$ by $x + y$
    **b.** $a^2 + 2ab + b^2$ by $a + b$
    **c.** $c^2 - 2cd + d^2$ by $c - d$
    **d.** $m^4 + 2m^2n^2 + n^4$ by $m^2 + n^2$
    **e.** $r^6 - 2r^3s^2 + s^4$ by $r^3 - s^2$

11. **a.** $4x^2 + 12xy + 9y^2$ by $2x + 3y$
    **b.** $3b^2 - 13bc + 4c^2$ by $b - 4c$
    **c.** $20a^2 - 2ad - 6d^2$ by $5a - 3d$
    **d.** $35c^2 + 37cx - 6x^2$ by $7c - x$
    **e.** $16m^4 - 26m^2n^2 - 35n^4$ by $8m^2 + 7n^2$

12. **a.** $9 + 6x + x^2$ by $3 + x$
    **b.** $12 + 7y + y^2$ by $4 + y$
    **c.** $24 - 10b - b^2$ by $2 - b$
    **d.** $21 - 4a - a^2$ by $7 + a$
    **e.** $20 + w - w^2$ by $5 - w$

13. **a.** $c^2d^2 - 2cdx + x^2$ by $cd - x$
    **b.** $4x^2 + 17xyz + 15y^2z^2$ by $4x + 5yz$
    **c.** $6b^2c^2 + 13bcmx - 28m^2x^2$ by $2bc + 7mx$
    **d.** $14a^4m^2 - 64a^2my + 32y^2$ by $7a^2m - 4y$
    **e.** $20d^6r^4 - 14d^3r^2tx^2 - 24t^2x^4$ by $5d^3r^2 + 4tx^2$

14. **a.** $2cd + d^2 + c^2$ by $c + d$
    **b.** $10 + x^2 - 7x$ by $x - 2$
    **c.** $14y - 8 + 15y^2$ by $4 + 3y$
    **d.** $15a^2 - 32b^2 - 28ab$ by $4b + 5a$
    **e.** $35y^2 - 62xy + 24x^2$ by $6x - 5y$
    **f.** $56a^2 - 45c^2 - 23ac$ by $8a - 9c$

15. **a.** $x^3 - 2x^2 + 2x - 1$ by $x - 1$
    **b.** $c^3 - 2c^2 - 13c + 20$ by $c - 4$
    **c.** $a^3 - 3a^2b + 3ab^2 - b^3$ by $a - b$
    **d.** $m^3 - 6m^2n + 2mn^2 + 3n^3$ by $m - n$
    **e.** $a^3b^3 + 3a^2b^2c - 6abc^2 + 2c^3$ by $ab - c$

16. **a.** $18r^3 - 57r^2 + 29r + 5$ by $6r - 5$
    **b.** $6m^3 - 4m^2 - 7m + 12$ by $3m + 4$
    **c.** $8b^3 - 18b^2c + 25bc^2 - 12c^3$ by $4b - 3c$
    **d.** $3a^3x^3 - 18a^2x^2y + 9axy^2 + 6y^3$ by $3ax - 3y$
    **e.** $4a^3x^2 + 3a^3x - a^2x^2 - 4a^2x^3 - 2ax^3$ by $a - x$

17. **a.** $c^2 + cx + cy + xy$ by $c + x$
    **b.** $m^2 - mr + ms - rs$ by $m - r$
    **c.** $bd + cd + bx + cx$ by $b + c$
    **d.** $x^2y^2 - x^2y - xy^2 + xy$ by $x^2 - x$
    **e.** $ab^2c - 4ac + 4d - b^2d$ by $ac - d$

18. **a.** $c^3 - 7c^2 + 3c + 14$ by $c^2 - 5c - 7$
    **b.** $12b^3 - 14b^2 + 13b - 6$ by $4b^2 - 2b + 3$
    **c.** $a^4 + 2a^3 - 4a^2 - 5a + 6$ by $a^2 + a - 2$
    **d.** $3x^4 - 14x^3 + 12x^2 - 26x - 7$ by $x^2 - 4x - 1$
    **e.** $8m^4 - 10m^3n - 13m^2n^2 + 13mn^3 - 10n^4$ by $2m^2 - mn - 5n^2$

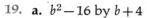

19. **a.** $b^2 - 16$ by $b + 4$      **b.** $x^2 - y^2$ by $x - y$
   **c.** $a^3 - 8$ by $a - 2$      **d.** $c^3 + d^3$ by $c + d$
   **e.** $x^4 - y^4$ by $x + y$      **f.** $a^4 - 81$ by $a - 3$
   **g.** $x^5 + 1$ by $x + 1$      **h.** $m^6 - n^6$ by $m - n$
   **i.** $r^6 - 64$ by $r + 2$      **j.** $a^8 - b^8$ by $a - b$

20. **a.** $x^2 + 3x + 7$ by $x + 2$      **b.** $x^2 + 6x - 9$ by $x + 3$
   **c.** $a^2 + ab + b^2$ by $a + b$      **d.** $3c^2 + 4c - 5$ by $c + 4$
   **e.** $5a^2 + 6a - 4$ by $a - 5$      **f.** $4x^2 - 4xy + 6y^2$ by $2x - 3y$
   **g.** $12c^2 - 4cd - 3d^2$ by $2c - 5d$      **h.** $5y^3 - 2y^2 + 5y - 9$ by $5y - 2$
   **i.** $6b^4 + 4b^3 - 4b^2 + 4b + 7$ by $3b^2 - 4b + 3$      **j.** $x^4 + 16$ by $x + 2$

## CUMULATIVE REVIEW

1. Add $6b^2 - 4bc - c^2$ and $-2b^2 - 4bc + c^2$.
2. Find the sum of $-10m - 8n$, $5m - n$, and $-6m + 12n$.
3. Combine: $3x^2 - 5x + x^2 - 9 - 2x + 7 - x - 8x^2 - 2x - 4$
4. From $-8y^2 - 3y + 6$ subtract $-3y^2 + 7y - 9$.
5. Subtract $3a^2 - 2b^2$ from $6ab$.
6. Find the difference between $4s^2 - 8s + 1$ and $2s^2 + 8s - 3$.
7. Multiply: $(-4m^7x^2y^4)(+5m^6xy^3)(-mx^4y^2)$
8. Find the product: $-10r^5s^8(-2r^6s^2 + 5r^4s^4 - r^2s^6)$
9. Multiply $6c^2 - 5d$ by $4c^2 - 3d$.

Divide:

10. $\dfrac{-30n^3}{-5}$    $\dfrac{-54a^7y^2}{9a^5y^2}$    $\dfrac{17c^4d^9}{-17c^4d^8}$    $\dfrac{-21x^3y^2z}{-1}$    $\dfrac{-72b^5d^7x^8}{8b^4x^4}$

11. $\dfrac{6mn - 18mx}{-6m}$    $\dfrac{24x^8y^6 - 16x^5y^9 + 36x^4y^{12}}{-4x^3y^6}$    $\dfrac{10a^8 - 25a^6 + 45a^3 - 5a^2}{5a^2}$

12. Divide $15a^2 + 23a - 28$ by $3a + 7$.
13. Divide $24x^2 - 82xy + 63y^2$ by $4x - 9y$.
14. Divide $n^4 - 16$ by $n + 2$.
15. Divide the quotient of $(-18r^6) \div (-3r^2)$ by the quotient of $(-8r^3) \div (-4r)$.
16. Subtract the sum of $3y^2 - 8y$ and $5y - 9$ from the quotient of $(16y^5 - 8y^4 - 12y^3) \div (4y^3)$.
17. Multiply the sum of $6t - 7$ and $2t + 5$ by the difference between $11t - 4$ and $-5t + 8$.
18. Divide the product of $9a^6b(8a^2b^5 - 4b^2)$ by the quotient of $(60a^5b^8) \div (-5a^3b^6)$.
19. Multiply the quotient of $(14x^2 + 15xy + 4y^2) \div (2x + y)$ by the quotient of $(36x^2 + 7xy - 15y^2) \div (4x + 3y)$.

# Removal of Parentheses

I. **Aim**   To remove parentheses.

II. **Procedure**

1. When the parentheses containing an expression are preceded by a plus (+) sign, proceed as in addition, or briefly remove the parentheses and rewrite all the terms which are within the parentheses without changing their signs. Combine like terms.

2. When the parentheses containing an expression are preceded by a minus (−) sign, proceed as in subtraction by adding the additive inverse of the expression given in the parentheses, or briefly remove the parentheses and rewrite all the terms which are within the parentheses but with their signs changed. Then combine like terms.

3. When there are parentheses within parentheses, remove one set of the parentheses at a time, starting with the innermost set of parentheses.

4. When the parentheses are immediately preceded by a term, remove the parentheses by multiplying each term in the parentheses by the term which precedes it. See sample solutions. Note that in $6x + 3(x − 5)$, the quantity $x − 5$ is thought of being immediately preceded by the term 3 and not by the term $6x$.

5. When the factors of a term are binomials, find their product and write it within parentheses. Then remove parentheses and combine like terms. See sample solution.

III. **Sample Solutions**

Remove parentheses and, if possible, add like terms:

1.   $4x + (3x − 5)$
   $= 4x + 3x − 5$
   $= 7x − 5$

*Answer: $7x − 5$*

2.   $8a + (−5a + 6)$
   $= 8a − 5a + 6$
   $= 3a + 6$

*Answer: $3a + 6$*

**189**

**3.**  $7x - (3x - 9)$
$= 7x + (-3x + 9)$
$= 7x - 3x + 9$
$= 4x + 9$
*Answer:* $4x + 9$

**4.**  $3a^2 - 2a - (2a^2 - 5a + 4)$
$= 3a^2 - 2a - 2a^2 + 5a - 4$
$= a^2 + 3a - 4$
*Answer:* $a^2 + 3a - 4$

**5.**  $3a - [4b + 5 - (-2a + 9b - 1) + a]$
$= 3a - [4b + 5 + 2a - 9b + 1 + a]$
$= 3a - [3a - 5b + 6]$
$= 3a - 3a + 5b - 6$
$= 5b - 6$
*Answer:* $5b - 6$

**6.**  $3a + 5(a + 2)$
$= 3a + (5a + 10)$
$= 3a + 5a + 10$
$= 8a + 10$
*Answer:* $8a + 10$

**7.**  $2b - 5b(3b - 2) + 18b^2$
$= 2b - (15b^2 - 10b) + 18b^2$
$= 2b - 15b^2 + 10b + 18b^2$
$= 3b^2 + 12b$
*Answer:* $3b^2 + 12b$

**8.**  $(x + 2)(x - 3) - (x - 5)(x - 1)$
$= (x^2 - x - 6) - (x^2 - 6x + 5)$
$= x^2 - x - 6 - x^2 + 6x - 5$
$= 5x - 11$
*Answer:* $5x - 11$

## DIAGNOSTIC TEST

Remove parentheses and, if possible, add like terms:

1. $6b + (5b + 7)$
2. $8x + (4x - 3y)$
3. $10 + (-6a + 3)$
4. $5m + (-8m - 4n)$
5. $4c^2 - cd + (d^2 - 6cd)$
6. $10y - 3 + (-8y^2 + y - 4)$
7. $9 - (7a + 5)$
8. $4a - (3a - 2x)$
9. $-2x^2 - (-x^2 + 4x)$
10. $3x - (-4y - 6x)$
11. $8b - 11c - (8b - 9c)$
12. $7n^2 - n - (6 - n + 10n^2)$
13. $(x - 3y) + (4x - 2y)$
14. $(c^2 - 3cx + x^2) - (2c^2 - 2cx + 3x^2)$
15. $(2a^3 - 5a^2 - 3a) - 2a^2$
16. $5a - (4a - 2x) + 3a$
17. $2x - (3x + 2) - 5 + (2x - 7)$
18. $[2c - (3c + 2d) - d]$
19. $-[4x^2 + (2x + 3) - 7]$
20. $4b + [3 - (2b - 4)]$
21. $3m - [m - (4m + 3n) + 5n]$
22. $2x^2 - [x^2 + 2x + (2x^2 - 5) - 4] - 4x$

**23.** $3a - 4b + [2a - (3b - 4a)] - (5a - 7b)$     **24.** $3(x + 4) + 2x$
**25.** $-5(4b - 3) + 6b - 4$                        **26.** $-a(a + 6) - 4a^2$
**27.** $4 + 6(y - 3)$                               **28.** $9 - 3(2c + 7)$
**29.** $8b - b(3b - 1)$                             **30.** $bc + 2c(b - 3c)$
**31.** $4x - 3x(5x - 4) + 20x^2$
**32.** $a^2 + 3a + 2a(a - 4) - 3a(a - 3) - 4a$
**33.** $b^2 - b - 4[2b + 3(b - 4)]$
**34.** $(x + 2)(x + 3) + (x - 1)(x - 2)$
**35.** $(m - 5)(m + 4) - (m - 6)(m - 2)$
**36.** $4x(x + 1) - 2x(x - 3) - (x + 2)(x - 8)$

## *RELATED PRACTICE EXAMPLES*

Remove parentheses and, if possible, add like terms:

**1. a.** $2x + (3x + 5)$
  **b.** $3y + (7c + 4y)$
  **c.** $6 + (3a + 9)$
  **d.** $-10 + (5b + 9)$
  **e.** $7a + (4c + 5d)$

**2. a.** $5m + (4m - 2)$
  **b.** $8x^2 + (3b^2 - 2x^2)$
  **c.** $15 + (9 - 3c)$
  **d.** $-2x + (5x^2 - 4x)$
  **e.** $4a^2 + (ab - 2b^2)$

**3. a.** $8 + (-4y + 6)$
  **b.** $5b + (-3b + 7c)$
  **c.** $2x + (-5y + 4x)$
  **d.** $-10m + (-3m + 5n)$
  **e.** $2x^2 + (-xy + 3y^2)$

**4. a.** $7r + (-3r - 4s)$
  **b.** $3m^2 + (-4n^2 - 3m^2)$
  **c.** $14 + (-6b - 8)$
  **d.** $-5d + (-8a - 4d)$
  **e.** $8c^2 + (-2c - 5)$

**5. a.** $4b + 6 + (3b + 7)$
  **b.** $5x^2 - 3x + (2x - 3x^2)$
  **c.** $2b^2 - 4bc + (5c^2 - 2bc)$
  **d.** $a^3 + a^2b + (-2a^2b - b^4)$
  **e.** $x - 3y + (-4x + 8y)$

**6. a.** $5a + (3a - 4b + 2c)$
  **b.** $3x + (-5x^2 - 8x + 7)$
  **c.** $6ab + (2a^2 + 3ab - b^2)$
  **d.** $25 - c + (-c^2 - 8c + 16)$
  **e.** $m^3 - 6m + (2m^2 - 5m + 1)$

**7. a.** $3a - (2a + 6)$
  **b.** $5m - (4y + 3m)$
  **c.** $9 - (4x + 7)$
  **d.** $-4c - (5c + 4d)$
  **e.** $8x - (3y + 9)$

**8. a.** $8c - (3c - 8)$
  **b.** $5d - (2a - 3d)$
  **c.** $4 - (7x - 9)$
  **d.** $-3s^2 - (2r^2 - 8s^2)$
  **e.** $4m - (5x - 3y)$

**9. a.** $10a - (-4a + 1)$
  **b.** $5xy - (-xy + 2y^2)$
  **c.** $6r^3 - (-2r^4 + 4r^3)$
  **d.** $-8b - (-3a + 4b)$
  **e.** $5m^2 - (-2m + 9)$

**10. a.** $8t - (-4t - 7)$
  **b.** $3ab - (-5ab - 2b^2)$
  **c.** $2a^2x - (-a^2x^2 - 2a^2x)$
  **d.** $-4c^3 - (-2c^2 - 5c^3)$
  **e.** $3x^2 - (-4xy - 3y^2)$

11. a. $3c + 2 - (c + 6)$    $3c + 2 - c - 6$
    b. $5d - 3s - (2d - 4s)$ $= 2c + 4$
    c. $4y^2 - 5y - (-2y + 9)$
    d. $8b^2 - 3x^2 - (-4xy - 6b^2)$
    e. $a^3 - a^2 + a - (2a^2 - a^3)$

12. a. $2y - (4a - 3b + 2y)$   $2y - 4a + 3b - 2y$
    b. $9x^2 - (-x^2 + 5x - 3)$ $= -4a + 3b$
    c. $4cd - (c^2 - 4cd - 7d^2)$
    d. $10b + b^2 - (-7b^2 - 9b + 8)$
    e. $x^3 - 5x^2 - (3x + 3 - x^3)$

13. a. $(3x - 4) + (2x + 3)$   $3x - 4 + 2x + 3$ $= 5x - 1$
    b. $(a - 3b) + (4a - 2b)$
    c. $(x^2 - 2x + 5) + (2x^2 - 3x - 9)$
    d. $(a^2 - 3ab + b^2) + (4b^2 - 2bc + c^2)$
    e. $(9 - 2c + 3c^2) + (-3c - c^2 + 4)$

14. a. $(x - 5) - (2x - 9)$   $x - 5 - 2x + 9$ $= -x + 4$
    b. $(3c - 9d) - (-4c + 2d)$
    c. $(x^2 + 2y^2) - (4y^2 - 3x^2)$
    d. $(b^2 - 5b + 3) - (3b - 7 - 2b^2)$
    e. $(m^2 - n^2) - (5mn - 3n^2)$

15. a. $(9x + 3) + 4x$ $= 13x + 3$
    b. $-(2ab - 3ab^2) + 4ab^2$
    c. $(10b^2 - 4b) - 3b$
    d. $-(49 - a^2) - 5a^2$
    e. $(r^2 - 2r + 6) + 3r - 7$

16. a. $8m + (6m + 4) + 6$   $8m + 6m + 4 + 6$ $= 14m + 10$
    b. $2x^2 + (3x^2 - 5x) + 2x$
    c. $3d - (4a - 7d) + 8d$
    d. $4n^2 - (n^2 - 8x) - 8x$
    e. $2a + b + (3a - 4b + 1) - (3b + 2)$

17. a. $2c + (4b - 3c) - (4c - 5b)$   $2c + 4b - 3c - 4c + 5b$ $= 9b - 5c$
    b. $(3x - 5y) - 2x + 6y - (4x - y)$
    c. $5a^2 - (2a + 6) - 3a + (a^2 - 1)$
    d. $-(x^3 - 4x^2 - 3x) + 2x^2 + (2x^3 - 4x)$
    e. $8m^4 - (2m^4 - m^2n^2) + (2m^2n^2 - n^4) - 2n^4$

18. a. $[5 + (3x + 2)]$
    b. $[2a + (3b - a) - b]$   $2a + 3b - a - b$ $= a + 2b$
    c. $(6x^2 - [2x^2 - y^2] - 2y^2)$
    d. $(m^2 + 1 - [2m + m^2] + 3$   $3x + 7$
    e. $[2a - 5c + (3b + 6c) - b]$

19. a. $-[2s + (3s - 5)]$   $-2s - 3s + 5$
    b. $-[-2b - (3b + 8c)]$   $= 5 + 5$
    c. $-(18 - [3x - 4x^2])$   $5s$
    d. $-(bx + [2bx - 9cx + 3] - 4)$
    e. $-[2x^3 - x^2 - (5x^3 - 3x) - 4x^3 + 2x]$

20. a. $8x + [2 + (4x + 3)]$   $8x + 2 + 4x + 3$
    b. $5d - [3m - (2d - m)]$   $12x + 5$
    c. $x^2 + (2x^2 - [x^2 + 2x - 3])$
    d. $a^2b - 5ab^2 + [2a^3 - (-3a^2b - ab^2)]$
    e. $m^3 - 2m^2 - [m^2 + (m^3 - 3m^2 - 4m)]$

21. a. $3t + [4t + (2t - 5) + 7]$   $3t + 4t + 2t - 5 + 7$
    b. $4r^2 - [2r^2 - r + (3r + 6) - 2]$   $= 9t + 6$
    c. $m^3 - [4m^2 - (2m^3 - m^2) + m^2]$   $+ 2$

**d.** $2ab + 5 + (3ab - 4 - [6 - 4ab] - 9)$

**e.** $8c^2d - 2cd^2 + [2c^2d - (cd^2 - 4c^2d) - 5cd^2]$

**22.** **a.** $3x + [6 + (4x - 1)] - 7$

**b.** $y^4 + [y^2 - (3y^4 - 2y^2)] - 5y^2$

**c.** $c^3 - d^3 - [2c^3 + (d^3 - 4c^3)] + 7d^3$

**d.** $5b^4 - 2b^2c^2 - (b^4 - [c^4 - 2b^2c^2] - c^4) + 2c^4$

**e.** $2x^2 - xy + [2y^2 + (3xy + x^2) - 4xy] - y^2$

**23.** **a.** $5x - 6 + (2x - 3) - [4x + (7 - 2x)]$

**b.** $3a - 9b - (4b - 3a) - [2a - (3a - 5b)]$

**c.** $(5c - 4d) - (4c - 5d) + [3d + (c - d)]$

**d.** $[2x - (4x - y)] - [-5x + (2x - 7y)]$

**e.** $x^2 - [x + (2x^2 - 1)] + 3x + [2x - (x^2 - 1)] - 2$

Simplify:

**24.** **a.** $2(a + 3) + 5$

**b.** $3(5b + 7) + 4b$

**c.** $9(x - 2) - 3x + 10$

**d.** $4(2a - 3y) - 7y - 3a$

**e.** $6(-5m + n) + 9m + 4n$

**25.** **a.** $-3(c + 6) + 10$

**b.** $-6(8 - 4c) - 5c$

**c.** $-5(-2b^2 - b) + 15b^2 - 3b$

**d.** $-7(4a - 5x) - 4x + 30a$

**e.** $-2(9m^2 - 3m + 2) + m^2 - 5m$

**26.** **a.** $x(x + 4) - 2x$

**b.** $b^2(b - 3) - 5b^2$

**c.** $-d(3a + 4c) + 7ad - 2cd$

**d.** $-a^3(2a^2 - a + 6) + 3a^4 - 4a^3$

**e.** $-ax(-6a - 5x) + 7ax^2 - 4a^2x$

**27.** **a.** $3 + 4(y + 3)$

**b.** $7 + 5(2x - 4)$

**c.** $9 + 3(-m + 6)$

**d.** $b + 7(-7a - 2b)$

**e.** $5c + 2(3x - 9c)$

**28.** **a.** $8 - 2(x + 6)$

**b.** $6 - 4(3x - 1)$

**c.** $5 - 5(-b^2 - 2)$

**d.** $7a - 2(3a - b)$

**e.** $4x - 6(2x - 3y)$

**29.** **a.** $6b + b(1 + b)$

**b.** $5x - x(3x - 5)$

**c.** $2y^2 - y(x - y)$

**d.** $6x^3 + x(4x - 5)$

**e.** $9b^3 - b^2(a - 2b)$

**30.** **a.** $ax + 2x(a + x)$

**b.** $4by - 3b(y - 2b)$

**c.** $3mn - 2m(-3m - n)$

**d.** $9x^2y + 5x(2xy - y)$

**e.** $4a^2b^2 - 4ab(2a^2b + 3ab^2)$

**31.** **a.** $2c + c(5c + 3) + 4c^2$

**b.** $9bx - x(2b - 5x) - 2x^2$

**c.** $6x + 3(7 - 4x) + 28$

**d.** $4m^2 - 2m(m - 3n) - 5mn$

**e.** $2c^2y - 5c(4cy - y) + 9cy$

**32.** **a.** $x^2 + 4x - 2(x + 3) + x(2x - 4)$

**b.** $2b^2 - b(b - 2) + 3b(2b - 5)$

**c.** $a^2 - 2a - (2a^2 - 5a) + 3a(a - 4)$

**d.** $3x^2 + (x^2 - 3x + 1) + 2x - x(x + 3) - 8$

**e.** $4a^2b - a(2ab + b^2) - 2ab^2 - 3b(4ab - 3a^2) - 2a^2b$

**33.** a. $4a+2[4a+2(3a-5)]$

$4a+2[12a-2a+6a-10]$

b. $x^2-5x-3[2x+x(x-3)]$

c. $3a^2b-ab^2-b[3a+b(a+1)]$

$4a+24a^2-4a^2+12a^2-20$

d. $5m^3-2m^2[5m-2(m-3)]+3m^2$

e. $4x^2y+3x[xy-3(y^2-2xy)]-4xy^2$

$4a+32a^2-20$

**34.** a. $(x+4)(x+2)+(x+2)(x+2)$

$x^2+2x+4x+8+x^2+2x+2x+4$

b. $(a-3)(a+1)+(a+6)(a+3)$

$=2x^2+10x+12$

c. $(3y-5)(4y-2)+(y+4)(y-3)$

d. $(x+y)(x+y)+(x-y)(x+y)$

e. $(c-3d)(c+3d)+(5c-4d)(2c+5d)$

**35.** a. $(c-2)(c+7)-(c+1)(c+1)$

$c^2+7c-2c-14+c^2+1c+1c+1c$

b. $(a-b)(a-b)-(a+b)(a-b)$

$2c^2 \quad 10c-c^2-14$

c. $(2x-3)(3x-2)-(4x-1)(3x+5)$

$3c-15$

d. $(3c-5d)(3c+2d)-(2c-7d)(5c-3d)$

e. $(x^2-y^2)(x^2+y^2)-(x^2-y^2)(x^2-y^2)$

**36.** a. $2a(a+2)+(a-3)(a-5)$

$2a^2+4a+(a^2-5a-3a+15)$

b. $(x-5)(x+5)-3x(x-3)+6x$

$2a^2+4a+a^2-8a+15$

c. $8y-(2y-3)(2y+5)-3y(y+2)$

$3a^2-4a+15$

d. $-5b(b+c)+3bc-(b-c)(b-2c)$

e. $(a-b)(a+b)-b(a-b)+a(a+b)-b^2$

## CUMULATIVE REVIEW

1. Add as indicated: $(6-2x+4x^2)+(8x-7x^2+3)$
2. Find the sum of: $3m^2-4m+5$, $6m-m^2$, and $10-2m+m^2$
3. Combine: $7y^3-y^2+4y-10-3y+2y^2-8y^3+9+y^3-2$
4. From $4n^2-11x^2$ subtract $5nx+12n^2$.
5. Subtract $6c^2-5cd+7d^2$ from $3c^2-cd-7d^2$.
6. Subtract as indicated: $(5x^2-xy)-(-4xy+8y^2)$

Multiply as indicated:

7. a. $(-2x^3)(-4x^2)(-3x)$  b. $-2y^3(-4y^2-3y)$  c. $5m^4n(2m^8-6mn+3n^3)$
8. a. $(4n-5)(7n-2)$    b. $(3c^2-6x)(5c^2+2x)$  c. $(2x^2-3x+7)(6x-1)$

Divide as indicated:

9. a. $(-80a^6x)\div(-16)$    b. $(-42x^{10}y^8)\div(6x^5y^8)$
10. a. $(81b^9-63b^4)\div(-9b^4)$    b. $(72a^2-66ax+5x^2)\div(6a-5x)$

Remove parentheses and, if possible, add like terms:

11. $4x+(5x^2-8x+2)$    12. $8a^2-6a-(4a^2-11a+3)$
13. $20b^2-(16b^2-5b)-[6b^2-(-b^2+8b)-9b]+11b$
14. $18n-12n(2n-3)+24n^2$

# Inclosing Terms within Parentheses

**I. Aim**    To inclose terms within a set of parentheses.

**II. Procedure**

1. If the set of parentheses is to be prefixed with a plus (+) sign, write within parentheses the terms which are to be inclosed without changing their signs.

2. If the set of parentheses is to be prefixed with a minus (−) sign, write within parentheses the terms which are to be inclosed but with their signs changed.

**III. Sample Solutions**

Rewrite, but inclose the last two terms within a set of parentheses prefixed with a plus (+) sign:

1. $4a - 3b + 6c - 2d$    *Answer:* $4a - 3b + (6c - 2d)$
2. $5b^3 - 2b^2 - 3b + 5$    *Answer:* $5b^3 - 2b^2 + (-3b + 5)$

Rewrite, but inclose the last two terms within a set of parentheses prefixed with a minus (−) sign:

3. $2b - 5c - 3x + 7y$    *Answer:* $2b - 5c - (3x - 7y)$
4. $3x^3 - 4x^2 + x - 5$    *Answer:* $3x^3 - 4x^2 - (-x + 5)$

## DIAGNOSTIC TEST

Rewrite each of the following expressions;
Inclose the last two terms of each expression within a set of parentheses and precede it by a plus (+) sign:

1. $2a + 3b + 4c + 5d$
2. $4x^4 - 5x^3 + 6x^2 - 3x$
3. $4b^3 + 5b^2 - 5b - 8$
4. $a^4 - a^3b + a^2b^2 - ab^3 + b^4$

Inclose the last three terms within a set of parentheses and precede it by a plus (+) sign:

5. $c^5 - 5c^4 - 10c^3 - 10c^2 + 5c - 1$

Inclose the last two terms of each expression within a set of parentheses and precede it by a minus (−) sign:

6. $2ab + 5ac - 3bc - 4cd$
7. $9a^2 + 4b^2 - 3c^2 + 5d^2$
8. $2y^3 - 3y^2 + 6y + 8$
9. $m^4 - 8m^3 + 12m^2 + 16m - 18$

10. Inclose the last three terms within a set of parentheses and precede it by a minus (−) sign: $4x^4 - 2x^3y + 3x^2y^2 - 2xy^3 + y^4$

## *RELATED PRACTICE EXAMPLES*

Rewrite each of the following expressions, but inclose the last two terms of each expression within a set of parentheses and precede it by a plus (+) sign:

1.  **a.** $a + x + b + y$
    **b.** $3c - 2d + 4m + 5x$
    **c.** $x^3 + 3x^2 + 2x + 7$
    **d.** $-b^4 - b^3 + b^2 + 3b$
    **e.** $d^3y + d^2y^2 + dy^3 + y^4$

2.  **a.** $c + d + m - n$
    **b.** $x^3 - 2x^2 + 4x - 8$
    **c.** $b^4 + 3b^3 + 5b^2 - 7b$
    **d.** $a^3b - a^2b^2 + ab^3 - b^4$
    **e.** $-m^4 - 2m^3 + 4m^2 - 3m$

3.  **a.** $b + d - x - y$
    **b.** $4ab - 3ac - 2ad - 5xy$
    **c.** $y^3 + 4y^2 - 3y - 9$
    **d.** $x^4 - x^3y + x^2y^2 - xy^3 - y^4$
    **e.** $-a^4 - 2a^3 - 5a^2 - 7a - 2$

4.  **a.** $m + r - s + x$
    **b.** $3a - 4b - 5c + 8d$
    **c.** $x^3 - 5x^2 - 8x + 2$
    **d.** $c^4 - 2c^3d + 2c^2d^2 - 5cd^3 + d^4$
    **e.** $-s^4 + 2s^3 - 4s^2 - 3s + 7$

Rewrite each of the following expressions, but inclose the last three terms of each expression within a set of parentheses and precede it by a plus (+) sign:

5.  **a.** $3a + 7b + 5c + 9d + 4e$
    **c.** $a^4 - 2a^3 - 5a^2 - 4a - 9$
    **e.** $x^4 - 2x^3y - 7x^2y^2 - 3xy^3 + 5y^4$

    **b.** $4x^4 - 3x^3 + 2x^2 - 3x + 4$
    **d.** $c^4 + 5c^3 - 6c^2 + 7c - 3$
    **f.** $8b^5 - 6b^4 + 2b^3 - b^2 + 9b - 10$

Rewrite each of the following expressions, but inclose the last two terms of each expression within a set of parentheses and precede it by a minus (−) sign:

6.  **a.** $c + d - m - x$
    **b.** $5a - 3b - 4c - 6d$
    **c.** $x^3 + x^2 - 2x - 5$
    **d.** $c^4 + c^3d + c^2d^2 - cd^3 - d^4$
    **e.** $b^4 - 2b^3 + 3b^2 - 5b - 7$

7.  **a.** $b + c - x + y$
    **b.** $4m - 3n - 2s + 3t$
    **c.** $c^3 - 5c^2 - 3c + 4$
    **d.** $-t^4 + 3t^3 - 2t^2 - 8t + 3$
    **e.** $a^4 - 5a^3b + 7a^2b^2 - 4ab^3 + b^4$

8.  **a.** $a + c + m + n$
    **b.** $3b - 5c + 9x + 3y$
    **c.** $d^3 + 2d^2 + 3d + 4$
    **d.** $-a^5 - a^4 + a^3 + a^2$
    **e.** $m^4 + 2m^3n - 8m^2n^2 + 3mn^3 + n^4$

9.  **a.** $r + s + t - x$
    **b.** $s^3 - 7s^2 + 3s - 6$
    **c.** $2a^3 + a^2 + 5a - 1$
    **d.** $x^4 - 4x^3y + 8x^2y^2 + 3xy^3 - 5y^4$
    **e.** $-x^4 + 5x^3 - 2x^2 + x - 2$

Rewrite each of the following expressions, but inclose the last three terms of each expression within a set of parentheses and precede it by a minus (−) sign:

10. **a.** $5b + 2d + 9m + 3x + y$
    **c.** $3c^4 - 2c^3 + 5c^2 - 8c + 2$

    **b.** $n^4 - 5n^3 - n^2 - 6n - 5$
    **d.** $x^4 + 9x^3 - 7x^2 + 3x - 8$

# REVIEW

*1—19 Parts A, c and e*

**1. a.** Add:
$$\begin{array}{r} -\ 9bx^3 \\ +\ 5bx^3 \\ \underline{-12bx^3} \end{array}$$

**b.** Add as indicated:
$$(-3a^2)+(-a^2)+(-9a^2)$$

**c.** Combine $p+.14p$

**2. a.** Add:
$$\begin{array}{r} 8x^2-5xy-\ y^2 \\ \underline{-3x^2-7xy+4y^2} \end{array}$$

**b.** Add as indicated:
$$(9b^2-4bc)+(7bc-4c^2)$$
$$+(8c^2-b^2)$$

**c.** Simplify: $4x^2-3x+7-11-3x-x^2+x-5+4x-2x^2$

**3. a.** Subtract the lower term from the upper term:
$$\begin{array}{r} -12a^4b^3 \\ \underline{-15a^4b^3} \end{array}$$

**b.** Subtract as indicated:
$$(-8x)-(-5y)$$

**c.** From $24c^2$ subtract $52c^2$

**4. a.** Subtract the lower expression from the upper expression:
$$\begin{array}{r} -6x^2-\ 9x+5 \\ \underline{x^2-11x-1} \end{array}$$

**b.** Subtract as indicated: $(2m^2-5mn-n^2)-(4m^2-7mn+8n^2)$

**c.** Subtract $2ab-5b^2$ from $6a^2-7ab$

Multiply:

**5.** $a^2b^7$
$a^9b$

**6. a.** $\begin{array}{r} -8x^2y^5z \\ -3x^7y^2z^6 \end{array}$

**b.** $(-12c)(9x)$

**c.** $(-2b^2)(-2b^2)(-2b^2)$

**7. a.** $-3b(-8a+2c)$

**b.** $7x^2y^5(6x^7-5x^4y^3+2y^{10})$

**c.** $\begin{array}{r} 10a^8n^7-9a^6n^5+a^4n^9 \\ \underline{-6a^4n^6} \end{array}$

**8. a.** $\begin{array}{r} 10x-11 \\ \underline{9x-12} \end{array}$

**b.** $\begin{array}{r} 4c^2-8cd+3d^2 \\ \underline{5c^2+2cd-\ d^2} \end{array}$

**c.** $(7a-5c)(8a+4b)$

Divide:

**9.** $\dfrac{a^{12}b^5}{a^9b^5}$

**10. a.** $\dfrac{-32x^{15}y^{10}}{-4x^8y^3}$

**b.** $\dfrac{63a^4b^{11}n^3}{7a^4b^5}$

**c.** $(-54c^3d^2)\div(9c^3d^2)$

**11. a.** $\dfrac{16b-8b^2}{-8b}$

**b.** $\dfrac{15x^8y^6-45x^6y^7+30x^4y^8}{5x^4y^6}$

**c.** $(12x^{10}-8x^9+14x^7-10x^6-2x^4)\div(-2x^4)$

*$a^2b^2 \quad a^9b$* **197**

*$7a$*

**12. a.** Divide $c^2 - 11c + 24$ by $c - 3$.
  **b.** Divide $14x^2 + 53xy - 45y^2$ by $2x + 9y$.
  **c.** Divide $10a^3 - 49a^2 + 48a - 12$ by $5a - 2$.

**13.** Remove parentheses and, if possible, add like terms:
  **a.** $12m + (-4m + n) - 5n$
  **b.** $15 - 6(7x - 9)$
  **c.** $4c^2 + [2c^2 + 3c - (8c^2 - 3) - 7] - 9c$
  **d.** $3a(a + 2) - 4a(a - 5) - (a - 3)(a + 7)$

**14.** Rewrite the following expression, but inclose the last two terms within a set of parentheses and precede it by:
  **a.** A plus sign: $6b^3 + 4b^2 - 7b + 8$
  **b.** A minus sign: $12x^4 - 5x^3y + 7x^2y^2 - xy^3 + 2y^4$

## COMPETENCY CHECK TEST

In each of the following select the letter corresponding to your answer:

**1.** Add: $(2x^2 - xy - 8y^2) + (-4x^2 + 3xy - y^2)$.

  The sum is:
  **a.** $6x^2 - 4xy - 9y^2$  **b.** $-2x^2 + 2xy - 9y^2$
  **c.** $-2x^2 + 2xy + 8y^2$  **d.** Answer not given

**2.** Subtract: $(-5b^2 + 9) - (3b^2 - 7b - 9)$.

  The difference is:
  **a.** $-2b^2 - 7b$  **b.** $8b^2 - 7b$
  **c.** $-8b^2 - 7b - 18$  **d.** $-8b^2 + 7b + 18$

**3.** Multiply: $(3x + 2y)(5x - 3y)$.

  The product is:
  **a.** $8y - y$  **b.** $15x^2 - 6y^2$
  **c.** $15x^2 - 19xy - 6y^2$  **d.** $15x^2 + xy - 6y^2$

**4.** Divide: $\dfrac{20m^6x^2 - 4m^3x^6}{-4m^3x^2}$.

  The quotient is:
  **a.** $-5m^3 + x^4$  **b.** $-5m^3 + x^3$
  **c.** $-5m^2 - x^3$  **d.** Answer not given

## ● 🔑    **Keyed Achievement Test**

The colored numeral indicates the Exercise where help may be found.

1. Which numerals name integers? Rational numbers? $\boxed{2\text{-}5}$

   $-3.7$      $50\%$      $-108$      $\frac{56}{8}$      $-4\frac{7}{12}$      $\sqrt{92}$      $-\frac{13}{25}$

2. Which of the following sentences are true? $\boxed{2\text{-}8}$
   - **(a)** $-12 > -7$       **(b)** $18 \div 2 \neq 2 \div 18$       **(c)** $.04 < .4$
   - **(d)** $.9 + 3.4 > 5 - .5$       **(e)** $9 \times 31 = 300 - 21$       **(f)** $\frac{11}{16} \not< \frac{12}{17}$

3. On a number line, draw the graph of: $-9, -4, -1, 0, 3, 5, 7$. $\boxed{2\text{-}7}$

4. Find the corresponding list of values of $x^2 - 5x + 2$ when the replacements for the variable are: $-3, -2, -1, \ldots, 3$. $\boxed{3\text{-}6}$

5. Find the value of $S$ when $n = 12$, $a = 6$, and $l = 39$, using the formula $S = \dfrac{n}{2}(a + l)$. $\boxed{3\text{-}7}$

6. Determine the missing numbers. Then write the word statement and the formula that expresses the relationship between the following rows of numbers.

   | $r$ | 1 | 2 | 3 | 4 | 9 |
   |-----|---|---|----|---|---|
   | $s$ | 3 | 7 | 11 |   |   |

7. Find the sum of $3x^2 - 7x - 5$, $8 - 9x$, and $4 - 2x - 6x^2$. $\boxed{5\text{-}3}$

8. Subtract $9c^2 - 8cd + d^2$ from $5c^2 - 8cd - 2d^2$. $\boxed{5\text{-}5}$

9. Multiply $4n^2 - 5n + 3$ by $6n - 9$. $\boxed{5\text{-}9}$

10. Divide $28a^2 + 43ab - 45b^2$ by $4a + 9b$. $\boxed{5\text{-}13}$

11. Remove parentheses and, if possible, add like terms: $\boxed{5\text{-}14}$
    $$8m - (4m - 7) + (-6m + 9) - 5$$

## *MAINTENANCE PRACTICE IN ARITHMETIC*

1. Add:
   479,686
   975,649
   889,578

2. Subtract:
   1,049,235
     289,196

3. Multiply:
   $\cdot$ 89,376
     8,694

4. Divide:
   $5,082\overline{)15,733,872}$

5. Add:
   $9\frac{5}{6} + 13\frac{4}{5}$

6. Subtract:
   $1\frac{1}{4} - \frac{29}{32}$

7. Multiply:
   $9\frac{3}{8} \times 3\frac{1}{5}$

8. Divide:
   $8\frac{2}{3} \div 6$

9. Add:
   $\$.29 + \$3.75 + \$.68$

10. Subtract:
    $.935 - .895$

11. Multiply:
    $.075 \times 940$

12. Divide: $\$.05\overline{)\$10}$

13. Find $6\frac{1}{4}\%$ of $\$7,600$.

14. What per cent of 125 is 250?

15. 120 is 4% of what number?

# INTRODUCTION

Students are already acquainted with statements like:

$$3 + 4 = 7 \qquad \square - 2 = 9 \qquad 5n = 40 \qquad \frac{x}{8} = 2$$

$$A = lw \qquad p = 2l + 2w \qquad i = prt$$

These are all *equations*—statements that two expressions are equal.

In this chapter we shall deal with equations that contain a variable that represents a number. We shall study how to solve the equation to determine this number.

An equation may be compared to balanced scales. To keep an equation in balance, any change on one side of the equality sign must be balanced by an equal change on the other side of the equality sign.

*To solve an equation,* we transform the given equation to the simplest equivalent equation which has only the variable itself as one member and a constant naming a number as the other member. To do this we use the axioms with inverse operations or with the additive or multiplicative inverse.

## The Four Basic Types of Equations:

| Exercise 6-4 | Exercise 6-5 | Exercise 6-6 | Exercise 6-7 |
|---|---|---|---|
| $n + 3 = 15$ | $n - 3 = 15$ | $3n = 15$ | $\frac{n}{3} = 15$ |

are solved both by inverse operations and by the additive or multiplicative inverse. When the inverse operation is used, the operation indicated in each case by the variable and its connected term is undone by performing the inverse operation on both sides of the equation.

The equation is considered one of the most powerful and practical tools of mathematics. It enables us to solve problems and to use formulas from science, mathematics, and business. This is illustrated in Exercise 6-10 as in Chapter Seven.

# Equations
# in One Variable

# Equations and Solutions

**I. Aim** To develop an understanding of the vocabulary used with equations and their solutions.

## II. Procedure

Study each of the following definitions and examine its related sample solution.

1. *To solve an equation in one variable* means to find the number represented by the variable which, when substituted for the variable, will make the sentence true.

2. Any number that makes the sentence true is said to *satisfy* the equation or to be the *root* or the *solution* of the equation.

3. *Checking an equation* is testing whether some number is a solution to this equation by substituting the number for the variable in the equation. If the resulting sentence is true, the number is a solution; if not, the number is not a solution.

4. The expressions at the left and at the right of the equality sign in an equation are called *members* or *sides* of the equation.

5. Equations which have exactly the same solutions are called *equivalent equations.*

6. An equation is considered to be in its *simplest form* when one member consists of only the variable itself and the other member is a constant term.

7. An equation in which the variable is raised to the first power only or, if it contains fractions in which the variable does not appear in the denominator, is called an *equation of the first degree* or a *simple equation.*
*Note:* In this chapter only equations of the first degree in one variable will be studied.

8. **a.** An equation that is true for some and not all replacements of the variable is called a *conditional equation.*

   **b.** An equation that is true for all replacements of the variable is called an *identical equation* or *identity.*

   **c.** An equation that is satisfied by no number is said to have no solution.

## III. Sample Solutions

**1.** Is the sentence $x + 2 = 10$ true: **a.** When $x = 5$? **b.** When $x = 8$?

| | |
|---|---|
| $x + 2 = 10$ | $x + 2 = 10$ |
| $5 + 2 \neq 10$ | $8 + 2 = 10$ |
| $7 \neq 10$ | $10 = 10$ |
| *Answer:* No | *Answer:* Yes |

    **c.** Which number is the root or solution of $x + 2 = 10$?

*Answer:* 8 is the root or solution.

**2.** Write the left member of the equation $n + 5 = 20$.   *Answer:* $n + 5$

**3. a.** Are the equations $y + 6 = 18$ and $y = 12$ equivalent equations.

*Answer:* Yes

    **b.** Is the equation $y = 12$ the simplest form of the equation $y + 6 = 18$?

*Answer:* Yes

**4.** Is $2n^2 + 3 = 5$   or   $3x + 5 = 11$ an equation of the first degree?

*Answer:* $3x + 5 = 11$

**5.** Which of the following equations are conditional equations? Which equations are satisfied by all numbers replacing the variable (identity)? Which equations have no solution?

$$3y - 11 = 4 \qquad 6 + x = x \qquad 8n + 4 = 4 + 8n \qquad 8 - x = x$$

*Answers:* Conditional equations: $3y - 11 = 4$ and $8 - x = x$
                Identity: $8n + 4 = 4 + 8n$
                No solution: $6 + x = x$

## DIAGNOSTIC TEST

**1.** Which of the following numbers will make $y - 5 = 10$ a true statement?

    **a.** $y = -2$     **b.** $y = 5$     **c.** $y = 15$     **d.** $y = 2$

**2.** Which of the following numbers is the root of the equation $\dfrac{r}{6} = 18$?

    **a.** $r = 3$     **b.** $r = \frac{1}{3}$     **c.** $r = 12$     **d.** $r = 108$

**3.** Which of the following is the solution of the equation $-3x = 12$?

    **a.** $-9$     **b.** $4$     **c.** $-4$     **d.** $-15$

**4.** Write the left member of the equation $4n - n = 2n + 17$.

5. Which of the following equations have $-2$ as the solution?
   **a.** $x + 7 = -5$   **b.** $x - 1 = -3$   **c.** $-6 = 12x$   **d.** $x = -2$
   Which of the above equations are equivalent equations? Which equation of the equivalent equations is the simplest form?
6. Which of the following equations are equations of the first degree?
   **a.** $6y - y^3 = 10$        **b.** $4x - 1 = 8$        **c.** $3n - 5 = 2n^2$
7. Which of the following equations are conditional equations? Which are satisfied by all numbers replacing the variable? Which have no solution?
   **a.** $2x + 5 = 5 + 2x$ **b.** $n + 7 = n$   **c.** $8t = t = 0$ **d.** $11 - x = x$

## *RELATED PRACTICE EXAMPLES*

1. Which of the following numbers will make the sentence:
   **a.** $n + 4 = 9$ true?
   | | | | |
   |---|---|---|---|
   | (1) $n = 13$ | (2) $n = 5$ | (3) $n = 36$ | (4) $n = -5$ |

   **b.** $x - 6 = 7$ true?
   | | | | |
   |---|---|---|---|
   | (1) $x = 1$ | (2) $x = -1$ | (3) $x = 13$ | (4) $x = 0$ |

   **c.** $12y = 84$ true?
   | | | | |
   |---|---|---|---|
   | (1) $y = 21$ | (2) $y = 96$ | (3) $y = 72$ | (4) $y = 7$ |

   **d.** $\dfrac{b}{3} = 6$ true?
   | | | | |
   |---|---|---|---|
   | (1) $b = 2$ | (2) $b = 3$ | (3) $b = 18$ | (4) $b = \frac{1}{2}$ |

   **e.** $5a + 8 = 53$ true?
   | | | | |
   |---|---|---|---|
   | (1) $a = -9$ | (2) $a = 7$ | (3) $a = 9$ | (4) $a = 10$ |

2. Which of the following numbers is the root of the equation:
   **a.** $x + 19 = 31$?
   | | | | |
   |---|---|---|---|
   | (1) $x = 50$ | (2) $x = -10$ | (3) $x = 22$ | (4) $x = 12$ |

   **b.** $y - 23 = 42$
   | | | | |
   |---|---|---|---|
   | (1) $y = 19$ | (2) $y = -8$ | (3) $y = 65$ | (4) $y = 75$ |

   **c.** $15m = 10$
   | | | | |
   |---|---|---|---|
   | (1) $m = 0$ | (2) $m = \frac{2}{3}$ | (3) $m = 1\frac{1}{2}$ | (4) $m = 150$ |

   **d.** $\dfrac{a}{9} = 8$
   | | | | |
   |---|---|---|---|
   | (1) $a = 45$ | (2) $a = -54$ | (3) $a = 0$ | (4) $a = 72$ |

   **e.** $6n + 2n = 56$
   | | | | |
   |---|---|---|---|
   | (1) $n = 9$ | (2) $n = -7$ | (3) $n = 7$ | (4) $n = -8$ |

3. Which of the following is the solution of the equation:
    **a.** $y + 24 = 96$?  (1) 120  (2) $-82$  (3) 72  (4) 4
    **b.** $n - 16 = 0$?  (1) $-16$  (2) 0  (3) 20  (4) 16
    **c.** $7x = -42$?  (1) 6  (2) $-6$  (3) $-35$  (4) 35
    **d.** $\dfrac{r}{20} = 5$?  (1) 4  (2) 60  (3) 100  (4) $\frac{1}{4}$
    **e.** $2d - 11 = 19$?  (1) 30  (2) 4  (3) $-4$  (4) 15

4. **a.** Write the left member of $x - 20 = 75$.
    **b.** Write the right member of $\dfrac{n}{17} = -4$.
    **c.** Write the left member of $8y + 26 = 53$.
    **d.** Write the right member of $7m - 14 = 5m - 47$.
    **e.** Write the left member of $6x - x + 10 = 26 - 2x$.

5. **a.** Which of the following equations have 5 as the solution?
    (1) $x + 1 = 6$  (2) $x = 5$  (3) $7x = 42$  (4) $x - 8 = 13$
    Which of the above equations are equivalent equations?
    Which equation of the equivalent equations is the simplest form?

    **b.** Which of the following equations have 3 as the solution?
    (1) $n - 2 = 1$  (2) $8n = 24$  (3) $n = 3$  (4) $4n + 5 = 18$
    Which of the above equations are equivalent equations?
    Which equation of the equivalent equations is the simplest form?

    **c.** Which of the following equations have $-4$ as the solution?
    (1) $y + 3 = 1$  (2) $2y - 7 = -15$  (3) $-y = 4$  (4) $y = -4$
    Which of the above equations are equivalent equations?
    Which equation of the equivalent equations is the simplest form?

    **d.** Which of the following equations have 0 as the solution?
    (1) $a - 5 = 5$  (2) $8a = 8$  (3) $a = 0$  (4) $3a + 17 = 17$
    Which of the above equations are equivalent equations?
    Which equation of the equivalent equations is the simplest form?

6. Which of the following equations are equations of the first degree?
    **a.** $x^2 + 1 = 7$          **b.** $2x - 3 = 9$
    **c.** $9y + 4 = 5y$          **d.** $3n^2 + 5n = 10$
    **e.** $8x^3 = 14$          **f.** $6(a + 9) = -10$
    **g.** $2c^2 - 4c + 7 = 0$          **h.** $5a + 7a - 3 = 0$

7. Which of the following equations are conditional equations? Which equations are satisfied by all numbers replacing the variable? Which equations have no solution?
    **a.** $x = x - 3$          **b.** $7n = -28$          **c.** $11y - 13 = 0$
    **d.** $5 - m = m$          **e.** $6x + 7 = 7 + 6x$          **f.** $-4 - 4y = y$
    **g.** $5(y - 3) = 5y - 15$          **h.** $2r + 19 = 2r$          **i.** $9b - 25 = 8(b + 6)$

# Inverse Operations

**I. Aim**　　To review inverse operations. They undo each other.

**II. Procedure**

1. Addition and subtraction are inverse operations.
   **a.** Use subtraction to undo addition.
   **b.** Use addition to undo subtraction.
2. Multiplication and division are inverse operations.
   **a.** Use division to undo multiplication.
   **b.** Use multiplication to undo division.
3. Observe that:
   **a.** When two numbers have zero (0) as their sum, each number is said to be the additive inverse of each other. See Exercise 2-4.
   **b.** When two numbers have one (1) as their product, each number is said to be the multiplicative inverse of the other. See Exercise 2-13. Also see Exercise 1-3.

**III. Sample Solutions**

Find the missing number and operation in each of the following:

1. $(n + 2) ? \square = n$　　$(n + 2) - 2 = n$　*Answer:* Subtract 2
2. $(x - 4) ? \square = x$　　$(x - 4) + 4 = x$　*Answer:* Add 4
3. $(5 \times n) ? \square = n$　　$(5 \times n) \div 5 = n$　*Answer:* Divide by 5
4. $(y \div 8) ? \square = y$　　$(y \div 8) \times 8 = y$　*Answer:* Multiply by 8
5. $\dfrac{m}{3} ? \square = m$　　$\dfrac{m}{3} \times 3 = m$　*Answer:* Multiply by 3

## DIAGNOSTIC TEST

Find the missing number or variable or operation:

1. **a.** $(12 + 2) ? \square = 12$　　　　**b.** $(x + 9) - \square = x$
2. **a.** $(39 - 19) + 19 = ?$　　　　**b.** $(n - 16) ? \square = n$
3. **a.** $(y \times 7) ? \square = y$　　　　**b.** $2d ? \square = d$
4. **a.** $(56 \div 8) ? \square = 56$　　　　**b.** $(b \div 4) \times \square = b$

**5. a.** $\frac{5}{6}?\square=5$      **b.** $\frac{r}{12}?\square=r$

**6.** Which operation is the inverse of multiplication? Which operation is the inverse of addition?

**7.** What number must be added to $-9.7$ to get 0 as the sum?

**8.** What is the additive inverse of: **a.** $-81$? **b.** $\frac{11}{16}$? **c.** $12.3$?

**9.** $-\frac{3}{8}$ multiplied by what number will give 1 as the product?

**10.** What is the multiplicative inverse of: **a.** $\frac{1}{24}$? **b.** $-13$? **c.** $\frac{22}{7}$?

## RELATED PRACTICE EXAMPLES

Find the missing number or variable or operation:

**1. a.** $(8+9)-\square=8$    **b.** $(12+3)-3=?$    **c.** $(10+15)?15=10$
   **d.** $(23+6)?\square=23$    **e.** $(x+12)-12=?$    **f.** $(n+7)?\square=n$

**2. a.** $(11-2)+\square=11$    **b.** $(22-8)?8=22$    **c.** $(30-16)+16=?$
   **d.** $(15-9)?\square=15$    **e.** $(y-2)+2=?$    **f.** $(x-5)?\square=x$

**3. a.** $(9\times7)\div\square=9$    **b.** $(25\times47)?\square=25$    **c.** $(a\times6)\div6=?$
   **d.** $(x\times10)?\square=x$    **e.** $(c\times29)?29=c$    **f.** $(k\times42)?\square=k$

**4. a.** $(27\div3)\times\square=27$    **b.** $(16\div2)?2=16$    **c.** $(42\div6)\times6=?$
   **d.** $(75\div15)?\square=75$    **e.** $(m\div4)\times4=?$    **f.** $(y\div9)?\square=y$

**5. a.** $\frac{3}{10}\times\square=3$    **b.** $\frac{22}{7}?7=22$    **c.** $\frac{5}{8}\times8=?$

   **d.** $\frac{2}{5}?\square=2$    **e.** $\frac{x}{4}\times4=?$    **f.** $\frac{n}{6}?\square=n$

**6. a.** Is the operation of division the inverse of subtraction?
   **b.** Is the operation of addition the inverse of multiplication?
   **c.** Is the operation of division the inverse of addition?
   **d.** Which operation is the inverse of subtraction?
   **e.** Which operation is the inverse of division?

**7.** To get 0 as the sum, what number must be added to:
   **a.** $+7$?    **b.** $+.8$?    **c.** $+\frac{3}{8}$?    **d.** $-11$?    **e.** $-\frac{3}{4}$?    **f.** $-1.7$?    **g.** $-25$?

**8.** What is the additive inverse of each of the following?
   **a.** $+18$    **b.** $\frac{1}{6}$    **c.** $-5$    **d.** $-\frac{7}{8}$    **e.** $+6.5$    **f.** $-.3$    **g.** $-1\frac{9}{16}$

**9.** To get 1 as the product, by what number must you multiply:
   **a.** $\frac{1}{3}$?    **b.** $\frac{5}{8}$?    **c.** $6$?    **d.** $-\frac{1}{4}$?    **e.** $-\frac{7}{12}$?    **f.** $-9$?    **g.** $\frac{5}{3}$?

**10.** What is the multiplicative inverse of each of the following?
   **a.** $12$    **b.** $\frac{1}{5}$    **c.** $-2$    **d.** $-\frac{2}{3}$    **e.** $\frac{9}{5}$    **f.** $-35$    **g.** $-3\frac{1}{2}$

# Properties of Equality—Axioms

**I. Aim**   To study the various properties of equality.

**II. Procedure**

1. The properties of equality which state that the results are equal when equals are increased or decreased or multiplied or divided by equals, with division by zero excluded, are called *axioms*.
   Use these axioms to obtain an equation in simplest form by
   a. Subtracting the same number from both sides of the given equation.
   b. Adding the same number to both sides of the given equation.
   c. Dividing both sides of the given equation by the same non-zero number.
   d. Multiplying both sides of the given equation by the same non-zero number.

2. Equality has a reflexive property, a symmetric property, and a transitive property.
   a. Reflexive property: Any number is equal to itself. In general, when $a$ is a real number, $a = a$.
   b. Symmetric property: If one number is equal to a second number, then the second number is equal to the first number.
      In general, when $a$ and $b$ are real numbers, and $a = b$, then $b = a$.
   c. Transitive property: If one number is equal to a second number and the second number is equal to a third number, then the first number is equal to the third.
      In general, when $a$, $b$, and $c$ are real numbers and $a = b$ and $b = c$, then $a = c$.

3. If two numbers are equal, then their opposites are equal.

**III. Sample Solutions**

1. What number should be subtracted from both sides of $x + 8 = 15$ to get $x = 7$?   *Answer:* Subtract 8 from both sides.

2. What number should be added to both sides of $n - 5 = 9$ to get $n = 14$?
   *Answer:* Add 5 to both sides.

3. By what number should you divide both sides of $4y = 12$ to get $y = 3$?
   *Answer:* Divide both sides by 4.

**4.** By what number should you multiply both sides of $\frac{s}{10} = 5$ to get $s = 50$?

*Answer:* Multiply both sides by 10.

**5.** If $-6 = w$, does $w = -6$?   *Answer:* Yes, by symmetric property of equality.

**6.** If $20 \div 4 = 5$ and $5 = 3 + 2$, does $20 \div 4 = 3 + 2$?

*Answer:* Yes, by transitive property of equality.

**7.** Write the sentence for $-b = 12$ which indicates that if two numbers are equal, then their opposites are equal.   *Answer:* $b = -12$

## DIAGNOSTIC TEST

1. What number should be subtracted from both sides of:
   **a.** $32 + 18 = 50$ to get the sentence $32 = 32$?
   **b.** $y + 3 = 14$ to get the sentence $y = 11$?
2. What number should be added to both sides of:
   **a.** $17 - 8 = 9$ to get the sentence $17 = 17$?
   **b.** $r - 12 = 42$ to get the sentence $r = 54$?
3. By what number should you divide both sides of:
   **a.** $8 \times 5 = 40$ to get the sentence $5 = 5$?
   **b.** $12y = 72$ to get the sentence $y = 6$?
4. By what number should you multiply both sides of:
   **a.** $63 \div 7 = 9$ to get the sentence $63 = 63$?
   **b.** $n \div 16 = 5$ to get the sentence $n = 80$?
5. Does equality have the reflexive property? Illustrate your answer.
6. Does equality have the symmetric property? Illustrate your answer.
7. Does equality have the transitive property? Illustrate your answer.
8. If two numbers are equal, are their opposites equal? Illustrate your answer.

## RELATED PRACTICE EXAMPLES

1. What number should be subtracted from both sides of:
   **a.** (1) $9 + 7 = 16$ to get the sentence $9 = 9$?
      (2) $n + 7 = 16$ to get the sentence $n = 9$?
   **b.** (1) $12 + 15 = 27$ to get the sentence $12 = 12$?
      (2) $x + 15 = 27$ to get the sentence $x = 12$?
   **c.** (1) $31 + 9 = 40$ to get the sentence $31 = 31$?
      (2) $y + 9 = 40$ to get the sentence $y = 31$?

2. What number should be added to both sides of:
   **a.** (1) $11 - 3 = 8$ to get the sentence $11 = 11$?
      (2) $n - 3 = 8$ to get the sentence $x = 11$?
   **b.** (1) $23 - 10 = 13$ to get the sentence $23 = 23$?
      (2) $r - 10 = 13$ to get the sentence $r = 23$?
   **c.** (1) $52 - 18 = 34$ to get the sentence $52 = 52$?
      (2) $x - 18 = 34$ to get the sentence $x = 52$?

3. By what number should you divide both sides of:
   **a.** (1) $6 \times 9 = 54$ to get the sentence $9 = 9$?
      (2) $6n = 54$ to get the sentence $n = 9$?
   **b.** (1) $17 \times 5 = 85$ to get the sentence $5 = 5$?
      (2) $17x = 85$ to get the sentence $x = 5$?
   **c.** (1) $9 \times 12 = 108$ to get the sentence $12 = 12$?
      (2) $9t = 108$ to get the sentence $t = 12$?

4. By what number should you multiply both sides of:
   **a.** (1) $36 \div 3 = 12$ to get the sentence $36 = 36$?
      (2) $n \div 3 = 12$ to get the sentence $n = 36$?
   **b.** (1) $\frac{75}{5} = 15$ to get the sentence $75 = 75$?

      (2) $\frac{n}{5} = 15$ to get the sentence $n = 75$?

   **c.** (1) $48 \div 6 = 8$ to get the sentence $48 = 48$?
      (2) $y \div 6 = 8$ to get the sentence $y = 48$?
   **d.** (1) $\frac{27}{9} = 3$ to get the sentence $27 = 27$?

      (2) $\frac{x}{9} = 3$ to get the sentence $x = 27$?

5. **a.** Does $8 = 8$?      **b.** Does $-\frac{5}{16} = -\frac{5}{16}$?      **c.** Does $4.3 = 4.3$?

6. **a.** If $8 + 2 = 10$, does $10 = 8 + 2$?
   **b.** If $9 \times 7 = 63$, does $63 = 9 \times 7$?
   **c.** If $30 \div 6 = 5$, does $5 = 30 \div 6$?      **d.** If $3 = x$, does $x = 3$?
   **e.** If $n = -7$, does $-7 = n$?      **f.** If $-12 = y$, does $y = -12$?

7. **a.** If $13 + 5 = 18$ and $18 = 6 \times 3$, does $13 + 5 = 6 \times 3$?
   **b.** If $20 - 11 = 9$ and $9 = 72 \div 8$, does $20 - 11 = 72 \div 8$?
   **c.** If $15 + 8 - 10 = (6 \times 2) + 1$ and $(6 \times 2) + 1 = (45 \div 3) - 2$, does
      $15 + 8 - 10 = (45 \div 3) - 2$?

8. Write the sentence for each of the following which indicates that the
   opposites of these equal numbers are equal:
   **a.** $12 = 12$      **b.** $-27 = -27$      **c.** $-m = 7$      **d.** $-x = -4$
   **e.** $-t = 16$      **f.** $-y = -41$      **g.** $-r = -39$      **h.** $-n = 0$

# Equations–Solution by Inverse Operation (Subtraction) or by Additive Inverse

I. **Aim**     To solve equations by the type $n + 3 = 15$
  (1) By the inverse operation (subtraction) or
  (2) By the additive inverse.

## II. Procedure

1. Solution by Inverse Operation
  Subtract from both sides of the equation the number that is on the same side of the equation as the variable so that only the variable remains on that side. See sample solution 1.a.
2. Solution of Additive Inverse
  Or add to both sides of the equation the additive inverse of the number that is on the same side of the equation as the variable.
3. When the variable appears on the right side of the equation, rewrite the equation in the final step, using the symmetric property of equality. See sample solution 2.
4. Note that the equation $n + 3 = 15$ may also appear in the form $3 + n = 15$ or $15 = n + 3$ or $15 = 3 + n$.
5. To check whether the number found is a solution, substitute this number for the variable in the given equation. If the resulting sentence is true, then this number is a solution. The ? mark and ✔ mark may be used. See check of sample solution 1.

## III. Sample Solutions

1. Solve and check: $n + 3 = 15$
  The indicated operation of $n + 3$ is addition. To eliminate the $+3$ from the left member use: a. the inverse operation of subtraction, or b. add its additive inverse.

**a.** *Solution:*   OR          **b.** *Solution:*                    *Check:*

$$n + 3 = 15 \qquad\qquad n + 3 = 15 \qquad\qquad n + 3 = 15$$
$$n + 3 - 3 = 15 - 3 \quad n + 3 + (-3) = 15 + (-3) \quad 12 + 3 = 15\,?$$
$$n = 12 \qquad\qquad\qquad n = 12 \qquad\qquad 15 = 15 \; ✔$$

The root is 12.
*Answer:* 12          *Answer:* 12

The question mark in the check indicates the uncertainty as to whether the resulting sentence is true before computation. The ✔ mark indicates that the sentence is true.

**2.** Solve and check: $10 = y + 2.7$
*Solution:*

$$10 = y + 2.7$$
$$10 - 2.7 = y + 2.7 - 2.7$$
$$7.3 = y$$
$$y = 7.3$$

*Answer:* 7.3

*Check:*
$$10 = y + 2.7$$
$$10 = 7.3 + 2.7 ?$$
$$10 = 10 \checkmark$$

**3.** Solve and check: $x + 7 = 2$
*Solution:*

$$x + 7 = 2$$
$$x + 7 - 7 = 2 - 7$$
$$x = -5$$

*Answer:* $-5$

*Check:*
$$x + 7 = 2$$
$$-5 + 7 = 2 ?$$
$$2 = 2 \checkmark$$

## DIAGNOSTIC TEST

Solve and check:
**1.** $x + 11 = 40$    **2.** $7 + a = 13$    **3.** $19 = b + 1$
**4.** $25 = 8 + y$    **5.** $c + 5\frac{3}{4} = 7$    **6.** $m + 1.8 = 9.4$
**7.** $4 + x = 4$    **8.** $y + 32 = 9$    **9.** $r + 15 = -17$

## *RELATED PRACTICE EXAMPLES*

Solve and check:

1. **a.** $x + 2 = 6$    **b.** $m + 5 = 9$    **c.** $r + 8 = 10$
   **d.** $x + 12 = 21$    **e.** $a + 26 = 28$    **f.** $b + 37 = 51$
   **g.** $n + 15 = 32$    **h.** $y + 20 = 109$    **i.** $x + 78 = 120$

2. **a.** $1 + x = 5$    **b.** $3 + y = 11$    **c.** $9 + n = 22$
   **d.** $36 + b = 43$    **e.** $17 + y = 70$    **f.** $58 + t = 91$
   **g.** $25 + x = 86$    **h.** $75 + c = 100$    **i.** $60 + b = 185$

3. **a.** $7 = x + 4$    **b.** $6 = y + 5$    **c.** $18 = a + 2$
   **d.** $38 = c + 13$    **e.** $47 = m + 32$    **f.** $40 = x + 18$
   **g.** $34 = b + 27$    **h.** $80 = x + 45$    **i.** $95 = y + 59$

4. **a.** $9 = 6 + b$    **b.** $10 = 5 + x$    **c.** $14 = 9 + a$
   **d.** $69 = 38 + n$    **e.** $42 = 25 + y$    **f.** $56 = 41 + b$
   **g.** $83 = 59 + x$    **h.** $110 = 37 + y$    **i.** $150 = 100 + g$

5. **a.** $x + \frac{1}{4} = \frac{1}{2}$  **b.** $n + \frac{1}{2} = 2\frac{1}{2}$  **c.** $a + 5\frac{1}{3} = 8\frac{2}{3}$
   **d.** $y + \frac{3}{4} = 5$  **e.** $x + 4\frac{1}{2} = 8$  **f.** $a + 6\frac{2}{3} = 9\frac{5}{6}$
   **g.** $\frac{3}{4} + x = 4\frac{1}{4}$  **h.** $7 = b + \frac{3}{8}$  **i.** $10\frac{1}{2} = 6\frac{7}{8} + x$

6. **a.** $x + .2 = .6$  **b.** $a + .4 = 7.2$  **c.** $b + .5 = 4$
   **d.** $1.9 + y = 6.3$  **e.** $20 = a + 4.9$  **f.** $9.7 = .04 + n$
   **g.** $\$10 = c + \$3.49$  **h.** $s + \$1.37 = \$5.20$  **i.** $\$7.98 = \$6.45 + m$

7. **a.** $x + 8 = 8$  **b.** $11 = b + 11$  **c.** $5 + y = 5$
   **d.** $n + \frac{3}{4} = \frac{3}{4}$  **e.** $b + .6 = .6$  **f.** $2 = x + 2$
   **g.** $4\frac{1}{2} + r = 4\frac{1}{2}$  **h.** $m + .02 = .02$  **i.** $9 = 9 + t$

8. **a.** $n + 6 = 4$  **b.** $x + 21 = 0$  **c.** $d + 53 = 46$
   **d.** $y + .21 = .18$  **e.** $8 + r = 3$  **f.** $a + 1 = \frac{11}{16}$
   **g.** $75 = 90 + t$  **h.** $2\frac{1}{3} + s = 1\frac{5}{6}$  **i.** $0 = l + 98$

9. **a.** $x + 2 = -6$  **b.** $h + 12 = -15$  **c.** $z + 78 = -29$
   **d.** $t + 3.4 = -5.2$  **e.** $-4 = a + 8$  **f.** $27 + c = -27$
   **g.** $-.95 = s + .15$  **h.** $-\frac{5}{8} = 1\frac{1}{2} + n$  **i.** $12 + y = -19$

## CUMULATIVE REVIEW

1. Which of the following numbers is the root of the equation $8x - 5 = 11$?
   **a.** $x = 3$  **b.** $x = 4$  **c.** $x = 2$  **d.** $x = -1$
2. Which of the following equations have 4 as the solution?
   **a.** $3n + 8 = 20$  **b.** $-n = 4$  **c.** $-15n = -60$  **d.** $n = 4$
   Which of the above equations are equivalent?
   Which equation of the equivalent equations is the simplest form?
3. Which of the following equations are conditional equations?
   Which equations are satisfied by all numbers replacing the variable?
   Which equations have no solution?
   **a.** $16 - n = n$  **b.** $9b - 11 = 11 - 9b$  **c.** $y - 7 = y$
   **d.** $20 + t = t$  **e.** $6n + 1 = 1 + 6n$  **f.** $5(x - 3) = 5x - 15$
4. What operation with what number should be used on both sides of:
   **a.** $18 + 8 = 26$ to get the sentence $18 = 18$?
   **b.** $n + 5 = 12$ to get the equation $n = 7$?
5. Solve and check:
   **a.** $x + 7 = 13$  **b.** $125 = y + 98$  **c.** $4\frac{1}{2} + n = 9$
   **d.** $6.8 = b + .29$  **e.** $47 = 47 + C$  **f.** $t + 19 = 0$
   **g.** $r + 10 = -8$  **h.** $m + 16 = -16$  **i.** $34 = r + 34$
   **j.** $n + \frac{2}{3} = 2\frac{5}{6}$  **k.** $c + \$2.59 = \$5.80$  **l.** $\$20 = \$6.50 + s$

# Equations–Solution by Inverse Operation (Addition) or by Additive Inverse

**I. Aim**   To solve equations of the type $n - 3 = 15$
 (1) By the inverse operation (addition) or
 (2) By the additive inverse

## II. Procedure

1. Solution by Inverse Operation
 Add to both sides of the equation the number that is on the same side of the equation as the variable so that only the variable remains on that side. See sample solution 1.a.
2. Solution by Additive Inverse
 Or add to both sides of the equation the additive inverse of the number that is on the same side of the equation as the variable.
3. When the variable appears on the right side of the equation, rewrite the equation in the final step, using the symmetric property of equality. See sample solution 2.
4. Note that the equation $n - 3 = 15$ may also appear in the form $15 = n - 3$ but not as $3 - n = 15$ or $15 = 3 - n$.
5. To check your answer, substitute it for the variable in the given equation. If the resulting sentence is true, then your answer is a solution.

## III. Sample Solutions

1. Solve and check: $n - 3 = 15$
 The indicated operation of $n - 3$ is subtraction. To eliminate the $-3$ from the left member use a. the inverse operation of addition, or b. add its additive inverse.

 **a.** *Solution:*         OR   **b.** *Solution:*

$$n - 3 = 15$$
$$n - 3 + 3 = 15 + 3$$
$$n = 18$$

The root is 18.

*Answer:* 18

$$n - 3 = 15$$
$$n + (-3) = 15$$
$$n + (-3) + (+3) = 15 + (+3)$$
$$n = 18$$

*Answer:* 18

*Check:*
$$n - 3 = 15$$
$$18 - 3 = 15\,?$$
$$15 = 15 \ ✔$$

**214**

**2.** Solve and check: $7 = s - 9\frac{1}{2}$

*Solution:*

$$7 = s - 9\frac{1}{2}$$
$$7 + 9\frac{1}{2} = s - 9\frac{1}{2} + 9\frac{1}{2}$$
$$16\frac{1}{2} = s$$
$$s = 16\frac{1}{2}$$

*Check:*

$$7 = s - 9\frac{1}{2}$$
$$7 = 16\frac{1}{2} - 9\frac{1}{2}?$$
$$7 = 7 \checkmark$$

*Answer:* $16\frac{1}{2}$

**3.** Solve and check:

*Solution:*

$$x - 9 = -15$$
$$x - 9 + 9 = -15 + 9$$
$$x = -6$$

*Check:*

$$x - 9 = -15$$
$$-6 - 9 = -15?$$
$$-15 = -15 \checkmark$$

*Answer:* $-6$

## DIAGNOSTIC TEST

Solve and check:

**1.** $x - 8 = 12$      **2.** $7 = b - 31$      **3.** $a - 5 = 5$

**4.** $c - 3 = 0$      **5.** $x - \frac{1}{8} = 2\frac{3}{8}$      **6.** $y - 4.5 = .16$

**7.** $x - 2 = -20$      **8.** $x - 11 = -3$

## RELATED PRACTICE EXAMPLES

Solve and check:

1. **a.** $x - 1 = 5$      **b.** $m - 8 = 4$      **c.** $y - 3 = 6$
   **d.** $n - 18 = 3$      **e.** $x - 7 = 9$      **f.** $a - 12 = 15$
   **g.** $c - 5 = 19$      **h.** $y - 53 = 42$      **i.** $x - 49 = 74$

2. **a.** $9 = x - 6$      **b.** $16 = y - 5$      **c.** $23 = a - 79$
   **d.** $36 = a - 93$      **e.** $28 = x - 18$      **f.** $100 = z - 47$
   **g.** $69 = b - 24$      **h.** $26 = y - 87$      **i.** $76 = n - 58$

3. **a.** $x - 4 = 4$      **b.** $a - 1 = 1$      **c.** $8 = n - 8$
   **d.** $b - 18 = 18$      **e.** $12 = x - 12$      **f.** $7 = x - 7$
   **g.** $25 = s - 25$      **h.** $x - 16 = 16$      **i.** $b - 20 = 20$

4. **a.** $x - 5 = 0$      **b.** $x - 20 = 0$      **c.** $0 = b - 8$
   **d.** $0 = y - 13$      **e.** $y - 4 = 0$      **f.** $a - 7 = 0$
   **g.** $0 = x - 14$      **h.** $0 = c - 2$      **i.** $d - 9 = 0$

5.  **a.** $x - \frac{1}{2} = \frac{1}{2}$      **b.** $b - \frac{2}{3} = 1\frac{2}{3}$      **c.** $y - \frac{7}{8} = 2\frac{3}{8}$
    **d.** $a - 4 = 6\frac{1}{6}$      **e.** $m - 1\frac{1}{2} = 5$      **f.** $c - 3\frac{1}{4} = 4\frac{11}{16}$
    **g.** $4\frac{1}{3} = x - 1\frac{5}{6}$      **h.** $10 = b - 7\frac{1}{4}$      **i.** $20\frac{1}{2} = x - 10\frac{7}{8}$

6.  **a.** $b - .2 = .6$      **b.** $x - 1.8 = 3.6$      **c.** $a - 7 = 10.4$
    **d.** $m - .8 = 5$      **e.** $7.2 = x - .04$      **f.** $25 = t - 1.3$
    **g.** $c - \$.59 = \$2$      **h.** $\$.67 = l - \$.19$      **i.** $a - \$1.25 = \$4.98$

7.  **a.** $x - 3 = -9$      **b.** $s - 5 = -13$      **c.** $-36 = n - 7$
    **d.** $m - 2.4 = -3.8$      **e.** $-.26 = x - .09$      **f.** $t - \frac{2}{3} = -\frac{5}{6}$
    **g.** $-57 = b - 29$      **h.** $y - 38 = -61$      **i.** $r - 4\frac{3}{5} = -5\frac{1}{2}$

8.  **a.** $x - 10 = -4$      **b.** $c - 9 = -8$      **c.** $-6 = d - 11$
    **d.** $-4.5 = b - 7.1$      **e.** $-58 = l - 96$      **f.** $p - 9\frac{1}{3} = -6\frac{1}{2}$
    **g.** $y - 5 = -5$      **h.** $r - 1.2 = -.84$      **i.** $-\frac{5}{8} = c - 2\frac{3}{4}$

## CUMULATIVE REVIEW

1.  Which of the following numbers is the root of the equation $7n + 9n = 80$?
    **a.** $n = 16$      **b.** $n = -5$       **c.** $n = 10$        **d.** $n = 5$

2.  Which of the following is the solution to the equation $\frac{x}{6} = 2$?

    **a.** 3        **b.** 12          **c.** 4          **d.** $\frac{1}{3}$

3.  Which of the following equations have $-7$ as the solution?
    **a.** $-x = 7$    **b.** $2x - 3 = 17$     **c.** $x = -7$      **d.** $x + 8 = 1$
    Which of the above equations are equivalent?
    Which equation of the equivalent equations is the simplest form?

4.  Which of the following equations are conditional equations?
    Which equations are satisfied by all numbers replacing the variable?
    Which equations have no solution?
    **a.** $y + 6 = y$  **b.** $3a - 9 = 9 - 3a$  **c.** $5x + 6 = 6 + 5x$  **d.** $10 - n = n$

5.  Find the missing number or variable or operation:
    **a.** $(n - 16) ? \square = n$         **b.** $(y + 21) ? \square = y$

6.  What operation with what number should be used on both sides of $y - 8 = 25$ to get the equation $y = 33$?

7.  Solve and check:
    **a.** $n + 85 = 101$      **b.** $2\frac{3}{8} + y = 7$      **c.** $4.9 = b + .23$
    **d.** $31 = 31 + r$      **e.** $m + 26 = 0$      **f.** $t + 4 = -3$

8.  Solve and check:
    **a.** $s - 9 = 13$      **b.** $n - 27 = 19$      **c.** $58 = r - 36$
    **d.** $24 = x - 42$      **e.** $b - 4\frac{3}{4} = 0$      **f.** $4.5 = x - 1.86$
    **g.** $m - 52 = -21$      **h.** $y - 7 = -17$      **i.** $b - 48 = 48$
    **j.** $p - .25 = .6$      **k.** $0 = x - 12.3$      **l.** $-6 = x - 10$

# Equations–Solution by Inverse Operation (Division) or by Multiplicative Inverse

**I. Aim**    To solve equations of the type $3n = 15$
     (1) By the inverse operation (division) or
     (2) By the multiplicative inverse.

## II. Procedure

1. Solution by Inverse Operation
   Divide both sides of the equation by the coefficient of the variable so that the variable remains with the coefficient one (1) understood. See sample solution 1.a.
2. Solution by Multiplicative Inverse
   Or multiply both sides of the equation by the multiplicative inverse of the coefficient of the variable.
3. When the variable appears on the right side of the equation, rewrite the equation in the final step, using the symmetric property of equality. See sample solution 2.
4. Note that the equation $3n = 15$ may also appear as $15 = 3n$.
5. To check your answer, substitute it for the variable in the given equation. If the resulting sentence is true, then your answer is a solution.
6. When the numerical coefficient of the variable is a fraction, multiply by the proper multiplicative inverse. See sample solution 7.
7. Observe in sample solution 9 that to determine the value of the variable, the opposite of both members of the equation are used.

## III. Sample Solutions

1. Solve and check: $3n = 15$
   The indicated operation of $3n$ is multiplication. To eliminate the 3 from the left member use a. the inverse operation of division, or b. multiply by its multiplicative inverse.

<table>
<tr><td>

**a.** *Solution:*
$$3n = 15$$
$$\frac{3n}{3} = \frac{15}{3}$$
$$n = 5$$
The root is 5.

</td><td>

OR   **b.** *Solution:*
$$3n = 15$$
$$\tfrac{1}{3} \cdot 3n = \tfrac{1}{3} \cdot 15$$
$$n = 5$$
*Answer:* 5

</td><td>

*Check:*
$$3n = 15$$
$$3 \cdot 5 = 15 \,?$$
$$15 = 15 ✔$$

</td></tr>
</table>

**2.** Solve and check: $12 = 4a$

*Solution:*     *Check:*

$12 = 4a$        $12 = 4a$

$\frac{12}{4} = \frac{4a}{4}$      $12 = 4 \cdot 3$ ?

           $12 = 12$ ✔

$3 = a$

$a = 3$   *Answer:* 3

**3.** Solve and check: $.02r = 6.4$

*Solution:*         *Check:*

$.02r = 6.4$        $.02r = 6.4$

$\frac{.02r}{.02} = \frac{6.4}{.02}$    $.02 \times 320 = 6.4$ ?

             $6.4 = 6.4$ ✔

$r = 320$   *Answer:* 320

**4.** Solve and check: $6x = 8$

*Solution:*     *Check:*

$6x = 8$       $6x = 8$

$\frac{6x}{6} = \frac{8}{6}$     $6 \cdot 1\frac{1}{3} = 8$ ?

          $8 = 8$ ✔

$x = 1\frac{1}{3}$

*Answer:* $1\frac{1}{3}$

**5.** Solve and check: $5m = 3$

*Solution:*     *Check:*

$5m = 3$       $5m = 3$

$\frac{5m}{5} = \frac{3}{5}$     $5 \cdot \frac{3}{5} = 3$ ?

          $3 = 3$ ✔

$m = \frac{3}{5}$

*Answer:* $\frac{3}{5}$

**6.** Solve and check: $12n = 4$

*Solution:*     *Check:*

$12n = 4$      $12n = 4$

$\frac{12n}{12} = \frac{4}{12}$    $12 \cdot \frac{1}{3} = 4$ ?

          $4 = 4$ ✔

$n = \frac{1}{3}$

*Answer:* $\frac{1}{3}$

**7.** Solve and check: $\frac{2}{3}c = 8$

*Solution:*        *Check:*

$\frac{2}{3}c = 8$       $\frac{2}{3}c = 8$

$\frac{3}{2} \cdot \frac{2}{3}c = \frac{3}{2} \cdot 8$    $\frac{2}{3} \cdot 12 = 8$ ?

$c = 12$         $8 = 8$ ✔

*Answer:* 12

**8.** Solve and check: $-6n = 72$

*Solution:*        *Check:*

$-6n = 72$       $-6n = 72$

$\frac{-6n}{-6} = \frac{72}{-6}$    $-6(-12) = 72$ ?

           $72 = 72$ ✔

$n = -12$

*Answer:* $-12$

**9.** Solve and check: $-x = 2$

**a.** Use principle that the opposite of equals are equal, or
**b.** divide both sides by $-1$ since $-x = -1x$.

**a.** *Solution:*    OR   **b.** *Solution:*

     $-x = 2$           $-x = 2$

       $x = -2$        $\frac{-x}{-1} = \frac{2}{-1}$

                      $x = -2$

*Check:*     $-x = 2$

        $-(-2) = 2$ ?

           $2 = 2$ ✔

*Answer:* $-2$

## DIAGNOSTIC TEST

Solve and check:
1. $9b = 63$
2. $32 = 8s$
3. $7a = 7$
4. $2y = 0$
5. $4x = 11$
6. $16c = 24$
7. $9r = 5$
8. $10m = 8$
9. $.06p = 18$
10. Find the root correct
    to the nearest
    hundredth:
    $12x = 14$
11. $\frac{3}{4}a = 27$
12. $6d = -30$
13. $-8x = 96$
14. $-15y = -60$
15. $-c = -7$

## RELATED PRACTICE EXAMPLES

Solve and check:

1. **a.** $5x = 25$  **b.** $3y = 21$  **c.** $7c = 28$
   **d.** $15a = 45$  **e.** $8b = 56$  **f.** $6x = 48$
   **g.** $9y = 54$  **h.** $10a = 70$  **i.** $12r = 36$

2. **a.** $18 = 6s$  **b.** $24 = 8w$  **c.** $56 = 7n$
   **d.** $32 = 4x$  **e.** $90 = 9m$  **f.** $22 = 11R$
   **g.** $72 = 24y$  **h.** $95 = 5x$  **i.** $100 = 20h$

3. **a.** $5b = 5$  **b.** $12y = 12$  **c.** $8a = 8$
   **d.** $17b = 17$  **e.** $9 = 9y$  **f.** $2 = 2b$
   **g.** $11a = 11$  **h.** $15 = 15t$  **i.** $21c = 21$

4. **a.** $9b = 0$  **b.** $6x = 0$  **c.** $7c = 0$
   **d.** $14y = 0$  **e.** $0 = 5x$  **f.** $0 = 4a$
   **g.** $10z = 0$  **h.** $0 = 18r$  **i.** $25t = 0$

5. **a.** $5n = 7$  **b.** $4y = 9$  **c.** $10a = 13$
   **d.** $6c = 25$  **e.** $3m = 4$  **f.** $9 = 5b$
   **g.** $23 = 7y$  **h.** $2x = 17$  **i.** $47 = 9d$

6. **a.** $8n = 10$  **b.** $12x = 16$  **c.** $9d = 15$
   **d.** $24c = 36$  **e.** $9 = 6y$  **f.** $14 = 10x$
   **g.** $15t = 20$  **h.** $26 = 4s$  **i.** $42 = 12d$

7. **a.** $3x = 1$  **b.** $5y = 2$  **c.** $9b = 4$
   **d.** $3 = 8n$  **e.** $10m = 7$  **f.** $21r = 4$
   **g.** $7c = 5$  **h.** $9 = 10a$  **i.** $15c = 13$

8. **a.** $6m = 2$     **b.** $8y = 4$     **c.** $12y = 10$
   **d.** $18s = 12$     **e.** $15 = 45b$     **f.** $16 = 36a$
   **g.** $88s = 33$     **h.** $26 = 39y$     **i.** $375n = 75$

9. **a.** $8x = .16$     **b.** $.5a = 3.5$     **c.** $1.6y = 4.8$
   **d.** $12 = .2a$     **e.** $.04p = 32$     **f.** $9c = \$.54$
   **g.** $1.06x = \$33.92$     **h.** $\$.07n = \$1.12$     **i.** $\$345 = 1.15c$

Find each root correct to nearest hundredth:

10. **a.** $6a = 8$     **b.** $11x = 25$     **c.** $7c = 31$
   **d.** $50 = 13b$     **e.** $30m = 25$     **f.** $18y = 15$
   **g.** $24x = 42$     **h.** $76 = 32c$     **i.** $1.7a = 81$

Solve and check:

11. **a.** $\frac{1}{4}n = 17$     **b.** $\frac{3}{8}x = 78$     **c.** $\frac{5}{6}c = 105$
   **d.** $\frac{7}{12}y = 3\frac{1}{2}$     **e.** $52 = \frac{4}{5}a$     **f.** $\$4.77 = \frac{3}{10}s$
   **g.** $1\frac{1}{2}c = 186$     **h.** $\frac{1}{5}s = \$.29$     **i.** $2\frac{3}{4}n = 3\frac{2}{3}$

12. **a.** $4x = -8$     **b.** $6y = -6$     **c.** $7b = -28$
   **d.** $5z = -23$     **e.** $8y = -4$     **f.** $9a = -1$
   **g.** $-21 = 3x$     **h.** $-54 = 6y$     **i.** $-11 = 4r$

13. **a.** $-5s = 35$     **b.** $-4x = 36$     **c.** $-9t = 12$
   **d.** $-3b = 5$     **e.** $-8y = 3$     **f.** $-12x = 8$
   **g.** $24 = -6x$     **h.** $2 = -3y$     **i.** $15 = -4b$

14. **a.** $-7b = -21$     **b.** $-3a = -36$     **c.** $-45x = -45$
   **d.** $-8z = -9$     **e.** $-4x = -1$     **f.** $-15r = -3$
   **g.** $-32 = -4y$     **h.** $-4 = -9z$     **i.** $-13 = -2x$

15. **a.** $-x = 4$     **b.** $-b = 3$     **c.** $-y = 1$
   **d.** $-a = -2$     **e.** $-x = -5$     **f.** $-m = -4$
   **g.** $6 = -x$     **h.** $-1 = -y$     **i.** $-9 = -n$

## *CUMULATIVE REVIEW*

Solve and check:

1. **a.** $x + 19 = 33$     **b.** $8 = y + 3\frac{2}{5}$     **c.** $0 = s + 25$
   **d.** $n + 4.1 = 18$     **e.** $r + 38 = 29$     **f.** $24 + x = -9$

2. **a.** $x - 15 = 10$     **b.** $9.7 = c - .28$     **c.** $n - 1\frac{1}{2} = 0$
   **d.** $y - 27 = -11$     **e.** $b - 5 = -18$     **f.** $a - 23 = 23$

3. **a.** $7x = 63$     **b.** $18y = 54$     **c.** $-12n = 6$
   **d.** $-10c = -80$     **e.** $15 = 25t$     **f.** $\frac{5}{6}a = 120$
   **g.** $-27y = 49$     **h.** $-9 = -n$     **i.** $.5c = \$.72$
   **j.** $-12t = -12$     **k.** $8y = 54$     **l.** $-9x = -45$
   **m.** $\frac{2}{3}c = \$1.46$     **n.** $-21d = 14$     **o.** $10n = 1$

# Equations–Solution by Inverse Operation (Multiplication) or by Multiplicative Inverse

EXERCISE
6-7

**I. Aim**   To solve equations of the type $\dfrac{n}{3} = 15$

   (1) By the inverse operation (multiplication) or
   (2) By the multiplicative inverse.

**II. Procedure**

1. Solution by Inverse Operation
   Multiply both sides of the equation by the denominator of the fractional coefficient of the variable so that the variable remains with the coefficient one (1) understood. See sample solution 1.a.
2. Solution by Multiplicative Inverse
   Or multiply both sides of the equation by the multiplicative inverse of the coefficient of the variable.
3. When the variable appears on the right side of the equation, rewrite the equation in the final step, using the symmetric property of equality. See sample solution 2.
4. Note that the equation $\dfrac{n}{3} = 15$ may also appear as $15 = \dfrac{n}{3}$ or $\frac{1}{3}n = 15$ or $15 = \frac{1}{3}n$. See sample solution 1.b.
5. To check your answer, substitute it for the variable in the given equation. If the resulting sentence is true, then your answer is a solution.

**III. Sample Solutions**

1. Solve and check: $\dfrac{n}{3} = 15$

   The indicated operation of $\dfrac{n}{3}$ is division. To eliminate the 3 from the left member use a. the inverse operation of multiplication, or b. multiply by its multiplicative inverse.

   **a.** *Solution:*   OR   **b.** *Solution:*           *Check:*

   $$\frac{n}{3}=15$$            $$\frac{n}{3}=15$$            $$\frac{n}{3}=15$$

   $$3\cdot\frac{n}{3}=3\cdot15$$            $$^*\tfrac{1}{3}n=15$$            $$\frac{45}{3}=15\,?$$
   $$15=15\ ✔$$

   $$n=45$$            $$3\cdot\tfrac{1}{3}n=3\cdot15$$            $$^*\frac{n}{3}\ \text{may be}$$
   The root is **45**.   *Answer:* 45            $$n=45$$            $$3\ \text{written as}\ \tfrac{1}{3}n.$$

221

**2.** Solve and check: $6 = \dfrac{a}{2}$

*Solution:*      *Check:*

$$6 = \dfrac{a}{2} \qquad 6 = \dfrac{a}{2}$$

$$2 \cdot 6 = 2 \cdot \dfrac{a}{2} \qquad 6 = \tfrac{12}{2}\,?$$

$$12 = a \qquad 6 = 6 \checkmark$$

$$a = 12$$

*Answer:* 12

**3.** Solve and check: $\dfrac{x}{4} = -3$

*Solution:*      *Check:*

$$\dfrac{x}{4} = -3 \qquad \dfrac{x}{4} = -3$$

$$4 \cdot \dfrac{x}{4} = 4 \cdot (-3) \qquad -\tfrac{12}{4} = -3\,?$$

$$x = -12 \qquad -3 = -3 \checkmark$$

*Answer:* $-12$

---

## DIAGNOSTIC TEST

Solve and check:

1. $\dfrac{x}{3} = 7$     2. $\dfrac{b}{5} = 4$     3. $\dfrac{a}{4} = 20$     4. $\dfrac{x}{12} = 6$

5. $\dfrac{c}{8} = 8$     6. $\dfrac{r}{7} = 1$     7. $\dfrac{m}{5} = 0$     8. $\dfrac{c}{6.3} = .5$

9. $2 = \dfrac{b}{9}$     10. $\dfrac{a}{3} = -12$     11. $\dfrac{t}{-6} = 4$     12. $\dfrac{x}{-7} = -21$

---

## *RELATED PRACTICE EXAMPLES*

Solve and check:

1. **a.** $\dfrac{x}{2} = 5$    **b.** $\dfrac{a}{7} = 10$    **c.** $\dfrac{b}{5} = 6$    **d.** $\dfrac{m}{4} = 9$    **e.** $\dfrac{b}{2} = 11$

2. **a.** $\dfrac{c}{7} = 3$    **b.** $\dfrac{b}{5} = 2$    **c.** $\dfrac{c}{9} = 4$    **d.** $\dfrac{x}{4} = 3$    **e.** $\dfrac{y}{6} = 5$

3. **a.** $\dfrac{x}{4} = 8$    **b.** $\dfrac{y}{2} = 4$    **c.** $\dfrac{c}{7} = 21$    **d.** $\dfrac{r}{3} = 12$    **e.** $\dfrac{b}{5} = 20$

4. **a.** $\dfrac{x}{4} = 2$    **b.** $\dfrac{b}{9} = 3$    **c.** $\dfrac{x}{12} = 3$    **d.** $\dfrac{b}{16} = 18$    **e.** $\dfrac{m}{15} = 10$

5. **a.** $\dfrac{b}{2} = 2$    **b.** $\dfrac{a}{5} = 5$    **c.** $\dfrac{m}{8} = 8$    **d.** $\dfrac{c}{6} = 6$    **e.** $\dfrac{x}{15} = 15$

6. **a.** $\dfrac{a}{5} = 1$    **b.** $\dfrac{y}{3} = 1$    **c.** $\dfrac{b}{4} = 1$    **d.** $\dfrac{d}{8} = 1$    **e.** $\dfrac{x}{20} = 1$

7. **a.** $\dfrac{b}{4}=0$  **b.** $\dfrac{m}{9}=0$  **c.** $\dfrac{x}{10}=0$  **d.** $\dfrac{y}{6}=0$  **e.** $\dfrac{b}{18}=0$

8. **a.** $\dfrac{x}{.2}=7$  **b.** $\dfrac{n}{1.5}=.8$  **c.** $\dfrac{a}{6}=3.9$  **d.** $\dfrac{x}{.1}=.02$  **e.** $\dfrac{c}{2.4}=5.7$

9. **a.** $4=\dfrac{x}{7}$  **b.** $9=\dfrac{y}{5}$  **c.** $8=\dfrac{c}{2}$  **d.** $10=\dfrac{n}{20}$  **e.** $1=\dfrac{b}{15}$

10. **a.** $\dfrac{n}{5}=-4$  **b.** $\dfrac{a}{10}=-12$  **c.** $\dfrac{r}{2}=-6$  **d.** $-24=\dfrac{d}{8}$

11. **a.** $\dfrac{x}{-6}=18$  **b.** $\dfrac{t}{-2}=35$  **c.** $\dfrac{c}{-9}=5$  **d.** $\dfrac{n}{-7}=7$

12. **a.** $\dfrac{b}{-4}=-20$  **b.** $\dfrac{n}{-3}=-3$  **c.** $\dfrac{d}{-4}=-28$  **d.** $-5=\dfrac{x}{-10}$

## CUMULATIVE REVIEW  *meen*

1. Which of the following numbers is the root of the equation $\dfrac{x}{12}=4$?

   **a.** 3  **b.** $\frac{1}{3}$  **c.** 8  **d.** 48

2. Which of the following equations have $-4$ as the solution?

   **a.** $-3=12x$  **b.** $\dfrac{x}{-2}=2$  **c.** $5x-1=3x-9$  **d.** $x=-4$

   Which of the above equations are equivalent?
   Which equation of the equivalent equations is the simplest form?

3. What operation with what number should be used on both sides of $\dfrac{r}{4}=20$ to get the equation $r=80$?

Solve and check:

4. **a.** $n+18=31$  **b.** $6+x=25$  **c.** $8.1=s+1.7$
   **d.** $6+n=9\frac{1}{6}$  **e.** $y+40=14$  **f.** $d+12=-12$

5. **a.** $r-7=3$  **b.** $x-2.4=9$  **c.** $0=b-16$
   **d.** $8\frac{2}{3}=n-9\frac{1}{2}$  **e.** $g-32=-13$  **f.** $-25=y-11$

6. **a.** $15y=90$  **b.** $-72=9s$  **c.** $-20b=8$
   **d.** $.08a=\$32$  **e.** $-6r=-54$  **f.** $\frac{3}{4}n=\$6.48$

7. **a.** $\dfrac{n}{8}=32$  **b.** $0=\dfrac{s}{5}$  **c.** $\dfrac{m}{23}=1$  **d.** $\dfrac{r}{14}=-7$

   **e.** $15=\dfrac{d}{15}$  **f.** $\dfrac{b}{-11}=5$  **g.** $-10=\dfrac{c}{-10}$  **h.** $\dfrac{y}{.6}=1.2$

# Solving Equations

## I. Aim    To solve simple equations.

## II. Procedure

1. Use either **a.** inverse operations, or **b.** the additive inverse and the multiplicative inverse. See sample solutions 3, 4, and 5.

2. When either or both members consist of two or more terms containing the variable, or two or more constant terms, combine like terms in the given members, then solve the resulting equivalent equation. See sample solutions 1 and 2.

3. When the variable has a fractional coefficient, solve the equation either by using the multiplicative inverse or by inverse operations. See sample solution 5.

4. When the equation contains terms with variables inclosed within parentheses, first remove the terms from the parentheses. See sample solutions 6 and 7.

5. To check your answer, substitute it for the variable in the given equation. If the resulting sentence is true, then your answer is a solution.

## III. Sample Solutions

1. Solve and check: $3n + 5n = 24$

   *Solution:*

   $3n + 5n = 28$

   $8n = 24$

   $$\frac{8n}{8} = \frac{24}{8}$$

   $n = 3$

   *Answer:* 3

   *Check:*

   $3n + 5n = 24$

   $(3 \cdot 3) + (5 \cdot 3) = 24\,?$

   $9 \;+\; 15 \;= 24$

   $24 = 24 \checkmark$

2. Solve and check: $7x = 39 - 4$

   *Solution*

   $7x = 39 - 4$

   $7x = 35$

   $$\frac{7x}{7} = \frac{35}{7}$$

   $x = 5$

   *Answer:* 5

   *Check:*

   $7x = 39 - 4$

   $7 \cdot 5 = 39 - 4\,?$

   $35 = 35 \checkmark$

**3.** Solve and check: $4x + 9 = 37$

**a.** *Solution:*

$$4x + 9 = 37$$
$$4x + 9 - 9 = 37 - 9$$
$$4x = 28$$
$$\frac{4x}{4} = \frac{28}{4}$$
$$x = 7$$

**b.** *Solution:*

$$4x + 9 = 37$$
$$4x + 9 + (-9) = 37 + (-9)$$
$$4x = 28$$
$$\frac{1}{4} \cdot 4x = \frac{1}{4} \cdot 28$$
$$x = 7$$

*Check:*

$$4x + 9 = 37$$
$$(4 \cdot 7) + 9 = 37?$$
$$28 + 9 = 37$$
$$37 = 37 ✔$$

*Answer:* 7

**4.** Solve and check: $6a + 8 = 62 - 3a$

**a.** *Solution:*

$$6a + 8 = 62 - 3a$$
$$6a + 8 - 8 = 62 - 8 - 3a$$
$$6a = 54 - 3a$$
$$6a + 3a = 54 - 3a + 3a$$
$$9a = 54$$
$$\frac{9a}{9} = \frac{54}{9}$$
$$a = 6$$

**b.** *Solution:*

$$6a + 8 = 62 - 3a$$
$$6a + 8 + (-8) = 62 + (-8) - 3a$$
$$6a = 54 - 3a$$
$$6a = 54 + (-3a)$$
$$6a + (3a) = 54 + (-3a) + (3a)$$
$$9a = 54$$
$$\frac{1}{9} \cdot 9a = \frac{1}{9} \cdot 54$$
$$a = 6$$

*Answer:* 6

*Check:*

$$6a + 8 = 62 - 3a$$
$$(6 \cdot 6) + 8 = 62 - 3 \cdot 6?$$
$$36 + 8 = 62 - 18$$
$$44 = 44 ✔$$

**5.** Solve and check: $\dfrac{2b}{3} = 10$

**a.** *Solution*

$$\frac{2b}{3} = 10$$
$$3 \cdot \frac{2b}{3} = 3 \cdot 10$$
$$2b = 30$$
$$\frac{2b}{2} = \frac{30}{2}$$
$$b = 15$$

**b.** *Solution:*

$$\frac{2b}{3} = 10$$
$$\frac{2}{3}b = 10$$
$$\frac{3}{2} \cdot \frac{2}{3}b = \frac{3}{2} \cdot 10$$
$$b = 15$$

*Check:*

$$\frac{2b}{3} = 10$$
$$\frac{2 \cdot 15}{3} = 10?$$
$$\frac{30}{3} = 10$$
$$10 = 10 ✔$$

*Answer:* 15

**6.** Solve and check:

$5x + 10(14 - x) = 95$

*Solution:*

$$5x + 10(14 - x) = 95$$
$$5x + 140 - 10x = 95$$
$$-5x + 140 = 95$$
$$-5x + 140 - 140 = 95 - 140$$
$$-5x = -45$$
$$\frac{-5x}{-5} = \frac{-45}{-5}$$
$$x = 9$$

*Answer:* 9

*Check:*

$$5x + 10(14 - x) = 95$$
$$5 \cdot 9 + 10(14 - 9) = 95?$$
$$45 + 10(5) = 95$$
$$45 + 50 = 95$$
$$95 = 95 ✔$$

**7.** Solve and check:

$(x + 6)(x - 4) = (x + 2)(x - 3)$

*Solution:*

$$(x + 6)(x - 4) = (x + 2)(x - 3)$$
$$x^2 + 2x - 24 = x^2 - x - 6$$
$$x^2 - x^2 + 2x + x - 24 =$$
$$x^2 - x^2 - x + x - 6$$
$$3x - 24 = -6$$
$$3x - 24 + 24 = -6 + 24$$
$$3x = 18$$
$$\frac{3x}{3} = \frac{18}{3}$$
$$x = 6$$

*Answer:* 6

*Check:* $(x + 6)(x - 4) = (x + 2)(x - 3)$
$$(6 + 6)(6 - 4) = (6 + 2)(6 - 3)?$$
$$(12)(2) = (8)(3)$$
$$24 = 24 ✔$$

## DIAGNOSTIC TEST

Solve and check:

1. $7x + 2x = 72$
2. $9y - 6y = 21$
3. $3x - x + 5x = 56$
4. $a + .02a = 510$
5. $\frac{7}{8}b - \frac{3}{8}b = 4$
6. $4m = 27 + 9$
7. $18x + 7 = 43$
8. $11 + 3b = 26$
9. $13 = 2w + 3$
10. $69 = 5 + 32t$
11. $8n + 15 = 25$
12. $3a - 4 = 11$
13. $8 = 5m - 12$
14. $6r - 24 = 0$
15. $1.8C + 32 = 158$
16. $\frac{3x}{4} = 12$
17. $5t = 15 + 2t$
18. $4x + 6 = 6x$
19. $8m - 35 = 3m$
20. $3z = 9z - 24$
21. $7x + 8 = 44 + 3x$
22. $10m - 25 = 2m - 9$
23. $3y + 15 = 67 - y$
24. $25 - 3s = 3s + 1$
25. $6y + 2 = 17 - 9$
26. $9t - 4t = 10 - 5t$
27. $x + x + 2 + x + 4 = 39$

28. $3x - 5x + 9 = 29 - 4x - 12$
29. $5n + 7n - 2 = 6 - 14n + 2n$
30. $8x + 6 - x = 4 + 3x + 2$
31. $4a + 33 = 12 - 3a$
32. $18 - (x - 4) = 9 + (x + 3)$
33. $30x + 40(15 - x) = 520$
34. $(b + 2)(b - 2) = (b + 8)(b - 3)$

## *RELATED PRACTICE EXAMPLES*

Solve and check:

**1.** **a.** $x + x = 10$    **b.** $4y + 7y = 66$    **c.** $8a + 2a = 50$
   **d.** $b + 4b = 15$    **e.** $8 = 2x + 2x$    **f.** $18 = 3b + 3b$
   **g.** $40 = 5y + 3y$    **h.** $28 = 6x + x$    **i.** $4x + 2x = 0$

**2.** **a.** $8x - 3x = 10$    **b.** $5y - y = 24$    **c.** $9b - 2b = 56$
   **d.** $3d - 2d = 4$    **e.** $6 = 10a - 4a$    **f.** $42 = 9x - 3x$
   **g.** $30 = 11y - 6y$    **h.** $1 = 8b - 7b$    **i.** $5 = 3r - r$

**3.** **a.** $2x + 3x + 4x = 36$   **b.** $5b + b + 2b = 64$   **c.** $21 = x + x + x$
   **d.** $10a - a - 2a = 35$   **e.** $48 = 21d - 3d - 2d$   **f.** $4x + x - 2x = 24$
   **g.** $9x - 3x + 2x = 96$   **h.** $2y + 2y - 3y = 10$   **i.** $60 = 7b - 3b + 8b$

**4.** **a.** $p + .15p = 69$    **b.** $c + .06c = \$74.20$    **c.** $a - .04a = 240$
   **d.** $2.3y + 1.6y = 117$    **e.** $l - .25l = \$.78$    **f.** $p + .03p = 61.8$

**5.** **a.** $\frac{1}{3}x + \frac{2}{3}x = 12$    **b.** $m + \frac{1}{4}m = 6\frac{1}{4}$    **c.** $\frac{7}{8}x = \frac{1}{8}x + 9$
   **d.** $10 = \frac{5}{6}b - \frac{1}{6}b$    **e.** $s - \frac{2}{5}s = 30$    **f.** $x + \frac{3}{4}x = 28$

**6.** **a.** $3x = 15 + 6$    **b.** $8y = 19 + 13$    **c.** $5 + 3 = 4c$
   **d.** $10 + 8 = 3b$    **e.** $6a = 18 - 6$    **f.** $8x = 17 - 9$
   **g.** $15y = 31 - 1$    **h.** $51 - 9 = 6m$    **i.** $8 - 6 = 4s$

**7.** **a.** $2x + 5 = 13$    **b.** $6b + 3 = 15$    **c.** $9d + 1 = 28$
   **d.** $12y + 8 = 56$    **e.** $5a + 4 = 29$    **f.** $8x + 7 = 15$
   **g.** $7b + 8 = 71$    **h.** $5m + 24 = 84$    **i.** $4r + 20 = 100$

**8.** **a.** $8 + 3r = 23$    **b.** $12 + 7x = 47$    **c.** $6 + 9y = 15$
   **d.** $5 + 6x = 35$    **e.** $32 + 5b = 47$    **f.** $45 + 2d = 69$
   **g.** $28 + 9x = 46$    **h.** $73 + 10b = 93$    **i.** $55 + 6m = 103$

**9.** **a.** $9 = 2x + 1$    **b.** $14 = 5a + 4$    **c.** $38 = 11y + 5$
   **d.** $56 = 7b + 21$    **e.** $19 = 4m + 15$    **f.** $35 = 8s + 11$
   **g.** $77 = 9r + 23$    **h.** $80 = 3x + 56$    **i.** $91 = 10d + 1$

**10.** **a.** $8 = 5 + 3x$    **b.** $44 = 9 + 7y$    **c.** $60 = 32 + 4a$
   **d.** $75 = 41 + 2m$    **e.** $82 = 27 + 11b$    **f.** $42 = 18 + 12d$
   **g.** $52 = 25 + 3r$    **h.** $61 = 36 + 5y$    **i.** $147 = 75 + 8s$

11. **a.** $2x + 7 = 10$    **b.** $3y + 9 = 16$    **c.** $5b + 12 = 15$
    **d.** $8a + 11 = 17$    **e.** $4x + 15 = 25$    **f.** $8 + 6m = 12$
    **g.** $45 = 12b + 24$    **h.** $64 = 4y + 59$    **i.** $48 = 18 + 9x$

12. **a.** $5y - 4 = 6$    **b.** $3a - 5 = 16$    **c.** $2b - 3 = 17$
    **d.** $8x - 16 = 24$    **e.** $7d - 32 = 31$    **f.** $9x - 57 = 15$
    **g.** $10y - 21 = 9$    **h.** $6x - 48 = 6$    **i.** $12m - 71 = 25$

13. **a.** $19 = 2x - 5$    **b.** $21 = 4b - 3$    **c.** $46 = 8y - 10$
    **d.** $53 = 5m - 7$    **e.** $33 = 2y - 5$    **f.** $23 = 6x - 67$
    **g.** $12 = 4a - 16$    **h.** $48 = 9r - 87$    **i.** $40 = 10x - 40$

14. **a.** $8x - 32 = 0$    **b.** $5r + 60 = 0$    **c.** $7a - 12 = 0$
    **d.** $9d - 6 = 0$    **e.** $4m + 3 = 0$    **f.** $18y - 12 = 0$
    **g.** $0 = 3b - 9$    **h.** $0 = 4x - 5$    **i.** $0 = 8d + 2$

15. **a.** $1.8c + 32 = 122$    **b.** $30 + .6t = 45$    **c.** $\$.05n - \$.12 = \$3.88$
    **d.** $1.8c + 32 = 194$    **e.** $200 + .25s = 325$    **f.** $.06p + 800 = 920$

16. **a.** $\dfrac{4d}{5} = 8$    **b.** $\dfrac{9a}{10} = 81$    **c.** $\dfrac{5x}{8} = 75$

    **d.** $56 = \dfrac{7c}{8}$    **e.** $72 = \dfrac{8n}{9}$    **f.** $\dfrac{3x}{16} = 96$

17. **a.** $4x = 8 + 2x$    **b.** $7a = 12 + a$    **c.** $6b = 36 - 3b$
    **d.** $9y = 4y + 30$    **e.** $8x = 65 - 5x$    **f.** $17n = 5n + 72$

18. **a.** $3a + 10 = 8a$    **b.** $5x + 6 = 7x$    **c.** $4b + 7 = 5b$
    **d.** $9m + 24 = 11m$    **e.** $240 + .4s = s$    **f.** $\$17 + .32s = s$

19. **a.** $6a - 20 = 2a$    **b.** $9x - 18 = 3x$    **c.** $5b - 48 = b$
    **d.** $12y - 35 = 7y$    **e.** $15x - 36 = 11x$    **f.** $21m - 63 = 12m$

20. **a.** $2b = 7b - 15$    **b.** $5a = 9a - 32$    **c.** $x = 7x - 78$
    **d.** $8d = 11d - 36$    **e.** $3x = 15x - 84$    **f.** $6y = 31y - 100$

21. **a.** $2x + 1 = 4 + x$    **b.** $5x + 3 = 15 + 2x$    **c.** $3 + 9y = 11 + 7y$
    **d.** $9 + 6z = z + 34$    **e.** $5x + 2 = 3x + 8$    **f.** $13n + 95 = 5n + 7$

22. **a.** $6r - 4 = 20 - 2r$    **b.** $5x - 2 = 28 - x$    **c.** $4d - 3 = 3d - 1$
    **d.** $10x - 22 = 8x - 10$    **e.** $6x - 3 = 15 - 3x$    **f.** $21y - 52 = 8y - 13$

23. **a.** $3x - 2 = 3 + 2x$    **b.** $9y - 5 = 7y + 3$    **c.** $z + 2 = 34 - 3z$
    **d.** $2x - 6 = x + 3$    **e.** $2 + 7x = 11 - 2x$    **f.** $7 + 11n = 97 - 7n$

24. **a.** $25 - 4x = 4x + 1$    **b.** $18 - 5x = 2x + 4$    **c.** $5n + 2 = 8n - 7$
    **d.** $8 + 2w = 5w - 13$    **e.** $-x + 17 = 4x + 2$    **f.** $40 - 8y = 13y - 23$

25. **a.** $2x + 3 = 7 + 6$    **b.** $5y - 2 = 17 - 4$    **c.** $8 + 7y = 12 + 24$
    **d.** $20 - 10 = 6b - 8$    **e.** $13 - 4 = 10x - 11$    **f.** $37 + 14 = 12n - 9$

26. **a.** $4r + 2r = 18 + 3r$    **b.** $7x + 2x = 8x + 4$    **c.** $-a + 3a = a + 6$
    **d.** $3x - 2x = 24 - x$    **e.** $4b - 3b = 5b - 28$    **f.** $27 + 7y = 6y + 4y$

27. **a.** $2x + 5x + 5 = 40$    **b.** $x + x + 10 = 20$
    **c.** $y + 2y + 3y + 30 = 360$    **d.** $x + x + 1 + x + 2 = 21$
    **e.** $77 - x - 2x - 6x - 5 = 0$    **f.** $s + s + 4 + s + s + 4 = 40$

g. $x + 3x + x + 3x = 32$

h. $0 = y + y + 2 - 3y + 5$

i. $a + a + 2 + a + 4 = 42$

j. $-5x + 8 - x + 9 - 2x - 11 = -34$

28. a. $3x + 5x + 3 = 5 + 7x$

b. $4y - y + 5 = 35 - 2y$

c. $8h + 5 - 2 = 6h + 17$

d. $-2x - 8 + 7x = 64 - 3x$

e. $4b + 2 + 3b - 4 = 21 + 5b - 17$

f. $6r - 4 + 2r + 8 = 18 - 3r - 8 + 9r$

g. $2y + 23 - 6 + y = 4y + 13 - 3y + 14$

h. $60 + 3x - 5x + 2 = 7x + 2x - 15 - 22$

i. $8y - 3y + 7 - 4y - 2$

$\quad = 15 - 9y - 5y + 5$

j. $-9n + 6 + 4n - 5$

$\quad = 10 - 2n - n + 7 - 3n$

29. a. $6x + 2 = 5$

b. $4r - 2r = 3$

c. $6b + 7 = 2b + 8$

d. $4 + 5m = 12m - 6$

e. $3x + x - 2 = 7 - 2x$

f. $5c - 3 + 4 = 9 - 2c + c$

g. $7y + y + 7 = 9 - y + 3y + 13$

h. $11a - 5a + 8 - 19 + 6a - 7 = 0$

i. $0 = 3x - 4 - x - 3 + 2x + 5 - 1$

j. $-7r + 9 - r - 1 = 2 - 2r - 9 + r$

30. a. $4m - 3 = 5m - 3$

b. $2 + 5x = x + 2$

c. $3b - 5 = 2 - 7$

d. $7x + 6 = 8 - x - 2$

e. $4x + 2 = 3 + 2x - 1$

f. $8 + 3m - 5 + m - 3 = 0$

g. $x + 3x - 2 = 2x - 8 + 3x + 6$

h. $2x + 7x + 5 - 3x = 3 + 4x + 2$

i. $9y - 4 + 2y - 3 + 2 - y + 5 = 0$

j. $-2n + 7 - 8n - 8 = -n + 6 + 2n - 7$

31. Solve and check.

a. $3x + 9 = 3$

b. $-d + 7 = 11$

c. $3a = 10 + 5a$

d. $-8n - 45 = 7n$

e. $9x - 2 = 3x - 10$

f. $3b + 12 = 9 - 6b$

g. $24 - 5t = 19 - 10t$

h. $5y = -6 - 4y$

i. $5 - m + 2m = 14 + 4m$

j. $3z + 6z + 25 = 4z - 31 - 2z$

32. a. $3x + (x + 2) = 10$

b. $y - (8 - y) = 2$

c. $(50 - b) - (3b + 2) = 0$

d. $12 + (4 - z) = 20 + z$

e. $x + (x + 1) + (x + 2) = 12$

f. $20 - (8 + y) - (y - 1) = 29$

g. $5m - (m + 3) = 7 + (m + 2)$

h. $x + (2x + 3) = (x + 3) + (x + 4)$

i. $(x - 1) - (x + 2) - (x - 3) = x$

j. $(3n - 5) + (-7n - 9)$

$\quad = (n - 8) - (-5n + 1)$

33. a. $4(2x + 6) = 48$

b. $2(x + 5) = x + 15$

c. $2(s + s + 3) = 14$

d. $5(x + 4) = 7(x + 2)$

e. $4(a - 1) = 5(a - 2)$

f. $5y + 10(8 - y) = 65$

g. $8 - 4(x - 1) = 2 + 3(4 - x)$

h. $3x + 2(x + 2) = 13 - (2x - 5)$

i. $4x + 5 - 3(2x - 5)$

$\quad = 9 - (8x - 1) + x$

j. $9m - 6(4m - 5) - 15$

$\quad = 5m + 7(-2m - 9)$

34. a. $x(x + 3) = x^2 + 9$

b. $x(x - 1) = x^2 - 2(x - 2)$

c. $2w(3w + 1) = 3w(2w + 1) - 2$

d. $(x + 2)(x + 2) = x^2 + 8$

e. $(a + 3)(a - 3) = a^2 + 2a - 35$

f. $(y - 4)(y - 1) - y^2 = 9$

g. $(x + 2)(x + 4) = x(x + 8)$

h. $(y - 6)(y - 2) = (y - 4)(y - 5)$

i. $(2n - 3)(3n + 1) = (6n - 3)(n - 1)$

j. $-2b(b - 5) = (b - 6)(b + 4) - 3b^2$

# Solving Equations with Absolute Values

**I. Aim**  To solve equations containing an absolute value of the variable.

**II. Procedure**

1. Use the equality axioms to simplify the given equation.
2. Since the absolute value of both a number and its opposite are the same arithmetic numbers, equations with absolute values of the variable will have two roots, except when the root is zero. See sample solution.
3. Check your answers by substituting each for the variable in the given equation. If the resulting sentence is true for the root, then it is a solution.

**III. Sample Solution**

Solve and check: $|y| + 6 = 14$

Solution:

$$|y| + 6 = 14$$
$$|y| + 6 - 6 = 14 - 6$$
$$|y| = 8$$
$$y = +8 \text{ or } -8$$

Answer: $+8, -8$

Check:

a. $|y| + 6 = 14$
$|+8| + 6 = 14$ ?
$8 + 6 = 14$
$14 = 14$ ✔

b. $|y| + 6 = 14$
$|-8| + 6 = 14$ ?
$8 + 6 = 14$
$14 = 14$ ✔

## DIAGNOSTIC TEST

1. Find the solutions of: **a.** $|x| = 65$  **b.** $|a| = 126$
2. Find the solutions of: **a.** $|y| - 9 = 17$  **b.** $56 = 7 \times |n|$

## RELATED PRACTICE EXAMPLES

Find the solutions of each of the following:

1. **a.** $|x| = 3$  **b.** $|y| = 7$  **c.** $|t| = 29$  **d.** $|r| = 0$
   **e.** $|n| = 31$  **f.** $46 = |d|$  **g.** $|v| = 100$  **h.** $|z| = 750$

2. **a.** $|p| + 5 = 11$  **b.** $|m| - 3 = 13$  **c.** $9 \times |b| = 72$

   **d.** $41 + |r| = 86$  **e.** $15 \times |c| = 105$  **f.** $\dfrac{|n|}{2} = 17$

   **g.** $25 = |x| - 32$  **h.** $288 = 9 \times |w|$  **i.** $6 = \dfrac{|y|}{24}$

230

# Evaluation of Formulas

**I. Aim**   To determine the value of any variable in a formula when the values of the other variables are known.

**II. Procedure**

**1.** Copy the formula.   **2.** Substitute the given values for the variables.
**3.** Perform the necessary operations.
**4.** Then solve the resulting equation for the value of the required variable.

**III. Sample Solutions**

**1.** Find the value of $t$ if $d = 80$ and $r = 20$, using the formula $d = rt$.

$$d = rt$$
$$80 = 20t$$
$$4 = t$$
$$t = 4 \qquad \textit{Answer: } t = 4$$

**2.** Find the value of $w$ if $V = 96$, $l = 8$, and $h = 4$, using the formula $V = lwh$.

$$V = lwh$$
$$96 = 8 \cdot w \cdot 4$$
$$96 = 32w$$
$$3 = w$$
$$w = 3 \qquad \textit{Answer: } w = 3$$

**3.** Find the value of $E$ if $I = 11$ and $R = 20$, using the formula $I = \dfrac{E}{R}$.

$$I = \frac{E}{R}$$
$$11 = \frac{E}{20}$$
$$220 = E$$
$$E = 220 \qquad \textit{Answer: } E = 220$$

**4.** Find the value of $r$ if $A = 54$, $p = 50$, and $t = 4$, using the formula $A = p + prt$.

$$A = p + prt$$
$$54 = 50 + 50 \cdot r \cdot 4$$
$$54 = 50 + 200r$$
$$54 - 50 = 50 - 50 + 200r$$
$$4 = 200r$$
$$.02 = r$$
$$r = .02 \text{ or } 2\% \quad \textit{Answer: } r = .02 \text{ or } 2\%$$

## DIAGNOSTIC TEST

Find the value of:

**1.** $w$ when $A = 40$ and $l = 8$, using the formula $A = lw$

**2.** $t$ when $i = 8$, $p = 50$, and $r = .04$, using the formula $i = prt$

**3.** $r$ when $d = 204$ and $t = 6$, using the formula $d = rt$

**4.** $l$ when $V = 672$, $w = 6$, and $h = 8$, using the formula $V = lwh$

**5.** $r$ when $i = 21$, $p = 60$, and $t = 7$, using the formula $i = prt$

**6.** $c$ when $s = 51$ and $g = 14$, using the formula $s = c + g$

**7.** $i$ when $A = 104$ and $p = 95$, using the formula $A = p + i$

**8.** $c$ when $l = 15$ and $s = 69$, using the formula $l = c - s$

**9.** $a$ when $l = 25$, $n = 8$, and $d = 3$, using the formula $l = a + (n - 1)d$

**10.** $l$ when $p = 38$ and $w = 7$, using the formula $p = 2l + 2w$

**11.** $t$ when $A = 88$, $p = 80$, and $r = .02$, using the formula $A = p + prt$

**12.** $E$ when $I = 4$ and $R = 55$, using the formula $I = \dfrac{E}{R}$

**13.** $a$ when $A = 18$ and $b = 9$, using the formula $A = \dfrac{ab}{2}$

**14.** $p$ when $A = 230$, $r = .05$, and $t = 3$, using the formula $A = p + prt$

**15.** $r$ when $T = 36$, $R = 150$, and $t = 24$, using the formula $TR = tr$

**16.** $C$ when $F = 4$, using the formula $F = 1.8C + 32$

**17.** $r$ when $a = 15$ and $S = 20$, using the formula $a = S(1 - r)$

**18.** $I$ when $E = 126$, $r = 6$, and $R = 12$, using the formula $E = I(r + R)$

## RELATED PRACTICE EXAMPLES

Find the value of:

1. **a.** $s$ when $p = 54$, using the formula $p = 6s$

   **b.** $s$ when $p = 60$, using the formula $p = 4s$

   **c.** $r$ when $d = 38$, using the formula $d = 2r$

   **d.** $s$ when $p = 21$, using the formula $p = 3s$

   **e.** $s$ when $p = 80$, using the formula $p = 5s$

   **f.** $w$ when $A = 48$ and $l = 8$, using the formula $A = lw$

   **g.** $d$ when $c = 21.98$ and $\pi = 3.14$, using the formula $c = \pi d$

   **h.** $b$ when $A = 65$ and $a = 13$, using the formula $A = ab$

   **i.** $s$ when $p = 132$ and $n = 12$, using the formula $p = ns$

    **j.**  $h$ when $V = 120$ and $B = 15$, using the formula $V = Bh$
    **k.**  $p$ when $c = 56$ and $n = 14$, using the formula $c = np$
    **l.**  $r$ when $p = 9$ and $b = 36$, using the formula $p = br$
    **m.** $t$ when $d = 72$ and $r = 18$, using the formula $d = rt$
    **n.**  $R$ when $E = 220$ and $I = 5$, using the formula $E = IR$
    **o.**  $E$ when $W = 300$ and $I = 4$, using the formula $W = IE$
    **p.**  $t$ when $v = 65$ and $a = 5$, using the formula $v = at$
    **q.**  $D$ when $P = 512$ and $H = 8$, using the formula $P = HD$
    **r.**  $R$ when $W = 270$ and $I = 3$, using the formula $W = I^2R$
    **s.**  $v$ when $P = 258$ and $F = 86$, using the formula $P = Fv$
    **t.**  $d$ when $W = 960$ and $F = 60$, using the formula $W = Fd$

**2.**  **a.**  $h$ when $V = 90$, $l = 6$, and $w = 3$, using the formula $V = lwh$
    **b.**  $t$ when $i = 12$, $p = 75$, and $r = .02$, using the formula $i = prt$
    **c.**  $r$ when $c = 43.96$ and $\pi = 3.14$, using the formula $c = 2\pi r$
    **d.**  r.p.s. when $T.S. = 748$, $\pi = \frac{22}{7}$, and $d = 7$, using the formula $T.S. = \pi d \times$ r.p.s.
    **e.**  $D$ when $F = 84$, $A = 3$, and $H = 4$, using the formula $F = AHD$
    **f.**  $b$ when $A = 75.36$, $\pi = 3.14$, and $a = 6$, using the formula $A = \pi ab$
    **g.**  $h$ when $V = 1{,}848$, $\pi = \frac{22}{7}$, and $r = 14$, using the formula $V = \pi r^2h$
    **h.**  $b$ when $A = 36$ and $a = 8$, using the formula $A = \frac{1}{2}ab$
    **i.**  $t$ when $i = 8.1$, $p = 54$, and $r = .03$, using the formula $i = prt$
    **j.**  $h$ when $V = 1{,}755$, $l = 15$, and $w = 13$, using the formula $V = lwh$

**3.**  **a.**  $l$ when $A = 85$ and $w = 5$, using the formula $A = lw$
    **b.**  $b$ when $A = 72$ and $h = 6$, using the formula $A = bh$
    **c.**  $n$ when $c = 112$ and $p = 14$, using the formula $c = np$
    **d.**  $I$ when $W = 55$ and $E = 220$, using the formula $W = IE$
    **e.**  $a$ when $A = 171$ and $b = 19$, using the formula $A = ab$
    **f.**  $I$ when $E = 110$ and $R = 5$, using the formula $E = IR$
    **g.**  $r$ when $d = 424$ and $t = 8$, using the formula $d = rt$
    **h.**  $B$ when $V = 108$ and $h = 9$, using the formula $V = Bh$
    **i.**  $n$ when $p = 56$ and $s = 7$, using the formula $p = ns$
    **j.**  $a$ when $v = 384$ and $t = 12$, using the formula $v = at$

**4.**  **a.**  $p$ when $i = 4$, $r = .05$, and $t = 2$, using the formula $i = prt$
    **b.**  $l$ when $V = 96$, $w = 2$, and $h = 6$, using the formula $V = lwh$
    **c.**  $A$ when $F = 14{,}400$, $H = 9$, and $D = 64$, using the formula $F = AHD$
    **d.**  $l$ when $V = 392$, $w = 8$, and $h = 7$, using the formula $V = lwh$
    **e.**  $p$ when $i = 33.60$, $r = .06$, and $t = 4$, using the formula $i = prt$

**5.**  **a.**  $w$ when $V = 270$, $l = 9$, and $h = 5$, using the formula $V = lwh$
    **b.**  $a$ when $A = 251.2$, $\pi = 3.14$, and $b = 8$, using the formula $A = \pi ab$
    **c.**  $r$ when $i = 60$, $p = 250$, and $t = 6$, using the formula $i = prt$

**d.** $H$ when $F = 875$, $A = 2$, and $D = 62\frac{1}{2}$, using the formula $F = AHD$

**e.** $d$ when $T.S. = 396$, $\pi = \frac{22}{7}$, and $r.p.s. = 21$, using the formula $T.S. = \pi d \times r.p.s.$

**6. a.** $p$ when $A = 46$ and $i = 9$, using the formula $A = p + i$

**b.** $c$ when $s = 54$ and $g = 21$, using the formula $s = c + g$

**c.** $C$ when $A = 312$, using the formula $A = C + 273$

**d.** $A$ when $B = 65$, using the formula $A + B = 90$

**e.** $D_w$ when $D_t = 98$ and $D_p = 36$, using the formula $D_t = D_w + D_p$

**f.** $L$ when $A = 7,540$ and $C = 3,850$, using the formula $A = L + C$

**g.** $M$ when $N = 93$, using the formula $M + N = 180$

**h.** $p$ when $m = 82$ and $e = 29$, using the formula $m = p + e$

**i.** $c$ when $s = 67$ and $m = 25$, using the formula $s = c + m$

**j.** $n$ when $l = 105$ and $d = 18$, using the formula $l = n + d$

**7. a.** $g$ when $s = 64$ and $c = 49$, using the formula $s = c + g$

**b.** $d$ when $l = 206$ and $n = 187$, using the formula $l = n + d$

**c.** $B$ when $A = 16$, using the formula $A + B = 90$

**d.** $i$ when $A = 342$ and $p = 315$, using the formula $A = p + i$

**e.** $c$ when $s = 570$ and $p = 491$, using the formula $s = p + c$

**f.** $e$ when $m = 43$ and $p = 28$, using the formula $m = p + e$

**g.** $C$ when $A = 9,138$ and $L = 1,252$, using the formula $A = L + C$

**h.** $D_p$ when $D_t = 116$ and $D_w = 79$, using the formula $D_t = D_w + D_p$

**i.** $N$ when $M = 135$, using the formula $M + N = 180$

**j.** $m$ when $s = 271$ and $c = 209$, using the formula $s = c + m$

**8. a.** $s$ when $g = 17$ and $c = 56$, using the formula $g = s - c$

**b.** $A$ when $i = 8$ and $p = 135$, using the formula $i = A - p$

**c.** $A$ when $C = 59$, using the formula $C = A - 273$

**d.** $l$ when $d = 26$ and $n = 184$, using the formula $d = l - n$

**e.** $A$ when $C = 3,575$ and $L = 962$, using the formula $C = A - L$

**f.** $A$ when $p = 400$ and $i = 18$, using the formula $p = A - i$

**g.** $c$ when $l = 33$ and $s = 114$, using the formula $l = c - s$

**h.** $m$ when $s = 820$ and $d = 65$, using the formula $s = m - d$

**i.** $s$ when $c = 46$ and $g = 13$, using the formula $c = s - g$

**j.** $s$ when $m = 394$ and $c = 52$, using the formula $m = s - c$

**9. a.** $b$ when $p = 19$ and $e = 7$, using the formula $p = b + 2e$

**b.** $V$ when $v = 216$, $g = 32$, and $t = 4$, using the formula $v = V + gt$

**c.** $a$ when $l = 23$, $n = 4$, and $d = 6$, using the formula $l = a + (n - 1)d$

**d.** $b$ when $p = 45$ and $e = 13$, using the formula $p = b + 2e$

**e.** $a$ when $l = 54$, $n = 10$, and $d = 5$, using the formula $l = a + (n - 1)d$

10.  **a.** $l$ when $p = 32$ and $w = 9$, using the formula $p = 2l + 2w$
     **b.** $C$ when $F = 77$, using the formula $F = 1.8C + 32$
     **c.** $e$ when $p = 63$ and $b = 17$, using the formula $p = b + 2e$
     **d.** $w$ when $p = 58$ and $l = 16$, using the formula $p = 2l + 2w$
     **e.** $t$ when $v = 169$, $V = 9$, and $g = 32$, using the formula $v = V + gt$
     **f.** $C$ when $F = 140$, using the formula $F = 1.8C + 32$
     **g.** $d$ when $l = 79$, $a = 25$, and $n = 10$, using the formula
       $l = a + (n - 1)d$
     **h.** $n$ when $S = 1,260$, using the formula $S = 180n - 360$
     **i.** $t$ when $v = 112$, $V = 16$, and $g = 32$, using the formula $v = V + gt$
     **j.** $l$ when $p = 32$ and $w = 9$, using the formula $p = 2l + 2w$

11.  **a.** $r$ when $A = 69$, $p = 60$, $t = 5$, using the formula $A = p + prt$
     **b.** $h$ when $A = 3,564$, $\pi = \frac{22}{7}$, and $r = 21$, using the formula
       $A = 2\pi r^2 + 2\pi rh$
     **c.** $l$ when $A = 440$, $\pi = \frac{22}{7}$, and $r = 7$, using the formula $A = \pi r^2 + \pi rl$
     **d.** $t$ when $A = 54$, $p = 40$, and $t = .05$, using the formula $A = p + prt$
     **e.** $R$ when $E = 75$, $I = 3$, and $r = 9$, using the formula $E = Ir + IR$

12.  **a.** $d$ when $r = 8$, using the formula $r = \dfrac{d}{2}$

     **b.** $s$ when $m = 6$, using the formula $m = \dfrac{s}{60}$

     **c.** $w$ when $H.P. = 3$, using the formula $H.P. = \dfrac{w}{746}$

     **d.** $c$ when $d = 13$, using the formula $d = \dfrac{c}{3.14}$

     **e.** $i$ when $m = 4$, using the formula $m = \dfrac{i}{39.37}$

     **f.** $d$ when $r = 8$ and $t = 5$, using the formula $r = \dfrac{d}{t}$

     **g.** $p$ when $r = .24$ and $b = 40$, using the formula $r = \dfrac{p}{b}$

     **h.** $m$ when $d = 8$ and $v = 12$, using the formula $d = \dfrac{m}{v}$

     **i.** $E$ when $R = 11$ and $I = 10$, using the formula $R = \dfrac{E}{I}$

     **j.** $F$ when $P = 50$ and $A = .3$, using the formula $P = \dfrac{F}{A}$

     **k.** $v$ when $a = 54$ and $t = 9$, using the formula $a = \dfrac{v}{t}$

l.  $W$ when $P = 4{,}000$ and $t = 5$,        m.  $s$ when $R = 8$ and $c = 6$,

   using the formula $P = \dfrac{W}{t}$              using the formula $R = \dfrac{s}{c}$

n.  $W$ when $L = 7$ and $A = 250$, using the formula $L = \dfrac{W}{A}$

o.  $W$ when $E = 110$ and $I = 1.5$, using the formula $E = \dfrac{W}{I}$

13.  a.  $h$ when $V = 28$ and $B = 7$,        b.  $b$ when $A = 16$ and $a = 8$,

   using the formula $V = \dfrac{Bh}{3}$              using the formula $A = \dfrac{ab}{2}$

c.  $V$ when $H.P. = 24$ and $T = 88$, using the formula $H.P. = \dfrac{TV}{550}$

d.  $N$ when $S = 4{,}400$, $\pi = \tfrac{22}{7}$, and $d = 21$, using the formula $S = \dfrac{\pi dN}{12}$

e.  $S$ when $L = 640$, $C_L = .64$, $d = .001$, and $V = 100$, using the formula
   $L = C_L \dfrac{d}{2} SV^2$

f.  $d$ when $P = 7{,}500$, $F = 250$, and $t = 10$, using the formula $P = \dfrac{Fd}{t}$

g.  $h$ when $V = 628$, $\pi = 3.14$, and $d = 8$, using the formula $V = \dfrac{\pi d^2 h}{4}$

h.  $W$ when $K.E. = 40$, $v = 16$, and $g = 32$, using the formula $K.E. = \dfrac{Wv^2}{2g}$

i.  $W$ when $F = 50$, $v = 20$, $g = 32$, and $r = 6$, using the formula
   $F = \dfrac{Wv^2}{gr}$

j.  $S$ when $D = 300$, $C_D = .025$, $d = .002$, and $V = 200$, using the formula
   $D = C_D \dfrac{d}{2} SV^2$

14.  a.  $p$ when $A = 186$, $r = .04$, and $t = 6$, using the formula $A = p + prt$
    b.  $I$ when $E = 52$, $r = 6$, and $R = 7$, using the formula $E = Ir + IR$
    c.  $S$ when $a = 16$ and $r = \tfrac{1}{2}$, using the formula $a = S - Sr$
    d.  $h$ when $A = 91$, $b = 18$, and $b' = 8$, using the formula $2A = bh + b'h$
    e.  $p$ when $A = 57.50$, $r = .05$, and $t = 3$, using the formula $A = p + prt$

15.  a.  $L$ when $W = 80$, $w = 60$, and $l = 12$, using the formula $WL = wl$
    b.  $P_1$ when $P_2 = 42$, $T_1 = 15$, and $T_2 = 35$, using the formula
       $P_1 T_2 = T_1 P_2$
    c.  $r$ when $D = 24$, $R = 250$, and $d = 20$, using the formula $DR = dr$
    d.  $V$ when $P = 42$, $P' = 36$, and $V' = 14$, using the formula $PV = P'V'$
    e.  $T_1$ when $V_1 = 18$, $V_2 = 8$, $T_2 = 16$, using the formula $V_1 T_2 = T_1 V_2$

16. **a.** $a$ when $v = -120$ and $t = 6$, using the formula $v = at$
   **b.** $C$ when $A = 84$, using the formula $A = C + 273$
   **c.** $C$ when $F = -76$, using the formula $F = 1.8C + 32$
   **d.** $A$ when $C = -350$ and $L = 925$, using the formula $C = A - L$
   **e.** $C$ when $F = 14$, using the formula $F = 1.8C + 32$

17. **a.** $l$ when $p = 174$ and $w = 39$, using the formula $p = 2(l + w)$
   **b.** $n$ when $l = 25$, $a = 5$, and $d = 2$, using the formula
   $l = a + (n - 1)d$
   **c.** $B$ when $A = 57$ and $C = 74$, using the formula $C = 180 - (A + B)$
   **d.** $R$ when $E = 78$, $I = 6$, and $r = 5$, using the formula $E = I(r + R)$
   **e.** $t$ when $A = 448$, $p = 350$, and $r = .04$, using the formula
   $A = p(1 + rt)$

18. **a.** $I$ when $E = 112$, $r = 5$, and $R = 9$, using the formula $E = I(r + R)$
   **b.** $S$ when $a = 6$ and $r = \frac{1}{2}$, using the formula $a = S(1 - r)$
   **c.** $p$ when $A = 592$, $r = .06$, and $t = 8$, using the formula $A = p(1 + rt)$
   **d.** $h$ when $A = 176$, $b = 18$, and $b' = 14$, using the formula
   $$A = h\left(\frac{b + b'}{2}\right)$$
   **e.** $p$ when $A = 812$, $r = .05$, and $t = 9$, using the formula $A = p(1 + rt)$

## CUMULATIVE REVIEW

Solve and check:
1. **a.** $8r + 19 = 67$    **b.** $4n = 11n - 35$    **c.** $p + .06p = 848$
   **d.** $0 = 12x + 9$    **e.** $9c - 8 = 10c$    **f.** $15d - 7 = 20 + 6d$
   **g.** $8b - 6 - 2b = 45 - b + 5$    **h.** $10t + 25(50 - t) = 980$
   **i.** $(11x - 8) - (20 - 3x) = 7(x - 3)$    **j.** $(n + 13)(n + 2) = (n + 8)(n - 8)$
2. **a.** $|x| = 8$    **b.** $|w| + 9 = 31$    **c.** $|r| - 17 = 23$    **d.** $16 \times |n| = 64$

Find the value of:
3. $E$ when $W = 550$ and $I = 110$, using the formula $W = IE$
4. $a$ when $l = 124$, $n = 15$, and $d = 8$, using the formula $l = a + (n - 1)d$
5. $V$ when $v = 413$, $g = 32$, and $t = 11$, using the formula $v = V + gt$
6. $m$ when $d = 3.2$ and $v = 7$, using the formula $d = \dfrac{m}{v}$
7. $r$ when $a = 12$ and $S = 16$, using the formula $a = S(1 - r)$
8. $W$ when $K.E. = 350$, $v = 20$, and $g = 32$, using the formula
   $$K.E. = \frac{Wv^2}{2g}$$
9. $C$ when $F = -49$, using the formula $F = 1.8C + 32$
10. $I$ when $E = 147$, $r = 8$, and $R = 13$, using the formula $E = Ir + IR$

# Graphing an Equation in One Variable on the Number Line

## I. Aim

To draw the graph of an equation in one variable on the number line.

## II. Procedure

1. Solve the given equation to find its solution or solutions.
2. Then on a number line locate the point or points whose coordinate or coordinates are numbers forming all the solutions to the given equation.
3. Indicate these points by solid or colored dots.
4. Observe that:
   a. The graph of an equation in one variable on the number line is the collection of all points on the number line whose coordinates are the numbers forming the solutions of the equation. Also see Exercise 2-7.
   b. The graph of an equation is the graph of its solutions.
5. a. Where the solution is one number, the graph will consist of one point.
   b. Where there are two solutions, the graph will consist of two points. See sample solution 4.
   c. When the solution of the given equation is an infinite number of numbers, its graph is the entire number line. It is indicated by a heavy solid or colored line with an arrowhead in each direction to show that it is endless. See sample solution 5.
6. When required to write an equation corresponding to a graph, use the simplest form of the equation or any other equivalent equation. See sample solution 6.

## III. Sample Solutions

1. Draw the graph of $x = 5$. The solution of the equation $x = 5$ is 5. The graph of $x = 5$ is the point whose coordinate is 5.

2. Draw the graph of $4x + 7 = 19$. The solution of the equation $4x + 7 = 19$ is 3. The graph of $4x + 7 = 19$ is the point whose coordinate is 3.

*Answer:*

*Answer:*

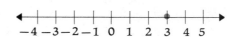

**3.** Draw the graph of $x = x + 2$. There is no solution of the equation $x = x + 2$.

**4.** Draw the graph of $|n| = 1$. The solutions of the equation $|n| = 1$ are $-1$ and 1. The graph of $|n| = 1$ are the points whose coordinates are $-1$ and 1.

*Answer:* We have no points to draw the graph.

*Answer:*

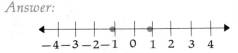

**5.** Draw the graph of $2x + 4 = 2(x + 2)$. The solutions of the equation $2x + 4 = 2(x + 2)$ are all the real numbers, since this equation (identity) is satisfied by every real number. The graph of $2x + 4 = 2(x + 2)$ is the entire number line.

*Answer:*

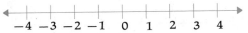

**6.** Write a corresponding equation which is pictured by the graph:

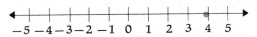

*Answer:* The above graph is a picture of the basic equation $x = 4$ or any equivalent equation, such as $5x = 20$, $x + 9 = 13$, $x - 1 = 3$, $\frac{x}{2} = 2$, etc.

## DIAGNOSTIC TEST

**1.** Write the coordinates of which the following is the graph:

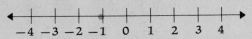

**2.** Draw the graph of each of the following groups of numbers on a number line.

    **a.** 1      **b.** $-4$      **c.** $-2, 2$      **d.** 6

For each of the following equations draw the appropriate number line, then graph its solution. The replacements for the variables are all the real numbers.

**3.** $y = -3$        **4.** $x - 7 = 0$        **5.** $6y + 19 = -5$

**6.** $|x| + 4 = 8$                **7.** $5(3x + 4) = 20 + 15x$

**8.** $n = n - 2$           **9.** $2(x - 3) = 2(3 - x)$

**10.** Write a corresponding equation which is pictured by the following graph:

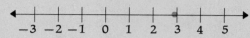

## RELATED PRACTICE EXAMPLES

**1.** Write the coordinates of which each of the following is the graph:

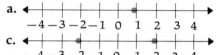

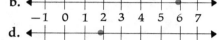

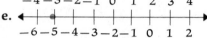

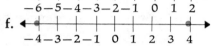

**2.** Draw the graph of each of the following groups of numbers on a number line.

     **a.** 4       **b.** $-2$       **c.** 0       **d.** $-7$       **e.** $-3, 3$
     **f.** 2       **g.** 8        **h.** $-9$      **i.** $-1$       **j.** $-5, 5$

For each of the following equations draw an appropriate number line, then graph its solution. The replacements for the variables are all the real numbers.

**3. a.** $n = 6$         **4. a.** $3y = 12$        **5. a.** $2n + 12 = 18$
    **b.** $y = -1$           **b.** $2 = x + 2$         **b.** $3y - 5 = 1$
    **c.** $b = 0$            **c.** $n - 5 = 3$         **c.** $x - 5x = -24$
    **d.** $r = -4$          **d.** $p + 8 = 2$         **d.** $0 = 21 + 7y$

    **e.** $x = -5$          **e.** $\dfrac{x}{-3} = 1$         **e.** $6w - 8 = -50$

**6. a.** $|x| = 4$           **b.** $|y| + 1 = 6$        **c.** $|t| - 3 = 0$

    **d.** $5 \times |n| = 5$       **e.** $\dfrac{|b|}{6} = 2$

**7. a.** $4x + 5 = 5 + 4x$           **b.** $2(3y + 1) = 6y + 2$
    **c.** $8w - 12 = 4(2w - 3)$       **d.** $6(4n + 3) = 3(6 + 8n)$
    **e.** $3(16 - 8x) = 8(6 - 3x)$      **f.** $5(7x - 14) = 7(5x - 10)$

**8. a.** $n + 2 = x$         **b.** $y - 3 = y$          **c.** $n = 7 + n$
    **d.** $r = r - 1$          **e.** $8 + w = w$         **f.** $z = z - 5$

9. **a.** $8x + 9 = 4x + 25$      **b.** $2n - 7n = 3n - 16$
   **c.** $4y - 8 - y = 20 - y$    **d.** $9m - 5 + m = 18 - 2m - 11$
   **e.** $25 - 2(3y - 8) = -7$    **f.** $(x - 1)(x + 9) = (x - 3)(x + 8)$

10. Write a corresponding equation which is pictured by each of the following graphs:

**a.**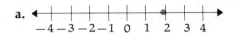

**b.**

**c.**

**d.**

**e.**

**f.**

## COMPETENCY CHECK TEST

In each of the following, select the letter corresponding to your answer:
1. The equation which has $-2$ as its solution is:
   **a.** $3x + 4 = 10$    **b.** $x - 2 = 4$      **c.** $1 = x - 1$      **d.** $4x + 11 = 3$
2. The equation that is equivalent to $n = 3$ is:
   **a.** $6n + 3 = 3$    **b.** $5n - 4 = 19$    **c.** $\dfrac{n}{-3} = -1$    **d.** $8n - 2n = 21$
3. The solution of $.08n = \$96$ is:
   **a.** $\$12$      **b.** $\$1.20$      **c.** $\$120$      **d.** $\$1,200$
4. The solution of $6x - 5 = 49$ is: **a.** $9$   **b.** $7$   **c.** $-5$   **d.** $60$
5. The solution of $p - .06p = 235$ is: **a.** $200$   **b.** $150$   **c.** $250$   **d.** $300$
6. The solution of $7x - 4(x - 6) = 36$ is: **a.** $2$   **b.** $-6$   **c.** $-8$   **d.** $4$
7. When the formula $p = 3s$ is used and $p = 27$, the value of $s$ is:
   **a.** $81$      **b.** $24$      **c.** $30$      **d.** $9$
8. When the formula $F = 1.8C + 32$ is used and $F = 14$, the value of $C$ is:
   **a.** $50$      **b.** $57.2$      **c.** $0$      **d.** $-10$
9. When the formula $v = V + gt$ is used and $v = 210$, $V = 18$, and $g = 32$, the value of $t$ is: **a.** $8$   **b.** $7$   **c.** $6$   **d.** $5$
10. When the formula $E = I(R + r)$ is used and $E = 108$, $I = 9$, and $r = 4$, the value of $R$ is: **a.** $36$   **b.** $8$   **c.** $13$   **d.** $121$
11. The graph of the equation $3x - 2 = 10$ is:

**a.**     **b.**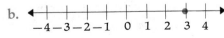

**c.** 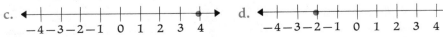    **d.**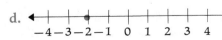

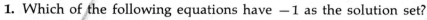

1. Which of the following equations have $-1$ as the solution set?
   **a.** $3x + 15 = 12$      **b.** $4x - 3 = 1$      **c.** $2x = x - 1$
   **d.** $x = -1$
   Which of the above equations are equivalent?
   Which equation of the equivalent equations is the simplest form?

Solve and check:

2. **a.** $x + 8 = 21$             **b.** $7 + p = 0$
   **c.** $3 = n + 11$

3. **a.** $y - 10 = 0$         **b.** $14 = w - 14$
   **c.** $x - 6 = -9$

4. **a.** $24m = -40$         **b.** $\frac{3}{5}c = 75$
   **c.** $-52 = -4r$

5. **a.** $\dfrac{b}{4} = 12$          **b.** $-5 = \dfrac{x}{8}$

   **c.** $\dfrac{t}{-2} = 10$

6. **a.** $11x + 36 = 3$      **b.** $6y - 17 = 8y + 9$
   **c.** $18 - (5 - 3n) = 2(6n - 7)$

7. Find the solution of:    **a.** $|x| = 15$    **b.** $|n| + 7 = 20$

Find the value of:

8. $w$ when $p = 278$ and $l = 83$, using the formula $p = 2l + 2w$

9. $F$ when $P = 36$, $d = 8$, and $t = 12$, using the formula $P = \dfrac{Fd}{t}$

10. $R$ when $E = 143$, $I = 11$, and $r = 4$, using the formula
    $E = Ir + IR$

11. $C$ when $F = -148$, using the formula $F = 1.8C + 32$

12. $n$ when $S = 540$, using the formula $S = 180(n - 2)$

13. $p$ when $A = 1{,}290$, $r = .06$, and $t = 12$, using the formula
    $A = p(1 + rt)$

14. For each of the following equations, draw an appropriate number
    line, then graph its solution. The replacements for the variable
    are all the real numbers.
    **a.** $12n = -36$               **b.** $5x - 8x = -15$
    **c.** $7r + 2 = 3r - 18$       **d.** $|x| - 3 = 1$
    **e.** $w + 8 = w$             **f.** $2x + 9 = 9 + 2x$

## Keyed Achievement Test

The colored numeral indicates the exercise where help may be found.

1. Find the sum of $-7a^2 + 3ab - b^2$, $4a^2 - 9b^2$, and $-9ab - 3b^2$. [5-3]
2. Simplify: $8n^2 - 4n + 6 - 2n - 7 - n^2 + 5n - 2n^2 - 3 - 4n^2$. [5-3]
3. Subtract $10x^2 - 11xy$ from $10xy - y^2$. [5-5]
4. Multiply $6b^2 - 4bc - c^2$ by $3b^2 - 5bc + 2c^2$. [5-9]
5. Divide $m^6 - 64n^6$ by $m - 2n$. [5-13]
6. Multiply the sum of $3b - 8$ and $-5b + 7$ by the difference between $-9b + 10$ and $-4b - 5$. [5-3, 5-5, 5-9]
7. Solve and check each of the following equations when the replacements for the variable are all the real numbers: [6-4 thru 6-8]
   **a.** $n + 15 = 81$   **b.** $8b = 56$   **c.** $x - 7 = -10$   **d.** $9x + x = 5$
   **e.** $\dfrac{x}{7} = 49$     **f.** $53 + 12s = 17$       **g.** $6x - 13 = 3x + 20$
   **h.** $25y + 5(18 - y) = 290$     **i.** $16 - (4n - 9) = 5 + (2n + 8)$
   **j.** $(x - 4)(x + 8) = (x + 2)(x + 1)$
8. Find the value of: [6-10]  **a.** $d$ when $l = 89$, $a = 19$, and $n = 11$, using the formula $l = a + (n - 1)d$.
   **b.** $W$ when $F = 216$, $v = 60$, $g = 32$, and $r = 25$, using the formula $F = \dfrac{Wv^2}{gr}$.
9. For each of the following equations draw an appropriate number line, then graph its solution. The replacements for the variable are all the real numbers. [6-11]
   **a.** $9n + 5 = 23$   **b.** $a = 10 - a$   **c.** $8x + 7 = 7 + 8x$
   **d.** $c = c + 4$     **e.** $|y| - 5 = 1$   **f.** $(4x - 1) - 2(5x - 4) = 19$

## MAINTENANCE PRACTICE IN ARITHMETIC

1. Add:
   839,175
   94,658
   478,966

2. Subtract:
   6,700,000
   3,947,098

3. Multiply:
   9,076
   8,065

4. Divide:
   $5,906 \overline{)35,932,104}$

5. Add:
   $6\frac{7}{10} + 9\frac{1}{6}$

6. Subtract: $7\frac{1}{4} - 3\frac{4}{5}$    7. Multiply: $3\frac{1}{7} \times 4\frac{3}{4}$      8. Divide: $\frac{13}{15} \div 1\frac{5}{8}$

9. Add: $.8 + .088 + 8.8$  10. Subtract: $\$25 - \$16.84$

11. Multiply: $1\frac{5}{8} \times \$4.56$  12. Divide: $3.5 \overline{).007}$       13. Find 17% of 629.

14. $74 is what per cent $185?       15. 12% of what amount is $30?

# INTRODUCTION

## *The Equation Method*

To solve verbal problems by algebraic methods, use the following general plan:

1. First read the problem carefully to determine what is required and any facts which are related to this unknown value.
2. When there is only one required value, represent it by a variable, usually a letter. When there are two required values, represent the smaller unknown value by some letter and express the other unknown value in terms of that letter.
3. Then find two facts, at least one fact involving the unknown value, which are equal to each other.
4. Translate these facts into algebraic expressions and form an equation by writing one algebraic expression equal to the other.
5. Solve the equation to find the required value.
6. Check the answer directly with the facts of the given problem.

In this chapter skills first are developed both in representing algebraic expressions by unique methods and in translating verbal sentences into equations. This preliminary instruction is the first part of a step-by-step development in the problem-solving techniques used in this text. Introductory problem representations and elementary problems are also provided with diagrams included where helpful. Problems involving the use of formulas and their solutions also are treated extensively.

# Problem Solving

**7**

# Algebraic Representation in Problems

**I. Aim**    To represent by algebraic expressions the quantity or quantities to be determined, often called the unknown.

**II. Procedure**

1. When only one unknown value requires representation, select a variable (any letter) to represent it.
2. When there are two related quantities that require algebraic representation, select a variable (any letter) to represent one of the quantities and represent the second quantity in related terms.
   Note that there are three types of representations:
   **a.** One quantity is a given number times the second quantity. See sample solution 1.
   **b.** One quantity is a given number more or less than the second quantity. See sample solution 2.
   **c.** The sum of the two quantities is given. See sample solution 3.

**III. Sample Solutions**

Represent the two unknown quantities in each of the following by related algebraic expressions.

1. One number is 8 times a second number.

   *Answer:* Let $x$ = second number
   and $8x$ = first number    OR    Let $x$ = first number
   and $\frac{1}{8}x$ = second number

2. One number is 8 more than a second number.

   *Answer:* Let $x$ = second number
   and $x + 8$ = first number    OR    Let $x$ = first number
   and $x - 8$ = second number

3. The sum of two numbers is 8.

   *Answer:* Let $x$ = first number
   and $8 - x$ = second number    OR    Let $x$ = second number
   and $8 - x$ = first number

## DIAGNOSTIC TEST

Represent the two unknown quantities in each of the following by related algebraic expressions:

**1.** There are 4 times as many nickels as dimes.

**2.** Steve weighs 18 pounds more than Charlotte.

**3.** Scott and Peter together have $100.

## *RELATED PRACTICE EXAMPLES*

1. Copy each table, then find the missing numbers or representations:

   **a.** One number is 5 times a second number.

   | Second number | 4 | 8 | 11 | 20 | $x$ |
   |---|---|---|---|---|---|
   | First number | | | | | |

   **b.** Larry is twice as old as Anthony.

   | Anthony's age | 3 yr. | 7 yr. | 12 yr. | 35 yr. | $n$ yr. |
   |---|---|---|---|---|---|
   | Larry's age | | | | | |

   **c.** There are 6 times as many nickels as dimes.

   | No. of dimes | 5 | 8 | $n$ | | |
   |---|---|---|---|---|---|
   | No. of nickels | | | | 24 | $n$ |

   **d.** Car $A$ travels $\frac{1}{4}$ as fast as Car $B$.

   | Car $B$ | 60 m.p.h. | $r$ m.p.h. | | | |
   |---|---|---|---|---|---|
   | Car $A$ | | | 18 m.p.h. | 13 m.p.h. | $r$ m.p.h. |

e. Represent the two unknown quantities in each of the following by related algebraic expressions:
   (1) One number is 7 times a second number.
   (2) David weighs 5 times as much as his baby brother.
   (3) There are 10 times as many 50¢ admission tickets as 25¢ admission tickets.
   (4) One side of a triangle is 8 times as long as a second side.
   (5) The speed of an airplane is 3 times that of a train.
   (6) Mrs. Jones is 4 times her daughter's age.
   (7) There are twice as many $5 bills as $1 bills.
   (8) The base of a rectangle is 4 times its altitude.
   (9) Bill is $\frac{1}{3}$ John's age.
   (10) There are $\frac{1}{8}$ as many 3¢ stamps as 2¢ stamps.

2. Copy each table, then find the missing numbers or representations:

   a. One number is 5 more than a second number.

| Second number | 1 | 4 | 12 | 25 | $n$ |
|---|---|---|---|---|---|
| First number | | | | | |

   b. Charlotte weighs 3 pounds more than Marilyn.

| Marilyn's weight | 80 lb. | $w$ lb. | | | |
|---|---|---|---|---|---|
| Charlotte's weight | | | 104 lb. | 126 lb. | $w$ lb. |

   c. Andrew had $25 less than Todd.

| Todd's money | $40 | $75 | $110 | $200 | $$m$ |
|---|---|---|---|---|---|
| Andrew's money | | | | | |

   d. Elaine is 2 years younger than Ed.

| Ed's age | 35 yr. | $x$ yr. | | | |
|---|---|---|---|---|---|
| Elaine's age | | | 29 yr. | 43 yr. | $x$ yr. |

   e. Represent the two unknown quantities in each of the following by related algebraic expressions:
      (1) The length of a rectangle is 7 inches greater than the width.

(2) There are 3 more dimes than quarters.
(3) Tony is 7 years older than Rosetta.
(4) One number is 15 more than a second number.
(5) The area of a rectangle is 8 sq. inches greater than the area of a square.
(6) There are 10 pounds more of the 60¢ candy than of the 50¢ candy.
(7) Bill has 4 cents more than Sue.
(8) Richard is 12 years younger than Carl.
(9) Jose weighs 40 pounds less than his father.
(10) Yale scored 25 points more than Harvard.

3. Copy each table, then find the missing numbers or representations:

   a. The sum of two numbers is 5.

   | First number | 2 | 4 | $n$ | | |
   |---|---|---|---|---|---|
   | Second number | | | | 1 | $n$ |

   b. A board 15 feet long is divided into two parts.

   | First part | 3 ft. | 8 ft. | $x$ ft. | | |
   |---|---|---|---|---|---|
   | Second part | | | | 9 ft. | $x$ ft. |

   c. There are 38 stamps; some are 15¢ stamps and the rest are 13¢ stamps.

   | No. 13¢ stamps | 7 | 18 | $x$ | | |
   |---|---|---|---|---|---|
   | No. 15¢ stamps | | | | 25 | $x$ |

   d. Represent the two unknown quantities in each of the following by related algebraic expressions:
      (1) The sum of two numbers is 25.
      (2) Two trains traveled a total distance of 150 miles.
      (3) There are 46 boys and girls in the algebra class.
      (4) Juanita and Carmella together weigh 210 pounds.
      (5) The sum of the ages of Joe and Tom is 33 years.
      (6) There were 200 pounds of $2.85 and $2.95 coffee.
      (7) A man left $1,500 to his son and daughter.
      (8) A 2-ton truck and 5-ton truck moved 40 tons.
      (9) The heights of Joan and Virginia add up to 128 inches.
      (10) Harry and Dick together sold 125 magazines.

# Generalized Representation

I. **Aim**     To generalize with representations.

II. **Procedure**

1. When variables are used instead of numerals in problems and a generalized answer is required, use these variables, proper symbols of operation, and given numerals to indicate the relationship involved. See sample solution 1.

2. When a variable represents a number of a specified number system, to determine whether a given algebraic expression containing the variable also represents a number in this number system:

   a. Select several numbers from the number system described in the problem. Use a wide range of values.

   b. Substitute each selected number for the variable in the given expression to determine whether the expression meets the requirements of the problem. See example solution 2.

III. **Sample Solutions**

1. Scott has $n$ dollars in the bank. If he deposits $x$ dollars, how many dollars does he then have in the bank?

   *Answer:* $n + x$ dollars

2. Does the expression $x - 5$ always represent a whole number when $x$ represents every whole number?
   *Solution:*
   When $x = 5$, then $x - 5 = 5 - 5 = 0$
   $\phantom{When }x = 6$, then $x - 5 = 6 - 5 = 1$, etc.
   Thus when $x$ is 5 or more, the expression $x - 5$ represents a whole number.
   When $x = 0$, then $x - 5 = 0 - 5 = -5$
   $\phantom{When }x = 4$, then $x - 5 = 4 - 5 = -1$
   Thus when $x$ is 0, 1, 2, 3, 4, the expression $x - 5$ is negative and does not represent a whole number.

   *Answer:* $x - 5$ does not always represent a whole number.

## DIAGNOSTIC TEST

1. Write the algebraic expression that represents the answer in the following problem:
   One of the two angles has a measure of $d$ degrees and the sum of their measures is $85°$. What is the measure of the second angle?

2. For all values of $x$ when $x$ represents every integer, which of the following expressions represent integers?

   **a.** $x + 6$ **b.** $8 - x$ **c.** $-4x$ **d.** $\dfrac{5x}{3}$

3. For all values of $s$ and $t$ when $s$ and $t$ each represent every whole number, which of the following expressions represent whole numbers?

   **a.** $s + t$ **b.** $s - t$ **c.** $3st$ **d.** $\dfrac{s + 5}{2s}$ (when $s \neq 0$)

4. For all values of $y$ when $y$ represents every negative integer, which of the following expressions represent positive integers?

   **a.** $y - 2$ **b.** $11 - 2y$ **c.** $-y$ **d.** $y + 12$

5. If $b$ represents an odd integer, which of the following expressions represent odd integers?

   **a.** $b + 1$ **b.** $b + 2$ **c.** $2b$ **d.** $3b - 1$

6. If $m$ represents an even integer, which of the following expressions represents the next consecutive even integer?

   **a.** $2m$ **b.** $m + 1$ **c.** $m + 2$ **d.** $2m - 2$

## RELATED PRACTICE EXAMPLES

1. In each of the following problems, write the algebraic expression that represents the answer:
   **a.** (1) Leon is $y$ years old. How old will he be in 6 years?
      (2) Julio is $r$ years old. How old will he be in $n$ years?
      (3) Lisa is 21 years old. How old was she $t$ years ago?
      (4) John is $b$ years old. How old was he $y$ years ago?
   **b.** (1) If one orange costs 5 cents, how much will $n$ oranges cost?
      (2) What is the cost of $m$ erasers when one eraser costs $c$ cents?
      (3) If $x$ books were bought for $d$ dollars, how much did one book cost?
      (4) If one pencil costs $p$ cents, how many pencils can you buy for $s$ cents?

**c.** (1) Debbie now has $10. If she earns *n* dollars, how many dollars does she then have?

    (2) Paul saves *d* dollars each week. How many weeks will it take him to save $200?

    (3) Teresa's brother earns *r* dollars per hour. If he works *h* hours per week, how much does he earn per week?

    (4) Ann earns *b* dollars per year. How much does she earn per month?

**d.** (1) Jerry's father receives a salary of *s* dollars per week. During a particular week he also received a commission of *c* dollars. How much did he earn in all during that week?

    (2) A merchant bought a radio for *t* dollars and sold it for *n* dollars. How much profit did he make?

    (3) Tom purchased a clock-radio at a sale for *d* dollars. It regularly sells for *m* dollars. How much did he save?

    (4) Cathy has *p* dollars in the bank. If the interest *i* is added to her bank account, how much money will she then have in her bank account?

**e.** (1) There are 30 pupils in the class. At the end of the school year *x* pupils were promoted. How many pupils were not promoted?

    (2) The school baseball team played *h* games. If the team lost *z* games, how many games did it win?

    (3) There are *n* students in the mathematics class. If there are *y* girls, how many boys are in the class?

    (4) If it snowed on *s* days during the month of January, on how many days during that month did it not snow?

**f.** (1) An automobile travels *r* miles per hour. How many miles can it go in 10 hours?

    (2) How many hours did it take an airplane to fly *m* miles if it averages 450 m.p.h.?

    (3) At what average rate of speed must a train travel to cover a distance of *d* miles in *t* hours?

    (4) How many miles can an automobile travel on *g* gallons of gasoline if it averages *m* miles per gallon?

**g.** (1) Joan has *x* books. Anne has 11 books more than 5 times as many books as Joan. How many books does Anne have?

    (2) Bill has *n* stamps. George has 6 stamps less than 4 times as many stamps as Bill. How many stamps does George have?

    (3) Angle *B* measures *x* degrees. Angle *A* measures 25° less than 7 times the measure of angle *B*. What is the measure of angle *A*?

    (4) Jane has *d* dollars. Her brother has $9 more than 3 times as many dollars as Jane. How many dollars does her brother have?

   **h.** (1) Elaine has $n$ \$10 bills and 7 times as many \$5 bills. How many \$10 and \$5 bills does she have in all?

   (2) Juan has $x$ 13¢ stamps and 9 times as many 15¢ stamps. How many stamps does he have in all?

   (3) Frank's mother bought $c$ oranges, twice as many apples as oranges, and 4 more peaches than oranges. How many pieces of fruit did she buy?

   (4) Tim has $y$ quarters, 6 more dimes than quarters, and 3 times as many nickels as quarters. How many coins (quarters, dimes, and nickels) does he have in all?

**2. a.** For all values of $n$ when $n$ represents every whole number, which of the following expressions represent whole numbers?

   (1) $n + 8$     (2) $n - 3$     (3) $4n$          (4) $\dfrac{n}{5}$

**b.** For all values of $x$ when $x$ represents every integer, which of the following expressions represent integers?

   (1) $x + 10$     (2) $x - 2$     (3) $9x$          (4) $\dfrac{x}{8}$

**c.** For all values of $y$ when $y$ represents every rational number, which of the following expressions represent rational numbers?

   (1) $y + 3$     (2) $y - 7$     (3) $6y$          (4) $\dfrac{y}{9}$

**d.** For all values of $a$ when $a$ represents every integer, which of the following expressions represent integers?

   (1) $a + 2$     (2) $9 - a$     (3) $-7a$          (4) $\dfrac{3a}{4}$

**3. a.** For all values of $b$ and $c$ when $b$ and $c$ each represent every whole number, which of the following expressions represent whole numbers?

   (1) $b + c$     (2) $b - c$     (3) $bc$          (4) $\dfrac{b}{c}$  (when $c \neq 0$)

**b.** For all values of $m$ and $n$ when $m$ and $n$ each represent every integer, which of the following expressions represent integers?

   (1) $mn$          (2) $2m + n$     (3) $m - 5n$     (4) $\dfrac{3m}{n}$  (when $n \neq 0$)

**c.** For all values of $r$ and $s$ when $r$ and $s$ each represent every rational number, which of the following expressions represent rational numbers?

   (1) $r + s$     (2) $rs$          (3) $2r - s$     (4) $\dfrac{r + 1}{3s}$  (when $s \neq 0$)

4. **a.** For all values of $y$ when $y$ represents every positive integer, which of the following expressions represent negative integers?
    (1) $-y$          (2) $y - 15$          (3) $6y$          (4) $3 - y$

   **b.** For all values of $x$ when $x$ represents every negative integer, which of the following expressions represent positive integers?
    (1) $4 - x$          (2) $x + 10$          (3) $-x$          (4) $x - 6$

    (5) $8x$          (6) $x^2$          (7) $-10x$          (8) $\dfrac{x}{2}$

   **c.** For all values of $n$ when $n$ represents every negative integer, which of the following expressions represent negative integers?
    (1) $-n$          (2) $2n + 5$          (3) $3n - 2$          (4) $4 - n$

   **d.** For all values of $t$ when $t$ represents every positive integer, which of the following expressions represent positive integers?

    (1) $t - 1$          (2) $6 - 2t$          (3) $(-7t)^2$          (4) $\dfrac{6t}{2}$

5. **a.** If $n$ represents an even integer, which of the following expressions represent even integers?
    (1) $n + 1$          (2) $5n$          (3) $3n + 2$          (4) $2n - 1$

   **b.** If $x$ represents an odd integer, which of the following expressions represent odd integers?
    (1) $2x$          (2) $x + 1$          (3) $x + 2$          (4) $x + 4$

   **c.** If $m$ represents an odd integer, which of the following expressions represent even integers?
    (1) $5m$          (2) $m + 3$          (3) $2m + 2$          (4) $3m + 5$

   **d.** If $r$ represents an even integer, which of the following expressions represent odd integers?
    (1) $r + 10$          (2) $r - 5$          (3) $2r + 7$          (4) $r + 1$

6. **a.** If $n$ represents an integer, which of the following expressions represent the next consecutive integer?
    (1) $n + 2$          (2) $n + 1$          (3) $n - 1$          (4) $2n$

   **b.** If $y$ represents an even integer, which of the following expressions represent the next consecutive even integer?
    (1) $2y$          (2) $y + 1$          (3) $y + 2$          (4) $2y + 2$

   **c.** If $s$ represents an odd integer, which of the following expressions represent the next consecutive odd integer?
    (1) $s + 1$          (2) $2s$          (3) $s + 2$          (4) $2s + 1$

   **d.** If $d$ represents an integer, which of the following pairs of expressions represent the next two consecutive integers?
    (1) $2d$ and $3d$                         (2) $d + 1$ and $2d + 2$
    (3) $d + 1$ and $d + 2$                   (4) $2d$ and $d + 3$

# Translating Verbal Sentences into Equations

## I. Aim    To translate a verbal sentence into an equation

## II. Procedure
1. Using the given algebraic representations for the unknowns, numerals if any, symbols of operation, and the equality sign, write the equation indicated by the verbal sentence.
2. Also review Exercise 1-7.

## III. Sample Solutions
1. Use $n$ = the number when writing as an equation:

$$\text{One third of some number is equal to 15.}$$
$$\tfrac{1}{3} \qquad n \qquad = \qquad 15$$

*Answer:* $\tfrac{1}{3}n = 15$

2. Use $x$ = Todd's age in years and $x + 3$ = Lisa's age in years when writing as an equation:

$$\text{The sum of Todd's age and Lisa's age is 21 years.}$$
$$x \quad + \quad x + 3 \quad = 21$$

*Answer:* $x + x + 3 = 21$

---

### DIAGNOSTIC TEST

1. Write an equation for each of the following verbal sentences, using $x$ to represent the number:
   a. Some number increased by itself is equal to forty-six.
   b. Six times some number minus ten is equal to eighty-one.
   c. Some number divided by twelve is equal to negative seven.
2. Write an equation for each of the following verbal sentences, using $x$ = first number and $x + 6$ = second number:
   a. Seven times the first number is equal to five times the second number.
   b. Three times the sum of the two numbers is equal to 22 more than the second number.
   c. The difference between eight times the first number and seven times the second number is eleven.

## *RELATED PRACTICE EXAMPLES*

Write an equation for each of the following verbal sentences:

1. **a.** Using $x$ to represent the number:
   (1) Some number increased by seven is equal to twenty.
   (2) One fourth of some number is seventeen.
   (3) Five times some number minus eight is equal to forty-three.
   **b.** Using $n$ to represent the number:
   (1) Twelve times some number decreased by seventy is equal to ninety-six.
   (2) Some number divided by six is equal to twenty-seven.
   (3) The sum of four times some number and three times this same number is forty-two.
   **c.** Using $c$ to represent the cost:
   (1) Twice the cost decreased by $3.50 is equal to $10.
   (2) Ninety-five per cent of the cost is $38.
   (3) The cost increased by two-fifths of the cost is equal to $5.60.
   **d.** Using $s$ to represent the selling price:
   (1) The difference between three times the selling price and $5 is equal to $25.
   (2) The selling price increased by one-half of the selling price is equal to $90.
   (3) The selling price decreased by twenty-five per cent of the selling price is equal to $4.50.

2. **a.** Using $x =$ first number and $2x =$ second number:
   (1) The sum of the first number and the second number is 15.
   (2) Three times the first number equals the second number increased by 12.
   (3) If four times the first number is subtracted from three times the second number, the result is 8.
   **b.** Using $x =$ John's age in years and $x + 3 =$ Mary's age in years:
   (1) The sum of John's and Mary's ages equals 17 years.
   (2) Twice John's age decreased by 5 years equals Mary's age increased by 2 years.
   (3) The difference between three times John's age and twice Mary's age is 10 years.
   **c.** Using $x =$ number of nickels and $2x + 5 =$ number of dimes:
   (1) Three times the number of nickels equals the number of dimes.
   (2) Twice the number of dimes plus the number of nickels equals 20.
   (3) The total number of nickels and dimes is 23.

**d.** Using $x =$ number of pounds of 99¢ grass seed and $50 - x =$ number of pounds of 79¢ grass seed:

(1) The number of pounds of 99¢ seed and twice the number of pounds of 79¢ seed equals 90 pounds.

(2) Twice the number of pounds of 99¢ seed less the number of pounds of 79¢ seed equals 40 pounds.

(3) Six times the number of pounds of 99¢ seed equals four times the number of pounds of 79¢ seed.

**e.** Using $x =$ width in feet and $2x + 4 =$ length in feet:

(1) Twice the length plus twice the width equals 38 feet.

(2) Five times the width is the same as twice the length plus 10 feet.

(3) The length minus the width is equal to double the width.

**f.** Using $3x + 7 =$ speed of train $A$ in m.p.h. and $2x - 5 =$ speed of train $B$ in m.p.h.:

(1) Train $A$ travels twice as fast as train $B$.

(2) The difference in the speeds of the two trains is the same as the sum of their speeds less 54 m.p.h.

(3) The speed of train $A$ increased by 14 m.p.h. is equal to 3 times the speed of train $B$.

## CUMULATIVE REVIEW

1. Represent the two unknown quantities in each of the following by related algebraic expressions:

   **a.** Ted is 3 inches taller than Bob.

   **b.** Margie has 25 coins consisting of dimes and nickels.

   **c.** Jim is twice as old as his brother.

2. Write the algebraic expression that represents the answer:
   Charlotte has $D$ dollars. Marilyn has $25 more than three times as many dollars as Charlotte. How many dollars does Marilyn have?

3. Write an equation for each of the following sentences, using $n$ to represent the number:

   **a.** Some number decreased by nine is equal to fifteen.

   **b.** One-sixth of some number is eleven.

   **c.** The sum of four times some number and eight times the same number is sixty.

4. Using $x =$ first number and $x + 6 =$ second number, write an equation for the sentence: Three times the sum of the two numbers is equal to twelve more than the second number.

# Number Problems

**I. Aim**    To solve number problems by using the equation method.

**II. Procedure**

1. Read the general directions given on page 244 how to solve verbal problems.

2. Use the chart below and on page 259 which illustrates how the equation method is used to find:

   Column *A*—A fractional part or decimal part of a number.
   Column *B*—What fractional part or what decimal part or what percent one number is of another.
   Column *C*—A number when a fractional part or decimal part or percent of it is known.

3. Review consecutive integers on page 52. See sample solution 4.

4. Check directly in the problem.

**III. Sample Solutions**

1. What number increased by 29 is equal to 67?

   | *Solution:* | *Check:* |
   |---|---|
   | Let $n$ = number | 38 |
   | $n + 29 = 67$ | $+29$ |
   | $n + 29 - 29 = 67 - 29$ | $67$ ✔ |
   | $n = 38$ | |

   *Answer:* 38

|  *A* | *B* | *C* |
|---|---|---|
| **2.** Find $\frac{3}{4}$ of 16. | What fractional part of 16 is 12? | $\frac{3}{4}$ of what number is 12? |
| $\frac{3}{4} \times 16 = n$ | $n \times 16 = 12$ | $\frac{3}{4} \times n = 12$ |
| $12 = n$ | $16n = 12$ | $\frac{3}{4}n = 12$ |
| $n = 12$ | $\frac{1}{16} \cdot 16n = \frac{1}{16} \cdot 12$ | $\frac{4}{3} \cdot \frac{3}{4}n = \frac{4}{3} \cdot 12$ |
| *Answer:* 12 | $n = \frac{3}{4}$ | $n = 16$ |
| | *Answer:* $\frac{3}{4}$ | *Answer:* 16 |

**258**

|            A            |            B            |            C            |
|-------------------------|-------------------------|-------------------------|

### A

Find .75 of 16.

$$.75 \times 16 = n$$
$$12 = n$$
$$n = 12$$

*Answer:* 12

### B

What decimal part of 16 is 12?

$$n \times 16 = 12$$
$$16n = 12$$
$$\tfrac{1}{16} \cdot 16n = \tfrac{1}{16} \cdot 12$$
$$n = \tfrac{3}{4}$$
$$n = .75$$

*Answer:* .75

### C

.75 of what number is 12?

$$.75 \times n = 12$$
$$.75n = 12$$
$$\frac{.75n}{.75} = \frac{12}{.75}$$
$$n = 16$$

*Answer:* 16

---

Find 75% of 16.

$$75\% \times 16 = n$$
$$.75 \times 16 = n$$
$$12 = n$$
$$n = 12$$

*Answer:* 12

What percent of 16 is 12?

$$n\% \times 16 = 12$$
$$\frac{n}{100} \times 16 = 12$$
$$\frac{16n}{100} = 12$$
$$16n = 1,200$$
$$n = 75$$
$$n\% = 75\%$$

*Answer:* 75%

75% of what number is 12?

$$75\% \times n = 12$$
$$.75n = 12$$
$$\frac{.75n}{.75} = \frac{12}{.75}$$
$$n = 16$$

*Answer:* 16

---

3. One number is five times a second number. The sum of the two numbers is 36. Find the numbers.

*Solution:* Let $n$ = second number
    Then $5n$ = first number

The equation is determined by the fact:
First number plus second number equals 36.

$$5n + n = 36$$
$$6n = 36$$
$$\frac{6n}{6} = \frac{36}{6}$$
$$n = 6, \quad \text{second number}$$
$$5n = 30, \quad \text{first number}$$

*Check:* $30 = 5 \times 6$ ✔
$\quad\quad\quad\;\, 30 + 6 = 36$ ✔

*Answer:* 30 and 6

**4.** Find two consecutive integers whose sum is 53.

*Solution:* Let $x$ = first integer
   Then $x + 1$ = next consecutive integer
The equation is determined by the fact:
The sum of the two integers is 53.

|  |  |
|---|---|
| $x + x + 1 = 53$ | *Check:* |
| $2x + 1 = 53$ | 26 and 27 are |
| $2x + 1 - 1 = 53 - 1$ | consecutive integers. |
| $2x = 52$ | $26 + 27 = 53$ |
| $x = 26$, first integer | |
| $x + 1 = 27$, next consecutive integer | |

*Answer:* 26 and 27

## DIAGNOSTIC TEST

Use the equation method to find the required number in each of the following:

**1.** The product of 37 and what number is 1,924?

**2. a.** Find $\frac{1}{16}$ of 144          **b.** Find $\frac{5}{8}$ of 784

**3. a.** Find .04 of 560          **b.** Find .245 of 8,000

**4. a.** Find 81% of 97          **b.** Find 6% of 128

**5. a.** What fractional part of 45 is 3?
   **b.** 42 is what fractional part of 63?

**6. a.** What decimal part of 25 is 18?
   **b.** 48 is what decimal part of 60?

**7. a.** What percent of 32 is 24?     **b.** 54 is what percent of 300?

**8. a.** $\frac{2}{3}$ of what number is 96?     **b.** 132 is $\frac{11}{16}$ of what number?

**9. a.** .47 of what number is 282?   **b.** 672 is 1.05 of what number

**10. a.** 6% of what number is 30?     **b.** 360 is 90% of what number?

**11.** If 119 students, or 14% of the student enrollment, were absent on a certain school day, how many students were present on that day?

**12.** One number is six more than a second number. Twice the second number increased by nine is sixteen less than three times the first number. What are the numbers?

**13.** The sum of the first and third integers of three consecutive even integers is 264. What are the integers?

## *RELATED PRACTICE PROBLEMS*

Use the equation method to find the required number in each of the following:

1. **a.** 15 times what number is equal to 90?
   **b.** 26 added to what number is equal to 103?
   **c.** 57 subtracted from what number is equal to 38?
   **d.** What number divided by 17 will give a quotient of 14?
   **e.** What number increased by 39 equals 64?
   **f.** The product of 23 and what number is 414?
   **g.** One-half of what number is 49?
   **h.** Five-sixths of what number is 60?
   **i.** What number increased by six times the same number equals 126?
   **j.** Scott asked Steve to guess his age. He said: "If you add 15 to four times my age, you get 43." How old is Scott?

Find:

2. **a.** $\frac{1}{4}$ of 28
   **b.** $\frac{3}{5}$ of 85
   **c.** $\frac{2}{3}$ of 150
   **d.** $\frac{7}{12}$ of 288
   **e.** $\frac{5}{8}$ of 76
   **f.** $\frac{9}{10}$ of 65

3. **a.** .7 of 48
   **b.** .25 of 59
   **c.** .04 of 350
   **d.** .581 of 2.7
   **e.** .375 of 9.1
   **f.** .065 of 700

4. **a.** 12% of 50
   **b.** 3% of 260
   **c.** 25% of 104
   **d.** $87\frac{1}{2}$% of 4,000
   **e.** 300% of 48
   **f.** 128% of 320

5. **a.** What fractional part of 18 is 12?
   **b.** What fractional part of 54 is 45?
   **c.** 30 is what fractional part of 48?
   **d.** 63 is what fractional part of 81?
   **e.** What fractional part of 91 is 39?
   **f.** 42 is what fractional part of 56?

6. **a.** What decimal part of 5 is 3?
   **b.** What decimal part of 16 is 12?
   **c.** 41 is what decimal part of 50?
   **d.** 65 is what decimal part of 125?
   **e.** What decimal part of 80 is 48?
   **f.** 18 is what decimal part of 40?

7. **a.** What percent of 10 is 7?
   **b.** What percent of 25 is 19?
   **c.** 54 is what percent of 72?
   **d.** 64 is what percent of 800?
   **e.** What percent of $5 is $.75?
   **f.** 49 is what percent of 56?

8. **a.** $\frac{1}{4}$ of what number is 16?
   **b.** $\frac{2}{3}$ of what number is 52?
   **c.** 75 is $\frac{5}{8}$ of what number?
   **d.** 56 is $\frac{7}{12}$ of what number?
   **e.** $\frac{9}{20}$ of what number is 63?
   **f.** $\frac{1}{3}$ of what number is 57?
   **g.** 96 is $\frac{4}{5}$ of what number?
   **h.** 81 is $\frac{3}{4}$ of what number?
   **i.** $\frac{19}{100}$ of what number is 114?
   **j.** $1\frac{1}{4}$ times what number is 375?

**9. a.** .03 of what number is 6?  **b.** .8 of what number is 20?
   **c.** 45 is .06 of what number?  **d.** 9 is .375 of what number?
   **e.** 1.04 of what number is 468?  **f.** 6.7 is .25 of what number?

**10. a.** 5% of what number is 21?  **b.** 9% of what number is 135?
   **c.** 12% of what number is 60.  **d.** 80% of what number is 24?
   **e.** 25% of what number is 13?  **f.** 104% of what number is 26?
   **g.** $37\frac{1}{2}$% of what number is 345?  **h.** 100% of what number is 53?
   **i.** 9.3% of what number is 620?  **j.** $2\frac{1}{2}$% of what number is 50?
   **k.** 81 is 18% of what number?  **l.** 14 is 4% of what number?
   **m.** 36 is 75% of what number?  **n.** 1,590 is 106% of what number?
   **o.** 47 is 10% of what number?  **p.** 59% of what number is 41.3?

**11. a.** Tom received $\frac{3}{4}$ of all votes cast in the election for school president. If he received 876 votes, how many students voted?
   **b.** Nancy bought a tennis racket at a sale paying $8.00 for it. If the price was reduced $\frac{1}{6}$, what was the regular price?
   **c.** If the school baseball team won 15 games, or $\frac{5}{8}$ of the games played, how many games were lost?
   **d.** If 118 students, or 40% of the graduating class, selected the college preparatory course, how many pupils were in the graduating class?
   **e.** The school athletic association sold 1,043 student membership tickets. If $87\frac{1}{2}$% of the school became members, what was the school enrollment?

**12. a.** One number is 4 times a second number. The sum of the two numbers is 30. What are the numbers?
   **b.** One number is 10 more than a second number. The sum of the two numbers is 16. What are the numbers?
   **c.** One number is twice a second number. Five more than the second number is the same as the first number less 3. What are the numbers?
   **d.** The sum of two numbers is 15. Twice the first number is the same as 4 times the second number. What are the numbers?
   **e.** The first number is 8 more than a second number. Three times the second number plus twice the first number is equal to 26. What is the first number?

**13. a.** Find two consecutive integers whose sum is 139.
   **b.** Find three consecutive integers whose sum is 81.
   **c.** What three consecutive odd integers have a sum of 45?
   **d.** Find three consecutive even integers such that 3 times the first equals the sum of the other two.
   **e.** Find four consecutive integers such that the sum of the first and fourth is 53.

# Investment, Business, Geometry, Age, Coin, Mixture and Motion Problems

**I. Aim**   To solve by the equation method: A. Investment Problems, B. Business Problems, C. Geometry Problems, D. Age Problems, E. Coin and Mixture Problems, and F. Motion Problems.

## II. Procedure

1. Read the general directions given on page 244 on how to solve verbal problems.
2. Study the basic information given in each topic.
3. Observe how the general directions are applied in the illustrated solution of the different types of problems.
4. Note that several topics include introductory problems and additional instructional procedures.
5. Also see decimal fractional equations in Exercise 11-3, on page 449.

## DIAGNOSTIC TEST

1. How much money must be invested at the annual rate of 6% to earn $7,500 per year?
2. Mr. Harris invests $8,900, part at 8% and the rest at 11% annual interest. How much does he invest at each rate if his total annual interest from both investments is $874?
3. What is the list price of a camera if its net price is $87.50 after a 30% discount is allowed?
4. Mr. Thompson received $168.40 commission last week. If he was paid at the rate of 2% of his sales, what was the amount of his sales?
5. At what price must a merchant sell a watch which costs $51.35 to make a profit of 35% on the selling price?
6. A rectangle is 21 feet longer than it is wide. What is its length and width if the perimeter is 554 feet?
7. Charlotte is 9 years older than Marilyn. In 1 year Charlotte will be four times as old as Marilyn. What are their ages now?
8. a. Pedro has 24 coins, some nickels and the rest dimes. If he has $1.50 in all, how many coins of each kind does he have?

**b.** How many pounds of cookies worth 96¢ a pound must be mixed with 60 pounds of cookies worth 70¢ a pound to make a mixture to sell at 90¢ a pound?

9. How long will it take an airplane, flying at an average speed of 360 m.p.h. to intercept an airplane, 1,075 miles away, flying directly toward it at an average speed of 285 m.p.h.?

## III. Sample Solutions
### A. *INVESTMENT*

*Interest* is money paid for the use of money. The money borrowed or invested on which interest is paid is called the *principal*. Interest paid on the principal only is called *simple interest*. When the interest is expressed as a percent of the principal, it is called the *rate of interest*. Unless specified otherwise, the rate of interest is the rate per year. The sum of the principal and the interest is called the *amount*.

*SAMPLE PROBLEMS*

1. How much money must be invested at an annual rate of 8% to earn $12,000 per year?

*Solution:*

Let $a$ = amount of money in dollars invested

8% of $a$ = 12,000

$.08a = 12,000$

$$\frac{.08a}{.08} = \frac{12,000}{.08}$$

$a = \$150,000$

*Check:*

$150,000

.08

$12,000✓

*Answer:* $150,000 is the required amount of investment.

2. Mrs. Ricci invested $2,400, part at 12% and the rest at 9% per year. If her total annual income from both investments was $261, how much did she invest at each rate?

*Solution:*

Let $x$ = amount invested in dollars at 12%.

Then $2,400 - x$ = amount invested in dollars at 9%.

$.12x$ = annual income from the 12% investment.

$.09(2,400 - x)$ = annual income from the 9% investment.

The equation is determined by the fact: The income from the 12% investment plus income from the 9% investment equals the total income.

$$.12x + .09(2,400 - x) = 261$$
$$.12x + 216 - .09x = 261$$
$$.12x - .09x + 216 = 261$$
$$.03x + 216 = 261$$
$$.03x + 216 - 216 = 261 - 216$$
$$.03x = 45$$
$$x = \$1,500 \text{ invested at } 12\%$$
$$2,400 - x = \$900 \text{ invested at } 9\%$$

*Answer:* $1,500 at 12% and $900 at 9%

| Check: | $1,500 | $900 | $1,500 | $180 |
|---|---|---|---|---|
| | .12 | .09 | +900 | +81 |
| | $180.00 | $81.00 | $2,400 invested | $261 income ✔ |

## RELATED PRACTICE PROBLEMS

1. **a.** How much money must be invested at the annual rate of 12% to earn $9,000 per year?
   **b.** How much money was borrowed at the annual rate of 9% if the interest charged was $513 per year?
   **c.** Find the principal on which the annual interest is:
   (1) $1,750 when the annual rate is 14%
   (2) $3,500 when the annual rate is $8\frac{3}{4}\%$
   (3) $840 when the annual rate is $10\frac{1}{2}\%$
   **d.** At what annual rate of interest must $75,000 be invested to earn $6,375 per year?
   **e.** What is the full amount to be repaid at the end of the year if, after 12% interest is deducted, the sum of $836 is obtained from the bank?

2. **a.** Mr. Bauman invested $2,700, part at 11% and the rest at 8% per year. How much did he invest at each rate if his total annual income from both investments was $264?
   **b.** A woman invested $6,000, part at 9% and the rest at 10% per year. If she received a total annual income of $562 from both investments, how much did she invest at each rate?
   **c.** A 8% investment brings an annual return of $34 less than a 6% investment. Find the two amounts invested if the total amount invested is $4,650.

**d.** A man invested $1,500 more at 8% than he did at 10% per year. If the total annual income from both investments was $462, how much did he invest at 10%?

**e.** Ms. Lopez invested a certain amount of money at 7% per year and twice as much at 12%. If her total annual income from both investments was $186, how much did she invest at each rate?

**f.** Mr. Smith invested $4,800, part at 8% and the rest at 6% per year. How much did he invest at 6% if his total annual income from both investments was $354?

**g.** A man invested $30,000, some of it at 9% and the rest at 11% per year. Find how much money he invested at each rate if he received the same amount of income on each investment.

**h.** *A* borrowed 12,000 from *B* and *C*. If he paid a total of $1,660 annual interest, paying *B* at the rate of 13% and *C* at the rate of 15%, how much money did he borrow from *B*?

## B. *BUSINESS PROBLEMS*

### Discount

Discount is the amount of reduction in the price of an article. The regular price or full price of the article is called the *list price* or *marked price*. The reduced price of the article after the discount is deducted is called the *net price* or *sale price*. The rate of discount is the percent the discount is of the list price.

### SAMPLE PROBLEM

The net price of a television set is $301 when a 14% discount is allowed. What is its list price?

*Solution:*    Let        $x$ = list price
              Then $.14x$ = amount of discount
The equation is determined by the fact:
The net price plus the discount equals the list price.

$$301 + .14x = x$$

*Check:*

$$301 + .14x - .14x = x - .14x$$

$350 list price

$$301 = .86x$$

$\times\ \ .14$

$$\frac{301}{.86} = \frac{.86x}{.86}$$

1400
350

$$350 = x$$

$49.00 discount

$$x = \$350, \text{ list price}$$

$350 list price
$-\ \ \ 49$ discount

*Answer:* $350 is the list price.

$301 net price ✔

## *RELATED PRACTICE PROBLEMS*

 **a.** What is the list price of a refrigerator if its net price is $204 after a 15% discount is allowed?

**b.** Find the regular price of a baseball glove that sold for $11.40 at a 25% reduction sale.

**c.** Find the list price when the:
(1) net price is $352 and the rate of discount is 12%.
(2) net price is $228 and the rate of discount is 5%.
(3) net price is $495 and the rate of discount is 40%.

**d.** Scott paid $35 for a watch at a sale. If he was allowed a 30% reduction, what was the regular price of the watch?

**e.** Find the list price of a mirror if its net price is $20.80 after a 35% discount is allowed.

Commission

The money that a person receives for selling goods for another person is called *commission*. The amount of money that remains after the commission is deducted from the selling price is called *net proceeds*. The rate of commission is the percent the commission is of the selling price.

*SAMPLE PROBLEM*

How much must the sales be for a salesperson to earn $175 weekly if paid a 7% commission on sales?

*Solution:*
Let    $a$ = amount of sales
Then $.07a$ = commission on sales
$.07a = 175$
$$\frac{.07a}{.07} = \frac{175}{.07}$$
$a = \$2,500$, sales

*Answer:* $2,500 is the amount of sales required.

Check:
$2,500 sales
   .07
———————
$175.00 commission ✔

## RELATED PRACTICE PROBLEMS

4. **a.** Marilyn received $112 commission last week. If she was paid at the rate of 8% of her sales, what was the amount of her sales?

**b.** Mr. Wilson is a car salesman. He earns 4% commission on his sales. What must his sales be for the year in order to earn $15,000 for the year?

**c.** Find the amount of sales when the:
(1) commission is $57 at the rate of 5% commission.
(2) commission is $420 at the rate of 6% commission.
(3) commission is $137 at the rate of 20% commission.

**d.** A lawyer charged her client 15% commission to collect a debt. If the client received $629 as his share, what was the amount of the debt that was collected?

**e.** A salesman earns $77 per week as salary and an additional 3% commission on his sales. If he received $380 as his total salary including commission, what was the amount of his sales for that week?

PROFIT AND LOSS

The *cost* of goods is the amount of money that the merchant pays for the goods. The *selling price* is the amount of money the merchant receives when the goods are sold. When the selling price is greater than the cost, the difference between them is called the *gross profit* or *margin* or *mark-up*. When the cost is greater than the selling price, the difference between them is called the *loss*. The rate of profit on the cost is the percent the profit is of the cost. The rate of profit on the selling price is the percent the profit is of the selling price.

SAMPLE PROBLEM

At what price must a dealer sell a radio which costs $27 to make a profit of 40% on the selling price?

*Solution:*

$$\text{Let} \quad s = \text{selling price}$$
$$\text{Then} \ .4s = \text{profit on selling price}$$

The equation is determined by the fact:
The cost plus the profit equals the selling price.

$$27 + .4s = s$$
$$27 + .4s - .4s = s - .4s$$

$$27 = .6s$$
$$\frac{27}{.6} = \frac{.6s}{.6}$$
$$45 = s$$
$$s = \$45, \text{ selling price}$$

*Answer:* $45 is the selling price.

*Check:*

| | |
|---|---|
| $45 selling price | $27 cost |
| .4 | + 18 profit |
| $18 profit | $45 selling price ✔ |

## RELATED PRACTICE PROBLEMS

5. **a.** A merchant bought a chair for $78. At what price must it be sold to make a 35% profit on the selling price?

**b.** Find the selling price of each of the following articles:
   (1) Electric heater, costing $38 and a 20% mark-up on the selling price.
   (2) Watch, costing $44 and a 45% profit on the selling price.
   (3) Washing machine, costing $192 and a 36% profit on the selling price.

**c.** A dealer sold a camera for $91, making a 30% profit on the cost. How much did the dealer pay for it?

## C. GEOMETRY PROBLEMS

*Procedure:*
1. If necessary, refer to page 281 for any required geometric facts.
2. Read the general directions given on page 244.
3. Include the required geometric figure. See sample problem below.

### SAMPLE PROBLEM

The length of a rectangle is 4 meters more than the width. The perimeter of the rectangle is 40 meters. What do the length and the width each measure?

*Solution:*
Let          $w$ = width in meters
Then $w + 4$ = length in meters
The equation is determined by the fact: The perimeter of a rectangle is equal to twice the length plus twice the width.

$$2(w + 4) + 2w = 40$$
$$2w + 8 + 2w = 40$$
$$2w + 2w + 8 = 40$$
$$4w + 8 = 40$$
$$4w + 8 - 8 = 40 - 8$$
$$4w = 32$$
$$w = 8 \quad \text{meters wide}$$
$$w + 4 = 8 \ + 4 \text{ or } 12 \text{ meters long}$$

*Answer:* 12 meters long and 8 meters wide

*Check:*

| 12 meters long | 24 meters (twice the length) |
|---|---|
| − 8 meters wide | +16 meters (twice the width) |
| 4 meters, difference | 40 meters, perimeter ✔ |

## RELATED PRACTICE PROBLEMS

6. Solve and check each of the following problems:
   **a.** The length of a rectangle is 14 meters more than the width. The perimeter of a rectangle is 264 meters. What do the length and the width each measure?
   **b.** The perimeter of a rectangle is 168 centimeters. Its length is five times its width. Find the length and the width.
   **c.** The width of a rectangle is 5 meters less than the length. Find the dimensions of the rectangle if its perimeter is 90 meters.
   **d.** The length of a rectangle is 8 centimeters more than six times its width. The perimeter of the rectangle is 156 centimeters. What do the length and the width each measure?
   **e.** If two opposite sides of a square are each increased by 12 millimeters, and the other two sides are each decreased by 4 millimeters, a rectangle is formed having the same area. How long is each side of the square?
   **f.** Side $a$ of a triangle is 2 centimeters longer than side $b$. Side $b$ is three times as long as side $c$. The perimeter of the triangle is 37 centimeters. Find the length of each side.
   **g.** The base of an isosceles triangle is 7 meters longer than each of the other equal sides. What does each side of the triangle measure if the perimeter is 58 meters?
   **h.** Complementary angles are two angles whose sum of measures is 90°.
   (1) If one of two complementary angles measures 40° more than the other, what is the measure of each angle?
   (2) If one of two complementary angles measures 30° less than twice the other, what is the measure of each angle?

    **i.** Supplementary angles are two angles whose sum of measures is 180°.
       (1) If one of two supplementary angles measures 52° less than the other, what is the measure of each angle?
       (2) If one of two supplementary angles measures 16° more than three times the other, what is the measure of each angle?
    **j.** The sum of the measures of the three angles of a triangle is 180°. Find the measure of each of the three angles of a triangle:
       (1) When the first angle measures four times the second angle and the third angle measures 12° more than the second angle.
       (2) When the first angle measures 31° more than the second angle and the third angle measures 5° less than twice the first angle.

## D. *AGE PROBLEMS*

INTRODUCTORY PROBLEMS

Solve each of the following problems:

    **a.** A man is twice as old as his son. Together the sum of their ages is 63 years. What are their ages?
    **b.** Ed is 5 years older than Jim. Four times Jim's age increased by 3 years equals three times Ed's age diminished by 2 years. Find Ed's age.
    **c.** The difference in ages of two girls is 1 year. The sum of their ages is 27 years. What are their ages?
    **d.** Neil is three times as old as Francis. Use this statement in the following:
       (1) If Francis is 12 years old now, how old is Neil now?
       (2) How old was Francis 4 years ago?
       If Francis is $x$ years old now, what algebraic expression represents:
       (4) Neil's age now?   (5) Francis's age 4 years ago?
       (6) Neil's age 4 years ago?
    **e.** Elaine is five times as old as Ann. Use this statement in the following:
       (1) If Ann is 8 years old now, how old is Elaine now?
       (2) How old will Ann be 6 years from now?
       (3) How old will Elaine be 6 years from now?
       If Ann is $x$ years old now, what algebraic expression represents:
       (4) Elaine's age now?   (5) Ann's age 6 years from now?
       (6) Elaine's age 6 years from now?
    **f.** Sara is 4 years older than Mary. Use this statement in the following:
       (1) If Mary is 10 years old now, how old is Sara now?
       (2) How old was Mary 3 years ago?
       (3) How old was Sara 3 years ago?
       (4) How old was Mary 7 years ago?
       (5) How old was Sara 7 years ago?

If Mary is $x$ years old now, what algebraic expression represents:
(6) Sara's age now?  (7) Mary's age 3 years ago?  (8) Sara's age 3 years ago?  (9) Mary's age 7 years ago?  (10) Sara's age 7 years ago?

g. Ralph is 5 years older than Walter. Use this statement in the following:
   (1) If Walter is 3 years old now, how old is Ralph now?
   (2) How old will Walter be 9 years from now?
   (3) How old will Ralph be 9 years from now?
   If Walter is $x$ years old now, what algebraic expression represents:
   (4) Ralph's age now?  (5) Walter's age 9 years from now?
   (6) Ralph's age 9 years from now?

h. The sum of Charlotte's present age and Marilyn's present age is 50 years. Use this statement in the following:
   (1) If Marilyn is 20 years old now, how old is Charlotte now?
   (2) How old was Marilyn 5 years ago?
   (3) How old was Charlotte 5 years ago?
   If Marilyn is $x$ years old now, what algebraic expression represents:
   (4) Charlotte's age now?  (5) Marilyn's age 5 years ago?  (6) Charlotte's age 5 years ago?

i. The sum of Steve's present age and Ronald's present age is 38 years. Use this statement in the following:
   (1) If Steve is 5 years old now, how old is Ronald now?
   (2) How old will Steve be 7 years from now?  (3) Ronald?
   If Steve is $x$ years old now, what algebraic expression represents:
   (4) Ronald's age now?  (5) Steve's age 7 years from now?
   (6) Ronald's age 7 years from now?

j. At the present time George is 8 years old and Joe is 2 years old.
   (1) How many years older is George than Joe now?
   (2) How many times as old is George compared to Joe now?
   Let us see whether these comparisons change as the boys get older.
   (3) One year from now, how old will George be?  (4) Joe?
   (5) One year from now how many years older will George be than Joe?
   (6) One year from now how many times as old will George be compared to Joe?
   (7) Four years from now, how old will George be?  (8) Joe?
   (9) Four years from now how many years older will George be than Joe?
   (10) Four years from now how many times as old will George be compared to Joe?

Mr. Kelly is four times as old as his son. Sixteen years from now he will be only twice as old. What are their ages now?

*Solution:*

Let $\quad x =$ son's present age in years

Then $4x =$ father's present age in years

$\quad x + 16 =$ son's age in years 16 years hence

$\quad 4x + 16 =$ father's age in years 16 years hence

The equation is determined by the fact: The father's age 16 years hence is equal to twice the son's age at that time.

$$4x + 16 = 2(x + 16)$$
$$4x + 16 = 2x + 32$$
$$4x + 16 - 16 = 2x + 32 - 16$$
$$4x = 2x + 16$$
$$4x - 2x = 2x - 2x + 16$$
$$2x = 16$$
$$x = 8 \text{ years, son's present age}$$
$$4x = 4 \times 8 \text{ or } 32 \text{ years, father's present age}$$

*Check:*

Father's age    Son's age

present

32 years $= 4 \times 8$ years

16 years hence

48 years $= 2 \times 24$ years ✔

*Answer:* son is 8 years old, father is 32 years old.

## RELATED PRACTICE PROBLEMS

7. Solve and check each of the following problems:

   **a.** Mr. Whitney is three times as old as his son. Twelve years from now he will be only twice as old. What are their ages now?

   **b.** Richard is twice as old as his brother. Four years ago he was four times as old. What are their ages now?

   **c.** Peggy is 6 years older than Rosa. Two years from now Peggy will be twice as old as Rosa. Find their present ages.

   **d.** Arthur is 12 years younger than Robert. Three years ago Robert was five times as old as Arthur. How old is Robert now?

   **e.** Jose is one fifth as old as Harry. Four years hence three times Jose's age will equal Harry's age. Find Harry's present age.

   **f.** Lucy is 5 years older than Dorothy. Four years ago eight times Dorothy's age equaled three times Lucy's age. What is Lucy's age now?

   **g.** Henry is four times as old as George. Six years ago he was ten times as old as George. Find Henry's present age.

   **h.** The sum of the ages of David and Bernice is 48 years. Eight years hence David will be three times Bernice's age. What are their ages now?

i. The sum of the ages of a mother and daughter is 45 years. Five years ago the mother was six times the daughter's age. What are their ages now?

 j. The sum of the ages of a father and son is 46 years. In two years four times the son's age will equal the father's age. Find their present ages.

k. Eight years ago Jim was three times as old as Tom. However, eight years from now Jim will only be twice as old as Tom. What are their present ages?

## E. COIN AND MIXTURE PROBLEMS

INTRODUCTORY PROBLEMS

Solve each of the following problems:

a. There are five times as many $1 bills as $2 bills. The total number of bills is 48. How many $1 bills are there?

b. Maria has 41 coins. She has 3 more nickels than pennies. How many nickels and how many pennies has she?

c. Among 580 admission tickets sold, there were three times as many 50¢ tickets sold as 25¢ tickets. Find the number of each.

d. There were 70 pounds of 97¢ cookies and 83¢ cookies sold. For each pound of 97¢ cookies, four times as many pounds of 83¢ cookies. How many pounds of 83¢ cookies were sold?

e. Among 35 coins there are 5 more dimes than nickels and 3 fewer nickels than pennies. Find the number of each kind of coin.

f. There are twice as many $2 bills as $5 bills. Use this statement in the following:
(1) If there are four $5 bills, how many $2 bills are there?
(2) What is the value of all the $5 bills in dollars?
(3) What is the value of all the $2 bills in dollars?
If there are $x$ $5 bills, write an algebraic expression that represents:
(4) The number of $2 bills.
(5) The value of all the $5 bills in dollars?
(6) The value of all the $2 bills in dollars?

g. There are three more nickels than dimes. Use this statement in the following:
(1) If there are 8 dimes, how many nickels are there?
(2) What is the value of all the dimes in cents?
(3) What is the value of all the nickels in cents?
If there are $x$ dimes, write an algebraic expression that represents:
(4) The number of nickels.   (5) The value of all the dimes in cents.
(6) The value of all the nickels in cents.

**h.** A grocer mixes cookies worth 98¢ a pound with cookies worth 79¢ a pound, mixing 70 pounds in all. Use this statement in the following:

(1) If 40 pounds of 98¢ cookies are used, how many pounds of 79¢ cookies are used?

(2) What is the cost in cents of all the 98¢ cookies being used?

(3) What is the cost in cents of all the 79¢ cookies being used?

If $x$ pounds of 98¢ cookies are used, write an algebraic expression that represents:

(4) The number of pounds of 79¢ cookies being used.

(5) The cost in cents of all the 98¢ cookies being used.

(6) The cost in cents of all the 79¢ cookies being used.

*Procedure:*

**1.** In coin problems:

  **a.** Represent the number of each kind of coin.

  **b.** Represent also their values in some common denomination.

  **c.** Read the general directions given on page 244.

**2.** In mixture problems:

  **a.** Represent the quantity of each commodity.

  **b.** Represent also their costs in some common denomination.

  **c.** Read the general directions given on page 244.

*SAMPLE PROBLEMS*

**1.** Mike has $1.55 in nickels and dimes. He has 7 more nickels than dimes. Find the number of each kind of coin.

*Solution:*   Let      $x$ = no. of dimes

Then $x + 7$ = no. of nickels

$10x$ = value of all dimes in cents

$5(x + 7)$ = value of all nickels in cents

The equation is determined by the fact: The number of cents in all dimes plus the number of cents in all nickels equals the total number of cents. $1.55 = 155 cents.

$$10x + 5(x + 7) = 155$$
$$10x + 5x + 35 = 155$$
$$15x + 35 = 155$$
$$15x + 35 - 35 = 155 - 35$$
$$15x = 120$$
$$x = 8 \text{ dimes}$$
$$x + 7 = 8 + 7 \text{ or } 15 \text{ nickels}$$

*Check:*

  15 nickels

  − 8 dimes

  7 difference

  $.75 value of nickels

  + .80 value of dimes

  $1.55 ✔

*Answer:* 8 dimes, 15 nickels

**2.** A grocer mixes cookies worth 80¢ a pound with cookies worth 95¢ a pound making a mixture to sell at 85¢ a pound.? If he mixes 60 pounds, how many pounds of each kind does he use?

*Solution:*

Let $n$ = number of pounds of 80¢ cookies.

Then $60 - n$ = number of pounds of 95¢ cookies.

$80n$ = cost of all the 80¢ cookies in cents.

$95(60 - n)$ = cost of all the 95¢ cookies in cents.

The equation is determined by the fact: The cost of all 80¢ cookies plus the cost of all 90¢ cookies equals the cost of 60 pounds of 85¢ cookies.

$$80n + 95(60 - n) = 85 \times 60$$
$$80n + 5700 - 95n = 5100$$
$$80n - 95n + 5700 = 5100$$
$$-15n + 5700 = 5100$$
$$-15n + 5700 - 5700 = 5100 - 5700$$
$$-15n = -600$$
$$n = 40 \text{ pounds of 80¢ cookies}$$
$$60 - n = 60 - 40 \text{ or 20 pounds of 95¢ cookies}$$

*Answer:* 40 pounds of 80¢ cookies, 20 pounds of 95¢ cookies

*Check:*   40 pounds + 20 pounds = 60 pounds

| 40 | 20 | 60 |
|---|---|---|
| $\times$ \$.80 | $\times$ \$.95 | $\times$ \$.85 |
| \$32.00 + | \$19.00 = | \$51.00 ✔ |

## RELATED PRACTICE PROBLEMS

**8** Solve and check each of the following problems:

**a.** James has \$1.25 in nickels and dimes. He has three times as many nickels as dimes. Find the number of each kind of coin.

**b.** Mr. Bennett had four times as many \$2 bills as \$1 bills, the total amounting to \$54. How many bills of each kind did he have?

**c.** Joe, in changing a \$2 bill, received 6 more dimes than quarters. How many dimes did he receive?

**d.** Harry has 4 more nickels than half-dollars; their value is \$1.85. Find the number of nickels.

**e.** Ester has 16 coins, some quarters and the rest nickels. The total value of all the coins is \$1.40. Find the number of each kind of coin.

**f.** Reba has \$2.34. She has three times as many dimes as nickels and 6 more pennies than dimes. Find the number of each kind of coin.

**g.** Stewart has 25 stamps; some 15¢ stamps and the rest 18¢ stamps. The value of all the stamps is $4.05. How many stamps of each kind does he have?

**h.** Sam bought stamps costing $4.20. He received the same number of 15¢ stamps as 18¢ stamps, but as many 1¢ stamps as the sum of the other two. How many stamps of each kind did he have?

**i.** There were 3,000 persons at a football game. Some paid $2 for their tickets while the rest paid $1. The total receipts amounted to $4,850. How many tickets of each kind were sold?

**j.** A grocer mixes cookies worth 89¢ a pound with cookies worth 99¢ a pound making a mixture to sell at 93¢ a pound. If he mixes 50 pounds, how many pounds of each kind does he use?

**k.** A confectioner wishes to make 80 pounds of mixed candy to sell at 50¢ a pound. If he mixes candy worth 38¢ a pound with candy worth 70¢ a pound, how many pounds of each kind does he use?

**l.** How many pounds of tea worth $2.31 a pound should be blended with tea worth $2.55 a pound to make 40 pounds of blended tea to sell at $2.40 a pound?

**m.** How many pounds of cookies worth 72¢ a pound must be mixed with 25 pounds of cookies worth 96¢ a pound to get a mixture worth 80¢ a pound?

**n.** Walnuts cost 60¢ a pound more than peanuts. If Mr. Carroll paid $15.60 for 4 pounds of peanuts and 6 pounds of walnuts, what did he pay per pound for each?

**o.** A farmer sent 500 bags of potatoes to a commission merchant; some at 90¢ a bag and the rest at 50¢ a bag. If he received $394 in payment, how many bags of each did he send?

**F.** *MOTION PROBLEMS*

INTRODUCTORY PROBLEMS

**a.** What is the distance in kilometers that a car travels:
  (1) At the rate of 44 km.p.h. in 4 hours?
  (2) At the rate of 64 km.p.h. in 9 hours?
  (3) At the rate of $x$ km.p.h. in 6 hours?
  (4) At the rate of 96 km.p.h. in $x$ hours?
  Write a formula for the distance ($d$) traveled in terms of the rate of speed ($r$) and the time ($t$) of travel.

**b.** What average rate of speed must a train travel to cover a distance of:
  (1) 264 kilometers in 3 hours?    (2) 208 kilometers in 2 hours?
  (3) 370 kilometers in $x$ hours?    (4) $x$ kilometers in 6 hours?
  Write a formula for the average rate of speed ($r$) required to travel a given distance ($d$) in a given time ($t$).

**c.** How many hours will it take an airplane to fly:
  (1) 1,920 kilometers at the rate of 480 km.p.h.?
  (2) 3,360 kilometers at the rate of 560 km.p.h.?
  (3) 1,520 kilometers at the rate of $x$ km.p.h.?
  (4) $x$ kilometers at the rate of 520 km.p.h.?
  Write a formula for the time ($t$) it takes to travel a given distance ($d$) at a given rate of speed ($r$).

*Procedure:*

**1.** First read the general directions given on page 244.
**2.** Use the motion formulas to relate the given facts and to form the required equation.
**3.** A diagram may also be used to picture the problem and aid in determining the required equation.
**4.** Observe that in some problems we have a starting point and move outward from that point. In other problems we move inward from two points toward a meeting point. In these problems the equation is usually developed by the fact that the sum of the distances traveled by the two vehicles is equal to the total distance.
**5.** Also observe that in problems dealing with one vehicle overtaking another vehicle, when starting from the same point, the distance traveled by each vehicle is the same.
**6.** The radius of action is the distance an airplane can fly on a given amount of fuel in a given direction under known conditions and still return to the point where it took off.

*SAMPLE PROBLEM*

Two trains leave a station, one traveling north at the rate of 60 km.p.h. and the other south at the rate of 54 km.p.h. In how many hours will they be 342 kilometers apart?

*Solution:*

Let $x$ = no. of hours each train traveled
  $60x$ = no. of kilometers traveled by train going north
  $54x$ = no. of kilometers traveled by train going south

The equation is determined by the fact: The distance traveled by one train plus the distance traveled by the other train equals the total distance.

$60x + 54x = 342$
$114x = 342$
$x = 3$ hours

*Check:*
60 km.p.h. $\times$ 3 hours = 180 kilometers
54 km.p.h. $\times$ 3 hours = 162 kilometers
342 kilometers ✔

*Answer:* 3 hours

### RELATED PRACTICE PROBLEMS

9. Solve and check each of the following problems:

   **a.** One car, traveling at an average speed of 55 m.p.h., leaves Philadelphia for Washington, a distance of 150 miles. At the same time another car, traveling on the same highway at an average speed of 45 m.p.h., leaves Washington for Philadelphia. In how many hours will they meet? How far from Philadelphia will they meet?

   **b.** Two boys on bicycles start from the same place but ride in opposite directions. Joe rides twice as fast as Tom, and in 4 hours they are 24 kilometers apart. What is the rate of speed of each boy?

   **c.** How fast did a car go to overtake a bus in 6 hours if the bus averaged 45 m.p.h. and left 3 hours before the car?

   **d.** Two hours after a plane left Newark for San Francisco at a speed of 320 km.p.h., another plane left traveling at a speed of 480 km.p.h. How many hours will it take the second plane to overtake the first plane?

   **e.** Two salesmen leave a hotel at the same time but travel by automobile in opposite directions. Mr. Smith's average rate of speed is 3 miles per hour faster than Mr. White's. Find their rates of speed if, after 5 hours, they are 435 miles apart.

   **f.** A passenger train traveling at the rate of 50 kilometers per hour leaves a station 6 hours after a freight train and overtakes it in 4 hours. Find the rate of speed of the freight train.

   **g.** Two groups of boy scouts, living in towns 17 miles apart, decide to pitch camp along the road joining the two towns. If the group from one town leaves at 7 A.M. and the group from the second town leaves at 8 A.M. but walks 1 mile per hour faster than the first group, they will meet at 10 A.M. How fast is each group walking?

   **h.** How long will it take an airplane, flying at an average speed of 380 m.p.h. to intercept an airplane, observed 84 miles away, flying directly toward it at an estimated speed of 250 m.p.h.?

   **i.** How far from an airport can an airplane fly and yet return (radius of action) in 5 hours if its outgoing speed is 300 m.p.h. and its return speed is 200 m.p.h.?

   **j.** If its outgoing speed is 270 km.p.h. and its return speed is 360 km.p.h., how far from an aircraft carrier can a plane fly and return in 7 hours?

   **k.** An unknown aircraft is sighted 194 miles from an airbase, flying at an estimated speed of 300 m.p.h. toward the base. If it takes an interceptor plane 2 minutes to take off, flying at an average speed of 420 m.p.h., how far from the base will it intercept the first airplane?

# Problems Solved by Formulas

I. **Aim**    To solve problems by the use of formulas and the equation method in the subject areas of: A. Geometry; B. Interest; C. Gears, Pulleys, and Sprockets; D. Electricity; E. Levers; F. Temperature; and G. Aviation.

II. **Procedure**

1. Study the basic information, formulas, and the sample solution given for each subject area.

2. To solve a problem:
   a. Read the problem carefully.
   b. Select the formula that relates the required value with the known values.
   c. Substitute the known values for the variables in the formula.
   d. Solve the resulting equation.

3. In some cases it may be necessary to use more than one formula to find the required answer.

## DIAGNOSTIC TEST

Solve each of the following problems, using the appropriate formulas:

1. a. How long is each side of a square garden if its perimeter is 236 meters?
   b. What is the area of a circular skating rink if its circumference is 132 meters?
      Use $\pi = \frac{22}{7}$.
   c. How high should a rectangular storage bin be to hold 31,800 cubic meters if the length is 53 meters and the width 24 meters?

2. How much did Fred borrow at 9% per year if at the end of 2 years he repaid $1,121, which included both the amount borrowed and the interest?

3. How many teeth are required on a gear if it is to run at 280 r.p.m. when driven by a gear with 76 teeth running at 210 r.p.m.?

4. A 1,350-watt electric heater is used on a 110-volt circuit. How much current does it take and what is the resistance of the heater?

5. How much downward pressure must be exerted on a 12-foot lever arm to raise a weight of 300 pounds on an 8-foot lever arm?

6. Arrange the following temperature readings in order of warmth, with the one recording the warmest temperature first: 47° Fahrenheit, 298° absolute, 10° Celsius.

7. An airplane has a wing span of 64 feet and a wing area of 512 square feet. It has a gross weight of 8,192 pounds. Find the mean chord, aspect ratio, and wing loading of the airplane.

## A. GEOMETRY

The following is a brief summary of geometric relationships that require the use of formulas.

The distance around a polygon is called *perimeter* (*p*).

The distance around a circle is called *circumference* (*c*).

The *area* (*A*) of any surface is the number of units of square measure contained in the surface.

The *volume* (*V*) is the number of units of cubic measure contained in a given space.

### RECTANGLE

The perimeter (*p*) of a rectangle is equal to twice the length (*l*) plus twice the width (*w*). Formula: $p = 2l + 2w$.

The area (*A*) of the interior of a rectangle is equal to the length (*l*) times the width (*w*).

Formula: $A = lw$.

### SQUARE

The perimeter (*p*) of a square is equal to 4 times the length of its side (*s*). Formula: $p = 4s$.

The area (*A*) of the interior of a square is equal to the length of its side (*s*) times itself (or the side squared).

Formula: $A = s^2$.

### CIRCLE

radius

diameter

The circumference ($c$) of a circle is equal to Pi ($\pi$) times the diameter ($d$). Formula: $c = \pi d$ where $\pi = \frac{22}{7}$ or 3.14. For greater accuracy 3.1416 is used.

The circumference ($c$) of a circle is equal to two times Pi ($\pi$) times the radius ($r$). Formula: $c = 2\pi r$.

The area ($A$) of the interior of a circle is equal to Pi ($\pi$) times the radius ($r$) squared. Formula: $A = \pi r^2$.

The area ($A$) of the interior of a circle is equal to one-fourth times Pi ($\pi$) times the diameter ($d$) squared. Formula: $A = \frac{1}{4}\pi d^2$.

### TRIANGLE

The perimeter ($p$) of a triangle is equal to the sum of the lengths of its sides.
Formula: $p = a + b + c$.

The area ($A$) of the interior of a triangle is equal to one-half the altitude ($a$) times the base ($b$).

Formula: $A = \frac{1}{2}ab$ or $A = \dfrac{ab}{2}$.

Sometimes the formula $A = \frac{1}{2}bh$ is used where $b$ represents the base and $h$ represents the height.

The perimeter ($p$) of an equilateral triangle is equal to three times the length of its side ($s$).
Formula: $p = 3s$.

### PARALLELOGRAM

The area ($A$) of the interior of a parallelogram is equal to the altitude ($a$) times the base ($b$).
Formula: $A = ab$.

Sometimes the formula $A = bh$ is used, where $b$ represents the base and $h$ represents the height.

### TRAPEZOID

The area ($A$) of the interior of a trapezoid is equal to the height ($h$) times the average of the two parallel sides (bases: $b_1$ and $b_2$).

Formula:      $A = h \times \dfrac{b_1 + b_2}{2}$ or $A = \dfrac{h}{2}(b_1 + b_2)$

### Rectangular Solid

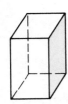

The volume ($V$) of a rectangular solid is equal to the length ($l$) times the width ($w$) times the height ($h$).
Formula: $V = lwh$.

The total area ($A$) of the outside surface of a rectangular solid is the area of its six faces.
Formula: $A = 2lw + 2lh + 2wh$.

### Cube

The volume ($V$) of a cube is equal to the length of its edge ($e$) times itself times itself (or the edge "cubed").
Formula: $V = e^3$.

The total area ($A$) of the outside surface of a cube is the area of its six congruent square faces.
Formula: $A = 6e^2$ or $A = 6s^2$.

### Right Circular Cylinder

The volume ($V$) of a right circular cylinder is equal to Pi ($\pi$) times the square of the radius ($r$) of the base times the height ($h$). Formula: $V = \pi r^2 h$.

The volume ($V$) of a right circular cylinder is equal to the area of the base ($B$) times the height. Formula: $V = Bh$. When the diameter is known, the formula $V = \frac{1}{4}\pi d^2 h$ may be used.

The lateral area or the area of the curved surface ($A$) of a right circular cylinder is equal to Pi ($\pi$) times the diameter ($d$) of the base times the height ($h$). Formula: $A = \pi dh$. Sometimes the formula $A = 2\pi rh$ is also used.

The total area ($A$) of a right circular cylinder, which includes the upper and lower circular bases, equals the lateral area plus the area of the two bases. Any one of the following formulas may be used: $A = 2\pi rh + 2\pi r^2$, or $A = 2\pi r(h + r)$, or $A = \pi dh + \frac{1}{2}\pi d^2$.

### Sphere

The volume ($V$) of a sphere is equal to $\frac{4}{3}$ times Pi ($\pi$) times the cube of the radius ($r$). Formula: $V = \frac{4}{3}\pi r^3$.

When the diameter is known, the formula $V = \dfrac{\pi d^3}{6}$ is used.

### Right Circular Cone

The volume ($V$) of a right circular cone is equal to $\frac{1}{3}$ times Pi ($\pi$) times the square of the radius ($r$) of the base times the height ($h$). Formula: $V = \frac{1}{3}\pi r^2 h$.

### Pyramid

The volume ($V$) of a pyramid is equal to $\frac{1}{3}$ times the area of the base ($B$) times the height ($h$). Formula: $V = \frac{1}{3}Bh$.

### Angles

Two angles whose sum of measures is 90° are called *complementary angles*. Each angle is said to be the complement of the other. The sum of measures of complementary angles $A$ and $B$ is 90°. Formula: $A + B = 90°$.

Two angles whose sum of measure is 180° are called *supplementary angles*. Each angle is said to be the supplement of the other. The sum of measures of supplementary angles $A$ and $B$ is 180°. Formula: $A + B = 180°$.

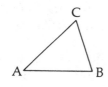

The sum of measures of the three angles of any triangle is 180°. In triangle $ABC$, the sum of measures of angles $A$, $B$, and $C$ is 180°. Formula: $A + B + C = 180°$.

## RELATED PRACTICE PROBLEMS

Solve each of the following problems, using the appropriate formulas:

1. **a.** Rectangle and Square
   (1) Find the perimeter of a rectangle 26 centimeters long and 19 centimeters wide.
   (2) What is the area of the interior of a rectangle 170 meters long and 96 meters wide?
   (3) How long is a rectangle when it is 35 millimeters wide and the perimeter is 200 millimeters?
   (4) If the length of a rectangular wing of an airplane is 50 meters and its area is 325 square meters, what is its width?
   (5) The width of a rectangular storage room is 48 meters and its floor area is 10,080 square meters. What is the perimeter of the room?
   (6) The perimeter of a rectangular airport site is 17,000 meters. If the length is 5,000 meters, find the area of the site in square meters.
   (7) What is the perimeter of a square whose side measures 57 millimeters?

(8) Find the area of the interior of a square whose side measures 31 centimeters.

(9) How long is a side of a square when its perimeter measures 76 meters?

(10) What is the area of a square garden if it takes 92 centimeters of fence to enclose the garden?

**b.** Circle

(1) Find the circumference of a circle whose diameter is 30 meters. ($\pi = 3.14$)

(2) What is the circumference of a circle whose radius is 84 centimeters? ($\pi = \frac{22}{7}$)

(3) Find the area of the interior of a circle with a radius of 17 millimeters. ($\pi = 3.14$)

(4) What is the area of the interior of a circle with a diameter of 28 meters? ($\pi = \frac{22}{7}$)

(5) What is the diameter of a circle when the circumference is 132 meters? ($\pi = \frac{22}{7}$)

(6) Find the radius of a circle when the circumference is 628 centimeters. ($\pi = 3.14$)

(7) Find the area of the interior of a circle when the circumference is 176 millimeters. ($\pi = \frac{22}{7}$)

(8) Find the area of the cross section of a tree trunk having a circumference of 3 meters 8 centimeters. ($\pi = \frac{22}{7}$)

(9) In making a complete turn about a pylon, an airplane travels $2\frac{3}{4}$ miles. How far away from the pylon did the plane fly? ($\pi = \frac{22}{7}$)

(10) What must the radius measure if a circular track is constructed so that the inside lane is 1 mile? How many square feet will be enclosed by this path of 1 mile? ($\pi = 3.14$)

**c.** Triangle, Parallelogram, Trapezoid

(1) Find the area of the interior of a triangle whose altitude is 26 meters and base is 19 meters.

(2) What is the area of the interior of a parallelogram with an altitude of 35 millimeters and a base of 40 millimeters?

(3) Find the area of the interior of a trapezoid with bases of 57 meters and 61 meters and a height of 66 meters.

(4) Find the perimeter of an equilateral triangle when its side measures 16 meters

(5) Find the base of a triangle if the area is 54 square meters and the altitude is 9 meters.

(6) The area of a triangle is 192 square centimeters and the base is 24 centimeters. What is the altitude?

(7) Find the base of a parallelogram if the area is 371 square centimeters and height is 7 centimeters.

(8) What is the height of a trapezoid if the area is 192 square meters, the upper base is 14 meters and lower base is 18 meters?

**d.** Rectangular Solid and Cube

(1) Find the volume of a rectangular solid 32 meters long, 25 meters wide, and 29 meters high.

(2) What is the volume of a cube whose edge measures 54 centimeters?

(3) What is the area of the outside surface of a rectangular solid 18 centimeters long, 14 centimeters wide, and 15 centimeters high?

(4) What is the area of the outside surface of a cube whose edge measures 11 meters?

(5) How long must a rectangular bin be to hold 1,800 cubic meters, if the width is 10 meters and height is 6 meters?

(6) A rectangular flower box is 18 millimeters long, 4 millimeters high and has a volume of 432 cubic millimeters. What is its width?

(7) How high must a packing crate be if it is required to hold 120 cubic meters and have a base of 24 square meters?

**e.** Right Circular Cylinder; Sphere; Cone; Pyramid

(1) Find the volume of a right circular cylinder with a radius of 14 meters and 20 meters high. ($\pi = \frac{22}{7}$)

(2) What is the volume of a right circular cylinder 6 centimeters in diameter and 8 centimeters high? ($\pi = 3.14$)

(3) Find the lateral area of a right circular cylinder $3\frac{1}{2}$ millimeters in diameter and 10 millimeters high. ($\pi = \frac{22}{7}$)

(4) What is the lateral area of a right circular cylinder with a radius of 30 meters and 19 meters high? ($\pi = 3.14$)

(5) What is the entire area of a right circular cylinder with a radius of 21 centimeters and 45 centimeters high? ($\pi = \frac{22}{7}$)

(6) What is the volume of a sphere whose radius is 7 millimeters? ($\pi = \frac{22}{7}$)

(7) Find the volume of a right circular cone 18 meters in diameter and 23 meters high. ($\pi = 3.14$)

(8) What is the volume of a square pyramid 125 meters on each side and 60 meters high?

(9) The lateral area of a cylindrical container is 132 square centimeters and the height is 7 centimeters. What is its volume? ($\pi = \frac{22}{7}$)

(10) A tin can is 6 millimeters high and its base has a diameter measuring 4 millimeters. If a second tin can has a base with a diameter

of 6 millimeters, how high must it be to have the same volume as the first can?

(11) 4,092 square meters of material is used to make a cylindrical tank including the two bases. How many cubic meters does the tank hold if the radius of the base is 21 meters? ($\pi = \frac{22}{7}$)

(12) The volume of a right circular cone is 12,474 cubic centimeters. Its base has a radius of 63 centimeters. Find its height. ($\pi = \frac{22}{7}$)

## B. *INTEREST*

The interest ($i$) equals the principal ($p$) times the rate of interest per year ($r$) times the time expressed in years ($t$). Formula: $i = prt$

The amount ($A$) is equal to the sum of the principal ($p$) and the interest ($i$). Formula: $A = p + i$ or $A = p + prt$

## *RELATED PRACTICE PROBLEMS*

2. Solve each of the following problems, using the appropriate formulas:

   a. How much money must be invested at 15% to earn an annual income of $1,800?

   b. How many years will it take $1,000 invested at 12% to earn $3,000 interest?

   c. Mr. Gomez borrowed $50 and at the end of 5 years paid $45 interest. Find the annual rate of interest.

   d. Mrs. Barr paid $210 interest at the end of 3 years. How much money did she borrow at 14% per year?

   e. Joan's brother borrowed $400 at 8% interest per year. How long was he in debt if he paid $128 interest?

   f. Camila repaid $192 of which $35 was interest. How much money did she borrow?

   g. Mr. Jensen borrowed $200 at 11% per year. He repaid the loan with a check in the amount of $248 which included the interest. How long did he borrow the money?

   h. Arnold borrowed $180. After 2 years he repaid $216, which included both the amount borrowed and the interest. Find the annual rate of interest.

   i. How much did Carol borrow at 9% per year if at the end of 6 years she repaid $2,002 which included both the amount borrowed and the interest?

   j. At the end of 4 years Patrick received $238, which included both the amount invested and the interest. How much money did he invest at $8\frac{1}{2}$% per year?

## C. *GEARS, PULLEYS, and SPROCKETS*

### (1) GEARS

Suppose one gear with 36 teeth is meshed (teeth of one gear fit the grooves of the other) with another gear of 12 teeth. When the gear with 12 teeth makes one complete turn, the gear with 36 teeth has made one third of a turn. When the larger gear makes one complete revolution, the smaller gear has turned 3 times. Since one gear drives the other, one of the meshed gears is called the driving gear and the other the driven gear. The number of teeth $(T)$ of the driving gear times its number of revolutions per minute $(R)$ equals the number of teeth $(t)$ of the driven gear times its number of revolutions per minute $(r)$. This relationship is expressed by the formula: $TR = tr$.

### SAMPLE SOLUTION

A gear with 48 teeth revolves at the rate of 150 r.p.m. (revolutions per minute) and drives a gear with 60 teeth. How many r.p.m. does the driven gear make?

$$
\begin{aligned}
\text{Given} \quad T &= 48 \text{ teeth} & TR &= tr \\
R &= 150 \text{ r.p.m.} & 48 \times 150 &= 60r \\
t &= 60 \text{ teeth} & 7{,}200 &= 60r \\
\text{Find } r. & & 120 &= r \\
& & r &= 120
\end{aligned}
$$

*Answer:* 120 r.p.m.

### (2) PULLEYS

If two pulleys (wheels mounted so that they turn on axles) are belted together and one pulley drives the other, a relationship similar to the meshed gears exists.

Suppose the driving pulley has an 8-inch diameter and the driven pulley a 4-inch diameter. The circumference of the driving pulley is twice the circumference of the driven pulley. Therefore, for each turn of the driving pulley there will be two turns of the driven pulley. The speed of a pulley is determined by the distance through which a point on the circumference moves in one minute. Consequently, the speed depends on the circumference of the pulley and the nuber of revolutions it makes per minute.

The circumference of the driving pulley times its number of revolutions per minute equals the circumference of the driven pulley times its number of revolutions per minute.

Or, since the diameters of the two pulleys are related exactly as the circumferences, the diameter ($D$) of the driving pulley times its number of revolutions per minute ($R$) equals the diameter ($d$) of the driven pulley times its number of revolutions per minute ($r$). This relationship is expressed by the formula: $DR = dr$.

### SAMPLE SOLUTION

Find the diameter of a driven pulley if a 24-inch pulley running at 125 r.p.m. is driving a pulley making 200 r.p.m.

$$\text{Given } D = 24 \text{ inches} \qquad DR = dr$$
$$R = 125 \text{ r.p.m.} \qquad 24 \times 125 = d \times 200$$
$$r = 200 \text{ r.p.m.} \qquad 3{,}000 = 200d$$
$$\text{Find } d. \qquad 15 = d$$
$$d = 15$$

*Answer:* 15 inches

### (3) SPROCKETS

If two sprockets are connected by a chain, a relationship similar to meshed gears and belted pulleys exists. The number of teeth ($T$) of the driving sprocket times the number of revolutions per minute ($R$) equals the number of teeth ($t$) of the driven sprocket times its number of revolutions per minute ($r$). Formula: $TR = tr$.

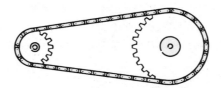

Also, the diameter ($D$) of the driving sprocket times its number of revolutions per minute ($R$) equals the diameter ($d$) of the driven sprocket times its number of revolutions per minute ($r$). Formula: $DR = dr$.

### RELATED PRACTICE PROBLEMS

3. Solve each of the following problems, using the appropriate formula:
   **a.** How many teeth are required on a gear if it is to run at 210 r.p.m. when driven by a gear with 60 teeth running at 140 r.p.m.?

   **b.** An 18-inch pulley turning 160 r.p.m. drives a 6-inch pulley. How many revolutions per minute is the 6-inch pulley turning?

**c.** The sprocket attached to the pedal of a bicycle has 32 teeth. The sprocket attached to the hub of the rear wheel has 8 teeth. If a boy turns the pedal 72 times per minute, how many times does the rear wheel turn in one minute?

**d.** What size pulley must be used on a motor running at 1,750 r.p.m. in order to drive a 16-inch pulley at 500 r.p.m.?

**e.** A tricycle with a chain drive has a sprocket with 15 teeth attached to the pedal and a sprocket with 5 teeth attached to the rear axle. If a youngster rotates the pedal 40 times per minute, how many times do the rear wheels turn in one minute?

**f.** Judy's bicycle has a sprocket with 30 teeth attached to the pedal and a sprocket with 10 teeth attached to the hub of the rear wheel. If she makes the pedal rotate at the average rate of 60 times per minute, how many times does the rear wheel turn in one minute? If the diameter of the rear wheel is 28 inches, how far can she go in one minute? At this rate how long will it take to ride a mile?

**g.** A 72-tooth gear running at 170 r.p.m. is required to drive a second gear at 255 r.p.m. How many teeth must there be in the driven gear?

**h.** A 28-inch pulley is belted to a 7-inch pulley. If the larger pulley makes 375 r.p.m., how many revolutions per minute does the 7-inch pulley make?

**i.** How many revolutions per minute is a 48-tooth gear running when driven by a 36-tooth gear running at 135 r.p.m.?

**j.** What size pulley is required on a machine if a 20-inch pulley on a shaft running at 240 r.p.m. is to drive the machine at 600 r.p.m.?

**D.** *ELECTRICITY*

The principles of electricity and related electrical formulas are employed in the fields of radio, television, and radar. Knowledge of some of the simpler elements is useful to the average person. Many technical terms such as watts, kilowatt-hours, volts, and amperes are commonly used.

Analysis and use of the following electrical formulas will enable pupils to understand some of the principles of electricity and the relationship between the technical terms.

(1) $E = IR$        (2) $W = IE$        (3) $H.P. = \dfrac{W}{746}$

(4) $K.W. = \dfrac{W}{1,000}$        (5) $K.W.Hr. = \dfrac{WT}{1,000}$

In these formulas $E$ represents the ~~electromotive force,~~ sometimes called the difference in pressure or potential. This force pushes the electric current through the conductor and is measured by the unit called the ~~volt.~~ It may be compared to the force a pump exerts in pushing water through a pipe.

$I$ represents the *intensity of current* or amount of electricity flowing and is measured by the unit called the *ampere*. This flow of electricity may be compared to the amount of water flowing through a pipe measured in gallons per minute.

$R$ represents the *resistance* of the conductor and is measured by the unit called the *ohm*. This resistance may be compared to the resistance a clogged pipe or a pipe with a small diameter may offer to the flow of water.

$W$ represents the *power* or *rate of work* done by the electric current and is measured by the unit called the *watt*. A *kilowatt* (K.W.) is 1,000 watts and is a unit measuring electric power. H.P. represents *horsepower* and is a unit measuring power. The *kilowatt-hour* is a unit measuring electrical energy. The amount of energy furnished may be found by multiplying the power in kilowatts by the time in hours ($T$).

The formula $E = IR$ shows the relationship between pressure, current, and resistance or between volts, amperes, and ohms. The number of *volts* equals the number of *amperes* times the number of *ohms*.

The formula $W = IE$ shows the relationship between power, current, and pressure or between watts, amperes, and volts. The number of *watts* equals the number of *amperes* times the number of *volts*.

## SAMPLE SOLUTION

How much current does a 660-watt electric iron take when operated on a 110-volt circuit? What is the resistance of the iron?

Given $W = 660$ watts     $W = IE$        $E = IR$

$E = 110$ volts       $660 = I \cdot 110$     $110 = 6R$

Find $I$ and $R$.        $660 = 110I$     $18\frac{1}{3} = R$

*Answers:* 6 amperes;     $6 = I$       $R = 18\frac{1}{3}$ ohms

$18\frac{1}{3}$ ohms       $I = 6$ amperes

## RELATED PRACTICE PROBLEMS

4. Solve each of the following problems, using the appropriate formulas:

   a. How many amperes of current are passing through a 110-volt circuit with a resistance of 30 ohms?

**b.** An electric iron has a resistance of 12 ohms. If it is used on a 110-volt circuit, how many amperes pass through it?

**c.** How many volts are required to ring a bell having a resistance of 40 ohms if a current of .2 amperes is used?

**d.** An electric broiler takes 3.5 amperes on a 110-volt circuit. What is the resistance of the broiler?

**e.** What is the resistance of an electric toaster if it takes 5 amperes at 110 volts?

**f.** How many amperes of current are flowing through a 40-watt lamp when operated on a 110-volt circuit?

**g.** A heating coil takes 20 amperes at 125 volts. How much power in watts is it consuming?

**h.** A 605-watt waffle iron is used on a 110-volt circuit. How much current does it take and what is the resistance of the waffle iron?

**i.** Find the voltage when a 100-watt lamp takes .4 amperes.

**j.** What is the resistance of a 25-watt, 110-volt lamp?

**k.** What power in kilowatts does a 110-volt lamp consume if it takes a current of .5 amperes?

**l.** If 75 amperes of current are used at 110 volts, what power in kilowatts is expended? In horsepower?

**m.** Find the horsepower consumed if a motor takes 12 amperes at 110 volts.

**n.** How much will it cost to burn five 40-watt lamps for 4 hours each evening for 30 days at the average price of 5 cents per kilowatt hour?

**o.** At 7 cents per kilowatt hour how much will it cost to operate for 8 hours a motor that uses 9 amperes of current at 125 volts?

**E.** *LEVERS*

The subject of the lever, a very common machine, is studied in physics. However, the simplest lever, the familiar seesaw or teeterboard, illustrates an important scientific principle which may be studied here.

In all likelihood you have seen two children at play with the seesaw. The heavier child sitting closer to the point of support balances a lighter child sitting further away from the point of support. The principle of balancing a teeterboard is the law of the lever.

The lever is a bar or board. Its point of support is called the *fulcrum*. The distance from the point where the weight is located to the fulcrum is called an *arm* of the lever.

The law of the lever states that one weight ($W$) times the length of its arm ($L$) is equal to the second weight ($w$) times the length of its arm ($l$). Expressed as a formula this relationship is: $WL = wl$.

Where must Tom, who weighs 112 pounds, sit on a seesaw to balance Richard, who weighs 126 pounds, and sits 4 feet from the fulcrum?

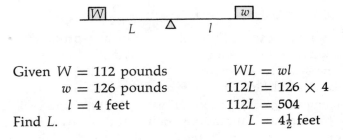

Given $W = 112$ pounds  $WL = wl$
$w = 126$ pounds  $112L = 126 \times 4$
$l = 4$ feet  $112L = 504$
Find $L$.  $L = 4\frac{1}{2}$ feet

*Answer:* $4\frac{1}{2}$ feet

## RELATED PRACTICE PROBLEMS

5. Solve each of the following problems, using the formula:
   a. When Marilyn sits 7 feet and Charlotte 6 feet from the fulcrum, they balance on the seesaw. How much does Charlotte weigh if Marilyn weighs 72 pounds?
   b. Frank weighs 90 pounds and sits on a teeterboard 5 feet from the fulcrum. If Ed weighs 105 pounds, how far from the fulcrum must he sit to just balance Frank?
   c. One arm of a lever is 8 inches long and the other is 10 inches long. If a weight of 4 pounds is placed on the 10-inch arm, how much weight is required on the 8-inch arm to balance the lever?
   d. Where must a 120-pound weight be hung to balance an 80-pound weight hung 5 feet from the fulcrum of a lever?
   e. Tony weighs 75 pounds and his cousin, Salvatore, weighs 50 pounds. If Tony sits 6 feet from the fulcrum of a teeterboard, where must Salvatore sit to balance the teeterboard?
   f. How much downward pressure must Richard exert on a 9-foot lever arm to raise a weight of 135 pounds on a 6-foot lever arm?
   g. Paul wants to lift a 60-pound rock with a 5-foot crowbar. If the fulcrum is 6 inches from the rock, how many pounds of pressure must he use to lift the rock?
   h. How much downward pressure must be exerted to raise a 210-pound rock with a 6-foot crowbar if the fulcrum is 1 foot from the rock?
   i. What downward force must be applied at a distance of 4 feet from the fulcrum to balance a downward force of 30 pounds applied 6 feet on the other side of the fulcrum?

**j.** What weight can Harry lift with a 5-foot crowbar when the fulcrum is 9 inches from the object and he exerts a downward pressure of 108 pounds?

## F. *TEMPERATURE*

In science, especially in the field of meteorology, we are often required to convert Fahrenheit, Celsius, and absolute temperature readings from one scale to another.

Gabriel Fahrenheit, a German scientist, invented the Fahrenheit scale using the freezing point of water (32°) and the boiling point (212°) as the basis of his calibrations. Between these two limits are 180 equal divisions called *degrees*. In the United States the Fahrenheit scale is in general use.

Most Europeans use the Celsius scale. It is used universally for scientific purposes. On the Celsius scale the freezing point of water is 0° and the boiling point is 100°. Between these two points there are 100 equal divisions which compare to the 180 smaller divisions on the Fahrenheit scale. The Kelvin temperature scale, used in SI measurement, is related to the Celsius scale. One degree Celsius is exactly equal to one *Kelvin* (the name used to mean degree Kelvin). The reading of a specific temperature on the Kelvin scale is approximately 273 Kelvins more than its reading on the Celsius scale.

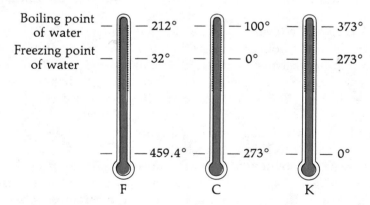

The following formulas show the relationships of the three temperature scales:

$$F = 1.8C + 32 \text{ or } C = \tfrac{5}{9}(F - 32)$$
$$K = C + 273$$

where $F$ represents Fahrenheit reading, $C$ represents Celsius reading, and $K$ represents Kelvin reading.

## Sample Solution

Find the Celsius temperature reading corresponding to a Fahrenheit temperature reading of 86°.

Given $F = 86$             $F = 1.8C + 32$
Find $C$.                $86 = 1.8C + 32$
           $86 + (-32) = 1.8C + 32 + (-32)$
                  $54 = 1.8C$
                  $30 = C$
                  $C = 30$

*Answer:* 30°

## *RELATED PRACTICE PROBLEMS*

6. Solve each of the following problems, using the appropriate formula:
   a. Change the following Fahrenheit readings to corresponding Celsius readings:
      (1) 176° F     (2) 41° F     (3) 320° F     (4) 185° F     (5) 77° F
   b. Change the following Celsius readings to corresponding Fahrenheit readings:
      (1) 50° C     (2) 75° C     (3) 110° C     (4) 18° C     (5) 31° C
   c. Change the following Celsius readings to corresponding Kelvin readings:
      (1) 28° C     (2) 43° C     (3) 87° C     (4) 9° C     (5) 112° C
   d. If the temperature of standard air at sea level is 59° F, what is the corresponding reading on the Celsius scale?
   e. If the temperature at an altitude of 1,000 feet is 29° C, what is the corresponding Fahrenheit reading?
   f. The normal room temperature is 68° F. What would it be on the Celsius scale?
   g. A French recipe calls for heating the oven at a temperature of 140° C. If the heat controls are calibrated in the Fahrenheit units, at what Fahrenheit temperature should the controls be set?
   h. The normal body temperature is 98.6° F. What would it be on the Celsius scale?
   i. Which temperature is warmer, 108° F or 40° C?
   j. The melting point of uranium is 1,690° C. What is the corresponding reading on the Fahrenheit scale?

## G. *AVIATION*

A great many technical problems in aeronautics involve the wing of the airplane. In this section a simple treatment of the airplane wing is undertaken, formulas are developed, and related problems are solved.

### AREA

The area of a rectangular wing is found by multiplying its length by its width. The length of an airplane wing is called the *span* and its width the *chord* so that, technically, the area equals the span times the chord. Expressed as a formula this relationship is: $A = sc$. If the wing is tapered, the average or mean chord is used.

### ASPECT RATIO

The ratio of the span to the chord of an airplane wing is called its *aspect ratio*.

Expressed as a formula this relationship is: $R = \frac{s}{c}$. The aspect ratio is an important factor in the construction of an airplane. If the wing is built too narrow, it may not be strong enough to carry the weight of the airplane.

### WING LOADING

The number of pounds of gross weight supported in flight by each square foot of the wing is called *wing loading*. The wing loading ($L_w$) of an airplane is found by dividing the gross weight ($W$) by the wing area ($A$). Expressed as a formula this relationship is:

$$L_w = \frac{W}{A}$$

### LIFT AND DRAG

The *lift* of an airplane is the force that supports the plane in flight. It is produced mostly by the effect the airflow has upon the wing of the plane. The *lift* ($L$) of the wing expressed in pounds depends on a number of factors including the *wing area* ($A$) usually expressed in square feet, the *air speed* or *velocity* ($V$) expressed in feet per second, the *air density* ($d$), and the airfoil and angle of attack together called the *coefficient of lift* ($C_L$). The relationship between these quantities is expressed by the formula:

$$L = \frac{C_L dAV^2}{2}$$

The *drag* of an airplane is the resistance of the air to the motion of the plane. The *drag* (D) of the wing expressed in pounds also depends on the *wing area* (A), the *air speed* or *velocity* of the plane (V), the *air density* (d), and a factor called *coefficient of drag* (C_D). Expressed as a formula this relationship is:

$$D = \frac{C_D d A V^2}{2}$$

## Sample Solution

Find the (a) mean chord, (b) aspect ratio, and (c) the wing loading of a plane with a wing area of 392 square feet, a wing span of 56 feet, and a gross weight of 5,880 pounds.

Given $A = 392$ square feet $\qquad A = sc \qquad R = \frac{s}{c} \qquad L_w = \frac{W}{A}$

$\qquad\qquad s = 56$ feet $\qquad\qquad 392 = 56c$

$\qquad\qquad W = 5,880$ pounds $\qquad\quad 7 = c \qquad R = \frac{56}{7} \quad L_w = \frac{5880}{392}$

$\qquad\qquad\qquad\qquad\qquad\qquad\qquad c = 7$ feet $\quad R = 8 \qquad L_w = 15$ pounds per square foot

Find $c$, $R$, and $L_w$. $\qquad$ (a) *Answer:* (b) *Answer:* 8 (c) *Answer:* 15
$\qquad\qquad\qquad\qquad\qquad\qquad\quad$ 7 feet $\qquad\qquad\qquad\qquad\qquad$ pounds per square foot

## *RELATED PRACTICE PROBLEMS*

7. Solve each of the following problems, using the appropriate formula:
   a. A plane with a wing span of 48 feet and a mean chord of 6 feet has a gross weight of 3,456 pounds. Find the aspect ratio and the wing loading.
   b. Find the mean chord, aspect ratio, and wing loading of a plane with a wing area of 486 square feet, a wing span of 54 feet, and a gross weight of 5,103 pounds.
   c. A plane has a wing span of 39 feet and a wing area of 234 square feet. It has a gross weight of 2,574 pounds. Find the mean chord, aspect ratio, and wing loading.
   d. Determine the lift and drag on the wing of an airplane flying at 150 m.p.h. where the air density is .002. The wing has an area of 180 square feet. The coefficient of lift is .56 and the coefficient of drag is .028.
   e. An airplane flying at 120 m.p.h. where the air density is .0018 has a wing area of 200 square feet. Find the lift and drag on the wing of the airplane if the coefficient of lift is .48 and the coefficient of drag is .024.

# REVIEW

1. Using $a$ = Steve's age in years, write an equation for the verbal sentence: Five times Steve's age is equal to 26 years more than three times his age.
2. Using $x$ = speed of car $A$ in m.p.h. and $x + 15$ = speed of car $B$ in m.p.h., write an equation for the verbal sentence: Twice the speed of car $A$ decreased by 20 m.p.h. is equal to the speed of car $B$ increased by 10 m.p.h.
3. Use the equation method to find the required number in each of the following:
   a. Seven-eighths of what number is 42?
   b. 40% of what number is 98?
   c. The first number is 3 more than a second number. Four times the first number increased by 9 is equal to seven times the second number. What are the numbers?
   d. What three consecutive even integers have a sum of 264?

Solve each of the following problems by the equation method:
4. A merchant purchased radios at $39 each. What price should each radio be marked so that a 25% profit on the selling price may be realized?
5. The length of a rectangle is seven times its width. If the perimeter is 208 inches, what is the length of the rectangle?
6. Mr. Williams cashed a check for $365, receiving $20 bills and $5 bills only. If he was given eight more $5 bills than $20 bills, how many $5 bills did he get?
7. How long will it take an airplane, flying at an average speed of 435 m.p.h. to intercept an airplane, 1,190 miles away, flying directly toward it at an average speed of 245 m.p.h.?
8. The sum of Scott's age and Peter's age is 20. Four years from now, Scott will be twice as old as Peter. What are their present ages?

Solve each of the following, using the appropriate formulas:
9. Find the area of the interior of a circle when the circumference is 132 inches. ($\pi = \frac{22}{7}$)
10. When the room temperature is 72° F, what would the temperature reading be on the Celsius scale?
11. A 660-watt electric iron is used on a 110-volt circuit. How much current does it take? What is the resistance of the iron?

## Keyed Achievement Test

The colored numeral indicates the exercise where help may be found.
1. Find the sum of $5m^2 - mn - 3n^2$, $6n^2 - 9m^2$, $5m^2 - 6mn$. [5-3]
2. From $3a^2 - 4b^2$ subtract $7a^2 - 6ab + b^2$. [5-5]
3. Multiply $6x^2 - 2xy + 7y^2$ by $5x - y$. [5-9]
4. Divide $28n^3 + 29n^2 - 80n + 42$ by $4n + 3$. [5-13]
5. Solve and check: [6-4 thru 6-8]                **a.** $7c + 11 = 53$
   **b.** $10y + 8 = 3$         **c.** $8n - 3n = -30$         **d.** $19 - x = x - 19$
   **e.** $11 - 3(5 - 2x) = 13x - 18$        **f.** $(r + 13)(r + 3) = (r + 10)(r - 1)$
6. Find the value of: [6-10]
   **a.** $n$ when $S = 1,620$, using the formula $S = 180n - 360$
   **b.** $d$ when $l = 103$, $a = 19$, and $n = 15$, using the formula $l = a + (n - 1)d$
   **c.** $N$ when $S = 3,300$, $\pi = \dfrac{22}{7}$, and $d = 14$, using the formula $S = \dfrac{\pi dN}{12}$

Solve each of the following problems by the equation method:
7. A dealer sold a car for $5,310, making an 18% profit on the cost. How much did the dealer pay for the car? [7-5]   4,500 $
8. Kim is now 3 years older than her brother. Nine years ago she was twice as old as her brother. What are their ages now? [7-5]   Kim 15 Brot 12

Solve each of the following, using the appropriate formulas:
9. The perimeter of a rectangular garden is 596 feet. If the width is 117 feet, what is the area of the garden? [7-6]   Area
10. How much did Tim borrow at 12% per year if at the end of 4 years he repaid $1,332 for the amount borrowed and the interest? [7-6]   900 $

## *MAINTENANCE PRACTICE IN ARITHMETIC*

**1.** Add:
53,296
88,739
45,687

**2.** Subtract:
527,406
85,297

**3.** Multiply:
605
867

**4.** Divide:
$704\overline{)68,992}$

**5.** Add:
$7\frac{2}{3} + 2\frac{3}{4}$

**6.** Subtract: $9\frac{4}{5} - 4$          **7.** Multiply: $12 \times 3\frac{5}{6}$          **8.** Divide: $5\frac{1}{4} \div \frac{15}{16}$
**9.** Add: $.66 + .6 + .666$ **10.** Subtract: $\$500 - \$387.49$ **11.** Multiply: $.87\frac{1}{2} \times \$200$
**12.** Divide: $\$.65\overline{)\$13.}$          **13.** Find $103\frac{1}{2}\%$ of $10,000.
**14.** $.45 is what percent of $1.20?          **15.** 9% of what amount is $558?

# INTRODUCTION

An *inequality* is a mathematical sentence that uses the symbol $<$, $>$, $\neq$, $\leq$, $\geq$, $\not<$, $\not>$, $\not\leq$, or $\not\geq$ as its verb. Sometimes $\leq$ is written as $\leqq$ and $\geq$ as $\geqq$. See Exercise 8-1 for the meanings of these symbols.

In this chapter we shall learn how to solve inequalities which contain a variable representing numbers. To solve an inequality in one variable means to find all the numbers each of which, when substituted for the variable will make the inequality a true sentence.

Inequalities, like equations, are solved by using transformations based on the axioms and on the additive and multiplicative inverses to get equivalent inequalities in simplest form, each of which has the same solutions as the given inequality. An *inequality in simplest form* is expressed with only the variable as one member of the inequality and a numeral for the other member. *Equivalent inequalities* are inequalities having exactly the same solutions.

The solutions of the basic types on inequalities in their simplest forms are studied first. See Exercise 8-3. Then we learn how to transform inequalities to these basic types in simplest form to determine their solutions. Since the solution of an inequality depends not only on the inequality but also on the replacements for the variable, number systems are reviewed. Also note the method of checking answers. See pages 314, 320, 325, 330, 335, and 340.

# Inequalities in One Variable

**8**

# Reading and Writing Inequalities

**I. Aim**    To learn to read inequalities and to write them symbolically.

**II. Procedure**

1. To read or write an inequality:
   **a.** Read or write each variable, numeral symbols of operation, and inequality symbol in the required order.
   **b.** The variable in an inequality is read as "each number and *name* the variable." The variable *n* in the inequality $n > 5$ is read: "each number *n*."
   **c.** Also review Exercises 1-6 and 1-7.

2. The verb symbols used with inequalities are:

   $\neq$ is read "is not equal to."

   $<$ is read "is less than."

   $\not<$ is read "is not less than."

   $>$ is read "is greater than."

   $\not>$ is read "is not greater than."

   $\leq$ is read "is less than or equal to."

   $\not\leq$ is read "is not less than or not equal to."

   $\geq$ is read "is greater than or equal to."

   $\not\geq$ is read "is not greater than or not equal to."

3. Observe that:
   **a.** Equivalent verb symbols

   $\not<$ (is not less than) and $\geq$ (is greater than or equal to) are equivalent.

   $\not>$ (is not greater than) and $\leq$ (is less than or equal to) are equivalent.

   $\not\leq$ (is not less than or not equal to) and $>$ (is greater than) are equivalent.

   $\not\geq$ (is not greater than or not equal to) and $<$ (is less than) are equivalent.

   **b.** Disjunction—(Two simple sentences connected by "or")
   (1) The verb symbols used are: $\leq$, $\geq$, $\not<$, and $\not>$.
   (2) The inequality $8n \leq 56$ is the shortened form of $8n < 56$ or $8n = 56$.
   The inequality $5x \geq 20$ is the shortened form of: $5x > 20$ or $5x = 20$.

302

    **c.** Conjunction—(Two simple sentences connected by "and")

        (1) The verb symbols are used in pairs with the variable written between the symbols. They include: $<<$, $>>$, $\leq\leq$, $\geq\geq$, $\leq<$, $<\leq$, $\geq>$, $>\geq$.

        (2) The sentence $10 < n < 18$ is the shortened form of: $10 < n$ and $n < 18$.

        The sentence $5 > y > 2$ is the shortened form of: $5 > y$ and $y > 2$.

**4. a.** Inequalities that use the same verb symbol or an equivalent verb symbol are said to be of the *same order*.

        The inequalities $n > 5$ and $x > 2$ are of the same order.

    **b.** Inequalities that use opposite verb symbols are said to be of the *reverse order*.

        The inequalities $n > 12$ and $n < 9$ are of the reverse order.

**5.** When required to rewrite an inequality so that the variable appears on the other side of the inequality symbol, reverse the order symbol.

        $8 < g$ and $g > 8$ are equivalent.

        But $8 < g$ and $g < 8$ cannot be true at the same time.

## III. Sample Solutions

**1.** Read: $n + 9 \neq 15$.

*Answer:* Each number plus nine is not equal to fifteen.

**2.** Read: $y < 7$.

*Answer:* Each number $y$ is less than seven.

**3.** Read: $c - 2 > 5$.

*Answer:* Each number $c$ minus two is greater than five.

**4.** Read: $a + 4 \not> 8$.

*Answer:* Each number $a$ plus four is not greater than eight.

**5.** Read: $8n \leq 56$.

*Answer:* Eight times each number $n$ is less than or equal to fifty-six.

**6.** Read: $5x \geq 20$.

*Answer:* Five times each number $x$ is greater than or equal to twenty.

**7.** Read: $b - 2 \not\geq 11$.

*Answer:* Each number $b$ minus two is not greater than or not equal to eleven.

**8.** Read: $a + 6 \not\leq 10$.

*Answer:* Each number $a$ plus six is not less than or not equal to ten.

**9.** Read: $10 < n < 18$.

*Answer:* Each number $n$ is greater than ten and less than eighteen.

**10.** Read: $5 > y > 2$.

*Answer:* Each number $y$ is less than five and greater than two.

**11.** Read: $4 \leq n \leq 14$.

*Answer:* Each number $n$ is greater than or equal to four and less than or equal to fourteen.

**12.** Read: $15 \geq d \geq 7$.

*Answer:* Each number $d$ is less than or equal to fifteen and greater than or equal to seven.

**13.** Write an inequality of the reverse order to that in $x > 8$.

*Answer:* $x < 8$

## DIAGNOSTIC TEST

**1.** Read, or write in words, each of the following:

　　**a.** $6x - 8 = 15$　　　　**b.** $t + 9 < 7$　　　　**c.** $25 > n + 2$

　　**d.** $11c \nless 50$　　　　**e.** $9a - 5 \neq 4a + 13$　　　　**f.** $\dfrac{3b}{2} \ngtr 10$

**2.** Write symbolically each of the following:

　　**a.** Some number $c$ decreased by one is equal to twelve.

　　**b.** Each number $y$ increased by six is greater than negative four.

　　**c.** Four times each number $x$ is not less than fifty-one.

　　**d.** Eight times each number $b$ is not greater than twenty-five.

　　**e.** Each number $t$ divided by four is less than thirteen.

　　**f.** Ten times each number $r$ is not equal to thirty-two.

**3.** Read, or write in words:

　　**a.** $6y \geq 48$　　　　　　　　**b.** $3x + 7 \leq -9$

　　**c.** $46 \nleq 7n - 11$　　　　　**d.** $\dfrac{m}{8} \ngeq 15$

**4.** Write symbolically each of the following:

　　**a.** Each number $x$ minus eleven is less than or equal to twenty.

　　**b.** Nine times each number $y$ is not greater than or not equal to negative six.

　　**c.** Twenty times each number $a$ increased by five is greater than or equal to forty-nine.

　　**d.** Twelve times each number $m$ decreased by two is not less than or not equal to fifty.

**5. a.** Is $8 \leq 12$ equivalent to $8 \ngtr 12$?

　　**b.** Is $10 \nless 4$ equivalent to $10 \ngeq 4$?

　　**c.** Is $5 > 1$ equivalent to $5 \nleq 1$?

　　**d.** Is $17 \ngeq 20$ equivalent to $17 < 20$?

6. Which of the following symbols is equivalent to:
   a. $<$?:          (1) $\geq$          (2) $\not>$          (3) $\nleq$          (4) $\ngeq$
   b. $\geq$?:          (1) $\nleq$          (2) $<$          (3) $\nless$          (4) $\ngtr$
7. Read, or write in words, each of the following:
   a. $6 < b < 16$          b. $2 > n > -3$
   c. $10 \geq x \geq 1$          d. $-11 \leq 5d \leq 0$
8. Write symbolically each of the following:
   a. Each number $y$ is greater than seven and less than fourteen.
   b. Each number $n$ is less than negative five and greater than negative twelve.
   c. Each number $r$ is less than or equal to zero and greater than or equal to negative seven.
   d. Twice each number $m$ is greater than or equal to fifteen and less than or equal to thirty.
9. Read, or write in words, each of the following:
   a. $-2 < x \leq 7$          b. $15 \geq h > 4$
10. Which of the following pairs of inequalities have the same order?
   a. $y < 6$ and $y > -2$          b. $n > -7$ and $n > 0$
11. Write an inequality of the reverse order to that indicated in each of the following:
   a. $c < 10$          b. $r > -5$          c. $x \not> 16$          d. $y \leq 2$
12. For each of the following rewrite the second inequality inserting the correct symbol:
   a. $10 < x$ and $x$ ? $10$ are equivalent.
   b. $-6 \geq y$ and $y$ ? $-6$ are equivalent.

## RELATED PRACTICE EXAMPLES

1. Read, or write in words, each of the following:
   a. $x > 9$          b. $0 = 16c - 8$          c. $3y < 21$
   d. $n + 4 \nless 15$          e. $c - 6 \neq 13$          f. $8b \not> 48$
   g. $m - 19 < 7$          h. $12a - 5 = 20$          i. $27 > b + 18$
   j. $14 \nless 5d - 6$          k. $10x - 8 \neq 2x + 27$          l. $\dfrac{n}{5} \not> 9$

2. Write symbolically each of the following:
   a. Some number $x$ increased by four is equal to twenty.
   b. Each number $n$ decreased by nine is less than two.
   c. Twelve times each number $c$ is not equal to zero.
   d. Seven times each number $a$ minus six is greater than forty-eight.

**e.** Sixteen times each number $m$ is not less than seventy.

**f.** Each number $n$ divided by five is not greater than four.

**g.** Each number $b$ plus nine is not less than negative five.

**h.** Some number $z$ minus six is equal to negative eighteen.

**i.** Six times each number $d$ plus fifteen is greater than forty.

**j.** Each number $g$ increased by twelve is not equal to negative ten.

**k.** Three times each number $h$ minus seven is less than twenty-one.

**l.** Eleven times each number $y$ decreased by one is not greater than eight.

3. Read, or write in words, each of the following:

**a.** $s \leq 7$      **b.** $10d \geq 50$      **c.** $4n \nleq -24$

**d.** $8 \ngeq r + 2$      **e.** $9t + 6 \leq 15$      **f.** $b - 10 \leq -23$

**g.** $18 \ngeq 6y$      **h.** $4 - x \geq 1$      **i.** $19 \ngeq 12a - 13$

**j.** $2m + 4 \leq m - 7$      **k.** $11x - 18 \geq 8x + 21$      **l.** $\dfrac{t}{7} \nleq 3$

4. Write symbolically each of the following:

**a.** Each number $b$ plus two is greater than or equal to twelve.

**b.** Five times each number $x$ is less than or equal to thirty.

**c.** Each number $n$ minus eight is not less than or not equal to four.

**d.** Fourteen times each number $y$ is not greater than or not equal to negative two.

**e.** Each number $g$ plus nineteen is less than or equal to ninety.

**f.** Seven times each number $c$ is not greater than or not equal to seventy-five.

**g.** Twice each number $p$ increased by seven is greater than or equal to negative nine.

**h.** Four times each number $x$ decreased by sixteen is not less than or not equal to zero.

5. **a.** Is $9 \ngtr 10$ equivalent to $9 \leq 10$?

**b.** Is $12 \nleq 15$ equivalent to $12 < 15$?

**c.** Is $4 \nless 2$ equivalent to $4 \ngeq 2$?      **d.** Is $8 \geq 3$ equivalent to $8 < 3$?

**e.** Is $1 < 7$ equivalent to $1 \ngeq 7$?      **f.** Is $5 > 0$ equivalent to $5 \leq 0$?

**g.** Is $11 \ngeq 16$ equivalent to $11 \nless 16$? **h.** Is $6 \leq 12$ equivalent to $6 > 12$?

6. Which of the following symbols is equivalent to:

**a.** $>$?:      (1) $\nless$      (2) $\ngtr$      (3) $\nleq$      (4) $\geq$

**b.** $\nless$?:      (1) $>$      (2) $\geq$      (3) $\leq$      (4) $\ngeq$

**c.** $\leq$?:      (1) $\geq$      (2) $<$      (3) $\ngtr$      (4) $\ngeq$

**d.** $\ngeq$?:      (1) $\leq$      (2) $\ngtr$      (3) $\nless$      (4) $<$

**e.** $\ngtr$?:      (1) $<$      (2) $\geq$      (3) $\leq$      (4) $\nleq$

**f.** $<$?:      (1) $\ngtr$      (2) $\geq$      (3) $\ngeq$      (4) $\nleq$

**g.** $\nleq$?:      (1) $\ngeq$      (2) $\nless$      (3) $\geq$      (4) $>$

**h.** $\geq$?:      (1) $\nleq$      (2) $<$      (3) $\ngtr$      (4) $\nless$

7. Read, or write in words, each of the following:
   **a.** $5 < n < 10$      **b.** $-3 < b < 7$      **c.** $18 > r > 0$
   **d.** $25 > c > -1$      **e.** $-8 < 3x < -4$      **f.** $7 \leq a \leq 17$
   **g.** $14 \geq d \geq -2$      **h.** $-6 \leq y \leq -1$      **i.** $27 \geq m \geq 12$

8. Write symbolically each of the following:
   **a.** Each number $x$ is greater than five and less than eighteen.
   **b.** Each number $d$ is less than nine and greater than zero.
   **c.** Each number $r$ is greater than negative twelve and less than negative two.
   **d.** Each number $h$ is less than or equal to twenty and greater than or equal to seventeen.
   **e.** Ten times each number $n$ is greater than or equal to seven and less than or equal to fifty.
   **f.** Four times each number $y$ is less than or equal to seventeen and greater than or equal to negative one.

9. Read, or write in words, each of the following:
   **a.** $5 \leq n < 8$      **b.** $6 < a \leq 13$      **c.** $14 > b \geq 3$
   **d.** $18 \geq y > 7$      **e.** $-3 < h \leq 0$      **f.** $-9 \geq t > -21$

10. Which of the following pairs of inequalities have the same order?
    **a.** $x < 6$ and $x > 6$      **b.** $y > 5$ and $y > 10$      **c.** $t < -3$ and $t < 0$
    **d.** $n \not< 3$ and $n \not> 11$      **e.** $x \leq 10$ and $x \leq -1$      **f.** $b > 14$ and $b < 9$

11. Write an inequality of the reverse order to that indicated in each of the following:
    **a.** $n < 6$      **b.** $b > 5$      **c.** $x > -8$      **d.** $r < -1$
    **e.** $t < 0$      **f.** $y \geq 2$      **g.** $s \not< -5$      **h.** $a \leq -4$

12. For each of the following rewrite the second inequality inserting the correct symbol:
    **a.** $12 > n$ and $n ? 12$ are equivalent.
    **b.** $-4 < y$ and $y ? -4$ are equivalent.
    **c.** $0 \not> t$ and $t ? 0$ are equivalent.
    **d.** $7 \not< x$ and $x ? 7$ are equivalent.
    **e.** $-1 \geq m$ and $m ? -1$ are equivalent.
    **f.** $9 \leq r$ and $r ? 9$ are equivalent.
    **g.** $-11 \not\geq x$ and $x ? -11$ are equivalent.
    **h.** $1.4 \not\leq n$ and $n ? 1.4$ are equivalent.
    **i.** $-\frac{3}{4} < b$ and $b ? -\frac{3}{4}$ are equivalent.
    **j.** $800 \not> p$ and $p ? 800$ are equivalent.
    **k.** $-2 \not< g$ and $g ? -2$ are equivalent.
    **l.** $-.47 > w$ and $w ? -.47$ are equivalent.
    **m.** $4\frac{1}{2} \neq y$ and $y ? 4\frac{1}{2}$ are equivalent.
    **n.** $-39 \leq x$ and $x ? -39$ are equivalent.

# Properties for Solving Inequalities

I. Aim    To develop the various properties used with inequalities

II. Procedure
1. Study the corresponding sample solutions and observe that:
    a. When the same number (positive or negative) is added to or subtracted from both sides of an inequality, another equivalent of the *same order* results.
    b. When both sides of an inequality are either multiplied or divided by the same *positive* number, another equivalent inequality of the *same order* results.
    c. When both sides of an inequality are either multiplied or divided by the same *negative* number, another equivalent inequality of the *reverse order* results.

2. Also observe that an inequality does not have a reflexive property or a symmetric property but does have a transitive property.
    An inequality has:
    a. No reflexive property—    A number cannot be greater than itself nor
       $a \not> a; a \not< a$    can it be less than itself.

    b. No symmetric property—If one number is greater than a second
       If $a > b$, then $b \not> a$    number, the second number cannot be greater than the first number.

       If $a < b$, then $b \not< a$    If one number is less than the second number, the second number cannot be less than the first number.

    c. Transitive property—    If one number is greater than a second
       If $a > b$ and $b > c$,    number and the second number is greater
       then $a > c$    than a third number, then the first number is greater than the third number.

       If $a < b$ and $b < c$,    If one number is less than a second num-
       then $a < c$    ber and the second number is less than a third number, then the first number is less than the third number.

## III. Sample Solutions

1. Using the true inequality $7 > 4$, determine whether a true inequality of the same order results when you:

   **a.** Add 2 to each side.

   $7 > 4$
   $7 + 2 > 4 + 2$
   $9 > 6$

   *Answer:* $9 > 6$ is true and has the same order as $7 > 4$.

   **b.** Add $-2$ to each side.

   $7 > 4$
   $7 + (-2) > 4 + (-2)$
   $5 > 2$

   *Answer:* $5 > 2$ is true and has the same order as $7 > 4$.

2. Using the true inequality $5 < 8$, determine whether a true inequality of the same order results when you:

   **a.** Subtract 4 from each side.

   $5 < 8$
   $5 - 4 < 8 - 4$
   $1 < 4$

   *Answer:* $1 < 4$ is true and has the same order as $5 < 8$.

   **b.** Subtract $-4$ from each side.

   $5 < 8$
   $5 - (-4) < 8 - (-4)$
   $5 + (4) < 8 + (4)$
   $9 < 12$

   *Answer:* $9 < 12$ is true and has the same order as $5 < 8$.

3. Using the true inequality $2 < 9$, determine whether a true inequality of the same order results when you:

   **a.** Multiply each side by 5.

   $2 < 9$
   $5 \times 2 < 5 \times 9$
   $10 < 45$

   *Answer:* $10 < 45$ is true and has the same order as $2 < 9$.

   **b.** Multiply each side by $-5$.

   $2 < 9$
   $(-5) \times 2 < (-5) \times 9$
   $-10 < -45$

   *Answer:* $-10 < -45$ is false, but the *reverse order* $(-10 > -45)$ is true.

4. Using the true inequality $8 > 6$, determine whether a true inequality of the same order results when you:

   **a.** Divide each side by 2.

   $8 > 6$
   $8 \div 2 > 6 \div 2$
   $4 > 3$

   *Answer:* $4 > 3$ is true and has the same order as $8 > 6$.

   **b.** Divide each side by $-2$.

   $8 > 6$
   $8 \div (-2) > 6 \div (-2)$
   $-4 > -3$

   *Answer:* $-4 > -3$ is false, but the *reverse order* $(-4 < -3)$ is true.

## DIAGNOSTIC TEST

Determine the resulting true inequality when:

1. $-6$ is added to both sides of $12 > 9$?

2. $8$ is subtracted from both sides of $-1 < 10$?

3. Both sides of $4 < 10$ are multiplied by $3$?

4. Both sides of $15 \not> 25$ are divided by $5$?

Does the resulting true inequality have the same order as the given inequality or the reverse order when:

5. Both sides of $9 > 8$ are multiplied by $-4$?

6. Both sides of $-18 < 24$ are divided by $-6$?

7. What number should be subtracted from both sides of $n + 6 < 13$ to get the sentence $n < 7$?

8. What number should be added to both sides of $x - 2 > 8$ to get the sentence $x > 10$?

9. By what number should both sides of $12y \leq 60$ be divided to get the sentence $y \leq 5$?

10. By what number should both sides of $\frac{x}{3} \not> 5$ be multiplied to get the sentence $x \not> 15$?

## *RELATED PRACTICE EXAMPLES*

In each of the following, determine the resulting true inequality when:

1. **a.** $8$ is added to both sides of $15 < 16$.
   **b.** $-8$ is added to both sides of $10 > 5$.
   **c.** $12$ is added to both sides of $-1 < 2$.
   **d.** $-3$ is added to both sides of $-5 \not> -3$.

2. **a.** $2$ is subtracted from both sides of $19 < 21$.
   **b.** $-2$ is subtracted from both sides of $0 > -4$.
   **c.** $11$ is subtracted from both sides of $7 \not< 2$.
   **d.** $-6$ is subtracted from both sides of $-8 < -5$.

3. **a.** Both sides of $5 > 4$ are multiplied by $3$.
   **b.** Both sides of $9 < 15$ are multiplied by $7$.
   **c.** Both sides of $-3 < 1$ are multiplied by $6$.
   **d.** Both sides of $-4 \not< -2$ are multiplied by $10$.

4. **a.** Both sides of $25 < 30$ are divided by 5.
  **b.** Both sides of $56 > 14$ are divided by 7.
  **c.** Both sides of $-36 \not> 48$ are divided by 12.
  **d.** Both sides of $-63 > -27$ are divided by 9.
Does the resulting true inequality have the same order as the given inequality or the reverse order when:

5. **a.** Both sides of $8 < 11$ are multiplied by $-8$?
  **b.** Both sides of $16 > 14$ are multiplied by $-3$?
  **c.** Both sides of $-15 < 7$ are multiplied by $-2$?
  **d.** Both sides of $-3 \not> -1$ are multiplied by $-12$?

6. **a.** Both sides of $18 < 24$ are divided by $-6$?
  **b.** Both sides of $30 > 21$ are divided by $-3$?
  **c.** Both sides of $96 \not< -72$ are divided by $-8$?
  **d.** Both sides of $-75 < -90$ are divided by $-15$?

7. What number should be subtracted from both sides of the inequality:
  **a.** (1) $6 + 4 > 9$ to get the sentence $6 > 5$?
     (2) $x + 4 > 9$ to get the sentence $x > 5$?
  **b.** (1) $11 + 19 < 43$ to get the sentence $11 < 24$?
     (2) $r + 19 < 43$ to get the sentence $r < 24$?
  **c.** (1) $8 + 13 \not< 0$ to get the sentence $8 \not< -13$?
     (2) $n + 13 \not< 0$ to get the sentence $n \not< -13$?
  **d.** (1) $x + 7 \geq 15$ to get the sentence $x \geq 8$?
     (2) $n + 16 \leq -3$ to get the sentence $n \leq -19$?

8. What number should be added to both sides of the inequality:
  **a.** (1) $12 - 5 < 8$ to get the sentence $12 < 13$?
     (2) $x - 5 < 8$ to get the sentence $x < 13$?
  **b.** (1) $9 - 11 > -15$ to get the sentence $9 > -4$?
     (2) $y - 11 > -15$ to get the sentence $y > -4$?
  **c.** (1) $7 - 2 \not> 29$ to get the sentence $7 \not> 31$?
     (2) $n - 2 \not> 29$ to get the sentence $n \not> 31$?
  **d.** (1) $w - 10 \leq 0$ to get the sentence $w \leq 10$?
     (2) $x - 6 \geq -9$ to get the sentence $x \geq -3$?

9. By what number should you divide both sides of the inequality:
  **a.** (1) $8 \times 4 > 24$ to get the sentence $4 > 3$?
     (2) $8x > 24$ to get the sentence $x > 3$?
  **b.** (1) $7 \times 9 < 98$ to get the sentence $9 < 14$?
     (2) $7n < 98$ to get the sentence $n < 14$?
  **c.** (1) $12 \times (-3) \not> 24$ to get the sentence $-3 \not> 2$?
     (2) $12y \not> 24$ to get the sentence $y \not> 2$?

   **d.** (1) $20t \le 4$ to get the sentence $t \le \frac{1}{5}$?

      (2) $5n \ge -30$ to get the sentence $n \ge -6$?

   **e.** (1) $(-6) \times 7 < 54$ to get the sentence $7 > -9$?

      (2) $-6n < 54$ to get the sentence $n > -9$?

10. By what number should you multiply both sides of the inequality:

   **a.** (1) $56 \div 8 < 9$ to get the sentence $56 < 72$?

      (2) $n \div 8 < 9$ to get the sentence $n < 72$?

   **b.** (1) $\frac{36}{4} > 3$ to get the sentence $36 > 12$?

      (2) $\frac{x}{4} > 3$ to get the sentence $x > 12$?

   **c.** (1) $\frac{-60}{15} \not< -7$ to get the sentence $-60 \not< -105$?

      (2) $\frac{a}{15} \not< -7$ to get the sentence $a \not< -105$?

   **d.** (1) $\frac{r}{8} \ge 1$ to get the sentence $r \ge 8$?

      (2) $\frac{n}{2} \le 0$ to get the sentence $n \le 0$?

   **e.** (1) $\frac{25}{-5} > -7$ to get the sentence $25 < 35$?

      (2) $\frac{x}{-5} > -7$ to get the sentence $x < 35$?

## CUMULATIVE REVIEW

1. Write symbolically:

   **a.** Six times each number $n$ is not less than twenty-five.

   **b.** Eleven times each number $x$ increased by nine is greater than or equal to negative twelve.

2. Write each of the following inequalities as an inequality of reverse order:

   **a.** $y > 14$       **b.** $m < -2$       **c.** $s \ge 0$       **d.** $x \not< 3$

3. What operation with what number should be used on both sides of the inequality:

   **a.** $w + 8 < 19$ to get the sentence $w < 11$?

   **b.** $-7x > -35$ to get the sentence $x < 5$?

   **c.** $y - 12 \ge -3$ to get the sentence $y \ge 9$?

   **d.** $15x \le -10$ to get the sentence $x \le -\frac{2}{3}$?

   **e.** $\frac{n}{-1} \not> 9$ to get the sentence $n \not< -9$?

# Solving Inequalities in Simplest Form

EXERCISE
8-3

**I. Aim**  To solve basic inequalities which are expressed in simplest form when the replacements for the variable are the:
a. natural numbers, b. whole numbers, c. integers, d. rational numbers, e. real numbers, f. prime numbers, g. odd integers, h. even integers, or i. a finite group of numbers.

**II. Procedure**

**1. a.** Select the numbers described by the given inequality that satisfies the inequality making it a true sentence.

**b.** List the solutions when possible. Study all the sample solutions.

**2.** Observe that:

**a.** To solve an inequality means to find all the numbers each of which, when substituted for the variable, will make the inequality a true sentence.

**b.** An inequality in simplest form is expressed with only the variable itself as one member of the inequality and a numeral for the other member.

**3.** The solution of an inequality depends not only on the inequality but also on the replacements for the variable. See all the sample solutions. Review each of the following number systems which may be used for the replacements:

**a.** The numbers, beginning with *zero*, named by the numerals 0, 1, 2, 3, 4, 5, 6, 7, 8, 9, 10, 11, 12, . . . are called the *whole numbers*.

**b.** The whole numbers, beginning with *one*, named by the numerals, 1, 2, 3, 4, 5, 6, 7, 8, 9, 10, 11, 12, . . . that are used in counting are called *natural numbers* or *counting numbers*.

**c.** All the whole numbers and their opposites are called *integers*.

**d.** All the fractional numbers and their opposites are called the *rational numbers*. The rational numbers include integers since each integer may be named in fraction form.

**e.** All the rational numbers and all the irrational numbers are called the *real numbers*. Also see page 50.

**f.** A *prime number* is a whole number, other than 0 and 1, which is divisible (is divided exactly) only by itself and by 1 and by no other whole number. It is any whole number greater than 1 whose only factors are 1 and itself.

**g.** The *even integers* are the integers that are divisible by two (2). Zero is considered to be an even integer.

    Numbers whose numerals end in 0, 2, 4, 6, or 8 are even integers.

**h.** The *odd integers* are the integers that are not divisible by two (2).

    Numbers whose numerals end in 1, 3, 5, 7, or 9 are odd integers.

**i.** When the replacements for the variable are limited, only numbers from that group can be used as solutions. See sample solutions 5*b.* and 7*b.*

**4.** Inequalities using equivalent symbols ($\not< $ and $\geq$; $\not> $ and $\leq$) have the same solution.

**5.** Check your answers by substituting one or more of these numbers for the variable and checking whether each will make the inequality a true sentence. Also select a number that is not a solution and check whether, when it is substituted for the variable, a false sentence results.

**6.** When there is no number that satisfies the inequality, there is no solution.

## III. Sample Solutions

**1.** Solve $n < 4$ when the replacements for the variable are:

  **a.** All the natural numbers.
                                     **b.** All the whole numbers.

      *Solution:* $n < 4$
                                *Solution:* $n < 4$

             $n = $ 1, 2, or 3
                            $n = $ 0, 1, 2, or 3

*Answers:* 1, 2, or 3
                          *Answers:* 0, 1, 2, or 3

**2.** Solve $n > 5$ when the replacements for the variable are:

  **a.** All the real numbers.
                                   **b.** All the even integers.

      *Solution:* $n > 5$
                              *Solution:* $n > 5$

             $n = $ all the real numbers greater than 5.
            $n = $ 6, 8, 10, . . .

*Answers:* All the real numbers greater than 5.
       *Answers:* 6, 8, 10, . . .

**3.** Solve $n \leq -2$ when the replacements for the variable are:

  **a.** All the integers.
                                     **b.** All the whole numbers.

      *Solution:* $n \leq -2$
                           Since there are no whole

      $n = $ . . . , $-4$, $-3$, $-2$
                    numbers less than or equal

*Answers:* . . . , $-4$, $-3$, $-2$
             to $-2$, there is no solution.

                                *Answer:* No solution

**4.** Solve $n \geq 6$ when the replacements for the variable are:

    **a.** All the real numbers.

      *Solution:* $n \geq 6$

      $n = 6$, all real
      numbers greater
      than 6.

  *Answer:* 6, all real numbers
        greater than 6.

    **b.** All the odd integers.

      *Solution:* $n \geq 6$

      $n = 7, 9, 11, \ldots$

  *Answer:* 7, 9, 11, . . .

**5.** Solve $n \not< 8$ when the replacements for the variable are:

    **a.** All the rational numbers.

      *Solution:* $n \not< 8$

      $n \geq 8$

      $n = 8$, all rational
      numbers greater than 8

  *Answer:* 8, all rational numbers
        greater than 8.

    **b.** 0, 1, 2, . . . , 10.

      *Solution:* $n \not< 8$

      $n \geq 8$

      $n = 8, 9,$ or $10$

  *Answer:* 8, 9, or 10.

**6.** Solve $n \not> 4$ when the replacements for the variable are:

    **a.** All the even integers.

      *Solution:* $n \not> 4$

      $n \leq 4$

      $n = \ldots, -2, 0, 2$

  *Answer:* . . . , −2, 0, 2

    **b.** All the prime numbers.

      *Solution:* $n \not> 4$

      $n \leq 4$

      $n = 2$ or $3$

  *Answer:* 2 or 3

**7.** Solve $n \neq 9$ when the replacements for the variable are:

    **a.** All the real numbers.

      *Solution:* $n \neq 9$

      $n =$ every real
      number except 9.

  *Answer:* Every real number except 9.

    **b.** 1, 2, 3, . . . , 9.

      *Solution:* $n \neq 9$

      $n = 1, 2, 3, \ldots, 8$

  *Answer:* 1, 2, 3, . . . , 8

## DIAGNOSTIC TEST

Find the solutions of each of the following inequalities when the replacements for the variable are:

**1.** All the whole numbers:   **a.** $x < 12$   **b.** $b < -6$   **c.** $g < 1$

**2.** All the integers:           **a.** $c > -3$   **b.** $x > 0$   **c.** $z > 2$

3. All the natural numbers:    **a.** $r \leq 1$    **b.** $n \leq 7$    **c.** $d \leq -5$
4. All the odd integers:    **a.** $w \geq 6$    **b.** $a \geq -9$    **c.** $c \geq -1$
5. All the real numbers:    **a.** $y \not< 10$    **b.** $r \not< 1$    **c.** $x \not< -2$
6. All the even integers:    **a.** $a \not> 7$    **b.** $p \not> -8$    **c.** $n \not> 0$
7. All the integers:    **a.** $m \neq -8$    **b.** $d \neq 11$    **c.** $a \neq 1$
8. 1, 2, 3, . . . , 16    **a.** $b < 5$    **b.** $x > 10$    **c.** $r \not\geq 7$
9. All the prime numbers:    **a.** $15 > n$    **b.** $10 \not< b$    **c.** $23 \geq m$

## *RELATED PRACTICE EXAMPLES*

Find the solutions of each of the following inequalities when the replacements for the variable are:

1. **a.** All the whole numbers:
   (1) $x < 8$      (2) $y < 0$      (3) $b < -5$
   **b.** All the natural numbers:
   (1) $n < 5$      (2) $t < -2$      (3) $x < 1$
   **c.** All the integers:
   (1) $z < 6$      (2) $m < -4$      (3) $r < 0$
   **d.** All the real numbers:
   (1) $r < 7$      (2) $x < 0$      (3) $g < -1$
   **e.** All the prime numbers:
   (1) $a < 16$      (2) $c < 25$      (3) $w < 40$
   **f.** All the rational numbers:
   (1) $y < 11$      (2) $n < -6$      (3) $t < \frac{2}{3}$
   **g.** All the odd integers:
   (1) $x < 8$      (2) $b < 1$      (3) $m < -10$

2. **a.** All the integers:
   (1) $y > 5$      (2) $b > 0$      (3) $c > -8$
   **b.** All the rational numbers:
   (1) $n > -.3$      (2) $x > 1\frac{3}{4}$      (3) $a > 0$
   **c.** All the whole numbers:
   (1) $d > 6$      (2) $r > -1$      (3) $n > -10$
   **d.** All the real numbers:
   (1) $m > -4$      (2) $z > 0$      (3) $x > 15$
   **e.** All the natural numbers:
   (1) $h > 12$      (2) $n > -6$      (3) $t > 1$

   **f.** All the even integers:
      (1) $x > -5$           (2) $s > 14$           (3) $v > 0$
   **g.** All the prime numbers:
      (1) $w > 7$            (2) $y > 0$            (3) $x > 21$

**3. a.** All the whole numbers:
      (1) $x \leq 4$           (2) $w \leq 0$           (3) $b \leq -3$
   **b.** All the integers:
      (1) $t \leq 11$          (2) $a \leq -6$         (3) $r \leq 1$
   **c.** All the natural numbers:
      (1) $m \leq 7$         (2) $x \leq 1$          (3) $y \leq -2$
   **d.** All the rational numbers:
      (1) $r \leq -5$        (2) $n \leq 0$          (3) $c \leq \frac{7}{8}$
   **e.** All the prime numbers:
      (1) $y \leq 17$         (2) $s \leq 2$          (3) $x \leq 35$
   **f.** All the real numbers:
      (1) $b \leq -8$        (2) $z \leq 12$         (3) $d \leq 0$
   **g.** All the even integers:
      (1) $c \leq 10$        (2) $g \leq -3$        (3) $y \leq 1$

**4. a.** All the integers:
      (1) $y \geq 8$           (2) $x \geq -1$        (3) $b \geq 0$
   **b.** All the prime numbers:
      (1) $c \geq 12$        (2) $n \geq 0$          (3) $r \geq 17$
   **c.** All the whole numbers:
      (1) $r \geq 9$          (2) $z \geq -3$        (3) $g \geq 14$
   **d.** All the real numbers:
      (1) $x \geq -4$        (2) $v \geq 25$        (3) $y \geq 0$
   **e.** All the natural numbers:
      (1) $t \geq 10$         (2) $m \geq 0$        (3) $x \geq -5$
   **f.** All the odd integers:
      (1) $w \geq -3$       (2) $p \geq 1$          (3) $a \geq 16$
   **g.** All the rational numbers:
      (1) $n \geq 7\frac{2}{5}$       (2) $x \geq -6$        (3) $d \geq 9.7$

**5. a.** All the real numbers:
      (1) $x \not< 8$          (2) $z \not< -5$       (3) $n \not< 0$
   **b.** All the rational numbers:
      (1) $y \not< -4$       (2) $r \not< 6.8$       (3) $c \not< 1\frac{1}{3}$
   **c.** All the integers:
      (1) $a \not< 6$          (2) $m \not< -1$       (3) $g \not< 14$
   **d.** All the natural numbers:
      (1) $c \not< 7$          (2) $x \not< 0$        (3) $v \not< -2$

   **e.** All the whole numbers:
     (1) $w \not< 15$          (2) $t \not< -7$          (3) $d \not< 1$
   **f.** All the positive even integers:
     (1) $n \not< -12$      (2) $v \not< 5$          (3) $a \not< 0$
   **g.** All the prime numbers:
     (1) $x \not< 9$          (2) $b \not< 19$         (3) $r \not< 39$

**6. a.** All the rational numbers:
     (1) $y \not> \frac{1}{2}$         (2) $w \not> -5$        (3) $t \not> 0$
   **b.** All the natural numbers:
     (1) $n \not> 8$          (2) $x \not> 0$         (3) $w \not> -7$
   **c.** All the integers:
     (1) $r \not> -2$       (2) $k \not> 6$         (3) $m \not> 1$
   **d.** All the real numbers:
     (1) $b \not> -1$       (2) $v \not> 0$         (3) $x \not> 9$
   **e.** All the whole numbers:
     (1) $x \not> 6$         (2) $c \not> -10$      (3) $n \not> 0$
   **f.** All the prime numbers:
     (1) $m \not> 11$       (2) $y \not> 30$       (3) $b \not> 49$
   **g.** All the negative odd integers:
     (1) $d \not> 3$         (2) $n \not> -3$       (3) $a \not> 0$

**7. a.** All the natural numbers:
     (1) $x \neq 7$         (2) $t \neq -3$        (3) $y \neq 9$
   **b.** All the rational numbers:
     (1) $c \neq 2.6$       (2) $m \neq 0$        (3) $b \neq -6\frac{3}{4}$
   **c.** All the whole numbers:
     (1) $w \neq 1$         (2) $z \neq -5$       (3) $r \neq 8$
   **d.** All the integers:
     (1) $a \neq -7$       (2) $x \neq 0$       (3) $n \neq 14$
   **e.** All the real numbers:
     (1) $y \neq 16$       (2) $v \neq -10$      (3) $t \neq 1$
   **f.** All the non-positive even integers:
     (1) $r \neq -9$       (2) $y \neq 6$       (3) $x \neq -4$
   **g.** All the odd numbers:
     (1) $x \neq 5$         (2) $b \neq 1$        (3) $c \neq 15$

**8. a.** 1, 2, 3, ... , 12    (1) $x > 7$    (2) $b < 8$    (3) $x \not> 6$
   **b.** 0, 2, 4, 6, 8, 10    (1) $y < 6$    (2) $n > 3$    (3) $r \geq 12$
   **c.** 1, 3, 5, 7, 9      (1) $n \geq 5$    (2) $c \not> 8$    (3) $m < 1$
   **d.** 2, 3, 5, 7, 11, 13, 17, 19    (1) $w \not< 10$    (2) $t \leq 13$    (3) $b > 3$
   **e.** $-10, -8, -6, \ldots , 0$    (1) $x < -2$    (2) $d \neq -4$    (3) $k \geq -6$
   **f.** 0, 2, 4, ... , 30    (1) $p > 15$    (2) $v > 20$    (3) $t \not> 15$

**g.** 1, 3, 5, ... , 25     (1) $r \leq 18$    (2) $b > 25$    (3) $x \not< 12$
**h.** $-8, -6, -4, -2, 0, 2, 4$    (1) $a \geq -4$    (2) $s < -6$    (3) $w \not> 2$
**i.** $-7, -5, -4, -1, 0, 1, 3, 6$    (1) $m \not< -5$    (2) $t > -1$    (3) $z \neq 4$
**j.** 0, 1, 2, 3, ... , 15    (1) $x > 6$    (2) $n \not\geq 12$    (3) $y \leq 9$

**9. a.** All the natural numbers:
     (1) $9 > x$        (2) $16 \leq w$        (3) $12 \not< r$
**b.** All the whole numbers:
     (1) $11 < y$        (2) $-5 \not> a$        (3) $8 \geq d$
**c.** All the integers:
     (1) $-7 > g$        (2) $12 \leq x$        (3) $0 < b$
**d.** All the rational numbers:
     (1) $-3 \not< m$        (2) $8\frac{1}{2} > c$        (3) $.4 \leq a$
**e.** All the real numbers:
     (1) $4 \not> b$        (2) $-9 < y$        (3) $2 \neq w$
**f.** All the even integers:
     (1) $5 < t$        (2) $10 > r$        (3) $-6 > x$
**g.** All the prime numbers:
     (1) $20 > x$        (2) $20 < n$        (3) $2 \neq y$
**h.** All the odd integers:
     (1) $-10 \geq d$        (2) $17 \not< g$        (3) $0 < n$
**i.** 0, 1, 2, 3, ... , 14:
     (1) $8 < x$        (2) $10 > t$        (3) $14 \not< b$
**j.** $-9, -8, -6, -3, -1, 2, 5, 7, 8$:
     (1) $-1 \not< n$        (2) $-9 < w$        (3) $8 \geq y$

## CUMULATIVE REVIEW

**1.** Write symbolically:
     **a.** Each number $s$ divided by eight is less than negative one.
     **b.** Seven times each number $n$ plus five is greater than or equal to sixteen.
     **c.** Each number $x$ is less than seven and greater than negative three.
**2.** Find the solutions of each of the following inequalities when the replacements for the variable are:

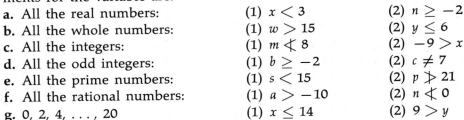

     **a.** All the real numbers:      (1) $x < 3$      (2) $n \geq -2$
     **b.** All the whole numbers:      (1) $w > 15$      (2) $y \leq 6$
     **c.** All the integers:      (1) $m \not< 8$      (2) $-9 > x$
     **d.** All the odd integers:      (1) $b \geq -2$      (2) $c \neq 7$
     **e.** All the prime numbers:      (1) $s < 15$      (2) $p \not> 21$
     **f.** All the rational numbers:      (1) $a > -10$      (2) $n \not< 0$
     **g.** 0, 2, 4, ... , 20      (1) $x \leq 14$      (2) $9 > y$

# Inequalities–Solution by Inverse Operation (Subtraction) or by Additive Inverse

**I. Aim**  To solve inequalities of the types: $n + 3 > 15$; $n + 3 < 15$; $n + 3 \geq 15$; $n + 3 \leq 15$; $n + 3 \neq 15$; $n + 3 \not> 15$; and $n + 3 \not< 15$.

(1) By the inverse operation (subtraction) or

(2) By the additive inverse

## II. Procedure

1. Solution by Inverse Operation:
   Subtract from both sides of the inequality the number that is on the same side of the inequality as the variable so that only the variable remains on that side. See sample solution 1.
2. Solution by Additive Inverse:
   Or add to both sides of the inequality the additive inverse of the number that is on the same side of the inequality as the variable.
3. When the variable appears on the right side of the inequality, rewrite in the final step by using the equivalent inequality. See sample solution 2.
4. See Exercise 8-3 for the solution of inequalities in simplest form.
5. Check your solutions. See page 314.

## III. Sample Solutions

1. Solve: $n + 3 > 15$ when the replacements for the variable are all the real numbers.

   **a.** *Solution:*           OR    **b.** *Solution:*

   $$n + 3 > 15$$
   $$n + 3 - 3 > 15 - 3$$
   $$n > 12$$
   $$n = \text{all real numbers}$$
   $$\text{greater than 12.}$$

   $$n + 3 > 15$$
   $$n + 3 + (-3) > 15 + (-3)$$
   $$n > 12$$
   $$n = \text{all real numbers}$$
   $$\text{greater than 12.}$$

   *Answer:* All real numbers greater than 12.

   *Check:* Is 8 a solution?           Is 17 a solution?

   $$n + 3 > 15$$                    $$n + 3 > 15$$
   $$8 + 3 > 15?$$                   $$17 + 3 > 15?$$
   $$11 \not> 15$$                   $$20 > 15 ✔$$
   $$\text{No}$$                      $$\text{Yes}$$

**320**

**2.** Solve: $9 \geq x + 6$ when the replacements are all the whole numbers.

*Solution:*

$$9 \geq x + 6$$
$$9 - 6 \geq x + 6 - 6$$
$$3 \geq x$$
$$x \leq 3$$

*Answer:* 1, 2, and 3

**3.** Solve: $x + 5 < 7$ when the replacements are all the integers.

*Solution:*

$$x + 5 < 7$$
$$x + 5 - 5 < 7 - 5$$
$$x < 2$$
$$x = \ldots, -2, -1, 1$$

*Answer:* $\ldots, -2, -1, 0, 1$

## DIAGNOSTIC TEST

Find the solutions of each of the following inequalities when the replacements for the variable are all the real numbers:

**1. a.** $x + 7 > 22$    **b.** $w + 14 > 10$    **c.** $16 + m > 16$

**2. a.** $y + 6 < -4$    **b.** $.9 + n < 1.8$    **c.** $x + 5 < 0$

**3. a.** $d + 1 \neq 8$    **b.** $r + 11 \neq 4$    **c.** $7 + t \neq -1$

**4. a.** $z + 10 \not> 3$    **b.** $2 + m \not> 9$    **c.** $a + 4\frac{2}{3} \not> 4\frac{2}{3}$

**5. a.** $c + 8 \not< 11$    **b.** $7 + y \not< 0$    **c.** $w + 1.2 \not< 2$

**6. a.** $s + 3 \geq 17$    **b.** $x + 5\frac{1}{2} \geq 1$    **c.** $16 + z \geq 0$

**7. a.** $x + 2 \leq 15$    **b.** $y + .18 \leq .18$    **c.** $5 + n \leq -5$

Find the solutions of each of the following inequalities when the replacements for the variable are:

**8. a.** All the integers:
(1) $x + 11 \leq 5$    (2) $p + 1 > -3$    (3) $10 + w \not< 0$

**b.** All the natural numbers:
(1) $r + 2 \not> 6$    (2) $12 + n \geq 12$    (3) $b + 9 < -1$

**c.** All the prime numbers:
(1) $m + 5 < 27$    (2) $y + 3 \neq 4$    (3) $17 + x > -2$

**9. a.** All the real numbers:
(1) $4 > t + 9$    (2) $-8 \not< x + 4$    (3) $7 \leq 7 + c$

**b.** All the even integers:
(1) $-5 < z + 12$    (2) $10 > h + 2$    (3) $-20 \neq 11 + d$

### RELATED PRACTICE EXAMPLES

Find the solutions of each of the following inequalities when the replacements for the variable are all the real numbers:

1. **a.** $c + 2 > 10$      $x + 17 > 85$      $40 + z > 40$
   **b.** $y + .3 > .8$      $5 + a > 6.7$      $g + .9 > 1$
   **c.** $v + 1\frac{1}{4} > 8\frac{1}{4}$      $m + \frac{2}{3} > 5\frac{2}{3}$      $x + 2\frac{1}{2} > 4\frac{7}{8}$
   **d.** $g + 18 > 6$      $c + 7 > -5$      $b + 10 > 0$

2. **a.** $b + 8 < 15$      $39 + d < 39$      $x + 12 < 61$
   **b.** $y + .7 < .7$      $n + 2.7 < 4.5$      $p + .5 < 2.1$
   **c.** $n + \frac{1}{2} < \frac{3}{4}$      $x + \frac{5}{6} < 10$      $d + 1\frac{5}{8} < 4$
   **d.** $y + 11 < 4$      $71 + t < 0$      $x + 16 < -2$

3. **a.** $d + 1 \neq 9$      $r + 14 \neq 23$      $28 + h \neq 55$
   **b.** $r + .5 \neq .9$      $y + .64 \neq 10.4$      $g + .08 \neq 5$
   **c.** $x + \frac{2}{3} \neq \frac{11}{12}$      $\frac{2}{5} + n \neq 1\frac{7}{10}$      $s + 1\frac{1}{6} \neq 7$
   **d.** $t + 4 \neq 2$      $w + 13 \neq -23$      $z + 39 \neq 18$

4. **a.** $x + 10 \not> 24$      $16 + y \not> 33$      $y + 17 \not> 17$
   **b.** $n + 6 \not> 7.4$      $r + .43 \not> .99$      $.4 + x \not> 1.2$
   **c.** $h + \frac{7}{8} \not> 1\frac{1}{2}$      $z + 2\frac{1}{2} \not> 5$      $n + 3\frac{2}{3} \not> 7\frac{1}{6}$
   **d.** $x + 15 \not> 11$      $6 + g \not> -17$      $r + 8 \not> -8$

5. **a.** $a + 6 \not< 18$      $r + 60 \not< 100$      $56 + t \not< 63$
   **b.** $e + .1 \not< 1$      $d + 2 \not< 4.2$      $a + .34 \not< 6.4$
   **c.** $n + 3\frac{4}{5} \not< 8\frac{4}{5}$      $x + \frac{11}{16} \not< \frac{5}{6}$      $\frac{3}{8} + y \not< 2\frac{1}{3}$
   **d.** $c + 8 \not< 4$      $h + 9 \not< -10$      $a + 28 \not< 0$

6. **a.** $x + 11 \geq 17$      $54 + a \geq 100$      $t + 37 \geq 91$
   **b.** $y + .2 \geq .5$      $c + 1.4 \geq 5$      $6 + x \geq 8.3$
   **c.** $n + \frac{1}{2} \geq 1\frac{7}{8}$      $x + 2\frac{2}{3} \geq 3\frac{1}{3}$      $y + 5\frac{3}{4} \geq 8\frac{5}{6}$
   **d.** $r + 19 \geq 4$      $3 + m \geq -5$      $z + 42 \geq 36$

7. **a.** $t + 3 \leq 10$      $b + 47 \leq 83$      $95 + g \leq 110$
   **b.** $s + 1.2 \leq 1.8$      $x + .09 \leq .4$      $n + 3.5 \leq 8$
   **c.** $x + 2\frac{1}{4} \leq 5\frac{1}{4}$      $m + \frac{13}{16} \leq 1$      $b + 3\frac{1}{4} \leq 5\frac{1}{3}$
   **d.** $y + 26 \leq 6$      $t + 7 \leq -2$      $6 + h \leq -6$

8. Find the solutions of each of the following inequalities:

   **a.** When the replacements for the variable are all the whole numbers:

     (1) $n + 9 < 14$          (2) $c + 2 > 11$        (3) $y + 6 \not< 12$

     (4) $x + 9 \not> 9$          (5) $m + 4 \neq 7$         (6) $r + 5 \leq 1$

     (7) $12 + g \not\leq 41$       (8) $t + 3 \geq 0$         (9) $x + 4 \not\geq -4$

   **b.** When the replacements for the variable are all the integers:

     (1) $a + 4 > 9$           (2) $r + 6 < 2$           (3) $m + 8 \not> 0$

     (4) $n + \frac{1}{2} \not< 3\frac{1}{2}$      (5) $y + 3 \leq -8$      (6) $x + 14 \not\geq 14$

     (7) $w + .5 \geq .7$       (8) $11 + r \neq 1$       (9) $v + 10 \not< -6$

   **c.** When the replacements for the variable are all the natural numbers:

     (1) $b + 1 < 7$           (2) $x + 5 > 14$        (3) $z + 2 \leq 2$

     (4) $y + 6 \geq 3$         (5) $m + 8 \neq 7$       (6) $r + 4 \not< 6.8$

     (7) $13 + k \not> 10$       (8) $x + 2\frac{1}{4} \not\leq 5$      (9) $n + .87 \not\geq 4.37$

   **d.** When the replacements for the variable are all the rational numbers:

     (1) $d + \frac{3}{4} > 1\frac{1}{2}$      (2) $x + 1\frac{2}{3} < \frac{5}{6}$      (3) $a + 2\frac{1}{2} \neq 2\frac{1}{2}$

     (4) $n + 7 \not< 5$         (5) $4 + c \not> 13$       (6) $y + 9 \not< -1$

     (7) $s + .6 \leq 1.5$       (8) $m + 8 \not\geq .74$      (9) $t + .25 \geq 1.25$

   **e.** When the replacements for the variable are all the prime numbers:

     (1) $x + 7 < 14$         (2) $n + 3 > 9$          (3) $b + 2 \geq 5$

     (4) $r + 8 \leq 21$        (5) $w + 5 \neq 22$       (6) $11 + d \not< 50$

     (7) $v + 6 \not> 32$       (8) $x + 10 \not< 10$      (9) $y + 4 \not\geq 42$

   **f.** When the replacements for the variable are all the odd integers:

     (1) $c + 6 > 16$         (2) $n + 4 < 23$       (3) $y + 13 \not> 41$

     (4) $x + 5 \not< 5$         (5) $a + 8 \not< 2$        (6) $1 + r \neq -9$

     (7) $p + 7 \geq 0$        (8) $x + \frac{2}{3} \not\geq 4\frac{5}{6}$      (9) $z + .25 \leq 8.57$

   **g.** When the replacements for the variable are all the even integers:

     (1) $w + 12 < 18$       (2) $x + 21 \not> 45$      (3) $c + 6 \neq 7$

     (4) $b + 11 \not< -3$      (5) $9 + m > 1$        (6) $y + 16 < 0$

     (7) $r + 1\frac{3}{8} > 2\frac{5}{8}$      (8) $w + 6.7 \leq 9.3$      (9) $x + 19 \geq 108$

   **h.** When the replacements for the variable are $-11, -8, -4, -1, 0, 2, 5,$ 6, 10:

     (1) $x + 1 \not< 7$         (2) $b + 4 \geq 3$        (3) $r + 6 \neq 2$

     (4) $m + 8 < 3$        (5) $a + 9 < 1$        (6) $c + 5 \leq 7$

     (7) $v + 3 > 0$         (8) $x + 7 > 12$       (9) $p + 10 \not> -2$

   **i.** When the replacements for the variable are $-12, -8, -4, \ldots, 12$:

     (1) $n + 2 < 9$         (2) $c + 1 > 1$        (3) $x + 6 \not< 3$

     (4) $t + 7 \not> 13$       (5) $d + 10 \neq 2$      (6) $y + 15 \leq 12$

     (7) $x + 5 \geq 18$       (8) $r + 14 > 21$      (9) $z + 8 < -3$

9. Find the solutions of each of the following inequalities:

   **a.** When the replacements for the variable are all the real numbers:

      (1) $17 > y + 7$          (2) $3 < c + 9$          (3) $6 \not> b + 6$

      (4) $28 \leq a + 11$       (5) $1 < x + 7$          (6) $5 \neq n + 8$

      (7) $10 \geq 18 + y$       (8) $0 \not< d + 15$       (9) $31 > 32 + t$

   **b.** When the replacements for the variable are all the whole numbers:

      (1) $8 < n + 4$           (2) $10 \geq b + 10$        (3) $5 \not> 11 + x$

   **c.** When the replacements for the variable are all the integers:

      (1) $-5 > d + 2$         (2) $0 \not< m + 1$          (3) $11 \leq 8 + y$

   **d.** When the replacements for the variable are all the prime numbers:

      (1) $10 \not> w + 3$        (2) $11 > 7 + x$          (3) $20 > m + 2$

   **e.** When the replacements for the variable are all the even integers:

      (1) $0 \geq x + 10$        (2) $1 < y + 1$          (3) $9 \not< 3 + c$

   **f.** When the replacements for the variable are $-9, -7, -5, \ldots, 9$

      (1) $13 > n + 8$         (2) $2 > b + 4$          (3) $21 > 15 + d$

## CUMULATIVE REVIEW

1. Write symbolically:

   **a.** Each number $x$ increased by nine is less than three.

   **b.** Twelve times each number $n$ is greater than negative seven.

   **c.** Twice each number $y$ plus six is not equal to seventeen.

2. Find the solutions of each of the following inequalities when the replacements for the variable are all the real numbers:

   **a.** $x \neq 13$            **b.** $y \geq -5$              **c.** $n + 4 < 11$

   **d.** $16 + w > 0$        **e.** $19 \not< r + 2$           **f.** $t + 5 \geq -5$

   **g.** $s + 2\frac{1}{2} < 6\frac{1}{4}$       **h.** $3.2 + m \geq 7$          **i.** $-27 \not> 17 + x$

3. Find the solutions of each of the following inequalities when the replacements for the variable are:

   **a.** All the whole numbers:

      (1) $n \not> 6$             (2) $b + 9 < 14$         (3) $2 + r > 15$

   **b.** All the integers:

      (1) $x \leq -3$           (2) $y + 10 > 6$        (3) $-5 \not< m + 7$

   **c.** All the prime numbers:

      (1) $y < 14$            (2) $n + 4 \not> 9$        (3) $32 > 11 + d$

   **d.** All the even whole numbers less than 19:

      (1) $m > 11$            (2) $x + 48 < 16$       (3) $12 \geq y + 3$

   **e.** $-9, -7, -5, \ldots, 9$:

      (1) $w \neq -1$           (2) $d + 1 > -2$        (3) $4 + n < 0$

# Inequalities–Solution by Inverse Operation (Addition) or by Additive Inverse

EXERCISE
8-5

**I. Aim**  To solve inequalities of the types: $n - 3 > 15$; $n - 3 < 15$; $n - 3 \geq 15$; $n - 3 \leq 15$; $n - 3 \neq 15$; $n - 3 \not> 15$; and $n - 3 \not< 15$.

(1) By the inverse operation (addition) or

(2) By the additive inverse

**II. Procedure**

**1.** Solution by Inverse Operation:

Add to both sides of the inequality the number that is on the same side of the equation as the variable so that only the variable remains on that side. See sample solution 1.a.

**2.** Solution by Additive Inverse:

Or add to both sides of the equation the additive inverse of the number that is on the same side of the equation as the variable.

**3.** When the variable appears on the right side of the inequality, rewrite in the final step, by using the equivalent inequality. See sample solution 3.

**4.** See Exercise 8-3 for the solution of inequalities in simplest form.

**5.** Check your solutions. See page 314.

**III. Sample Solutions**

**1.** Solve $n - 3 < 15$ when the replacements for the variable are all the real numbers.

**a.** *Solution:*

$$n - 3 < 15$$
$$n - 3 + 3 < 15 + 3$$
$$n < 18$$

$n =$ all real numbers less than 18.

**b.** *Solution:*

$$n - 3 < 15$$
$$n + (-3) + (+3) < 15 + (+3)$$
$$n < 18$$

$n =$ all real numbers less than 18.

*Answer:* All real numbers less than 18.

*Check:* Is 10 a solution?

$$n - 3 < 15$$
$$10 - 3 < 15 ?$$
$$7 < 15 ✔$$
Yes

Is 20 a solution?

$$n - 3 < 15$$
$$20 - 3 < 15 ?$$
$$17 \not< 15$$
No

**2.** Solve $4 > n - 2$ when the replacements for the variable are all whole numbers.

*Solution:*

$$4 > n - 2$$
$$4 + 2 > n - 2 + 2$$
$$6 > n$$
$$n < 6$$
$$n = 0, 1, 2, 3, 4, 5$$

*Answer:* 0, 1, 2, 3, 4, 5

**3.** Solve $x - 1 \leq 2$ when the replacements for the variable are all the integers.

*Solution:*

$$x - 1 \leq 2$$
$$x \leq 2 + 1$$
$$x \leq 3$$
$$x = \ldots, -1, 0, 1, 2, 3$$

*Answer:* $\ldots, -1, 0, 1, 2, 3$

## DIAGNOSTIC TEST

Find the solutions of each of the following inequalities when the replacements for the variable are all the real numbers:

**1. a.** $x - 7 > 21$  **b.** $m - 4\frac{2}{3} > 2$  **c.** $d - 31 > -16$

**2. a.** $n - 11 < 5$  **b.** $r - 3.9 < 0$  **c.** $x - 23 < -42$

**3. a.** $w - 12 \neq 19$  **b.** $p - \frac{1}{4} \neq 1\frac{1}{2}$  **c.** $w - 9 \neq -9$

**4. a.** $b - 37 \not> 28$  **b.** $v - 7.5 \not> 8.2$  **c.** $t - 18 \not> -6$

**5. a.** $y - 54 \not< 0$  **b.** $a - 18 \not< -24$  **c.** $z - 41 \not< 57$

**6. a.** $x - 17 \geq 22$  **b.** $g - 4\frac{1}{8} \geq 2\frac{3}{4}$  **c.** $d - 10 \geq 16$

**7. a.** $r - 11 \leq -3$  **b.** $w - .25 \leq 8$  **c.** $x - 21 \leq 0$

Find the solutions of each of the following inequalities when the replacements for the variable are:

**8. a.** All the whole numbers:

 (1) $n - 6 \leq 7$     (2) $t - 2 > 0$     (3) $a - 5 \not> -5$

 **b.** All the prime numbers:

 (1) $b - 4 < 3$     (2) $y - 1 \neq 1$     (3) $x - 7 < -9$

 **c.** All the integers:

 (1) $x - 3 \geq 10$     (2) $d - 2\frac{1}{2} \not> 3\frac{3}{4}$     (3) $m - 11 \not< -20$

**9. a.** All the real numbers:

 (1) $14 < n - 8$     (2) $-9 \not> a - 6$     (3) $0 \geq r - 10$

 **b.** All the odd integers:

 (1) $1 > x - 3$     (2) $-10 \leq c - 13$     (3) $-6 \not< w - 5$

## *RELATED PRACTICE EXAMPLES*

Find the solutions of each of the following inequalities when the replacements for the variable are all the real numbers:

1. **a.** $x - 8 > 15$     $s - 16 > 4$     $r - 9 > 0$
   **b.** $r - 2.9 > 7.4$     $n - 4 > .9$     $x - .8 > 6$
   **c.** $x - \frac{5}{8} > \frac{1}{8}$     $y - \frac{3}{4} > \frac{3}{4}$     $z - 1\frac{2}{3} > 2\frac{1}{3}$
   **d.** $d - 10 > -3$     $m - 9 > -25$     $s - 28 > -28$

2. **a.** $z - 9 < 26$     $y - 23 < 11$     $h - 7 < 7$
   **b.** $n - .7 < .4$     $x - 6 < .25$     $a - .18 < 1.2$
   **c.** $t - 1\frac{1}{2} < 2\frac{1}{2}$     $h - \frac{3}{8} < \frac{3}{8}$     $m - 2\frac{5}{6} < \frac{1}{3}$
   **d.** $a - 17 < -7$     $y - 5 < -5$     $x - 4 < -12$

3. **a.** $b - 2 \neq 7$     $c - 10 \neq 0$     $a - 56 \neq 20$
   **b.** $a - .1 \neq .5$     $b - 3.8 \neq .02$     $r - 5.4 \neq .6$
   **c.** $n - \frac{2}{3} \neq 5\frac{1}{3}$     $x - \frac{7}{12} \neq \frac{11}{16}$     $t - 3\frac{1}{8} \neq 2\frac{1}{2}$
   **d.** $b - 25 \neq -12$     $n - 16 \neq -17$     $r - 83 \neq -68$

4. **a.** $t - 1 \not> 10$     $l - 25 \not> 25$     $p - 84 \not> 0$
   **b.** $x - .8 \not> 3$     $h - 7.1 \not> .9$     $z - 3.9 \not> 7.1$
   **c.** $w - \frac{3}{5} \not> 3\frac{2}{5}$     $c - 4\frac{7}{10} \not> 1\frac{3}{10}$     $d - 2\frac{5}{8} \not> \frac{1}{2}$
   **d.** $n - 70 \not> -25$     $b - 31 \not> -50$     $y - 46 \not> -63$

5. **a.** $s - 12 \not< 21$     $x - 100 \not< 57$     $s - 51 \not< 79$
   **b.** $g - .3 \not< 1.7$     $s - 4 \not< .5$     $b - .68 \not< 0$
   **c.** $w - 2\frac{1}{6} \not< \frac{5}{6}$     $d - \frac{5}{12} \not< \frac{5}{12}$     $x - 4\frac{2}{3} \not< 7\frac{5}{6}$
   **d.** $y - 36 \not< -19$     $x - 52 \not< -61$     $b - 108 \not< -108$

6. **a.** $x - 11 \geq 20$     $y - 27 \geq 14$     $t - 40 \geq 40$
   **b.** $r - .7 \geq 4.1$     $c - 3 \geq 10.5$     $b - 2.6 \geq 8$
   **c.** $n - \frac{3}{8} \geq \frac{1}{2}$     $a - 1\frac{1}{3} \geq 2\frac{2}{3}$     $x - 5\frac{7}{12} \geq 5\frac{7}{12}$
   **d.** $g - 33 \geq -42$     $m - 10 \geq -6$     $s - 18 \geq -18$

7. **a.** $y - 25 \leq 19$     $x - 8 \leq 21$     $a - 16 \leq 0$
   **b.** $m - .17 \leq .7$     $b - 3.4 \leq 1.9$     $y - .6 \leq .6$
   **c.** $t - \frac{4}{5} \leq 1\frac{3}{10}$     $d - 7\frac{2}{3} \leq 2\frac{5}{6}$     $r - 2\frac{3}{4} \leq 1\frac{7}{12}$
   **d.** $p - 9 \leq -24$     $n - 36 \leq -29$     $w - 51 \leq -49$

8. Find the solutions of each of the following inequalities:

   **a.** When the replacements for the variable are all the natural numbers:

     (1) $x - 5 > 21$          (2) $y - 7 \leq 0$          (3) $b - 10 \not< 10$

     (4) $c - 1 < 2$          (5) $r - 2 < -5$       (6) $w - 8 \neq -3$

     (7) $n - 2\frac{1}{2} \not> 3\frac{1}{2}$       (8) $h - 6 \geq 4$        (9) $z - 1.6 > 2.7$

   **b.** When the replacements for the variable are all the integers:

     (1) $n - 3 < 9$          (2) $y - 8 \not> 2$          (3) $t - 1 \geq 0$

     (4) $b - 6 \neq 8$         (5) $c - 10 > -4$       (6) $x - 5 > -9$

     (7) $v - .26 \leq .74$      (8) $a - 7\frac{1}{4} < -3\frac{3}{4}$    (9) $m - 19 \not< -16$

   **c.** When the replacements for the variable are all the whole numbers:

     (1) $b - 11 \not< 5$        (2) $r - 4 \geq 4$          (3) $d - 2 < 7$

     (4) $g - 3\frac{5}{8} \leq 6\frac{3}{5}$      (5) $t - 6 \neq 5$        (6) $y - 9 \not> -10$

     (7) $x - 7 > -2$       (8) $p - 4.6 > .05$      (9) $n - 3 < 0$

   **d.** When the replacements for the variable are all the prime numbers:

     (1) $h - 5 \leq 2$         (2) $c - 1 \not> 15$       (3) $x - 6 \neq -1$

     (4) $n - 1\frac{1}{2} > 3$       (5) $b - 11 < 0$       (6) $w - 2.8 < 7.5$

     (7) $r - 12 > -20$      (8) $y - 2 \geq 2$        (9) $s - 7 \not< -7$

   **e.** When the replacements for the variable are all the even integers:

     (1) $x - 4 \not> 12$       (2) $m - 9 < 5$        (3) $d - 3 \geq -3$

     (4) $y - 7 > -4$       (5) $v - 1 \neq -8$       (6) $b - 2 > 0$

     (7) $a - 4.3 \not< 5.2$      (8) $n - 15 \leq 2$      (9) $z - 3\frac{1}{3} < 4\frac{5}{6}$

   **f.** When the replacements for the variable are all the rational numbers:

     (1) $n - 8 < 5$         (2) $c - 1 \not> -1$      (3) $x - 5\frac{7}{8} > 1\frac{1}{4}$

     (4) $a - 6 \neq 0$        (5) $x - 9 > -3$       (6) $y - 4.6 < -5.7$

     (7) $r - 11 \leq -13$      (8) $m - 8 \not< 7$       (9) $w - 2 \geq -3$

   **g.** When the replacements for the variable are all the odd integers:

     (1) $a - 6 > 4$         (2) $x - 1 \not< -2$       (3) $p - 7 < 8$

     (4) $n - 10 > 0$       (5) $c - 1\frac{3}{5} \leq 4\frac{1}{2}$     (6) $t - 2 \neq 5$

     (7) $k - 3.4 < -1.5$     (8) $x - 11 \geq 1$      (9) $y - 5 \not> -5$

   **h.** When the replacements for the variable are $-8, -7, -6, \ldots, 8$:

     (1) $x - 5 > 2$         (2) $m - 3 < 1$        (3) $a - 1 \leq -2$

     (4) $n - 2 \not> 0$        (5) $c - 6 \geq -8$      (6) $x - 12 \not< -7$

     (7) $d - 4 < -4$      (8) $b - 11 \neq -3$      (9) $y - 6 > 6$

   **i.** When the replacements for the variable are $-16, -14, -12, \ldots, 20$:

     (1) $t - 9 < 4$         (2) $x - 11 > -5$      (3) $n - 7 \not> 3$

     (4) $w - 8 \leq -20$      (5) $y - 1 \not< 15$      (6) $a - 5 < 6$

     (7) $r - 5 \neq 0$       (8) $m - 7 \geq 14$      (9) $x - 8 > -8$

9. Find the solutions of each of the following inequalities:

   **a.** When the replacements for the variable are all the real numbers:

     (1) $11 < c - 3$          (2) $1 > x - 9$          (3) $0 \not< y - 4$

     (4) $\frac{7}{8} \geq n - 1\frac{3}{4}$       (5) $-3 \neq b - 12$       (6) $1.45 \not> r - .75$

     (7) $-11 < d - 11$      (8) $17 \leq t - 10$       (9) $8 > g - 7$

   **b.** When the replacements for the variable are all the natural numbers:

     (1) $2 > x - 5$          (2) $-3 \leq w - 9$       (3) $-8 \not> y - 8$

   **c.** When the replacements for the variable are all the integers:

     (1) $14 < t - 11$        (2) $-15 < b - 7$       (3) $4 \neq x - 12$

   **d.** When the replacements for the variable are all the prime numbers:

     (1) $15 \not> n - 23$       (2) $5 \geq x - 6$        (3) $-1 < a - 1$

   **e.** When the replacements for the variable are all the odd integers:

     (1) $-7 > c - 16$       (2) $3 > m - 10$       (3) $-20 \not< y - 13$

   **f.** When the replacements for the variable are $-11, -9, -7, \ldots, 11$.

     (1) $1 < n - 8$          (2) $-7 \geq x - 2$       (3) $3 \not> b - 3$

## CUMULATIVE REVIEW

Find the solutions of each of the following inequalities when the replacements for the variable are all the real numbers:

**1. a.** $t > 5$               **b.** $x < -7$              **c.** $y \geq .6$

**2. a.** $n + 10 < 23$      **b.** $15 \not> w - 18$      **c.** $-6d \neq 12$

   **d.** $6 + s \geq 0$         **e.** $-9n < -27$         **f.** $\frac{x}{4} \leq -5$

**3. a.** $r - 7 < 16$        **b.** $k - 8 > -5$       **c.** $-3 \not< y - 12$

   **d.** $x - 9.3 \geq 4.8$     **e.** $c - 1 \neq 14$       **f.** $p - 3\frac{1}{4} \not> 2\frac{3}{4}$

   **g.** $b - 2 > -11$      **h.** $0 \leq g - 6$         **i.** $x - 9 < -9$

Find the solutions of each of the following inequalities when the replacements for the variable are:

**4.** All the whole numbers:

   **a.** $b - 3 < 4$         **b.** $w - 5 \not> 1$        **c.** $6 \geq x - 6$

**5.** All the even integers:

   **a.** $n - 1 > -2$       **b.** $0 \not< t - 7$         **c.** $a - 10 \leq -10$

**6.** All the prime numbers:

   **a.** $r - 4 < 12$        **b.** $b - 9 \geq 2$         **c.** $m - 1 \not> 1$

**7.** All the odd whole numbers greater than 6 and less than 19:

   **a.** $x - 8 > 3$         **b.** $a - 2 \neq 11$       **c.** $-8 \not< y - 20$

**8.** $-9, -7, -5, \ldots, 9$:

   **a.** $n - 5 \not> 0$         **b.** $x - 3 < -4$        **c.** $-4 \leq c - 1$

# Inequalities–Solution by Inverse Operation (Division) or by Multiplicative Inverse

### I. Aim

To solve inequalities of the types: $3n > 15$; $3n < 15$; $3n \geq 15$; $3n \leq 15$; $3n \neq 15$; $3n \not> 15$; and $3n \not< 15$.
(1) By the inverse operation (division) or
(2) By the multiplicative inverse

### II. Procedure

1. Solution by Inverse Operation:
   Divide both sides of the inequality by the coefficient of the variable so that the variable remains with the coefficient one (1) understood.
   a. If the number you divide by is a positive number, *keep* the order of the resulting inequality the *same* as the given inequality.
   b. If the number you divide by is a negative number, *reverse* the order of the resulting inequality. See sample solutions 4 and 5.
2. Solution by Multiplicative Inverse:
   Or multiply both sides of the inequality by the multiplicative inverse of the coefficient of the variable.
   a. If the number you multiply by is a positive number, keep the order of the resulting inequality the *same* as the given inequality.
   b. If the number you multiply by is a negative number, *reverse* the order of the resulting inequality.
3. When the variable appears on the right side of the inequality, rewrite in the final step by using the equivalent inequality. See sample solution 2.
4. See Exercise 8-3 for the solutions of inequalities in simplest form.
5. Check your solutions. See page 314.

### III. Sample Solutions

1. Solve $3n < 15$ when the replacements for the variable are all the real numbers.

   **a.** *Solution:*

   $3n < 15$

   $\dfrac{3n}{3} < \dfrac{15}{3}$

   $n < 5$

   $n = $ all real numbers less than 5.

   **b.** *Solution:*

   $3n < 15$

   $\frac{1}{3} \cdot 3n < \frac{1}{3} \cdot 15$

   $n < 5$

   *Check:*

   Is 6 a solution?

   $3n < 15$

   $3 \cdot 6 < 15$?

   $18 \not< 15$

   No

   Is 2?

   $3n < 15$

   $3 \cdot 2 < 15$?

   $6 < 15$ ✔

   Yes

   *Answer:* All real numbers less than 5.

330

**2.** Solve $24 > 6x$ when the replacements for the variable are all the whole numbers.

*Solution:*

$24 > 6x$

$\frac{24}{6} > \frac{6x}{6}$

$4 > x$

$x < 4$

$x = 0, 1, 2, 3$

*Answer:* 0, 1, 2, 3

**3.** Solve $4y \not< -28$ when the replacments for the variable are all the integers.

*Solution:*

$4y \not< -28$

$\frac{4y}{4} \not< \frac{-28}{4}$

$y \not< -7$

$y \geq -7$

$y = -7, -6, -5, \ldots$

*Answer:* $-7, -6, -5, \ldots$

**4.** Solve $-9n > 54$ when the replacements for the variable are all the integers.

*Solution:*

$-9n > 54$

$\frac{-9n}{-9} < \frac{54}{-9}$

$n < -6$

$n = \ldots, -9, -8, -7$

*Answer:* $\ldots, -9, -8, -7$

**5.** Solve $-2x \leq -8$ when the replacements for the variable are all the integers.

*Solution:*

$-2x \leq -8$

$\frac{-2x}{-2} \leq \frac{-8}{-2}$

$x \geq 4$

$x = 4, 5, 6, \ldots$

*Answer:* 4, 5, 6, $\ldots$

## DIAGNOSTIC TEST

Find the solutions of each of the following inequalities when the replacements for the variable are all the real numbers.

**1. a.** $4x > 76$   **b.** $.05m > 20$   **c.** $\frac{2}{3}r > 12$   **d.** $7w > -21$

**2. a.** $9y < 45$   **b.** $8s < 3.2$   **c.** $\frac{1}{4}p < 15$   **d.** $17t < -68$

**3. a.** $14w \neq 56$   **b.** $.3x \neq 2.7$   **c.** $\frac{3}{5}n \neq 42$   **d.** $90k \neq -54$

**4. a.** $5t \not> 11$   **b.** $4.5c \not> 9$   **c.** $\frac{1}{8}h \not> 24$   **d.** $16r \not> -80$

**5. a.** $6m \not< 72$   **b.** $.01y \not< 7$   **c.** $\frac{5}{6}x \not< 60$   **d.** $3y \not< -19$

**6. a.** $12p \geq 8$   **b.** $.24x \geq 9.6$   **c.** $\frac{7}{12}n \geq 49$   **d.** $8w \geq -5$

**7. a.** $2n \leq 17$   **b.** $.7z \leq .49$   **c.** $\frac{4}{7}a \leq 56$   **d.** $15z \leq -20$

**8. a.** $-7t > 63$   **b.** $-3a > -36$

**9. a.** $-10n < 5$   **b.** $-15p < -75$

**10. a.** $-11x \neq 110$        **b.** $-t \neq -8$
**11. a.** $-5v \not> 0$        **b.** $-2n \not> -6$
**12. a.** $-12z \not< 36$        **b.** $-\frac{4}{5}w \not< -80$
**13. a.** $-8x \geq 10$        **b.** $-19r \geq -57$
**14. a.** $-9m \leq 7.2$        **b.** $-6v \leq -84$

Find the solutions of each of the following inequalities when the replacements for the variable are:

**15. a.** All the natural numbers:
    (1) $10b < 90$        (2) $-5x \geq 45$        (3) $\frac{2}{5}n \not< -20$
    **b.** All the integers:
    (1) $7a \not> -42$        (2) $-13y \leq -13$        (3) $.06b > 4.8$
    **c.** All the prime numbers:
    (1) $6n \leq 108$        (2) $4t \not\geq 27$        (3) $-11c \not< -77$
**16. a.** All the real numbers:
    (1) $54 > 3b$        (2) $-69 \leq 12d$        (3) $-48 \not> -\frac{3}{8}y$
    **b.** All the even integers:
    (1) $-12 < 4x$        (2) $105 \not< 15y$        (3) $54 \geq -9g$

## RELATED PRACTICE EXAMPLES

Find the solutions of each of the following inequalities when the replacements for the variable are all the real numbers:

**1. a.** $8n > 56$        $6x > 84$        $15s > 25$        $20z > 10$
    **b.** $.02a > 6$        $1.8d > .9$        $.7h > .14$        $.05s > 4.9$
    **c.** $\frac{2}{3}x > 28$        $\frac{5}{6}t > 30$        $\frac{1}{4}r > 67$        $\frac{7}{15}b > 91$
    **d.** $2x > -48$        $.03n > -.18$        $\frac{2}{3}b > -16$        $12n > -42$

**2. a.** $4c < 24$        $16a < 12$        $5k < 0$        $24p < 60$
    **b.** $3g < 3.6$        $1.8x < .54$        $2.5b < 50$        $.01n < .8$
    **c.** $\frac{9}{10}y < 45$        $\frac{5}{4}d < 65$        $\frac{3}{8}y < 81$        $\frac{7}{12}c < 84$
    **d.** $8t < -64$        $\frac{7}{8}w < -28$        $16a < -12$        $7x < -5.6$

**3. a.** $9x \neq 72$        $15r \neq 90$        $2n \neq 3$        $10g \neq 6$
    **b.** $1.5c \neq 3$        $6a \neq .09$        $.98t \neq 4.9$        $.02p \neq 20$
    **c.** $\frac{3}{8}m \neq 81$        $\frac{7}{10}r \neq 42$        $\frac{1}{4}t \neq 13$        $\frac{3}{5}y \neq 15$
    **d.** $19b \neq -57$        $.95t \neq -190$        $14x \neq -35$        $\frac{5}{12}z \neq -100$

**4. a.** $7y \not> 35$        $12p \not> 8$        $4x \not> 25$        $9y \not> 4$
    **b.** $.8y \not> .2$        $4.2s \not> .7$        $.03a \not> 7.8$        $.6x \not> 18$
    **c.** $\frac{7}{12}b \not> 56$        $\frac{1}{3}x \not> 23$        $\frac{5}{8}n \not> 80$        $\frac{3}{10}r \not> 27$
    **d.** $9g \not> -54$        $.03p \not> -7.2$        $\frac{2}{5}b \not> -39$        $25n \not> -10$

5. **a.** $14b \not< 56$    $6z \not< 0$    $7n \not< 31$    $6x \not< 2$
   **b.** $3.2n \not< 6.4$    $9x \not< 1.8$    $.12c \not< 36$    $.004y \not< 5$
   **c.** $\frac{7}{9}t \not< 63$    $\frac{2}{5}d \not< 74$    $\frac{5}{6}x \not< 42$    $\frac{8}{5}d \not< 56$
   **d.** $10x \not< -200$    $.1h \not< -10$    $\frac{7}{10}y \not< -147$    $36r \not< -24$

6. **a.** $3x \geq 81$    $9y \geq 6$    $2c \geq 31$    $15t \geq 0$
   **b.** $6a \geq 2.4$    $.7n \geq .63$    $1.3y \geq 7.8$    $.04r \geq 10$
   **c.** $\frac{1}{8}n \geq 19$    $\frac{2}{3}b \geq 30$    $\frac{4}{5}t \geq 72$    $\frac{11}{7} \geq 121$
   **d.** $5c \geq -65$    $.06t \geq -48$    $\frac{7}{12}x \geq -21$    $10b \geq -105$

7. **a.** $12g \leq 96$    $6b \leq 14$    $8r \leq 2$    $10c \leq 3$
   **b.** $.7x \leq 4.9$    $15y \leq 7.5$    $2.3c \leq .69$    $.08d \leq 80$
   **c.** $\frac{1}{5}n \leq 17$    $\frac{3}{8}x \leq 39$    $\frac{10}{9}t \leq 70$    $\frac{5}{6}a \leq 85$
   **d.** $9g \leq -54$    $\frac{7}{10}d \leq -28$    $1.2c \leq -144$    $24x \leq -18$

8. **a.** $-11x > 22$    $-5a > 90$    $-6m > 2$    $-\frac{1}{6}t > 2\frac{2}{3}$
   **b.** $-14y > -42$    $-s > -2$    $-8b > -18$    $-.9x > -4.5$

9. **a.** $-15n < 75$    $-9b < 12$    $-x < 20$    $-.04t < 16$
   **b.** $-7c < -49$    $-21r < -7$    $-8p < -9.6$    $-\frac{7}{8}d < -28$

10. **a.** $-12d \neq 84$    $-10c \neq 6$    $-.1v \neq .5$    $-\frac{3}{4}a \neq 15$
    **b.** $-20n \neq -160$    $-6d \neq -20$    $-\frac{1}{3}s \neq -17$    $-.02h \neq -100$

11. **a.** $-3c \not> 81$    $-18r \not> 40$    $-1.9x \not> .95$    $-\frac{5}{8}w \not> 30$
    **b.** $-5m \not> -105$    $-12n \not> -8$    $-\frac{2}{5}g \not> -4$    $-1.5r \not> -90$

12. **a.** $-4c \not< 52$    $-24x \not< 18$    $-.125n \not< 72$    $-\frac{4}{5}m \not< 48$
    **b.** $-16n \not< -144$    $-25t \not< -65$    $-\frac{9}{10}z \not< 18$    $-.18y \not< -.198$

13. **a.** $-6x \geq 48$    $-y \geq 19$    $-.4b \geq 8$    $-\frac{1}{6}a \geq 25$
    **b.** $-15c \geq -135$    $-8n \geq -6$    $-\frac{2}{3}t \geq -16$    $-5d \geq -3.5$

14. **a.** $-18x \leq 162$    $-9r \leq 21$    $-\frac{1}{2}g \leq 75$    $-.48y \leq 24$
    **b.** $-25n \leq -400$    $-72z \leq -36$    $-.4y \leq -92$    $-\frac{7}{12}m \leq -119$

15. Find the solutions of each of the following inequalities:

   **a.** When the replacements for the variable are all the integers:
     (1) $8n > 40$      (2) $-9y < 72$      (3) $10b \not> -70$
     (4) $2x \not< 69$      (5) $6w \neq -54$      (6) $.4t \leq 20$
     (7) $\frac{3}{5}y > 75$      (8) $-12b \geq -27$      (9) $18m < 12$

   **b.** When all the replacements for the variable are all the natural numbers:
     (1) $15x < 75$      (2) $20c \geq 200$      (3) $-3b > -21$
     (4) $7n \not> 0$      (5) $2w > -12$      (6) $8t \neq 4$
     (7) $.05d < 10$      (8) $9x \leq -21$      (9) $\frac{2}{3}n \not< 40$

   **c.** When the replacements for the variable are all the whole numbers:
     (1) $16w \leq 96$    (2) $5a \not> -15$    (3) $9d \neq 63$    (4) $4n > 50$
     (5) $7r \not< -28$    (6) $8m < 32.8$    (7) $\frac{5}{6}x > 20$    (8) $-3y \geq 84$

**d.** When the replacements for the variable are all the even integers:

(1) $6p \not< 42$             (2) $11y > -22$         (3) $7x \leq 7$

(4) $-15b < -90$        (5) $14d \neq 42$           (6) $9c \not< 99$

(7) $-\frac{7}{9}x \geq 21$         (8) $.2y < -16$            (9) $13n > 54$

**e.** When the replacements for the variable are all the prime numbers:

(1) $10x < 60$     (2) $4t \neq 68$     (3) $6y > 57$     (4) $20c \not< -50$

(5) $\frac{3}{4}n \geq 69$     (6) $11b \leq 11$     (7) $8g > 120$     (8) $.9x \not> 270$

**f.** When the replacements for the variable are all the rational numbers:

(1) $.7b > .14$         (2) $12d \not< 6$            (3) $\frac{5}{8}x \leq 70$

(4) $-11y < 132$       (5) $-4w < -63$      (6) $.08x \not> -.45$

(7) $16n > -2$        (8) $1.05c \geq 525$      (9) $\frac{6}{7}y \neq -20$

**g.** When the replacements for the variable are all the odd integers:

(1) $4n \geq 72$     (2) $-27z < 0$     (3) $14b \neq -14$     (4) $-21y > -35$

(5) $.6p \not> -7.8$     (6) $\frac{1}{4}c > 24$     (7) $9b \not< -5$     (8) $25n \leq 175$

**h.** When the replacements for the variable are $-7, -6, -5, \ldots, 7$.

(1) $14y < 28$     (2) $3x \geq -3$     (3) $-8b \not> 40$     (4) $4r \leq -32$

(5) $\frac{7}{8}x \neq 21$     (6) $.5t > 10$     (7) $19n < 0$     (8) $12c \not< -54$

**i.** When the replacements for the variable are $-14, -12, -10, \ldots, 20$.

(1) $-22x \leq 88$       (2) $10y \neq -140$      (3) $51c \not> 0$

(4) $-7d \geq -63$        (5) $12r \leq 240$       (6) $21b < -84$

(7) $18y < -18$        (8) $.2z > 3.6$        (9) $\frac{5}{12}w \not< 5$

**16.** Find the solutions of each of the following inequalities:

**a.** When the replacements for the variable are all the real numbers:

(1) $45 > 9r$        (2) $6 < 10t$         (3) $.28 \not\leq .4n$

(4) $32 \geq 14c$       (5) $18 \neq \frac{2}{3}y$        (6) $-40 \not< 5x$

(7) $12 \not> -2b$       (8) $-42 \leq -6n$      (9) $23 \not\geq -3w$

**b.** When the replacements for the variable are all the natural numbers:

(1) $63 < 7b$            (2) $0 \not> -8x$          (3) $-72 \geq 9z$

**c.** When the replacements for the variable are all the integers:

(1) $54 > 6a$            (2) $11 \leq -y$           (3) $-30 \not< -10t$

**d.** When the replacements for the variable are all the even integers:

(1) $64 \leq -16x$        (2) $-56 \not\geq 14b$       (3) $150 > 25r$

**e.** When the replacements for the variable are all the prime numbers:

(1) $132 \not< 12y$        (2) $-10 > -2w$         (3) $35 \leq -5n$

**f.** When the replacements for the variable are $-8, -6, -4, \ldots, 10$:

(1) $-16 \geq 8c$         (2) $12 < -3x$          (3) $-1 \not> -m$

# Inequalities—Solution by Inverse Operation (Multiplication) or by Multiplicative Inverse

**I. Aim** To solve inequalities of the types: $\frac{n}{3} > 15$; $\frac{n}{3} < 15$; $\frac{n}{3} \geq 15$;

$\frac{n}{3} \leq 15$; $\frac{n}{3} \neq 15$; $\frac{n}{3} \not> 15$; and $\frac{n}{3} \not< 15$.

(1) By the inverse operation (multiplication) or
(2) By the multiplicative inverse

## II. Procedure

1. Solution by Inverse Operation:
   Multiply both sides of the inequality by the denominator of the fractional coefficient of the variable so that the variable remains with coefficient one (1) understood. See sample solution 1.a.
   **a.** If the number you multiply by is a positive number, keep the order of the resulting inequality the *same* as the given inequality.
   **b.** If the number you multiply by is a negative number, reverse the order of the resulting inequality. See sample solution.
2. Solution by Multiplicative Inverse:
   Or multiply both sides of the inequality by the multiplicative inverse of the coefficient of the variable. See 1.a. and 1.b. above.
3. When the variable appears on the right side of the inequality, rewrite in the final step by using the equivalent inequality. See sample solution 2.
4. See Exercise 8-3 for the solutions of inequalities in simplest form.
5. Check your solutions. See page 314.

## III. Sample Solutions

1. Solve $\frac{n}{3} > 15$ when the replacements for the variable are all the real numbers.

   **a.** *Solution:*

   $$\frac{n}{3} > 15$$

   $$3 \cdot \frac{n}{3} > 3 \cdot 15$$

   $$n > 45$$
   $n =$ all real numbers greater than 45.

   **b.** *Solution:*

   $$\frac{n}{3} > 15$$

   $$\tfrac{1}{3}n > 15$$
   $$3 \cdot \tfrac{1}{3}n > 3 \cdot 15$$

   $$n > 45$$

   *Check:*
   Is 60 a
   solution?        Is 24?

   $$\frac{n}{3} > 15 \qquad \frac{n}{3} > 15$$

   $$\tfrac{60}{3} > 15\,? \qquad \tfrac{24}{3} > 15\,?$$
   $$20 > 15 ✔ \qquad 8 \not> 15$$
   Yes              No

   *Answer:* All real numbers greater than 45.

335

**2.** Solve $2 \geq \dfrac{n}{5}$ when the replacements for the variable are all the natural numbers.

*Solution:*

$$2 \geq \frac{n}{5}$$

$$5 \cdot 2 \geq 5 \cdot \frac{n}{5}$$

$$10 \geq n$$
$$n \leq 10$$
$$n = 1, 2, 3, \ldots, 10$$

*Answer:* 1, 2, 3, ..., 10

**4.** Solve $\dfrac{n}{-6} < 3$ when the replacements for the variable are all the integers.

*Solution:*

$$\frac{n}{-6} < 3$$

$$-6 \cdot \frac{n}{-6} > (-6)(3)$$

$$n > -18$$
$$n = -17, -16, -15, \ldots$$

*Answer:* $-17, -16, -15, \ldots$

**3.** Solve $\dfrac{n}{2} < 3$ when the replacements for the variable are all the whole numbers:

*Solution:*

$$\frac{n}{2} < 3$$

$$2 \cdot \frac{n}{2} < 2 \cdot 3$$

$$n < 6$$
$$n = 0, 1, 2, 3, 4, 5$$

*Answer:* 0, 1, 2, 3, 4, 5

**5.** Solve $\dfrac{b}{-2} \geq -4$ when the replacements for the variable are all the real numbers.

*Solution:*

$$\frac{b}{-2} \geq -4$$

$$-2 \cdot \frac{b}{-2} \leq (-2)(-4)$$

$$b \leq 8$$
$$b = 8 \text{ and all the real numbers less than 8.}$$

*Answer:* 8 and all the real numbers less than 8.

## DIAGNOSTIC TEST

Find the solutions of each of the following inequalities when the replacements for the variable are all the real numbers:

**1.** $\dfrac{n}{5} > 11$     **2.** $\dfrac{x}{8} < 6$     **3.** $\dfrac{b}{3} \neq 1$     **4.** $\dfrac{r}{.2} \not> 9$

**5.** $\dfrac{m}{7} \not< 0$     **6.** $\dfrac{p}{4} \geq 12$     **7.** $\dfrac{v}{9} \leq 20$     **8.** $\dfrac{x}{-3} > 15$

**9.** $\dfrac{a}{-2} < -4$     **10.** $\dfrac{c}{-11} \neq 6$     **11.** $\dfrac{z}{-4} \not> -16$

**12.** $\dfrac{m}{-24} \not< 8$ **13.** $\dfrac{r}{-15} \geq 0$ **14.** $\dfrac{y}{-9} \leq -1$

Find the solutions of each of the following inequalities when the replacements for the variable are all the:

**15. a.** Natural numbers: **(1)** $\dfrac{n}{7} < 2$ **(2)** $\dfrac{x}{-5} < -6$ **(3)** $\dfrac{b}{4} \not> 10$

**b.** Integers: **(1)** $\dfrac{w}{-2} > 6$ **(2)** $\dfrac{m}{8} > 0$ **(3)** $\dfrac{x}{-7} \neq -3$

**c.** Prime numbers: **(1)** $\dfrac{r}{3} \leq 5$ **(2)** $\dfrac{a}{7} \not> 4$ **(3)** $\dfrac{t}{-1} \geq -11$

**16. a.** Real numbers: **(1)** $8 > \dfrac{c}{6}$ **(2)** $-9 \leq \dfrac{y}{-4}$ **(3)** $-10 \not< \dfrac{x}{7}$

**b.** Odd integers: **(1)** $3 < \dfrac{n}{2}$ **(2)** $-5 > \dfrac{b}{5}$ **(3)** $-12 \not< \dfrac{w}{-1}$

## RELATED PRACTICE EXAMPLES

Find the solutions of each of the following inequalities when the replacements for the variable are all the real numbers:

**1. a.** $\dfrac{x}{2} > 3$ $\quad \dfrac{n}{4} > -8$ $\quad \dfrac{d}{15} > 0$ $\quad \dfrac{a}{.6} > 2.7$

**b.** $\dfrac{c}{3} > 5$ $\quad \dfrac{y}{9} > 6$ $\quad \dfrac{g}{8} > -1$ $\quad \dfrac{r}{10} > 10$

**2. a.** $\dfrac{n}{6} < 2$ $\quad \dfrac{v}{8} < 11$ $\quad \dfrac{x}{15} < -15$ $\quad \dfrac{b}{9} < 1$

**b.** $\dfrac{r}{12} < 0$ $\quad \dfrac{a}{6} < 7$ $\quad \dfrac{t}{.2} < -.9$ $\quad \dfrac{m}{4} < 4$

**3. a.** $\dfrac{b}{4} \neq 10$ $\quad \dfrac{t}{3} \neq 1$ $\quad \dfrac{m}{8} \neq 4$ $\quad \dfrac{c}{12} \neq -20$

**b.** $\dfrac{h}{21} \neq 7$ $\quad \dfrac{r}{1.5} \neq 12$ $\quad \dfrac{b}{8} \neq -9$ $\quad \dfrac{d}{7} \neq 0$

**4. a.** $\dfrac{y}{8} \not> 4$ $\quad \dfrac{s}{7} \not> 15$ $\quad \dfrac{a}{9} \not> 4.2$ $\quad \dfrac{b}{11} \not> -100$

**b.** $\dfrac{x}{16} \not> -1$ $\quad \dfrac{n}{2} \not> 32$ $\quad \dfrac{w}{3.4} \not> .17$ $\quad \dfrac{m}{8} \not> 13$

**5. a.** $\dfrac{m}{5} \not< 30$ $\quad \dfrac{x}{12} \not< 1$ $\quad \dfrac{r}{6} \not< 0$ $\quad \dfrac{n}{3} \not< -14$

**b.** $\dfrac{y}{13} \not< 4$ $\quad \dfrac{t}{6.2} \not< 10$ $\quad \dfrac{a}{19} \not< -19$ $\quad \dfrac{g}{20} \not< 11$

**6. a.** $\dfrac{x}{4} \geq 6$ $\qquad$ $\dfrac{a}{9} \geq -3$ $\qquad$ $\dfrac{m}{21} \geq 21$ $\qquad$ $\dfrac{v}{63} \geq 1$

**b.** $\dfrac{d}{18} \geq 0$ $\qquad$ $\dfrac{y}{5.4} \geq .25$ $\qquad$ $\dfrac{t}{6} \geq -36$ $\qquad$ $\dfrac{n}{2} \geq 17$

**7. a.** $\dfrac{c}{3} \leq 16$ $\qquad$ $\dfrac{x}{5} \leq 2$ $\qquad$ $\dfrac{z}{22} \leq 11$ $\qquad$ $\dfrac{d}{12} \leq -12$

**b.** $\dfrac{x}{95} \leq 1$ $\qquad$ $\dfrac{n}{10.4} \leq -.3$ $\qquad$ $\dfrac{a}{38} \leq 0$ $\qquad$ $\dfrac{p}{100} \leq 25$

**8.** $\dfrac{x}{-5} > 8$ $\qquad$ $\dfrac{b}{-4} > 1$ $\qquad$ $\dfrac{t}{-7} > -3$ $\qquad$ $\dfrac{m}{-10} > -15$

**9.** $\dfrac{n}{-3} < 12$ $\qquad$ $\dfrac{a}{-9} < 0$ $\qquad$ $\dfrac{r}{-6} < -7$ $\qquad$ $\dfrac{x}{-8} < -2$

**10.** $\dfrac{c}{-7} \neq -1$ $\qquad$ $\dfrac{b}{-8} \neq 5$ $\qquad$ $\dfrac{a}{-2} \neq 11$ $\qquad$ $\dfrac{n}{-9} \neq -9$

**11.** $\dfrac{y}{-5} \not> 4$ $\qquad$ $\dfrac{x}{-10} \not> -8$ $\qquad$ $\dfrac{w}{-3} \not> -12$ $\qquad$ $\dfrac{t}{-11} \not> 0$

**12.** $\dfrac{a}{-12} \not< -3$ $\qquad$ $\dfrac{c}{-5} \not< 3$ $\qquad$ $\dfrac{n}{-9} \not< 14$ $\qquad$ $\dfrac{h}{-1} \not< -1$

**13.** $\dfrac{x}{-2} \geq 20$ $\qquad$ $\dfrac{s}{-15} \geq -5$ $\qquad$ $\dfrac{p}{-14} \geq 6$ $\qquad$ $\dfrac{r}{-6} \geq -16$

**14.** $\dfrac{d}{-4} \leq -32$ $\qquad$ $\dfrac{n}{-7} \leq 2$ $\qquad$ $\dfrac{x}{-28} \leq 0$ $\qquad$ $\dfrac{y}{-20} \leq -4$

**15.** Find the solutions of each of the following inequalities:
**a.** When the replacements for the variable are all the whole numbers:

(1) $\dfrac{x}{3} > 4$ $\quad$ (2) $\dfrac{b}{5} < 3$ $\quad$ (3) $\dfrac{n}{8} \neq 1$ $\quad$ (4) $\dfrac{a}{2} \not> 6$ $\quad$ (5) $\dfrac{r}{-3} \not< -3$

(6) $\dfrac{t}{-1} \geq 4$ $\quad$ (7) $\dfrac{p}{2} \leq 1$ $\quad$ (8) $\dfrac{w}{-5} < -9$ $\quad$ (9) $\dfrac{d}{-7} > 0$

**b.** When the replacements for the variable are all the integers:

(1) $\dfrac{t}{3} \leq 8$ (2) $\dfrac{m}{-2} > 4$ (3) $\dfrac{y}{-6} \not< -1$ (4) $\dfrac{c}{4} \neq 4$ (5) $\dfrac{v}{-7} \geq -2$

(6) $\dfrac{x}{5} > -3$ $\quad$ (7) $\dfrac{b}{2} < 0$ $\quad$ (8) $\dfrac{n}{-8} \not> 3$ $\quad$ (9) $\dfrac{a}{-10} < 1$

**c.** When the replacements for the variable are all the prime numbers:

(1) $\dfrac{n}{6} \not> 3$ $\quad$ (2) $\dfrac{v}{5} \geq 7$ $\quad$ (3) $\dfrac{x}{-2} > -1$ $\quad$ (4) $\dfrac{b}{7} \not< 4$ $\quad$ (5) $\dfrac{w}{-3} \neq 8$

(6) $\dfrac{c}{4} < 6$      (7) $\dfrac{t}{9} > 5$      (8) $\dfrac{r}{-2} > -9$      (9) $\dfrac{y}{10} \le 0$

**d.** When the replacements for the variable are all the even integers:

(1) $\dfrac{m}{4} > 5$      (2) $\dfrac{w}{11} \neq -2$      (3) $\dfrac{p}{8} \le 7$

(4) $\dfrac{x}{9} \ngtr 1$      (5) $\dfrac{r}{-3} < 11$      (6) $\dfrac{z}{-4} < -3$

(7) $\dfrac{h}{10} \nless 2$      (8) $\dfrac{a}{15} \ge 0$      (9) $\dfrac{b}{-1} > 6$

**e.** When the replacements for the variable are $-10, -9, -8, \ldots, 12$:

(1) $\dfrac{r}{5} > 2$      (2) $\dfrac{x}{8} < -1$      (3) $\dfrac{y}{-3} \ngtr 3$

(4) $\dfrac{n}{-2} < -2$      (5) $\dfrac{w}{7} \le 0$      (6) $\dfrac{x}{-5} \nless 4$

(7) $\dfrac{d}{-1} > 7$      (8) $\dfrac{m}{3} \neq -4$      (9) $\dfrac{c}{-8} \ge -3$

**16.** Find the solutions of each of the following inequalities:

**a.** When the replacements for the variable are all the real numbers:

(1) $6 > \dfrac{x}{2}$      (2) $1 < \dfrac{b}{9}$      (3) $4 \neq \dfrac{n}{4}$

(4) $-10 \ngtr \dfrac{y}{3}$      (5) $2 \nless \dfrac{t}{-8}$      (6) $-7 \ge \dfrac{r}{-6}$

(7) $0 \le \dfrac{s}{12}$      (8) $5 < \dfrac{w}{-1}$      (9) $8 > \dfrac{p}{9}$

**b.** When the replacements for the variable are all the natural numbers:

(1) $3 < \dfrac{x}{4}$      (2) $1 \ge \dfrac{n}{7}$      (3) $-2 \ngtr \dfrac{a}{-5}$

**c.** When the replacements for the variable are all the prime numbers:

(1) $1 < \dfrac{r}{2}$      (2) $5 > \dfrac{x}{4}$      (3) $-7 \le \dfrac{y}{-1}$

**d.** When the replacements for the variable are all the integers:

(1) $10 \nless \dfrac{t}{8}$      (2) $0 < \dfrac{m}{-6}$      (3) $-9 \ge \dfrac{c}{3}$

**e.** When the replacements for the variable are $-10, -8, -6, \ldots, 10$:

(1) $-2 > \dfrac{d}{2}$      (2) $-3 \le \dfrac{x}{-1}$      (3) $0 \nless \dfrac{b}{-4}$

# Solving Inequalities

**I. Aim**    To solve inequalities in one variable.

**II. Procedure**

1. Transform the given inequality to the simplest equivalent inequality by using inverse operations or the additive or multiplicative inverse, and by following the properties of order of an inequality. See the sample solutions.

2. When either or both members consist of two or more terms containing the variable or two or more constant terms, combine like terms in the members, then solve the resulting inequality.

3. When the inequality contains terms with variables inclosed within parentheses, first remove the terms from the parentheses.

4. See Exercise 8-3 for the solutions of inequalities in simplest form.

5. Check your solutions. See page 314.

**III. Sample Solutions**

1. Solve $8x + 9 > 41$ when the replacements for the variable are all the real numbers:

   *Solution by Inverse Operations:*

$$8x + 9 > 41$$
$$8x + 9 - 9 > 41 - 9$$
$$8x > 32$$
$$\frac{8x}{8} > \frac{32}{8}$$
$$x > 4$$
$$x = \text{all real numbers}$$
$$\text{greater than 4.}$$

*Answer:* All the real numbers greater than 4.

*Check:*

| Is 5 a solution? | Is 2 a solution? |
|---|---|
| $8x + 9 > 41$ | $8x + 9 > 41$ |
| $8 \cdot 5 + 9 > 41\,?$ | $8 \cdot 2 + 9 > 41\,?$ |
| $40 + 9 > 41$ | $16 + 9 > 41$ |
| $49 > 41 \; ✔$ | $25 \not> 41$ |
| Yes | No |

**340**

**2.** Solve $4n - 5 \leq 7 + 2n$ when the replacements for the variable are all the whole numbers.

*Solution by Additive and Multiplicative Inverses:*
$$4n - 5 \leq 7$$
$$4n + (-2n) - 5 \leq 7 + 2n + (-2n)$$
$$2n - 5 \leq 7$$
$$2n + (-5) \leq 7$$
$$2n + (-5) + (+5) \leq 7 + (+5)$$
$$2n \leq 12$$
$$\tfrac{1}{2} \cdot 2n \leq \tfrac{1}{2} \cdot 12$$
$$n \leq 6$$
$$n = 0, 1, 2, 3, 4, 5, 6$$

*Answer:* 0, 1, 2, 3, 4, 5, 6

*Check:*

Is 3 a solution?
$$4n - 5 \leq 7 + 2n$$
$$4 \cdot 3 - 5 \leq 7 + 2 \cdot 3 ?$$
$$12 - 5 \leq 7 + 6$$
$$7 \leq 13 ✔$$
Yes

Is 8 a solution?
$$4n - 5 \leq 7 + 2n$$
$$4 \cdot 8 - 5 \leq 7 + 2 \cdot 8 ?$$
$$32 - 5 \leq 7 + 16$$
$$27 \not\leq 23$$
No

Sole and graph each solution set
P. A, b, and C

## DIAGNOSTIC TEST

Find the solutions of each of the following inequalities when the replacements for the variable are all the real numbers:

**1. a.** $2n + 9 > 23$
  **b.** $3 + 9c < 48$
  **c.** $10b + 7 \neq 37$
  **d.** $6 + 12y \not> 44$
  **e.** $8d + 23 \not< 65$
  **f.** $7x + 41 \geq -8$
  **g.** $2n + 19 \leq 7$

**2. a.** $4y - 11 > 17$
  **b.** $7a - 25 < 10$
  **c.** $2 \neq 6m - 2$
  **d.** $15s - 75 \not> 0$
  **e.** $9c - 6 \not< 3$
  **f.** $12x - 4 \geq 5$
  **g.** $8y - 11 \leq -25$

**3. a.** $d + d > 56$
  **b.** $r - 13r < 84$
  **c.** $7b + 9b \neq 80$
  **d.** $4n + 4n \not> 72$

**4. a.** $9n > 10 + 4n$
  **b.** $11y - 24 < 3y$
  **c.** $7w + 16 \neq 5w$
  **d.** $2x \not> 9x - 42$

**e.** $3y - 18y \not< -105$
**f.** $3c - c + 2c \geq -28$
**g.** $y + y - 9y \leq 35$

**e.** $12a - 15 \not< 17a$
**f.** $8y \geq y - 63$
**g.** $3x + 32 \leq 7x$

**5. a.** $6y + 27 > 2y - 1$
**b.** $3n - 5 < 11n - 21$
**c.** $1 + a + 2a + 3a \neq 19$
**d.** $9d - 8 + d \not> 16 + 4d$
**e.** $x + x + 1 + x + 2 \not< -24$
**f.** $7t - 11 + 6t - 15 \geq 0$
**g.** $5c - 9 + c \leq 25 - 3c - 7$

**6. a.** $8n - (2n + 7) > 29$
**b.** $t + 2(9 - t) < 15$
**c.** $10 + (y - 6) \neq 4 - (5 - 2y)$
**d.** $5c - 4 \not> c + (c + 2)$
**e.** $b(b + 5) \not< b^2 + 20$
**f.** $(x + 1)(x - 1) \geq (x + 3)(x - 2)$
**g.** $(3n - 5) - (n - 7) \leq 6(2n - 3)$

**7.** Find the solutions of each of the following inequalities:

**a.** $8x + 3 < 27 + 5x$ when the placements for the variable are all the natural numbers.

**b.** $2w - 9 + 8w \geq 14w + 35$ when the replacements for the variable are all the even integers.

**c.** $y(y - 5) \not> (y + 3)(y - 6)$ when the replacements for the variable are $-7, -5, -3, \ldots, 7$.

## RELATED PRACTICE EXAMPLES

Find the solutions of each of the following inequalities when the replacements for the variable are all the real numbers:

**1. a.** $10x + 3 > 63$     $7c + 8 > -20$     $76 + 3d > 58$
**b.** $9 + 5z < 44$     $12t + 32 < 8$     $7a + 15 < 15$
**c.** $13 + 4r \neq 21$     $9w + 65 \neq 2$     $33 \neq 24 + 12x$
**d.** $8s + 25 \not> 57$     $51 + 14y \not> 9$     $2n + 37 \not> 42$
**e.** $6m + 62 \not< 98$     $11 + 20d \not< -91$     $18y + 64 \not< 79$
**f.** $7x + 9 \geq -5$     $12c + 19 \geq 25$     $13 + 9y \geq 40$
**g.** $3a + 17 \leq 32$     $15 + 16n \leq -33$     $24t + 21 \leq 35$

**2. a.** $9x - 5 > 67$     $2d - 9 > -3$     $25r - 27 > 8$
**b.** $11b - 8 < 25$     $10m - 6 < -45$     $14h - 2 < 5$
**c.** $5y - 15 \neq 0$     $6s - 13 \neq -55$     $13 \neq 9b - 17$
**d.** $8t - 40 \not> 16$     $18w - 11 \not> -83$     $24h - 19 \not> 41$
**e.** $3r - 7 \not< 29$     $15g - 25 \not< -130$     $12x - 6 \not< 2$
**f.** $14x - 13 \geq 57$     $40y - 17 \geq -37$     $33n - 53 \geq 112$
**g.** $6t - 9 \leq 0$     $63n - 19 \leq 16$     $20w - 3 \leq -19$

3. **a.** $7y - 6y > 17$      $2c - 7c > 45$      $x + x - 5x > -12$
   **b.** $9d - d < 48$      $b - 2b < 10$      $8n - 13n < -35$
   **c.** $10x + 3x - 8x \neq 75$      $5y - 9y \neq 32$      $12c - 21c \neq -81$
   **d.** $6s - s - s \not> 84$      $10d - 16d \not> 54$      $y - 5y \not> -48$
   **e.** $18m + 7m \not< 150$      $13z - 21z \not< 96$      $3a - 2a - 8a \not< -42$
   **f.** $8a - 3a \geq -35$      $3x - 6x + x \geq 12$      $10b - 15b \geq -45$
   **g.** $11y + y \leq 60$      $c - 9c \leq 64$      $5a + 7a - a \leq 0$

4. **a.** $7x > 8 + 5x$      $11n + 5 > 6n$
   **b.** $4y < 9y - 20$      $3x + 19 < 2x$
   **c.** $5a - 40 \neq a$      $21c \neq 72 - 3c$
   **d.** $12t - 55 \not> 23t$      $16y \not> 21y + 30$
   **e.** $9x \not< 3x + 54$      $9 + 10d \not< 7d$
   **f.** $y \geq 4y - 81$      $14b - 3 \geq 17b$
   **g.** $12a \leq 5a + 63$      $5x - 19 \leq 6x$

5. **a.** $9y - 5 > 4y + 25$      $15 + 9b - 8 > 13 + 11b$
   **b.** $10b - 24 < 3b - 3$      $5a + a + 7 < 24 + 3a - 6$
   **c.** $6a + 7 \neq 91 - 5a$      $17 + 11n - 13 \neq 4n + 4 + 2n$
   **d.** $8x - 2x \not> 2x - 52$      $7d - 10d - 5 \not> d + 3$
   **e.** $12c - 9 \not< 35 - 8$      $1 + y - 2y - 3y \not< y - 1$
   **f.** $9t + 38 \geq 4t + 13$      $3x + 21 + 5x \geq 23 + 2x - 8$
   **g.** $3w + 25 \leq 17 - w$      $x + x + 2 + x + 4 \leq -21$

6. **a.** $2n + 2(n + 6) > 84$      $(y + 6)(y + 5) - y^2 > 52$
   **b.** $(4a + 3) - (2a - 5) < 16$      $8c - (c - 7) < c + 19$
   **c.** $5(x - 2) \neq 4(x + 1)$      $7(4 - d) \neq 5(6 - d)$
   **d.** $b + (2b + 3) \not> (b + 3) + (b + 4)$      $(h - 2)(h - 1) \not> h(h + 1)$
   **e.** $(x + 4)(x + 3) \not< x^2 + 26$      $(b + 8)(b - 2) \not< (b - 4)(b + 13)$
   **f.** $3y + (y - 12) \geq 20$      $c^2 - 12 \geq (c - 6)(c + 2)$
   **g.** $x(x + 2) \leq x^2 + 18$      $a - (10 - a) \leq 6$

7. Find the solutions of each of the following inequalities:
   **a.** When the replacements for the variable are all the natural numbers:
       (1) $3a + 7 \geq 1$            (2) $5c - 2 < 12$
       (3) $11x - 4x \not> 49$        (4) $15b - 10 > 20b$
       (5) $6y - 7y \neq -13$       (6) $10n + 6 > 6 + 5n$
       (7) $3n - (n - 9) \leq 15$     (8) $(x + 4)(x + 5) \not< x(x + 10)$
   **b.** When the replacements for the variable are all the integers:
       (1) $9x + 3 \neq 66$           (2) $7y - 5 > 30$
       (3) $8n + 2 < 3n + 17$      (4) $13t - 9 \leq 9t + 7$
       (5) $7x - 14 \geq 5x - 2$      (6) $4c + 9c \not> 10 - 36$
       (7) $11d - 2d \not< 3d + 18$    (8) $(y - 3)(y + 2) > y^2 - 2$

c. When the replacements for the variable are 0, 2, 4, 6, 8:
  (1) $5a - 4 \not< 6$
  (2) $9d - 3d \neq 12$
  (3) $10y + 51 < 3y + 51$
  (4) $6t - 4 \leq 5t - 8$
  (5) $b - 3b + 12 \not\geq 5b - 16$
  (6) $x(x - 2) > (x - 4)(x + 6)$

d. When the replacements for the variable are $-3, -2, -1, \ldots, 3$:
  (1) $3x + 10 \not< 1$
  (2) $7y - y > -6$
  (3) $8s - 2 \neq 6s + 6$
  (4) $2x + 3 < x + 4$
  (5) $4x - 5x - 8 \geq x - 2$
  (6) $3y - (y - 2) \not> 6(y - 1)$

e. When the replacements for the variable are $0, 1, 2, \ldots, 9$:
  (1) $x + x < 12$
  (2) $4y - 1 \not> 15$
  (3) $6x - 4 > 4x + 16$
  (4) $7g + 9 \neq 9$
  (5) $12a - a \geq 3a + 26$
  (6) $(x - 2)(x - 3) > (x + 1)(x - 4)$

f. When the replacements for the variable are all the non-negative integers:
  (1) $8b + 6 > 22$
  (2) $5n + 2 \neq 4n - 3$
  (3) $11y - 28 \leq -3y$
  (4) $14d - 4 \not< 11d - 4$
  (5) $9 - 4(a - 6) < (2a + 5) - (3a - 1)$
  (6) $(c - 6)(c + 2) - c^2 < 20$

g. When the replacements for the variable are all the prime numbers:
  (1) $10n + 8n \neq 54$
  (2) $2b + 14 < 7$
  (3) $4x + 5 \not> 8 - 3$
  (4) $9t + 15 < 2t + 36$
  (5) $7x - 15 - x \geq 25 + x$
  (6) $x(x - 1) > (x + 5)(x - 3)$

h. When the replacements for the variable are all the even integers less than 20:
  (1) $12n - 45 > 27$
  (2) $b + b - 3b \not< -10$
  (3) $6a + 7 > 4a + 7$
  (4) $19y - 2 \not> 3y + 30$
  (5) $10c - c + 3 \leq 40 - 3c + 59$
  (6) $4y - 6(y - 2) < 16 - y$

i. When the replacements for the variable are all integers greater than $-7$ and less than $-2$:
  (1) $w + 3 < -1$
  (2) $16d - d \neq -45$
  (3) $11c - 7 \not> 16c$
  (4) $2m + 3 \not< -1$
  (5) $4n + n < 2n - 9$
  (6) $8x - 11 + 3x \geq 2x - 55$

j. When the replacements for the variable are all the odd integers greater than $-10$ and less than 9:
  (1) $3m + 8 \not> 8$
  (2) $10y - 12y < 17$
  (3) $6a - 9 \neq 4a - 3$
  (4) $7s - 5 > s + 8$
  (5) $16r + 18 - 9r \leq 6r + 17$
  (6) $(x + 3) - (2x - 1) < (3x - 2) + (x + 1)$

# Graphing an Inequality in One Variable on the Number Line

**I. Aim**  To draw the graph of an inequality in one variable on the number line.

**II. Procedure**

1. Solve the given inequality to find its solutions.

2. Then on a number line draw the graph of these solutions.

3. Observe that:

   **a.** The *graph of an inequality in one variable on a number line* is the collection of all points on the number line whose coordinates are the numbers forming the solutions of the inequality.

   **b.** The basic graph of an inequality in one variable expressed in simplest form may be a *half line* (sample solutions 1 and 2), a *ray* (sample solutions 3 and 4), *two half-lines* (sample solution 5), and *interval* (sample solution 6), or a *line segment* or *segment* (sample solution 7).

   **c.** When the given inequality is not in simplest form, transform it to its simplest form and draw the graph of the numbers described. See sample solution 8.

   **d.** When there are no solutions as in $x > x + 3$, there are no points to draw the graph.

   **e.** When there is an infinite number of solutions as in $x < x + 3$, its graph is the complete number line and is indicated by a solid or a colored line with an arrowhead in each direction to show that it is endless. See sample solution 10.

4. When required to write an inequality corresponding to a graph, use the simplest form of the inequality or any other inequality having the same solutions. See sample solution 11.

5. Verb symbols $\not>$ and $\leq$ are equivalent and verb symbols $\not<$ and $\geq$ are equivalent. The graphs of inequalities with equivalent verb symbols are the same.
   The graph of $x \not> 2$ is the same as the graph of $x \leq 2$. See sample solution 4.
   The graph of $x \not< 2$ is the same as the graph of $x \geq 2$. See sample solution 3.

**345**

## III. Sample Solutions

Draw the graph of each of the following inequalities on the number line when the replacements for the variable are all the real numbers.

**1.** $x > 2$   *Answer:*

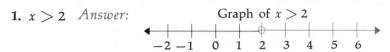

Graph of $x > 2$

The graph of all real numbers greater than a given number is a half-line extending along the number line to the right of a point whose coordinate is the given number.

    A *line* is a collection of points. A point on the line separates the line into two *half-lines*. The half-line does not include the endpoint separating it from the half-line. An open dot is used to indicate that this point is not included in the graph. The half-line extends indefinitely in one direction. The arrowhead indicates the direction and that the half-line is endless.

**2.** $x < 2$   *Answer:*

Graph of $x < 2$

The graph of all real numbers less than a given number is a half-line extending along the number line to the left of the point whose coordinate is the given number.

**3.** $x \geq 2$   *Answer:*

Graph of $x \geq 2$

The graph of all real numbers greater than or equal to a given number is a ray extending along the number line to the right of the point whose coordinate is the given number.

    A *ray* is a half-line which includes one end point (the point that separates the line into two half-lines). A heavy solid or colored dot is used to indicate the inclusion of the endpoint.

**4.** $x \leq 2$   *Answer:*

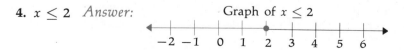

Graph of $x \leq 2$

The graph of all real numbers less than or equal to a given number is a ray extending along the number line to the left of the point whose coordinate is the given number.

**5.** $x \neq 2$   *Answer:*

Graph of $x \neq 2$

The graph of all real numbers with the exception of one number is the entire real number line excluding the point whose coordinate is the given number. The open dot is used to indicate that this point is not included in the graph. This graph consists of two half-lines.

**6.** $2 < x < 5$
or $5 > x > 2$   *Answer:*

The graph of $2 < x < 5$ is the same as the graph of $5 > x > 2$

The graph of all real numbers between two given numbers but not including these two given numbers is an *interval* which is a definite part of a line excluding the endpoints (the two points whose coordinates are the two given numbers). Open dots are used to indicate that these points are not included in the graph.

**7.** $2 \leq x \leq 5$
or $5 \geq x \geq 2$   *Answer:*

The graph of $2 \leq x \leq 5$ is the same as the graph of $5 \geq x \geq 2$

The graph of all real numbers between two given numbers and including these two given numbers is a *line segment* which is a definite part of a line including the endpoints (the two points whose coordinates are the two given numbers). Heavy solid or colored dots are used to indicate that these points are included in the graph.

**8.** $3x + 2 < 14$
*Solution:*

$$3x + 2 < 14$$
$$3x + 2 - 2 < 14 - 2$$
$$3x < 12$$
$$x < 4$$

*Answer:*

Graph of $3x + 2 < 14$

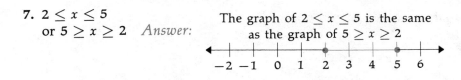

**9.** $x > x + 3$

There are no solutions to $x > x + 3$. Therefore there are no points to draw a graph.

*Answer:* $x > x + 3$ has no graph.

**10.** $x < x + 3$
  $x =$ all the real    *Answer:*
    numbers.

Graph of $x < x + 3$

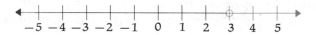

$$-4 \ -3 \ -2 \ -1 \ \ 0 \ \ 1 \ \ 2 \ \ 3 \ \ 4$$

**11.** Write a corresponding inequality which is pictured by the graph:

$$-5 \ -4 \ -3 \ -2 \ -1 \ \ 0 \ \ 1 \ \ 2 \ \ 3 \ \ 4 \ \ 5$$

*Answer:* $x > 3$ or any other equivalent inequality as $4x > 12$, $x + 1 > 4$, $x - 2 > 1$, or $\frac{x}{3} > 1$.

## DIAGNOSTIC TEST

For each of the following inequalities draw an appropriate number line, then draw the graph of its solutions. The replacements for the variable are all the real numbers.

**1. a.** $x > 1$    **b.** $n > -4$    **2. a.** $y < 3$    **b.** $w < -5$
**3. a.** $b \neq -3$    **b.** $r \neq 4$    **4. a.** $x \geq -1$    **b.** $m \geq 5$
**5. a.** $z \leq 6$    **b.** $t \leq -2$    **6. a.** $p \not> 7$    **b.** $z \not> -8$
**7. a.** $v \not< -6$    **b.** $x \not< 5$    **8.** $-5 < x < 4$
**9.** $0 > y > -3$    **10.** $-7 \leq n \leq -1$
**11.** $9 \geq t \geq 1$    **12.** $-6 \leq x < 0$
**13.** $4 > m \geq -4$    **14. a.** $n < n - 4$  **b.** $x + 3 > x$
**15.** $y + 6 \leq 4$    **16.** $t - 2 \not> 3$
**17.** $-4x > 16$    **18.** $\frac{x}{3} \neq -1$
**19.** $9y + 7 < 52$    **20.** $3b - 11 < -20$
**21.** $2x - 3x \geq -5$    **22.** $5w \not< 7w + 18$
**23.** $10n - 17 > 3n + 25$    **24.** $16 - y - 2y \geq 18 + y - 2$
**25.** $6x(x - 6) < (3x - 4)(2x - 7)$
**26.** Write a corresponding inequality which is pictured by the following graph:

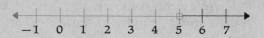

$$-1 \ \ 0 \ \ 1 \ \ 2 \ \ 3 \ \ 4 \ \ 5 \ \ 6 \ \ 7$$

## *RELATED PRACTICE EXAMPLES*

For each of the following inequalities draw the appropriate number line, then draw the graph of its solutions. The replacements for the variable are all the real numbers.

1. **a.** $x > 3$ **b.** $y > 5$ **c.** $x > -2$ **d.** $z > -7$
   **e.** $n > 0$ **f.** $4 < b$ **g.** $-1 < x$ **h.** $b > -6$

2. **a.** $x < 4$ **b.** $z < 6$ **c.** $n < -3$ **d.** $w < -1$
   **e.** $x < 5$ **f.** $y < 0$ **g.** $-7 > a$ **h.** $1 > n$

3. **a.** $x \neq 5$ **b.** $n \neq 0$ **c.** $p \neq -4$ **d.** $-1 \neq x$
   **e.** $w \neq 3$ **f.** $a \neq -2$ **g.** $10 \neq n$ **h.** $r \neq -5$

4. **a.** $n \geq 1$ **b.** $x \geq 7$ **c.** $w \geq -3$ **d.** $y \geq -6$
   **e.** $m \geq 0$ **f.** $4 \leq x$ **g.** $-2 \leq b$ **h.** $g \geq 5$

5. **a.** $z \leq 3$ **b.** $x \leq 0$ **c.** $r \leq -1$ **d.** $v \leq -5$
   **e.** $-6 \geq d$ **f.** $a \leq 7$ **g.** $10 \geq x$ **h.** $y \leq -9$

6. **a.** $w \not> 3$ **b.** $t \not> 8$ **c.** $y \not> -1$ **d.** $x \not> -5$
   **e.** $n \not> 4$ **f.** $-6 \not< b$ **g.** $1 \not< x$ **h.** $z \not> 0$

7. **a.** $p \not< 5$ **b.** $x \not< 0$ **c.** $s \not< -7$ **d.** $m \not< -2$
   **e.** $4 \not> a$ **f.** $n \not< -9$ **g.** $y \not< 10$ **h.** $-3 \not> y$

8. **a.** $1 < x < 7$ **b.** $0 < y < 9$ **c.** $-4 < n < 5$
   **d.** $-9 < b < -1$ **e.** $-2 < a < 0$ **f.** $-8 < x < 4$

9. **a.** $3 > n > 0$ **b.** $8 > w > 1$ **c.** $-1 > x > -6$
   **d.** $-3 > y > -11$ **e.** $5 > r < -2$ **f.** $0 > p > -7$

10. **a.** $4 \leq y \leq 8$ **b.** $-5 \leq x \leq -1$ **c.** $-4 \leq w \leq 0$
    **d.** $-2 \leq d \leq 2$ **e.** $-9 \leq m \leq 1$ **f.** $0 \leq n \leq 10$

11. **a.** $12 \geq n \geq 5$ **b.** $-6 \geq z \geq -11$ **c.** $4 \geq a \geq -5$
    **d.** $0 \geq x \geq -8$ **e.** $6 \geq t \geq 0$ **f.** $3 \geq y \geq -4$

12. **a.** $1 \leq n < 5$ **b.** $-4 < x \leq 0$ **c.** $-7 < b \leq -2$
    **d.** $0 \leq r < 9$ **e.** $2 \leq w < 10$ **f.** $-6 < y \leq 3$

13. **a.** $9 > d \geq 4$ **b.** $-1 \geq z > -8$ **c.** $5 \geq x > -5$
    **d.** $11 > t \geq 0$ **e.** $3 \geq a > -3$ **f.** $0 > x \geq -4$

14. **a.** $x < x - 2$ **b.** $y > y + 4$ **c.** $n < n + 2$
    **d.** $w > w - 1$ **e.** $b + 5 > b$ **f.** $m + 6 < m$
    **g.** $t - 3 < t$ **h.** $x - 5 > x$ **i.** $y < y - 7$

Find the solutions of each of the following inequalities and draw its graph on an appropriate number line. The replacements for the variable are all the real numbers.

15. **a.** $x+4<4$     **b.** $n+2>7$     **c.** $y+1\nless-5$
    **d.** $y+5\geq-2$     **e.** $b+3\ngtr0$     **f.** $n+6\neq3$
    **g.** $a+7\leq11$     **h.** $10<x+8$     **i.** $4+y>9$

16. **a.** $n-2>4$     **b.** $t-3\nless1$     **c.** $x-5\leq0$
    **d.** $a-1<-4$     **e.** $x-8\neq-3$     **f.** $5>c-6$
    **g.** $b-12\geq-12$     **h.** $r-4<3$     **i.** $t-7\ngtr-10$

17. **a.** $5n\ngtr20$     **b.** $7b<56$     **c.** $2r>-10$
    **d.** $11m\neq11$     **e.** $-8x>-24$     **f.** $-15c<0$
    **g.** $-5\geq-y$     **h.** $3m\nless-21$     **i.** $-10a\leq40$

18. **a.** $\dfrac{x}{5}>2$     **b.** $\dfrac{a}{3}\leq4$     **c.** $\dfrac{m}{8}\neq1$

    **d.** $\dfrac{c}{2}\nless-3$     **e.** $\dfrac{n}{-1}<7$     **f.** $5\geq\dfrac{w}{4}$

    **g.** $\dfrac{b}{-2}<-2$     **h.** $\dfrac{r}{-3}>0$     **i.** $\dfrac{d}{6}\ngtr-1$

19. **a.** $3x+8\leq14$     **b.** $8m+5>5$     **c.** $2c+7\nless-3$
    **d.** $15y+6\geq51$     **e.** $11d+9\neq-2$     **f.** $4+5t<39$
    **g.** $19>6x+1$     **h.** $7t+25\ngtr4$     **i.** $10b+17<-23$

20. **a.** $5n-3<12$     **b.** $8d-11\ngtr29$     **c.** $13y-54\leq37$
    **d.** $2p-5\nless-3$     **e.** $7y-9\geq-23$     **f.** $3t-18\neq0$
    **g.** $4c-25>-1$     **h.** $9x-10>-55$     **i.** $n-16<-16$

21. **a.** $9n+2n\geq44$     **b.** $6x-x\nless-10$     **c.** $-3w+w>16$
    **d.** $8r-11r>-12$     **e.** $10m-3m\ngtr49$     **f.** $7c-8c\leq-6$
    **g.** $2x-x+3x\neq20$     **h.** $n+n+n<-21$     **i.** $3y-4y-y<18$

22. **a.** $14d+45>5d$     **b.** $5m-12\geq4m$     **c.** $7b>2b+15$
    **d.** $8r\ngtr2r-24$     **e.** $10n<21+13n$     **f.** $19c-35\neq26c$
    **g.** $x\leq3x+20$     **h.** $10-y\nless-6y$     **i.** $9t<17t-40$

23. **a.** $10c+7\nless8c+21$     **b.** $9n-4\geq5n+24$     **c.** $11t-8<21t-22$
    **d.** $8w+19\neq7-4w$     **e.** $y-6y<2y-21$     **f.** $7x-8\leq15x-56$

24. **a.** $4n-3+n>7-2n$           **b.** $t+5t-9>t-4t$
    **c.** $11-w-2\ngtr6w+9-8w$     **d.** $n-4n+6\leq15-n+7$
    **e.** $16-3a-19\neq21-2a+5a$     **f.** $13x-8x+9-x<5x+16$

25. **a.** $6r+(2r-11)<7$          **b.** $19-3(7x-5)\geq3x-14$
    **c.** $7(4y-9)\nless5(6y-7)$     **d.** $(3x-8)-(2x+7)$
$$>(7x+12)+(3-x)$$

    **e.** $(2y+1)(5y+6)\neq10(y^2+4)$     **f.** $(x-3)(x-4)>(x+5)(x+12)$

**26.** Write a corresponding inequality which is pictured by each of the following graphs:

**a.**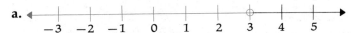

(number line from −3 to 5, open circle at 3)

**b.** (number line from −5 to 3, open circle at −2)

**c.** (number line from −4 to 4, open circle at −1)

**d.** (number line from −2 to 6, closed dot at 4)

**e.** (number line from −4 to 4, open circles at −3 and 4)

**f.** (number line from −4 to 4, closed dots at −1 and 2)

**g.** (number line from −4 to 4, open circle at 1)

**h.** (number line from −7 to 1, closed dot at −4)

**i.** (number line from −10 to −2, closed dot at −5)

**j.** (number line from 6 to 14, closed dot at 9)

**k.** (number line from −15 to −7, open circle at −11)

**l.** (number line from −12 to −4, closed dot at −11, open circle at −6)

**m.** (number line from −11 to −3, open circle at −7)

**n.** (number line from −1 to 7, closed dot at 4)

**o.** (number line from −12 to −4, open circle at −9)

**p.** (number line from −4 to 4, closed dots at −3 and 4)

# REVIEW

1. Write symbolically:
   a. Four times each number $x$ is not greater than negative six.
   b. Nine times each number $n$ decreased by two is less than or equal to forty-nine.

Find the solutions of each of the following inequalities when the replacements for the variable are all the:

2. a. Whole numbers:   (1) $x < 8$   (2) $n + 4 > 19$
   b. Even integers:   (1) $n \le 7$   (2) $-8r \not< 0$
   c. One-digit prime numbers:   (1) $12 > w + 5$   (2) $-10n \not< -30$
   d. Real numbers:   (1) $m \not< 4.5$   (2) $\dfrac{w}{-7} > -2$

3. a. Whole numbers:   (1) $4c - c \ge 8c - 50$   (2) $8x + 11 \ne 29 - x$
   b. Real numbers:   (1) $3w + 7 > 9$   (2) $4x - 11x < -63$
   c. Whole numbers $< 11$:   (1) $2x - 7x \not\ge -35$
                                   (2) $9 + 5y < y + 33$
   d. Integers $-4, -3, -2, \ldots, 4$:   (1) $x > 20 + 6x$
                                   (2) $(3x - 2) - 4(2 - x) \not> -17$

4. Draw on a number line the graph of each inequality. The replacements are all the real numbers.
   a. $x > 6$        b. $w \le -2$     c. $y \not< -1$      d. $a \ne -3$
   e. $7 \ge m \ge -1$  f. $w - 4 < w$   g. $-5x \ge -20$   h. $r - 2 \not> 1$

5. Write a corresponding inequality which is pictured by each graph:

a.
   $-3 \;-2\; -1 \quad 0 \quad 1 \quad 2 \quad 3$

b.
   $-3 \;-2\; -1 \quad 0 \quad 1 \quad 2 \quad 3$

c.
   $-5 \;-4\; -3 \;-2\; -1 \quad 0 \quad 1$

d.
   $-7 \;-6\; -5 \;-4\; -3\; -2\; -1$

## COMPETENCY CHECK TEST

In each of the following, select the letter corresponding to your answer:
When the replacements for the variable are:

1. All the whole numbers, the solutions of $3x < 15$ are:
   a. 5          b. 5, 6, 7, . . .          c. 0, 1, 2, 3, 4, 5   d. 0, 1, 2, 3, 4

2. All the prime numbers, the solutions of $4n - 1 \not> 35$ are:
   a. 2, 3, 5, 7      b. 1, 3, 5, 7, 9      c. 3, 5, 7          d. 2, 4, 6, 8

3. 1, 2, 3, . . . , 8, the solutions of $2s + 3 \le 11$ are:
   a. 1, 2, 3      b. 1, 2, 3, 4      c. 4, 5, 6, 7, 8    d. 5, 6, 7, 8

## Keyed Achievement Test

The numeral at the end of each problem indicates the section where help may be found.

1. Solve and check: 6-4 thru 6-8
   a. $16y - 9 = 39$

   b. $11x + 7 = 21 - x$

   c. $38 - (5n + 2) =$ $7(2n - 9)$

   d. $(y - 1)(y - 4) =$ $(y + 2)(y - 5)$

2. Draw on a number line the graph of each of the following equations: 6-11

   a. $17y = -34$

   b. $2x - 9x = -28$

   c. $|n| - 2 = 0$

3. Find the value of: 6-10
   a. $C$ when $F = -63$, using the formula $F = 1.8C + 32$.
   b. $p$ when $A = 234$, $r = .04$, and $t = 9$, using the formula $A = p + prt$.

4. a. Andrew has $4.65 in dimes and quarters. He has eight more dimes than quarters. Find the number of each kind of coin. 7-5
   b. At what price must a merchant sell a camera which costs $59.50 to make a profit of 30% on the selling price? 7-5
   c. A rectangle is 32 meters longer than it is wide. What is its length and width if its perimeter is 300 meters? 7-5

5. Find the solution of: 8-4 thru 8-8
   a. $y - 9 < 15$ when the variable replacements are all the real numbers.
   b. $14x > 56$ when the variable replacements are all the whole numbers.
   c. $25 \geq 6n - 5$ when the variable replacements are all the real numbers.
   d. $5x - 9x \not< 20$ when the variable replacements are: $-9$, $-7$, $-5$, $\ldots$, $0$.
   e. $\dfrac{n}{-3} < 3$ when the variable replacements are all the real numbers.

6. Draw on a number line the graph of each inequality. The replacements for the variable are all the real numbers. 8-9
   a. $n - 7n > 30$

   b. $3x + 48 \leq 15x$

   c. $-4 \leq x \leq 6$

# INTRODUCTION

In order to save time when multiplying binomials, special rules are developed so that products may be found mentally. The study of finding products of binomials by short methods is called *special products*.

A complete study of algebraic fractions, fractional equations, formulas, and quadratic equations involves the process of finding two or more expressions whose product is the given expression. This process is called *factoring*.

Factoring is the direct reverse of multiplication. In multiplication the factors are given and we are required to find the product. In factoring the product is given and we are required to find the factors. Division and factoring are not the same process. In division the product and a factor are given and we are required to find the other factor. In factoring only the product is given.

When we solve the equation $2x + 3x = 15$, we combine $2x$ and $3x$ to obtain $5x$. However, when solving the equation $ax + bx = c$, we cannot add $ax$ and $bx$ to get a single term. Thus, to solve for $x$ we factor the expression $ax + bx$, selecting $x$ as one of the factors.

This process of finding the common monomial factor is also useful when dealing with formulas like $A = p + prt$ and $E = IR + Ir$. Students will find the process of factoring the difference of two squares helpful when formulas like $a = \sqrt{c^2 - b^2}$, $b = \sqrt{c^2 - a^2}$ and $A = \pi R^2 - \pi r^2$ occur. On page 572 factoring is used in the solution of quadratic equations. It is also used to develop the quadratic formula and in the solution of quadratic equations by completing the square.

# Special Products and Factoring

# Product of Two Binomials— Sight Method

**I. Aim**    To find at sight the product of two binomials.

**II. Procedure**

1. **a.** Multiply the first terms of the given binomials to obtain the first term of the product.

   **b.** Find the algebraic sum of the products of the first term of each given binomial by the second term of the other given binomial to obtain the second term of the product.

   **c.** Multiply the second terms of the given binomials to obtain the third term of the product.

2. Follow the same procedure as above when both terms of the binomials contain a variable or when higher powers of the variables are used.

3. Sometimes the given binomials contain a common term. Also see Exercise 8-5 for a special method.

4. Sometimes the binomials are identical. Also see Exercise 8-4 for a special method.

5. When the algebraic sum of the products of the first term of each given binomial multiplied by the second term of the other given binomial is zero, the final product contains only two terms. Also see Exercise 8-3 for a special method.

6. When the resulting products of the first term of each given binomial multiplied by the second term of the other given binomial are unlike terms, the final product contains four terms.

**III. Sample Solutions**

1. Find at sight the product of: $(4n - 3)(5n + 2)$

   **a.** $(4n - 3)(5n + 2) = 20n^2$

   **b.** $(4n - 3)(5n + 2) = 20n^2 + 8n - 15n = 20n^2 - 7n$

   **c.** $(4n - 3)(5n + 2) = 20n^2 - 7n - 6$

   However, all three steps should be combined and written in one step at sight as follows:

   $(4n - 3)(5n + 2) = 20n^2 - 7n - 6$

   *Answer:* $20n^2 - 7n - 6$

**2.** Find at sight the product of: $(6b - 7x)(3b - 4x)$
$(6b - 7x)(3b - 4x) = 18b^2 - 45bx + 28x^2$
*Answer:* $18b^2 - 45bx + 28b^2$

**3.** Find at sight the product of: $(b + 8)(b + 2)$
$(b + 8)(b + 2) = b^2 + 10b + 16$
*Answer:* $b^2 + 10b + 16$

**4.** Find at sight the product of: $(n + 5)(n + 5)$
$(n + 5)(n + 5) = n^2 + 10n + 25$
*Answer:* $n^2 + 10n + 25$

**5.** Find at sight the product of: $(c + 7)(c - 7)$
$(c + 7)(c - 7) = c^2 - 7c + 7c - 49$
*Answer:* $c^2 - 49$

**6.** Find at sight the product of: $(2a + 3b)(5a + 4c)$
$(2a + 3b)(5a + 4c) = 10a^2 + 15ab + 8ac + 12bc$
*Answer:* $10a^2 + 15ab + 8ac + 12bc$

## DIAGNOSTIC TEST

Find the product of each of the following at sight:

**1.** $(9n + 5)(6n + 7)$
**2.** $(4b - 8)(3b - 4)$
**3.** $(7a + 10)(2a - 1)$
**4.** $(12m + 13)(5m - 12)$
**5.** $(6w - 11)(4w + 9)$
**6.** $(2y - 19)(y + 3)$
**7.** $(5c + 8d)(3c - 5d)$
**8.** $(4bx + 1)(9bx + 7)$
**9.** $(10bc - 2ax)(5bc - 3ax)$
**10.** $(8x^2 - 9)(2x^2 + 5)$
**11.** $(7a^2 - 3x^3y)(6a^2 - x^3y)$
**12.** $(12c - 5d)(12c + 6d)$
**13.** $(3a + 11x)(3a + 11x)$
**14.** $(2x + 9y)(2x - 9y)$
**15.** $(5b - 2x)(4c + 7y)$

## *RELATED PRACTICE EXAMPLES*

**1.** Find the product of each of the following at sight:

**a.** $(2x + 9)(6x + 1)$
**b.** $(7b + 2)(4b + 5)$
**c.** $(5m - 6)(4m + 11)$
**d.** $(4r + 7)(r + 3)$
**e.** $(5 + 8t)(4 + 9t)$
**f.** $(10n + 13)(3n + 8)$

**2.** **a.** $(3x - 4)(2x - 5)$
**b.** $(7a - 1)(4a - 9)$
**c.** $(8t - 3)(5t - 12)$
**d.** $(12n - 7)(3n - 8)$
**e.** $(9 - 10n)(6 - n)$
**f.** $(11y - 1)(15y - 1)$

3. **a.** $(5c + 8)(3c - 2)$      **b.** $(9r + 19)(r - 1)$
  **c.** $(6x + 9)(4x - 5)$      **d.** $(11 + 12b)(7 - 2b)$
  **e.** $(5n + 10)(7n - 3)$      **f.** $(8a + 6)(15a - 4)$

4. **a.** $(6x + 5)(8x - 9)$      **b.** $(7y + 4)(3y - 12)$
  **c.** $(11a + 2)(5a - 8)$      **d.** $(4t + 3)(9t - 10)$
  **e.** $(10b + 1)(9b - 5)$      **f.** $(x + 9)(2x - 21)$

5. **a.** $(2x - 3)(4x + 7)$      **b.** $(8n - 5)(3n + 2)$
  **c.** $(8 - 5d)(1 + 10d)$      **d.** $(9r - 2)(6r + 9)$
  **e.** $(7y - 4)(y + 12)$      **f.** $(15m - 8)(3m + 6)$

6. **a.** $(5d - 9)(2d + 3)$      **b.** $(6y - 4)(8y + 1)$
  **c.** $(6 - 11b)(12 + 7b)$      **d.** $(4y - 21)(y + 5)$
  **e.** $(12c - 5)(10c + 2)$      **f.** $(15x - 9)(9x + 4)$

7. **a.** $(3a + 7b)(5a + 4b)$      **b.** $(4m - 5n)(8m - 6n)$
  **c.** $(9x + 4y)(3x - 8y)$      **d.** $(6r - 11s)(r + 2s)$
  **e.** $(12x - 15y)(4x + 3y)$      **f.** $(7c + 3d)(14c - 5d)$

8. **a.** $(2xy + 3)(4xy - 9)$      **b.** $(8ab - 5)(11ab - 7)$
  **c.** $(3rs - 5)(10rs + 12)$      **d.** $(12 - 4cd)(1 + 8cd)$
  **e.** $(9abc + 4)(16abc - 6)$      **f.** $(8 - xyz)(15 + 2xyz)$

9. **a.** $(9ab + 8c)(2ab + 11c)$      **b.** $(7xy - 3z)(5xy + 4z)$
  **c.** $(6b - 5dx)(b - 2dx)$      **d.** $(4ab - 3xy)(7ab + 5xy)$
  **e.** $(5ax + 9by)(8ax - 4by)$      **f.** $(3bd - 5cm)(7bd + 11cm)$

10. **a.** $(6a^2 - 1)(4a^2 - 10)$      **b.** $(3x^2 + 9)(5x^2 - 9)$
  **c.** $(7 - 4b^3)(8 + 3b^3)$      **d.** $(2a^2b + 1)(9a^2b + 5)$
  **e.** $(4m^3x^4 - 2)(5m^3x^4 - 3)$      **f.** $(10c^5d^6 + 9)(6c^5d^6 - 12)$

11. **a.** $(4c^2 - 7d^2)(3c^2 - 5d^2)$      **b.** $(a^2 - 6b^2)(8a^2 + 7b^2)$
  **c.** $(5x^3 - 2y^2)(11x^3 + 2y^2)$      **d.** $(9m^4 + n^5)(4m^4 - 3n^5)$
  **e.** $(6a^3x^2 + 2bc^4)(9a^3x^2 + 7bc^4)$      **f.** $(3c^6n - 8a^3y^5)(4c^6n - a^3y^5)$

12. **a.** $(c + 5)(c + 9)$      **b.** $(n - 6)(n + 7)$
  **c.** $(12 - y)(8 - y)$      **d.** $(4a - 5b)(4a + 3b)$
  **e.** $(3x^2 + 10y^2)(3x^2 - 7y^2)$      **f.** $(6ab^3 - 1)(6ab^3 + 11)$

13. **a.** $(x + 8)(x + 8)$      **b.** $(y - 3)(y - 3)$
  **c.** $(10 + d)(10 + d)$      **d.** $(n^2 - 9)(n^2 - 9)$
  **e.** $(5r - 4t)(5r - 4t)$      **f.** $(8a^2b^5 + c^2)(8a^2b^5 + c^2)$

14. **a.** $(x + 4)(x - 4)$      **b.** $(n - 15)(n + 15)$
  **c.** $(6 + m)(6 - m)$      **d.** $(a + c)(a - c)$
  **e.** $(12b - 5c)(12b + 5c)$      **f.** $(7d^2 + xy)(7d^2 - xy)$

15. **a.** $(b + c)(x + y)$      **b.** $(2a - x)(3d - 2y)$
  **c.** $(9m - 5n)(4n - 11r)$      **d.** $(x^2 - 6z)(3y - 4)$
  **e.** $(5ab - 2c)(7ad - 8b)$      **f.** $(s + 6a)(2r - t)$

# Squaring a Monomial

**I. Aim**    To square a monomial.

**II. Procedure**

1. **a.** Multiply the numerical coefficient by itself.
   **b.** Multiply the exponent of each of its variables by 2.
2. Observe:
   **a.** The square of a monomial is the product when the monomial is multiplied by itself.
   **b.** The square of any positive or negative number is a positive number. Therefore the numerical coefficient of the product is always positive.
   **c.** The square of a rational number is sometimes called a *perfect square*.
   **d.** The square of 0 is 0.

**III. Sample Solutions**

1. The square of 7, indicated as $(7)^2$, is $7 \times 7$ or 49.
2. The square of $-5$, indicated as $(-5)^2$, is $(-5)(-5)$ or $+25$.
3. The square of $\frac{3}{8}$, indicated as $(\frac{3}{8})^2$, is $\frac{3}{8} \times \frac{3}{8}$ or $\frac{9}{64}$.
4. The square of $-.6$, indicated as $(.6)^2$, is $(-.6)(-.6)$ or $+.36$.
5. The square of $x$, indicated as $(x)^2$, is $x \cdot x$ or $x^2$.
6. The square of $y^3$, indicated as $(y^3)^2$, is $y^3 \cdot y^3$ or $y^6$.
7. The square $a^2b^5c^4$, indicated as $(a^2b^5c^4)^2$, is $(a^2b^5c^4)(a^2b^5c^4)$ or $a^4b^{10}c^8$.
8. The square of $6x^6$, indicated as $(6x^6)^2$, is $(6x^6)(6x^6)$ or $36x^{12}$.
9. The square of $-3a^2b$, indicated as $(-3a^2b)(-3a^2b)$ or $9a^4b^2$.

## DIAGNOSTIC TEST

Find the square of each of the following:

   **1.** 9          **2.** $-11$          **3.** $-\frac{7}{8}$          **4.** 3.5

Find the product of each of the following:

   **5.** $(-25)^2$          **6.** $(\frac{7}{16})^2$          **7.** $(-2.4)^2$
   **8.** Find the square of $x^2yz^5$.
   **9.** Find the product of $(a^4b^3c^9)^2$.
   **10.** Find the square of $-8n^2x^3$.
   **11.** Find the product of $(-10d^8r^6s^7)^2$.

## *RELATED PRACTICE EXAMPLES*

Find the square of each of the following:

1. **a.** 11      **b.** 6      **c.** $+2$      **d.** $+8$
   **e.** $+10$      **f.** 14      **g.** $+16$

2. **a.** $-8$      **b.** $-1$      **c.** $-12$      **d.** $-15$
   **e.** $-20$      **f.** $-17$      **g.** $-9$

3. **a.** $\frac{4}{5}$      **b.** $\frac{5}{8}$      **c.** $-\frac{1}{3}$      **d.** $-\frac{3}{4}$
   **e.** $-\frac{4}{7}$      **f.** $\frac{5}{9}$      **g.** $-\frac{11}{6}$

4. **a.** .7      **b.** .02      **c.** $-.1$      **d.** $+.3$
   **e.** $-.18$      **f.** 6.2      **g.** $-5.37$

Find the product of each of the following:

5. **a.** $(3)^2$      **b.** $(-9)^2$      **c.** $(+1)^2$
   **d.** $(-8)^2$      **e.** $(-11)^2$      **f.** $(16)^2$

6. **a.** $(\frac{1}{2})^2$      **b.** $(\frac{2}{3})^2$      **c.** $(-\frac{1}{4})^2$
   **d.** $(-\frac{4}{9})^2$      **e.** $(\frac{8}{15})^2$      **f.** $(\frac{22}{7})^2$

7. **a.** $(.5)^2$      **b.** $(-.2)^2$      **c.** $(.34)^2$
   **d.** $(-1.6)^2$      **e.** $(.111)^2$      **f.** $(-25.4)^2$

8. Find the square of each of the following:
   **a.** $m$      **b.** $xy$      **c.** $a^2$      **d.** $b^3$
   **e.** $x^4y^6$      **f.** $ab^3c^9$      **g.** $m^2y^7$      **h.** $r^8st^5$
   **i.** $b^{10}c^3d^9$      **j.** $a^4m\,x^7y^{12}$

9. Find the product of each of the following:
   **a.** $(c)^2$      **b.** $(n^4)^2$      **c.** $(bd)^2$      **d.** $(m^3y^2)^2$
   **e.** $(r^8t^{12})^2$      **f.** $(bc^2d^7)^2$      **g.** $(w^3x^9y^{10})^2$      **h.** $(m^6n^2x)^2$
   **i.** $(hm^4s^3)^2$      **j.** $(b^5c^{11}x^6y^{20})^2$

10. Find the square of each of the following:
   **a.** $8b$      **b.** $11c^3$      **c.** $4a^5x^4z^7$
   **d.** $\frac{2}{3}n$      **e.** $\frac{4}{5}d^2$      **f.** $.3m^3x^6y^9$
   **g.** $-2a$      **h.** $-nx^2z^4$      **i.** $-\frac{5}{8}a^5cd^3$
   **j.** $-2.6mn^6$      **k.** $9r^3st^8$      **l.** $-5bx^2y^{10}$

11. Find the product of each of the following:
   **a.** $(7r)^2$      **b.** $(-4x^4)^2$      **c.** $(-\frac{1}{6}b^3)^2$
   **d.** $(12ad^2n^5)^2$      **e.** $(-.05m^6x^2)^2$      **f.** $(\frac{9}{16}w^3y^9)^2$
   **g.** $(-\frac{3}{10}at^2)^2$      **h.** $(3.4c^4d^8r)^2$      **i.** $(-5b^6x^3y^7)^2$
   **j.** $(1.7m^2n^9)^2$      **k.** $(-31r^4s^5t^{10})^2$      **l.** $(-\frac{9}{4}ab^3c^8d^{12})^2$

# Product of the Sum and Difference of the Same Two Terms–Special Method

**I. Aim**  To find the product of the sum and the difference of the same two terms.

**II. Procedure**
1. Square the first term to obtain the first term of the product.
2. Square the second term to obtain the second term of the product.
3. Indicate the difference of these squares by inserting a subtraction symbol ($-$) between the two squares.

   Note: The product of the sum and the difference of the same two terms is a binomial consisting of the difference of the squares of these terms.

**III. Sample Solutions**

Find at sight each of the following products:

1. $(a + 6)(a - 6)$  *Answer:* $a^2 - 36$
2. $(4 - x)(4 + x)$  *Answer:* $16 - x^2$
3. $(2b + 3c)(2b - 3c)$  *Answer:* $4b^2 - 9c^2$
4. $(\frac{1}{2}ab - x^2)(\frac{1}{2}ab + x^2)$  *Answer:* $\frac{1}{4}a^2b^2 - x^4$
5. $(-5 + b)(5 + b) = (b - 5)(b + 5)$  *Answer:* $b^2 - 25$
6. $(x + 3)(x - 3)(x^2 + 9)$  *Solution:*  $(x + 3)(x - 3)(x^2 + 9)$
   $$= (x^2 - 9)(x^2 + 9)$$
   $$= x^4 - 81 \quad \text{\textit{Answer:} } x^4 - 81$$

---

### DIAGNOSTIC TEST

Find the product of each of the following at sight:

1. $(r+9)(r-9)$
2. $(y-4)(y+4)$
3. $(w+v)(w-v)$
4. $(s-a)(s+a)$
5. $(2-n)(2+n)$
6. $(x^2-y^2)(x^2+y^2)$
7. $(ab^2+c)(ab^2-c)$
8. $(3x+2)(3x-2)$
9. $(7a-6b)(7a+6b)$
10. $(\frac{2}{3}cd-x)(\frac{2}{3}cd+x)$
11. $(.9x+.5y)(.9x-.5y)$
12. $(-3+n)(3+n)$
13. $(b+2)(b-2)(b^2+4)$

## RELATED PRACTICE EXAMPLES

Find the product of each of the following at sight:

1. **a.** $(x+2)(x-2)$    **b.** $(n+1)(n-1)$    **c.** $(c+10)(c-10)$

2. **a.** $(n-5)(n+5)$    **b.** $(d-8)(d+8)$    **c.** $(y-9)(y+9)$

3. **a.** $(x+y)(x-y)$    **b.** $(c+d)(c-d)$    **c.** $(w+r)(w-r)$

4. **a.** $(m-n)(m+n)$    **b.** $(b-a)(b+a)$    **c.** $(s-c)(s+c)$

5. **a.** $(8-b)(8+b)$    **b.** $(6+n)(6-n)$    **c.** $(4-t)(4+t)$

6. **a.** $(a^2+x^2)(a^2-x^2)$    **b.** $(b^3-c^3)(b^3+c^3)$    **c.** $(r^4-s^4)(r^4+s^4)$

7. **a.** $(xy+6)(xy-6)$    **b.** $(a-mn)(a+mn)$    **c.** $(bx^2+cy^2)(bx^2-cy^2)$

8. **a.** $(2n+5)(2n-5)$      **b.** $(6b-9)(6b+9)$
    **c.** $(4r+1)(4r-1)$      **d.** $(3-11y)(3+11y)$
    **e.** $(z^2-10)(z^2+10)$      **f.** $(8+13t)(8-13t)$
    **g.** $(9xy-20)(9xy+20)$      **h.** $(ab^2+7)(ab^2-7)$
    **i.** $(5c^3-12)(5c^3+12)$      **j.** $(2a^4-5)(2a^4+5)$
    **k.** $(7r^2t^3+9)(7r^2t^3-9)$      **l.** $(10-3m^5x^2)(10+3m^5x^2)$

9. **a.** $(3x+2y)(3x-2y)$      **b.** $(2r-5t)(2r+5t)$
    **c.** $(7y+z)(7y-z)$      **d.** $(10b+11c)(10b-11c)$
    **e.** $(4x^2-9y)(4x^2+9y)$      **f.** $(6cy+7z)(6cy-7z)$
    **g.** $(8n^3+5x^2)(8n^3-5x^2)$      **h.** $(9c^2-11d)(9c^2+11d)$
    **i.** $(13m-2n)(13m+2n)$      **j.** $(5a^2b-x^4)(5a^2b-x^4)$
    **k.** $(11cx^3+2dy)(11cx^3-2dy)$      **l.** $(8r^2x^5-3y^6)(8r^2x^5+3y^6)$
    **m.** $(2g^2-c^3x)(2g^2+c^3x)$      **n.** $(5ax+3by)(5ax-3by)$

10. **a.** $(x+\frac{1}{2})(x-\frac{1}{2})$      **b.** $(\frac{3}{5}rs-t)(\frac{3}{5}rs+t)$
    **c.** $\left(\frac{a}{4}+\frac{b}{9}\right)\left(\frac{a}{4}-\frac{b}{9}\right)$      **d.** $\left(\frac{4}{a}-\frac{9}{b}\right)\left(\frac{4}{a}+\frac{9}{b}\right)$

11. **a.** $(.4a+.6)(.4a-.6)$      **b.** $(1.5m-n)(1.5m+n)$
    **c.** $(.3x^2y+.7z^2)(.3x^2y-.7z^2)$      **d.** $(.02x^5-3.1y^3)(.02x^5+3.1y^3)$

12. **a.** $(1+t)(-1+t)$      **b.** $(-6+b)(6+b)$
    **c.** $(-8+n^2)(8+n^2)$      **d.** $(-y+x)(y+x)$
    **e.** $(4c+3a)(-4c+3a)$      **f.** $(-10mn+b)(10mn+b)$

13. Find the product of each of the following:
    **a.** $(x+1)(x-1)(x^2+1)$      **b.** $(b-5)(b+5)(b^2+25)$
    **c.** $(c+d)(c-d)(c^2+d^2)$      **d.** $(x^2+4y^2)(x-2y)(x+2y)$
    **e.** $(4m-9n)(4m+9n)(16m^2+81n^2)$      **f.** $(2a-3x)(4a^2+9x^2)(2a+3x)$

# The Square of a Binomial— Special Method

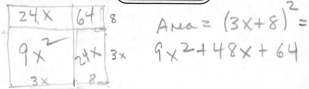

**I. Aim**   To square a binomial.

**II. Procedure**

1. To find the first term, square the first term of the binomial.
2. To find the second term, find the product of the two terms of the binomial. Then multiply this product by 2.
3. To find the third term, square the second term of the binomial.
4. Observe:

   **a.** The signs of the first and third terms are always positive.

   **b.** The sign of the numerical coefficient of the middle term (second term) of the product is negative when the given binomial contains *one* term which is either a negative number or has a negative numerical coefficient. Otherwise the sign is positive.

**III. Sample Solutions**

Find the product of each of the following:

1. $(3x + 8)^2 = (3x)^2 + 2(3x)(8) + (8)^2 = 9x^2 + 48x + 64$
   *Answer:* $9x^2 + 48x + 64$

2. $(7a + 4b)^2 = (7a)^2 + 2(7a)(4b) + (4b)^2 = 49b^2 + 56ab + 16b^2$
   *Answer:* $49a^2 + 56ab + 16b^2$

3. $(5n - 2)^2 = (5n + [-2])^2 = (5n)^2 + 2(5n)(-2) + (-2)^2 =$
   $$25n^2 - 20n + 4$$
   *Answer:* $25n^2 - 20n + 4$

4. $(6c - 7d)^2 = (6c + [-7d])^2 = (6c)^2 + 2(6c)(-7d) + (-7d)^2 =$
   $$36c^2 - 84cd + 49d^2$$
   *Answer:* $36c^2 - 84cd + 49d^2$

5. $(x + 2)(x - 2)(x^2 - 4)$
   *Solution:*   $(x + 2)(x - 2)(x^2 - 4)$
   $= (x^2 - 4)(x^2 - 4)$
   $= x^4 - 8x^2 + 16$   *Answer:* $x^4 - 8x^2 + 16$

6. Indicate the square of the binomial $x + 5$ in two ways.
   *Answer:* $(x + 5)(x + 5)$ or briefly $(x + 5)^2$.

**363**

## PRELIMINARY EXAMPLES

Finding the middle term.

| | First term | Second term | Twice the product of two terms | | | First term | Second term | Twice the product of two terms |
|---|---|---|---|---|---|---|---|---|
| 1. | $x$ | 3 | ? | 2. | | $b$ | 1 | ? |
| 3. | $a$ | $b$ | ? | 4. | | $x$ | $y$ | ? |
| 5. | $5a$ | 4 | ? | 6. | | $7x$ | $2y$ | ? |
| 7. | $9c$ | $7d$ | ? | 8. | | $10r$ | $3s$ | ? |
| 9. | $x$ | $\frac{1}{2}$ | ? | 10. | | $4m$ | .3 | ? |

## DIAGNOSTIC TEST

**1.** Express $(7x - 3y)(7x - 3y)$ in brief form, using exponents.

**2.** Write $(4s - 15)^2$ in another way.

Find the product of each of the following at sight:

**3.** $(t + 6)(t + 6)$          **4.** $(3 + n)(3 + n)$

**5.** $(d + r)(d + r)$          **6.** $(b^2 + d^2)(b^2 + d^2)$

**7.** $(a + my)(a + my)$        **8.** $(9x + 5)(9x + 5)$

**9.** $(w + 12)^2$              **10.** $(15 + y)^2$

**11.** $(x + z)^2$               **12.** $(c^3x + dy^2)^2$

**13.** $(8a + 3b)^2$            **14.** $(s - 7)(s - 7)$

**15.** $(t - w)(t - w)$          **16.** $(10r - 6t)(10r - 6t)$

**17.** $(5a^3 - 8cx)(5a^3 - 8cx)$     **18.** $(11b - 4)^2$

**19.** $(m - x)^2$              **20.** $(15w - 3v)^2$

**21.** $(6n^2 - 2rx)^2$          **22.** $(\frac{2}{5}x + 9y)^2$

**23.** $(.9a - 1.2b)^2$         **24.** Square the binomial $12b - 11x$.

**25.** Find the product of $(3c + d)(3c - d)(9c^2 - d^2)$.

### *RELATED PRACTICE EXAMPLES*

1. Express each of the following in brief form, using exponents:
   **a.** $(n+9)(n+9)$
   **b.** $(4a+7c)(4a+7c)$
   **c.** $(3m^2+y^2)(3m^2+y^2)$
   **d.** $(b-3)(b-3)$
   **e.** $(9ab-1)(9ab-1)$
   **f.** $(12-a)(12-a)$
   **g.** $(10c+3d)(10c+3d)$
   **h.** $(5a^2-2b^2)(5a^2-2b^2)$

2. Write each of the following in another way:
   **a.** $(b+6)^2$
   **b.** $(9+x)^2$
   **c.** $(n-20)^2$
   **d.** $(4x^2-5)^2$
   **e.** $(7cd-8)^2$
   **f.** $(3r+10s)^2$
   **g.** $(\frac{3}{5}b-12)^2$
   **h.** $(14-3y)^2$
   **i.** $(.5m+.2n)^2$

Find the product of each of the following at sight:

3. **a.** $(x+3)(x+3)$    **b.** $(n+1)(n+1)$    **c.** $(c+10)(c+10)$

4. **a.** $(2+a)(2+a)$    **b.** $(7+y)(7+y)$    **c.** $(11+n)(11+n)$

5. **a.** $(x+y)(x+y)$    **b.** $(b+c)(b+c)$    **c.** $(w+v)(w+v)$

6. **a.** $(a^2+c^2)(a^2+c^2)$    **b.** $(s^3+t^3)(s^3+t^3)$    **c.** $(b^4+x)(b^4+x)$

7. **a.** $(bc+d)(bc+d)$    **b.** $(mx+ny)(mx+ny)$    **c.** $(b^3c^3+d^6)(b^3c^3+d^6)$

8. **a.** $(4r+9)(4r+9)$    **b.** $(12s+7)(12s+7)$
   **c.** $(8x+18y)(8x+18y)$    **d.** $(2c+11d)(2c+11d)$
   **e.** $(5m+3n^2)(5m+3n^2)$    **f.** $(6ax+8y)(6ax+8y)$

9. **a.** $(t+8)^2$    **b.** $(m+2)^2$    **c.** $(x+15)^2$

10. **a.** $(6+w)^2$    **b.** $(11+y)^2$    **c.** $(20+n)^2$

11. **a.** $(r+t)^2$    **b.** $(c+d)^2$    **c.** $(g+h)^2$

12. **a.** $(b^3+x^3)^2$    **b.** $(x+yz)^2$    **c.** $(a^2b+cd^2)^2$

13. **a.** $(5m+3)^2$    **b.** $(9c+11)^2$    **c.** $(2c+7n)^2$
    **d.** $(6x+13y)^2$    **e.** $(8rs+t^3)^2$    **f.** $(3b^3c^2+10x^2y^3)^2$

Find the product of each of the following at sight:

14. **a.** $(c-8)(c-8)$    **b.** $(v-16)(v-16)$    **c.** $(x-9)(x-9)$

15. **a.** $(a-b)(a-b)$    **b.** $(r-s)(r-s)$    **c.** $(c-d)(c-d)$

16. **a.** $(4x-5)(4x-5)$    **b.** $(7b-8n)(7b-8n)$    **c.** $(15a-11x)(15a-11x)$

17. **a.** $(6a^2-2c^2)(6a^2-2c^2)$    **b.** $(9xy-z)(9xy-z)$
    **c.** $(10b^3-4cd)(10b^3-4cd)$

18. **a.** $(b-1)^2$    **b.** $(3m-8)^2$    **c.** $(2y-13)^2$

19. **a.** $(x-y)^2$    **b.** $(b-a)^2$    **c.** $(n-t)^2$

20. **a.** $(8a-9c)^2$    **b.** $(7w-20x)^2$    **c.** $(10d-3y)^2$

**21. a.** $(m^2 - n^2)^2$          **b.** $(a^2b - xy^3)^2$          **c.** $(5rt - 8s^2)^2$

Find the product of each of the following at sight:

**22. a.** $(\frac{1}{4}a + 2)(\frac{1}{4}a + 2)$          **b.** $(\frac{2}{3}b - 6c)^2$          **c.** $(\frac{5}{6}x^2 - 12y)^2$

**23. a.** $(1.2r + .5s)(1.2r + .5s)$          **b.** $(.7x - .3y)^2$          **c.** $(.6a - 2.4b^2)(.6a - 2.4b^2)$

**24.** Square each of the following binomials:

| | | | |
|---|---|---|---|
| **a.** $r + 8$ | **b.** $m - 9$ | **c.** $y + z$ | **d.** $6b - 1$ |
| **e.** $ac - 4$ | **f.** $10r + 9$ | **g.** $8m - 3n$ | **h.** $5ab - 12c$ |
| **i.** $11x^2 - 8$ | **j.** $3 - 20rt$ | **k.** $15n^4 - 7x^2$ | **l.** $6ax + 9by$ |
| **m.** $b^2c - xy^2$ | **n.** $.7m - .8x$ | **o.** $\frac{3}{4}c + 12$ | **p.** $4g^2 + 25h^3$ |

**25.** Find the product of each of the following:

**a.** $(x + 3)(x - 3)(x^2 - 9)$          **b.** $(m - n)(m + n)(m^2 - n^2)$

**c.** $(5b + 4)(5b - 4)(25b^2 - 16)$          **d.** $(2x - 3y)(4x^2 - 9y^2)(2x + 3y)$

**e.** $(4a + bc)(4a - bc)(16a^2 - b^2c^2)$          **f.** $(3r^2 - 5s^2)(3r^2 + 5s^2)(9r^4 - 25s^4)$

## CUMULATIVE REVIEW

Find the square of each of the following:

**1. a.** 7     **b.** $+4$     **c.** $-10$     **d.** 25     **e.** $-30$     **f.** $-11$     **g.** $+.8$
   **h.** $-\frac{5}{8}$     **i.** $+\frac{2}{3}$     **j.** $-1.2$     **k.** $+.01$     **l.** $-\frac{3}{5}$     **m.** $-.19$     **n.** $+\frac{11}{6}$

**2. a.** $mx$          **b.** $r^3$          **c.** $x^2y^5$          **d.** $cx^8$          **e.** $x^3y^7z^2$

**3. a.** $-5c^4$          **b.** $2n^3y$          **c.** $-.4a^3b^5$          **d.** $\frac{3}{4}rs^2t^8$

Find the product of each of the following:

**4. a.** $(5)^2$     **b.** $(-1)^2$     **c.** $(-\frac{5}{9})^2$     **d.** $(.3)^2$     **e.** $(-1.7)^2$  **f.** $(-\frac{7}{3})^2$

**5. a.** $(-s)^2$   **b.** $(x^2z^3)^2$   **c.** $(ab^5)^2$   **d.** $(x^4yz^3)^2$     **e.** $(-c^6d^8)^2$

**6. a.** $(6t)^2$     **b.** $(-\frac{2}{3}r^2)^2$   **c.** $(.5m^3x^4)^2$   **d.** $(-9x^6yz^7)^2$

Find the product of each of the following at sight:

**7. a.** $(x - 5)(x + 5)$                              **b.** $(.2m^2n + .3y^3)(.2m^2n - .3y^3)$

**8. a.** $(y - 5)(y + 8)$                              **b.** $(6a - 11)(6a + 7)$

**9. a.** $(2n + 7)(2n + 7)$                            **b.** $(3c - 8d)^2$

**10. a.** $(5x + 2)(3x - 9)$                           **b.** $(8a - 3x)(3a + 4x)$

**11. a.** $(n - 12)(n + 2)$                            **b.** $(6 - 2x)(5 - 3x)$
   **c.** $(10m + n)^2$                                 **d.** $(\frac{1}{6}b + 8)(\frac{1}{6}b - 2)$
   **e.** $(3a + 5b)(3a - 5b)$                          **f.** $(2r - 7)(4r - 3)$
   **g.** $(s + 9)(s - 10)$                             **h.** $(9b^2 - 5c^2)(9b^2 - 5c^2)$
   **i.** $(5cd^2 - x^3)(5cd^2 - x^3)$                  **j.** $(.7w - 2.1y)(.8w - .6y)$
   **k.** $(9a + 7y)(9a - 7y)$                          **l.** $(15b^2 - 2c^3)^2$
   **m.** $(4x - 3y)(4x + 3y)(16x^2 + 9y^2)$            **n.** $(b - 2c)(2a - b)$

# Product of Two Binomials Having a Common Term— Special Method

**I. Aim**   To find the product of two binomials having a common term.

**II. Procedure**

**1. a.** To find the first term of the product, square the common term.
   **b.** To find the second term of the product, find the algebraic sum of the second terms of the binomials and multiply it by the common term.
   **c.** To find the third term of the product, multiply the second terms of the binomials.

**2.** Sometimes the common term is just a number. See sample solution 2.

**3.** Sometimes the given binomials consist of two terms both containing variables.

**4.** When the common term of the given binomials is the second term in each of the given binomials, the square of the common term is the third term of the product.

**III. Sample Solutions**

**1.** Find the product of: $(9b - 3)(9b + 5)$

**Step a.**

$(9b - 3)(9b + 5) = (9b)^2 = 81b^2$

**Step b.**

$$(9b - 3)(9b + 5) = 81b^2 + (-3 + 5)(9b)$$
$$= 81b^2 + (2)(9b)$$
$$= 81b^2 + 18b$$

**Step c.**

$$(9b - 3)(9b + 5) = 81b^2 + 18b + (-3)(+5)$$
$$= 81b^2 + 18b - 15$$

However, all three steps should be combined and written in one step at sight as follows:

$$(9b - 3)(9b + 5) = 81b^2 + 18b - 15$$
$$\textit{Think: } (9b)^2 + (-3 + 5)(9b) + (-3)(+5)$$

*Answer:* $81b^2 + 18b - 15$

2. Find the product of: $(3 + 5a)(3 + 2a)$

$(3 + 5a)(3 + 2a) = (3)^2 + (7a \cdot 3) + (5a \cdot 2a) = 9a + 21a + 10a^2$

*Answer:* $9 + 21a + 10a^2$

3. Find the product of: $(4x + 3y)(4x + 5y)$

$$(4x + 3y)(4x + 5y) = (4x)^2 + (3y + 5y)(4x) + (3y)(5y)$$
$$= 16x^2 + (8y)(4x) + 15y^2$$
$$= 16x^2 + 32xy + 15y^2$$

*Answer:* $16x^2 + 32xy + 15y^2$

4. Find the product of: $(11 - 4c)(1 - 4c)$

$$(11 - 4c)(1 - 4c) = (11 \cdot 1) + (11 + 1)(-4c) + (-4c)^2$$
$$= 11 + (12)(-4c) + 16c^2$$
$$= 11 - 48c + 16c^2$$

*Answer:* $11 - 48c + 16c^2$

## PRELIMINARY EXAMPLES

**A.** Add the second and third terms algebraically and then multiply the sum by the first term:

|     | First | Second | Third |
|-----|-------|--------|-------|
| 1.  | $x$   | $+2$   | $+3$  |
| 3.  | $b$   | $+6$   | $+4$  |
| 5.  | $y$   | $-4$   | $+3$  |
| 7.  | $x$   | $+2$   | $-6$  |
| 9.  | $b$   | $-8$   | $-9$  |
| 11. | $3x$  | $+5$   | $+3$  |
| 13. | $4a$  | $-3$   | $+2$  |
| 15. | $w$   | $+6$   | $-6$  |
| 17. | $b$   | $-3x$  | $-4x$ |
| 19. | $3a$  | $+5m$  | $-9m$ |

|     | First  | Second  | Third   |
|-----|--------|---------|---------|
| 2.  | $a$    | $+7$    | $+1$    |
| 4.  | $m$    | $-2$    | $+5$    |
| 6.  | $d$    | $+7$    | $-4$    |
| 8.  | $r$    | $-4$    | $-3$    |
| 10. | $x$    | $-5$    | $+11$   |
| 12. | $8y$   | $-4$    | $-7$    |
| 14. | $s$    | $-5$    | $-5$    |
| 16. | $x$    | $+2y$   | $+y$    |
| 18. | $c$    | $-8d$   | $+7d$   |
| 20. | $x^2$  | $-3y^2$ | $-5y^2$ |

**B.** Find the product of the second and third terms, using examples in Part A.

## DIAGNOSTIC TEST

Find the product of each of the following at sight:

**1.** $(x+6)(x+4)$          **2.** $(n-11)(n-5)$          **3.** $(t+10)(t-2)$

**4.** $(b+1)(b-12)$         **5.** $(y-9)(y+3)$           **6.** $(c-4)(c+7)$

**7.** $(n-10)(n+9)$         **8.** $(x+9)(x-9)$           **9.** $(3x+8)(3x-5)$

**10.** $(m^5-15)(m^5-1)$    **11.** $(ab^2-3)(ab^2+7)$    **12.** $(\frac{5}{6}xy+12)(\frac{5}{6}xy-18)$

**13.** $(7a+.4)(7a+1.9)$    **14.** $(8-9cd)(8+6cd)$      **15.** $(5x-11y)(5x-4y)$

**16.** $(13+6b)(13+10b)$

## *RELATED PRACTICE EXAMPLES*

Find the product of each of the following at sight:

**1. a.** $(x+1)(x+2)$          **b.** $(a+5)(a+8)$          **c.** $(y+9)(y+7)$

**2. a.** $(c-4)(c-9)$          **b.** $(d-6)(d-1)$          **c.** $(a-3)(a-3)$

**3. a.** $(n+8)(n-3)$          **b.** $(x+10)(x-5)$         **c.** $(y+12)(y-2)$

**4. a.** $(a+5)(a-8)$          **b.** $(r+1)(r-4)$          **c.** $(w+10)(w-12)$

**5. a.** $(r-3)(r+1)$          **b.** $(s-5)(s+2)$          **c.** $(x-18)(x+3)$

**6. a.** $(x-2)(x+8)$          **b.** $(y-7)(y+10)$         **c.** $(n-5)(n+15)$

**7. a.** $(b+4)(b-3)$          **b.** $(x-2)(x+1)$          **c.** $(a+8)(a-9)$

**8. a.** $(n-10)(n+10)$        **b.** $(c+8)(c-8)$          **c.** $(r-12)(r+12)$

**9. a.** $(2a+3)(2a+1)$        **b.** $(3x+8)(3x+2)$        **c.** $(6n-5)(6n-7)$

   **d.** $(8b+9)(8b-3)$        **e.** $(10d-6)(10d+5)$      **f.** $(11y+8)(11y-12)$

**10. a.** $(c^2+5)(c^2+4)$     **b.** $(x^3-10)(x^3+9)$     **c.** $(a^4-5)(a^4-8)$

**11. a.** $(xy+6)(xy+9)$       **b.** $(c^3x^2-8)(c^3x^2-4)$  **c.** $(5m^2n+1)(5m^2n-2)$

**12. a.** $(\frac{3}{4}n+4)(\frac{3}{4}n+8)$  **b.** $(\frac{1}{3}cd-12)(\frac{1}{3}cd+15)$  **c.** $(\frac{7}{8}b^2+24)(\frac{7}{8}b^2-16)$

**13. a.** $(b+2.3)(b-.6)$      **b.** $(3x+.8)(3x+1.8)$     **c.** $(.5y-.9)(.5y-.2)$

**14. a.** $(5+4b)(5+7b)$       **b.** $(3-x)(3-2x)$         **c.** $(9+6n)(9-8n)$

   **d.** $(20-7c^2)(20+5c^2)$  **e.** $(15-7t^2)(15+7t^2)$  **f.** $(10+2ab)(10-ab)$

**15. a.** $(c+2d)(c+3d)$                                   **b.** $(2x-y)(2x-5y)$

   **c.** $(a+8b)(a-9b)$                            **d.** $(3m-2x)(3m+5x)$

   **e.** $(7a^2+4t)(7a^2-6t)$                      **f.** $(10ab+5cd)(10ab-8cd)$

**16. a.** $(4+x)(5+x)$         **b.** $(1-c)(3-c)$          **c.** $(7+3y)(4+3y)$

   **d.** $(3a+3)(a+3)$         **e.** $(9b-4)(7b-4)$        **f.** $(6n+5x)(4n+5x)$

   **g.** $(2m-8n^2)(7m-8n^2)$  **h.** $(10c+9ab)(5c+9ab)$   **i.** $(11r-2s)(5r-2s)$

# Factoring Polynomials Having a Common Monomial Factor

**I. Aim**  To factor a polynomial whose terms have a common monomial factor.

**II. Procedure**

1. To determine the greatest common factor of two or more monomials,
   a. Express the numerical coefficient of each given monomial as a product of prime numbers.

   A *prime number* is a whole number other than 0 and 1 which can be divided exactly by itself and by 1 and by no other whole number.
   b. Select the highest power of each prime number common to the given monomials and the highest power of each variable common to the given monomials. The product of these is the greatest common factor.

2. To factor a polynomial whose terms have a common monomial factor,
   a. Find the greatest common factor of its terms.
   b. Divide the given polynomial by the greatest common factor.
   c. Write the indicated product of these two factors.

3. When one of the terms of the given polynomial is exactly divisible by the common monomial factor, use this quotient 1. There should be as many terms in the quotient factor as there are in the given polynomial. See sample solution 5.

4. If all the terms of the given polynomial have a negative coefficient, use a common monomial factor that has a negative coefficient. See sample solution 6.

5. Sometimes a binomial may be a common factor.

**III. Sample Solutions**

1. Find the greatest common factor of: $12a^3$ and $8a^2$

   $12a^3 = 2 \cdot 2 \cdot 3 \cdot a \cdot a \cdot a = 2^2 \cdot 3 \cdot a^3$

   $8a^2 = 2 \cdot 2 \cdot 2 \cdot a \cdot a = 2^3 \cdot a^2$

   Highest power of 2 common to both $12a^3$ and $8a^2$ is $2^2$

   Highest power of a common to both $12a^3$ and $8a^2$ is $a^2$

   The factor 3 is not common to both $\underline{12a^3 \text{ and } 8a^2}$

*Answer:* Greatest common factor is $2^2 \cdot a^2$ or $4a^2$

Factor each of the following:

**2.** $5x - 5y$  *Answer:* $5(x - y)$

**3.** $ax - dx + rx$  *Answer:* $x(a - d + r)$

**4.** $8c^5d^4 - 16c^3d^7$  *Answer:* $8c^3d^4(c^2 - 2d^3)$

**5.** $ab + ac - a$  *Answer:* $a(b + c - 1)$

**6.** $-2ax^4 - 6b^2x^5 - 4c^2x^3$  *Answer:* $-2x^3 (ax + 3b^2x^2 + 2c^2)$

**7.** $x(x + 2) + 4 (x + 2)$  *Answer:* $(x + 2)(x + 4)$

## DIAGNOSTIC TEST

**1.** Find the greatest common factor of 72 and 48.

**2.** Find the greatest common factor of $r^6s^5$ and $r^4s^{10}$.

**3.** Find the greatest common factor of $75m^7n^3$, $50m^6n^8$, $100m^5n^9$.

**4.** Since $24x^4y^2 - 30x^3y^3 = 6x^3y^2 (4x - 5y)$, what is the given common monomial factor of $24x^4y^2 - 30x^3y^3$?

**5.** Find the second factor, given the polynomial $32b^3 - 48b^2 + 64b$ and the factor $16b$.

Factor each of the following polynomials:

**6.** $11m + 11n$

**7.** $cx - cy + cz$

**8.** $ar - rl$

**9.** $ax - bx + x$

**10.** $8x + 16$

**11.** $a^4 + 2a^3b$

**12.** $a^2bx + ab^2x - abx^2$

**13.** $12x^8y^6 - 8x^6y^3 + 24x^4y^5$

**14.** $-3b^3 - 6b^2 - 9b$

**15.** Write $r(r - 3) + 4 (r - 3)$ in factor form.

## *RELATED PRACTICE EXAMPLES*

$(x - y + z)$

**1.** Find the greatest common factor of each of the following:

  **a.** 6 and 7      **b.** 18 and 24      **c.** 90 and 63

  **d.** 75 and 45      **e.** 8, 12, and 18      **f.** 54, 72, and 108

**2.** Find the greatest common factor of each of the following:

  **a.** $n^4$ and $n^2$      **b.** $a^3b$ and $a^5$      **c.** $x^3y^2$ and $x^2y^3$

  **d.** $ar^2s^4$ and $r^6s^3t$      **e.** $m^7n^4, m^5n^{10}, m^4n^9$      **f.** $b^9c^3x^2, b^6c^5x^3,$ and $b^4c^4x^4$

**3.** Find the greatest common factor of each of the following:

  **a.** $10m^5$ and $35m^3$             **b.** $9c^4d$ and $14c^2d^2$

  **c.** $28x^6y^5$ and $70x^5y^8z^2$      **d.** $16a^5b^2, 40a^4b^6,$ and $56b^8c^3$

  **e.** $15n^8y^3, 29n^6y^2,$ and $48n^7y^5$      **f.** $8r^2s^3l^5, 18r^4l^8,$ and $12s^2l^{10}$

$11(m + n)$             $-3b^1 (b^2 + b + 3)$

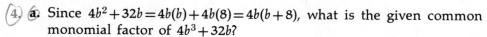

4. a. Since $4b^2+32b=4b(b)+4b(8)=4b(b+8)$, what is the given common monomial factor of $4b^3+32b$?

   b. Since $12x^5-18x^4=6x^4(2x-3)$, what is the given common monomial factor of $12x^5-18x^4$?

   c. Since $25ab^4+20a^3b^2-15a^4b^3=5ab^2(5b^2+4a^2-3a^3b)$, what is the given common monomial factor of $25ab^4+20a^3b^2-15a^4b^3$?

5. Find the second factor, given the polynomial:

   a. $18c^4+24c^6$ and the factor $6c^4$.

   b. $30a^2-75a+90$ and the factor 15.

   c. $72m^5n^3+96m^4n^5-48m^2n^6$ and the factor is $24m^2n^3$.

   d. $27b^6d^5-81b^6d^6+54b^3d^4$ and the factor $27b^3d^4$.

Factor each of the following polynomials:

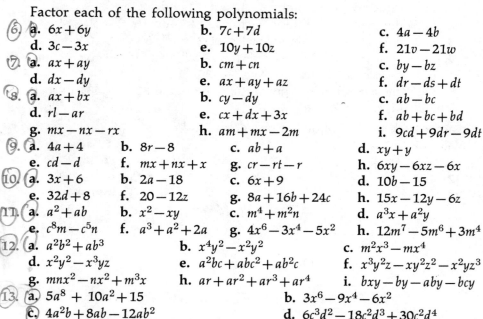

6. a. $6x+6y$          b. $7c+7d$          c. $4a-4b$
   d. $3c-3x$          e. $10y+10z$        f. $21v-21w$

7. a. $ax+ay$          b. $cm+cn$          c. $by-bz$
   d. $dx-dy$          e. $ax+ay+az$       f. $dr-ds+dt$

8. a. $ax+bx$          b. $cy-dy$          c. $ab-bc$
   d. $rl-ar$          e. $cx+dx+3x$       f. $ab+bc+bd$
   g. $mx-nx-rx$       h. $am+mx-2m$       i. $9cd+9dr-9dt$

9. a. $4a+4$      b. $8r-8$       c. $ab+a$         d. $xy+y$
   e. $cd-d$      f. $mx+nx+x$    g. $cr-rt-r$      h. $6xy-6xz-6x$

10. a. $3x+6$     b. $2a-18$      c. $6x+9$         d. $10b-15$
    e. $32d+8$    f. $20-12z$     g. $8a+16b+24c$   h. $15x-12y-6z$

11. a. $a^2+ab$        b. $x^2-xy$        c. $m^4+m^2n$      d. $a^3x+a^2y$
    e. $c^8m-c^5n$     f. $a^3+a^2+2a$    g. $4x^6-3x^4-5x^2$ h. $12m^7-5m^6+3m^4$

12. a. $a^2b^2+ab^3$         b. $x^4y^2-x^2y^2$          c. $m^2x^3-mx^4$
    d. $x^2y^2-x^3yz$        e. $a^2bc+abc^2+ab^2c$      f. $x^3y^2z-xy^2z^2-x^2yz^3$
    g. $mnx^2-nx^2+m^3x$     h. $ar+ar^2+ar^3+ar^4$      i. $bxy-by-aby-bcy$

13. a. $5a^8+10a^2+15$                      b. $3x^6-9x^4-6x^2$
    c. $4a^2b+8ab-12ab^2$                   d. $6c^3d^2-18c^2d^3+30c^2d^4$
    e. $2d^2r^2+10d^3r^3-8d^4r^4$           f. $20a^5d^3+15a^4d^7+10a^2d^9$
    g. $9a^3c^3-15a^3c+3a^3$                h. $30bx+10b^2-20b^3$
    i. $4a^3b^4-2a^3b^3c^2$                 j. $27b^6c^3d^5-36b^4c^4d^6-18b^3c^5d^8$

14. a. $-ab-ac-ad$                          b. $-x^3-x^2-x$
    c. $-a^2y^2-a^2y-ay^2$                  d. $-2c^3-4c^2-6c$
    e. $-15bx^3-10b^2x^2-25b^3x^2$          f. $-63m^4n^4-18m^3n^5-54m^2n^6$

15. Write each of the following expressions in factored form:

    a. $a(a-b)+b(a-b)$                      b. $x(x+y)-y(x+y)$
    c. $c(c+3)+3(c+3)$                      d. $b(a-2)-2(a-2)$
    e. $x^2(x-5)+6(x-5)$                    f. $n^2(n-8)-7(n-8)$

# Square Root of a Monomial

**I. Aim** To find the square root of a monomial.

**II. Procedure**

**1. a.** Take the square root of its numerical coefficient.
   **b.** Divide the exponent of each of its variables by 2.

**2.** Observe:
   **a.** The *square root of a monomial* is that monomial which, when multiplied by itself, produces the given monomial.
   **b.** Every positive number has two square roots, one positive and the other negative. Each is the opposite of the other.
   **c.** The positive square root of a number is called the *principal square root*.
   **d.** Zero has only one square root, which is zero.

**3.** When a perfect square number is named by a common fraction, determine its square root by finding the square root of both the numerator and the denominator of the fraction.

**III. Sample Solutions**

**1.** The square roots of 9 are $+3$ and $-3$.

**2.** The square roots of $\frac{4}{25}$ are $+\frac{2}{5}$ and $-\frac{2}{5}$.

**3.** The square roots of $x^6$ are $x^3$ and $-x^3$.

**4.** The square roots of $a^4b^2c^{10}$ are $a^2bc^5$ and $-a^2bc^5$.

**5.** The square roots of $64m^6n^{14}x^2$ are $8m^3n^7x$ and $-8m^3n^7x$.

**6.** The principal square root of 49 is 7.

---

## DIAGNOSTIC TEST

Find the square roots of each of the following:

**1.** 36      **2.** 81      **3.** $\frac{16}{25}$      **4.** $a^4$

**5.** $r^6s^2t^{10}$      **6.** $144c^8d^2$      **7.** $.04w^4x^{12}y^6$      **8.** $\frac{36}{49}a^{10}b^2c^4$

**9.** Find the principal square root of 121.

## RELATED PRACTICE EXAMPLES

Find the square roots of each of the following:

1. **a.** 25　　　**b.** 64　　　**c.** 49　　　**d.** 81　　　**e.** 100
   **f.** 169　　**g.** 225　　**h.** 400　　**i.** 121　　**j.** 576

2. **a.** .01　　**b.** .16　　**c.** .09　　**d.** .64　　**e.** .36
   **f.** .25　　**g.** .81　　**h.** .04　　**i.** .49　　**j.** 1.44

3. **a.** $\frac{1}{4}$　　**b.** $\frac{16}{25}$　　**c.** $\frac{4}{9}$　　**d.** $\frac{36}{49}$　　**e.** $\frac{25}{81}$
   **f.** $\frac{49}{100}$　**g.** $\frac{64}{121}$　**h.** $\frac{9}{196}$　**i.** $\frac{225}{361}$　**j.** $\frac{169}{256}$

4. **a.** $x^2$　　**b.** $n^2$　　**c.** $b^4$　　**d.** $y^6$　　**e.** $c^{10}$
   **f.** $r^{18}$　**g.** $r^8$　　**h.** $t^{12}$　**i.** $a^{20}$　**j.** $n^{100}$

5. **a.** $x^2y^2$　　**b.** $n^2x^2$　　**c.** $a^4b^2$　　**d.** $c^2d^6$　　**e.** $x^8y^4$
   **f.** $m^{12}n^6$　**g.** $a^{10}b^4c^2$　**h.** $r^{14}s^8t^6$　**i.** $b^{12}x^2y^8$　**j.** $m^4n^{10}r^{16}$

6. **a.** $9c^2$　　　　**b.** $100x^2$　　　**c.** $81x^4$　　　**d.** $121x^8$
   **e.** $36a^2b^2$　**f.** $25x^4y^{10}$　**g.** $4m^4n^2x^8$　**h.** $144a^6x^{12}y^{16}$

7. **a.** $.04b^2$　　　**b.** $.25x^2$　　　**c.** $.36a^2c^2$　　**d.** $.81c^2x^4$
   **e.** $.49a^2b^4y^6$　**f.** $.09b^{12}c^8d^{14}$　**g.** $.64r^8x^2y^4$　**h.** $.16a^2b^6c^{10}d^8$

8. **a.** $\frac{1}{4}n^2$　　　**b.** $\frac{9}{25}a^2$　　　**c.** $\frac{4}{81}x^6$　　　**d.** $\frac{49}{144}b^{12}$
   **e.** $\frac{25}{64}a^8n^{18}$　**f.** $\frac{100}{121}m^4x^{10}$　**g.** $\frac{81}{100}b^2c^{12}d^{16}$　**h.** $\frac{169}{225}r^8s^6t^{14}$

9. Find the principal square root of each of the following:
   **a.** 100　　　**b.** 25　　　**c.** 81　　　**d.** 256
   **e.** 900　　　**f.** 169　　　**g.** 144　　　**h.** 400

## CUMULATIVE REVIEW

1. Find the product of each of the following at sight:
   **a.** $(n - 8)(n - 8)$　　　　　　**b.** $(25 - n)(25 + n)$

   **c.** $(6w - 5)(6w + 4)$　　　　**d.** $(11ad - 8bc)(4ad + 2bc)$

   **e.** $(12b^2c + \frac{1}{2}d^3)(12b^2c - \frac{1}{2}d^3)$　　**f.** $(.4a + .5b)(.4a - 1.2b)$

   **g.** $(10x + 2y)(3x + 9y)$　　　**h.** $(m + n)(m - n)(m^2 - n^2)$

2. Factor each of the following polynomials:
   **a.** $2m^4 - 6m^3 + 3m^2$　　　　**b.** $9y^2 + 81$

   **c.** $10b^{10} + 25b^6 - 20b^4$　　　**d.** $18c^5d^3 - 12c^3d^4 + 24c^2d^5$

   **e.** $x(x + 9) + 5(x + 9)$　　　　**f.** $-4x^4y - 2x^2y^2 - 6x^2y^4$

3. Find the square roots of each of the following:

   **a.** 121　　　**b.** $n^{16}$　　　**c.** $25x^8y^2$　　　**d.** $.09a^{10}b^6c^{12}$

# Factoring the Difference of Two Squares

EXERCISE 9-8

I. **Aim**    To factor the difference of two squares.

II. **Procedure**

1. Find the positive square root of each of the given squares.
2. Write the sum of two square roots as one factor of an indicated product.
3. Write the difference of the same two square roots as the second factor of the indicated product.
4. Note that the sum of the two squares cannot be factored by this method.
5. Sometimes a resulting factor is factorable. If so, continue factoring until each factor cannot be factored any further.

III. **Sample Solutions**

1. Factor: $x^2 - 36$

   $x^2 - 36 = (x)^2 - (6)^2 = (x + [6])(x - [6])$

   In brief, $x^2 - 36 = (x + 6)(x - 6)$

   *Answer:* $(x + 6)(x - 6)$

2. Factor $\frac{16}{25}m^2 - 121$

   *Answer:* $(\frac{4}{5}m + 11)(\frac{4}{5}m - 11)$

3. Factor: $b^4 - x^4$

   $b^4 - x^4 = (b^2 + x^2)(b^2 - x^2)$
   $= (b^2 + x^2)(b + x)(b - x)$

   *Answer:* $(b^2 + x^2)(b + x)(b - x)$

---

### DIAGNOSTIC TEST

Factor each of the following binomials:

1. $n^2 - 4$
2. $25 - y^2$
3. $81x^2 - 16$
4. $a^2x^4 - y^6$
5. $49c^2 - 144d^2$
6. $\frac{9}{100}t^2 - 64z^2$
7. $c^2 - .36x^2$
8. $a^4 - b^4$

## RELATED PRACTICE EXAMPLES

Factor each of the following binomials:

1. **a.** $c^2 - 9$     **b.** $x^2 - 25$     **c.** $a^2 - 49$
   **d.** $b^2 - 100$     **e.** $d^2 - 144$     **f.** $n^2 - 1$
   **g.** $y^2 - 16$     **h.** $r^2 - 121$     **i.** $m^2 - 64$

2. **a.** $4 - t^2$     **b.** $100 - r^2$     **c.** $81 - y^2$
   **d.** $36 - x^2$     **e.** $16 - c^2$     **f.** $121 - b^2$
   **g.** $49 - x^2$     **h.** $64 - a^2$     **i.** $1 - d^2$

3. **a.** $4x^2 - 25$     **b.** $49a^2 - 1$     **c.** $9y^2 - 16$
   **d.** $100 - 81c^2$     **e.** $121 - 64b^2$     **f.** $x^2y^2 - 4$
   **g.** $b^4 - 36$     **h.** $9 - a^4x^2$     **i.** $144 - n^6$

4. **a.** $x^2 - y^2$     **b.** $r^2 - s^2$     **c.** $b^4 - a^2$
   **d.** $c^6 - d^4$     **e.** $a^2b^2 - y^2$     **f.** $c^2d^2 - x^2$
   **g.** $b^2 - x^2y^2$     **h.** $a^4x^2 - w^2$     **i.** $b^2c^6 - x^4y^8$

5. **a.** $9a^2 - 4b^2$     **b.** $16x^2 - 25y^2$     **c.** $49m^2 - 4n^2$
   **d.** $144c^2 - 121d^2$     **e.** $64a^2b^2 - x^2$     **f.** $4a^4x^2 - 49b^2y^4$
   **g.** $121b^6 - 16m^4$     **h.** $81m^4x^2y^2 - 25a^2$     **i.** $a^2b^2x^2 - 4c^2d^2y^2$

6. **a.** $a^2 - \frac{1}{9}$     **b.** $\frac{1}{4}n^2 - 25$     **c.** $\frac{4}{25}m^2 - 64x^2$

   **d.** $\dfrac{x^2}{36} - \dfrac{y^2}{49}$     **e.** $\dfrac{36}{a^2} - \dfrac{25}{b^2}$

7. **a.** $b^2 - .16$     **b.** $x^2 - .01$     **c.** $.64 - a^2$
   **d.** $49 - .81m^2n^4$     **e.** $.25c^2 - .36d^2$     **f.** $.04b^4c^6 - .09x^8y^2$

8. **a.** $m^4 - n^4$     **b.** $a^4 - 81$     **c.** $x^4 - 16y^4$
   **d.** $c^8 - 1$     **e.** $a^8 - b^8$     **f.** $x^{16} - 1$

## CUMULATIVE REVIEW

1. Find the product at sight:
   **a.** $(m + x)(m + x)$
   **b.** $(3a - 4)(3a + 4)$
   **c.** $(x - 3)(x - 7)$
   **d.** $(4a + 9b)^2$
   **e.** $(6m - 7)(9m + 5)$
   **f.** $(9xy + 8z^2)(9xy - 8z^2)$
   **g.** $(2a - b)(2a + b)(4a^2 + b^2)$

2. Factor:
   **a.** $12c^2 - 9c^3 + 6c^4$
   **b.** $x^2 - 81$
   **c.** $-8m^6y^3 + 4m^5y^4 - 6m^4$
   **d.** $25c^2d^4 - 36x^2$
   **e.** $54xy^3z^2 + 18x^2z^5 - 36x^4y^2z$
   **f.** $100 - 9c^2$
   **g.** $b^2(b - 7) + 8(b - 7)$

EXERCISE
**9-9**

**I. Aim** To factor a perfect square trinomial.

**II. Procedure**

1. First determine whether the given trinomial is a *perfect square trinomial* that fits the following description:
   a. Two terms (usually the first and third terms of the given trinomial) must be perfect squares containing positive numerical coefficients or positive numbers.
   b. The remaining term (usually the second term of the given trinomial) must be twice the product of the square roots of the two perfect square terms.

2. If the given trinomial is found to be a perfect square trinomial, factor it as follows:
   a. Find the positive square root of each of the two perfect square terms.
   b. Connect the two square roots with the sign of the numerical coefficient of the remaining term of the given trinomial.
   c. Indicate that this binomial is used twice as a factor.

**III. Sample Solutions**

1. Which of the following are perfect square trinomials?
   **a.** $x^2 + 6x + 9$   **b.** $x^2 - 6x + 9$   **c.** $x^2 + 6x - 9$   **d.** $x^2 + 5x + 9$

   *Answer:* $x^2 + 6x + 9$ and $x^2 - 6x + 9$ are perfect square trinomials.

    $x^2 + 6x - 9$ is not a perfect square trinomial because of the negative sign preceding the 9.

   $x^2 + 5x + 9$ is not a perfect square because 5 is not twice the product of the square roots of $x^2$ and 9.

Factor each of the following:

2. $x^2 + 6x + 9$
   *Answer:* $(x + 3)(x + 3)$ or $(x + 3)^2$
3. $c^2 - 2cd + d^2$
   *Answer:* $(c - d)(c - d)$ or $(c - d)^2$
4. $4x^2 + 20xy + 25y^2$
   *Answer:* $(2x + 5y)(2x + 5y)$
5. $49a^6 - 56a^3xy^2 + 16x^2y^4$
   *Answer:* $(7a^3 - 4xy^2)^2$
6. $m^2 + 3m + \frac{9}{4}$
   *Answer:* $(m + \frac{3}{2})(m + \frac{3}{2})$

## DIAGNOSTIC TEST

1. Which of the following are perfect square trinomials?
   a. $m^2 + 6m + 36$
   b. $n^2 - 22n + 121$
   c. $9c^2 + 30cd - 25d^2$
   d. $64b^2 + 144bc + 81c^2$

2. Find the term required in each of the following to make a perfect square trinomial:
   a. $r^2 + 10r + ?$
   b. $a^2 - ? + 49$
   c. $? + 40n + 100$
   d. $121s - 88st + ?$

Factor each of the following trinomials:

3. $x^2 + 8x + 16$
4. $t^2 - 24t + 144$
5. $9 + 6w + w^2$
6. $m^2 + 2mn + n^2$
7. $a^4 + 2a^2b^2 + b^4$
8. $16x^2 + 40x + 25$
9. $81 - 36x + 4x^2$
10. $9a^2b^2 + 12ab + 4$
11. $49x^2 - 84xy + 36y^2$
12. $64b^2c^2 + 144bcx + 81x^2$
13. $n^2 + 9n + \frac{81}{4}$
14. $a^2 - 1.2a + 0.36$

## RELATED PRACTICE EXAMPLES

1. Which of the following are perfect square trinomials?
   a. $b^2 + 14b - 49$
   b. $n^2 + 10m + 16$
   c. $x^2 - 5x + 4$
   d. $a^2 + 10a + 25$
   e. $c^2 - 2cd - d^2$
   f. $36d^2 + 13dx + x^2$
   g. $81a^2 - 72ab + 16b^2$
   h. $4x^2 + 4x + 1$
   i. $49m^2 + 140mn + 100n^2$
   j. $16b^2 - 20by + 25y^2$

2. Find the term required in each of the following to make a perfect square trinomial:
   a. $a^2 + ? + 9$
   b. $x^2 - ? + 25$
   c. $b^2 + 8b + ?$
   d. $m^2 - 18m + ?$
   e. $r^2 + ? + s^2$
   f. $? + 2ax + x^2$
   g. $81c^2 + ? + 36$
   h. $9b^2 - 42b + ?$
   i. $64a^2b^2 + 80abx + ?$
   j. $? - 240a^2b^2xy + 144b^4y^2$

Factor each of the following trinomials:

3. **a.** $b^2 + 2b + 1$
   **b.** $x^2 + 10x + 25$
   **c.** $x^2 + 16x + 64$
   **d.** $c^2 + 20c + 100$

4. **a.** $y^2 - 8y + 16$
   **b.** $x^2 - 12x + 36$
   **c.** $d^2 - 18d + 81$
   **d.** $n^2 - 22n + 121$

5. **a.** $25 + 10b + b^2$
   **b.** $144 + 24x + x^2$
   **c.** $1 - 2m + m^2$
   **d.** $64 - 16y + y^2$

6. **a.** $c^2 + 2cd + d^2$
   **b.** $x^2 - 2xy + y^2$
   **c.** $d^2 - 2dy + y^2$
   **d.** $a^2 + 2ab + b^2$

7. **a.** $x^4 + 2x^2y^2 + y^4$
   **b.** $r^2 - 2rs^2 + s^4$
   **c.** $a^4 + 2a^2t^3 + t^6$
   **d.** $x^6 - 2x^3y^5 + y^{10}$

8. **a.** $9b^2 + 12b + 4$
   **b.** $49y^2 - 56y + 16$
   **c.** $100a^2 - 180a + 81$
   **d.** $36m^2 + 60m + 25$

9. **a.** $25 + 40x + 16x^2$
   **b.** $64 - 176b + 121b^2$
   **c.** $121 - 66c + 9c^2$
   **d.** $100 + 140a + 49a^2$

10. **a.** $c^2x^2 + 16cx + 64$
    **b.** $25b^2x^2 + 70bx + 49$
    **c.** $144c^4y^2 - 120c^2y + 25$
    **d.** $121x^2y^4 - 132xy^2 + 36$

11. **a.** $16a^2 + 8ab + b^2$
    **b.** $49c^2 - 28rd + 4d^2$
    **c.** $64m^2 + 80mn + 25n^2$
    **d.** $100a^2 - 220ar + 121r^2$

12. **a.** $25a^2d^2 + 20acd + 4c^2$
    **b.** $100x^4 - 140x^2y^2 + 49y^4$
    **c.** $4a^4x^2 + 12a^2b^3x + 9b^6$
    **d.** $81b^6x^4 - 90b^3c^2x^2y + 25c^4y^2$

13. **a.** $x^2 + x + \frac{1}{4}$
    **b.** $a^2 - 5a + \frac{25}{4}$
    **c.** $9b^2 + 4b + \frac{4}{9}$
    **d.** $36d^2 + 15d + \frac{25}{16}$

14. **a.** $c^2 + 0.8c + 0.16$
    **b.** $m^2 - 0.2m + 0.01$
    **c.** $4s^2 + 2.8s + 0.49$
    **d.** $0.16a^2 - 0.72a + 0.81$

## CUMULATIVE REVIEW

1. Find the product at sight:
   **a.** $(5c - 6)(5c - 6)$
   **b.** $(11 + r)(11 - r)$
   **c.** $(x - 10)(x + 9)$
   **d.** $(9x + 6y)(4x - 7y)$
   **e.** $(c^2 + d^2)^2$

2. Factor:
   **a.** $m^2 + 16m + 64$
   **b.** $4c^2 + 16$
   **c.** $9a^2 - 30ab + 25b^2$
   **d.** $49x^2 - 36y^2$
   **e.** $d^6 + 2d^3y^3 + y^6$

# Factoring–Factors Are Binomials Having a Common Term

## I. Aim

To factor a trinomial whose factors are two binomials having a common term with the numerical coefficient 1.

## II. Procedure

1. First find this common term by taking the square root of the perfect square term (usually the first term of the given trinomial).
2. Then find two numbers whose product is one of the terms of the trinomial (usually the third term of the given trinomial) and whose algebraic sum is the numerical coefficient of the remaining term (usually the second term) of the given trinomial. To do this, first think of all the pairs of factors whose product is the given number and then select the pair whose sum is the other given number.
3. Write as one factor the binomial consisting of the common term and one of the two numbers found in step 2, and as the second factor the binomial consisting of the common term and the second of these numbers.

## III. Sample Solutions

1. What two numbers have the product of $+6$ and the sum of $+5$? Each of the following pairs of factors has a product $+6$:
   $$(+6)(+1) = +6 \quad (-6)(-1) = +6 \quad (+3)(+2) = +6 \quad (-3)(-2) = +6$$
   but only $+3$ and $+2$ has the sum $+5$.
   *Answer:* $+3$ and $+2$

   Factor each of the following:
2. $x^2 + 9x + 20$
   *Answer:* $(x + 5)(x + 4)$

3. $b^2 - 11b + 18$
   *Answer:* $(b - 9)(b - 2)$

4. $m^2 - 6m - 16$
   *Answer:* $(m - 8)(m + 2)$

5. $s^2 + s - 12$
   *Answer:* $(s + 4)(s - 3)$

## DIAGNOSTIC TEST

Find two numbers which, when added, give the following sum, and which, when multiplied, give the following product:

1. Sum, $+11$; Product, $+18$    2. Sum, $-22$; Product, $+40$
3. Sum, $+4$; Product, $-96$     4. Sum, $0$; Product, $-81$
5. Sum, $-10$; Product, $-56$

Factor each of the following:

6. $a^2 + 7a + 12$      7. $x^2 - 6x + 8$       8. $b^2 + 10b - 24$
9. $y^2 - 7y - 18$     10. $x^2 + x - 20$

## RELATED PRACTICE EXAMPLES

Find two numbers which, when added, give the following sum, and which, when multiplied, give the following product:

1. a. Sum, $+6$; Product, $+8$       b. Sum, $+8$; Product, $+7$
   c. Sum, $+22$; Product, $+72$     d. Sum, $+25$; Product, $+84$

2. a. Sum, $-3$; Product, $+2$       b. Sum, $-10$; Product, $+21$
   c. Sum, $-11$; Product, $+18$     d. Sum, $-20$; Product, $+91$

3. a. Sum, $+4$; Product, $-5$       b. Sum, $+10$; Product, $-24$
   c. Sum, $+1$; Product, $-42$      d. Sum, $+11$; Product, $-60$

4. a. Sum, $-3$; Product, $-18$      b. Sum, $-8$; Product, $-20$
   c. Sum, $-1$; Product, $-12$      d. Sum, $-29$; Product, $+54$

5. a. Sum, $0$; Product, $-16$       b. Sum, $0$; Product, $-81$
   c. Sum, $0$; Product, $-100$      d. Sum, $0$; Product, $-49$

Factor each of the following trinomials:

6. a. $x^2 + 4x + 3$      b. $b^2 + 7b + 10$      c. $y^2 + 16y + 28$
   d. $m^2 + 12m + 27$    e. $s^2 + 18s + 45$     f. $c^2 + 16c + 39$

7. a. $x^2 - 5x + 6$      b. $x^2 - 9x + 8$       c. $a^2 - 8a + 12$
   d. $d^2 - 30d + 56$    e. $r^2 - 18r + 72$     f. $b^2 - 52b + 100$

8. a. $b^2 + 2b - 24$     b. $x^2 + 5x - 6$       c. $n^2 + 9n - 10$
   d. $y^2 + 13y - 48$    e. $x^2 + 11x - 80$     f. $b^2 + 7b - 98$

9. a. $x^2 - 2x - 8$      b. $n^2 - 5n - 24$      c. $b^2 - 19b - 20$
   d. $s^2 - 9s - 90$     e. $x^2 - 15x - 54$     f. $m^2 - 11m - 42$

10. a. $a^2 + a - 6$      b. $x^2 - x - 2$        c. $y^2 + y - 56$
    d. $x^2 - x - 30$     e. $b^2 - b - 42$       f. $r^2 - r - 72$

# Factoring–General Method

**I. Aim**   To factor a trinomial using the general method. This method excludes common factoring.

**II. Procedure**

**1.** Find two terms whose product is a term of the trinomial (usually the first term). These terms are the first terms of the two binomial factors.

**2.** Find two terms whose product is another term of the trinomial (usually the third term). These terms are the second terms of the two binomial factors.

**3.** From all the possible combinations, select the pair of binomials whose algebraic sum of the cross products of terms found in steps 1 and 2 matches the remaining term of the given trinomial (usually the second term).

**4.** Use the above procedures when all three terms of the given trinomial contain variables.

**5.** The given terms may have more than one variable or higher powers of the variables.

**6.** Perfect square trinomials may be factored by this method.

**7.** The difference of two squares may be factored by this general method. Write it as a trinomial with zero (0) as the numerical coefficient of the middle term. Thus $4x^2 - 9$ is considered as $4x^2 - 0x - 9$.

**III. Sample Solutions**

**1.** Factor $8x^2 + 2x - 3$

The factors of $8x^2$ are $8x \cdot x$ or $2x \cdot 4x$.
The factors of $-3$ are $(+1)(-3)$ or $(-1)(+3)$.

The possible combinations are:

| | |
|---|---|
| $(8x + 1)(x - 3)$ | $(2x + 1)(4x - 3)$ |
| $(8x - 3)(x + 1)$ | $(2x - 3)(4x + 1)$ |
| $(8x - 1)(x + 3)$ | $(2x - 1)(4x + 3)$ |
| $(8x + 3)(x - 1)$ | $(2x + 3)(4x - 1)$ |

**382**

Checking these combinations of binomials, you will find that only $(2x-1)(4x+3)$ has $+2x$ as the middle term. Sometimes all combinations fail, in which case the trinomial cannot be factored.

*Answer:* $(2x-1)(4x+3)$ or $(4x+3)(2x-1)$

Factor each of the following:

**2.** $20x^2+21x+4$   *Answer:* $(5x+4)(4x+1)$

**3.** $12a^2-52a+35$   *Answer:* $(2a-7)(6a-5)$

**4.** $32b^2+28b-15$   *Answer:* $(4b+5)(8b-3)$

**5.** $6y^2-7y-24$   *Answer:* $(3y-8)(2y+3)$

**6.** $30x^2-xy-42y^2$   *Answer:* $(6x+7y)(5x-6y)$

## DIAGNOSTIC TEST

Factor each of the following:

**1.** $8x^2+26x+15$    **2.** $15c^2-41c+28$    **3.** $12a^2+17a-5$
**4.** $24b^2-50b-9$    **5.** $6-23a+20a^2$    **6.** $x^2+8xy+15y^2$
**7.** $16a^2-16a-5$    **8.** $6x^2+5xy-21y^2$    **9.** $24b^2d^2-2bcd-15c^2$
**10.** $16d^4-42d^2+5$    **11.** $9n^2-48n+64$    **12.** $144c^2-25$

## *RELATED PRACTICE EXAMPLES*

Factor each of the following trinomials:

**1. a.** $6b^2+23b+7$        **b.** $24y^2+47y+20$      **c.** $4d^2+21d+27$
**2. a.** $15x^2-26x+8$       **b.** $8n^2-26n+21$       **c.** $12s^2-44s+35$
**3. a.** $10c^2+3c-4$        **b.** $84d^2+d-15$        **c.** $5x^2+33x-56$
**4. a.** $2t^2-3t-14$        **b.** $16x^2-6x-27$       **c.** $56a^2-33a-108$
**5. a.** $28+71b+18b^2$      **b.** $2-11t+12t^2$       **c.** $20+7m-6m^2$
**6. a.** $x^2+6xy+5y^2$      **b.** $a^2+12ab+20b^2$    **c.** $a^2-6ax+8x^2$
**7. a.** $4x^2+16xy+15y^2$   **b.** $25c^2+10cd-8d^2$   **c.** $49n^2-21nx-10x^2$
**8. a.** $40b^2+31bc+6c^2$   **b.** $12x^2-19xy-21y^2$  **c.** $70m^2-83mn+18n^2$
**9. a.** $6a^2b^2-abc-40c^2$   **b.** $35m^2+43mrs+12r^2s^2$
   **c.** $36c^2-37cdx+7d^2x^2$
**10. a.** $18x^6-35x^3+12$     **b.** $21c^4+19c^2d-12d^2$
   **c.** $45b^4c^2+71b^2cx+28x^2$
**11. a.** $y^2+16y+64$       **b.** $4a^2-20a+25$       **c.** $144r^2+128rs+49s^2$
**12. a.** $n^2-81$           **b.** $121b^2-36$         **c.** $49x^2-225y^2$

# Complete Factoring

## I. Aim     To factor a polynomial completely.

## II. Procedure

1. Factor by some method, replacing the given polynomial by its equivalent factors. Try the common factor method first.

2. If possible, factor the resulting factors.

3. Some expressions require extended factoring. Repeat step 2 until the resulting factors cannot be factored further.

4. Sometimes the given polynomial has a common binomial factor. See sample solution 5.

5. Sometimes by grouping the terms of a given polynomial you may find it possible to factor the polynomial as the difference of two squares. See sample solution 4.

## III. Sample Solutions

Factor each of the following:

**1.**  $cx^2 - c$
$= c(x^2 - 1)$
$= c(x + 1)(x - 1)$

*Answer:* $c(x + 1)(x - 1)$

**2.**  $ar^5 - a^5r^5$
$= ar^5(1 - a^4)$
$= ar^5(1 + a^2)(1 - a^2)$
$= ar^5(1 + a^2)(1 + a)(1 - a)$

*Answer:* $ar^5(1 + a^2)(1 + a)(1 - a)$

**3.**  $2x^2y - 4xy - 30y$
$= 2y(x^2 - 2x - 15)$
$= 2y(x - 5)(x + 3)$

*Answer:* $2y(x - 5)(x + 3)$

**4.**  $x^2(x^2 - 9) - 4(x^2 - 9)$
$= (x^2 - 9)(x^2 - 4)$
$= (x + 3)(x - 3)(x + 2)(x - 2)$

*Answer:* $(x + 3)(x - 3)(x + 2)(x - 2)$

**5.**  $a^2 + 2ab + b^2 - c^2$
$= (a + b)^2 - c^2$
$= (a + b + c)(a + b - c)$

*Answer:* $(a + b + c)(a + b - c)$

## DIAGNOSTIC TEST

Factor each of the following polynomials completely:

1. $ax^2 - a$       2. $5x^2 - 20y^2$

3. $ab^4 - ac^4$       4. $3x^2 + 12x + 12$

5. $2b^3 - 8b^2 - 10b$       6. $a^2(a + 2) - 9(a + 2)$

7. $b^2 - 2bc + c^2 - d^2$

## *RELATED PRACTICE EXAMPLES*

Factor each of the following polynomials completely:

1. **a.** $bx^2 - b$    **b.** $9y^3 - 9y$    **c.** $s^3 - s$
   **d.** $ar - a^3r^3$    **e.** $8t^2 - 98$    **f.** $36x^2 - 100$

2. **a.** $11x^2 - 11y^2$    **b.** $3a^2 - 75d^2$    **c.** $25m^2 - 100n^2$
   **d.** $2a^2 - 72b^2$    **e.** $\pi R^2 - \pi r^2$    **f.** $20c^3 - 45ca^3$

3. **a.** $ab^4 - ax^4$    **b.** $6m^5 - 6m$    **c.** $2b^4 - 32x^4$
   **d.** $5x^8 - 5y^8$    **e.** $a^5r^5 - ar^5$    **f.** $64x^4 - 625y^4$

4. **a.** $2x^2 + 12x + 18$      **b.** $a^3 - 8a^2 + 16a$
   **c.** $9a^4 - 18ab + 9b^2$      **d.** $2ab^2 + 20ab + 50a$
   **e.** $4bx^2 - 8bxy + 4by^2$      **f.** $3x^4 + 42x^3 + 147x^2$
   **g.** $27x^3 - 72x^2 + 48x$      **h.** $20a^4b - 60a^3b^2 + 45a^2b^3$

5. **a.** $3b^2 + 21b + 36$      **b.** $y^3 - y^2 - 2y$
   **c.** $5c^2 - 25c + 30$      **d.** $3cd^2 - 3cd - 126c$
   **e.** $4a^2b + 12ab - 72b$      **f.** $30x^3y + 55x^2y^2 - 50xy^3$
   **g.** $9n^4 - 27n^3 - 90n^2$      **h.** $36r^6s^2 + 78r^5s^3 - 30r^4s^4$

6. **a.** $x(x^2 - 4) + 2(x^2 - 4)$      **b.** $x^2(x - 3) - 4(x - 3)$
   **c.** $a^3(a^2 - c^2) + b^2(a^2 - c^2)$      **d.** $x^2(x^2 - y^2) - y^2(x^2 - y^2)$
   **e.** $b^2(b^2 - 9) - 16(b^2 - 9)$      **f.** $n^2(n^4 - 16) - 25(n^4 - 16)$
   **g.** $25(1 - c^2) - c^6(1 - c^2)$      **h.** $m^4x^4(m^2 - n^2) - y^2(m^2 - n^2)$

7. **a.** $x^2 + 2xy + y^2 - 4$      **b.** $a^2 - 6a + 9 - b^2$
   **c.** $4x^2 + 12xy + 9y^2 - 25z^2$      **d.** $16r^2 - 64rs + 64s^2 - t^2$
   **e.** $a^2 - b^2 - 2bc - c^2$      **f.** $9 - m^2 - 2mn - n^2$
   **g.** $36a^2 - 9b^2 + 24bc - 16c^2$      **h.** $16c^2 - r^2 + 2rs - s^2$

# Applications–Special Products and Factoring

**I. Aim**     To compute by using special products and to rewrite the formula in factored form.

## II. Procedure

1. Use special products to multiply two numbers when they may be expressed as the sum and difference of two numbers, one of which is usually some multiple of 10.
2. Use special products to multiply two numbers when they may be expressed as either the sum of two numbers or the difference of two numbers, one of which is some multiple of 10.
3. Sometimes common monomial factoring may be used in computation.
4. The right member of some formulas may be arranged in factored form.

## III. Sample Solutions

1. Using the method of the difference of two squares, find the product of: $42 \times 38$

   Since $42 = 40 + 2$ and $38 = 40 - 2$,
   then $42 \times 38 = (40 + 2)(40 - 2) = 40^2 - 2^2 = 1{,}600 - 4 = 1{,}596$

   *Answer: 1,596*

2. Using the method of squaring a binomial, find the product of: $(23)^2$
   Since $23 = 20 + 3$,
   then $23^2 = (20 + 3)^2 = 20^2 + 2(20)(3) + 3^2 = 400 + 120 + 9 = 529$
   Or since $23 = 30 - 7$,
   then $23^2 = (30 - 7)^2 = 30^2 - 2(30)(7) + 7^2 = 900 - 420 + 49 = 529$

   *Answer: 529*

3. Compute: $18 \times 26 + 12 \times 26$
   $18 \times 26 + 12 \times 26 = 26(18 + 12) = 26 \times 30 = 780$    *Answer: 780*

4. Rewrite the formula $b = \sqrt{c^2 - a^2}$ with the right member in factored form.
   $b = \sqrt{c^2 - a^2}$
   $b = \sqrt{(c + a)(c - a)}$    *Answer: $b = \sqrt{(c + a)(c - a)}$*

## DIAGNOSTIC TEST

1. Using the method of the difference of two squares, find the product of $44 \times 56$.
2. Using the method of squaring a binomial, find the product of $(87)^2$.
3. Compute $37 \times 89 + 13 \times 89$
4. Rewrite the formula $A = \pi R^2 - \pi r^2$ with the right member arranged in factored form.
5. Using the factored form of the formula $A = p + prt$, find the value of $A$ when $p = 60$, $r = .03$, and $t = 8$.

## RELATED PRACTICE EXAMPLES

1. Find the product of each of the following, using the method of the difference of two squares:

   a. $21 \times 19$     b. $52 \times 48$     c. $37 \times 43$     d. $18 \times 22$
   e. $63 \times 57$     f. $84 \times 76$     g. $92 \times 88$     h. $49 \times 51$
   i. $26 \times 34$     j. $73 \times 67$     k. $96 \times 104$     l. $510 \times 490$

2. Find the product of each of the following, using the method of squaring a binomial:

   a. $(34)^2$    b. $(42)^2$    c. $(21)^2$    d. $(63)^2$    e. $(2\frac{1}{2})^2$    f. $(1.5)^2$
   g. $(7\frac{1}{2})^2$    h. $(3.5)^2$    i. $(96)^2$    j. $(57)^2$    k. $(105)^2$    l. $(410)^2$

3. Compute:

   a. $7 \times 19 + 3 \times 19$     b. $24 \times 35 + 16 \times 35$     c. $18 \times 14 + 32 \times 14$
   d. $71 \times 27 + 29 \times 27$     e. $42 \times 58 + 28 \times 58$     f. $27 \times 45 - 7 \times 45$
   g. $89 \times 33 - 19 \times 33$     h. $13 \times \frac{1}{2} + 17 \times \frac{1}{2}$     i. $56 \times 9 - 16 \times 9$
   j. $35 \times 51 + 55 \times 51$

4. Rewrite each of the following formulas with the right member arranged in factored form:

   a. $p = 21 + 2w$     b. $A = \pi r^2 + \pi r l$     c. $A = S^2 - s^2$
   d. $A = 2\pi rh + 2\pi r^3$     e. $A = \pi R^2 - \pi r^2$     f. $E = Ir + IR$
   g. $A = p + prt$     h. $a = \sqrt{c^2 - b^2}$     i. $A = \frac{1}{4}\pi D^2 - \frac{1}{4}\pi d^2$

5. Using the factored form of the formula, find the value of:

   a. $S$ when $n = 7$, using the formula $S = 180n - 360$.
   b. $p$ when $l = 29$ and $w = 15$, using the formula $p = 2l + 2w$.
   c. $E$ when $I = 4$, $r = 67$, and $R = 23$, using the formula $E = Ir + IR$.
   d. $A$ when $S = 18$ and $s = 12$, using the formula $A = S^2 - s^2$.
   e. $A$ when $\pi = 3.14$, $R = 24$, and $r = 8$, using the formula $A = \pi R^2 - \pi r^2$.

Find the product of each of the following at sight:

1. **a.** $(6b + 7)(4b + 3)$ **b.** $(9x - 5)(8x + 1)$
   **c.** $(2c - 9d)(5c - 4d)$ **d.** $(8ab + 11c^2)(6ab - 3c^2)$
2. **a.** $(-\frac{7}{8})^2$ **b.** $(r^2s^3t^5)^2$ **c.** $(-9x^4y)^2$ **d.** $(.3a^6t^2)^2$
3. **a.** $(b - 12)(b + 12)$ **b.** $(10 + 9m)(10 - 9m)$
   **c.** $(4a^2b + 15xy^2)(4a^2b - 15xy^2)$ **d.** $(2m - n)(2m + n)(4m^2 + n^2)$
4. **a.** $(r - 9)(r - 9)$ **b.** $(7x + 10)^2$
   **c.** $(4b - 11c)^2$ **d.** $(6m^2 + 5n^3)(6m^2 + 5n^3)$
5. **a.** $(c - 3)(c + 10)$ **b.** $(d + 4)(d - 5)$
   **c.** $(7m + 6n)(7m + 8n)$ **d.** $(12 - 3a)(5 - 3a)$

6. Find the greatest common factor of $24a^5c^2$, $18a^4c^6$, and $36a^3c^8$.

7. Factor each of the following polynomials:
   **a.** $21b^3 - 14b^7$ **b.** $16r^6s^4 - 12r^5s^8 + 4r^4s^3$
   **c.** $-8a^4 - 5a^3b - 4a^2b^2$ **d.** $x(x + 6) - 3(x + 6)$

8. Find the square roots of each of the following:
   **a.** $100$ **b.** $81b^{12}c^6d^2$ **c.** $.04a^{10}x^4$ **d.** $\frac{16}{25}r^8s^2$

9. Factor each of the following binomials:
   **a.** $49c^2 - y^2$ **b.** $9 - 121b^4c^6$ **c.** $m^4 - x^4$

10. Find the term required in each of the following to make a perfect square trinomial:
    **a.** $25a^2 + 80a + ?$ **b.** $144c^2 + ? + 49d^2$

11. Factor each of the following trinomials:
    **a.** $121b^2 + 88bc + 16c^2$ **b.** $81x^4y^6 - 90x^2y^3 + 25$

12. Factor each of the following:
    **a.** $w^2 - 10w - 24$ **b.** $t^2 + t - 72$
    **c.** $64x^2 + 96xy + 27y^2$ **d.** $21r^2 - 65rs + 50s^2$

13. Factor each of the following polynomials completely:
    **a.** $50c^4 - 200d^4$ **b.** $7a^3 - 21a^2 - 28a$
    **c.** $m^2 - 2mn + n^2 - y^2$ **d.** $x^2(x^2 - 25) - 16(x^2 - 25)$

14. Using special products, find the product of:
    **a.** $38 \times 42$ **b.** $(96)^2$

15. Using the factored form of the formula $A = S^2 - s^2$, find the value of $A$ when $S = 75$ and $s = 25$.

## Keyed Achievement Test

The colored numeral indicates the exercise where help may be found.
1. Solve and check: 6-8
   a. $6n - 7 = -31$
   b. $4c + 8 = 9c - 22$
   c. $10x + 5(30 - x) = 235$
   d. $(x + 6)(x - 3) = (x - 5)(x + 4)$
2. Draw on a number line the graph of $3x + 7 = 1$. 6-11
3. Find the solutions of: 8-8
   a. $3y - 4 < 2$ when the variable replacements are all the whole numbers.
   b. $6x + 5 > 37 - 2x$ when the variable replacements are all the even integers.
4. Draw on a number line the graph of $11n + 20 < 6n$ when the variable replacements are all the real numbers. 8-9

Find the product of each of the following at sight:
5. a. $(9m - 8)(12m + 3)$     b. $(10c - 8d)(5c - 9d)$ 9-1
6. a. $(8a - 5)(8a + 5)$     b. $(m - 2n)(m + 2n)(m^2 + 4n^2)$ 9-3
7. a. $(4r - 9s)(4r - 9s)$     b. $(3ab - c^2)^2$ 9-4
8. a. $(t + 11)(t - 8)$     b. $(3w - 7)(3w - 10)$ 9-5

Factor each of the following completely:
9. a. $36b^2x^2 - 81b^5x^7 + 63b^4x^{10}$     b. $n(n - 2) + 7(n - 2)$ 9-6
10. a. $64c^2 - 81$     b. $x^4 - 16$ 9-8
11. a. $9w^2 + 30w + 25$     b. $121x^4y^2 - 154x^2yz + 49z^2$ 9-9
12. a. $m^2 - m - 56$     b. $r^2 + 7r - 18$ 9-10
13. a. $36a^2 + 47a + 15$     b. $42x^2 - 47x - 9$ 9-11
14. a. $4 - 4a^2$     b. $9n^2 - 63n + 54$ 9-12

## MAINTENANCE PRACTICE IN ARITHMETIC

1. Add:
948,329
77,594
497,868

2. Subtract:
27,600,050
9,807,195

3. Multiply:
486,093
70,080

4. Divide:
$576\overline{)3,863,808}$

5. Add:
$7\frac{5}{12} + 6\frac{3}{8} + 2\frac{7}{10}$

6. Subtract:
$15\frac{1}{4} - 14\frac{5}{6}$

7. Multiply:
$12\frac{2}{3} \times 1\frac{4}{5}$

8. Divide:
$6 \div 2\frac{1}{4}$

9. Add:
$4.83 + .556 + 70.8$

10. Subtract:
$1.803 - .9$

11. Multiply:
$24.8 \times .006$

12. Divide:
$6.1\overline{).305}$

13. Find 30% of $12.60.

14. 25 is what percent of 125?

15. 36 is 12% of what number?

# INTRODUCTION

A number that is named by the indicated quotient of any integer divided by a non-zero integer is called a *rational number*.

An algebraic expression that is the indicated quotient of one polynomial divided by a polynomial having a non-zero value is called a *rational algebraic expression* or simply an *algebraic fraction*. Since we do not divide by zero, the denominator of the algebraic fraction cannot have a zero value.

Expressions such as $\dfrac{x}{5}$, $\dfrac{12}{y}$, $\dfrac{r}{s}$, $\dfrac{x+7}{x-3}$, $\dfrac{b}{b+8}$, and $\dfrac{x^2 + 6x + 5}{x^2 - 4}$ are algebraic fractions provided their denominators do not have a zero value.

Since algebraic fractions are sometimes found in formulas and equations, it is necessary to learn to operate with them. We shall find the principles used in the study of arithmetic fractions are generalized in the work with algebraic fractions.

# Operations with Algebraic Fractions

# Rational Algebraic Expressions–Algebraic Fractions

**I. Aim**    To learn to write algebraic fractions and to determine the value of variables which make the denominator in these fractions equal to zero.

## II. Procedure

**1.** Use an algebraic fraction to represent a part of a whole.

**2.** Use algebraic fractions to indicate a ratio.

**3.** Use algebraic fractions to indicate division.

**4.** Observe that for each value of the variables a rational algebraic expression (algebraic fraction) is a real number provided the denominator is not zero. An algebraic fraction is meaningless or undefined when its denominator is zero.

**5.** To determine the number that makes a denominator equal zero when it replaces the variable, let the given denominator equal zero and solve the resulting equation. See sample solutions 4a and 4b.

**6.** Note that the values that make the denominator of an algebraic fraction zero must be excluded from the replacements for the variable.

For example, in $\dfrac{x}{x-4}$, the variable $x$ cannot equal 4 since the denominator $x - 4$ would then equal $4 - 4$ or 0.

This exclusion is indicated as $x \neq 4$.

**7.** Consider the value of an algebraic fraction to be 1 when its numerator and denominator are the same provided the replacements for the variables that make the denominator zero are excluded.

## III. Sample Solutions

**1.** Write 3 parts of $x$ equal parts as a fraction.    *Answer:* $\dfrac{3}{x}$

**2.** Write the ratio $x$ to $y$.    *Answer:* $x:y$ or $\dfrac{x}{y}$

**3.** Write $a \div b$ as a fraction.    *Answer:* $\dfrac{a}{b}$

**4.** For which values of the variable is the denominator of each of the following algebraic fractions equal to zero?

   **a.** $\dfrac{x}{x-5}$    First let $x - 5 = 0$, which has the solution $x = 5$. Thus, $x - 5 = 5 - 5 = 0$.

   *Answer:* when $x = 5$, the denominator $x - 5$ is equal to 0.

**b.** $\dfrac{n + 6}{n^2 + 3n - 28}$  First factor the denominator to get: $\dfrac{n + 6}{(n - 4)(n + 7)}$.

Let the denominator $(n - 4)(n + 7) = 0$.
If the product of two factors is zero, then one of its factors is zero or both factors are zero. Thus,

(1) When $n - 4 = 0$, then $n = 4$.
and $(n - 4)(n + 7)$
becomes $(4 - 4)(4 + 7) = 0(11) = 0$

(2) When $n + 7 = 0$, then $n = 7$.
and $(n - 4)(n + 7)$
becomes $(-7 - 4)(-7 + 7) = (-11)(0) = 0$

*Answers:* When $n = 4$ or $n = -7$, the denominator is equal to 0.

**5.** Which of the following name the number 1, assuming that the replacements of the variable that make the denominator zero are excluded:

**a.** $\dfrac{3n}{3 + n}$?   **b.** $\dfrac{3 + n}{3 + n}$?   *Answer:* $\dfrac{3 + n}{3 + n}$

## DIAGNOSTIC TEST

Write each of the following as a rational algebraic expression:
**1.** $w$ parts of $w + 9$ equal parts
**2.** The ratio of $x - y$ to $x + y$     **3.** $(c + d) \div m$

**4.** Find the number that is represented by $\dfrac{n}{n - 12}$ when $n = 18$.

**5.** For which values of the variable is the denominator of

**a.** $\dfrac{20}{n(n - 5)}$ zero?     **b.** $\dfrac{x - 5}{x^2 - 3x - 18}$ zero?

**6.** Which replacements of the variable must be excluded in:

**a.** $\dfrac{b - 9}{9b}$?     **b.** $\dfrac{13}{2c(c + 8)}$?     **c.** $\dfrac{n + 1}{n^2 - 10n + 21}$?

**7.** What is the value of $\dfrac{x - 11}{x - 11}$ when $x = 20$?

**8.** Which of the following name the number 1, assuming that the replacements of the variable that make the denominator zero are excluded?

**a.** $\dfrac{6 - n}{-6n}$     **b.** $\dfrac{19c^3y}{19c^3y}$     **c.** $\dfrac{(a + 3)^2}{a^2 + 6a + 9}$

## RELATED PRACTICE EXAMPLES

1. Write each of the following as a rational algebraic expression:
   **a.** $n$ parts of 9 equal parts
   **b.** 8 parts of $c$ equal parts
   **c.** $x$ parts of $y$ equal parts
   **d.** $b$ parts of $b + 1$ parts
   **e.** $x - 4$ parts of $x + 7$ parts

2. Write each of the following as a rational algebraic expression:
   **a.** The ratio of 7 to $d$
   **b.** The ratio of $n$ to 12
   **c.** The ratio of $g$ to $h$
   **d.** The ratio of $t$ to $t^2 + 7$
   **e.** The ratio of $w + y$ to $w - y$

3. Write each of the following as a rational algebraic expression:
   **a.** $y \div 4$       **b.** $10 \div r$       **c.** $b \div d$
   **d.** $(x - y) \div x$       **e.** $m \div (m + n)$       **f.** $(c^2 + d^2) \div (c^2 - d^2)$

4. Find the number that each of the following rational algebraic expressions represents:

   **a.** $\dfrac{x}{5}$ when $x = 10$      **b.** $\dfrac{6}{y}$ when $y = 18$

   **c.** $\dfrac{s}{t}$ when $s = 6$ and $t = 2$      **d.** $\dfrac{n^2}{7}$ when $n = 4$

   **e.** $\dfrac{a + 3}{8}$ when $a = -11$      **f.** $\dfrac{x}{x + 2}$ when $x = 16$

   **g.** $\dfrac{n + 9}{n}$ when $n = 3$      **h.** $\dfrac{b - 2}{b + 4}$ when $b = -10$

   **i.** $\dfrac{x + y}{x - y}$ when $x = 2$ and $y = 6$      **j.** $\dfrac{c^2 + 5c + 4}{c^2 - 16}$ when $c = 5$

5. For which values of the variable is the denominator of each of the following rational algebraic expressions equal to zero?

   **a.** $\dfrac{6}{x}$      **b.** $\dfrac{c}{n^2}$      **c.** $\dfrac{w + 6}{5w}$

   **d.** $\dfrac{7}{y - 2}$      **e.** $\dfrac{m}{m + 8}$      **f.** $\dfrac{n - 9}{n + 1}$

   **g.** $\dfrac{16}{x(x - 10)}$      **h.** $\dfrac{4b}{3b^2 - 9b}$      **i.** $\dfrac{7s}{(3s + 2)(2s - 7)}$

   **j.** $\dfrac{10}{y^2 - 25}$      **k.** $\dfrac{b - 4}{b^2 + 5b - 24}$      **l.** $\dfrac{7c - 5}{c^2 - 21c - 72}$

6. Which replacements of the variable must be excluded in each of the following rational expressions?

a. $\dfrac{12}{r}$

b. $\dfrac{x + 6}{6x}$

c. $\dfrac{m + 1}{2m^2}$

d. $\dfrac{r}{r - 3}$

e. $\dfrac{c - 11}{c + 12}$

f. $\dfrac{10}{y(y - 4)}$

g. $\dfrac{25}{4n(n + 9)}$

h. $\dfrac{w + 7}{(2w - 1)(6w + 11)}$

i. $\dfrac{9r}{r^2 - 100}$

j. $\dfrac{t + 6}{t^2 + 12t + 35}$

k. $\dfrac{21x}{x^2 - 15x - 54}$

l. $\dfrac{y - 9}{y^2 - 22y + 96}$

7. What is the value of the fraction:

a. $\dfrac{5n}{5n}$ when $n$ is replaced by 1? by 4? by 10?

b. $\dfrac{b^3}{b^3}$ when $b$ is replaced by 2? by 3? by 5?

c. $\dfrac{-7rs^2}{-7rs^2}$ when $r$ is replaced by 6 and $s$ by 4?

d. $\dfrac{x + 6}{x + 6}$ when $x$ is replaced by 4? by 10? by 36?

e. $\dfrac{(c - 5)(c + 7)}{(c - 5)(c + 7)}$ when $c$ is replaced by 10? by 6? by $-2$?

8. Which of the following name the number 1, assuming that the replacements of the variable that make the denominator zero are excluded?

a. $\dfrac{6x^2y}{6xy^2}$

b. $\dfrac{5n}{5 + n}$

c. $\dfrac{-21b^3x^4}{-21b^3x^4}$

d. $\dfrac{2 + t}{t + 2}$

e. $\dfrac{x^2 - 36}{x^2 - 36}$

f. $\dfrac{4(r + 5)}{4r + 20}$

g. $\dfrac{(x - 3)(x + 9)}{(x - 9)(x + 3)}$

h. $\dfrac{(b - 10)^2}{b^2 - 10^2}$

## MAINTENANCE PRACTICE IN ARITHMETIC

1. Reduce to lowest terms:  a. $\frac{10}{25}$  b. $\frac{9}{54}$  c. $\frac{52}{91}$  d. $\frac{60}{200}$

2. Raise to higher terms:  a. $\frac{1}{8} = \dfrac{?}{32}$  b. $\frac{3}{5} = \dfrac{?}{100}$  c. $\frac{11}{12} = \dfrac{?}{84}$

3. Multiply:  a. $\frac{5}{6} \times \frac{4}{5}$  b. $\frac{3}{4} \times \frac{7}{8}$  c. $\frac{2}{3} \times 72$  d. $4 \times \frac{9}{16}$

4. Divide:  a. $\frac{7}{8} \div \frac{2}{3}$  b. $\frac{5}{12} \div \frac{3}{16}$  c. $\frac{2}{5} \div 8$  d. $18 \div \frac{27}{32}$

5. Add:  a. $\frac{2}{5} + \frac{1}{5}$  b. $\frac{1}{2} + \frac{3}{8}$  c. $\frac{3}{4} + \frac{2}{5}$  d. $\frac{5}{6} + \frac{7}{8}$

6. Subtract:  a. $\frac{13}{16} - \frac{7}{16}$  b. $\frac{11}{12} - \frac{1}{6}$  c. $\frac{1}{4} - \frac{1}{10}$  d. $\frac{4}{5} - \frac{2}{3}$

# Simplifying Algebraic Fractions

**I. Aim**    To change algebraic fractions to lowest terms.

**II. Procedure**

**1. a.** Find the greatest common factor of both numerator and denominator. If the numerator or denominator or both are polynomials, first factor them completely when possible.

   **b.** Divide both the numerator and denominator by this greatest common factor.

**2. a.** Or factor the numerator and denominator, writing their greatest common factor as one of the factors.

   **b.** Then write this pair of greatest common factors as a separate fraction using one (1 is sometimes called the multiplicative identity) as its quotient. See sample solutions 1.b., 2.b., 3.b., 5., 6., and 7.

     When any number, except zero, is divided by itself, the quotient is 1. Replacements for the variables that make the denominator zero are excluded.

**3.** Follow the rule of signs for division when either the numerator or the denominator or both have negative coefficients.

**4.** Sometimes the common factor is a binomial.

**5.** Do not  cancel (divide) individual terms of polynomials. See sample solutions 6 and 7.

**6.** Observe that:

   **a.** A rational algebraic expression is *in simplest form* when its numerator and denominator have no common factor other than 1 or $-1$. When an algebraic fraction is expressed in lowest terms, it is in simplest form.

     An expression such as $\dfrac{xy}{-xy}$ is not in simplest form since the numerator and denominator have the common factor $xy$.

     An expression such as $\dfrac{x+y}{x-y}$ is in simplest form since the numerator and denominator have no common factor other than 1.

   **b.** Rational algebraic expressions which have the same simplest form are *equivalent algebraic fractions*.

## III. Sample Solutions

1. Reduce $\frac{24}{36}$ to lowest terms:

   **a.** $\frac{24}{36} = \dfrac{24 \div 12}{36 \div 12} = \frac{2}{3}$   or   **b.** $\frac{24}{36} = \dfrac{12 \cdot 2}{12 \cdot 3} = \frac{12}{12} \cdot \frac{2}{3} = 1 \cdot \frac{2}{3} = \frac{2}{3}$

2. Write $\dfrac{16a^4b^5}{20a^7b^2}$ in simplest form.   *Answer:* $\frac{2}{3}$

   **a.** $\dfrac{16a^4b^5}{20a^7b^2} = \dfrac{16a^4b^5 \div 4a^4b^2}{20a^7b^2 \div 4a^4b^2} = \dfrac{4b^3}{5a^3}$   or

   **b.** $\dfrac{16a^4b^5}{20a^7b^2} = \dfrac{4a^4b^2 \cdot 4b^3}{4a^4b^2 \cdot 5a^3} = \dfrac{4a^4b^2}{4a^4b^2} \cdot \dfrac{4b^3}{5a^3} = 1 \cdot \dfrac{4b^3}{5a^3} = \dfrac{4b^3}{5a^3}$   *Answer:* $\dfrac{4b^3}{5a^3}$

3. Write $\dfrac{a^5}{a^8}$ in simplest form.

   **a.** $\dfrac{a^5}{a^8} = \dfrac{a^5 \div a^5}{a^8 \div a^5} = \dfrac{1}{a^3}$   or   **b.** $\dfrac{a^5}{a^8} = \dfrac{a^5 \cdot 1}{a^5 \cdot a^3} = \dfrac{a^5}{a^5} \cdot \dfrac{1}{a^3} = 1 \cdot \dfrac{1}{a^3} = \dfrac{1}{a^3}$

   *Answer:* $\dfrac{1}{a^3}$

4. Write $\dfrac{-9c^3}{27c}$ in simplest form.

   $\dfrac{-9c^3}{27c} = -\dfrac{c^2}{3}$   *Answer:* $-\dfrac{c^2}{3}$

5. Write $\dfrac{b(x-y)}{c(x-y)}$ in simplest form.

   $\dfrac{b(x-y)}{c(x-y)} = \dfrac{(x-y)}{(x-y)} \cdot \dfrac{b}{c} = 1 \cdot \dfrac{b}{c} = \dfrac{b}{c}$   *Answer:* $\dfrac{b}{c}$

6. Write $\dfrac{3x}{3(x+y)}$ in simplest form.

   $\dfrac{3x}{3(x+y)} = \dfrac{3}{3} \dfrac{x}{x+y} = 1 \cdot \dfrac{x}{x+y} = \dfrac{x}{x+y}$   *Answer:* $\dfrac{x}{x+y}$

   Note in $\dfrac{x}{x+y}$ that $x$ is not a common factor.

7. Write $\dfrac{x^2+6x+9}{x^2-9}$ in simplest form.

   $\dfrac{x^2+6+9}{x^2-9} = \dfrac{(x+3)(x+3)}{(x+3)(x-3)} = \dfrac{(x+3)(x+3)}{(x+3)(x-3)} = 1 \cdot \dfrac{(x+3)}{(x-3)} = \dfrac{x+3}{x-3}$

   *Answer:* $\dfrac{x+3}{x-3}$

## DIAGNOSTIC TEST

1. Which of the following fractions are in simplest form: $\frac{12}{51}$? $\frac{19}{34}$? $\frac{13}{65}$?

2. Reduce $\frac{48}{60}$ to lowest terms.

3. Which of the following algebraic fractions are in simplest form?

    **a.** $\dfrac{4abc}{5cxy}$               **b.** $\dfrac{m+n}{mn}$               **c.** $\dfrac{6a+9b}{8a+10b}$

Write each of the following algebraic fractions in simplest form:

4. $\dfrac{ax}{ay}$            5. $\dfrac{a^2b^3c^4}{ab^2c^6}$          6. $\dfrac{x^4y^2z}{x^2y^2z^2}$

7. $\dfrac{a^3x^2}{ay}$          8. $\dfrac{18a^4b}{30ab^2}$          9. $\dfrac{14c^2d^3}{cd^4}$

10. $\dfrac{c}{c^3}$           11. $\dfrac{6b}{18ab}$           12. $\dfrac{-9x^2y}{18xy^2}$

13. $\dfrac{5a^2(x+y)}{15a(x+y)}$      14. $\dfrac{12ab^2}{6b^4(a-b)}$      15. $\dfrac{3x^2y^3(a-3)(a+7)}{6xy^4(a+4)(a-3)}$

16. $\dfrac{b+3}{(b+3)^2}$      17. $\dfrac{ax-a^2x^2}{bx-abx^2}$      18. $\dfrac{2ab^2c}{2ab+4ac}$

19. $\dfrac{2x^2+2xy}{x^2-y^2}$      20. $\dfrac{a^2-16}{a^2-8a+16}$      21. $\dfrac{x^2+4x-5}{x^2+8x+15}$

22. $\dfrac{2n^2+4n-30}{3n^2+21n+30}$

23. Write $\dfrac{x^2-x-30}{x^2+13x+40}$ in simplest form. Determine which replacements of the variables must be excluded. Use any other replacements to check whether the given expression and the expression in simplest form name the same number.

### RELATED PRACTICE EXAMPLES

1. Which of the following fractions are in simplest form?

    **a.** $\frac{4}{9}$     **b.** $\frac{56}{63}$     **c.** $\frac{16}{80}$     **d.** $\frac{51}{75}$     **e.** $\frac{65}{91}$     **f.** $\frac{17}{68}$     **g.** $\frac{53}{97}$

2. Reduce each of the following fractions to lowest terms.

    **a.** $\frac{6}{20}$          **b.** $\frac{30}{48}$          **c.** $\frac{36}{54}$          **d.** $\frac{21}{56}$

    **e.** $\frac{28}{42}$          **f.** $\frac{54}{72}$          **g.** $\frac{96}{120}$          **h.** $\frac{96}{160}$

**i.** $\frac{63}{144}$  **j.** $\frac{80}{112}$  **k.** $\frac{225}{400}$  **l.** $\frac{105}{140}$

**m.** $\frac{65}{156}$  **n.** $\frac{104}{256}$

3. Which of the following algebraic fractions are in simplest form?

**a.** $\frac{3x}{5x^7}$   **b.** $\frac{8+c}{8c}$   **c.** $\frac{a+b}{b+a}$   **d.** $\frac{x+1}{x-1}$

**e.** $\frac{21}{48(n+5)}$   **f.** $\frac{t+6}{(t+4)(t+2)}$   **g.** $\frac{5(c+d)^2}{3(c-d)^3}$   **h.** $\frac{4x-8y}{10c-5d}$

Write each of the following algebraic fractions in simplest form:

4. **a.** $\frac{bc}{bd}$   **b.** $\frac{5ax}{8ax}$   **c.** $\frac{7cd}{4ac}$   **d.** $\frac{3a}{10a}$

5. **a.** $\frac{c^4d^3}{c^2d^6}$   **b.** $\frac{a^8b^3c^5}{a^3b^7c^2}$   **c.** $\frac{m^5n^2x^5}{m^2nx^6}$   **d.** $\frac{c^8xy^9}{c^{10}x^4y^5}$

6. **a.** $\frac{m^3}{m^3}$   **b.** $\frac{ab^4}{a^2b^4}$   **c.** $\frac{c^2d^3x^2}{c^2d^3x}$   **d.** $\frac{m^5xy^6}{m^3xy^8}$

7. **a.** $\frac{a^5b^2c}{a^4b^3}$   **b.** $\frac{amx^2}{m^3xy^2}$   **c.** $\frac{r^9s^2t}{m^2s^8x^2}$   **d.** $\frac{a^2b^2c^3d^8}{b^5cd^4e}$

8. **a.** $\frac{6a^2b^5}{12ab^2}$   **b.** $\frac{5x^4y^3}{15xy}$   **c.** $\frac{4a^3x^8}{10a^6x^2}$   **d.** $\frac{12c^8d^4}{3cd^9}$

9. **a.** $\frac{8c^4}{c^7}$   **b.** $\frac{b^3c^5}{3b^5c^2}$   **c.** $\frac{a^4x^3y^2}{5x^2y^4z}$   **d.** $\frac{4\pi r^2}{\pi r^3}$

10. **a.** $\frac{x}{x^2}$   **b.** $\frac{cd}{c^4d^5}$   **c.** $\frac{c^4}{3c^8}$   **d.** $\frac{m^4x^6}{m^4x^{10}}$

11. **a.** $\frac{5c}{10bc}$   **b.** $\frac{16a^2x}{48a^4x^5}$   **c.** $\frac{25b^3c^2}{50b^9c^5}$   **d.** $\frac{2x^4z^2}{12x^5z^{11}}$

12. **a.** $\frac{-8a^3}{-4a^5}$   **b.** $\frac{-2c^3d^7}{6c^2d^9}$   **c.** $\frac{-15c^3m^{12}}{-5c^4m^5}$   **d.** $\frac{18a^5xy^2}{-6a^2xy^6}$

13. **a.** $\frac{5(c+d)}{6(c+d)}$   **b.** $\frac{4r(s-2t)}{2r^4(s-2t)}$   **c.** $\frac{a(a+c)}{b(a+c)}$   **d.** $\frac{3x^2(x-y)}{9x(x-y)}$

14. **a.** $\frac{a(x-y)}{ax}$   **b.** $\frac{9m}{9(m+n)}$   **c.** $\frac{8a^6b^3(a+b)}{24ab^4}$   **d.** $\frac{10m^2x^5}{5x^2(m-x)}$

15. **a.** $\frac{(a+b)(a+b)}{(a+b)(a-b)}$   **b.** $\frac{(c+2)(c-3)}{(c-3)(c+8)}$

**c.** $\frac{2x(x+y)(x-y)}{6x(x+y)}$   **d.** $\frac{6a^2b(c-4)(c+2)}{8ab^2(c+2)(c+1)}$

16. a. $\dfrac{c+2}{(c+2)(c-2)}$    b. $\dfrac{a-b}{(a-b)^2}$    c. $\dfrac{8b^2(x-y)^4}{16b(x-y)^7}$    d. $\dfrac{10x^3y^2(x+y)^3}{15xy^3(x+y)^4}$

17. a. $\dfrac{3x+3y}{3x-3y}$    b. $\dfrac{bc-bd}{cx-dx}$    c. $\dfrac{4a+4b}{8a+8b}$    d. $\dfrac{abc+2ab}{acm+2am}$

    e. $\dfrac{2x+4y}{6x+12y}$    f. $\dfrac{cx-c^2x^2}{dx-cdx^2}$    g. $\dfrac{ax+bx}{cx+dx}$    h. $\dfrac{ac+ad+ax}{bc+bd+bx}$

18. a. $\dfrac{12x}{6x^2+9x^3}$    b. $\dfrac{27abc}{18ab-63ac}$    c. $\dfrac{6b^3c^2-4bc^3}{8bc^2}$    d. $\dfrac{15m^2n^2x}{6mnx-18m^2nx}$

19. a. $\dfrac{y^2-1}{2y+2}$    b. $\dfrac{3x+15}{x^2-25}$    c. $\dfrac{2a+10}{a^2+10a+25}$    d. $\dfrac{x^2-4x}{x^2-6x+8}$

20. a. $\dfrac{a^2-b^2}{(a-b)^2}$    b. $\dfrac{x^2-1}{x^2+2x+1}$    c. $\dfrac{x^2-14x+49}{x^2-49}$    d. $\dfrac{x^2+2xy+y^2}{x^2-y^2}$

21. a. $\dfrac{a^2+5a+6}{a^2+6a+9}$    b. $\dfrac{b^2-b-12}{b^2+2b-15}$    c. $\dfrac{x^2-12x+36}{x^2+4x-48}$    d. $\dfrac{a^2-2a+1}{a^3+3a-4}$

22. a. $\dfrac{x^3-xy^2}{xy(x-y)^2}$          b. $\dfrac{4c^2+16c+16}{6c^2+18c+12}$

   c. $\dfrac{3ab^3-27ab}{6ab^2+6ab-72a}$ $\dfrac{3ab(b^2-9)}{6a(b^2+b-12)}$    d. $\dfrac{8b^3+24b^2-32b}{4b^3-16b^2+12b}$

*(handwritten at left: $16$, $6\overline{)72}$)*

23. First determine the simplest form of each of the following rational algebraic expressions. Then check whether both the given expression and
the expression in simplest form name the same number when the variable is replaced by any integers you select except those which make the
value of the denominator zero. Also determine for each given rational
algebraic expression the replacements that must be excluded.

   a. $\dfrac{b^2}{b^7}$           b. $\dfrac{mx}{nx}$           c. $\dfrac{8(a+b)}{12(a+b)}$

   d. $\dfrac{2b(c+d)}{4b^2}$       e. $\dfrac{(x+3)(x-5)}{(x-2)(x-5)}$       f. $\dfrac{x-4}{6(x-4)^3}$

   g. $\dfrac{4a^2-8a}{16a}$        h. $\dfrac{4x-4y}{8x-8y}$        i. $\dfrac{b^2+3b-18}{b^2-10b+21}$

   j. $\dfrac{n^2-8n+16}{n^2+n-20}$      k. $\dfrac{a^2+12a+36}{a^2+5a-6}$      l. $\dfrac{b^2+b-6}{b^2-9}$

*(handwritten at bottom: $= \dfrac{3ab(b+3)(b-3)}{6a(b+4)(b-3)} =$)*

# Multiplication of Algebraic Fractions

## I. Aim    To multiply algebraic fractions.

## II. Procedure
**1. a.** Where possible, first divide any numerator and any denominator by their greatest common factor.
   **b.** Then multiply the resulting numerator to find the numerator of the answer.
   **c.** Also divide the resulting denominator to find the denominator of the answer.
**2.** Where possible, factor the numerator and denominator completely before you multiply.
**3.** Follow the same procedure with three or more fractions.
**4.** Every polynomial may be considered an algebraic fraction with the denominator 1. See sample solution 6.
**5.** Replacements for the variables that make the denominator zero are excluded.

## III. Sample Solutions
Multiply in each of the following:

**1.** $\frac{2}{3} \cdot \frac{4}{5} = \frac{2 \cdot 4}{3 \cdot 5} = \frac{8}{15}$   *Answer:* $\frac{8}{15}$

**2.** $\frac{a}{x} \cdot \frac{b}{c} = \frac{a \cdot b}{x \cdot c} = \frac{ab}{cx}$   *Answer:* $\frac{ab}{cx}$

**3.** $\frac{x^2}{y^3} \cdot \frac{y^4}{x^5} = \frac{x^2 y^4}{x^5 y^3} = \frac{x^2 y^3}{x^2 y^3} \cdot \frac{y}{x^3} = 1 \cdot \frac{y}{x^3} = \frac{y}{x^3}$   *Answer:* $\frac{y}{x^3}$

**4.** $\frac{9m^3}{5nx} \cdot \frac{10nx^2}{63m^5} = \frac{\overset{1}{\cancel{9m^3}}}{\underset{1}{\cancel{5nx}}} \cdot \frac{\overset{2x}{\cancel{10nx^2}}}{\underset{7m^2}{\cancel{63m^5}}} = \frac{2x}{7m^2}$   *Answer:* $\frac{2x}{7m^2}$

**5.** $\dfrac{(a + x)^2}{a^2 - x^2} \cdot \dfrac{a^2 - 2ax + x^2}{ab - bx} = \dfrac{(a + x)(a + x)}{(a + x)(a - x)} \cdot \dfrac{(a - x)(a - x)}{b(a - x)}$

$= \dfrac{(a + x)(a + x)(a - x)(a - x)}{b(a + x)(a - x)(a - x)}$

$= \dfrac{(a + x)(a - x)(a - x)}{(a + x)(a - x)(a - x)} \cdot \dfrac{(a + x)}{b}$

$= 1 \cdot \dfrac{a + x}{b} = \dfrac{a + x}{b}$   *Answer:* $\dfrac{a + x}{b}$

6. $2a^2b \cdot \dfrac{3a-4b}{4a} = \dfrac{2a^2b}{1} \cdot \dfrac{3a-4b}{4a} = \dfrac{ab(3a-4b)}{2} = \dfrac{3a^2b-4ab^2}{2}$

$\qquad\qquad\qquad$ *Answer:* $\dfrac{3a^2b-4ab^2}{2}$

## DIAGNOSTIC TEST

Multiply as indicated:

1. $\dfrac{9}{16} \cdot \dfrac{3}{4}$

2. $\dfrac{2a}{5x} \cdot \dfrac{4b}{3y}$

3. $\dfrac{c^4d^3}{a^6b} \cdot \dfrac{a^2b^7}{c^5d}$

4. $\dfrac{7b^{10}}{6x^8} \cdot \dfrac{9x^3}{14b^6}$

5. $\dfrac{4a^6y^4}{5b^3x^2} \cdot \dfrac{15b^4x^2}{16a^8y}$

6. $\dfrac{14x^2}{5y} \cdot \dfrac{y^3}{x^3} \cdot \dfrac{15x}{4y^2}$

7. $\dfrac{x+y}{x-y} \cdot \dfrac{x-y}{x+y}$

8. $\dfrac{c+d}{10cd} \cdot \dfrac{5c^3d}{(c+d)^2}$

9. $\dfrac{(a-2)(a-3)}{(a+2)(a+4)} \cdot \dfrac{(a-1)(a+2)}{(a+1)(a-3)}$

10. $\dfrac{10a(a+2)}{3b(b-3)} \cdot \dfrac{6b(b-3)(b+3)}{25a^2(a-2)(a+2)}$

11. $\dfrac{x^2-4}{x^2+4x+4} \cdot \dfrac{2x+4}{x^2+x-6}$

12. $\dfrac{5x^5y^2}{12} \cdot 24x^2y^3$    13. $9n^2 \cdot \dfrac{n+3}{6n}$

## *RELATED PRACTICE EXAMPLES*

Multiply as indicated;

1. a. $\dfrac{3}{8} \cdot \dfrac{3}{5}$     b. $\dfrac{5}{6} \cdot \dfrac{8}{15}$     c. $\dfrac{11}{12} \cdot \dfrac{1}{3}$     d. $\dfrac{7}{10} \cdot \dfrac{15}{16}$

2. a. $\dfrac{m}{x} \cdot \dfrac{n}{y}$     b. $\dfrac{r^2}{b} \cdot \dfrac{a}{s^3}$     c. $\dfrac{2d}{5w} \cdot \dfrac{3t}{7z}$     d. $\dfrac{4a^5b}{9x^2} \cdot \dfrac{5c^2d^4}{3y^2z}$

3. a. $\dfrac{c}{d} \cdot \dfrac{d}{c}$     b. $\dfrac{a}{b^2} \cdot \dfrac{b}{a^2}$     c. $\dfrac{m^2y^3}{ny^4} \cdot \dfrac{n^6y^2}{m^3x^2}$     d. $\dfrac{c^5x^7}{d^2y^4} \cdot \dfrac{d^6z}{c^8x^5}$

4. a. $\dfrac{ab}{3} \cdot \dfrac{9}{c}$     b. $\dfrac{14}{x} \cdot \dfrac{xy}{4}$     c. $\dfrac{5x}{4a^2} \cdot \dfrac{6a}{10x^3}$     d. $\dfrac{16x^2y^3}{9a^3} \cdot \dfrac{27a^4}{12y^2}$

5. a. $\dfrac{2c}{5d} \cdot \dfrac{10d}{6c}$     b. $\dfrac{4a}{3c} \cdot \dfrac{3cd}{4b}$     c. $\dfrac{14r^3}{15t^7} \cdot \dfrac{40t^4}{21r^2}$

   d. $\dfrac{3b^2c^5}{10x^3y} \cdot \dfrac{15x^2y^4}{16b^3c^2}$     e. $\dfrac{7b^6x^3}{12a^2y^5} \cdot \dfrac{24a^7z}{35c^4x^6}$     f. $\dfrac{4a^4y^9}{9b^8d^4} \cdot \dfrac{5b^5d^8}{7a^2y^{11}}$

**6. a.** $\dfrac{2x}{5a} \cdot \dfrac{3a^2}{4x} \cdot \dfrac{5}{6ax}$

**b.** $\dfrac{16a^2b^2}{3ac^4} \cdot \dfrac{25c^2}{32ab^4} \cdot \dfrac{9ab^3}{5c}$

**c.** $\dfrac{2b^2m^4}{9ad^2} \cdot \dfrac{3a^2c^3}{8bx^3} \cdot \dfrac{6dx^2}{7c^6m}$

**d.** $\dfrac{12r^3t^8}{25a^6m^2} \cdot \dfrac{10s^6x^5}{21m^3t^8} \cdot \dfrac{7a^8}{8r^4x^3}$

**7. a.** $\dfrac{a+b}{a-b} \cdot \dfrac{a-b}{a+b}$

**b.** $\dfrac{x+6}{x-3} \cdot \dfrac{x-3}{x+6}$

**c.** $\dfrac{m-1}{m+5} \cdot \dfrac{m-5}{m-1}$

**8. a.** $\dfrac{a+x}{3ax} \cdot \dfrac{6ax}{(a+x)^2}$

**b.** $\dfrac{(b-2)^5}{4b} \cdot \dfrac{12b^3}{(b-2)^3}$

**c.** $\dfrac{3xy}{x-2y} \cdot \dfrac{(x-2y)^4}{4y^2}$

**9. a.** $\dfrac{(x-1)(x+5)}{(x-3)(x+5)} \cdot \dfrac{(x-3)(x-2)}{(x+4)(x-3)}$

**b.** $\dfrac{(a+c)(b+c)}{(c+d)(a+b)} \cdot \dfrac{(c+d)(a+2b)}{(b-c)(a+c)}$

**10. a.** $\dfrac{3a(x+2)}{5x(a+5)} \cdot \dfrac{15x(a-5)}{4a(x+2)}$

**b.** $\dfrac{12xy(x+y)}{7x^2(x-y)} \cdot \dfrac{35y^3(x-y)}{48x^2y(x-y)(x+y)}$

**11. a.** $\dfrac{6x+6y}{x-y} \cdot \dfrac{5x-5y}{12}$

**b.** $\dfrac{4a^2+10}{a-3} \cdot \dfrac{a^2-9}{6a^2-15}$

**c.** $\dfrac{(a+b)^2}{a^2-b^2} \cdot \dfrac{ax-bx}{ay+by}$

**d.** $\dfrac{4x+8}{6x-24} \cdot \dfrac{9x-36}{2x+4}$

**e.** $\dfrac{a^2-8a+16}{a^2+3a-10} \cdot \dfrac{a^2+2a-8}{a^2-16}$

**f.** $\dfrac{b^2-b-12}{b^2-6b+8} \cdot \dfrac{b^2-4}{b^2+5b+6}$

**g.** $\dfrac{x^2+x-2}{x^2-7x} \cdot \dfrac{x^2-13x+42}{x^2+2x}$

**h.** $\dfrac{m^2-7m+12}{m^2-m-6} \cdot \dfrac{m^2+7m+10}{m^2+m-20}$

**i.** $\dfrac{x^2+5xy+6y^2}{x^2+4xy-5y^2} \cdot \dfrac{x^2+3xy-10y^2}{x^2+xy-6y^2}$

**j.** $\dfrac{m^2+7mn+10n^2}{m^2+mn-2n^2} \cdot \dfrac{m^2-5mn+4n^2}{m^2+mn-20n^2}$

**12. a.** $3x \cdot \dfrac{y}{3}$

**b.** $\dfrac{b}{a^3} \cdot 4a^2$

**c.** $6c^4 \cdot \dfrac{a}{6c^4}$

**d.** $\dfrac{8m^3n}{15x^2} \cdot 10x^3$

**13. a.** $\dfrac{x+y}{2} \cdot 10$

**b.** $4 \cdot \dfrac{r-d}{3}$

**c.** $3x^2y^3 \cdot \dfrac{4x-y}{9x}$

**d.** $(r+7) \cdot \dfrac{5r}{(r-7)}$

**e.** $(m+6)(m-2) \cdot \dfrac{12m}{(m+6)(m-2)}$

**f.** $16(b+3)(b-5) \cdot \dfrac{b+9}{8(b+3)}$

# Division of Algebraic Fractions

### I. Aim   To divide algebraic fractions.

### II. Procedure
1. Find the reciprocal (also called the multiplicative inverse) of the given divisor by interchanging its numerator and denominator.
2. Then multiply the given dividend by this reciprocal of the divisor, following the procedure for the multiplication of algebraic fractions. See Exercise 9-3.
3. Where possible, factor the numerators and denominators completely.
4. Replacements for the variables that make the denominator zero are excluded.
5. Observe that:
   a. Dividing by a number is the same as multiplying by its reciprocal. See Exercise 2-12.
   b. If the product of two numbers is one (1), then each factor is said to be the *multiplicative inverse* or *reciprocal* of the other. Each non-zero rational algebraic expression has a reciprocal.
      (1) 4 (or $\frac{4}{1}$) and $\frac{1}{4}$ are reciprocals of each other.
      (2) $\frac{5}{6}$ and $\frac{6}{5}$ are reciprocals of each other.
      (3) $\dfrac{x}{5}$ and $\dfrac{5}{x}$ are reciprocals of each other with $x \neq 0$.
      (4) $\dfrac{n-3}{n}$ and $\dfrac{n}{n-3}$ are reciprocals of each other with $n \neq 0$ and $n \neq 3$, since the denominator cannot be zero.

### III. Sample Solutions
1. $\frac{2}{3} \div \frac{7}{8} = \frac{2}{3} \times \frac{8}{7} = \frac{16}{21}$   *Answer:* $\frac{16}{21}$

2. $\dfrac{2a^2c}{3bd^2} \div \dfrac{4ac^2}{6b^2d} = \dfrac{2a^2c}{3bd^2} \cdot \dfrac{6b^2d}{4ac^2} = \dfrac{ab}{cd}$   *Answer:* $\dfrac{ab}{cd}$

3. $\dfrac{x^2 + 2x - 8}{x^2 - 16} \div \dfrac{x^2 - 6x + 8}{x^2 - 8x + 16} = \dfrac{x^2 + 2x - 8}{x^2 - 16} \times \dfrac{x^2 - 8x + 16}{x^2 - 6x + 8}$

$$= \dfrac{(x+4)(x-2)}{(x+4)(x-4)} \cdot \dfrac{(x-4)(x-4)}{(x-4)(x-2)}$$

$$= \dfrac{(x+4)(x-2)(x-4)(x-4)}{(x+4)(x-2)(x-4)(x-4)}$$

$$= 1 \quad \textit{Answer: 1}$$

## DIAGNOSTIC TEST

1. Write the reciprocal of **a.** $\frac{7}{10}$  **b.** $-5$
2. Assuming that the denominators are not zero, write the reciprocal of $\frac{m}{m-12}$.

Divide as indicated:

3. $\frac{3}{10} \div \frac{2}{5}$

4. $\frac{x^7}{y^3} \div \frac{x^2}{y^8}$

5. $\frac{4a^9c^3}{5b^2x^5} \div \frac{8a^{10}c^2}{15b^2x^4}$

6. $\frac{7m^2n^2}{8a^3b^2} \div 21mn^4$

7. $\frac{a^2-b^2}{9x^3} \div \frac{(a-b)^2}{27x^3}$

8. $\frac{a^2+6a+9}{a^2+2a-3} \div \frac{a^2-9}{a^2-a-6}$

## RELATED PRACTICE EXAMPLES

1. Write the reciprocals of each of the following:
   **a.** $7$   **b.** $\frac{1}{8}$   **c.** $\frac{5}{12}$   **d.** $\frac{15}{7}$   **e.** $-3$   **f.** $-\frac{1}{4}$   **g.** $-\frac{9}{16}$

2. Write the reciprocals of each of the following, assuming that the denominators are not zero.

   **a.** $\frac{n}{3}$   **b.** $\frac{8}{c}$   **c.** $\frac{r}{s}$   **d.** $\frac{a-b}{a+b}$

   **e.** $\frac{b^2}{b-7}$   **f.** $\frac{x+2}{x^2-9}$   **g.** $\frac{w-10}{6w}$   **h.** $\frac{m^2-13m+40}{m^2-4m-21}$

Divided as indicated:

3. **a.** $\frac{2}{3} \div \frac{3}{4}$   **b.** $\frac{15}{16} \div \frac{9}{10}$   **c.** $\frac{5}{6} \div \frac{7}{8}$   **d.** $\frac{3}{16} \div \frac{5}{12}$

4. **a.** $\frac{1}{a^2} \div \frac{1}{a}$   **b.** $\frac{b^2}{x^3} \div \frac{b}{x^2}$   **c.** $\frac{x}{a} \div \frac{y}{b}$   **d.** $\frac{m^4}{n^5} \div \frac{m^2}{n^6}$

5. **a.** $\frac{ax}{by} \div \frac{x}{y}$   **b.** $\frac{3a^2c}{8bd^2} \div \frac{6ac^3}{bd}$   **c.** $\frac{5x^2y}{6ad} \div \frac{10xy^2}{3a^2d}$

   **d.** $\frac{2m^3x}{11ny^4} \div \frac{4mx^2}{33ny^6}$   **e.** $\frac{5b^2c^3d}{7a^2x^2y^2} \div \frac{15bc^4}{28a^2xy^3}$   **f.** $\frac{15r^2tx^3}{16dy^7} \div \frac{3t^5x^8}{4r^3d^2y^6}$

6. **a.** $\frac{3a^2c^4}{4b^2} \div 6ac^2$   **b.** $\frac{2x^4y^2}{5m^2n} \div 8xy^2$   **c.** $14a^2m^3 \div \frac{7a^2m^4}{8bx^2}$

   **d.** $4abc \div \frac{2a^2b}{3d^2}$   **e.** $\frac{9c^2m^3x^4}{10by^2} \div 27cx^7$   **f.** $25x^5y^3z^2 \div \frac{5x^2z^5}{6a^6y^2}$

**7. a.** $\dfrac{4a - 4c}{5a + 5c} \div \dfrac{4}{15}$ 　　　**b.** $\dfrac{c^2 - d^2}{4c^2d} \div \dfrac{c^2 - 2cd + d^2}{2cd^2}$

**c.** $\dfrac{4m^2 + 20m + 25}{5m^4n^2} \div \dfrac{4m^2 - 25}{4m^2n^5}$ 　　**d.** $\dfrac{a^2 + a - 20}{8a^3b} \div \dfrac{a^2 - 8a + 16}{6ac^4}$

**8. a.** $\dfrac{5x - 5}{x^2 - 25} \div \dfrac{x - 1}{x - 5}$ 　　　**b.** $\dfrac{a^2 - b^2}{(a + b)^2} \div \dfrac{a - b}{4a + 4b}$

**c.** $\dfrac{x + y}{x^2 - xy} \div \dfrac{1}{x^2 - y^2}$ 　　　**d.** $\dfrac{ab^2 - a^2b}{ab^2 - ab} \div \dfrac{b^2 - a^2}{ab}$

**e.** $\dfrac{x^2 - xy}{cx^2 - cy^2} \div \dfrac{x^3 - x^2}{cx^2 - cx}$ 　　**f.** $\dfrac{b^2 + 4b - 12}{b^2 + 9b + 18} \div \dfrac{3b + 12}{6b + 6}$

**g.** $\dfrac{c^2 + 14c + 49}{c^2 + 2c - 35} \div \dfrac{c^2 + 9c + 14}{c^2 - 3c - 10}$ 　　**h.** $\dfrac{x^2 + xy}{x^2 + y^2} \div \dfrac{x^3y + 2x^2y^2 + xy^3}{x^4 - y^4}$

## CUMULATIVE REVIEW

**1.** For which values of the variable is the denominator of each of the following algebraic expressions equal to zero?

**a.** $\dfrac{15}{2a}$ 　　**b.** $\dfrac{3t}{(t + 6)(2t - 3)}$ 　　**c.** $\dfrac{x - 6}{x^2 - 7x - 18}$

**2.** Write each of the following algebraic expressions in simplest form:

**a.** $\dfrac{6r^4t^2}{12r^{10}t^2}$ 　　**b.** $\dfrac{18b^6x^2y}{28a^4b^3x^7}$ 　　**c.** $\dfrac{54a^6b^3c^9}{27a^4b^6c^2 - 72a^3b^2c^{10}}$

**d.** $\dfrac{y^2 - 36}{y^2 - 5y - 6}$ 　　**e.** $\dfrac{n^3 - 5n^2 - 36n}{5n^2 + 10n - 40}$ 　　**f.** $\dfrac{x^2 - a^2}{a^2 - 2ax + x^2}$

**3.** Multiply as indicated:

**a.** $\dfrac{32m^9n^2}{25a^5c^8} \cdot \dfrac{35a^7n^3}{48c^2m^{10}}$ 　　**b.** $\dfrac{3c^2d^9}{7x^2y^6} \cdot \dfrac{8dy}{15c^2x} \cdot \dfrac{5x^3y^4}{2cd^6}$

**c.** $42(w + 8)(w - 6) \cdot \dfrac{w - 5}{7(w + 8)}$ 　　**d.** $\dfrac{6m - 18}{m^2 - 6m + 9} \cdot \dfrac{m^2 - 9}{m^2 - 3m - 18}$

**4.** Divide as indicated:

**a.** $\dfrac{27a^{12}c}{16b^9d^2} \div \dfrac{9a^5c^6}{32b^4d^2}$ 　　**b.** $\dfrac{21m^8x^5}{16y^7} \div 14m^6x^4$

**c.** $\dfrac{(r - s)^2}{8r^2} \div \dfrac{r^2 - s^2}{12s}$ 　　**d.** $\dfrac{x^3 + 4x^2 - 21x}{x^2 + 14x + 49} \div \dfrac{x^2 - 10x + 21}{2x^2 - 98}$

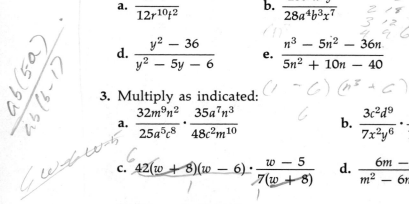

# Addition and Subtraction of Algebraic Fractions–with Like Denominators

**EXERCISE 10-5**

### I. Aim

To add and subtract algebraic fractions with like (same) denominators.

### II. Procedure

1. To add, form a fraction.
   a. For its denominator, write the common denominator.
   b. For its numerator, write first the indicated sum of the given numerators and then its sum. See sample solution.
2. To subtract, form a fraction.
   a. For its denominator, write the common denominator.
   b. For its numerator, write first the indicated difference of the given numerators and then its difference. See sample solutions 1–7.
3. If possible, combine any similar terms found in the numerator.
4. Also, if possible, change the resulting fraction to lowest terms. Factoring may be required to simplify the result. See sample solutions 8–11.
5. When the numerators are binomials or trinomials, observe in sample solutions 13 and 14 how the indicated addition and subtraction are written and simplified.
6. The variables in the denominators must have no replacement which makes the value of any denominator zero.

### III. Sample Solutions

Add or subtract as indicated in the following:

1. $\dfrac{7}{15} + \dfrac{4}{15} = \dfrac{7+4}{15} = \dfrac{11}{15}$  *Answer:* $\dfrac{11}{15}$

2. $\dfrac{1}{8} + \dfrac{3}{8} = \dfrac{1+3}{8} = \dfrac{4}{8} = \dfrac{1}{2}$  *Answer:* $\dfrac{1}{2}$

3. $\dfrac{x}{9} + \dfrac{7x}{9} = \dfrac{x+7x}{9} = \dfrac{8x}{9}$  *Answer:* $\dfrac{8x}{9}$

4. $\dfrac{2c}{5x} + \dfrac{3d}{5x} = \dfrac{2c+3d}{5x}$  *Answer:* $\dfrac{2c+3d}{5x}$

5. $\dfrac{5a}{6x} + \dfrac{7a}{6x} = \dfrac{5a+7a}{6x} = \dfrac{12a}{6x} = \dfrac{2a}{x}$  *Answer:* $\dfrac{2a}{x}$

6. $\dfrac{9b}{7} + \dfrac{4b}{7} = \dfrac{9b+4b}{7} = \dfrac{13b}{7}$  *Answer:* $\dfrac{13b}{7}$

**407**

7. $\dfrac{5a}{a+b} + \dfrac{5b}{a+b} = \dfrac{5a+5b}{a+b} = \dfrac{5(a+b)}{(a+b)} = \dfrac{(a+b)}{(a+b)} \cdot \dfrac{5}{1} = 1 \cdot 5 = 5$   *Answer:* 5

8. $\frac{4}{5} - \frac{2}{5} = \dfrac{4-2}{5} = \frac{2}{5}$   *Answer:* $\frac{2}{5}$

9. $\dfrac{5m}{8} - \dfrac{3n}{8} = \dfrac{5m-3n}{8}$   *Answer:* $\dfrac{5m-3n}{8}$

10. $\dfrac{7x}{2} - \dfrac{x}{2} = \dfrac{7x-x}{2} = \dfrac{6x}{2} = 3x$   *Answer:* $3x$

11. $\dfrac{11a}{5b} - \dfrac{3a}{5b} = \dfrac{11a-3a}{5b} = \dfrac{8a}{5b}$   *Answer:* $\dfrac{8a}{5b}$

12. $\dfrac{c^2}{c-2} - \dfrac{4}{c-2} = \dfrac{c^2-4}{c-2} = \dfrac{(c+2)(c-2)}{(c-2)} = \dfrac{(c-2)}{(c-2)} \cdot \dfrac{(c+2)}{1}$
$= 1(c+2) = c+2$   *Answer:* $c+2$

13. $\dfrac{3x+2y}{5} + \dfrac{2x+y}{5} = \dfrac{(3x+2y)+(2x+y)}{5} = \dfrac{3x+2y+2x+y}{5} = \dfrac{5x+3y}{5}$

*Answer:* $\dfrac{5x+3y}{5}$

14. $\dfrac{2a^2+4a+17}{a^2+5a+6} - \dfrac{a^2-4a+2}{a^2+5a+6} = \dfrac{(2a^2+4a+17)-(a^2-4a+2)}{a^2+5a+6}$

$= \dfrac{(2a^2+4a+17)+(-a^2+4a-2)}{a^2+5a+6}$

$= \dfrac{2a^2+4a+17-a^2+4a-2}{a^2+5a+6}$

$= \dfrac{a^2+8a+15}{a^2+5a+6} = \dfrac{(a+3)(a+5)}{(a+3)(a+2)} = \dfrac{a+5}{a+2}$

*Answer:* $\dfrac{a+5}{a+2}$

## DIAGNOSTIC TEST

Add as indicated:

1. $\dfrac{9}{16} + \dfrac{5}{16}$

2. $\dfrac{5s}{8} + \dfrac{2s}{8}$

3. $\dfrac{3b}{4} + \dfrac{5b}{4}$

4. $\dfrac{x}{6} + \dfrac{3x}{6}$

5. $\dfrac{x}{3} + \dfrac{y}{3}$

6. $\dfrac{b}{5y} + \dfrac{2b}{5y}$

7. $\dfrac{3}{4a} + \dfrac{1}{4a}$

8. $\dfrac{5r}{10b} + \dfrac{3r}{10b}$

9. $\dfrac{c}{c-d} + \dfrac{d}{c-d}$

10. $\dfrac{5xy}{x+y} + \dfrac{7xy}{x+y}$

11. $\dfrac{3c}{c+d} + \dfrac{3d}{c+d}$

Subtract as indicated:

12. $\dfrac{11}{12} - \dfrac{7}{12}$

13. $\dfrac{3x}{5} - \dfrac{2x}{5}$

14. $\dfrac{17w}{6} - \dfrac{5w}{6}$

15. $\dfrac{7a}{8} - \dfrac{3a}{8}$

16. $\dfrac{m}{2} - \dfrac{n}{2}$

17. $\dfrac{5}{c} - \dfrac{3}{c}$

18. $\dfrac{12b}{5x} - \dfrac{2b}{5x}$

19. $\dfrac{7n^2}{12x^2} - \dfrac{n^2}{12x^2}$

20. $\dfrac{a}{c-d} - \dfrac{b}{c-d}$

21. $\dfrac{4x}{x+y} - \dfrac{2x}{x+y}$

22. $\dfrac{b^2}{b-4} - \dfrac{16}{b-4}$

Add or subtract as indicated:

23. $\dfrac{3x-y}{5} + \dfrac{4x+2y}{5}$

24. $\dfrac{a+2b}{3} + \dfrac{2a+b}{3}$

25. $\dfrac{2c-3d}{4} - \dfrac{c+3d}{4}$

26. $\dfrac{4r-s}{6} - \dfrac{2r-7s}{6}$

27. $\dfrac{c+3d}{c+d} + \dfrac{c-d}{c+d}$

28. $\dfrac{3x}{(x-y)(x-y)} - \dfrac{2y}{(x-y)(x-y)}$

29. $\dfrac{a}{(a+b)(a+b)} + \dfrac{b}{(a+b)(a+b)}$

30. $\dfrac{c^2-cd}{(c+d)(c-d)} - \dfrac{cd-d^2}{(c+d)(c-d)}$

31. $\dfrac{b^2}{b^2+2b-15} + \dfrac{25}{b^2+2b-15}$

32. $\dfrac{x^2}{x^2-9} - \dfrac{6x-9}{x^2-9}$

### RELATED PRACTICE EXAMPLES

Add as indicated:

1. a. $\dfrac{2}{5} + \dfrac{1}{5}$

   b. $\dfrac{3}{7} + \dfrac{2}{7}$

   c. $\dfrac{1}{6} + \dfrac{1}{6}$

   d. $\dfrac{5}{16} + \dfrac{7}{16}$

   e. $\dfrac{7}{15} + \dfrac{4}{15}$

   f. $\dfrac{1}{12} + \dfrac{7}{12}$

   g. $\dfrac{13}{32} + \dfrac{15}{32}$

   h. $\dfrac{21}{64} + \dfrac{27}{64}$

2. a. $\dfrac{c}{4} + \dfrac{2c}{4}$

   b. $\dfrac{x}{5} + \dfrac{3x}{5}$

   c. $\dfrac{3a}{7} + \dfrac{6a}{7}$

3. **a.** $\dfrac{5t}{8} + \dfrac{3t}{8}$     **b.** $\dfrac{9b}{6} + \dfrac{3b}{6}$     **c.** $\dfrac{7x}{2} + \dfrac{5x}{2}$

4. **a.** $\dfrac{7n}{16} + \dfrac{5n}{16}$     **b.** $\dfrac{3m}{8} + \dfrac{3m}{8}$     **c.** $\dfrac{5c}{12} + \dfrac{3c}{12}$

5. **a.** $\dfrac{c}{4} + \dfrac{d}{4}$     **b.** $\dfrac{2a}{5} + \dfrac{3b}{5}$     **c.** $\dfrac{4x}{9} + \dfrac{5y}{9}$

6. **a.** $\dfrac{5}{x} + \dfrac{3}{x}$     **b.** $\dfrac{4m}{5ad} + \dfrac{3m}{5ad}$     **c.** $\dfrac{2a}{3x^2} + \dfrac{2a}{3x^2}$

7. **a.** $\dfrac{7}{5a} + \dfrac{3}{5a}$     **b.** $\dfrac{7d}{8x} + \dfrac{9d}{8x}$     **c.** $\dfrac{5ab}{4c^2} + \dfrac{7ab}{4c^2}$

8. **a.** $\dfrac{1}{8x} + \dfrac{5}{8x}$     **b.** $\dfrac{3x}{10y} + \dfrac{x}{10y}$     **c.** $\dfrac{3m^2}{16x^2} + \dfrac{9m^2}{16x^2}$

9. **a.** $\dfrac{4a^2}{y+1} + \dfrac{3b^2}{y+1}$     **b.** $\dfrac{m}{m-n} + \dfrac{n}{m-n}$     **c.** $\dfrac{3x}{x+y} + \dfrac{2y}{x+y}$

10. **a.** $\dfrac{5}{b+c} + \dfrac{4}{b+c}$     **b.** $\dfrac{x}{a+b} + \dfrac{2x}{a+b}$     **c.** $\dfrac{7dx}{d-x} + \dfrac{5dx}{d-x}$

11. **a.** $\dfrac{a}{a+b} + \dfrac{b}{a+b}$     **b.** $\dfrac{2c}{c+d} + \dfrac{2d}{c+d}$     **c.** $\dfrac{4m}{m+2n} + \dfrac{8n}{m+2n}$

Subtract as indicated:

12. **a.** $\dfrac{4}{5} - \dfrac{2}{5}$     **b.** $\dfrac{6}{7} - \dfrac{5}{7}$     **c.** $\dfrac{8}{9} - \dfrac{1}{9}$     **d.** $\dfrac{11}{15} - \dfrac{4}{15}$
    **e.** $\dfrac{9}{16} - \dfrac{3}{16}$     **f.** $\dfrac{5}{6} - \dfrac{1}{6}$     **g.** $\dfrac{9}{10} - \dfrac{3}{10}$     **h.** $\dfrac{31}{32} - \dfrac{15}{32}$

13. **a.** $\dfrac{7n}{3} - \dfrac{5n}{3}$     **b.** $\dfrac{4a}{6} - \dfrac{3a}{6}$     **c.** $\dfrac{9c}{11} - \dfrac{4c}{11}$

14. **a.** $\dfrac{6a}{5} - \dfrac{a}{5}$     **b.** $\dfrac{25h}{8} - \dfrac{9h}{8}$     **c.** $\dfrac{23x}{7} - \dfrac{2x}{7}$

15. **a.** $\dfrac{9x}{10} - \dfrac{3x}{10}$     **b.** $\dfrac{7m}{12} - \dfrac{m}{12}$     **c.** $\dfrac{17b}{32} - \dfrac{5b}{32}$

16. **a.** $\dfrac{13a}{6} - \dfrac{7b}{6}$     **b.** $\dfrac{14b}{3} - \dfrac{5c}{3}$     **c.** $\dfrac{3bc}{5} - \dfrac{2ad}{5}$

17. **a.** $\dfrac{8}{x} - \dfrac{6}{x}$     **b.** $\dfrac{4c}{xy} - \dfrac{2c}{xy}$     **c.** $\dfrac{10bx}{ad^2} - \dfrac{7bx}{ad^2}$

18. **a.** $\dfrac{9}{2s} - \dfrac{3}{2s}$     **b.** $\dfrac{3g}{2z} - \dfrac{g}{2z}$     **c.** $\dfrac{16a}{5c^2} - \dfrac{6a}{5c^2}$

19. **a.** $\dfrac{4d}{6x} - \dfrac{d}{6x}$     **b.** $\dfrac{9x}{10r^2} - \dfrac{5x}{10r^2}$     **c.** $\dfrac{23c^2d^2}{24x^2y} - \dfrac{15c^2d^2}{24x^2y}$

20. **a.** $\dfrac{5b^2c}{y+3} - \dfrac{2a^2x}{y+3}$    **b.** $\dfrac{a}{a-b} - \dfrac{3}{a-b}$    **c.** $\dfrac{x}{x+y} - \dfrac{y}{x+y}$

21. **a.** $\dfrac{7}{a+d} - \dfrac{5}{a+d}$    **b.** $\dfrac{3x}{x-y} - \dfrac{x}{x-y}$    **c.** $\dfrac{8c^2}{c-d} - \dfrac{3c^2}{c-d}$

22. **a.** $\dfrac{c}{c-d} - \dfrac{d}{c-d}$    **b.** $\dfrac{5m}{m-n} - \dfrac{5n}{m-n}$    **c.** $\dfrac{b^2}{b+3} - \dfrac{9}{b+3}$

Add or subtract as indicated:

23. **a.** $\dfrac{5c-3d}{2} + \dfrac{3c+2d}{2}$    **b.** $\dfrac{4a+3b}{3a} + \dfrac{a+b}{3a}$

24. **a.** $\dfrac{2x+3y}{5} + \dfrac{3x+2y}{5}$    **b.** $\dfrac{2a-b}{4} + \dfrac{2a+5b}{4}$

   **c.** $\dfrac{2c+d}{9} + \dfrac{c+2d}{9}$    **d.** $\dfrac{5m-2n}{6} + \dfrac{7m-4n}{6}$

25. **a.** $\dfrac{8x-2y}{3} - \dfrac{4x-5y}{3}$    **b.** $\dfrac{7d+4m}{10} - \dfrac{2d+3m}{10}$

26. **a.** $\dfrac{10s+4t}{7} - \dfrac{3s-3t}{7}$    **b.** $\dfrac{6x+5y}{2x} - \dfrac{4x+y}{2x}$

   **c.** $\dfrac{7m-2n}{3} - \dfrac{m+8n}{3}$    **d.** $\dfrac{4a+3ab}{a^2b^2} - \dfrac{a-2ab}{a^2b^2}$

27. **a.** $\dfrac{4a-3x}{a+x} - \dfrac{2a+9x}{a+x}$    **b.** $\dfrac{5c-4}{c-2} - \dfrac{2c+2}{c-2}$

28. **a.** $\dfrac{5a}{(a+b)(a+b)} + \dfrac{4b}{(a+b)(a+b)}$    **b.** $\dfrac{2x-3}{(x-2)(x+1)} - \dfrac{x}{(x-2)(x+1)}$

29. **a.** $\dfrac{x}{(x+y)(x+y)} + \dfrac{y}{(x+y)(x+y)}$    **b.** $\dfrac{2r-3n}{(r-n)(r-n)} - \dfrac{r-2n}{(r-n)(r-n)}$

30. **a.** $\dfrac{3x}{x(x+2)} + \dfrac{5x}{x(x+2)}$    **b.** $\dfrac{b^2+6b+2}{(b+3)(b-2)} - \dfrac{2b-1}{(b+3)(b-2)}$

   **c.** $\dfrac{c^2+7c+3}{(c+4)(c+5)} - \dfrac{2c+3}{(c+4)(c+5)}$    **d.** $\dfrac{x^2+x-7}{(x+5)(x-2)} + \dfrac{x^2+5x-13}{(x+5)(x-2)}$

31. **a.** $\dfrac{a^2}{a^2+2ab+b^2} + \dfrac{b^2}{a^2+2ab+b^2}$    **b.** $\dfrac{3m^2}{m^2-8m+16} - \dfrac{16}{m^2-8m+16}$

32. **a.** $\dfrac{x}{x^2-16} + \dfrac{4}{x^2-16}$    **b.** $\dfrac{x^2+4x}{x^2-x-6} - \dfrac{x-6}{x^2-x-6}$

   **c.** $\dfrac{4b^2-5b-20}{b^2+2b-15} - \dfrac{3b^2-5b+5}{b^2+2b-15}$    **d.** $\dfrac{a^2+ab-b^2}{a^2+2ab+b^2} + \dfrac{a^2+ab+b^2}{a^2+2ab+b^2}$

# Least Common Denominator

I. **Aim**     To determine the least common denominator of two or more algebraic fractions.

II. **Procedure**

1. Find the smallest possible algebraic expression that can be divided exactly by all the denominators of the given algebraic fraction. To do this:

   a. Select the highest power of each prime numerical factor appearing in the factored forms of all the denominators and form a product of these prime numbers. Find the numerical value of this product. A prime number is a whole number other than 0 and 1 which can be divided exactly by itself and by 1 and by no other whole number.

   b. Select the highest power of each variable factor appearing in the denominators.

   c. Multiply the numerical factor (step a.) and the variable factor (step b.).

2. Observe that:

   a. The *least common denominator* (*LCD*) of two or more algebraic fractions is (1) the smallest possible algebraic expression that can be divided exactly by all the denominators of the given algebraic fractions or (2) the *least common multiple* (*LCM*) of the given denominators.

   b. A *multiple* of a given whole number is a product of the given number and another whole number factor. A multiple of a given number is divisible by the given number.

   c. The *least common multiple* (*LCM*) of two or more numbers is the smallest number, zero excluded, which is a multiple of all of them. It is the smallest number that can be divided exactly by all the given numbers.

3. Sometimes the denominators of the algebraic fractions are binomials or trinomials. See sample solutions 5b and 5c.

4. Sometimes the binomial or trinomial denominators must be factored before the least common denominator may be determined. See sample solution 5c.

III. **Sample Solutions**

1. What are the first seven multiples of: **a.** 4? **b.** 6?

*Answers:* **a.** 0, 4, 8, 12, 16, 20, 24

          **b.** 0, 6, 12, 18, 24, 30, 36

**2.** What is the least common multiple of 4 and 6?
The multiples listed in sample solution 1 indicate that the common multiples are 0, 12, 24. Since 0 is excluded, 12 is the least common multiple.
*Answer:* 12

**3.** Find the least possible algebraic expression that can be divided exactly by $6a^3b$, $10a^2b^4$, and $8ab^5$.
$6a^3b = 2 \cdot 3 \cdot a^3 \cdot b$
$10a^2b^4 = 2 \cdot 5 \cdot a^2 \cdot b^4$
$8ab^5 = 2^3 \cdot a \cdot b$
Using the highest power of each factor, we get the product $2^3 \cdot 3 \cdot 5a^3 \cdot b^5$ or
$8 \cdot 3 \cdot 5a^3b^5 = 120a^3b^5$
*Answer:* $120a^3b^5$

**4.** Find the least possible algebraic expression that can be divided exactly by
  **a.** $x + 2$ and $x - 5$    *Answer:* $(x + 2)(x - 5)$
  **b.** $n - 6$ and $3(n - 6)$    *Answer:* $3(n - 6)$
  **c.** $2(t + 7)$ and $(t + 7)^2$    *Answer:* $2(t + 7)^2$
  **d.** $6(r + 2)$ and $4(r - 8)$    *Answer:* $12(r + 2)(r - 8)$
  **e.** $mn$ and $(m + n)$    *Answer:* $mn(m + n)$
  **f.** $x^2 + 6x + 9$ and $x^2 + 5x + 6$    $x^2 + 6x + 9 = (x + 3)^2$
                                        $x^2 + 5x + 6 = (x + 3)(x + 2)$
*Answer:* $(x + 3)^2(x + 2)$

**5.** Find the least common denominator of each of the following:

  **a.** $\dfrac{1}{8x^2y^3}$ and $\dfrac{1}{12x^4y^2}$      $8x^2y^3 = 2^3x^2y^3$      $\text{LCD} = 2^3 \cdot 3x^4y^3$
                                     $12x^4y^2 = 2^2 \cdot 3x^4y^2$         $= 8 \cdot 3x^4y^3$
                                                              $= 24x^4y^3$
*Answer:* $24x^4y^3$

  **b.** $\dfrac{1}{n - 5}$ and $\dfrac{1}{n + 6}$    $\text{LCD} = (n - 5)(n + 6)$
*Answer:* $(n - 5)(n + 6)$

  **c.** $\dfrac{1}{3x - 12}$ and $\dfrac{1}{x^2 - 16}$      $3x - 12 = 3(x - 4)$
                                          $x^2 - 16 = (x + 4)(x - 4)$
                                      $\text{LCD} = 3(x - 4)(x + 4)$
*Answer:* $3(x - 4)(x + 4)$

## DIAGNOSTIC TEST

1. Write the first eight multiples of 9.
2. Write the first six common multiples of 8 and 12.
3. Find the least common multiple of 4, 12, and 20.

Find the least possible algebraic expression that can be divided exactly by:

4. $10c^9d^3$ and $15c^6d^4$       5. $x - 4$ and $x + 6$
6. $12(a - 7)$ and $6(a - 7)^3$     7. $(n + 3)(n - 9)$ and $(n + 4)(n + 3)$
8. $r^2 - 10r - 25$ and $r^2 - 7r + 10$

Find the least common denominator of:

9. $\frac{9}{16}$ and $\frac{3}{10}$

10. $\frac{1}{x^5y^3}$, $\frac{1}{x^6y^2}$, and $\frac{1}{x^4y^4}$

11. $\frac{1}{c^4}$ and $\frac{1}{d^5}$

12. $\frac{1}{18b^8x^2}$ and $\frac{1}{24b^5x^6}$

13. $\frac{1}{4a - b}$, $\frac{1}{a - 4b}$, and $\frac{1}{4ab}$

14. $\frac{1}{6x - 4y}$, $\frac{1}{9x^2 - 4y^2}$, and $\frac{1}{9x^2 - 12xy + 4y^2}$

## RELATED PRACTICE EXAMPLES

1. Write the first eight multiples of each of the following:
   **a.** 5     **b.** 9     **c.** 12     **d.** 17
   **e.** 45     **f.** 70     **g.** 100
2. Write the first six common multiples of each of the following:
   **a.** 3 and 2     **b.** 7 and 14     **c.** 8 and 10
   **d.** 25 and 15     **e.** 6, 8, and 12     **f.** 10, 24, and 30
3. Find the least common multiple of each of the following:
   **a.** 2 and 5     **b.** 12 and 18     **c.** 32 and 48
   **d.** 4, 6, and 10     **e.** 6, 8, and 12     **f.** 18, 45, and 54

Find the least possible algebraic expression that can be divided exactly by:
4. **a.** $x^3y^2$ and $xy^5$     **b.** $3a^6b^4$ and $4a^2b^3$
   **c.** $12m^9n^5$ and $18m^7n^{10}$     **d.** $5a$ and $6b$
   **e.** $20c^3d^5$ and $25c^2x^6$     **f.** $16r^4x^7$ and $24s^8y^2$
   **g.** $4t^9$, $8t^7$, and $12t^5$     **h.** $2m^8n^2$, $3m^5n^5$, and $5mn^6$
   **i.** $18b^6x^5$, $27c^4x^6$, and $45x^4y^3$

5. **a.** $n+3$ and $n+6$     **b.** $a+1$ and $a+4$     **c.** $x+7$ and $x-7$
  **d.** $ab$ and $a+b$       **e.** $r-8$ and $r-9$       **f.** $b-4$, $b-1$, and $4$

6. **a.** $m+7$ and $2(m+7)$            **b.** $(c-2)^2$ and $c-2$
  **c.** $6(m-n)$ and $3(m-n)$       **d.** $a+b$ and $(a+b)^3$
  **e.** $4(c-2)$ and $6(c-2)$        **f.** $8(n-5)^4$ and $n(n-5)^2$

7. **a.** $x+6$ and $(x+6)(x-2)$
  **b.** $(m+7)(m-2)$ and $(m+7)(m+4)$
  **c.** $(b+2)(b-5)$ and $(b-5)(b+3)$
  **d.** $3(b-5)$ and $4(b+2)$
  **e.** $8(c+9)$ and $12(c-9)$
  **f.** $6n^2(n-6)$ and $(n-4)(n-1)$

8. **a.** $b+3$ and $3b+9$            **b.** $5m-5n$ and $4m-4n$
  **c.** $4b+4c$ and $6b-6c$        **d.** $m-6$ and $m^2-36$
  **e.** $c+9$ and $c^2-18c+81$     **f.** $b+1$, $b-1$, and $b^2-1$
  **g.** $r-2s$, $r^2-4s^2$, and $r+2s$    **h.** $x^2+7x+10$ and $x^2+8x+15$
  **i.** $s^2-5s+4$ and $s^2-2s-8$     **j.** $a^2-16$ and $a^2+6a+8$
  **k.** $m^2+2mn+n^2$ and $m^2-n^2$    **l.** $b^2-9$, $b^2+6b+9$, and $b^2-b-6$

9. Find the least common denominator of each of the following sets of fractions:
  **a.** $\frac{1}{2}$ and $\frac{7}{16}$    **b.** $\frac{2}{3}$ and $\frac{3}{5}$    **c.** $\frac{5}{6}$ and $\frac{5}{8}$    **d.** $\frac{3}{4}$ and $\frac{7}{10}$
  **e.** $\frac{2}{5}$ and $\frac{11}{12}$    **f.** $\frac{13}{18}$ and $\frac{9}{16}$    **g.** $\frac{4}{5}$, $\frac{1}{4}$, and $\frac{5}{8}$    **h.** $\frac{7}{12}$, $\frac{13}{16}$, and $\frac{9}{10}$

Find the least common denominator of each of the following groups of algebraic fractions:

10. **a.** $\dfrac{1}{a^4}$ and $\dfrac{1}{a^2}$     **b.** $\dfrac{1}{m^3}$ and $\dfrac{1}{m^4}$     **c.** $\dfrac{1}{a^3b^2}$ and $\dfrac{1}{a^2b^4}$

  **d.** $\dfrac{1}{x^5y^2}$ and $\dfrac{1}{x^3y^7}$     **e.** $\dfrac{1}{c^2d}$, $\dfrac{1}{cd^2}$, and $\dfrac{1}{c^3d}$     **f.** $\dfrac{1}{m^2x^3}$, $\dfrac{1}{m^5x^2}$, and $\dfrac{1}{m^3x^4}$

11. **a.** $\dfrac{1}{a}$ and $\dfrac{1}{x}$     **b.** $\dfrac{1}{b^2}$ and $\dfrac{1}{y^3}$     **c.** $\dfrac{1}{a^2c}$ and $\dfrac{1}{ad^2}$

  **d.** $\dfrac{1}{r^4t^3}$ and $\dfrac{1}{r^8t^5}$     **e.** $\dfrac{1}{m^4n^2}$, $\dfrac{1}{m^3x^2}$, and $\dfrac{1}{m^2y^3}$     **f.** $\dfrac{1}{c^3d}$, $\dfrac{1}{d^2x}$, and $\dfrac{1}{x^2y}$

12. **a.** $\dfrac{1}{5x}$ and $\dfrac{1}{3y}$     **b.** $\dfrac{1}{6c^2}$ and $\dfrac{1}{10c^3}$     **c.** $\dfrac{1}{8a^2d}$ and $\dfrac{1}{4ad^2}$

  **d.** $\dfrac{1}{16b^9x^2}$ and $\dfrac{1}{24b^4x^8}$     **e.** $\dfrac{1}{2a}$, $\dfrac{1}{5a^2}$, and $\dfrac{1}{6a^3}$     **f.** $\dfrac{1}{12m}$, $\dfrac{1}{8n}$, and $\dfrac{1}{10x}$

13. **a.** $\dfrac{1}{b+8}$ and $\dfrac{1}{b+9}$           **b.** $\dfrac{1}{n-2}$ and $\dfrac{1}{n-3}$

  **c.** $\dfrac{1}{rs}$ and $\dfrac{1}{r+s}$           **d.** $\dfrac{1}{8(a+b)}$ and $\dfrac{1}{6(a+b)}$

**e.** $\dfrac{1}{5(x-2)^2}$ and $\dfrac{1}{10(x-2)}$    **f.** $\dfrac{1}{a-9}$ and $\dfrac{1}{(a+7)(a-9)}$

**g.** $\dfrac{1}{(w+6)(w-5)}$ and $\dfrac{1}{(w-8)(w+6)}$    **h.** $\dfrac{1}{16(c-3d)}$ and $\dfrac{1}{12(3c+d)}$

**i.** $\dfrac{1}{c-5}$, $\dfrac{1}{c+8}$, and $\dfrac{1}{2c}$

**j.** $\dfrac{1}{x-2y}$, $\dfrac{1}{(x+2y)(x-2y)}$, and $\dfrac{1}{x+2y}$

**14. a.** $\dfrac{1}{r-8}$ and $\dfrac{1}{8r-64}$

**b.** $\dfrac{1}{9c+18}$ and $\dfrac{1}{6c-12}$

**c.** $\dfrac{1}{n^2-4n+4}$ and $\dfrac{1}{n^2-2n}$

**d.** $\dfrac{1}{b^2-100}$ and $\dfrac{1}{b^2+15b+50}$

**e.** $\dfrac{1}{r-s}$, $\dfrac{1}{r^2-s^2}$, and $\dfrac{1}{r+s}$

**f.** $\dfrac{1}{x^2+16x+64}$, $\dfrac{1}{x^2-2x-80}$, and $\dfrac{1}{x^2-64}$

## CUMULATIVE REVIEW

**1.** Write each of the following algebraic fractions in simplest form:

**a.** $\dfrac{27b^9x^2y^6}{18a^5b^3x^2}$    **b.** $\dfrac{10a^4b^3}{10a^4b^3-15a^2b^5}$    **c.** $\dfrac{6x^2+5x-6}{3x^2+10x-8}$

**2.** Multiply as indicated:

**a.** $\dfrac{6b^3}{5d^2x^3}\cdot\dfrac{x^4y}{14b^2}\cdot\dfrac{35d^3}{9bx}$    **b.** $\dfrac{9x^2+81}{x^2-9}\cdot\dfrac{6x-18}{x^4+18x^2+81}$

**3.** Divide as indicated:

**a.** $\dfrac{63b^8d^9}{8c^7}\div 7b^4c^3$    **b.** $\dfrac{4n-20}{n^2-2n-15}\div\dfrac{n^2-9}{n^2+6n+9}$

**4.** Add or subtract as indicated:

**a.** $\dfrac{3x}{10}+\dfrac{x}{10}$    **b.** $\dfrac{m}{m-n}-\dfrac{n}{m-n}$    **c.** $\dfrac{r^2}{r+s}-\dfrac{s^2}{r+s}$

**5.** Find the least common denominator of:

**a.** $\dfrac{1}{6x^5y^2}$, $\dfrac{1}{4x^4y^5}$, and $\dfrac{1}{10x^3y^8}$    **b.** $\dfrac{1}{n^2-25}$ and $\dfrac{1}{n^2-3n-10}$

# Changing Algebraic Fractions to Equivalent Fractions

**I. Aim**  To change an algebraic fraction to an equivalent fraction having a specified new denominator.

## II. Procedure

1. Divide the specified new denominator by the denominator of the given fraction.
2. Then multiply both the numerator and the denominator of the given fraction by this quotient.
   This is the same as multiplying the fraction by one (1).
3. Observe that:
   **a.** Fractions that have the same simplest form are called *equivalent algebraic fractions*.
   **b.** When the numerator and denominator of a fraction are each multiplied by the same number, except by zero, the result is an equivalent fraction.

## III. Sample Solutions

In each of the following, find the equivalent fraction with the specified denominator.

**1.** $\dfrac{5}{8} = \dfrac{?}{24}$  $\qquad \dfrac{5}{8} = \dfrac{5}{8} \cdot 1 = \dfrac{5}{8} \cdot \dfrac{3}{3} = \dfrac{15}{24}$  *Answer:* $\dfrac{15}{24}$

**2.** $\dfrac{2a}{5x} = \dfrac{?}{20bx}$  $\qquad \dfrac{2a}{5x} = \dfrac{2a}{5x} \cdot 1 = \dfrac{2a}{5x} \cdot \dfrac{4b}{4b} = \dfrac{8ab}{20bx}$

Observe in the given fraction $x \neq 0$; in the new fraction both $x \neq 0$ and $b \neq 0$.

*Answer:* $\dfrac{8ab}{20bx}$

**3.**
$$\frac{4x}{(x+3)(x-2)} = \frac{?}{(x+3)(x-2)(x+5)}$$

$$\frac{4x}{(x+3)(x-2)} = \frac{4x}{(x+3)(x-2)} \cdot 1 = \frac{4x}{(x+3)(x-2)} \cdot \frac{(x+5)}{(x+5)}$$

$$= \frac{4x(x+5)}{(x+3)(x-2)(x+5)} = \frac{4x^2 + 20x}{(x+3)(x-2)(x+5)}$$

Observe in the given fraction $x \neq -3$ and $x \neq 2$; in the new fraction, $x \neq -3$, $x \neq 2$, and $x \neq -5$.  *Answer:* $\dfrac{4x^2 + 20x}{(x+3)(x-2)(x+5)}$

## DIAGNOSTIC TEST

In each of the following, find the equivalent fraction with the specified denominator:

**1.** $\dfrac{7}{12} = \dfrac{?}{72}$        **2.** $\dfrac{3b^4}{5a^4x^3} = \dfrac{?}{30a^5b^6x^7}$

**3.** $\dfrac{c-6}{8c(c-5)} = \dfrac{?}{24c(c-5)(c+4)}$

### *RELATED PRACTICE EXAMPLES*

In each of the following, find the equivalent fraction with the specified denominator:

1. **a.** $\dfrac{2}{3} = \dfrac{?}{12}$    **b.** $\dfrac{4}{5} = \dfrac{?}{45}$    **c.** $\dfrac{7}{16} = \dfrac{?}{96}$    **d.** $\dfrac{11}{20} = \dfrac{?}{100}$

    **e.** $\dfrac{21}{32} = \dfrac{?}{64}$    **f.** $\dfrac{9}{13} = \dfrac{?}{65}$    **g.** $\dfrac{5}{12} = \dfrac{?}{108}$    **h.** $\dfrac{13}{18} = \dfrac{?}{144}$

2. **a.** $\dfrac{c}{d} = \dfrac{?}{d^2}$    **b.** $\dfrac{8}{x^3} = \dfrac{?}{x^3y}$    **c.** $\dfrac{5b}{9} = \dfrac{?}{18cd}$

    **d.** $\dfrac{2x^4y^2}{3a^2b^3} = \dfrac{?}{15a^2b^6}$    **e.** $\dfrac{5a^6}{8b^2x^2} = \dfrac{?}{16b^7c^3x^2}$    **f.** $\dfrac{11bx}{12ay} = \dfrac{?}{60abxy}$

    **g.** $\dfrac{8w^5x^2}{7r^2t} = \dfrac{?}{49r^4t^2}$    **h.** $\dfrac{13y^2}{16a^5x^2} = \dfrac{?}{144a^6x^2y^5}$    **i.** $\dfrac{4a^4z^9}{15x^3y^4} = \dfrac{?}{120a^2x^3y^5z^2}$

3. **a.** $\dfrac{8}{x-5} = \dfrac{?}{(x-5)(x+2)}$

    **b.** $\dfrac{6b}{b+7} = \dfrac{?}{4(b+7)}$

    **c.** $\dfrac{x-4}{2x(x+4)} = \dfrac{?}{2x(x+4)(x-4)}$

    **d.** $\dfrac{y-10}{(y+8)(y-6)} = \dfrac{?}{(y+8)(y-6)(y-1)}$

    **e.** $\dfrac{n-9}{5n} = \dfrac{?}{10n(n-2)(n+2)}$

    **f.** $\dfrac{5t-7}{4t(t-1)} = \dfrac{?}{12t(t+6)(t-1)(t-3)}$

# Addition and Subtraction of Algebraic Fractions—with Unlike Monomial Denominators

**I. Aim**   To add and subtract algebraic fractions with unlike monomial denominators.

## II. Procedure

1. Determine the least common denominator of the given algebraic fractions. See Exercise 10-6.
2. Change each of the given algebraic fractions to an equivalent fraction having the least common denominator as its denominator. See Exercise 10-7.
3. Then add or subtract as you do with fractions having the same denominator. See Exercise 10-5.
4. Simplify the answer whenever possible.
5. Note that the problems in this exercise are assumed to have denominators only with non-zero values.
6. When combining two or more terms, add successively or subtract successively or add and subtract in the same steps.

## III. Sample Solutions

**1.** Add: $\dfrac{5}{12} + \dfrac{3}{8}$.

The LCD is 24.

$$\frac{5}{12} + \frac{3}{8} = \frac{5}{12} \cdot \frac{2}{2} + \frac{3}{8} \cdot \frac{3}{3} = \frac{10}{24} + \frac{9}{24} = \frac{10+9}{24} = \frac{19}{24}$$

*Answer:* $\dfrac{19}{24}$

**2.** Add: $\dfrac{a}{5} + \dfrac{2a}{3}$.

The LCD is 15.

$$\frac{a}{5} + \frac{2a}{3} = \frac{a}{5} \cdot \frac{3}{3} + \frac{2a}{3} \cdot \frac{5}{5} = \frac{3a}{15} + \frac{10a}{15} = \frac{3a+10a}{15} = \frac{13a}{15}$$

*Answer:* $\dfrac{13a}{15}$

**3.** Subtract: $\dfrac{1}{a} - \dfrac{1}{b}$

The LCD is $ab$.

**419**

$$\frac{1}{a} - \frac{1}{b} = \frac{1}{a} \cdot \frac{b}{b} - \frac{1}{b} \cdot \frac{a}{a} = \frac{b}{ab} - \frac{a}{ab} = \frac{b-a}{ab} \quad \textit{Answer: } \frac{b-a}{ab}$$

4. Subtract: $\dfrac{c+d}{2cd^2} - \dfrac{c-d}{6c^2d}$

The LCD is $6c^2d^2$.

$$\frac{c+d}{2cd^2} - \frac{c-d}{6c^2d} = \frac{c+d}{2cd^2} \cdot \frac{3c}{3c} - \frac{c-d}{6c^2d} \cdot \frac{d}{d} = \frac{3c(c+d)}{6c^2d^2} - \frac{d(c-d)}{6c^2d^2}$$

$$= \frac{3c(c+d) - d(c-d)}{6c^2d^2} = \frac{3c^2 + 3cd - cd + d^2}{6c^2d^2}$$

$$= \frac{3c^2 + 2cd + d^2}{6c^2d^2} \quad \textit{Answer: } \frac{3c^2 + 2cd + d^2}{6c^2d^2}$$

## DIAGNOSTIC TEST

Add or subtract as indicated:

1. $\dfrac{7}{12} + \dfrac{5}{16}$       2. $\dfrac{5}{6} - \dfrac{1}{2}$       3. $\dfrac{a}{3} + \dfrac{a}{6}$

4. $\dfrac{3x}{5} - \dfrac{x}{10}$       5. $\dfrac{5c}{6} + \dfrac{2c}{5}$       6. $\dfrac{c}{2} - \dfrac{d}{3}$

7. $\dfrac{3b}{4} + \dfrac{5c}{6}$       8. $\dfrac{5m}{12} - \dfrac{7m}{30}$       9. $\dfrac{7}{10a} + \dfrac{3}{2a}$

10. $\dfrac{2a}{3b} - \dfrac{a}{4b}$       11. $\dfrac{5a}{6c} + \dfrac{7b}{8c}$       12. $\dfrac{3}{x} + \dfrac{4}{x^2}$

13. $\dfrac{b}{2a^2} + \dfrac{c}{3a^3}$       14. $\dfrac{2}{m} - \dfrac{5}{n}$       15. $\dfrac{3c}{4x} + \dfrac{5c}{8y}$

16. $\dfrac{5a}{12x^2y} - \dfrac{3b}{10xy^2}$       17. $\dfrac{x+4}{6} + \dfrac{1}{2}$       18. $\dfrac{x+2}{4} + \dfrac{x+3}{2}$

19. $\dfrac{x+y}{6} + \dfrac{x+y}{2}$       20. $\dfrac{a+5}{3} - \dfrac{a+2}{4}$       21. $\dfrac{2d-5}{2} - \dfrac{d-4}{5}$

22. $\dfrac{2a+3}{a} + \dfrac{a+2}{2}$       23. $\dfrac{3x+6}{2x} + \dfrac{5y+4}{2y}$       24. $\dfrac{c-d}{cd^2} - \dfrac{3c-3d}{c^2d}$

25. Combine $\dfrac{a+b}{2a} - \dfrac{a-b}{3b} + \dfrac{b-c}{c}$

## *RELATED PRACTICE EXAMPLES*

1. Add as indicated:

   **a.** $\frac{1}{2} + \frac{1}{6}$      **b.** $\frac{2}{5} + \frac{1}{4}$      **c.** $\frac{3}{4} + \frac{5}{32}$      **d.** $\frac{7}{10} + \frac{3}{16}$

   **e.** $\frac{5}{6} + \frac{3}{8}$      **f.** $\frac{11}{24} + \frac{9}{16}$      **g.** $\frac{1}{3} + \frac{1}{5} + \frac{1}{6}$      **h.** $\frac{13}{16} + \frac{7}{12} + \frac{5}{8}$

2. Subtract as indicated:

   **a.** $\frac{1}{3} - \frac{1}{6}$      **b.** $\frac{11}{12} - \frac{2}{3}$      **c.** $\frac{7}{8} - \frac{3}{10}$      **d.** $\frac{15}{16} - \frac{23}{32}$

   **e.** $\frac{3}{4} - \frac{2}{5}$      **f.** $\frac{5}{8} - \frac{1}{12}$      **g.** $\frac{2}{3} - \frac{7}{16}$      **h.** $\frac{17}{20} - \frac{9}{25}$

Add or subtract as indicated:

**3. a.** $\dfrac{3x}{4} + \dfrac{5x}{8}$      **b.** $\dfrac{3b}{10} + \dfrac{2b}{5}$      **c.** $\dfrac{5a}{3} + \dfrac{4b}{9}$

**4. a.** $\dfrac{a}{2} - \dfrac{a}{4}$      **b.** $\dfrac{3x^2}{5} - \dfrac{x^2}{10}$      **c.** $\dfrac{5c^2}{6} - \dfrac{c}{3}$

**5. a.** $\dfrac{4b}{5} + \dfrac{3b}{2}$      **b.** $\dfrac{7a}{8} + \dfrac{5x}{9}$      **c.** $\dfrac{2cd}{3} + \dfrac{3xy}{4}$

**6. a.** $\dfrac{7y}{2} - \dfrac{y}{3}$      **b.** $\dfrac{2ab}{3} - \dfrac{ab}{5}$      **c.** $\dfrac{3bx}{4} - \dfrac{2cx}{3}$

**7. a.** $\dfrac{5a}{6} + \dfrac{a}{4}$      **b.** $\dfrac{3xy}{8} + \dfrac{5xy}{12}$      **c.** $\dfrac{9a^2}{16} + \dfrac{5b^2}{18}$

**8. a.** $\dfrac{3b}{4} - \dfrac{7b}{10}$      **b.** $\dfrac{5ad}{6} - \dfrac{3ad}{8}$      **c.** $\dfrac{11x^2y}{12} - \dfrac{9xy^2}{16}$

**9. a.** $\dfrac{7}{6x} + \dfrac{5}{2x}$      **b.** $\dfrac{7a}{4b} - \dfrac{7a}{8b}$      **c.** $\dfrac{9a}{8m} - \dfrac{11a}{16m}$

**10. a.** $\dfrac{3}{2c} + \dfrac{8}{3c}$      **b.** $\dfrac{7x}{4a} - \dfrac{3x}{5a}$      **c.** $\dfrac{5c}{2x} + \dfrac{d}{9x}$

**11. a.** $\dfrac{5x}{4a} + \dfrac{7x}{6a}$      **b.** $\dfrac{15b}{8d} - \dfrac{3b}{10d}$      **c.** $\dfrac{7a}{12m} + \dfrac{3b}{8m}$

**12. a.** $\dfrac{2}{b^2} + \dfrac{5}{b}$      **b.** $\dfrac{a}{m^2} - \dfrac{2a}{m^5}$      **c.** $\dfrac{b^2}{x^4} - \dfrac{bd}{x^3}$

**13. a.** $\dfrac{3a}{4b^2} + \dfrac{5a}{2b^3}$      **b.** $\dfrac{4b}{5x^4} - \dfrac{5c}{3x}$      **c.** $\dfrac{13ab^2}{6y^4} - \dfrac{7a^2b}{8y^2}$

**14. a.** $\dfrac{1}{x} + \dfrac{1}{y}$      **b.** $\dfrac{4}{a} - \dfrac{3}{b}$      **c.** $\dfrac{2a}{c} + \dfrac{5b}{d}$

**15. a.** $\dfrac{8a}{5x} + \dfrac{2b}{3y}$      **b.** $\dfrac{3c}{8mx} - \dfrac{5d}{16ny}$      **c.** $\dfrac{a}{x} + \dfrac{3}{5}$

**16. a.** $\dfrac{7x}{a^2b} + \dfrac{x^2}{ab^2}$  **b.** $\dfrac{2a}{3c^2} - \dfrac{3b}{4cd}$  **c.** $\dfrac{9r^2s}{10x^3y^2} + \dfrac{3st^2}{8x^2y^4}$

**17. a.** $\dfrac{5b+4}{12} + \dfrac{3}{16}$  **b.** $\dfrac{3a+5}{4} - \dfrac{7}{8}$  **c.** $\dfrac{2b-3}{5} + \dfrac{9}{10}$

**18. a.** $\dfrac{r+2}{2} + \dfrac{r+6}{4}$  **b.** $\dfrac{m-4}{8} + \dfrac{m+2}{4}$  **c.** $\dfrac{y+6}{2} + \dfrac{y-3}{3}$

**d.** $\dfrac{4b+3}{5} + \dfrac{6y-5}{20}$  **e.** $\dfrac{c+d}{4} + \dfrac{3c-2d}{2}$  **f.** $\dfrac{5a+3x}{2} + \dfrac{2a+3y}{5}$

**19. a.** $\dfrac{2b}{5} + \dfrac{b+5d}{10}$  **b.** $\dfrac{2b+2x}{3} + \dfrac{b+x}{12}$  **c.** $\dfrac{3x-5y}{6} + \dfrac{3x+y}{3}$

**20. a.** $\dfrac{a-b}{4} - \dfrac{a+b}{8}$  **b.** $\dfrac{b+3}{3} - \dfrac{b+2}{5}$  **c.** $\dfrac{x-2}{6} - \dfrac{x+3}{2}$

**21. a.** $\dfrac{2a+4}{7} - \dfrac{3a-2}{3}$  **b.** $\dfrac{4x+2y}{3} - \dfrac{x-2y}{10}$  **c.** $\dfrac{3a-2b}{4} - \dfrac{3a-2b}{20}$

**22. a.** $\dfrac{3}{a} + \dfrac{a+2}{5}$  **b.** $\dfrac{r+3}{r} + \dfrac{r+2}{2}$  **c.** $\dfrac{c-4}{4} - \dfrac{c-4}{c}$

**23. a.** $\dfrac{a+2}{3a} + \dfrac{b+3}{3b}$  **b.** $\dfrac{ax+1}{a^2} + \dfrac{x+a}{a}$  **c.** $\dfrac{2x-3y}{4x} - \dfrac{3x-2y}{6y}$

**24. a.** $\dfrac{2a-2b}{a^2b} + \dfrac{a-b}{ab^2}$  **b.** $\dfrac{b-x}{bx} - \dfrac{c-x}{cx}$  **c.** $\dfrac{b-2x}{12b^2x} - \dfrac{2b+x}{10bx^2}$

**25.** Combine:

**a.** $\dfrac{5a}{2b^3} - \dfrac{2a^2}{3b^2} - \dfrac{5a^3}{8b}$  **b.** $\dfrac{x+y}{2xy^2} - \dfrac{2x-y}{4x^2y} + \dfrac{x^2-2y^2}{3x^2y^2}$

**c.** $\dfrac{4a-3b}{6ab} - \dfrac{a-4c}{8ac} - \dfrac{3b-c}{4bc}$  **d.** $\dfrac{x^2-y^2}{xy} + \dfrac{y(x-3)}{4x} - \dfrac{x(y-2)}{6y}$

*CUMULATIVE REVIEW*

**1.** Multiply as indicated:

$$\dfrac{3r-6}{r^2-r-6} \cdot \dfrac{r^2-4}{r^2-4r+4}$$

**2.** Divide as indicated:

$$6x^3yz^4 \div \dfrac{3x^2y}{4z^5}$$

**3.** Add as indicated:

$$\dfrac{2x-y}{4x^2y} + \dfrac{8x-y}{6xy^2}$$

**4.** Subtract as indicated:

$$\dfrac{1}{c} - \dfrac{1}{d}$$

# Addition and Subtraction of Algebraic Fractions–with Binominal and Trinomial Denominators

**I. Aim** To add and subtract algebraic fractions with binomial or tri-nominal denominators which cannot be factored.

## II. Procedure

**1.** Determine the least common denominator of the given algebraic fractions. See Exercise 10-6.
   **a.** Sometimes the least common denominator is the product of the denominators of the given algebraic fractions.
   **b.** Sometimes the least common denominator is the denominator of one of the given algebraic fractions.
**2.** Change each of the given algebraic fractions to an equivalent fraction having the least common denominator as its denominator. See Exercise 10-7.
**3.** Then add or subtract as you do with fractions having the same denominators. See Exercise 10-5.
**4.** Simplify the answer whenever possible.
**5.** Note that the problems in this exercise are assumed to have denominators only with non-zero values.

## III. Sample Solution

Subtract: $\dfrac{3x}{x-5} - \dfrac{x-2}{x+5}$     The LCD is $(x-5)(x+5)$.

$$\frac{3x}{x-5} - \frac{x-2}{x+5} = \frac{3x}{x-5} \cdot \frac{x+5}{x+5} - \frac{x-2}{x+5} \cdot \frac{x-5}{x-5}$$

$$= \frac{3x(x+5)}{(x-5)(x+5)} - \frac{(x-2)(x-5)}{(x+5)(x-5)}$$

$$= \frac{3x(x+5)-(x-2)(x-5)}{(x+5)(x-5)} = \frac{(3x^2+15x)-(x^2-7x+10)}{(x+5)(x-5)}$$

$$= \frac{3x^2+15x-x^2+7x-10}{(x+5)(x-5)} = \frac{2x^2+22x-10}{(x+5)(x-5)} = \frac{2(x^2+11x-5)}{(x+5)(x-5)}$$

*Answer:* $\dfrac{2(x^2+11x-5)}{(x+5)(x-5)}$ or $\dfrac{2x^2+22x-10}{x^2-25}$

## DIAGNOSTIC TEST

Add or subtract as indicated:

1. $\dfrac{2}{a+2} + \dfrac{3}{a+3}$

2. $\dfrac{7}{x+1} - \dfrac{5}{x-1}$

3. $\dfrac{5}{x-3} - \dfrac{3}{x+3}$

4. $\dfrac{6}{x} + \dfrac{2}{x+2}$

5. $\dfrac{5m}{m-2} + \dfrac{2m}{m+4}$

6. $\dfrac{3x}{x-3} + \dfrac{x-4}{x-5}$

7. $\dfrac{2}{c+d} + \dfrac{c+d}{cd}$

8. $\dfrac{3a}{a-5} - \dfrac{a+2}{a+6}$

9. $\dfrac{b-x}{b+x} - \dfrac{b+x}{b-x}$

10. $\dfrac{3}{a+3} + \dfrac{2}{a+2} + \dfrac{2}{3}$

11. $\dfrac{5}{b+2} + \dfrac{3}{2(b+2)}$

12. $\dfrac{2x}{x-2} - \dfrac{6x}{(x+1)(x-2)}$

13. $\dfrac{3a}{(a-3)^2} - \dfrac{3}{a-3}$

14. $\dfrac{x+2}{3(x+5)} - \dfrac{x}{5(x+5)}$

15. $\dfrac{a}{2(a+2b)} + \dfrac{b}{4(2a-b)}$

16. $\dfrac{b-2}{(b+5)(b+1)} - \dfrac{b-2}{(b+5)(b+2)}$

## RELATED PRACTICE EXAMPLES

Add or subtract as indicated:

1. **a.** $\dfrac{6}{x+5} + \dfrac{3}{x+4}$  **b.** $\dfrac{4}{a+4} + \dfrac{5}{a-2}$  **c.** $\dfrac{9}{c-2} + \dfrac{2}{c-3}$

2. **a.** $\dfrac{8}{c+4} - \dfrac{2}{c-6}$  **b.** $\dfrac{9}{b+2} - \dfrac{4}{b-3}$  **c.** $\dfrac{3}{m+1} - \dfrac{5}{m-2}$

3. **a.** $\dfrac{7}{y-6} - \dfrac{1}{y+6}$  **b.** $\dfrac{10}{x-3} - \dfrac{2}{x+4}$  **c.** $\dfrac{4}{x-5} - \dfrac{7}{x+2}$

4. **a.** $\dfrac{5}{2} + \dfrac{6}{a+3}$  **b.** $\dfrac{3}{y+1} - \dfrac{3}{y}$  **c.** $\dfrac{3}{x} + \dfrac{4}{x+4}$

5. **a.** $\dfrac{2c}{c-2} + \dfrac{c}{c-1}$  **b.** $\dfrac{x}{x-y} - \dfrac{y}{x+y}$  **c.** $\dfrac{4}{d+7} - \dfrac{3d}{d-2}$

6. **a.** $\dfrac{a + 2}{a - 4} + \dfrac{2a}{a - 2}$   **b.** $\dfrac{x + 5}{x - 6} + \dfrac{3x}{x + 5}$   **c.** $\dfrac{2b}{b + 2} + \dfrac{b + 2}{b - 2}$

7. **a.** $\dfrac{4x}{x - 3} + \dfrac{x + 3}{2x}$   **b.** $\dfrac{x - 2}{x - 4} - \dfrac{5}{3x}$   **c.** $\dfrac{a}{a - b} + \dfrac{a + b}{ab}$

8. **a.** $\dfrac{2x}{x - 1} - \dfrac{x + 3}{x + 2}$   **b.** $\dfrac{3b}{b + 2} - \dfrac{b + 4}{b - 3}$   **c.** $\dfrac{5c}{c - 2} - \dfrac{c - 1}{c + 2}$

9. **a.** $\dfrac{a + 5}{a - 5} + \dfrac{a - 5}{a + 5}$   **b.** $\dfrac{c - d}{c + d} - \dfrac{c + d}{c - d}$   **c.** $\dfrac{m - 2n}{m + 2n} - \dfrac{m + 2n}{m - 2n}$

10. **a.** $\dfrac{1}{b - 3} - \dfrac{4}{b - 6} + \dfrac{5}{6}$     **b.** $\dfrac{3}{x} + \dfrac{2}{x - y} - \dfrac{1}{x + y}$

11. **a.** $\dfrac{b}{5} + \dfrac{4x^2}{5(b + 4x)}$     **b.** $\dfrac{2a}{a + b} - \dfrac{8b}{3(a + b)}$

12. **a.** $\dfrac{6}{(x + 5)(x + 5)} + \dfrac{2}{x + 5}$     **b.** $\dfrac{2b - 7}{(b - 2)(b + 5)} + \dfrac{b + 5}{b - 2}$

13. **a.** $\dfrac{8}{a + x} + \dfrac{4x}{(a + x)^2}$     **b.** $\dfrac{x}{(x - 4)^2} - \dfrac{1}{x - 4}$

    **c.** $\dfrac{9}{(a - 6)^2} + \dfrac{a + 4}{a - 6}$     **d.** $\dfrac{5}{(c - 5)^2} - \dfrac{c + 5}{c - 5}$

14. **a.** $\dfrac{5}{2(a + 1)} + \dfrac{2}{3(a + 1)}$     **b.** $\dfrac{4d}{5(d - 2)} - \dfrac{3d}{10(d - 2)}$

15. **a.** $\dfrac{3a}{4(a + 3)} + \dfrac{3}{5(a - 3)}$     **b.** $\dfrac{7c}{10(c + d)} - \dfrac{5d}{16(c - d)}$

16. **a.** $\dfrac{2x}{(x - 1)(x + 3)} - \dfrac{x}{(x - 1)(x + 2)}$

    **b.** $\dfrac{x - 2}{2(x + 3)(x - 3)} + \dfrac{x + 3}{3(x - 3)(x + 2)}$

## CUMULATIVE REVIEW

1. Multiply as indicated:

$\dfrac{6x - 7}{12x^2y} \cdot 8y^6$

2. Divide as indicated:

$\dfrac{b^2 + 2b - 24}{b^2 + 2b - 8} \div \dfrac{b^2 - 6b + 8}{b^2 - 36}$

3. Add as indicated:

$\dfrac{a - b}{b} + \dfrac{2a}{a - b}$

4. Subtract as indicated:

$\dfrac{5}{x - 8} - \dfrac{x - 5}{x + 8}$

# Addition and Subtraction of Algebraic Fractions–Binomial and Trinomial Denominators Requiring Factoring

EXERCISE **10-10**

**I. Aim** To add and subtract algebraic fractions with binomial and trinomial denominators which can be factored.

**II. Procedure**
1. Factor the denominators of the given fractions.
2. Then follow the procedure outlined in Exercise 10-9.

**III. Sample Solution**

Add: $\dfrac{3}{x+1} + \dfrac{3}{x^2+x}$

The LCD is $x(x+1)$.

$$\frac{3}{x+1} + \frac{3}{x^2+x} = \frac{3}{x+1} + \frac{3}{x(x+1)} = \frac{3}{x+1}\cdot\frac{x}{x} + \frac{3}{x(x+1)}$$

$$= \frac{3x}{x(x+1)} + \frac{3}{x(x+1)} = \frac{3x+3}{x(x+1)} = \frac{3(x+1)}{x(x+1)} = \frac{3}{x}$$

*Answer,* $\dfrac{3}{x}$

---

## DIAGNOSTIC TEST

Add or subtract as indicated:

1. $\dfrac{b}{b+2} + \dfrac{b-3}{2b+4}$

2. $\dfrac{6s}{s^2-100} - \dfrac{3}{s-10}$

3. $\dfrac{4m}{m-5} - \dfrac{4m-8}{m^2-10m+25}$

4. $\dfrac{2a+3b}{2a+2b} + \dfrac{2a}{3a+3b}$

5. $\dfrac{a-2x}{4a+2x} - \dfrac{2a-x}{2a+4x}$

6. $\dfrac{a+3}{a^2+9a+14} + \dfrac{a-2}{a^2+4a-21}$

### *RELATED PRACTICE EXAMPLES*

Add or subtract as indicated:

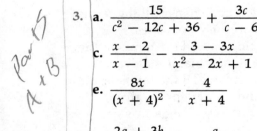

**1. a.** $\dfrac{2}{x+3} + \dfrac{3}{2x+6}$

**b.** $\dfrac{2}{a^2+a} + \dfrac{2}{a+1}$

**c.** $\dfrac{c^2}{3c-3d} - \dfrac{c+d}{3}$

**d.** $\dfrac{x+3}{x-3} - \dfrac{x+3}{4x-12}$

**2. a.** $\dfrac{5}{x^2-4} + \dfrac{3}{x+2}$

**b.** $\dfrac{t+7}{t-7} - \dfrac{14t}{t^2-49}$

**c.** $\dfrac{3a}{a^2+6a-16} + \dfrac{3}{a+8}$

**d.** $\dfrac{m-4}{m-6} - \dfrac{3m}{m^2-3m-18}$

**e.** $\dfrac{2x^2+5x}{x^2-8x-20} - \dfrac{x-2}{x-10}$

**f.** $\dfrac{2t^2+7}{t^2+t-2} - \dfrac{t-3}{t+2}$

**g.** $\dfrac{6a^2}{a^2-9} - \dfrac{3a}{a+3} - \dfrac{2a}{a-3}$

**h.** $\dfrac{c-2}{c+2} + \dfrac{10c-4}{c^2-4} - \dfrac{c+2}{c-2}$

**3. a.** $\dfrac{15}{c^2-12c+36} + \dfrac{3c}{c-6}$

**b.** $\dfrac{4-b}{b^2-8b+16} + \dfrac{2}{b-4}$

**c.** $\dfrac{x-2}{x-1} - \dfrac{3-3x}{x^2-2x+1}$

**d.** $\dfrac{2}{a+2} + \dfrac{4}{(a+2)^2}$

**e.** $\dfrac{8x}{(x+4)^2} - \dfrac{4}{x+4}$

**f.** $\dfrac{c+4}{(c-4)^2} - \dfrac{c+4}{c-4}$

**4. a.** $\dfrac{2a+3b}{2a+2b} + \dfrac{a}{3a+3b}$

**b.** $\dfrac{c+2d}{4c-4d} - \dfrac{d}{3c-3d}$

**5. a.** $\dfrac{2}{2x-8} + \dfrac{3}{2x-2}$

**b.** $\dfrac{b-2x}{4b+2x} - \dfrac{2b-x}{2b+4x}$

**6. a.** $\dfrac{5}{3x+15} + \dfrac{4}{x^2-25}$

**b.** $\dfrac{c+5d}{c^2-d^2} - \dfrac{2d}{c^2-cd}$

**c.** $\dfrac{a-4}{2a-10} + \dfrac{a}{a^2-10a+25}$

**d.** $\dfrac{4x}{3x^2-3y^2} - \dfrac{x-y}{x^2+2xy+y^2}$

**e.** $\dfrac{x}{x^2+5x+4} - \dfrac{1}{x^2+2x+1}$

**f.** $\dfrac{2n}{5n^2-5n-30} + \dfrac{3n}{4n^2+20n+24}$

**g.** $\dfrac{2}{x^2-4} - \dfrac{3}{x^2-4x+4} + \dfrac{4}{x^2+x-2}$

**h.** $\dfrac{6}{b^2-10b+25} - \dfrac{2}{b^2-2b-15} + \dfrac{3}{b^2-9b+20}$

**I. Aim**    To change signs in algebraic fractions in order to simplify operations.

**II. Procedure**

**1.** Observe that:

**a.** $-\dfrac{+6}{+8}$, $-\dfrac{-6}{-8}$, $+\dfrac{-6}{+8}$, and $+\dfrac{+6}{-8}$ in simplified form are all equivalent to $-\frac{3}{4}$.

**b.** Every fraction has signs in three locations: (1) preceding the fraction, (2) numerator, and (3) denominator.

**c.** But a change of signs to their opposites in any two locations does not change the value of the fraction.

**2.** Where necessary:

**a.** Retain the given sign preceding the fraction and change all the terms of both the numerator and denominator to their opposites.

**b.** Change the given sign preceding the fraction and all the terms of the numerator to their opposites. Retain the denominator as given.

**c.** Change the given sign preceding the fraction and all the terms of the denominator to their opposites. Retain the numerator as given.

**3.** Terms of a sum may be interchanged without any change in sign.

$b+a=a+b$    or    $-y+x=x-y$

**III. Sample Solutions**

**1.** Write $\dfrac{y-x}{x^2-y^2}$ in simplest form.

$$\dfrac{y-x}{x^2-y^2}=\dfrac{y-x}{(x+y)(x-y)}=-\dfrac{(x-y)}{(x+y)(x-y)}=-\dfrac{1}{x+y}\qquad Answer:\ -\dfrac{1}{x+y}$$

**2.** Multiply as indicated:

$$\dfrac{x^2-8x+15}{x^2-10x+25}\cdot\dfrac{10+3x-x^2}{15-2x-x^2}=\dfrac{x^2-8x+15}{x^2-10x+25}\cdot\dfrac{x^2-3x-10}{x^2+2x-15}$$

$$=\dfrac{(x-5)(x-3)}{(x-5)(x-5)}\cdot\dfrac{(x-5)(x+2)}{(x+5)(x-3)}=\dfrac{x+2}{x+5}$$

*Answer:* $\dfrac{x+2}{x+5}$

**428**

3. Combine: $\dfrac{3}{x+3} - \dfrac{2}{3-x} - \dfrac{4x}{x^2-9}$

$$\dfrac{3}{x+3} - \dfrac{2}{3-x} - \dfrac{4x}{x^2-9} = \dfrac{3}{x+3} + \dfrac{2}{x-3} - \dfrac{4x}{(x+3)(x-3)}$$

$$= \dfrac{3(x-3)}{(x+3)(x-3)} + \dfrac{2(x+3)}{(x-3)(x+3)} - \dfrac{4x}{(x+3)(x-3)}$$

$$= \dfrac{3(x-3)+2(x+3)-4x}{(x+3)(x-3)} = \dfrac{3x-9+2x+6-4x}{(x+3)(x-3)}$$

$$= \dfrac{x-3}{(x+3)(x-3)} = \dfrac{x-3}{x-3} \cdot \dfrac{1}{x+3} = 1 \cdot \dfrac{1}{x+3} = \dfrac{1}{x+3} \qquad \textit{Answer: } \dfrac{1}{x+3}$$

## DIAGNOSTIC TEST

Which of the following are true?

1. a. $\dfrac{-25}{+5} = \dfrac{+25}{-5}$  b. $\dfrac{-m^2}{+m} = +\dfrac{+m^2}{-m}$  c. $+\dfrac{-30}{-6} = -\dfrac{+30}{-6}$

2. a. $\dfrac{r-s}{s-r} = \dfrac{s-r}{r-s}$  b. $\dfrac{7-t}{8-5t+t^2} = \dfrac{t-7}{t^2-5t+8}$  c. $\dfrac{n-5}{25-n^2} = +\dfrac{n-5}{n^2-25}$

3. Write $\dfrac{a-b}{b^2-a^2}$ in simplest form.

4. Multiply: $\dfrac{a^2-49}{a^2+4a+4} \cdot \dfrac{2-a-a^2}{49-a^2}$        5. Divide: $\dfrac{c+d}{c-d} \div \dfrac{d+c}{d-c}$

6. Combine: $\dfrac{7}{y+2} + \dfrac{5}{2-y} + \dfrac{y+18}{y^2-4}$

## RELATED PRACTICE EXAMPLES

Which of the following are true?

1. a. $\dfrac{+6}{+3} = \dfrac{-6}{-3}$        b. $\dfrac{+10}{-2} = \dfrac{-10}{+2}$        c. $\dfrac{-15}{+3} = \dfrac{-15}{-3}$

   d. $\dfrac{+x^5}{+x^2} = \dfrac{-x^5}{+x^2}$        e. $\dfrac{-x^2}{+a} = \dfrac{+x^2}{-a}$        f. $+\dfrac{+20}{+5} = -\dfrac{-20}{+5}$

   g. $-\dfrac{+18}{+2} = -\dfrac{-18}{-2}$        h. $-\dfrac{+a^2}{-a} = +\dfrac{-a^2}{-a}$        i. $-\dfrac{-d^6}{+d^2} = -\dfrac{+d^6}{+d^2}$

**2. a.** $\dfrac{3}{m-n} = \dfrac{-3}{n-m}$

**b.** $\dfrac{a-b}{b-a} = \dfrac{b-a}{a-b}$

**c.** $\dfrac{5}{s^2-r^2} = \dfrac{5}{r^2-s^2}$

**d.** $\dfrac{a+2}{a^2-3a+7} = \dfrac{2+a}{7-3a+a^2}$

**e.** $\dfrac{b-2}{b^2+3b-5} = \dfrac{2-b}{5-3b-b^2}$

**f.** $+\dfrac{c-d}{d^2-c^2} = -\dfrac{d-c}{d^2-c^2}$

**g.** $-\dfrac{3a}{b-a} = +\dfrac{3a}{b-a}$

**h.** $+\dfrac{m}{m+n} = -\dfrac{m}{n+m}$

**i.** $-\dfrac{a-3}{9-a^2} = +\dfrac{a-3}{a^2-9}$

**j.** $+\dfrac{x-w}{w^2-x^2} = -\dfrac{w-x}{w^2-x^2}$

**3.** Write each of the following algebraic fractions in simplest form:

**a.** $\dfrac{2a+b}{b+2a}$

**b.** $\dfrac{3x-y}{y-3x}$

**c.** $\dfrac{a+3}{9-a^2}$

**d.** $\dfrac{2a-2b}{b-a}$

**e.** $\dfrac{2x-8}{12-3x}$

**f.** $\dfrac{c^2-b^2}{b^2+2bc+c^2}$

**g.** $\dfrac{x+3}{6-x-x^2}$

**h.** $\dfrac{2ab-a^2-b^2}{5a^2-5b^2}$

**i.** $\dfrac{3c^2-3d^2}{2cd-c^2-d^2}$

**4.** Multiply as indicated:

**a.** $\dfrac{9-a^2}{a-5} \cdot \dfrac{25-a^2}{a-3}$

**b.** $\dfrac{x^2-16}{x^2+2x+1} \cdot \dfrac{3+2x-x^2}{16-x^2}$

**c.** $\dfrac{w^2+9w+20}{w^2+7w+12} \cdot \dfrac{15+2w-w^2}{45+4w-w^2}$

**d.** $\dfrac{d^2-36}{8-2d-d^2} \cdot \dfrac{d^2-4d+4}{12+4d-d^2}$

**5.** Divide as indicated:

**a.** $\dfrac{3}{x-y} \div \dfrac{6}{y-x}$

**b.** $\dfrac{b^2-64}{b^2-10b+25} \div \dfrac{64-b^2}{10b-b^2-25}$

**c.** $\dfrac{a^2-2ax+x^2}{4a+4x} \div \dfrac{x^2-a^2}{8a+8x}$

**d.** $\dfrac{81-c^2}{c^2+2c-15} \div \dfrac{c^2-18c+81}{9-c^2}$

**6.** Combine:

**a.** $\dfrac{2}{x-y} - \dfrac{1}{y-x}$

**b.** $\dfrac{3}{c^2-d^2} + \dfrac{5}{d-c}$

**c.** $\dfrac{a-2}{4-a^2} + \dfrac{2-a}{a^2-4}$

**d.** $\dfrac{x}{2x-8} + \dfrac{4x}{16-x^2}$

**e.** $\dfrac{8b}{3-b} + \dfrac{3b}{b+3}$

**f.** $\dfrac{2}{a+3} + \dfrac{4}{3-a} + \dfrac{3a+15}{a^2-9}$

**g.** $\dfrac{m+n}{m-n} - \dfrac{m-n}{m+n} + \dfrac{m^2}{n^2-m^2}$

**h.** $\dfrac{x}{64-x^2} + \dfrac{2}{3x-24} - \dfrac{3}{2x+16}$

# Mixed Expressions and Complex Fractions

**I. Aim**  To simplify (1) mixed expressions involving multiplication or division and (2) complex fractions.

**II. Procedure**

**1. a.** A *mixed expression* consists of the sum or difference of terms with integral coefficients and an algebraic fraction such as:

$$n + \frac{5}{n} \quad \text{or} \quad x + 3 - \frac{6}{x - 2}$$

**b.** To change a mixed expression to a single algebraic fraction, write the integral term as a fraction with the denominator 1 and then combine it with the given algebraic fraction.

**c.** To simplify mixed expressions involving multiplication or division, change the mixed expressions to algebraic fractions and then multiply or divide as indicated. See sample solution 2.

**2. a.** A *complex fraction* is a fraction that has one or more fractions in either its numerator or denominator or both.

**b.** To simplify a complex fraction, express it first as an indicated division, using the ÷ symbol, and then divide as indicated. See sample solutions 3 and 4.

**c.** When there is a mixed expression in either the numerator or denominator or in both, express it as a fraction. Any polynomial may be considered an algebraic fraction with the denominator 1. See sample solution 5.

**III. Sample Solutions**

**1.** Write $\dfrac{a^2 - 3}{4a} + 3a$ as an algebraic fraction.

$$\frac{a^2 - 3}{4a} + 3a = \frac{a^2 - 3}{4a} + \frac{3a}{1} = \frac{a^2 - 3}{4a} + \frac{3a}{1} \cdot \frac{4a}{4a} = \frac{a^2 - 3}{4a} + \frac{12a^2}{4a}$$

$$= \frac{a^2 - 3 + 12a^2}{4a} = \frac{13a^2 - 3}{4a}$$

*Answer:* $\dfrac{13a^2 - 3}{4a}$

**431**

**2.** Simplify $\left(\dfrac{a}{b}+2\right)\left(\dfrac{a}{b}+3\right)$

$$\left(\frac{a}{b}+2\right)\left(\frac{a}{b}+3\right)=\frac{a+2b}{b}\cdot\frac{a+3b}{b}=\frac{(a+2b)(a+3b)}{b^2}\text{ or }\frac{a^2+5ab+6b^2}{b^2}$$

*Answer:* $\dfrac{a^2+5ab+6b^2}{b^2}$

Simplify each of the following:

**3.** $\dfrac{\frac{2}{3}}{\frac{5}{7}}=\frac{2}{3}\div\frac{5}{7}$

$$=\frac{2}{3}\times\frac{7}{5}=\frac{14}{15}$$

**4.** $\dfrac{\dfrac{a+b}{a}}{\dfrac{a-b}{a^2}}=\dfrac{a+b}{a}\div\dfrac{a-b}{a^2}$

$$=\frac{a+b}{a}\times\frac{a^2}{a-b}=\frac{a^2(a+b)}{a(a-b)}=\frac{a(a+b)}{a-b}$$

*Answer:* $\dfrac{14}{15}$  ·  *Answer:* $\dfrac{a(a+b)}{a-b}$

**5.** $\dfrac{x+\dfrac{12}{x-7}}{x-3}=\left(\dfrac{x}{1}+\dfrac{12}{x-7}\right)\div\dfrac{x-3}{1}=\left(\dfrac{x^2-7x}{x-7}+\dfrac{12}{x-7}\right)\div\dfrac{x-3}{1}$

$$=\left(\frac{x^2-7x+12}{x-7}\right)\div\frac{x-3}{1}=\left(\frac{x^2-7x+12}{x-7}\right)\times\frac{1}{x-3}$$

$$=\frac{(x-4)(x-3)}{x-7}\times\frac{1}{x-3}=\frac{x-4}{x-7}$$

*Answer:* $\dfrac{x-4}{x-7}$

## DIAGNOSTIC TEST

**1.** Express $\dfrac{4x^2-7}{3x}+2x$ as an algebraic fraction.

**2.** Simplify: $\left(\dfrac{c}{d}-1\right)\left(1+\dfrac{d}{c}\right)$

**3.** Simplify: $\left(\dfrac{b^2}{4}-4x^2\right)\div\left(\dfrac{b}{2}+2x\right)$

Express each of the following as an algebraic fraction:

**4.** $\dfrac{8m}{m-4} - m$

**5.** $x - y + \dfrac{y^2}{x+y}$

Simplify:

**6.** $\dfrac{\frac{3}{8}}{\frac{9}{16}}$

**7.** $\dfrac{\dfrac{x^2-y^2}{x}}{\dfrac{x+y}{x^2}}$

**8.** $\dfrac{b+2}{b-\dfrac{b}{b+3}}$

**9.** $\dfrac{c-\dfrac{d^2}{c}}{1+\dfrac{d}{c}}$

**10.** $\dfrac{\dfrac{n}{n-2}+\dfrac{n}{n+2}}{\dfrac{n}{n-2}-\dfrac{n}{n+2}}$

## RELATED PRACTICE EXAMPLES

Express each of the following as an algebraic fraction in simplest form:

**1. a.** $10 - \dfrac{b}{a}$

**b.** $\dfrac{ax+2}{ax} + 4$

**c.** $\dfrac{4cd^2-3}{2cd} - 5d$

**d.** $n + 8 + \dfrac{12}{n}$

**e.** $1 + \dfrac{r-s}{3r}$

**f.** $y - \dfrac{y-6}{4y} + 7$

Simplify as indicated:

**2. a.** $\left(2 + \dfrac{3}{a}\right)\left(2 - \dfrac{3}{a}\right)$

**b.** $\left(\dfrac{a}{x} + 3\right)\left(\dfrac{a}{x} + 2\right)$

**c.** $\left(1 + \dfrac{y^2}{x^2}\right)\left(1 - \dfrac{y^2}{x^2}\right)$

**d.** $\left(\dfrac{c}{d} - \dfrac{d}{c}\right)\left(\dfrac{c^2}{d} + d\right)$

**3. a.** $\left(4 + \dfrac{1}{x}\right) \div \left(4 - \dfrac{1}{x}\right)$

**b.** $\left(1 - \dfrac{b}{a}\right) \div \left(a - \dfrac{b^2}{a}\right)$

**c.** $\left(\dfrac{a-b}{b} - \dfrac{a-b}{a}\right) \div \left(\dfrac{a}{b} - \dfrac{b}{a}\right)$

**d.** $\left(\dfrac{m^2}{9} - 9n^2\right) \div \left(\dfrac{m}{3} + 3n\right)$

Express each of the following as an algebraic fraction:

**4. a.** $a + \dfrac{a^2}{a-5}$

**b.** $\dfrac{3}{x+y} + 4$

**c.** $a + b - \dfrac{2ab}{a-b}$

**5. a.** $1 - \dfrac{x+y}{x-y}$

**b.** $b - 5 + \dfrac{3b-7}{b-3}$

**c.** $a + 3b - \dfrac{a^2+b^2}{a+3b}$

Simplify:

6. **a.** $\dfrac{\frac{1}{2}}{\frac{3}{4}}$    **b.** $\dfrac{\frac{5}{3}}{\frac{10}{9}}$    **c.** $\dfrac{2\frac{1}{2}}{1\frac{3}{4}}$    **d.** $\dfrac{4+\frac{1}{2}}{\frac{5}{8}}$    **e.** $\dfrac{3+\frac{3}{4}}{2-\frac{1}{2}}$

7. **a.** $\dfrac{\frac{a}{x}}{\frac{a^2}{x}}$    **b.** $\dfrac{\frac{3b^2}{8c^3}}{\frac{7b}{16c^2}}$    **c.** $\dfrac{\frac{a}{a+b}}{\frac{b}{a+b}}$    **d.** $\dfrac{\frac{c^2-d^2}{c}}{\frac{c-d}{c^2}}$    **e.** $\dfrac{\frac{a^2-b^2}{c+d}}{\frac{a+b}{c^2-d^2}}$

8. **a.** $\dfrac{1}{1-\frac{b}{a}}$    **b.** $\dfrac{x}{x-\frac{x}{2}}$    **c.** $\dfrac{\frac{a+2}{2a}}{a^2-4}$    **d.** $\dfrac{x-\frac{35}{x-2}}{x+5}$

9. **a.** $\dfrac{\frac{m^2}{2}-2}{\frac{m}{2}-1}$    **b.** $\dfrac{a^2-\frac{b^2}{9}}{a+\frac{b}{3}}$    **c.** $\dfrac{a-\frac{b^2}{a}}{\frac{a}{b}-\frac{b}{a}}$    **d.** $\dfrac{\frac{x}{y}-\frac{y}{x}}{\frac{x}{y}+\frac{y}{x}}$

10. **a.** $\dfrac{x+5+\frac{6}{x}}{x-\frac{9}{x}}$    **b.** $\dfrac{\frac{x}{x+y}}{1-\frac{y}{x+y}}$    **c.** $\dfrac{c-\frac{cd}{c+d}}{c+\frac{cd}{c-d}}$    **d.** $\dfrac{\frac{x-y}{x}+\frac{x-y}{y}}{\frac{x-y}{y}-\frac{x-y}{x}}$

## COMPETENCY CHECK TEST

In each of the following, select the letter corresponding to your answer:

1. The factors of $x^2-10x-24$ are:

   **a.** $(x-6)(x-4)$    **b.** $(x-8)(x+3)$    **c.** $(x-4)(x+6)$    **d.** $(x+2)(x-12)$

2. The simplest form of $\dfrac{12c^6d^2}{18c^2d^8}$ is:

   **a.** $\dfrac{2c^3}{3d^4}$      **b.** $\dfrac{2c^4}{3d^6}$      **c.** $\frac{2}{3}c^3d^4$      **d.** answer not given

3. The sum of $\dfrac{m}{m+n}+\dfrac{n}{m+n}$ is:

   **a.** $m+n$       **b.** $m$       **c.** $1$       **d.** $n$

4. The product of $10x^2y^2\cdot\dfrac{3x-y}{2xy^2}$ is:

   **a.** $15x^2y$      **b.** $15x-y$      **c.** $15x^2y^2$      **d.** $15x^2-5xy$

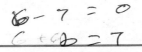
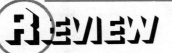

1. Which replacements of the variables must be excluded in:

   **a.** $\dfrac{9}{11x}$?
   **b.** $\dfrac{4n}{(n-5)(3n+2)}$?
   **c.** $\dfrac{b-10}{b^2-b-42}$?

Assume that the denominators in each of the following problems have a non-zero value.

2. Write each of the following algebraic fractions in simplest form:

   **a.** $\dfrac{32b^3x^9}{72x^4y^2}$
   **b.** $\dfrac{15a-40}{9a^2-64}$
   **c.** $\dfrac{64x^2-48xy+9y^2}{16x^2+34xy-15y^2}$

3. Multiply as indicated:

   **a.** $\dfrac{8r^2s^3}{3x^6y} \cdot \dfrac{18x^5}{25r^3s}$
   **b.** $\dfrac{n^2-36}{n^2+7n+10} \cdot \dfrac{n^2-n-6}{n^2-4n-12}$

4. Divide as indicated:

   **a.** $\dfrac{7a-21b}{9} \div 14a^2b^3$
   **b.** $\dfrac{3c-6d}{15cd} \div \dfrac{c^2-4d^2}{c^2+4cd+4d^2}$

5. Add or subtract as indicated:

   **a.** $\dfrac{6x}{5y} + \dfrac{9x}{5y}$
   **b.** $\dfrac{4t}{t-2} - \dfrac{8}{t-2}$

   **c.** $\dfrac{n^2-7n}{n^2-25} - \dfrac{3n-25}{n^2-25}$

6. Find the least possible algebraic expression that can be divided exactly by:

   **a.** $18r^3x^7$ and $24r^2x^9$
   **b.** $mn$, $m^2+6m+9n^2$, and $m^3-9mn$

7. Find the least common denominator of:

   **a.** $\dfrac{1}{xy}$ and $\dfrac{1}{x-y}$
   **b.** $\dfrac{1}{c^2-64}$, $\dfrac{1}{c^2-16c+64}$,

   and $\dfrac{1}{c^2-10c+16}$

8. Find the equivalent fraction with the specified denominator:

   **a.** $\dfrac{2a^4}{15c^3d^7} = \dfrac{?}{45a^2c^4d^7}$
   **b.** $\dfrac{x-4}{6x(x+5)} = \dfrac{?}{12x^2(x+5)(x-8)}$

9. Add or subtract as indicated:

   **a.** $\dfrac{6r-5s}{12r} - \dfrac{3r+s}{18s}$
   **b.** $\dfrac{5c}{14d^4} + \dfrac{11c^2}{42d^3} - \dfrac{6c^3}{63d^2}$

**10.** Express $n + 5 - \dfrac{n - 8}{3n}$ as an algebraic fraction.

**11.** Simplify:

$$\left(\frac{1}{y^2} - \frac{1}{x^2}\right)\left(\frac{xy}{x + y}\right)$$

**12.** Simplify:

$$\left(\frac{1}{c} + \frac{1}{d}\right) \div \left(c - \frac{d^2}{c}\right)$$

**13.** Add or subtract as indicated:

a. $\dfrac{3m}{m + 4n} + \dfrac{12m}{m + 4n}$

b. $\dfrac{a - 3x}{2a + 4x} - \dfrac{x}{4a + 2x}$

c. $\dfrac{5}{x} + \dfrac{3}{x + y} - \dfrac{6}{x - y}$

d. $\dfrac{4}{n^2 - 1} - \dfrac{7}{n^2 + 2n + 1} + \dfrac{2}{n^2 - 5n + 4}$

**14.** Combine:

$$\frac{2x}{2 - x} + \frac{3x - 2}{x^2 - x - 2} - \frac{x + 2}{x + 1}$$

**15.** Simplify:

$$\frac{c + d - \dfrac{cd}{c + d}}{c - \dfrac{d^2}{c + d}}$$

## MAINTENANCE PRACTICE IN ARITHMETIC

**1.** Add:
896,745
87,586
375,669

**2.** Subtract:
15,000,000
9,601,877

**3.** Multiply:
46,598
8,739

**4.** Divide:
$809\overline{)4,051,472}$

**5.** Add:
$6\frac{5}{12} + \frac{7}{8} + 1\frac{9}{10}$

**6.** Subtract:
$14\frac{1}{2} - \frac{7}{10}$

**7.** Multiply:
$8\frac{1}{3} \times 7\frac{2}{5}$

**8.** Divide:
$14 \div 1\frac{1}{8}$

**9.** Add:
.8 + 3.7 + .584

**10.** Subtract:
1.6 − 1.599

**11.** Multiply:
.067 × 6.5

**12.** Divide:
$.05\overline{)4}$

**13.** Find $8\frac{1}{2}\%$ of $1,500.

**14.** What percent of 64 is 96?

**15.** $60 is 5% of what amount?

## Keyed Achievement Test

1. **a.** $(6w-5)(9w-7)$   **b.** $(8x-3y)(9x+8y)$  9-1
2. **a.** $(n-7)(n-7)$   **b.** $(10t+9x)(10t-9x)$  9-3
3. **a.** $(r-9)(r-9)$   **b.** $(7x-3y^2)^2$  9-4
4. **a.** $(a+10)(a-15)$   **b.** $(4c+11)(4c+5)$  9-5

Factor each of the following completely:

5. **a.** $25n^2+100y^2$   **b.** $54a^7c^3-81a^4c^5+72a^3c^9$  9-6
6. **a.** $16x^2-9y^2$   **b.** $25a^4c^2-49d^6$  9-8
7. **a.** $81s^2-72s+16$   **b.** $49m^2+154mn+121n^2$  9-9
8. **a.** $r^2-18r-63$   **b.** $w^2-w-2$  9-10
9. **a.** $18b^2+61b-90$   **b.** $48r^2-86r+35$  9-11
10. **a.** $64a^2y^2-16a^2$   **b.** $x^5-81x$  9-12

Write each of the following algebraic fractions in simplest form:

11. **a.** $\dfrac{24x^5y^4z^2}{64a^2x^6y^3}$   **b.** $\dfrac{b-c}{c-b}$   **c.** $\dfrac{6n-4}{9n^2-12n+4}$  10-2

12. Multiply as indicated:  10-3

   **a.** $\dfrac{25r^4s^8}{12t^5}\cdot\dfrac{18s^2t^6}{35r^6x^7}\cdot\dfrac{21x^3}{54s^3t}$   **b.** $\dfrac{4x^2-24x}{x^2-9}\cdot\dfrac{x^2-9x+18}{x^2+12x+36}$

13. Divide as indicated:  10-4

   **a.** $36b^9c^6\div\dfrac{9a^2b^5c^7}{5d^2}$   **b.** $\dfrac{n^2-5n+6}{4n^2-64}\div\dfrac{n^2-n-6}{n^2-9n+20}$

14. Add or subtract as indicated:  10-5 thru 10-11

   **a.** $\dfrac{9c^2}{3c-d}-\dfrac{d^2}{3c-d}$

   **b.** $\dfrac{r-4s}{6r^2s}-\dfrac{4r-s}{8rs^2}$

   **c.** $\dfrac{b-2}{b-9}-\dfrac{b}{b+2}+\dfrac{72}{b^2-7b-18}$

   **d.** $\dfrac{8}{y^2-5y}+\dfrac{4}{y^2-10y+25}+\dfrac{5}{25-y^2}$

15. Simplify:  10-12

   $$\dfrac{1+\dfrac{3b}{a}}{a-\dfrac{9b^2}{a}}$$

# INTRODUCTION

$M$any algebraic problems and applications using formulas involve the solution of a fractional equation with fractional coefficients. Sometimes these equations are expressed as proportions.

A *fractional equation* is one which contains a variable in one or more denominators. Equations with fractional coefficients are also included in this chapter because the principles used in their solution are the same as those used in the solution of fractional equations.

The equations $\frac{2}{3}x = 10$ and $\frac{2x}{3} = 10$ are equivalent and each is considered to be an equation with fractional coefficients.

An equation such as $\frac{2}{x} = 6$ is a fractional equation but $\frac{x}{2} = 6$ or $\frac{1}{2}x = 6$ is an equation with fractional coefficients.

A *proportion* is a statement that two ratios are equal. Since ratios generally are written in fraction form, a proportion may be treated as a fractional equation. Also see Exercise 16-1.

When solving a fractional equation, we shall find it necessary to eliminate the fractions which the equation contains. To transform the given equation to a simple equation we multiply both sides of the equation by the least common denominator of the given fractions, employing the principle that the products are equal when equals are multiplied by equals.

# Fractional Equations and Problem Solving

**11**

# Equations with Fractional Coefficients

## I. Aim

To solve an equation in one variable with fractional coefficients.

## II. Procedure

1. Transform the given equation to an equivalent equation with no fractions.
   a. Determine the least common denominator (LCD) of all the given fractions in the equation. See Exercise 10-6.
   b. Multiply both sides of the given equation by the least common denominator.

2. Solve the resulting equation.

3. Check by substituting the root for the variable in the given equation.

4. When the given equation is in the form of a proportion, either the above method or the method of cross-products (see Exercise 16-1) may be used.

5. Note that a fractional part of a quantity may be expressed as an indicated product of a fractional coefficient and this quantity or as a fraction.

   For example: One-fourth of an amount ($a$) may be indicated as $\frac{1}{4}a$ or $\frac{a}{4}$.

## III. Sample Solutions

1. Solve and check: $\dfrac{2a - 3}{6} = \dfrac{2a + 3}{10}$

   The LCD is 30.

   *Solution:*

   $$\frac{2a - 3}{6} = \frac{2a + 3}{10}$$

   $$30 \cdot \frac{2a - 3}{6} = 30 \cdot \frac{2a + 3}{10}$$

   $$5(2a - 3) = 3(2a + 3)$$

   $$10a - 15 = 6a + 9$$

   $$10a - 6a - 15 + 15 = 6a - 6a + 9 + 15$$

   $$4a = 24$$

   $$a = 6$$

   *Check:*

   $$\frac{2a - 3}{6} = \frac{2a + 3}{10}$$

   $$\frac{2 \cdot 6 - 3}{6} = \frac{2 \cdot 6 + 3}{10}?$$

   $$\frac{12 - 3}{6} = \frac{12 + 3}{10}$$

   $$\frac{9}{6} = \frac{15}{10}$$

   $$1\tfrac{1}{2} = 1\tfrac{1}{2} ✔$$

   *Answer:* 6

**440**

**2.** Solve and check: $\dfrac{x}{2} + \dfrac{x}{3} = 5$

The LCD is 6.

*Solution:* $\dfrac{x}{2} + \dfrac{x}{3} = 5$

$$6\left(\dfrac{x}{2} + \dfrac{x}{3}\right) = 6(5)$$

$$\left(6 \cdot \dfrac{x}{2}\right) + \left(6 \cdot \dfrac{x}{3}\right) = 6(5)$$

$$3x + 2x = 30$$
$$5x = 30$$
$$x = 6 \quad \text{Answer: 6}$$

*Check:*

$$\dfrac{x}{2} + \dfrac{x}{3} = 5$$
$$\dfrac{6}{2} + \dfrac{6}{3} = 5\,?$$
$$3 + 2 = 5$$
$$5 = 5 \checkmark$$

**3.** Solve and check: $\dfrac{x + 6}{2} - \dfrac{2x - 3}{5} = \dfrac{3x + 4}{4}$

The LCD is 20.

*Solution:* $\dfrac{x + 6}{2} - \dfrac{2x - 3}{5} = \dfrac{3x + 4}{4}$

$$20\left(\dfrac{x + 6}{2} - \dfrac{2x - 3}{5}\right) = 20\left(\dfrac{3x + 4}{4}\right)$$

$$20\left(\dfrac{x + 6}{2}\right) - 20\left(\dfrac{2x - 3}{5}\right) = 20\left(\dfrac{3x + 4}{4}\right)$$

$$10(x + 6) - 4(2x - 3) = 5(3x + 4)$$
$$10x + 60 - 8x + 12 = 15x + 20$$
$$2x + 72 = 15x + 20$$
$$2x - 15x + 72 - 72 = 15x - 15x + 20 - 72$$
$$-13x = -52$$
$$x = 4$$

*Check:*

$$\dfrac{x + 6}{2} - \dfrac{2x - 3}{5} = \dfrac{3x + 4}{4}$$

$$\dfrac{4 + 6}{2} - \dfrac{(2 \cdot 4) - 3}{5} = \dfrac{(3 \cdot 4) + 4}{4}\,?$$

$$\dfrac{10}{2} - \dfrac{8 - 3}{5} = \dfrac{12 + 4}{4}$$

$$\dfrac{10}{2} - \dfrac{5}{5} = \dfrac{16}{4}$$
$$5 - 1 = 4$$
$$4 = 4 \checkmark \quad \text{Answer: 4}$$

## DIAGNOSTIC TEST

1. Express in two ways: Four-fifths of the number $y$.
2. Rewrite $\frac{7}{12}x$ as a fraction without the fractional coefficient.

Solve and check:

3. $\frac{1}{5}r = 4$

4. $\frac{b}{6} = 8$

5. $\frac{7}{8}y = 63$

6. $\frac{3t}{5} = 6$

7. $\frac{a}{16} = \frac{5}{4}$

8. $\frac{b}{2} = \frac{2}{3}$

9. $\frac{3}{10} = \frac{m}{15}$

10. $\frac{9}{2} = \frac{3y}{4}$

11. $\frac{x-5}{12} = \frac{5}{6}$

12. $\frac{3b+3}{4} = \frac{b}{2}$

13. $\frac{5c-2}{3} = \frac{8-3c}{5}$

14. $\frac{x}{3} + \frac{x}{4} = 14$

15. $\frac{y}{2} = \frac{5}{6} - \frac{y}{3}$

16. $x + \frac{x}{2} + \frac{x}{3} + \frac{x}{4} + \frac{x}{6} = 27$

17. $\frac{b-1}{3} + \frac{b+5}{5} = 6$

18. $\frac{m+2}{2} - \frac{m-6}{8} = \frac{m+11}{4}$

19. $\frac{5x-1}{2} - \frac{3x+5}{5} = \frac{7x+13}{6}$

## *RELATED PRACTICE EXAMPLES*

1. Express each of the following in two ways:
   a. One-tenth of the number $n$    b. One-ninth of the number $r$
   c. Two-fifths of the number $x$    d. Seven-eighths of the number $b$
   e. Five-twelfths of the number $t$  f. Eleven-sixteenths of the number $y$
2. Rewrite each as a fraction without the fractional coefficient:

   a. $\frac{1}{5}c$    b. $\frac{1}{7}y$    c. $\frac{1}{10}m$    d. $\frac{3}{4}b$    e. $\frac{5}{9}x$    f. $\frac{15}{4}w$    g. $\frac{22}{7}d$

Solve and Check:

3. a. $\frac{1}{7}x = 4$    b. $\frac{1}{3}n = 2$    c. $\frac{1}{6}b = 12$    d. $\frac{1}{4}y = 20$    e. $24 = \frac{1}{8}t$

4. a. $\frac{n}{5} = 3$    b. $\frac{c}{9} = 6$    c. $\frac{w}{8} = 16$    d. $45 = \frac{r}{5}$    e. $\frac{x}{18} = 12$

5. a. $\frac{3}{4}b = 21$    b. $\frac{2}{3}s = 4$    c. $35 = \frac{7}{10}n$    d. $\frac{5}{8}c = 2$    e. $\frac{9}{5}b = 10$

6. a. $\frac{2a}{3} = 4$    b. $\frac{3d}{4} = 6$    c. $\frac{5m}{8} = 8$    d. $21 = \frac{7x}{12}$    e. $\frac{11c}{4} = 1$

7. a. $\frac{a}{2} = \frac{1}{2}$    b. $\frac{c}{15} = \frac{3}{5}$    c. $\frac{5}{6} = \frac{a}{30}$    d. $\frac{11}{72} = \frac{n}{8}$    e. $\frac{7}{3} = \frac{m}{21}$

8. a. $\frac{b}{4} = \frac{2}{5}$    b. $\frac{x}{8} = \frac{10}{7}$    c. $\frac{r}{10} = \frac{4}{3}$    d. $\frac{1}{5} = \frac{a}{2}$    e. $\frac{9}{10} = \frac{b}{11}$

**9. a.** $\dfrac{r}{8} = \dfrac{5}{6}$    **b.** $\dfrac{a}{25} = \dfrac{12}{10}$    **c.** $\dfrac{3}{16} = \dfrac{y}{6}$    **d.** $\dfrac{b}{4} = \dfrac{1}{6}$    **e.** $\dfrac{15}{8} = \dfrac{x}{10}$

**10. a.** $\dfrac{7m}{9} = \dfrac{14}{3}$    **b.** $\dfrac{15}{2} = \dfrac{5a}{12}$    **c.** $\dfrac{3}{4} = \dfrac{3b}{16}$    **d.** $\dfrac{9b}{10} = \dfrac{3}{4}$    **e.** $\dfrac{3x}{9} = \dfrac{1}{6}$

**11. a.** $\dfrac{n+4}{9} = \dfrac{5}{9}$    **b.** $\dfrac{1}{4} = \dfrac{x+1}{8}$    **c.** $\dfrac{a+2}{2} = 2$    **d.** $\dfrac{t-5}{6} = 7$

   **e.** $\dfrac{x-2}{6} = \dfrac{3}{4}$    **f.** $\dfrac{4x+5}{6} = \dfrac{7}{2}$    **g.** $\dfrac{4}{5} = \dfrac{5x-9}{20}$    **h.** $7 = \dfrac{y-2}{4}$

**12. a.** $\dfrac{x+2}{4} = \dfrac{x}{2}$    **b.** $\dfrac{a+8}{3} = \dfrac{a}{5}$    **c.** $b = \dfrac{5b+3}{6}$    **d.** $\dfrac{c-2}{6} = \dfrac{c}{4}$

   **e.** $\dfrac{x}{5} = \dfrac{x-3}{2}$    **f.** $\dfrac{a}{8} = \dfrac{2a-5}{6}$    **g.** $\dfrac{3b+7}{5} = \dfrac{2b}{3}$    **h.** $\dfrac{5a-6}{5} = \dfrac{2a}{5}$

**13. a.** $\dfrac{x+2}{2} = \dfrac{x+6}{4}$    **b.** $\dfrac{b-2}{8} = \dfrac{b+4}{24}$    **c.** $\dfrac{x-5}{4} = \dfrac{x-2}{3}$

   **d.** $\dfrac{5-x}{6} = \dfrac{2+x}{8}$    **e.** $\dfrac{3x+1}{5} = \dfrac{2x+5}{5}$    **f.** $\dfrac{6x+7}{10} = \dfrac{2x+9}{6}$

   **g.** $\dfrac{3b+34}{14} = \dfrac{b+44}{21}$    **h.** $\dfrac{2a-5}{15} = \dfrac{3a-20}{10}$    **i.** $\dfrac{6-c}{2} = \dfrac{2c+9}{4}$

**14. a.** $\dfrac{x}{3} + \dfrac{x}{2} = 10$    **b.** $\dfrac{a}{4} - \dfrac{a}{8} = 3$    **c.** $4 = \dfrac{b}{6} + \dfrac{b}{3}$

   **d.** $2 = \dfrac{x}{3} - \dfrac{x}{5}$    **e.** $\dfrac{r}{9} + 4 = 6$    **f.** $\dfrac{n}{10} = 9 - \dfrac{n}{5}$

   **g.** $x = \dfrac{x}{3} + 4$    **h.** $6 = c - \dfrac{c}{7}$    **i.** $9 - \dfrac{3x}{4} = 0$

**15. a.** $\dfrac{x}{10} + \dfrac{x}{5} = \dfrac{9}{10}$    **b.** $\dfrac{n}{6} + \dfrac{n}{3} = \dfrac{1}{2}$    **c.** $\dfrac{x}{3} - \dfrac{x}{4} = \dfrac{1}{12}$

   **d.** $\dfrac{a}{8} - \dfrac{a}{12} = \dfrac{1}{8}$    **e.** $\dfrac{35}{48} = \dfrac{x}{16} + \dfrac{x}{12}$    **f.** $\dfrac{9}{40} = \dfrac{c}{8} - \dfrac{c}{10}$

   **g.** $\dfrac{b}{8} + \dfrac{3}{4} = \dfrac{b}{5}$    **h.** $\dfrac{5y}{6} + \dfrac{2y}{3} = \dfrac{9}{2}$    **i.** $\dfrac{4x}{5} - \dfrac{3x}{8} = \dfrac{17}{4}$

**16. a.** $x + \dfrac{x}{2} + \dfrac{x}{3} = 22$      **b.** $a - \dfrac{a}{2} + \dfrac{a}{4} = 6$

   **c.** $y + \dfrac{y}{2} + \dfrac{y}{3} + \dfrac{y}{4} = 50$      **d.** $b - \dfrac{b}{3} + \dfrac{b}{5} = 26$

   **e.** $x = 1 + \dfrac{x}{2} + \dfrac{x}{4} + \dfrac{x}{8} + \dfrac{x}{16}$      **f.** $\dfrac{x}{2} + \dfrac{2x}{3} + \dfrac{3x}{4} = 32$

   **g.** $\dfrac{5n}{6} + \dfrac{n}{4} + \dfrac{2n}{3} = 42$      **h.** $\dfrac{3x}{2} - \dfrac{5x}{16} - \dfrac{3x}{8} = 52$

**17. a.** $\dfrac{x}{2} + \dfrac{x+2}{3} = 9$  **b.** $\dfrac{y+1}{4} + \dfrac{y+5}{2} = 5$

**c.** $\dfrac{b+5}{6} + \dfrac{b+3}{5} = 4$  **d.** $\dfrac{c-2}{6} + \dfrac{c-4}{8} = \dfrac{3}{2}$

**e.** $\dfrac{b+1}{5} + \dfrac{b+2}{2} = 4$  **f.** $\dfrac{a+1}{3} + \dfrac{a+3}{4} = \dfrac{a+3}{2}$

**g.** $\dfrac{x+2}{4} + \dfrac{x+4}{6} = \dfrac{x+1}{2}$  **h.** $\dfrac{x+1}{2} + \dfrac{x+9}{4} = \dfrac{x-11}{5}$

**18. a.** $\dfrac{d+7}{2} - \dfrac{d+3}{4} = 4$  **b.** $\dfrac{n+4}{3} - \dfrac{n-5}{6} = 4$

**c.** $\dfrac{x+5}{4} - \dfrac{x+3}{8} = \dfrac{7}{4}$  **d.** $\dfrac{y-1}{12} - \dfrac{y+1}{16} = \dfrac{1}{24}$

**e.** $\dfrac{x-4}{2} - 2 = \dfrac{x-6}{4}$  **f.** $\dfrac{x-5}{2} - \dfrac{x-3}{4} = \dfrac{x-7}{10}$

**g.** $\dfrac{a+8}{3} - \dfrac{a-4}{5} = a$  **h.** $\dfrac{m+10}{4} - \dfrac{m-10}{6} = \dfrac{m+10}{2}$

**19. a.** $\dfrac{5b+3}{8} + \dfrac{3b+5}{2} = 5$  **b.** $6 = \dfrac{8x-1}{3} + \dfrac{2x+1}{5}$

**c.** $\dfrac{7x-2}{11} - \dfrac{2x-7}{3} = 1$  **d.** $\dfrac{7a+5}{8} - 2 = \dfrac{3a+15}{10}$

**e.** $\dfrac{3x-5}{3} + \dfrac{5x+3}{6} = \dfrac{13}{3}$  **f.** $\dfrac{4b-3}{6} - \dfrac{2b-5}{8} = \dfrac{7b}{16}$

**g.** $\dfrac{7m-6}{10} + \dfrac{4m+3}{5} = \dfrac{16m-13}{15}$  **h.** $\dfrac{4-3x}{8} + 2 = \dfrac{x-5}{4} - x$

## CUMULATIVE REVIEW

Solve and Check:

**1.** $\dfrac{n}{6} + \dfrac{n}{4} = 20$  **2.** $\dfrac{2x}{3} - \dfrac{5x}{12} = \dfrac{5}{2}$

**3.** $\dfrac{5a}{8} + \dfrac{a}{3} - \dfrac{7a}{8} = 4$  **4.** $\dfrac{x-3}{6} + \dfrac{x+3}{3} = 5$

**5.** $\dfrac{x}{2} - \dfrac{x+2}{5} = 2$  **6.** $\dfrac{x-5}{10} = \dfrac{x+4}{4}$

**7.** $a - \dfrac{a}{2} - \dfrac{a}{4} - \dfrac{a}{8} - \dfrac{a}{16} = 6$  **8.** $\dfrac{4y+3}{4} - \dfrac{5y}{6} = \dfrac{13}{12}$

**9.** $\dfrac{3b-4}{6} - \dfrac{5b-4}{8} = \dfrac{b+3}{12}$  **10.** $\dfrac{2x+3}{5} + \dfrac{2x+3}{6} = \dfrac{3x+4}{4}$

# Fractional Equations

**I. Aim**   To solve a fractional equation in one variable.

**II. Procedure**

1. Transform the given equation to an equivalent equation with no fractions.
   **a.** Determine the least common denominator (LCD) of all the given fractions in the equation. See exercise 10-6.
   **b.** Multiply both sides of the given equation by the least common denominator.
2. Solve the resulting equation.
3. Check by substituting the root for the variable in the given equation.
   **a.** Sometimes when both sides are multiplied by the least common denominator the new equation may have roots which are not roots of the given equation. Always check in the original (given) equation. Only an answer that satisfies the given equation is a solution.
   **b.** Solutions that make the denominator zero are excluded since the denominator in a fractional equation cannot have the value zero.
4. When the given equation is in a form of a proportion, either the above method or the method of cross-products (see Exercise 16-1) may be used.
5. Sometimes it is necessary to change the terms of the denominator to their additive inverses so that the least common denominator may be found. See Exercise 10-11.

**III. Sample Solutions**

1. Solve and check: $\dfrac{3}{4} + \dfrac{2}{3x} = \dfrac{14}{6x} - \dfrac{1}{12}$

   The LCD is $12x$.

   *Solution:* $\dfrac{3}{4} + \dfrac{2}{3x} = \dfrac{14}{6x} - \dfrac{1}{12}$

   $12x\left(\dfrac{3}{4} + \dfrac{2}{3x}\right) = 12x\left(\dfrac{14}{6x} - \dfrac{1}{12}\right)$

   $\left(12x \cdot \dfrac{3}{4}\right) + \left(12x \cdot \dfrac{2}{3x}\right) = \left(12x \cdot \dfrac{14}{6x}\right) - \left(12x \cdot \dfrac{1}{12}\right)$

   $(3x \cdot 3) + (4 \cdot 2) = (2 \cdot 14) - (x \cdot 1)$

   $9x + 8 = 28 - x$

   $9x + x + 8 - 8 = 28 - 8 - x + x$

   $10x = 20$

   $x = 2$   *Answer:* 2

   *Check:* $\dfrac{3}{4} + \dfrac{2}{3x} = \dfrac{14}{6x} - \dfrac{1}{12}$

   $\dfrac{3}{4} + \dfrac{2}{3 \cdot 2} = \dfrac{14}{6 \cdot 2} - \dfrac{1}{12}?$

   $\dfrac{3}{4} + \dfrac{2}{6} = \dfrac{14}{12} - \dfrac{1}{12}$

   $\dfrac{3}{4} + \dfrac{1}{3} = \dfrac{13}{12}$

   $\dfrac{13}{12} = \dfrac{13}{12}$ ✔

**445**

**2.** Solve and check: $\dfrac{x}{x-3} = \dfrac{2x^2+x}{x^2-9} - \dfrac{x-2}{x+3}$ if $x \neq 3$ and $x \neq -3$.

*Solution:*

$$\frac{x}{x-3} = \frac{2x^2+x}{x^2-9} - \frac{x-2}{x+3}$$

$$\frac{x}{x-3} = \frac{2x^2+x}{(x+3)(x-3)} - \frac{x-2}{x+3}$$

The LCD is $(x+3)(x-3)$.

$$(x+3)(x-3) \cdot \frac{x}{x-3} = (x+3)(x-3)\left(\frac{2x^2+x}{(x+3)(x-3)} - \frac{x-2}{x+3}\right)$$

$$(x+3)(x-3) \cdot \frac{x}{(x-3)} = (x+3)(x-3) \cdot \frac{2x^2+x}{(x+3)(x-3)}$$

$$- (x+3)(x-3) \cdot \frac{x-2}{(x+3)}$$

$$x(x+3) = 2x^2 + x - (x-3)(x-2)$$
$$x^2 + 3x = 2x^2 + x - (x^2 - 5x + 6)$$
$$x^2 + 3x = 2x^2 + x - x^2 + 5x - 6$$
$$x^2 + 3x = x^2 + 6x - 6$$
$$x^2 - x^2 + 3x - 6x = x^2 - x^2 + 6x - 6x - 6$$
$$-3x = -6$$
$$x = 2$$

*Answer:* 2

*Check:*

$$\frac{x}{x-3} = \frac{2x^2+x}{x^2-9} - \frac{x-2}{x+3}$$

$$\frac{2}{2-3} = \frac{2 \cdot 2^2 + 2}{2^2 - 9} - \frac{2-2}{2+3} \,?$$

$$\frac{2}{-1} = \frac{2 \cdot 4 + 2}{4 - 9} - \frac{0}{5}$$

$$-2 = \frac{8+2}{-5} - 0$$

$$-2 = \frac{10}{-5} - 0$$

$$-2 = -2 - 0$$

$$-2 = -2 \checkmark$$

## DIAGNOSTIC TEST

Solve and check:

1. $\dfrac{18}{r} = 6$

2. $\dfrac{3}{x} + \dfrac{5}{3} = \dfrac{19}{3x}$

3. $\dfrac{4}{5} + \dfrac{7}{4x} = \dfrac{13}{2x} - \dfrac{3}{20}$

4. $\dfrac{4}{m-3} = \dfrac{6}{m+3}$

5. $\dfrac{n-2}{n+4} = \dfrac{n+1}{n+10}$

6. $\dfrac{2}{a-3} + \dfrac{3a+1}{a+3} = 3$

7. $\dfrac{5}{x+4} + \dfrac{3}{x-1} = \dfrac{9x+4}{x^2+3x-4}$

8. $\dfrac{y}{y+2} = \dfrac{2y^2+6}{y^2-4} - \dfrac{y+1}{y-2}$

9. $\dfrac{2a-1}{2a+1} + \dfrac{1}{4a^2-1} = 2 - \dfrac{2a}{2a-1}$

10. $\dfrac{3}{5-x} + \dfrac{1}{3-x} = \dfrac{7x+3}{15-8x+x^2}$

11. $\dfrac{5}{r+4} - \dfrac{4}{4-r} = \dfrac{7r+8}{r^2-16}$

## RELATED PRACTICE EXAMPLES

Solve and check:

1. **a.** $\dfrac{8}{x} = 4$  **b.** $9 = \dfrac{6}{n}$  **c.** $\dfrac{5}{x} = \dfrac{5}{6}$  **d.** $\dfrac{2}{5} = \dfrac{10}{x}$  **e.** $16 = \dfrac{8}{3a}$

2. **a.** $\dfrac{1}{3} + \dfrac{1}{6} = \dfrac{1}{x}$  **b.** $\dfrac{2}{b} + \dfrac{1}{2} = \dfrac{5}{2b}$  **c.** $\dfrac{4}{x} + \dfrac{15}{2x} = \dfrac{23}{4}$

  **d.** $\dfrac{11}{2y} - \dfrac{2}{3y} = \dfrac{1}{6}$  **e.** $\dfrac{9}{4a} - \dfrac{1}{8} = 1$  **f.** $\dfrac{13}{2x} - \dfrac{5}{9} = \dfrac{29}{6x}$

3. **a.** $\dfrac{4}{x} + 2 = \dfrac{14}{x} - 3$  **b.** $\dfrac{2}{3y} + \dfrac{1}{4} = \dfrac{11}{6y} - \dfrac{1}{3}$

  **c.** $\dfrac{5}{2b} + \dfrac{1}{6} = \dfrac{3}{5b} + \dfrac{4}{5}$  **d.** $\dfrac{1}{2} - \dfrac{3}{2x} = \dfrac{4}{x} - \dfrac{5}{12}$

  **e.** $1 + \dfrac{1}{2x} + \dfrac{1}{3x} = \dfrac{13}{6x}$  **f.** $1 - \dfrac{5}{x} + \dfrac{5}{2x} - \dfrac{5}{3x} + \dfrac{5}{4x} = \dfrac{17}{24}$

4. **a.** $\dfrac{5}{x+1} = \dfrac{5}{8}$  **b.** $\dfrac{5}{y+2} = \dfrac{10}{y+3}$  **c.** $\dfrac{7}{x-5} = \dfrac{4}{x-1}$

  **d.** $\dfrac{3}{2x} = \dfrac{7}{5x-2}$  **e.** $\dfrac{5}{2x-5} = \dfrac{10}{3x+5}$  **f.** $\dfrac{8}{5y+6} = \dfrac{4}{5y-2}$

**5.6 a.** $\dfrac{x-3}{x+1} = \dfrac{x-6}{x-5}$    **b.** $\dfrac{a-2}{a-4} = \dfrac{a-7}{a+1}$

**c.** $\dfrac{b+1}{b+5} = \dfrac{b-4}{b-1}$    **d.** $\dfrac{a+4}{a-4} = \dfrac{a-4}{a+4}$

**e.** $\dfrac{2b}{b+2} = \dfrac{2b+8}{b+7}$    **f.** $\dfrac{2x-5}{3x-4} = \dfrac{2x-3}{3x-2}$

**6. a.** $\dfrac{15}{x} + \dfrac{9x-7}{x+2} = 9$    **b.** $\dfrac{6a-12}{a+3} + \dfrac{5}{a-2} = 6$

**c.** $\dfrac{3b-2}{b+1} = 4 - \dfrac{b+2}{b-1}$    **d.** $\dfrac{2s-4}{s-4} - 2 = \dfrac{20}{s+4}$

**7. a.** $\dfrac{4}{x+2} + \dfrac{2}{x-4} = \dfrac{30}{x^2-2x-8}$    **b.** $\dfrac{2}{y-3} - \dfrac{4}{y+3} = \dfrac{8}{y^2-9}$

**c.** $\dfrac{4}{x+2} + \dfrac{3x-2}{x^2-4} = \dfrac{4}{x-2}$    **d.** $\dfrac{3}{d+3} + \dfrac{5}{d+4} = \dfrac{12d+19}{d^2+7d+12}$

**8. a.** $\dfrac{c+2}{c+1} - \dfrac{c}{c+2} = \dfrac{4c+1}{c^2+3c+2}$

**b.** $\dfrac{m+1}{m+3} + \dfrac{m-3}{m-2} = \dfrac{2m^2-15}{m^2+m-6}$

**c.** $\dfrac{6x^2+14}{4x^2-9} - \dfrac{2x+1}{2x-3} = \dfrac{x+1}{2x+3}$

**d.** $\dfrac{y+5}{y^2-4} - \dfrac{3}{2y-4} = \dfrac{1}{2y+4}$

**9. a.** $\dfrac{5x-7}{2x-3} + \dfrac{x+2}{2x+3} - \dfrac{6}{4x^2-9} = 3$

**b.** $\dfrac{2d}{d-1} - \dfrac{3}{d^2-1} = 4 - \dfrac{2d-1}{d+1}$

**c.** $\dfrac{5t-2}{5t-3} = 1 + \dfrac{3}{5t+3} + \dfrac{2t}{25t^2-9}$

**d.** $2 - \dfrac{3x-2}{3x-1} = \dfrac{2x}{3x+1} + \dfrac{3x^2+20}{9x^2-1}$

**10. a.** $\dfrac{2}{4-x} + \dfrac{3}{4+x} = \dfrac{17}{16-x^2}$    **b.** $\dfrac{26}{1-n^2} = \dfrac{3}{1-n} - \dfrac{2}{1+n}$

**c.** $\dfrac{3}{3-x} + \dfrac{1}{2-x} = \dfrac{2+3x}{6-5x+x^2}$    **d.** $\dfrac{1+b}{5-b} - \dfrac{4}{5+b} = \dfrac{15+b^2}{25-b^2}$

**11. a.** $\dfrac{7x+5}{x+2} + \dfrac{x+1}{2-x} = 6$    **b.** $\dfrac{3}{y+3} - \dfrac{5}{3-y} = \dfrac{9y+1}{y^2-9}$

**c.** $\dfrac{4x+1}{x^2-x-6} + \dfrac{2}{3-x} = \dfrac{5}{x+2}$    **d.** $\dfrac{2m^2-25}{m^2-3m+2} + \dfrac{m-3}{2-m} = \dfrac{m+2}{m-1}$

# Equations with Decimal Fractions

**I. Aim**   To solve equations with decimal fractions.

**II. Procedure**

Solve by either of the following methods:
1. Transform the given equation to an equation with no fractions by multiplying both sides of the given equation by the power of 10 which will make the smallest fraction a whole number. Then solve this equation.
2. Or solve the given equation using the arithmetic of decimal fractions (see alternate method).
3. Check by substituting the root in the given equation.

**III. Sample Solutions**

1. Solve and check: $.08n = 9.6$

    *Solution:*

    $$.08n = 9.6$$
    $$100 \times .08n = 100 \times 9.6$$
    $$8n = 960$$
    $$n = 120$$

    *Check:*

    $$.08n = 9.6$$
    $$.08(120) = 9.6 ?$$
    $$9.6 = 9.6 ✔$$

    *Alternate Solution:*

    $$.08n = 9.6$$
    $$\frac{.08n}{.08} = \frac{9.6}{.08}$$
    $$n = 120 \quad \textit{Answer: } 120$$

2. Solve and check: $x + .05x = 42$

    *Solution:*

    $$x + .05x = 42$$
    $$100(x + .05x) = 100(42)$$
    $$100x + 5x = 4,200$$
    $$105x = 4,200$$
    $$x = 40$$

    *Check:*

    $$x + .05x = 42$$
    $$40 + .05(40) = 42 ?$$
    $$40 + 2 = 42$$
    $$42 = 42 ✔$$

    *Alternate Solution:*

    $$x + .05x = 42$$
    $$1.05x = 42$$
    $$x = 40 \quad \textit{Answer: } 40$$

## DIAGNOSTIC TEST

Solve and check:

1. $.4x = .36$                    2. $b + .3 = 1.2$

3. $1.5x - .4 = 4.1$              4. $a + .06a = 21.2$

5. $x - .25x = .35x + 24$         6. $.03(c - 10) = .07(c - 50)$

## *RELATED PRACTICE EXAMPLES*

Solve and check:

**1.**   **a.** $2x = .8$      **b.** $.3b = .6$      **c.** $.8m = 5.68$
    **d.** $.09a = 7.2$     **e.** $.04x = 100$     **f.** $1.25b = 625$

**2.**   **a.** $x + .5 = 3.6$    **b.** $c + 2 = 4.7$     **c.** $x - .4 = 6.25$
    **d.** $n - .02 = 2.5$    **e.** $.8 = t + .3$     **f.** $y + 5.1 = 9$

**3.**   **a.** $.3x + .2 = .8$    **b.** $2.4y + .15 = .87$    **c.** $3.2x - .3 = 6.1$
    **d.** $.6h - .8 = 4$     **e.** $1.02x + 4.8 = 55.8$    **f.** $.24 + 8y = .64$

**4.**   **a.** $1.2x + .3x = 4.5$   **b.** $4.2s - 3.6s = 9.6$   **c.** $.92x - 124 = .3x$
    **d.** $a + .06a = 636$    **e.** $1.2y - .08 = .8y$    **f.** $40 - 3.5x = .5x$

**5.**   **a.** $.6x + .3 = .3x + .9$          **b.** $.06r - .25 = .03r + .35$
    **c.** $x + .05x + .02x = 321$     **d.** $2x + 1.08x - 30.6 = .02x$
    **e.** $1.8m + .5m - .48 = .7m$    **f.** $1.2x + .05x = .15x + 5.5$

**6.**   **a.** $.04(x - 5) = 4$            **b.** $.02(m + 6) = .04m$
    **c.** $.05(x - 4) = .06(x - 5)$    **d.** $.25(x + 60) = 6 + x$
    **e.** $.04x + .05(500 - x) = 23$   **f.** $.03x + .02(800 - x) = 19$

## *CUMULATIVE REVIEW*

Solve and check:

1. $\dfrac{x - 3}{x + 7} = \dfrac{x - 2}{x + 10}$        2. $\dfrac{12}{t + 5} + \dfrac{5t - 1}{t - 2} = 5$

3. $\dfrac{a + 1}{a + 2} - \dfrac{a}{a + 5} = \dfrac{6a - 1}{a^2 + 7a + 10}$   4. $\dfrac{3x + 2}{3x + 4} + \dfrac{3x - 3}{3x - 4} = 2 + \dfrac{5x - 4}{9x^2 - 16}$

5. $\dfrac{m + 3}{m + 7} + \dfrac{m - 6}{7 - m} = \dfrac{2m + 7}{m^2 - 49}$   6. $.04x + .06(900 - x) = 46$

# Evaluation of Formulas

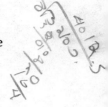

**I. Aim**  To determine the value of any variable in a formula when the values of all the other variables are known.

**II. Procedure**
1. Rewrite the formula.
2. Substitute the given values for the variables.
3. Perform the necessary operations.
4. Solve the resulting equation for the value of the required variable.
5. Check by substituting the given value and answer in the formula.
6. Be careful with the subscripts in the formula. They are *not* exponents.

**III. Sample Solutions**

1. Find the value of $n$ if $a = 40$, using the formula $a = \dfrac{360}{n}$.

   *Solution:*
   $$a = \frac{360}{n}$$
   $$40 = \frac{360}{n}$$
   $$n \cdot 40 = n \cdot \frac{360}{n}$$
   $$40n = 360$$
   $$n = 9$$
   *Answer: $n = 9$*
   *Check:*
   $$a = \frac{360}{n}$$
   $$40 = \frac{360}{9}\ ?$$
   $$40 = 40\ \checkmark$$

2. Find the value of $d$ if $f = 2$ and $D = 6$, using the formula $\dfrac{1}{f} = \dfrac{1}{d} + \dfrac{1}{D}$.

   *Solution:*
   $$\frac{1}{f} = \frac{1}{d} + \frac{1}{D}$$
   $$\frac{1}{2} = \frac{1}{d} + \frac{1}{6}$$
   $$6d \cdot \frac{1}{2} = 6d\left(\frac{1}{d} + \frac{1}{6}\right)$$
   $$6d \cdot \frac{1}{2} = \left(6d \cdot \frac{1}{d}\right) + \left(6d \cdot \frac{1}{6}\right)$$
   $$3d = 6 + d$$
   $$3d - d = 6 + d - d$$
   $$2d = 6$$
   $$d = 3$$
   *Answer: $d = 3$*

   *Check:* $\dfrac{1}{f} = \dfrac{1}{d} + \dfrac{1}{D}$
   $$\tfrac{1}{2} = \tfrac{1}{3} + \tfrac{1}{6}\ ?$$
   $$\tfrac{1}{2} = \tfrac{1}{2}\ \checkmark$$

## DIAGNOSTIC TEST

Find the value of:

**1.** $t$ when $r = 15$ and $d = 405$, using the formula $r = \dfrac{d}{t}$

**2.** $V$ when $V' = 72$, $P' = 12$, and $P = 8$, using the formula $\dfrac{V}{V'} = \dfrac{P'}{P}$

**3.** $R_4$ when $R_1 = 18$, $R_2 = 42$, and $R_3 = 27$, using the formula $\dfrac{R_1}{R_2} = \dfrac{R_3}{R_4}$

**4.** $D$ when $f = 8$ and $d = 24$, using the formula $\dfrac{1}{f} = \dfrac{1}{d} + \dfrac{1}{D}$

**5.** $F$ when $C = 10$, using the formula $C = \frac{5}{9}(F - 32)$

**6.** $r$ when $I = 4$, $n = 4$, $E = 40$, and $R = 16$, using the formula
$I = \dfrac{nE}{R + nr}$

## *RELATED PRACTICE EXAMPLES*

Find the value of:

1. **a.** $R$ when $I = 22$ and $E = 110$, using the formula $I = \dfrac{E}{R}$

   **b.** $v$ when $d = .8$ and $m = 15$, using the formula $d = \dfrac{m}{v}$

   **c.** $t$ when $v = 20$ and $s = 5$, using the formula $v = \dfrac{s}{t}$

   **d.** $c$ when $R = 8$ and $s = 96$, using the formula $R = \dfrac{s}{c}$

2. **a.** $W_1$ when $W_2 = 18$, $W_3 = 8$, and $W_4 = 12$, using the formula $\dfrac{W_1}{W_2} = \dfrac{W_3}{W_4}$

   **b.** $P'$ when $V = 9$, $V' = 21$, and $P = 35$, using the formula $\dfrac{V}{V'} = \dfrac{P'}{P}$

   **c.** $h$ when $F = 28$, $W = 24$, and $d = 42$, using the formula $\dfrac{F}{W} = \dfrac{h}{d}$

   **d.** $V$ when $V' = 3$, $T = 4$, and $T' = 6$, using the formula $\dfrac{V}{V'} = \dfrac{T}{T'}$

   **e.** $R_3$ when $R_1 = 25$, $R_2 = 45$, and $R_4 = 54$, using the formula $\dfrac{R_1}{R_2} = \dfrac{R_3}{R_4}$

3. **a.** $W$ when $F = 35$, $h = 15$, and $d = 40$, using the formula $\dfrac{F}{W} = \dfrac{h}{d}$

**b.** $L_1$ when $W_1 = 63$, $W_2 = 49$, and $L_2 = 18$, using the formula $\dfrac{W_1}{W_2} = \dfrac{L_2}{L_1}$

**c.** $R_2$ when $R_1 = 104$, $R_3 = 32$, and $R_4 = 24$, using the formula $\dfrac{R_1}{R_2} = \dfrac{R_3}{R_4}$

**d.** $P$ when $V = 72$, $V' = 96$, and $P' = 180$, using the formula $\dfrac{V}{V'} = \dfrac{P'}{P}$

**e.** $T'$ when $P = 9$, $V = 6$, $T = 3$, $P' = 24$, and $V' = 3$, using the formula

$$\dfrac{PV}{T} = \dfrac{P'V'}{T'}$$

**4. a.** $R_1$ when $R = 3$ and $R_2 = 4$, using the formula $\dfrac{1}{R} = \dfrac{1}{R_1} + \dfrac{1}{R_2}$

    **b.** $f$ when $d = 8$ and $D = 24$, *using the formula* $\dfrac{1}{f} = \dfrac{1}{d} + \dfrac{1}{D}$

**c.** $R_3$ when $R = 4$, $R_1 = 24$, and $R_2 = 6$, using the formula

$$\dfrac{1}{R} = \dfrac{1}{R_1} + \dfrac{1}{R_2} + \dfrac{1}{R_3}$$

    **d.** $R$ when $R_1 = 5$, $R_2 = 10$, and $R_3 = 2$, using the formula

$$\dfrac{1}{R} = \dfrac{1}{R_1} + \dfrac{1}{R_2} + \dfrac{1}{R_3}$$

**e.** $D$ when $f = 12$ and $d = 48$, using the formula $\dfrac{1}{f} = \dfrac{1}{d} + \dfrac{1}{D}$

**5. a.** $l$ when $S = 55$, $n = 5$, and $a = 3$, using the formula $S = \dfrac{n}{2}(a + l)$

**b.** $b$ when $A = 72$, $h = 6$, and $b' = 10$, using the formula

$$A = \dfrac{h}{2}(b + b')$$

    **c.** $a$ when $S = 110$, $n = 10$, and $I = 20$, using the formula

$$S = \dfrac{n}{2}(a + l)$$

**d.** $F$ when $C = 40$, using the formula $C = \tfrac{5}{9}(F - 32)$

**e.** $F$ when $C = -25$, using the formula $C = \tfrac{5}{9}(F - 32)$

**6. a.** $a$ when $n = 10$, $S = 250$, and $l = 45$, using the formula $n = \dfrac{2S}{a + l}$

**b.** $l$ when $p = 400$, $A = 480$, and $r = .04$, using the formula $p = \dfrac{A}{1 + rl}$

**c.** $r$ when $S = 8$ and $a = 4$, using the formula $S = \dfrac{a}{1 - r}$

**d.** $r$ when $S = 363$, $l = 243$, and $a = 3$, using the formula $S = \dfrac{rl - a}{r - 1}$

**e.** $n$ when $I = 2$, $E = 7$, $R = 2$, and $r = 3$, using the formula $I = \dfrac{nE}{R + nr}$

**I. Aim**   To solve by the equation method: A. Number Problems, B. Age Problems, C. Work Problems, and D. Mixture Problems.

## II. Procedure

1. Read the general directions on page 244 to help solve verbal problems.
2. Study the basic information given in each topic.
3. Observe how the general directions are applied in the illustrated solution of the different types of problems.

---

### DIAGNOSTIC TEST

1. What number should be both added to the numerator and subtracted from the denominator of the fraction $\frac{7}{13}$ so that the answer in simplest form is $\frac{2}{3}$?
2. Pedro is one-seventh as old as his cousin. In 8 years he will be one-third as old. What are their ages now?
3. Joan can clean her house in 6 hours but it takes 9 hours for her sister to do it. How long will it take the two girls to clean the house if they work together?
4. How many ounces of water should be added to 20 ounces of a 10% boric acid solution to reduce it to a 4% solution?
5. How many liters of pure anti-freeze should be added to 12 liters of a 25% anti-freeze solution to make a 40% anti-freeze solution?

---

**A.** Number Problems

*Sample Problem*

The sum of one-fourth of a certain number and three-eighths of the same number is 15. Find the number.

*Solution:* Let $n =$ the number

The equation is determined by the fact: One-fourth of the number plus three-eighths of this number is equal to 15.

$$\tfrac{1}{4}n + \tfrac{3}{8}n = 15$$

$$\text{or } \frac{n}{4} + \frac{3n}{8} = 15$$

$$8\left(\frac{n}{4} + \frac{3n}{8}\right) = 8 \cdot 15$$

$$\left(8 \cdot \frac{n}{4}\right) + \left(8 \cdot \frac{3n}{8}\right) = 8 \cdot 15$$

$$2n + 3n = 120$$

$$5n = 120$$

$$n = 24$$

*Check:*

$$\tfrac{1}{4} \times 24 = 6$$

$$\tfrac{3}{8} \times 24 = 9$$

$$6 + 9 = 15 \checkmark$$

*Answer:* 24

### RELATED PRACTICE PROBLEMS

1. **a.** If one-half of a certain number is added to one-third of the same number, the sum is 30. Find the number.
   **b.** What number increased by $\frac{5}{8}$ of itself equals 26?
   **c.** One number is 24 more than another. If the larger is divided by the smaller, the quotient is 5. Find the numbers.
   **d.** What number should be added to both the numerator and denominator of the fraction $\frac{5}{11}$ so that the answer in simplest form is equal to $\frac{3}{5}$?
   **e.** What is the value of an estate if a man wills $\frac{1}{3}$ of it to his daughter, $\frac{1}{4}$ of it to his son, and the remainder, $10,000, to his wife?

**B.** Age Problems

*Sample Problem*

Elaine is one third as old as her sister. In 6 years she will be one half as old. How old is Elaine now?

*Solution:*

$$\text{Let } x = \text{sister's present age in years}$$

$$\text{Then } \tfrac{1}{3}x \text{ or } \frac{x}{3} = \text{Elaine's present age in years}$$

$$x + 6 = \text{sister's age in years 6 years hence}$$

$$\frac{x}{3} + 6 = \text{Elaine's age in years 6 years hence}$$

The equation is determined by the fact: Elaine's age 6 years hence is equal to one half her sister's age at that time.

$$\frac{x}{3} + 6 = \tfrac{1}{2}(x + 6)$$

or

$$\frac{x}{3} + 6 = \frac{x + 6}{2}$$

$$6 \cdot \left(\frac{x}{3} + 6\right) = 6 \cdot \left(\frac{x + 6}{2}\right)$$

$$\left(6 \cdot \frac{x}{3}\right) + (6 \cdot 6) = 6 \cdot \left(\frac{x + 6}{2}\right)$$

$$2x + 36 = 3(x + 6)$$
$$2x + 36 = 3x + 18$$
$$2x - 3x + 36 - 36 = 3x - 3x + 18 - 36$$
$$-x = -18$$
$$x = 18 \text{ years, sister's present age}$$
$$\tfrac{1}{3}x = \tfrac{1}{3} \cdot 18 = 6 \text{ years, Elaine's present age}$$

*Check:*

| Elaine's age | Sister's age |
|---|---|
| | present |
| 6 years = $\frac{1}{3}$ of 18 years | |
| 6 years hence | |
| 12 years = $\frac{1}{2}$ of 24 years ✔ | |

*Answer:* Elaine is 6 years old now.

## *RELATED PRACTICE PROBLEMS*

2. **a.** John is one-fifth as old as his father. The sum of their ages is 42 years. How old is John?

   **b.** Lisa is one-sixth as old as her mother. In 12 years she will only be one-third as old. What are their ages now?

   **c.** Scott is one-fourth as old as his dad. Six years ago he was one-tenth as old. Find Scott's present age.

   **d.** Todd is one-half as old as his dad. He is now 20 years old. In how many years will he be three-fourths as old as his father?

   **e.** Steve's age four years ago was two-thirds of what his age will be one year from now. How old is he now?

## C. Work Problems

In these work problems the equation is determined by the fact that the amount of work done by one person (or one machine) in a certain time plus the amount of work done by a second person (or second machine) in the same time is equal to the total amount of work done by both persons (or both machines) in that time.

*Sample Problem*

Peter can wash a car in 20 minutes and Andrew can do it in 30 minutes. How long will it take the two boys to wash the car if they work together?

*Solution:*       Let $x$ = no. of minutes it takes both boys together

$$\text{Then } \frac{1}{x} = \text{part of work done by both boys in 1 minute}$$

$$\tfrac{1}{20} = \text{part of work done by Peter in 1 minute}$$

$$\tfrac{1}{30} = \text{part of work done by Andrew in 1 minute}$$

$$\frac{1}{20} + \frac{1}{30} = \frac{1}{x}$$

$$60x \cdot \left( \frac{1}{20} + \frac{1}{30} \right) = 60x \cdot \frac{1}{x}$$

$$\left( 60x \cdot \frac{1}{20} \right) + \left( 60x \cdot \frac{1}{30} \right) = 60x \cdot \frac{1}{x}$$

$$3x + 2x = 60$$

$$5x = 60$$

$$x = 12 \text{ minutes}$$

*Answer:* 12 minutes

*Check:*

Peter does $12 \times \dfrac{1}{20} = \dfrac{12}{20} = \dfrac{3}{5}$ of all work

Andrew does $12 \times \dfrac{1}{20} = \dfrac{12}{30} = \dfrac{2}{5}$ of all work

Together they do $\dfrac{3}{5} + \dfrac{2}{5} = \dfrac{5}{5}$ or all the work ✔

## *RELATED PRACTICE PROBLEMS*

3. **a.** Luis can mow a lawn in 24 minutes, and Roberto can do it in 48 minutes. How long does it take the two boys to mow the lawn together if they use two lawn mowers?

   **b.** A farmer can plow a field in 12 hours, and his son can plow it in 36 hours. If they work together, using two plows, how long will it take them to plow the field?

   **c.** Sara can paint a house in 8 days, but her father can do it in 6 days. How long does it take both to paint the house if they work together?

   **d.** Maria can address 75 envelopes in 60 minutes, but Ann can do the same number in 30 minutes. If they work together, how long will it take them to address 75 envelopes?

   **e.** A water tank can be filled by one pipe in 3 hours and by a second pipe in 2 hours. How many hours will it take the two pipes together to fill the tank?

   **f.** Joan can mow the lawn twice as quickly as Phil. Together they can do it in 4 hours. How long will it take each person to mow the lawn alone?

   **g.** A farmer can plow a field in 9 hours, and his son can do it in 18 hours? After the farmer had plowed for 3 hours, the boy joined him and together they completed the plowing of the field. How long did it take them?

   **h.** Charlotte can type a certain number of form letters in a third of the time it takes Harriet. If they do it together, they can complete the task in 12 hours. How long will it take Charlotte to do all the typing?

   **i.** Paul can paint a house in 8 days, but he and his brother can do the same job in 6 days. How many days will it take the brother to paint the house alone?

   **j.** A tank can be filled by one pipe in 4 hours, and can be emptied by another pipe in 6 hours. If the two pipes are open, how long will it take to fill the tank?

## D. Mixture Problems

*Sample Problems*

Observe in the following two sample problems that the solution is weakened by the addition of water (see first model solution) and is strengthened by the addition of acid (see second model solution). Analyze the representations in the model solutions.

1. How many ounces of water should be added to 16 ounces of a 25% nitric acid solution to reduce it to a 10% solution?

*Solution:*   Let  $x$ = no. of ounces of water to be added.

16 = total no. of ounces in given solution.

Then 16 + $x$ = total no. of ounces in new solution.

4 = no. of ounces of pure nitric acid (25% of 16).

$\frac{1}{10}(16 + x)$ or $\frac{16 + x}{10}$ = 10% of no. of ounces in new solution.

The equation is determined by the fact: 10% of the total number of ounces in the new solution equals 4 ounces of nitric acid full strength.

*Solution:*

$$\frac{16x + x}{10} = 4$$

$$10 \cdot \frac{16 + x}{10} = 10 \cdot 4$$

$$16 + x = 40$$

$$16 - 16 + x = 40 - 16$$

$$x = 24 \text{ ounces of water}$$

*Check:*

16 ounces in given solution

+24 ounces of water added

40 ounces in new solution

$\frac{4}{40}$ = 10% nitric acid solution ✔

*Answer:* 24 ounces of water

2. How many liters of alcohol should be added to 40 liters of a 20% solution to make a $33\frac{1}{3}$% solution?

*Solution:*     Let $x$ = no. of liters of alcohol to be added.

8 = no. of liters of pure alcohol in given solution (20% of 40).

8 + $x$ = no. of liters of pure alcohol in new solution.

40 = total no. of liters in given solution.

40 + $x$ = total no. of liters in new solution.

$\frac{1}{3}(40 + x)$ or $\frac{40 + x}{3}$ = $33\frac{1}{3}$% of no. of liters in new solution.

The equation is determined by the fact: $33\frac{1}{3}$% of the total number of liters in the new solution equals the number of liters of pure alcohol in the new solution.

*Solution:*

$$\frac{40 + x}{3} = 8 + x$$

$$3 \cdot \frac{40 + 3}{3} = 3(8 + x)$$

$$40 + x = 24 + 3x$$

$$40 - 40 + x - 3x = 24 - 40 + 3x - 3x$$

$$-2x = -16$$

$$x = 8 \text{ liters of alcohol}$$

*Answer:* 8 liters          to be added

*Check:*

40 L in given solution

+ 8 L alcohol to be added

48 L in new solution

8 L alcohol in given solution

8 L alcohol to be added

16 L alcohol in new solution

$\frac{16}{48}$ = $33\frac{1}{3}$% solution ✔

## RELATED PRACTICE PROBLEMS

4. **a.** How many milliliters of water should be added to 20 milliliters of a 30% sulphuric acid solution to reduce it to a 15% solution?

   **b.** How many ounces of water should be added to 28 ounces of a 25% hydrocholoric acid solution to reduce it to a 5% solution?

   **c.** A dairy has 400 quarts of milk containing 5% butter fat. How many quarts of milk containing no butter fat should be added to produce milk containing 4% butter fat?

   **d.** If there are 2 pounds of salt in a 20-pound salt solution, how many pounds of water should be added to make an 8% salt solution?

   **e.** A chemist has 6 liters of a $33\frac{1}{3}$% acetic acid solution. How many liters of water should be added to make a 20% solution?

5. **a.** How many liters of pure alcohol should be added to 25 liters of a 20% solution to make a 60% solution?

   **b.** How many ounces of pure sulphuric acid should be added to 32 ounces of a 25% solution to make a 40% solution?

   **c.** How many pounds of salt should be added to 20 pounds of a 10% salt solution to make it a $33\frac{1}{3}$% solution?

   **d.** How many milliliters of vinegar should be added to 40 milliliters of a 10% solution to make it a 25% solution?

   **e.** How many quarts of pure anti-freeze should be added to 10 quarts of a 20% anti-freeze solution to make a 50% anti-freeze solution?

## COMPETENCY CHECK TEST

In each of the following select the letter corresponding to your answer:

1. Subtract: $\dfrac{r-t}{rt} - \dfrac{s-t}{st}$   The answer is:

   a. $\dfrac{r-s}{rs}$     b. $\dfrac{r-s+2t}{rst}$     c. $\dfrac{r-s-t}{rt-st}$     d. Answer not given.

2. The solution of $x - \dfrac{x}{2} + \dfrac{x}{3} - \dfrac{x}{4} = 21$ is:

   a. 12         b. 48              c. 36              d. Answer not given.

3. When the formula $\dfrac{R_1}{R_2} = \dfrac{R_3}{R_4}$ is used and $R_1 = 800$, $R_2 = 480$, and $R_3 = 750$, the value of $R_4$ is:     a. 600     b. 900     c. 450     d. 300

4. The number of quarts of water to be added to 5 quarts of a 60% anti-freeze solution to reduce it to a 25% anti-freeze solution is:
   a. 3         b. 5              c. 7              d. 9

# REVIEW

Solve and check:

1. a. $\dfrac{x}{8} + \dfrac{x}{6} = 28$

   b. $n - \dfrac{n}{2} + \dfrac{n}{4} - \dfrac{n}{8} = 10$

   c. $\dfrac{3x}{5} + \dfrac{x - 5}{20} = 29$

   d. $\dfrac{5y - 4}{8} - \dfrac{y - 2}{5} = \dfrac{y + 8}{4}$

2. a. $\dfrac{11}{3x} - \dfrac{5}{8} = \dfrac{19}{8x}$

   b. $\dfrac{y - 6}{y - 7} = \dfrac{y + 9}{y + 3}$

   c. $\dfrac{3t - 1}{t + 6} - 2 = \dfrac{t + 2}{t - 9}$

   d. $\dfrac{b - 5}{b - 6} - \dfrac{b - 3}{b + 8} = \dfrac{5b - 2}{b^2 + 2b - 48}$

3. a. $n + .61 = .8$

   b. $c - .3c = 56$

   c. $4.8y - .82 = .44y + 7.9$

   d. $.08x + .07(1{,}500 - x) = 114$

4. Find the value of:

   a. $d$ when $F = 108$, $W = 81$, and $h = 36$, using the formula $\dfrac{F}{W} = \dfrac{h}{d}$

   b. $b'$ when $A = 243$, $h = 18$, and $b = 11$, using the formula
   $$A = \dfrac{h}{2}(b + b')$$

   c. $f$ when $d = 28$ and $D = 84$, using the formula $\dfrac{1}{f} = \dfrac{1}{d} + \dfrac{1}{D}$

   d. $R$ when $I = 4$, $E = 5$, $n = 16$, and $r = 3$, using the formula
   $$I = \dfrac{nE}{R + nr}$$

   e. $r$ when $S = 124$, $l = 4$, and $a = 64$, using the formula
   $$S = \dfrac{rl - a}{r - 1}$$

5. a. When two-thirds of a number is subtracted from three-fourths of this same number, the remainder is 6. Find the number.

   b. A certain wading pool can be filled with water in 16 minutes but it takes 20 minutes to empty it. How long will it take to fill the pool if the water is permitted to run in and out at the same time?

   c. How many ounces of pure acid should a chemist add to 25 ounces of a 24% acid solution to make a 50% acid solution?

## Keyed Achievement Test

The colored numeral indicates where help may be found.
1. Find the product of each of the following at sight: $\boxed{\text{9-1 thru 9-5}}$
   **a.** $(9t+6)(4t-5)$                  **b.** $(m-11)(m-16)$
   **c.** $(12bc^2-x^3)^2$                 **d.** $(3x^2y-4z)(3x^2y+4z)$
2. Factor each of the following completely:
   **a.** $35r^9s^3t^8+63r^6s^3t^5-14s^5t^4$ $\boxed{\text{9-6}}$    **b.** $100c^4-81d^2$ $\boxed{\text{9-8}}$
   **c.** $64n^2-80n+25$ $\boxed{\text{9-9}}$    **d.** $m^2+10m-56$ $\boxed{\text{9-10}}$
   **e.** $35x^2+51x+18$ $\boxed{\text{9-11}}$    **f.** $ab-ab^5$ $\boxed{\text{9-12}}$
3. Write $\dfrac{30b^2x^2-25bx^3}{40b^3x^2}$ in simplest form. $\boxed{\text{10-2}}$
4. Multiply: $\dfrac{4x^2}{4x^2-100}\cdot\dfrac{3x+15}{18x}$    5. Divide: $\dfrac{6a^2y}{25cx^2}\div\dfrac{12a^4x}{35cy}$
   $\boxed{\text{10-3}}$                                   $\boxed{\text{10-4}}$
6. Combine: **a.** $\dfrac{c-4}{3}+\dfrac{8-c}{8}-\dfrac{2c+5}{4}$  **b.** $\dfrac{3}{x}+\dfrac{5}{x-2}-\dfrac{7}{x-3}$  $\boxed{\text{10-9}}$
   $\boxed{\text{10-8}}$
7. Solve and check: $\boxed{\text{11-1 thru 11-3}}$
   **a.** $\dfrac{3x}{10}-\dfrac{2x}{3}=\dfrac{11}{15}$    **b.** $\dfrac{d+8}{d-4}-\dfrac{d-18}{d-6}=\dfrac{48}{d^2-10d+24}$
   **c.** $n-.08n=59.8$    **d.** $.05x+.07(800-x)=52$
8. Find the value of $n$ when $I=3$, $E=12$, $R=10$, and $r=2$, using the formula $I=\dfrac{nE}{R+nr}$. $\boxed{\text{11-4}}$
9. A tank can be filled by one pipe in 4 hours and by a second pipe in 8 hours. It can be emptied by a third pipe in 6 hours. If the three pipes are open, how long will it take to fill the tank? $\boxed{\text{11-5}}$

## MAINTENANCE PRACTICE IN ARITHMETIC

**1.** Add: 674,527    **2.** Subtract:    **3.** Multiply:    **4.** Divide:    **5.** Add:
         989,798      1,040,683      683,507      $526\overline{)4,212,734}$   $8\frac{2}{3}+5\frac{5}{16}+4\frac{1}{4}$
         824,789        361,595        7,089

**6.** Subtract: $16\frac{3}{8}-10\frac{11}{32}$    **7.** Multiply: $13\frac{1}{3}\times2\frac{1}{4}$    **8.** Divide: $6\frac{3}{4}\div18$
**9.** Add: $.89+.9+5.7$    **10.** Subtract: $8.7-.54$    **11.** Multiply: $.32\times.15$
**12.** Divide: $\$.10\overline{)\$18}$    **13.** Find 175% of 684
**14.** What percent of 72 is 54?    **15.** 4% of what number is 900?

# INTRODUCTION

**W**hen a vertical number line and a horizontal number line are drawn perpendicular (each angle formed measures 90°) to each other, a number plane is determined. If the number lines are real number lines (see page 54), the number plane consists of an infinite number of points and is called the *real number plane*.

These two perpendicular number lines form a pair of *coordinate axes* or guide lines so that points in this plane may be located or plotted.

The horizontal number line is the horizontal axis or *x-axis*. The vertical number line is the vertical axis or *y-axis*.

We have found on page 55 that only a single number (coordinate) is required to locate a point on a number line. However, an ordered pair of numbers is needed to locate a point in the number plane because the point is located with respect to both the *x*-axis and *y*-axis. An *ordered pair of numbers* is a pair of numbers expressed in a definite order so that one number is first (*first coordinate*) and the other number is second (*second coordinate*).

The graph of an equation in two variables is the graph of its solutions. See page 472. When the replacements for the variables are all the real numbers, there is an unlimited number of pairs of real numbers which satisfy the equation. Each ordered pair of real numbers corresponds to only one point of the real number plane.

462

# Graphing Linear Equations

**12**

# Points in the Number Plane

**I. Aim**    To locate and plot points in the real number plane.

**II. Procedure**

**1.** To express an *ordered pair of numbers*, write within parentheses the numerals for these numbers with the numeral for the first component first, followed by a comma, and then the numeral for the second component.

**2.** *To locate a point* in a number plane,
   **a.** Draw a perpendicular line from the point to the $x$-axis.
   **b.** Use the number corresponding to the point on the $x$-axis where this perpendicular line intersects the $x$-axis as the $x$-*coordinate* or *abscissa* of the point.
   
   The $x$-coordinate of the point in the illustration below is $+4$ or $4$.
   The $x$-coordinate indicates the horizontal distance, measured parallel to the $x$-axis, that a point is located to the left or to the right of the $y$-axis. This distance and direction is represented to the left of the $y$-axis by a negative number and to the right by a positive number.
   **c.** Then draw a perpendicular line from the point to the $y$-axis.
   **d.** Use the number corresponding to the point on the $y$-axis where this perpendicular line intersects the $y$-axis as the $y$-*coordinate* or *ordinate* of the point.
   
   The $y$-coordinate indicates the vertical distance, measured parallel to the $y$-axis, that a point is located up from the $x$-axis (by a positive number) and down (by a negative number).
   
   The $y$-coordinate of the point in the illustration below is $+3$ or $3$.

**3.** Represent the coordinates of the point by an ordered pair of numbers in which the $x$-coordinate is expressed first and the $y$-coordinate is second.

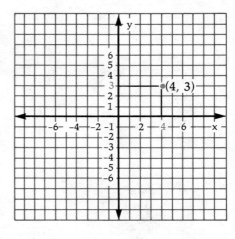

The coordinates of the point in the illustration to the right are the ordered pair of numbers named by (4,3).

**464**

**4.** Observe that:

**a.** The $x$-axis and the $y$-axis intersect at a point called the *origin,* where the coordinates are (0, 0).

**b.** Any point on the $x$-axis has 0 for its $y$-coordinate.

**c.** Any point on the $y$-axis has 0 for its $x$-coordinate.

**d.** The real number plane is divided into four regions, each of which is called a *quadrant.*

(1) In the quadrant marked I, each point has a positive $x$-coordinate and a positive $y$-coordinate.

(2) In the quadrant marked II, each point has a negative $x$-coordinate and a positive $y$-coordinate.

(3) In the quadrant marked III, each point has a negative $x$-coordinate and a negative $y$-coordinate.

(4) In the quadrant marked IV, each point has a positive $x$-coordinate and a negative $y$-coordinate.

**e.** An equation indicates the relationship between the abscissa (described by the variable $x$) and the ordinate (described by the variable $y$).

The equation $y = 3x$ indicates that the ordinate is three times the abscissa.

**5.** *To plot a point* that corresponds to a given ordered pair of numbers, locate this point in the number plane and indicate its position by a dot.

**a.** Locate the point by using the first component of the given ordered pair of numbers as the $x$-coordinate and the second component as the $y$-coordinate.

**b.** Then use a solid or colored dot to indicate the position of the point.

## III. Sample Solutions

**1.** Write the ordered pair of numbers which has $-7$ as the first component and 3 as the second component. *Answer:* $(-7, 3)$

**2.** What are the coordinates of point $A$? Point $B$? Point $C$? Point $D$? Point $E$? Point $F$? Point $G$? Point $H$? Point $I$?

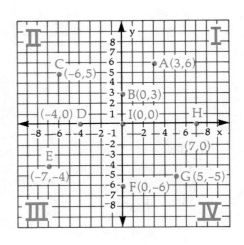

*Answers:* Point $A$, $(3, 6)$; Point $B$, $(0, 3)$; Point $C$, $(-6, 5)$; Point $D$, $(-4, 0)$; Point $E$, $(-7, -4)$; Point $F$, $(0, -6)$; Point $G$, $(5, -5)$; Point $H$, $(7, 0)$; Point $I$, $(0, 0)$

**3.** Describe the relationship of $x$ and $y$ in the equation $y = x + 2$, using the words abscissa and ordinate.

*Answer:* The ordinate is 2 more than the abscissa.

**4.** Plot the points:
   A whose coordinates are (6, 2)
   B whose coordinates are (−3, 4)
   C whose coordinates are (0, −5)

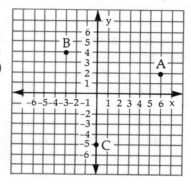

## DIAGNOSTIC TEST

**1.** Write the ordered pair of numbers which has −5 as the first component and −1 as the second component.

**2.** Write all the ordered pairs of numbers which have 0, 1, or 2 as the first component and −2 or −1 as the second component.

**3.** What are the coordinates of each of the points shown in the figure below?

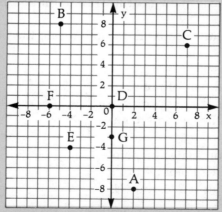

**4.** On graph paper draw the axes and plot the points that have the following coordinates:

   **a.** (−7, 1)     **b.** (0, −8)     **c.** (−9, −4)
   **d.** (5, 10)     **e.** (6, −6)     **f.** (−2, 0)

5. Using the notation I, II, III, and IV, as shown on page 465, in which quadrant is a point located when its corresponding coordinates are:
   **a.** $(-5, -4)$      **b.** $(-6, 3)$      **c.** $(8, 2)$
   **d.** $(4, -5)$      **e.** $(-9, 1)$      **f.** $(-1, -7)$
6. Describe the relationship of $x$ and $y$ in the equation $y = x - 3$, using the words abscissa and ordinate.
7. Write as an equation in two variables ($x$ and $y$): The ordinate is 7 more than the abscissa.

### *RELATED PRACTICE EXAMPLES*

1. Write the ordered pair of numbers which has:
   **a.** 4 as the first component and 9 as the second component.
   **b.** 5 as the first component and 0 as the second component.
   **c.** 2 as the first component and $-4$ as the second component.
   **d.** $-6$ as the first component and 3 as the second component.
   **e.** $-1$ as the first component and $-7$ as the second component.
2. Write all the ordered pairs of numbers which have:
   **a.** 1 or 2 as the first component and 1 as the second component.
   **b.** 0 or 1 as the first component and $-1$ or 0 as the second component.
   **c.** 1, 2, or 3 as the first component and 0 or 1 as the second component.
   **d.** $-3$ or $-1$ as the first component and $-1$, 0, or 1 as the second component.
   **e.** $-2$, 0, or 2 as the first component and $-2$, 0, or 2 as the second component.
3. What are the coordinates of each of the following indicated points?

**a.**

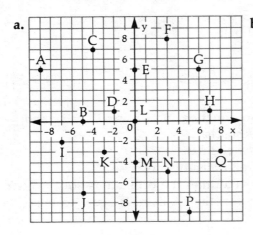

**b.**

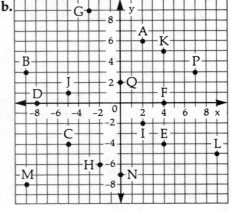

**c.** 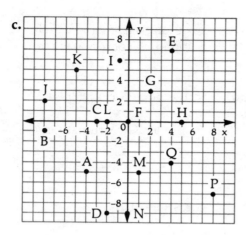   **d.**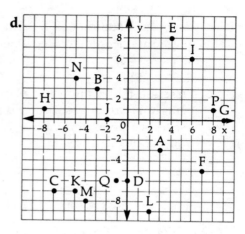

4. On graph paper draw the axes and plot the points that have the following coordinates:

| | | | |
|---|---|---|---|
| **a.** (4, 1) | **b.** (−2, 3) | **c.** (5, −2) | **d.** (−1, −7) |
| **e.** (3, 0) | **f.** (−5, −4) | **g.** (6, −8) | **h.** (0, −2) |
| **i.** (−9, 5) | **j.** (−3, −3) | **k.** (2, −7) | **l.** (5, 6) |
| **m.** (0, 0) | **n.** (−4, 8) | **o.** (−6, −1) | **p.** (−9, 0) |
| **q.** (−1, −1) | **r.** (8, −4) | **s.** (7, 7) | **t.** (−8, −5) |

5. Using the notation I, II, III, and IV, in which quadrant is a point located when its corresponding coordinates are:

| | | | |
|---|---|---|---|
| **a.** (−2, 5) | **b.** (3, −1) | **c.** (6, 3) | **d.** (−9, −7) |
| **e.** (1, −2) | **f.** (−4, 4) | **g.** (5, 8) | **h.** (−3, 11) |
| **i.** (−6, −9) | **j.** (2, −6) | **k.** (−8, 8) | **l.** (−1, −4) |

6. Describe the relationship of $x$ and $y$ in each of the following equations, using the words abscissa and ordinate:

| | | |
|---|---|---|
| **a.** $y = x$ | **b.** $y = 5x$ | **c.** $y = \frac{1}{3}x$ |
| **d.** $y = x + 9$ | **e.** $y = x - 6$ | **f.** $x = -7y$ |
| **g.** $x + y = 0$ | **h.** $5x + 3y = 9$ | **i.** $4x - 9y = 11$ |

7. Write each of the following as an equation in two variables ($x$ and $y$):
   **a.** The ordinate is four times the abscissa.
   **b.** The abscissa is 5 more than the ordinate.
   **c.** The ordinate is 1 less than three times the abscissa.
   **d.** Four times the abscissa plus seven times the ordinate is equal to ten.
   **e.** Six times the abscissa minus twice the ordinate is equal to negative five.

# Solutions of an Equation in Two Variables

### I. Aim

To determine ordered pairs of numbers that satisfy an equation in two variables.

### II. Procedure

1. Write the *y-form* of an equation in two variables by solving the equation for $y$ in terms of $x$. See sample solution 2.
2. Then select several values for the variable $x$ from the real numbers and substitute each of them for the variable $x$ in the *y*-form of the equation.
3. For each, compute the corresponding value of the $y$ variable.
4. Write the ordered pairs of numbers.
5. To check whether a number pair satisfies an equation in two variables, substitute the $x$-value for the variable $x$ and the $y$ value for the variable $y$ in the given equation. If a true sentence results, then the number pair satisfies the equation and is a solution.

   Thus, the solutions of an equation in two variables consists of all the ordered pairs or numbers which make the sentence true.

### III. Sample Solutions

1. Which of the following ordered pairs of numbers $(2, -3)$ or $(-1, 6)$ satisfies the equation $x + y = 5$?

   *Solution:*      *Check:*

$$x + y = 5 \qquad\qquad x + y = 5$$
$$(2) + (-3) = 5\,? \qquad (-1) + (6) = 5\,?$$
$$-1 \neq 5 \qquad\qquad\quad 5 = 5\;✔ \quad \textit{Answer: } (-1, 6)$$

2. Write the *y*-form of: **a.** $2x + y = 0$    **b.** $2x + 3y = 9$

   **a.** *Solution:*               **b.** *Solution:*

$$2x + y = 0 \qquad\qquad\qquad 2x + 3y = 9$$
$$2x - 2x + y = 0 - 2x \qquad 2x - 2x + 3y = 9 - 2x$$
$$y = -2x \qquad\qquad\qquad\qquad 3y = 9 - 2x$$

$$\textit{Answer: } y = -2x \qquad\qquad \frac{3y}{3} = \frac{9 - 2x}{3}$$

$$y = \frac{9 - 2x}{3}$$

$$y = 3 - \tfrac{2}{3}x \text{ or } y = -\tfrac{2}{3}x + 3$$

$$\textit{Answer: } y = -\tfrac{2}{3}x + 3$$

**469**

3. Find ordered pairs of numbers that satisfy $x + y = 5$ when $x = 0, 1,$ and 6.

*Solution:*

First, find the $y$-form of the equation

$$x + y = 5$$
$$x + (-x) + y = 5 + (-x)$$
$$y = 5 - x$$

Then, when $x = 0,$            when $x = 1,$            when $x = 6,$

$y = 5 - x$            $y = 5 - x$            $y = 5 - x$

$y = 5 - 0$            $y = 5 - 1$            $y = 5 - 6$

$y = 5$            $y = 4$            $y = 5 + (-6)$

The ordered pair is    The ordered pair is            $y = -1$

(0, 5)            (1, 4)            The ordered pair is

(6, -1)

*Answer:* $(0, 5), (1, 4),$ and $(6, -1)$

---

## DIAGNOSTIC TEST

1. Which of the following ordered pairs of numbers satisfies the equation $3x - 2y = 6$?
   **a.** $(4, 3)$      **b.** $(0, -3)$      **c.** $(3, 1)$      **d.** $(5, 4)$      **e.** $(-6, -12)$
2. Write the $y$-form of the equation: $6x + 3y = 15$
3. Find the ordered pairs of numbers that satisfy the equation $5x - 2y = 18$ when $x = -2, 0,$ and 4.

---

### RELATED PRACTICE EXAMPLES

1. Which of the following ordered pairs of numbers satisfies the equation:
   **a.** $x + y = 11$?
      (1) $(8, 3)$            (2) $(4, 9)$            (3) $(-1, 12)$            (4) $(11, 0)$

   **b.** $x - y = 5$?
      (1) $(-3, -2)$            (2) $(11, 8)$            (3) $(0, 5)$            (4) $(4, -1)$

   **c.** $3x + y = 15$?
      (1) $(2, 9)$            (2) $(3, 7)$            (3) $(6, -3)$            (4) $(-4, 3)$

   **d.** $2x - 5y = 4$?
      (1) $(7, 2)$            (2) $(2, 0)$            (3) $(5, 1)$            (4) $(\frac{1}{2}, -1)$

    e. $x + 4y = -12$?
       (1) $(-4, -2)$      (2) $(-10, 1)$      (3) $(0, -3)$      (4) $(-8, -1)$

    f. $5x - 3y = 12$?
       (1) $(0, -4)$      (2) $(1, 3)$      (3) $(3, 1)$      (4) $(6, 6)$

    g. $x - 8y = 24$?
       (1) $(16, 1)$      (2) $(8, -2)$      (3) $(-16, -5)$      (4) $(0, -3)$

    h. $4x + 3y = 18$?
       (1) $(0, -6)$      (2) $(3, 1)$      (3) $(2, 2)$      (4) $(-3, 10)$

    i. $6x - y = 0$?
       (1) $(1, 6)$      (2) $(0, 0)$      (3) $(-2, 12)$      (4) $(-1, 6)$

    j. $2x + 6y = -8$?
       (1) $(2, -2)$      (2) $(-1, -1)$      (3) $(-4, 0)$      (4) $(8, 4)$

2. Write the $y$-form for each of the following equations:

    a. $x + y = 7$                b. $x + y = -2$
    c. $3x + y = 12$            d. $x - y = 5$
    e. $x - y = 0$               f. $2x - y = -3$
    g. $x + 2y = 9$            h. $5x + 3y = 6$
    i. $x - 4y = 2$             j. $3x - 7y = 10$
    k. $24x = 6y$              l. $3y = 7x$
    m. $9x - 5y = 0$           n. $12x + 2y = 6$
    o. $11x - 4y = -5$

3. Find the ordered pairs of numbers that satisfy:

    a. $x + y = 9$ when $x = 2, 5$, and $10$.
    b. $x - y = 3$ when $x = 0, 4$, and $7$.
    c. $7x + y = -4$ when $x = -1, 0$, and $3$.
    d. $x + 9y = 15$ when $x = -3, 2$, and $6$.
    e. $2x - y = 0$ when $x = -4, 0$, and $5$.
    f. $3x - 2y = 6$ when $x = -6, 0$, and $2$.
    g. $9x + 5y = 10$ when $x = -1, 0$, and $5$.
    h. $3x - 4y = -2$ when $x = -2, 0$, and $4$.
    i. $18x + 6y = 12$ when $x = -4, 0$, and $3$.
    j. $6x - 7y = -7$ when $x = -7, 0$, and $2$.

## CUMULATIVE REVIEW

1. On graph paper draw the axes and plot the points whose coordinates are:
    a. $(-5, 6)$     b. $(0, -7)$     c. $(4, -3)$     d. $(8, 5)$     e. $(-3, -9)$

2. Which ordered pairs of numbers satisfy the equation $5x - 2y = 16$?
    a. $(2, -3)$     b. $(-4, -2)$     c. $(0, -8)$     d. $(4, 2)$     e. $(-2, -13)$

3. Find the ordered pairs of numbers that satisfy the equation
    $4x - 5y = 2$ when $x = -2, 0$, and $8$.

# Graphing an Equation in Two Variables in the Real Number Plane

**I. Aim**    To draw the graph of a linear equation in two variables.

**II. Procedure**

The graph of an equation in two variables in the real number plane is the collection of all points in the real number plane whose coordinates are the ordered pairs of numbers which are the solutions of the equation.

To draw the graph of the equation:

1. Determine three ordered pairs of numbers that are solutions of the equation. To do this:

    **a.** Transform the given equation to its $y$-form.

    **b.** Select three values for the $x$-coordinate. Substitute each of them for the $x$-variable in the $y$-form of the equation to compute the corresponding values of the $y$-variable, each of which becomes the corresponding $y$-coordinate.

2. Plot the points by locating their coordinates with respect to the $x$-axis and $y$-axis.

3. Draw a line through the three plotted points. Although two points determine a straight line, three points are used for greater accuracy.

    The graph of an equation in two variables of the first degree (see page 473) is a straight line. This kind of equation is usually called a *linear equation*.

4. To check whether the line is correct, select any other point on the line and find its coordinates. If the line is the graph of the given equation, the coordinates of the selected point should satisfy the equation. Thus, the coordinates of a point not on the line will not satisfy the equation whose graph is the line.

5. Observe that:

    **a.** If the equation does not contain a constant term, the line passes through the origin $(0, 0)$.

**b.** When a linear equation has only one variable, its graph is parallel to an axis.

(1) If the equation contains only the $y$-variable, the line is parallel to the $x$-axis.

(2) If the equation contains only the $x$-variable, the line is parallel to the $y$-axis.

6. The $x$-coordinate of any point on the $y$-axis is 0. To find the corresponding $y$-coordinate of the point at which the graph of an equation crosses the $y$-axis, substitute 0 for the $x$-variable in the given equation to obtain the corresponding value of $y$.

7. The $y$-coordinate of any point on the $x$-axis is 0. To find the corresponding $x$-coordinate of the point at which the graph of an equation crosses the $x$-axis, substitute 0 for the $y$-variable in the given equation to obtain the corresponding value of $x$.

## III. Sample Solutions

1. Draw the graph of
   $x + y = 5$
   *Solution:*
   $x + y = 5$
   $\quad y = 5 - x$
   This chart describes the coordinates: (0, 5), (3, 2), and (5, 0)

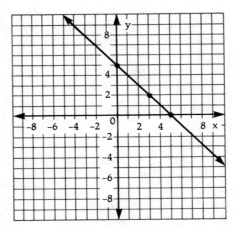

| $x$ | 0 | 3 | 5 |
|---|---|---|---|
| $y$ | 5 | 2 | 0 |

2. Draw the graph of $2y = 8x$
   *Solution:*
   $2y = 8x$
   $\quad y = 4x$

| $x$ | 0 | 2 | $-1$ |
|---|---|---|---|
| $y$ | 0 | 8 | $-4$ |

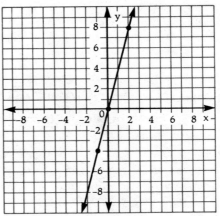

**3.** Draw the graph of $3y = 12$

*Solution:*

$$3y = 12$$
$$0x + 3y = 12$$
$$3y = 12 - 0x$$
$$y = 4 - 0x$$

| $x$ | 2 | 5 | $-3$ |
|---|---|---|---|
| $y$ | 4 | 4 | 4 |

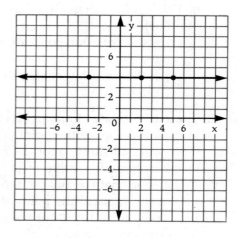

No matter what value is assigned to $x$, here the value of $y$ is always equal to 4. The line is parallel to the $x$-axis.

**4.** Draw the graph of $5x = -15$

*Solution:*

$$5x = -15$$
$$5x + 0y = -15$$
$$x + 0y = -3$$
$$x = -3 - 0y$$

| $x$ | $-3$ | $-3$ | $-3$ |
|---|---|---|---|
| $y$ | 1 | 2 | $-4$ |

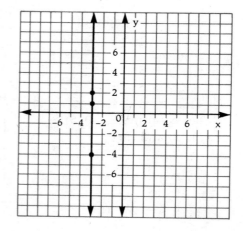

No matter what value is assigned to $y$, here the value of $x$ is always equal to $-3$. The line is parallel to the $y$-axis.

## DIAGNOSTIC TEST

Draw the graph of each of the following equations:

1. $y = x + 3$
2. $y = 6 - x$
3. $x + y = 4$
4. $x - y = 1$
5. $y = -2x$
6. $y - 1 = x$
7. $3y = -12x$
8. $2y = x$
9. $4x - 6y = 0$
10. $3x + y = 10$
11. $5x - 2y = -2$
12. $x = 6$

**13.** Without drawing the graph, determine which points corresponding to the following given coordinates are on the line which forms the graph of the equation $3x - 4y = 4$.
    **a.** $(-4, -4)$   **b.** $(8, -5)$   **c.** $(4, 2)$   **d.** $(-8, 5)$   **e.** $(0, -1)$

**14.** Without drawing the graph, determine which of the following equations have as a graph a line that passes through the origin:
    **a.** $x - y = 5$         **b.** $x + y = 0$         **c.** $4y = -5x$

**15.** Without drawing the graph, determine the coordinates of the point at which the graph of the equation $2x + 3y = 6$ crosses the Y-axis.

**16.** Without drawing the graph, determine the coordinates of the point at which the graph of the equation $5x - 4y = -15$ crosses the X-axis.

## RELATED PRACTICE EXAMPLES

Draw the graph of each of the following equations:

**1. a.** $y = x + 1$
   **b.** $y = x - 2$
   **c.** $y = 2 + x$
   **d.** $y = 2x + 3$
   **e.** $y = 3x - 4$

**2. a.** $y = -x + 4$
   **b.** $y = 7 - x$
   **c.** $y = 1 - 3x$
   **d.** $y = -x - 2$
   **e.** $y = -5 - 2x$

**3. a.** $x + y = 3$
   **b.** $x + y = 6$
   **c.** $x + y = -4$
   **d.** $x + y = 2$
   **e.** $x + y = -1$

**4. a.** $x - y = 2$
   **b.** $x - y = 5$
   **c.** $x - y = -1$
   **d.** $x - y = -6$
   **e.** $x - y = 7$

**5. a.** $y = x$
   **b.** $y = 3x$
   **c.** $y = -4x$
   **d.** $y = -\frac{2}{3}x$
   **e.** $y = \dfrac{x}{4}$

**6. a.** $y - 2 = x$
   **b.** $y - x = -5$
   **c.** $4 - y = x$
   **d.** $-y = -2x$
   **e.** $3 - x = -y$

**7. a.** $3y = 6x$
   **b.** $4y = 12x$
   **c.** $-2y = 6x$
   **d.** $-6y = -24x$
   **e.** $5y = -35x$

**8. a.** $3y = x$
   **b.** $4y = 7x$
   **c.** $-2y = x$
   **d.** $-5y = -4x$
   **e.** $x = \dfrac{y}{3}$

**9. a.** $x + y = 0$
   **b.** $x - 2y = 0$
   **c.** $4x + 2y = 0$
   **d.** $8y + 2x = 0$
   **e.** $3x - 2y = 0$

**10. a.** $2x + y = 5$
    **b.** $y + 3x = 3$
    **c.** $4x = 15 - y$
    **d.** $10 = 3x + y$
    **e.** $2x - y = -4$

**11. a.** $2x + 2y = 6$
    **c.** $8x - 2y = 10$
    **e.** $3x = 6 - 4y$
    **g.** $x - 3y + 12 = 0$
    **i.** $9y + 12x = 15$

    **b.** $6x + 3y = 9$
    **d.** $4x - 3y = 12$
    **f.** $5x - 2y = 20$
    **h.** $4x - 5y = 6$
    **j.** $10x - 4y - 9 = 0$

12. **a.** $y = 2$       **b.** $x = 4$       **c.** $y = -6$       **d.** $x = -5$
    **e.** $y = 0$       **f.** $x = 0$       **g.** $7y = 21$       **h.** $-5y = 5$
    **i.** $10x = 80$       **j.** $-9y = -81$       **k.** $12x = -72$       **l.** $15y = 0$

13. Without drawing the graph, determine which points corresponding to the following given coordinates are on the line which forms the graph of the equation:

    **a.** $x + y = 6$:
    (1) $(4, 2)$       (2) $(-1, 7)$       (3) $(3, 4)$       (4) $(-3, -3)$
    **b.** $x - y = 1$:
    (1) $(5, 4)$       (2) $(-4, 3)$       (3) $(0, -1)$       (4) $(-2, -1)$
    **c.** $3x + y = -7$:
    (1) $(2, 1)$       (2) $(-2, -1)$       (3) $(1, -10)$       (4) $(-4, 5)$
    **d.** $4x + 3y = 0$:
    (1) $(3, 4)$       (2) $(6, -8)$       (3) $(0, 0)$       (4) $(-9, 12)$
    **e.** $x - 5y = 4$:
    (1) $(-1, -1)$       (2) $(-6, -2)$       (3) $(9, 1)$       (4) $(4, 0)$
    **f.** $2x + 7y = 18$:
    (1) $(2, 2)$       (2) $(-9, 0)$       (3) $(16, -2)$       (4) $(5, 1)$
    **g.** $5x - 2y = -3$:
    (1) $(-3, -6)$       (2) $(5, 11)$       (3) $(-1, -1)$       (4) $(7, 16)$
    **h.** $6x + 8y = 16$:
    (1) $(0, 2)$       (2) $(-4, 5)$       (3) $(4, -1)$       (4) $(8, 4)$
    **i.** $9x + 4y = -36$:
    (1) $(4, 0)$       (2) $(-3, -2)$       (3) $(0, -9)$       (4) $(-8, 9)$
    **j.** $4x - 6y = 48$:
    (1) $(9, -2)$       (2) $(6, 4)$       (3) $(-5, -16)$       (4) $(0, -8)$

14. Without drawing the graphs, determine which of the following equations have as a graph a line that passes through the origin:
    **a.** $y = 3x$       **b.** $x + y = 7$       **c.** $y = -x$
    **d.** $4x + 9y = 3$       **e.** $y = -2$       **f.** $6x - 5y = 0$
    **g.** $x = 4$       **h.** $11x - y = 8$       **i.** $8y = 3x$

15. Without drawing the graph, determine the coordinates of the point at which the graph of each of the following equations crosses the Y-axis:
    **a.** $x + y = 3$       **b.** $x - y = 8$       **c.** $2y = x$
    **d.** $3x + y = 7$       **e.** $2x + 5y = 10$       **f.** $6x - y = -3$
    **g.** $5x - 9y = 3$       **h.** $4x + 8y = -12$       **i.** $8x - 7y = 5$

16. Without drawing the graph, determine the coordinates of the point at which the graph of each of the following equations crosses the X-axis:
    **a.** $x + y = 6$       **b.** $x - y = 8$       **c.** $x - y = -3$
    **d.** $4y = 3x$       **e.** $2x + 3y = 12$       **f.** $5x - 2y = -10$
    **g.** $7x + 3y = 0$       **h.** $9x - 11y = 3$       **i.** $4x + 5y = -9$

# Slope of a Line

**I. Aim**    To measure the slope or slant of a line directly from the graph and directly from the related equation.

**II. Procedure**

**1.** To measure the slope of a line directly from the graph:

   **a.** Find the ratio of the vertical change to the corresponding horizontal change when moving to the right from one point on the line to another point on the line. To do this:

     (1) Count from any point on the line a convenient number of units to the right. This measure will be the horizontal change.

     (2) If the line slants upward (positive direction), count vertically up until you meet the line at the second point. Prefix this numerical measure for the vertical change with a positive sign.

     (3) If the line slants downward (negative direction), count vertically down until you meet the line at a second point. Prefix this numerical measure for the vertical change with a negative sign.

     (4) Then divide the vertical change by the horizontal change to find the required number measuring the slope.

   **b.** Observe that slope is measured by a number, positive if the line slants upward to the right; negative if the line slants downward to the right; and zero if the line is parallel to the $x$-axis (horizontal line). There is no number that measures the slope of a line that is parallel to the $y$-axis (vertical line).

**2.** To draw a line that has a given slope indicated by a numeral:

   **a.** Select a point in the number plane.

   **b.** Then use the horizontal change and the corresponding vertical change indicated by the absolute value of this numeral to locate a second point.

   **c.** Draw the line through the two points.

**3.** When the coordinates of two points $(x_1, y_1)$ and $(x_2, y_2)$ are known, find the slope ($m$) by substituting the coordinates in the formula:

$$m = \frac{y_2 - y_1}{x_2 - x_1} \quad \text{or} \quad m = \frac{y_1 - y_2}{x_1 - x_2}$$

**477**

4. The coordinates of the endpoints may be used to determine the slope of a line segment by following the procedure of (3.) on the previous page.

5. Examine the following composite graph:

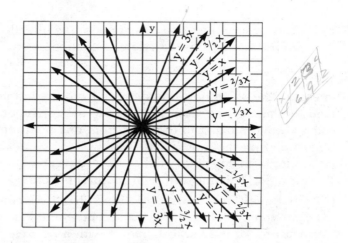

Observe that:

The slope of the line related to the equation $y = 3x$ is 3; to $y = x$ is 1; to $y = \frac{1}{3}x$ is $\frac{1}{3}$; to $y = \frac{2}{3}x$ is $\frac{2}{3}$; to $y = \frac{3}{2}$ is $\frac{3}{2}$. Each of these lines rises when drawn to the right and the slope for each is positive.

The slope of the line related to the equation $y = -3x$ is $-3$; to $y = -x$ is $-1$; to $y = -\frac{1}{3}x$ is $\frac{1}{3}$; to $y = -\frac{2}{3}$ is $-\frac{2}{3}$; to $y = -\frac{3}{2}$ is $-\frac{3}{2}$. Each of these lines falls when drawn to the right and the slope for each is negative. This comparison of each specific line to its related equation reveals an interesting mathematical fact. The *numerical measure of the slope* of the line and the *numerical coefficient of x* are identical when the related *equation is expressed in the* y-*form*.

## III. Sample Solutions

1. Determine the slope of the line at the right by selecting two points and comparing the vertical change to the corresponding horizontal change.

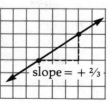

*Solution:*

   **a.** Select a point. Count 3 units to the right. The horizontal change is 3.

   **b.** Count 2 units up to meet the line. The vertical change is $+2$.

   **c.** The ratio of the vertical change ($+2$) to the horizontal change (3) is the slope, $+\frac{2}{3}$ or just $\frac{2}{3}$.

2. Draw a line with the slope of $+\frac{3}{4}$.

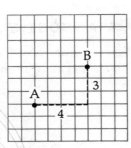

   **a.** Select on the graph paper a point ($A$) where two grid lines intersect.

   **b.** Count 4 units horizontally to the right and 3 units up to locate point ($B$). The ratio $\frac{3}{4}$ means that for every 4 units of horizontal change there are 3 units of vertical change.

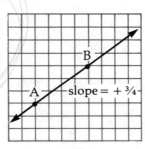

   **c.** Draw a straight line through points $A$ and $B$.

3. What is the slope of a line that passes through the points (1, 5) and (3, 9)?

   *Solution:*

   Given:

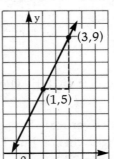

   $y_1 = 5$

   $y_2 = 9$

   $x_1 = 1$

   $x_2 = 3$

   $$m = \frac{y_2 - y_1}{x_2 - x_1}$$

   $$m = \frac{9 - 5}{3 - 1}$$

   $$m = \frac{4}{2}$$

   $$m = 2$$

   *Answer:* Slope is 2.

4. What is the slope of a line segment whose endpoints have the coordinates (2, 4) and (5, −8)?

   *Solution:*

   Given:

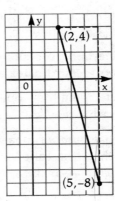

   $y_1 = 4$

   $y_2 = -8$

   $x_1 = 2$

   $x_2 = 5$

   $$m = \frac{y_2 - y_1}{x_2 - x_1}$$

   $$m = \frac{-8 - 4}{5 - 2}$$

   $$m = \frac{-12}{3}$$

   $$m = -4$$

   *Answer:* Slope is −4.

## DIAGNOSTIC TEST

1. Which of the following lines has a slope that is measured by a positive number? By a negative number? By zero?

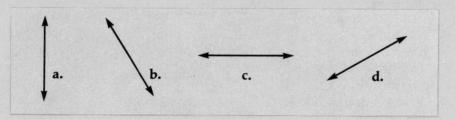

2. Determine the slope of each of the following lines:

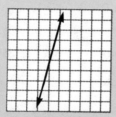     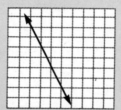     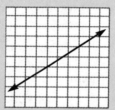

3. Does a line rise or fall when drawn to the right if its slope is:
   a. $-8$?   b. $+\frac{9}{4}$?

4. How does the vertical change compare to the corresponding horizontal change when a line has the slope:
   a. $+6$?   b. $-9$?   c. $-\frac{5}{8}$?   d. $+\frac{11}{3}$?

5. Draw a line with the slope of:
   a. $+5$   b. $-4$   c. $+\frac{2}{5}$   d. $-\frac{3}{2}$

6. Determine the slope of the line that passes through two points whose coordinates are $(-5, 4)$ and $(1, -8)$.

7. Determine the slope of the line segment if its endpoints have the coordinates $(-2, -3)$ and $(0, 9)$.

8. What number measures the slope of the line that is the graph of each of the following equations?
   a. $y = -9x$   b. $4x - 7y = 0$   c. $6x - 3y = 24$

## *RELATED PRACTICE EXAMPLES*

1. Which of the following lines has a slope that is measured by a positive number? By a negative number? By zero?

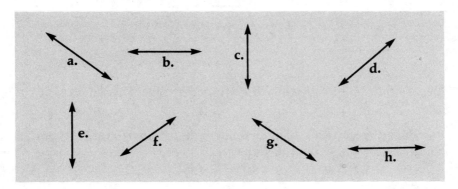

2. Determine the slope of each of the following lines by selecting for each line two convenient points and comparing the vertical change to the corresponding horizontal change.

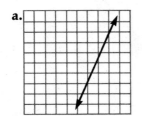

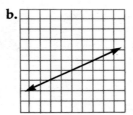

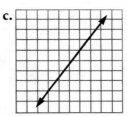

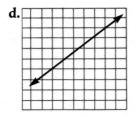

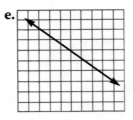

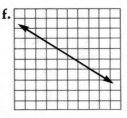

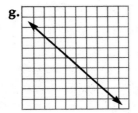

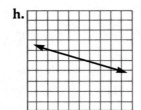

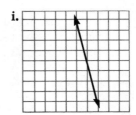

j. 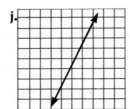   k. 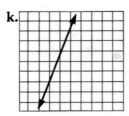   l.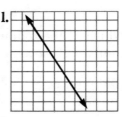

3. Does a line rise or fall when drawn to the right if its slope is:
   **a.** $-5$?    **b.** $+7$?    **c.** $-\frac{4}{3}$?    **d.** $+\frac{2}{5}$?    **e.** $0$?    **f.** $-\frac{1}{4}$?

4. How does the vertical change compare to the corresponding horizontal change when a line has the slope:
   **a.** $+2$?    **b.** $+5$?    **c.** $-4$?    **d.** $-8$?    **e.** $+\frac{3}{5}$?    **f.** $+\frac{7}{2}$?
   **g.** $-\frac{9}{4}$?    **h.** $-\frac{1}{6}$?    **i.** $+1$?    **j.** $-10$?    **k.** $-\frac{1}{3}$?    **l.** $+\frac{9}{7}$?

5. Draw a line with the slope of:
   **a.** $+3$    **b.** $+1$    **c.** $4$    **d.** $-2$    **e.** $-5$    **f.** $-1$
   **g.** $+\frac{2}{3}$    **h.** $+\frac{1}{6}$    **i.** $\frac{5}{4}$    **j.** $-\frac{3}{5}$    **k.** $-\frac{1}{2}$    **l.** $-\frac{10}{7}$
   **m.** $\frac{7}{3}$    **n.** $-\frac{3}{8}$    **o.** $\frac{8}{9}$    **p.** $-\frac{9}{4}$    **q.** $-\frac{2}{7}$    **r.** $\frac{5}{2}$

6. Determine the slope of the line that passes through two points whose coordinates are:
   **a.** $(1, 1)$ and $(5, 9)$
   **c.** $(2, 7)$ and $(4, 1)$
   **e.** $(2, -2)$ and $(8, -2)$
   **g.** $(-5, -7)$ and $(3, 5)$
   **i.** $(-6, 8)$ and $(0, -7)$

   **b.** $(3, 4)$ and $(7, 6)$
   **d.** $(1, 10)$ and $(5, 2)$
   **f.** $(-3, 10)$ and $(4, -4)$
   **h.** $(-8, -7)$ and $(-2, -4)$
   **j.** $(-4, 6)$ and $(-1, -3)$

7. Determine the slope of the line segment if its endpoints have the coordinates:
   **a.** $(2, 10)$ and $(4, 2)$
   **c.** $(-1, 7)$ and $(-2, 1)$
   **e.** $(6, 0)$ and $(9, -6)$
   **g.** $(-4, 6)$ and $(8, -3)$
   **i.** $(-4, 11)$ and $(8, 11)$

   **b.** $(-5, -1)$ and $(0, 9)$
   **d.** $(-8, 3)$ and $(2, 7)$
   **f.** $(1, 7)$ and $(4, 5)$
   **h.** $(3, 12)$ and $(9, -2)$
   **j.** $(0, -8)$ and $(7, 6)$

8. What number measures the slope of the line that is the graph of each of the following equations?
   **a.** $y = 7x$    **b.** $y = 11x$    **c.** $y = 4x$    **d.** $y = -\frac{5}{6}x$
   **e.** $y = -\frac{8}{3}x$    **f.** $4y = 12x$    **g.** $5y = -2x$    **h.** $3x - 9y = 0$
   **i.** $y = -\frac{3}{4}x + 8$    **j.** $7x - y = 10$    **k.** $4x + 2y = 12$    **l.** $5x - 3y = 2$

# *Y*-Intercept

**I. Aim** To determine directly from the linear equation the *y*-intercept number and the coordinates of the point where the graph crosses the *y*-axis.

## II. Procedure

**1. a.** Use the *y*-coordinate (ordinate) of the point where the line (graph) crosses the *y*-axis as the *y*-intercept.

**b.** The *y*-intercept is positive if this point is above the *x*-axis, negative if this point is below the *x*-axis, and 0 if the point is at the origin.

**c.** Sometimes the *y*-intercept is thought of as the distance from the origin to the point where the graph crosses the *y*-axis.

**2. a.** Examine this graph:

**b.** Observe that the lines are parallel since they all have the same slope but the *y*-intercepts differ. For equation $y = x + 7$, the *y*-intercept number is $+7$; for $y = x + 4$, it is $+4$; for $y = x + 2$, it is $+2$; for $y = x$, it is 0; for $y = x - 2$, it is $-2$; and for $y = x - 5$, it is $-5$.

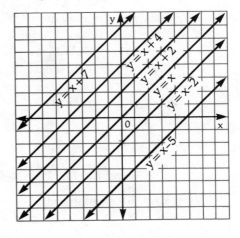

**c.** Comparison of each equation and its corresponding graph reveals that the *y-intercept number* in each case corresponds to the *constant term* (the term without *x* or *y*) found in the *y-form of the equation*.

**3.** Write as the coordinates of a point where the graph crosses the *y*-axis the number pair (0, *y*-intercept number) since the *x*-coordinate of any point on the *y*-axis is 0.

**483**

4. To determine quickly from a given linear equation in two variables the slope and the $y$-intercept number of its related line graph:
   **a.** Rearrange the linear equation in the $y$-form: $y = mx + b$.
   **b.** Use the coefficient of $x$, represented by $m$ as the measure of the *slope*.
   **c.** Use the constant term, represented by $b$, as the *y-intercept number*.

## III. Sample Solutions

1. What are the coordinates of the point where the graph of $y = 5x - 3$ crosses the $y$-axis?   *Answer:* $(0, -3)$

2. Without drawing the graph, determine the slope and the $y$-intercept number of the equation $2x + y = 8$.

   *Solution:* $2x + y = 8$            *Answer:* $-2$, slope
   $\qquad\qquad\quad y = -2x + 8$                  $+8$, $y$-intercept number

## DIAGNOSTIC TEST

1. Determine the $y$-intercept number when the equation is:
   **a.** $y = -8x - 7$        **b.** $4x - y = 6$        **c.** $3x + 5y = -25$
2. What are the coordinates of the point where the graph crosses the Y-axis when the $y$-intercept number is: **a.** $-7$? **b.** $+\frac{2}{3}$?
3. What are the coordinates of the point where the graph of the equation $3x - 7y = 28$ crosses the Y-axis?
4. Without drawing the graph, determine the slope and the $y$-intercept number of the graph and the coordinates of the point where the graph of the equation $5x + 2y = -6$ crosses the Y-axis.

## RELATED PRACTICE EXAMPLES

1. Determine the $y$-intercept number when the equation is:

   **a.** $y = x - 4$           **b.** $y = -5x + 3$          **c.** $y = 9x$
   **d.** $y = \frac{3}{4}x - 6$          **e.** $y = \frac{7}{8} + 10x$          **f.** $3x + y = 15$
   **g.** $2x - y = 12$          **h.** $4x + 2y = 8$           **i.** $16x = 4y$
   **j.** $7x - 3y = -6$         **k.** $9x + 18y = 12$         **l.** $5x - 10y = 20$
   **m.** $3x + 5y = -2$         **n.** $x = -4y - 18$          **o.** $8x - 7y = -24$

2. What are the coordinates of the point where the graph crosses the Y-axis when the $y$-intercept number is:
   **a.** $+6$?   **b.** $-2$?   **c.** $+8$?   **d.** $-1$?   **e.** $+\frac{1}{2}$?   **f.** $-2\frac{3}{4}$?   **g.** $0$?

3. What are the coordinates of the point where the graph of each of the following equations crosses the Y-axis?

**a.** $y = x + 7$      **b.** $y = x - 5$      **c.** $y = \frac{2}{3}x + \frac{3}{4}$

**d.** $x + y = 6$      **e.** $x - y = 1$      **f.** $4x - y = 4$

**g.** $3x - 4y = 0$      **h.** $7x = 10y$      **i.** $4x + 9y = 45$

**j.** $x - 8y = -24$      **k.** $6x + 3y = -3$      **l.** $5x + 2y = 7$

**m.** $x - 4y = 11$      **n.** $3x + 5y = -3$      **o.** $10x - 8y = 6$

4. Without drawing the graph, determine the slope and the $y$-intercept number of each graph and the coordinates of the point where the graph of each of the following equations crosses the Y-axis:

**a.** $y = 8x - 6$      **b.** $y = -3x + 5$      **c.** $y = \frac{1}{2}x - 2$

**d.** $y = -7 - 2x$      **e.** $x + y = 9$      **f.** $2x - y = 15$

**g.** $3x = 6y$      **h.** $8x - 2y = 0$      **i.** $6x + 5y = 30$

**j.** $7x + 21y = -35$      **k.** $10x - 5y = -20$      **l.** $3x + 4y = -6$

**m.** $x = 6 - 3y$      **n.** $8x - 7y = 2$      **o.** $9x - 4y = -5$

## CUMULATIVE REVIEW

1. On graph paper draw the axes and plot the points whose coordinates are:
**a.** $(-6, 2)$    **b.** $(0, -3)$    **c.** $(1, 8)$      **d.** $(-7, -2)$    **e.** $(6, -4)$

2. Which ordered pairs of numbers satisfy the equation $4x - 3y = 12$?
**a.** $(6, 4)$      **b.** $(1, -2)$      **c.** $(0, -4)$      **d.** $(2, -1)$      **e.** $(-3, -8)$

3. Write the $y$-form of the equation $6x - 3y = 11$.

4. Find the ordered pairs of numbers that satisfy the equation $7x - 3y = -9$ when $x = -3$, 0, and 2.

5. Draw on the number plane the graph of each of the following equations:
**a.** $y = 2x - 3$         **b.** $y = -3x$         **c.** $y = \frac{5}{2}x$
**d.** $x - y = -7$         **e.** $4x + 5y = 10$       **f.** $-7y = 28$

6. Without drawing the graph, determine which points corresponding to the following given coordinates are on the line which forms the graph of:
**a.** $3x + y = -15$:   (1) $(0, -5)$   (2) $(-4, -3)$   (3) $(-5, 0)$   (4) $(-6, 3)$
**b.** $5x - 2y = 20$:    (1) $(-4, 0)$   (2) $(2, 5)$      (3) $(6, -5)$   (4) $(0, -10)$

7. Determine the slope of the line that passes through two points whose coordinates are: **a.** $(-5, 6)$ and $(1, -6)$   **b.** $(-3, -4)$ and $(4, 1)$

8. Determine the slope of the line segment if its endpoints have the coordinates: **a.** $(-2, -3)$ and $(2, 7)$   **b.** $(-6, 2)$ and $(2, -4)$

9. Without drawing a graph, determine the slope and the $y$-intercept number of each graph and the coordinates of the point where the graph of each of the following equations crosses the $y$-axis:
**a.** $y = -\frac{2}{3}x + 6$      **b.** $16x - 4y = 8$      **c.** $5x + 6y = -3$

# Graphing by the Slope-Intercept Method

## I. Aim
To draw a graph of a linear equation using the slope and the $y$-intercept.

## II. Procedure
1. Locate two points which will be on the graph of the given equation by:
   a. First expressing the given equation in the form $y = mx + b$.
   b. Use the value of $b$, the constant term, as the $y$-intercept number to locate the point where the required line will cross the $y$-axis. This is the first point.
   c. Use the slope, represented by $m$ (the coefficient of the variable $x$) to locate the second point, counting to the right a convenient number of units from the first point, then counting the required number of units up or down depending on whether the slope is positive or negative.

2. Draw a line through these two points.

3. When the coordinates of a point and the slope of the line are known, use the above procedure to draw the graph.

4. Check whether the graph is correct by selecting a third point on the line and testing whether the coordinates satisfy the given equation.

## III. Sample Solution

Draw a graph of $2x + y = 5$ by the slope—intercept method.

*Solution:* $2x + y = 5$
$$y = -2x + 5$$

The $y$-intercept is 5. Locate point $(0, 5)$. The slope is $-2$. Count from point $(0, 5)$ one unit to the right and 2 units down, reaching point $(1, 3)$. Draw a straight line through points $(0, 5)$ and $(1, 3)$.

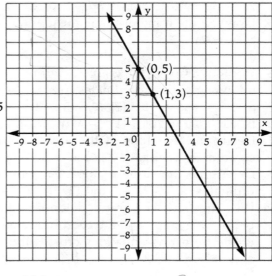

## DIAGNOSTIC TEST

1. Draw the graph of the equation $5x - 2y = 8$ by the slope-intercept method.

2. Draw the line that passes through the point $(4, -1)$ and has the slope $-3$.

## RELATED PRACTICE EXAMPLES

1. Draw the graph of each of the following equations by the slope-intercept method:

   **a.** $y = 5x + 1$       **b.** $y = 2x - 4$       **c.** $y = 6 - 4x$
   **d.** $y = -3x - 2$       **e.** $y = \frac{3}{4}x + 2$       **f.** $y = \frac{1}{2}x - 3$
   **g.** $y = 5x$       **h.** $y = -\frac{4}{3}x$       **i.** $x + y = 7$
   **j.** $x - y = -2$       **k.** $3x + y = 6$       **l.** $x - 3y = 9$
   **m.** $4x + 2y = 10$       **n.** $15x + 3y = -12$       **o.** $2x - 4y = 0$
   **p.** $5x - 6y = 30$       **q.** $3x - 2y = -6$       **r.** $28x - 7y = 21$
   **s.** $9y + 6x = -27$       **t.** $8x - 5y = -10$       **u.** $10x + 8y = 4$
   **v.** $4x - 3y = 2$       **w.** $x = -5y + 15$       **x.** $3y - 6x + 9 = 0$

2. Draw the line that passes through the point:

   **a.** $(3, 1)$ and has the slope 2.       **b.** $(0, -2)$ and has the slope 4.
   **c.** $(0, 0)$ and has the slope $\frac{1}{2}$.       **d.** $(-3, -4)$ and has the slope 1.
   **e.** $(2, -1)$ and has the slope $-5$.       **f.** $(-7, -3)$ and has the slope $-2$.
   **g.** $(4, 5)$ and has the slope $-\frac{3}{4}$.       **h.** $(1, -2)$ and has the slope $-\frac{5}{2}$.
   **i.** $(-5, 4)$ and has the slope $\frac{3}{8}$.       **j.** $(-3, 0)$ and has the slope $-\frac{5}{6}$.

## CUMULATIVE REVIEW

1. Without drawing the graph, determine the slope and $y$-intercept number of each graph and the coordinates of the point where the graph of each of the following equations crosses the $y$-axis:

   **a.** $y = 3x + 4$       **b.** $2x + y = 5$       **c.** $12x - 2y = 8$

2. Draw the graph of each equation by the slope-intercept method:

   **a.** $y = 6 - x$       **b.** $2x - y = 0$       **c.** $5x - 4y = 16$

3. Draw the line that passes through the point:

   **a.** $(2, -3)$ and has the slope 3       **b.** $(-5, 6)$ and has the slope $-\frac{4}{3}$

# Graph of a Formula

### I. Aim   To draw the graph of a formula containing two variables.

### II. Procedure

1. Construct a table of values showing the relationship between the two variables of the formula.
2. Draw two guide lines (axes) on graph paper, one at the bottom of the paper for the horizontal scale and the other at the left side of the paper for the vertical scale.
3. Select convenient scales for each variable so that the points corresponding to the number pairs in the table may be plotted without difficulty. Label scales.
4. Calculate two number pairs or, for checking purposes, three number pairs and plot their corresponding points.
5. Draw a line through these points. This line is the graph of the formula.
6. Print the formula and the title.

### III. Sample Solution

Draw the graph of the formula $i = .05p$ where $i$ represents the annual interest, $p$ represents the principal, and .05 indicates a 5% annual rate of interest.

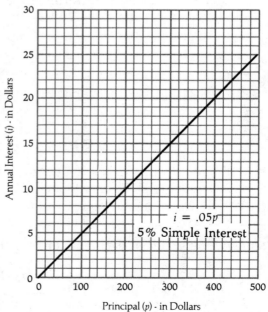

*Solution:*

1. Substitute for $p$: $100; $200; $240; $300; $360; $400 and write the related values in tabular form.

| $p$ | $100 | $200 | $240 | $300 | $360 | $400 |
|---|---|---|---|---|---|---|
| $i$ | $5 | $10 | $12 | $15 | $18 | $20 |

2. Draw guide lines and label scales, selecting each horizontal unit to represent a principal of $20 and each vertical unit to represent an annual interest of $1.

**488**

3. Plot points corresponding to ($100, $5), ($200, $10), and ($400, $20).

4. Draw the line through these points. This line is the graph of the formula $i = .05p$.

---

## DIAGNOSTIC TEST

1. Draw the graph of the formula $M = 180 - N$ showing the relationship between the supplementary angles $M$ and $N$.

2. A ship travels at an average speed of 25 knots. Develop the formula showing the relationship between the distance traveled and the time of travel. Draw the graph of this formula.

---

## RELATED PRACTICE EXAMPLES

1. Draw the graph of each of the following formulas:
   a. Perimeter and length of side of a square: $p = 4s$

   b. Absolute-Celsius temperature readings: $A = C + 273$

   c. Complementary angles: $B = 90 - A$

   d. Radius and diameter of a circle: $r = \dfrac{d}{2}$

   e. Fahrenheit-Celsius temperature readings: $F = 1.8C + 32$

2. Develop the formula from the given facts, then draw the graph of the formula in each of the following:
   a. Money is invested at 8% interest. Draw the graph showing the relationship between the annual interest and the principal.

   b. An airplane flies at an average ground speed of 300 m.p.h. Draw the graph showing the relationship between distance and flying time.

   c. The sum of $100 is borrowed at 13% interest. Draw the graph showing the relationship between the total interest and the time (in years) for which the money is borrowed.

   d. The wage scale for a certain type of work is $3.50 per hour. Draw the graph showing the relationship between wages earned and the number of hours of work (40 hours maximum).

   e. The speed in feet per second equals $\frac{22}{15}$ times the speed in miles per hour. Draw the graph showing the relationship between speeds expressed in feet per second and in miles per hour.

# Writing Linear Equations from Graphs

I. **Aim**   To write the linear equation from certain given facts such as: (1) the slope and $y$-intercept number, (2) coordinates of one point on the line and slope, (3) coordinates of two points on the line, and (4) directly from its graph.

II. **Procedure**

1. When the slope and $y$-intercept number are known:
   a. Write the equation $y = mx + b$.

   b. Substitute the given slope for $m$ and the $y$-intercept number for $b$. See sample solution 1.

2. When the coordinates $(x_1, x_2)$ of one point on the line and the slope $(m)$ are known:
   a. Write the formula $y - y_1 = m(x - x_1)$.

   b. Substitute the given values for $x_1$, $y_1$, and $m$ in the formula.

   c. Simplify the resulting equation. See sample solution.

3. When the coordinates of two points $(x_1, y_1)$ and $(x_2, y_2)$ are known:
   a. Write the formula $y - y_1 = \left( \dfrac{y_1 - y_2}{x_1 - x_2} \right)(x - x_1)$.

   b. Substitute the given coordinates for $x_1$, $y_1$, $x_2$, and $y_2$ in the formula.

   c. Compute as required and simplify the resulting equation. See sample solution.

4. Directly from the graph:
   Use any one of the following methods:
   a. Select two points on the given line. Then follow method (3) above.

   b. Find the slope and the $y$-intercept number on the given line. Then follow method (1) above.

   c. Find the coordinates of a point on the given line and the slope of the line. Then follow method (2) above.

5. Check whether your resulting equation is correct by selecting a point on the given line and testing whether its coordinates satisfy the equation.

## III. Sample Solutions

**1.** Write the linear equation when the slope is $-4$ and the $y$-intercept number is 7.

*Solution:*

Given:

$m = -4$

$b = 7$

$$y = mx + b$$
$$y = -4x + 7$$

*Answer:* $y = -4x + 7$

**2.** Write an equation of the straight line that passes through the point whose coordinates are $(5, 6)$ and has the slope $-3$.

*Solution:*

Given:

$x_1 = 5$

$y_1 = 6$

$m = -3$

$$y - y_1 = m\,(x - x_1)$$
$$y - 6 = -3\,(x - 5)$$
$$y - 6 = -3x + 15$$
$$y - 6 + 6 + 3x = -3x + 3x + 15 + 6$$
$$3x + y = 21$$

*Answer:* $3x + y = 21$

**3.** Write an equation of the straight line that passes through the points whose coordinates are $(4, 8)$ and $(2, 2)$.

*Solution:*

Given:

$x_1 = 4$

$y_1 = 8$

$x_2 = 2$

$y_2 = 2$

$$y - y_1 = \left(\frac{y_1 - y_2}{x_1 - x_2}\right)(x - x_1)$$

$$y - 8 = \left(\frac{8 - 2}{4 - 2}\right)(x - 4)$$

$$y - 8 = \left(\frac{6}{2}\right)(x - 4)$$

$$y - 8 = 3(x - 4)$$
$$y - 8 = 3x - 12$$
$$y - 8 + 8 - 3x = 3x - 3x - 12 + 8$$
$$-3x + y = -4$$

*Answer:* $-3x + y = -4$

## DIAGNOSTIC TEST

1. For each of the following, write an equation of the straight line having the:
   **a.** slope 3 and $y$-intercept number 2.
   **b.** slope $-\frac{2}{5}$ and $y$-intercept number $-7$.

2. For each of the following, write an equation of the straight line that passes through the point whose coordinates are:
   **a.** $(1, 5)$ with the slope 2          **b.** $(-4, 2)$ with the slope $-\frac{5}{6}$

3. For each of the following, write the equation of the straight line that passes through two points having the coordinates:
   **a.** $(3, 8)$ and $(11, 4)$                  **b.** $(-6, -5)$ and $(-2, 7)$

Write the equation of the following lines and check your equations:

4. By selecting two points on the line and using their coordinates:

5. By finding the slope and the $y$-intercept number:

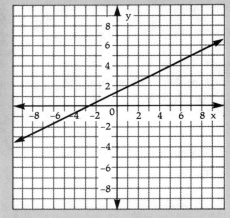

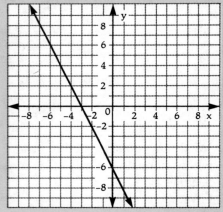

6. By finding the slope and selecting a point on the line, using its coordinates:

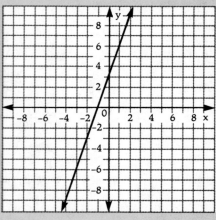

## RELATED PRACTICE EXAMPLES

1. For each of the following, write an equation of the straight line having the following given slope and $y$-intercept number:

|  | Slope | $y$-intercept number |  | Slope | $y$-intercept number |
|---|---|---|---|---|---|
| **a.** | 5 | 1 | **b.** | 6 | 3 |
| **c.** | $-2$ | $-7$ | **d.** | 9 | $-2$ |
| **e.** | 4 | 0 | **f.** | $-3$ | 6 |
| **g.** | 0 | $-1$ | **h.** | $\frac{2}{3}$ | 8 |
| **i.** | 7 | $-\frac{1}{2}$ | **j.** | $-\frac{5}{8}$ | $-\frac{3}{4}$ |

2. For each of the following, write an equation of the straight line that passes through the point whose coordinates are:
   **a.** (3, 5) with the slope 4
   **c.** $(-4, 1)$ with the slope $-2$
   **e.** (0, 4) with the slope $-1$
   **g.** (0, 0) with the slope $\frac{7}{4}$
   **i.** (2, 2) with the slope $-\frac{5}{3}$
   **b.** $(6, -2)$ with the slope 5
   **d.** $(-2, -3)$ with the slope $-3$
   **f.** (8, 3) with the slope $\frac{2}{3}$
   **h.** $(-7, 6)$ with the slope $-\frac{1}{2}$
   **j.** $(-5, -6)$ with the slope $\frac{3}{8}$

3. For each of the following, write the equation of the straight line that passes through two points having the coordinates:
   **a.** (2, 3) and (5, 9)
   **c.** (0, 4) and (6, 10)
   **e.** $(-1, -4)$ and (8, 3)
   **g.** $(-9, 4)$ and $(3, -8)$
   **i.** $(-7, -10)$ and $(-1, -2)$
   **k.** $(-5, 3)$ and $(0, -2)$
   **b.** (1, 8) and (3, 2)
   **d.** $(-2, 5)$ and (4, 7)
   **f.** $(-6, 10)$ and $(-2, -1)$
   **h.** (0, 0) and $(2, -6)$
   **j.** (8, 0) and $(-2, 4)$
   **l.** (6, 9) and (9, 6)

4. Write the equation of each of the following lines by selecting two points on the line and using their coordinates. Check your equations.

   **a.**

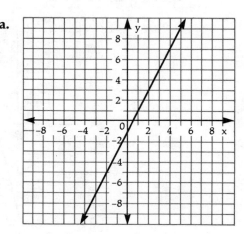

**b.** 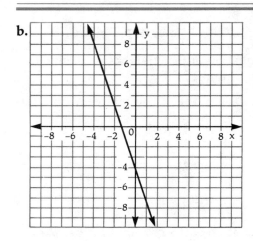    **c.**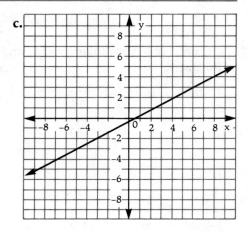

5. Write the equation of each of the following lines by finding the slope and the *y*-intercept number. Check your equations.

**a.** 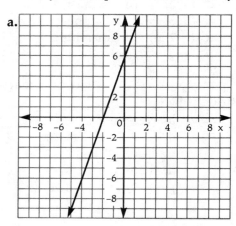    **b.**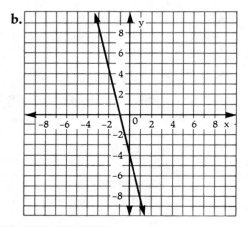

**c.**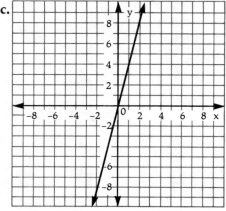

6. Write the equation of each of the following lines by finding the slope and selecting a point on the line, using its coordinates. Check your equations.

a.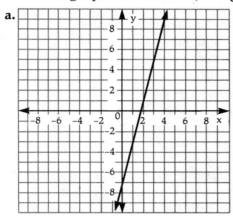

b.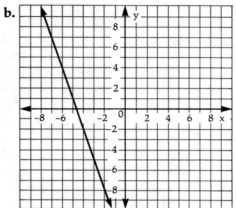

c.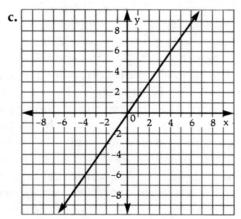

## COMPETENCY CHECK TEST

In each of the following select the letter corresponding to your answer.

1. The ordered pair of numbers that satisfies $3x - 2y = -9$ is:
   a. $(0, 3)$      b. $(-2, -3)$      c. $(5, -1)$      d. $(-1, 3)$

2. The slope of the graph of $8x - 2y = 12$ is measured by:
   a. 12      b. 8      c. 4      d. 2

3. The coordinates of the point where the graph $x + y = 3$ crosses the $x$-axis are: a. $(3, 0)$      b. $(0, 3)$      c. $(3, 3)$      d. answer not given

4. The slope of a line with points whose coordinates are $(4, 4)$ and $(6, 8)$ is:
   a. 6      b. 8      c. 2      d. 4

5. One of the points in the graph of $4x - 5y = 0$ has the coordinates:
   a. $(0, 0)$      b. $(5, -4)$      c. $(-10, 8)$      d. $(-3, -2)$

1. Draw the axes and plot the points of these coordinates:
   **a.** $(3, -5)$  **b.** $(0, 7)$  **c.** $(-2, -4)$  **d.** $(6, -1)$  **e.** $(-8, 3)$
2. Write as an equation in two variables ($x$ and $y$):
   The ordinate is 9 more than the abscissa.
3. Which ordered pair of numbers satisfies $6x - y = 7$?
   **a.** $(-1, 1)$  **b.** $(0, 7)$  **c.** $(1, -1)$  **d.** $(-2, -5)$  **e.** $(12, 5)$
4. Write the $y$-form of the equation $5x - 3y = -2$.
5. Find the ordered pairs of numbers that satisfy the equation
   $3x - y = 4$ when $x = -1$, 0, and 4.
6. Draw in the number plane the graph of each of the following
   equations: **a.** $y = 9 - x$  **b.** $3x - 4y = 6$  **c.** $2x + 3y = 0$
7. Without drawing the graph, find which points corresponding to
   these coordinates are on the line which forms the graph of:
   **a.** $5x - y = -3$:  (1) $(1, 2)$  (2) $(-1, 8)$  (3) $(0, 3)$
   **b.** $4x + 7y = 8$:  (1) $(-5, 4)$  (2) $(2, 0)$  (3) $(9, -6)$
8. Without drawing the graph, find the coordinates of the point at
   which the graph of each equation crosses the $x$-axis:
   **a.** $x - y = -5$  **b.** $4x + 3y = 12$  **c.** $6x - 5y = 3$
9. Determine the slope of a line that passes through the two points:
   **a.** $(-2, -8)$ and $(4, 10)$  **b.** $(-1, 9)$ and $(-5, 0)$
10. Determine the slope of a line segment if its endpoints have the
    coordinates: **a.** $(3, 0)$ and $(7, 8)$  **b.** $(-4, -5)$ and $(-3, 6)$
11. Without drawing the graph, determine the slope and the $y$-inter-
    cept number of each graph and the coordinates of the point
    where the graph of each of the following equations crosses the
    $y$-axis: **a.** $y = -4x + 7$  **b.** $5x - 8y = -16$  **c.** $6x + 4y = 3$
12. Draw the graph of each of the following by the slope-intercept
    method: **a.** $y = 3x - 2$  **b.** $4x - 5y = 10$  **c.** $2x + 6y = -12$
13. Draw the line that passes through point $(-1, 4)$ and has slope
    $-\frac{5}{6}$.
14. Draw the graph of the formula $i = .08p$ where $i$ represents the
    annual interest and $p$ represents the principal.
15. Write an equation of the straight line that:
    **a.** Has the slope $-5$ and $y$-intercept number $-4$.
    **b.** Has the slope 2 and passes through the point $(2, -3)$.
    **c.** Passes through 2 points $(-1, 6)$ and $(-7, 0)$.

## Keyed Achievement Test

The colored numeral indicates the section where help may be found.
1. Draw on the number line the graph of each equation: 6-11
   **a.** $6n = -24$  **b.** $2x - 9x = -42$  **c.** $r + 3 = r$
   **d.** $|a| - 1 = 0$  **e.** $7b - 35 = 35 - 7b$  **f.** $7b + 35 = 35 + 7b$
2. Draw in the number plane the graph of each equation: 12-3
   **a.** $x - 2y = 0$  **b.** $5x + y = 10$  **c.** $3x = 8y + -6$
3. Without drawing the graph, determine which points corresponding to the following given coordinates are on the line which forms
   the graph of: 12-3
   **a.** $4x + y = 5$:  (1) $(-2, 3)$  (2) $(-1, -1)$  (3) $(3, -7)$
   **b.** $2x - 3y = -3$:  (1) $(3, 1)$  (2) $(9, 5)$  (3) $(-3, -1)$
4. Determine the slope of a line that passes through two points
   whose coordinates are: **a.** $(-1, 3)$ and $(3, 15)$ **b.** $(-8, 10)$ and
   $(-2, -5)$ 12-4
5. Without drawing the graph, determine the slope and the $y$-inter-
   cept number of each graph and the coordinates of the point where
   the graph of each equation crosses the $y$-axis. 12-4 and 12-5
   **a.** $y = 6x - 2$  **b.** $4x - 2y = 10$  **c.** $9x - 6y = -5$
6. Draw the graph of each equation by the slope-intercept method:
   **a.** $y = -2x - 3$  **b.** $6x - 3y = -9$  **c.** $4x - 5y = 15$ 12-6
7. Draw the line that passes through the point: 12-6
   **a.** $(-3, 4)$ and has the slope $-3$ **b.** $(0, -2)$ and has the slope $\frac{3}{4}$
8. Write the equation of the straight line that passes through two
   points whose coordinates are $(3, -6)$ and $(-2, 5)$. 12-8

## MAINTENANCE PRACTICE IN ARITHMETIC

**1.** Add: 638,527  **2.** Subtract:  **3.** Multiply:  **4.** Divide:
        793,895      6,146,200        70,509       $483\overline{)1,727,208}$
        469,478      5,248,397        83,056

**5.** Add: $2\frac{5}{6} + 3\frac{1}{4} + 1\frac{3}{8}$  **6.** Subtract: $12 - 6\frac{5}{16}$  **7.** Multiply: $4\frac{5}{6} \times \frac{7}{8}$
**8.** Divide: $14\frac{3}{4} \div 3$  **9.** Add: $.09 + .999 + .9$  **10.** Subtract: $.82 - .567$
**11.** Multiply: $6.09 \times .103$  **12.** Divide: $.125\overline{)3}$  **13.** Find 19% of 43.6
**14.** What percent of 30 is 40?  **15.** 6% of what amount is $7.20?

# INTRODUCTION

Frequently in problems dealing with two related quantities, it may be difficult to represent the quantities in terms of one variable. However, if two variables are used, two equations containing the variables must be formulated and must be solved together. This pair of equations forms a *system of linear equations in two variables* (sometimes called *simultaneous equations*). Each of these equations has the same two variables, each of the first degree. Systems of linear equations in two variables are solved both algebraically and graphically.

In the algebraic solutions one of the variables is eliminated so that a simple equation in only the other variable remains. See pages 508 and 513.

In the graphic solution the coordinates of the point of intersection is found. When the graphs of two linear equations in the same two variables are drawn on the same set of axes, only one of the following possibilities exists at any one time:

a. The two lines intersect.
b. The two lines are parallel.
c. The two lines are one and the same line, they coincide.

When the two lines intersect, the point of intersection is obviously on both lines. Consequently the number pair associated with this point of intersection satisfies both equations.

When the two lines are parallel, they do not intersect and therefore have no common point. The equations have no common solution.

When the lines coincide, they each have exactly the same points. There is an infinite number of number pairs (common solutions) that satisfy both equations. Equations of this type are equivalent.

# Systems of Linear Equations in Two Variables and Problem Solving

# Systems of Linear Equations and Their Graphs

### I. Aim

To determine without drawing the graphs, whether the graphs of two linear equations are: (a) lines that intersect, (b) lines that are parallel, or (c) lines that coincide.

### II. Procedure

**1.** Express the given equations in their *y*-forms.

**(a)** If the *y*-forms of the two equations show that the numerical coefficients of *x* are different indicating different slopes, then the lines intersect.

The system of linear equations in two variables whose graphs are intersecting lines and so having only one common solution is said to be *independent* and *consistent*.

**(b)** If the *y*-forms of the two equations have different *y*-intersect numbers but the same numerical coefficients of *x*, indicating that the slopes of both lines are the same, then the lines are parallel.

The system of linear equations whose graphs are parallel lines is said to be *inconsistent*.

**(c)** If the *y*-forms of the two equations are exactly the same, then the lines coincide.

The system of linear equations whose graphs are lines that coincide is said to be *dependent*.

**2.** To express a linear equation in *standard form*, write it in the form of $ax + by = c$ where $x$ and $y$ are variables and $a$, $b$, and $c$ are rational numbers but $a$ and $b$ cannot both be equal to zero (0).

### III. Sample Solutions

Using their *y*-forms, determine which of the following systems of equations has as graphs a pair of intersecting lines? Parallel lines? Lines that coincide?

**1.** $2x + 2y = 7$

$5x - 4y = 8$

*Solution:* The *y*-form of $2x + 2y = 7$ is $y = -2x + 7$

The *y*-form of $5x - 4y = 8$ is $y = \frac{5}{4}x - 2$

These *y*-forms indicate that the related line graphs have different slopes ($-2$ and $\frac{5}{4}$).

*Answer:* The lines intersect.

**2.** $3x - y = 4$
$6x - 2y = 10$

*Solution:* The *y*-form of $3x - y = 4$ is $y = 3x - 4$
The *y*-form of $6x - 2y = 10$ is $y = 3x - 5$
These *y*-forms indicate that the related line graphs have the same slope (3) but different *y*-intercept numbers ($-4$ and $-5$).
*Answer:* The lines are parallel.

**3.** $6x + 3y = 9$
$4x + 2y = 6$

*Solution:* The *y*-form of $6x + 3y = 9$ is $y = -2x + 3$
The *y*-form of $4x + 2y = 6$ is $y = 2x + 3$
The given equations have exactly the same *y*-form.
*Answer:* The lines coincide. They form one and the same line.

**4.** Write each of the following equations in standard form:
  **a.** $5y = 9 - 7x$      *Answer:* $7x + 5y = 9$
  **b.** $3x = 2$            $3x + 0y = 2$
  **c.** $8y = -13$          $0x + 8y = -13$

## DIAGNOSTIC TEST

**1.** Write the equation $2x - 5 = 7y - 9$ in standard form.
**2.** Rewrite the equation $8y = 3$ in standard form as an equation in $x$ and $y$.
**3.** Find the coordinates of the point where the two lines intersect.

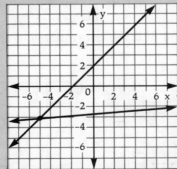

**4.** Which of the following number pairs satisfy both equations
$5x + 2y = 2$    **a.** $(4, -9)$    **b.** $(0, 1)$    **c.** $(2, -4)$    **d.** $(6, 14)$
$3x - 7y = 34$?

Using their $y$-forms determine which of the following systems of equations have as graphs a pair of lines:

**5.** That coincide: **a.** $6x+9y=15$    **b.** $8x-4y=24$    **c.** $9x-12y=18$
                          $10x+15y=20$       $10x-5y=30$       $15x-18y=24$

**6.** That are parallel: **a.** $2x-7y=3$    **b.** $4x-2y=8$    **c.** $15x+10y=16$
                        $2x+7y=3$        $6x-3y=12$        $12x+8y=20$

**7.** That intersect: **a.** $3x-y=7$      **b.** $12x+9y=18$    **c.** $25x-20y=35$
                     $6x-2y=13$       $9x+6y=15$        $5x-4y=7$

**8.** Which of the following systems of linear equations are inconsistent? Which are dependent? Which are independent and consistent?

    **a.** $7x-5y=-4$            **b.** $8x+12y=15$           **c.** $4x-9y=7$
       $14x-10y=-8$            $2x+3y=5$             $4x+9y=3$

## RELATED PRACTICE EXAMPLES

**1.** Write each of the following equations in standard form:

   **a.** $y=x+7$           **b.** $9x=8y$             **c.** $11y-5-3x=0$

   **d.** $2x+5=5y-6$      **e.** $12-x=2y$        **f.** $x-y=3x-1$

   **g.** $5(x-1)=3(y+2)$    **h.** $(4x-7y)-(2y-5)=0$    **i.** $10-2(6-x)=8y$

**2.** Rewrite each of the following equations in standard form as an equation in $x$ and $y$:

   **a.** $7x=18$                  **b.** $11x=25$               **c.** $4y=9$

   **d.** $14y=56$                **e.** $12x=-84$            **f.** $3y=0$

**3.** In each of the following find the coordinates of the point where the two lines intersect:

   **a.**

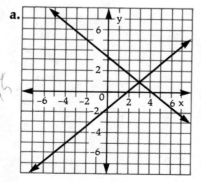

   **b.**

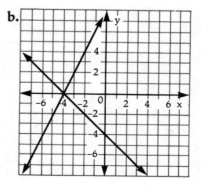

**c.** 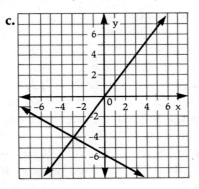    **d.**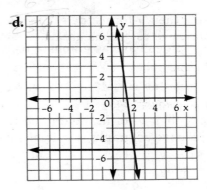

**e.** 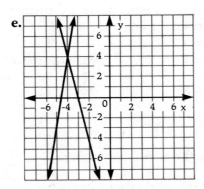    **f.**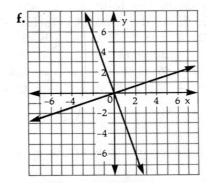

4. Which of the following number pairs satisfy both equations in each of the following systems?

   **a.** $x+y=13$　　(1) (5, 8)　　(2) (9, 4)　　(3) (8, 5)　　(4) (11, 2)
   　　$x-y=3$

   **b.** $4x+y=2$　　(1) $(-1, 6)$　(2) $(2, -6)$　(3) $(\frac{1}{2}, 0)$　　(4) $(1, -2)$
   　　$x-2y=14$

   **c.** $8x-4y=-16$　(1) (1, 6)　　(2) $(0, -4)$　(3) $(-3, -2)$　(4) (2, 8)
   　　$8x+7y=-10$

   **d.** $3x+5y=2$　　(1) $(-1, 1)$　(2) $(-16, 10)$　(3) $(14, -8)$　(4) $(4, -2)$
   　　$2x-6y=20$

   **e.** $5x+3y=11$　(1) (1, 2)　　(2) $(4, -3)$　(3) $(-2, 7)$　　(4) $(2, -7)$
   　　$7x+2y=0$

5. Using their $y$-forms, determine which of the following systems of equations have as graphs a pair of lines that coincide:

   **a.** $12x+8y=16$　　　　**b.** $4x-8y=18$　　　　**c.** $20x+24y=40$
   　　$8x+6y=12$　　　　　　$2x+4y=9$　　　　　　$15x+18y=30$

   **d.** $35x-30y=20$　　　　**e.** $10x+15y=-25$　　**f.** $18x-45y=54$
   　　$42x-36y=24$　　　　　$15x+20y=-35$　　　　$14x-35y=42$

6. Using their $y$-forms, determine which of the following systems of equations have as graphs a pair of parallel lines?

   **a.** $x + 3y = 6$           **b.** $9x - y = 10$        **c.** $4x + 10y = 60$
        $x + 3y = 5$               $9x + y = 7$            $2x - 5y = 30$
   **d.** $8x + 16y = -48$      **e.** $7x - y = 9$         **f.** $9x + 15y = 21$
        $4x + 8y = 24$            $21x - 3y = 15$       $6x + 10y = 14$

7. Using their $y$-forms, determine which of the following systems of equations have as graphs a pair of intersecting lines:

   **a.** $4x + y = 21$         **b.** $9x - 12y = 18$     **c.** $3x + 8y = -9$
        $x + 4y = 1$              $6x - 8y = 12$        $3x + 8y = 16$
   **d.** $x + 5y = 9$          **e.** $x - 3y = 7$          **f.** $2x - y = 4$
        $3x - 15y = -3$        $5x - 15y = 35$       $6x - 3y = 10$

8. Which of the following systems of linear equations are inconsistent? Which are dependent? Which are independent and consistent?

   **a.** $2x + y = 9$          **b.** $5x - 2y = 12$      **c.** $10x - 16y = 28$
        $4x + 3y = 13$          $5x - 2y = 8$          $5x + 8y = 14$
   **d.** $3x + 7y = 9$         **e.** $16x - 36y = 24$     **f.** $12x + 7y = -6$
        $9x + 21y = 27$         $4x - 9y = 6$          $24x + 14y = 12$

## CUMULATIVE REVIEW

1. Find the coordinates of the point where the two lines intersect:

   **a.** 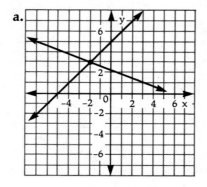       **b.**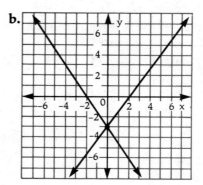

2. Using their $y$-forms, without drawing the graphs, determine which of the following systems of equations have as graphs a pair of lines that are parallel. A pair of intersecting lines. A pair of lines that coincide. Compare the three given systems of equations and explain how you can tell directly from each pair of equations what their graphs will be.

   **a.** $4x + y = 2$          **b.** $x + 3y = 6$        **c.** $20x + 24y = 40$
        $x - 2y = 14$          $x + 3y = 5$          $15x + 18y = 30$

# Solving Two Linear Equations in Two Variables by Graphing

**I. Aim**    To solve a system of two linear equations by graphing.

**II. Procedure**

1. Draw the graph of each equation on the same axes.
2. If the lines intersect, locate the point of intersection and determine its coordinates.
3. This number pair is the common solution of the two equations.
4. Check by substituting the coordinates of this point of intersection in the two given equations. If the two resulting sentences are true, then this number pair is the required solution.
5. Note that if the two lines are parallel, there is no common solution. If the two lines coincide, they are one and the same line and there are an infinite number of solutions.

**III. Sample Solution**

1. Solve graphically, then check:    $x + y = 0$
$$3x - 2y = 10$$

*Solution:*

$$x + y = 0$$
$$x - x + y = 0 - x$$
$$y = -x$$

| $x$ | 0 | 3 | $-4$ |
|---|---|---|---|
| $y$ | 0 | $-3$ | 4 |

$$3x - 2y = 10$$
$$3x - 2y + 2y = 10 + 2y$$
$$3x = 10 + 2y$$
$$3x - 10 = 10 - 10 + 2y$$
$$3x - 10 = 2y$$
$$\tfrac{1}{2}(3x - 10) = \tfrac{1}{2} \cdot 2y$$
$$\tfrac{3}{2}x - 5 = y$$
$$y = \tfrac{3}{2}x - 5$$

| $x$ | 0 | 4 | 6 |
|---|---|---|---|
| $y$ | $-5$ | 1 | 4 |

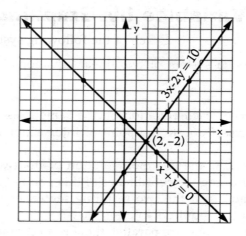

Check:

$x + y = 0$

$2 + (-2) = 0?$

$0 = 0 ✔$

$3x - 2y = 10$

$(3 \cdot 2) - 2(-2) = 10?$

$6 - (-4) = 10$

$6 + (4) = 10$

$10 = 10 ✔$

Answer: $x = 2$, $y = -2$, or $(2, -2)$

## DIAGNOSTIC TEST

Using all the real numbers, solve each of the following systems of equations graphically, then check:

1. $y = x + 1$
   $x + y = 5$

2. $y = -2x$
   $3x - y = 5$

3. $x - y = -7$
   $4x + 3y = 7$

4. $x - 6y = 2$
   $2x = 5y - 3$

5. $5x + 3y = 12$
   $x + 2y = 8$

6. $x + y = 0$
   $y = -3x$

7. $4x = 20$
   $3y = -18$

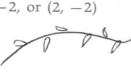

### RELATED PRACTICE EXAMPLES

Using all the real numbers, solve each of the following systems of equations graphically, then check:

**1. a.** $y = x$
$x + y = 4$

**b.** $y = 4x$
$x + y = 5$

**c.** $x = 2y$
$x - y = 2$

**d.** $x = 5y$
$x - y = 4$

**e.** $y = 3x$
$x + y = 8$

**f.** $y = x + 3$
$x + y = 9$

**g.** $y = x - 1$
$x + y = 11$

**h.** $x = y + 5$
$2x + y = 13$

**i.** $x = y - 2$
$3x - y = 8$

**j.** $y = x - 3$
$2x + 3y = 16$

**k.** $x + y = 6$
$x - y = 2$

**l.** $x + y = 5$
$x - 2y = -4$

**m.** $2x - y = 0$
$3y - 2x = 4$

**n.** $2x = 5 + y$
$3x - 4y = 0$

**o.** $3x - 5y = 2$
$4x - 6 = 5y$

**2. a.** $y = -x$
$x - y = 2$

**b.** $x = -4y$
$2x + 3y = 10$

**c.** $x = 2$
$2x - 3y = 13$

**d.** $3x = -5y$
$x + 4y = -7$

**e.** $y = x - 4$
$x + y = -2$

**f.** $x = y + 7$
$y = 2x - 10$

**g.** $x + y = -1$
$x - y = 3$

**h.** $x - y = 8$
$2x + 3y = 6$

**i.** $3x + 4y = -6$
$4x - 2y = 14$

**3. a.** $y = -2x$
$x + y = 1$

**b.** $2x = -3y$
$x + y = -2$

**c.** $x = -3$
$x + 2y = 5$

**d.** $3x + y = 0$
$2x + y = 2$

**e.** $y = 6 - x$
$x - y = -10$

**f.** $x + y = 1$
$x - y = -7$

**g.** $2x + y = -5$
$x + 2y = 2$

**h.** $5x + y = -3$
$2x = 3y - 8$

**i.** $4x - 3y = -11$
$2x + 5y = 1$

**4. a.** $x = 3y$
$2x - y = -5$

**b.** $3x = 2y$
$x - 3y = 7$

**c.** $y = -5$
$x - y = 2$

**d.** $4x - 2y = 0$
$3x + 2y = -7$

**e.** $y = x + 1$
$x + y = -5$

**f.** $x + y = -6$
$x - y = 4$

**g.** $3x + y = -7$
$x - 3y = 1$

**h.** $5x - 2y = -7$
$5 = y - 3x$

**i.** $2x - 7y = 6$
$5x - 8y = -4$

**5. a.** $x = y - 1$
$x + 4y = 4$

**6. a.** $y = x$
$3x + y = 0$

**7. a.** $x = 6$
$y = 3$

**b.** $x + y = 3$
$2x - y = 6$

**b.** $x + y = 0$
$x - 2y = 0$

**b.** $x = -5$
$y = 4$

**c.** $x - 3y = 6$
$3x - y = 2$

**c.** $y = -2x$
$3y + x = 0$

**c.** $3x = 12$
$4y = -8$

**d.** $3x - y = -12$
$2x + 8 = y$

**d.** $2x - 5y = 0$
$3x = 2y$

**d.** $8x = -16$
$5y = -20$

**e.** $4x - 5y = -4$
$3x = 2y - 3$

**e.** $x - y = 0$
$3y = 2x$

**e.** $-2x = 6$
$-8y = -24$

# Solving Two Linear Equations in Two Variables by Addition or Subtraction

**I. Aim**  To solve a system of two linear equations in two variables by the addition or subtraction method of elimination.

**II. Procedure**

1. Reduce the two equations to a single equation in one variable.

   **a.** When the two given equations, arranged in standard form, both have one variable with the same absolute value for its numerical coefficient, select this variable for elimination by:

   (1) Adding the two equations (left member to left member and right member to right member) when the numercial coefficients of this variable have opposite signs.

   (2) Subtracting the two equations (left member from left member and right member from right member) when the numerical coefficients of this variable have the same signs.

   **b.** When neither variable has a coefficient of equal absolute value, select a variable for elimination.

   (1) Multiply one equation by a number, or if necessary, the second equation by another number so that the absolute values of the numerical coefficient of the variable to be eliminated is the least common multiple of these absolute values.

   (2) Then follow step **a.** above.

2. Solve the resulting equation for the value of the remaining variable.

3. Substitute this value in one of the given equations to find the value of the other variable.

4. The resulting number pair is the common solution.

5. Check by substituting the found values for the variables in the two given equations.

6. Not always is the variable $y$ selected for elimination. When the numerical coefficients of the variable $x$ in the two equations have the same absolute value, select it for elimination. See sample solution 3.

7. Variables may be represented by letters other than $x$ and $y$. Two kinds of frames like $\triangle$ and $\square$ may also be used.

508

## III. Sample Solutions

1. Solve and check: $5x + y = 14$
$$3x - y = 2$$

Solution:
$$5x + y = 14$$
$$3x - y = 2$$
$$8x \quad\quad = 16$$
$$x = 2$$
$$5x + y = 14$$
$$(5 \cdot 2) + y = 14$$
$$10 + y = 14$$
$$10 - 10 + y = 14 - 10$$
$$y = 4 \quad Answer:\ x = 2,\ y = 4,\ or\ (2, 4)$$

Check:
$$5x + y = 14$$
$$(5 \cdot 2) + 4 = 14?$$
$$10 + 4 = 14$$
$$14 = 14\ \checkmark$$

$$3x - y = 2$$
$$(3 \cdot 2) - 4 = 2?$$
$$6 - 4 = 2$$
$$2 = 2\ \checkmark$$

2. Solve and check: $5x + 2y = 23$
$$8x + 2y = 32$$

Solution:
$$5x + 2y = 23$$
$$8x + 2y = 32$$
$$5x + 2y = 23$$
$$-8x - 2y = -32$$
$$-3x \quad\quad = -9$$
$$x = 3$$
$$5x + 2y = 23$$
$$(5 \cdot 3) + 2y = 23$$
$$15 + 2y = 23$$
$$15 - 15 + 2y = 23 - 15$$
$$2y = 8$$
$$y = 4 \quad Answer:\ x = 3,\ y = 4,\ or\ (3, 4)$$

Check:
$$5x + 2y = 23$$
$$(5 \cdot 3) + (2 \cdot 4) = 23?$$
$$15 + 8 = 23$$
$$23 = 23\ \checkmark$$

$$8x + 2y = 32$$
$$(8 \cdot 3) + (2 \cdot 4) = 32?$$
$$24 + 8 = 32$$
$$32 = 32\ \checkmark$$

3. Solve and check: $3x + y = 10$
$$3x - 2y = 16$$

Solution:
$$3x + y = 10$$
$$3x - 2y = 16$$
$$3x + y = 10$$
$$-3x + 2y = -16$$
$$3y = -6$$
$$y = -2$$

$$3x + y = 10$$
$$3x + (-2) = 10$$
$$3x - 2 + 2 = 10 + 2$$
$$3x = 12$$
$$x = 4$$

Check:
$$3x + y = 10$$
$$(3 \cdot 4) + (-2) = 10?$$
$$12 + (-2) = 10$$
$$10 = 10\ \checkmark$$

$$3x - 2y = 16$$
$$(3 \cdot 4) - 2(-2) = 16?$$
$$12 - (-4) = 16$$
$$12 + (4) = 16$$
$$16 = 16\ \checkmark$$

$Answer:\ x = 4,\ y = -2,\ or\ (4, -2)$

**4.** Solve and check: $2x + 3y = 6$
$3x + 4y = 7$

*Solution:*

$$2x + 3y = 6$$
$$3x + 4y = 7$$

$$2x + 3y = 6$$
$$4(2x + 3y) = 4(6)$$
$$8x + 12y = 24$$

$$3x + 4y = 7$$
$$3(3x + 4y) = 3(7)$$
$$9x + 12y = 21$$

$$8x + 12y = 24$$
$$9x + 12y = 21$$

$$8x + 12y = 24$$
$$-9x - 12y = -21$$

$$-x \qquad = 3$$
$$x = -3$$

$$2x + 3y = 6$$
$$2(-3) + 3y = 6$$

$$(-6) + 3y = 6$$
$$-6 + 6 + 3y = 6 + 6$$

$$3y = 12$$
$$y = 4$$

*Answer:* $x = -3$, $y = 4$, or $(-3, 4)$

*Check:*

$$2x + 3y = 6$$
$$2(-3) + (3 \cdot 4) = 6 ?$$
$$(-6) + (12) = 6$$
$$6 = 6 ✔$$

$$3x + 4y = 7$$
$$3(-3) + (4 \cdot 4) = 7 ?$$
$$(-9) + (16) = 7$$
$$7 = 7 ✔$$

## DIAGNOSTIC TEST

**1.** To find the common solution of the system $\begin{aligned} -5x + 7y &= 18 \\ 5x - 9y &= 21 \end{aligned}$ which variable, $x$ or $y$, is easier to eliminate?

Solve by the addition or subtraction method and check:

**2.** $x + y = 9$
$x - y = 3$

**3.** $a - 3b = 1$
$2a + 3b = 20$

**4.** $5m + n = 15$
$3m + n = 11$

**5.** $2c - 5d = 7$
$4c - 5d = 19$

**6.** $2x - 3y = -4$
$2x + 5y = 12$

**7.** $-5b - 4x = 23$
$2b - 4x = 2$

**8.** Transform the given equations in the system $\begin{aligned} 6x - 9y &= 12 \\ 8x + 5y &= 7 \end{aligned}$ so that one of the variables will have a numerical coefficient of the same absolute value. Do not solve this system of equations.

Solve by the addition or subtraction method and check:

**9.** $2x + 6y = 18$
$4x - 3y = 6$

**10.** $4x + 2y = 12$
$5x + 3y = 17$

**11.** $6x - 4y = 20$
$8x - 10y = 36$

**12.** $3x + 7y = 5$
$6x + 8y = 7$

### *RELATED PRACTICE EXAMPLES*

1. To find the common solution of each of the following systems of equations, which variable, $x$ or $y$, is easier to eliminate?

**a.** $5x + 9y = 16$
  $2x + 9y = 10$

**b.** $3x - 5y = 8$
  $3x - 6y = 5$

**c.** $7x - 8y = 9$
  $-7x + y = -3$

**d.** $x + 4y = 5$
  $2x - 4y = 6$

**e.** $x - y = -15$
  $x + y = 7$

**f.** $4x - 5y = 10$
  $4x + 5y = 21$

Solve and check:

2. **a.** $x + y = 12$
  $x - y = 2$

**b.** $x - y = 5$
  $x + y = 11$

**c.** $m - n = 4$
  $m + n = 6$

**d.** $c + d = 13$
  $c - d = 5$

**e.** $x - y = -1$
  $x + y = 9$

**f.** $a + b = 8$
  $a - b = -4$

**g.** $u - v = -3$
  $u + v = 5$

**h.** $r + s = 10$
  $r - s = 0$

**i.** $d - t = 0$
  $d + t = 6$

3. **a.** $3x + y = 10$
  $4x - y = 4$

**b.** $5b - c = 33$
  $7b + c = 51$

**c.** $x + 2y = 15$
  $x - 2y = -9$

**d.** $4x - 5y = -2$
  $2x + 5y = 14$

**e.** $a + 3b = 8$
  $6a - 3b = 27$

**f.** $5b - 6d = -1$
  $3b + 6d = 9$

**g.** $12x - 9y = 12$
  $x + 9y = 79$

**h.** $6r - 7s = 0$
  $4r + 7s = 70$

**i.** $4m + 3n = 15$
  $4m - 3n = 9$

4. **a.** $2x + y = 10$
  $x + y = 6$

**b.** $2c + d = 13$
  $5c + d = 28$

**c.** $5x + 3y = 30$
  $3x + 3y = 18$

**d.** $4x + 5y = 41$
  $7x + 5y = 53$

**e.** $3a + 2b = 17$
  $9a + 2b = 35$

**f.** $7m + 4n = 80$
  $6m + 4n = 72$

**g.** $-8b + 7c = 4$
  $4b + 7c = 40$

**h.** $5x + 8y = 23$
  $-3x + 8y = -1$

**i.** $2x + 9y = 47$
  $-5x + 9y = 40$

5. **a.** $3x - y = 5$
  $x - y = 1$

**b.** $2m - n = 2$
  $7m - n = 12$

**c.** $4a - 2b = 2$
  $3a - 2b = -1$

**d.** $9x - 5y = -22$
  $6x - 5y = -28$

**e.** $5c - 4d = 0$
  $8c - 4d = 12$

**f.** $4x - 7y = 12$
  $8x - 7y = 24$

**g.** $-5m - 8n = -65$
  $m - 8n = -35$

**h.** $4y - 6x = 12$
  $-2y - 6x = -24$

**i.** $-7r - 10s = -96$
  $-9r - 10s = -112$

6. **a.** $x + 5y = 7$
  $x + 3y = 5$

**b.** $b - 3c = -2$
  $b - 4c = -4$

**c.** $3c + 4d = 26$
  $3c - 5d = 8$

**d.** $-2y - 3z = -13$
  $-2y + 6z = 14$

**e.** $-5r + 4s = 0$
  $-5r + 6s = 10$

**f.** $-4m - n = -16$
  $-4m - 5n = -32$

**g.** $x + 3y = 9$
  $-x + 7y = 21$

**h.** $-9b - 4c = -71$
  $9b - 2c = 59$

**i.** $8a + 6b = 56$
  $-8a - 4b = -48$

**7.** **a.** $x - y = 4$  
   $x + y = 0$

**b.** $4x + y = 3$  
   $x + y = 0$

**c.** $5a - b = -13$  
   $2a - b = -7$

**d.** $m + 6n = 19$  
   $m - 2n = -13$

**e.** $8r + 6s = -54$  
   $3r - 6s = -12$

**f.** $2s + 3t = 13$  
   $4s + 3t = 5$

**g.** $8r - 7d = 69$  
   $6r - 7d = 57$

**h.** $9x + 10y = 32$  
   $-3x + 10y = 56$

**i.** $-5x - 8y = 26$  
   $-7x + 8y = -2$

**8.** Transform the given equations in each of the following systems of equations so that one of the variables will have a numerical coefficient of the same absolute value. Do not solve these systems of equations.

**a.** $4x + y = 8$  
   $x - 5y = 2$

**b.** $6x - 4y = 9$  
   $2x - 8y = 3$

**c.** $7x + 2y = 11$  
   $5x + 3y = 7$

**d.** $9x + 4y = -5$  
   $8x - 5y = 9$

**e.** $10x + 6y = 21$  
   $8x + 14y = 18$

**f.** $-4x + 12y = 15$  
   $6x - 8y = -3$

**g.** $3x - 6y = 10$  
   $-x + y = 5$

**h.** $2x + 7y = 12$  
   $5x - 3y = -4$

**i.** $12x - 14y = 7$  
   $9x + 21y = -5$

Solve and check:

**9.** **a.** $2a - b = 8$  
   $a + 2b = 9$

**b.** $4c + d = 15$  
   $c + 4d = 15$

**c.** $3x + 8y = -6$  
   $6x + 4y = -12$

**d.** $4x + 2y = 12$  
   $2x + 6y = -4$

**e.** $5m + 3n = 39$  
   $m - n = 3$

**f.** $2x - 9y = 40$  
   $6x - 3y = 24$

**g.** $5x - 2y = 3$  
   $2x - y = 0$

**h.** $5x - 2y = -4$  
   $3x + 4y = 34$

**i.** $-4a - 6b = 36$  
   $2a - 9b = 78$

**10.** **a.** $2x + 3y = 5$  
   $3x - 2y = 1$

**b.** $3x + 5y = 27$  
   $2x + 3y = 17$

**c.** $5x - 3y = -2$  
   $3x - 4y = -10$

**d.** $7a + 4b = 51$  
   $6a - 5b = 10$

**e.** $4x - 3y = -7$  
   $3x + 7y = 4$

**f.** $6x + 2y = 18$  
   $5x - 9y = 47$

**g.** $11c - 4d = 74$  
   $9c - 7d = 68$

**h.** $5r - 3s = 12$  
   $8r - 2s = 8$

**i.** $-4b + 6c = 2$  
   $3b - 7c = 6$

**11.** **a.** $4x + 6y = 10$  
   $10x - 8y = 2$

**b.** $12x - 16y = 20$  
   $8x + 6y = 30$

**c.** $10a + 4b = 28$  
   $6a + 18b = 48$

**d.** $20m - 6n = 72$  
   $8m - 4n = 32$

**e.** $8a + 6b = 4$  
   $6a - 10b = -26$

**f.** $9b - 4d = 30$  
   $6b + 14d = -30$

**g.** $4x - 12y = -20$  
   $6x + 15y = 102$

**h.** $12x - 9y = 21$  
   $10x - 6y = 10$

**i.** $21x + 12y = -99$  
   $14x - 8y = -18$

**12.** **a.** $2x + 4y = 4$  
   $x - 2y = 0$

**b.** $6x - 4y = -6$  
   $3x + y = 3$

**c.** $4x + 7y = 19$  
   $6x - 2y = -9$

**d.** $9x - 2y = 10$  
   $3x + 6y = -10$

**e.** $5x + 7y = 4$  
   $9x + 3y = 4$

**f.** $x - 3y = 0$  
   $4x - 2y = 5$

**g.** $3x - 4y = 3$  
   $12x - 2y = 5$

**h.** $6x - 4y = 0$  
   $5x + 2y = 4$

**i.** $7x + 5y = 15$  
   $5x - 9y = 17$

# Solving Two Linear Equations in Two Variables by Substitution

### I. Aim
To solve a system of two linear equations in two variables by the substitution method of elimination.

### II. Procedure
1. If neither one of the given equations is already arranged so that it expresses one variable in terms of the other, then select the simpler equation and solve it for one variable in terms of the second variable.
2. To eliminate the variable that is expressed in terms of the other, use the other given equation and replace this variable by the algebraic expression that is equal to it.
3. Solve the resulting equation to find the value of the variable.
4. Substitute the value of this variable in the simplest equation to find the value of the other variable. The resulting number pair is the common solution.
5. Check by substituting this number pair in the two given equations.
6. Sometimes one of the given equations has one variable missing. If necessary, solve this equation and substitute the value of this variable in the equation with the two variables to find the value of the other value.

### III. Sample Solutions
1. Solve and check:
$$y = x + 5$$
$$4x + 3y = 29$$

*Solution:*
$$y = x + 5$$
$$\underline{\underline{4x + 3y = 29}}$$

$$4x + 3(x + 5) = 29$$
$$4x + 3x + 15 = 29$$
$$7x + 15 = 29$$
$$7x + 15 - 15 = 29 - 15$$
$$7x = 14$$
$$x = 2$$
$$y = x + 5$$
$$y = 2 + 5$$
$$y = 7$$

*Check:*
$$y = x + 5$$
$$7 = 2 + 5?$$
$$7 = 7 \checkmark$$
$$4x + 3y = 29$$
$$(4 \cdot 2) + (3 \cdot 7) = 29?$$
$$8 + 21 = 29$$
$$29 = 29 \checkmark$$

*Answer:* $x = 2$, $y = 7$, or $(2, 7)$

**513**

**2.** Solve and check: $y = -3x$
$$5x - y = 8$$

*Solution:*

$$y = -3x$$
$$\underline{\underline{5x - y = 8}}$$
$$5x - (-3x) = 8$$
$$5x + (3x) = 8$$
$$8x = 8$$
$$x = 1$$
$$y = -3x$$
$$y = -3 \cdot 1$$
$$y = -3$$

*Check:*

$$y = -3x$$
$$-3 = -3 \cdot 1?$$
$$-3 = -3 \text{ ✔}$$

$$5x - y = 8$$
$$(5 \cdot 1) - (-3) = 8?$$
$$5 + 3 = 8$$
$$8 = 8 \text{ ✔}$$

*Answer:* $x = 1$, $y = -3$, or $(1, -3)$

**3.** Solve and check: $x + 2y = 10$
$$2x - y = 5$$

*Solution:*

$$x + 2y = 10$$
$$\underline{\underline{2x - y = 5}}$$
$$x + 2y = 10$$
$$x + 2y - 2y = 10 - 2y$$
$$\underline{x = 10 - 2y}$$
$$2x - y = 5$$
$$2(10 - 2y) - y = 5$$
$$20 - 4y - y = 5$$
$$20 - 5y = 5$$
$$20 - 20 - 5y = 5 - 20$$
$$-5y = -15$$
$$y = 3$$
$$x = 10 - 2y$$
$$x = 10 - (2 \cdot 3)$$
$$x = 10 - 6$$
$$x = 4$$

*Check:*

$$x + 2y = 10 \qquad\qquad 2x - y = 5$$
$$4 + (2 \cdot 3) = 10? \qquad (2 \cdot 4) - 3 = 5?$$
$$4 + 6 = 10 \qquad\qquad 8 - 3 = 5$$
$$10 = 10 \text{ ✔} \qquad\qquad 5 = 5 \text{ ✔}$$

*Answer:* $x = 4$, $y = 3$, or $(4, 3)$

## DIAGNOSTIC TEST

Solve by the substitution method and check:

**1.** $y = x$
$x + y = 4$

**2.** $y = x + 2$
$2x + 3y = 21$

**3.** $x = -3y$
$2x + 7y = 1$

**4.** $y = 2x$
$x - 3y = 10$

**5.** $x = y - 3$
$4x + 2y = 24$

**6.** $y = -4x$
$x - 2y = 27$

**7.** $x = y + 3$
$4y - 2x = 2$

**8.** $y = 10 - 7x$
$5x - 6y = 34$

**9.** $y = 3$
$8x + 5y = 31$

**10.** $x + y = 6$
$3x - 4y = 4$

## *RELATED PRACTICE EXAMPLES*

Solve by the substitution method and check:

**1.** **a.** $y = x$
$x + y = 6$

**b.** $x = y$
$3x + y = -8$

**c.** $y = 4x$
$x + 2y = 36$

**d.** $y = 7x$
$4x + 3y = -75$

**e.** $x = 3y$
$2x + 5y = 33$

**f.** $6x = y$
$3x + 4y = -27$

**2.** **a.** $y = x + 3$
$2x + 4y = 24$

**b.** $y = x + 1$
$5x + 3y = 51$

**c.** $x = y + 4$
$3x + 7y = -18$

**d.** $y = 6x + 3$
$2x + 6y = -20$

**e.** $y = 3 + x$
$3x + 8y = 46$

**f.** $x = 5 + 2y$
$4x + y = 2$

**3.** **a.** $y = -x$
$3x + y = 4$

**b.** $y = -2x$
$2x + 5y = -8$

**c.** $x = -4y$
$3x + 2y = -30$

**d.** $x = -y$
$4x + 3y = 6$

**e.** $y = -3x$
$6x + 5y = 18$

**f.** $-5y = x$
$2x + 7y = 12$

**4.** **a.** $y = x$
$4x - y = 6$

**b.** $y = 3x$
$2x - 3y = -7$

**c.** $x = 2y$
$7y - 4x = -3$

**d.** $y = 6x$
$3x - 5y = 27$

**e.** $x = 4y$
$5y - x = -2$

**f.** $y = 5x$
$6x - 3y = -45$

**5.** **a.** $y = x - 2$
$x + 5y = 20$

**b.** $y = x - 5$
$2x + 4y = 10$

**c.** $x = y - 1$
$3x + 2y = -13$

**d.** $y = 5 - x$
$6x + 5y = -23$

**e.** $x = 3 - y$
$4x + 7y = 18$

**f.** $y = x - 6$
$5x + 3y = -34$

6. **a.** $y = -x$
   $x - y = 2$
   **d.** $y = -7x$
   $x - 5y = -72$

   **b.** $x = -y$
   $y - x = 6$
   **e.** $y = -2x$
   $6x - 4y = -70$

   **c.** $y = -3x$
   $2x - 3y = 22$
   **f.** $x = -8y$
   $9x - 8y = 80$

7. **a.** $y = x + 2$
   $2x - y = 1$
   **d.** $y = 6 + x$
   $4x - 5y = -28$

   **b.** $y = 2x + 5$
   $x - 3y = -20$
   **e.** $y = 3x + 2$
   $9x - 7y = 22$

   **c.** $x = y + 3$
   $2y - x = -5$
   **f.** $y = 4x + 1$
   $3x - 4y = 35$

8. **a.** $y = x - 1$
   $4x - y = 19$
   **d.** $y = 8 - 3x$
   $4x - 5y = -2$

   **b.** $y = 3x - 2$
   $6x - 4y = -16$
   **e.** $x = 7 - 2y$
   $3y - 8x = 20$

   **c.** $x = 4y - 5$
   $5y - x = 4$
   **f.** $y = 5x - 3$
   $7x - 2y = 0$

9. **a.** $x = 5$
   $2x + 3y = 19$
   **d.** $3x = -9$
   $4x + 7y = 2$

   **b.** $6y = 12$
   $3x - 5y = -1$
   **e.** $x = 0$
   $8x - 3y = 6$

   **c.** $y = -7$
   $3x + 6y = -27$
   **f.** $4y = -16$
   $5x - 9y = -14$

10. **a.** $x + y = 5$
    $2x + 5y = 16$
    **d.** $x - y = 5$
    $3x - 2y = 14$
    **g.** $2x + 4y = 14$
    $5x - 6y = 3$

    **b.** $2x + y = 7$
    $4x - 2y = -6$
    **e.** $4x - y = 0$
    $6x - 3y = 12$
    **h.** $12x + 3y = -6$
    $7x - 2y = -11$

    **c.** $x + 3y = 16$
    $7x + 4y = 10$
    **f.** $x - 5y = 3$
    $8x - 7y = -9$
    **i.** $10x - 5y = 0$
    $4x + 9y = -66$

## CUMULATIVE REVIEW

Solve each of the following systems of equations, then check:

1. Solve graphically:
   **a.** $y = 8 - x$
   $x - y = 4$

   **b.** $y = -2x$
   $3x - 2y = -7$

   **c.** $2x + y = -7$
   $x - 3y = 7$

2. Solve by addition or subtraction method:
   **a.** $8x - 3y = 12$
   $4x - 6y = -12$

   **b.** $3m + 5n = 31$
   $2m - 3n = 8$

   **c.** $11x - 9y = 27$
   $8x + 14y = -36$

3. Solve by the substitution method:
   **a.** $y = 4x$
   $5x + y = 18$

   **b.** $y = x - 5$
   $4x - 3y = 17$

   **c.** $x + y = 7$
   $6x - 2y = 18$

# Miscellaneous Systems of Linear Equations

## I. Aim

To solve a system of linear equations when either given equation or both given equations are not in standard form.

## II. Procedure

1. Transform the given equations to equivalent equations in standard form.
2. Then solve by either the addition and subtraction method or the substitution method. See Exercises 13-3 and 13-4.
3. If either or both equations contain parentheses, remove the parentheses and transform the equations to the standard form before solving the system of equations.
4. If either or both equations contain decimals, multiply each decimal equation by some power of ten to clear the equation of decimals. Transform each equation to the standard form before solving the system of equations.
5. If either or both equations have fractional coefficients, clear each equation of all fractions by multiplying each equation by the least common denominator of its fractional coefficients. Transform each equation to the standard form before solving the system of equations.
6. If the given equations are fractional equations of the reciprocal type (see sample solution), first eliminate one of the variables by addition or subtraction before clearing the resulting equation of fractions. If necessary, multiply the equations by required numbers so that the numerators of the fractional terms to be eliminated are the same.
7. Check by substituting the found values for the variables in the two given (original) equations.

## III. Sample Solutions

1. Solve and check: $2x + 3 = 7 - y$
$$5x - 2y = 4x - 3$$

Transform given equations:

$$2x + 3 = 7 - y$$
$$2x + y + 3 - 3 = 7 - 3 - y + y$$
$$\underline{2x + y = 4}$$

$$5x - 2y = 4x - 3$$
$$5x - 4x - 2y = 4x - 4x - 3$$
$$\underline{x - 2y = -3}$$

*Solution:*

$$2x + y = 4$$
$$\underline{x - 2y = -3}$$

$$4x + 2y = 8$$
$$\underline{x - 2y = -3}$$
$$5x \quad\quad = 5$$
$$x = 1$$

$$2x + y = 4$$
$$(2 \cdot 1) + y = 4$$
$$2 + y = 4$$
$$2 - 2 + y = 4 - 2$$
$$y = 2$$

*Answer:* $x = 1$, $y = 2$, or $(1, 2)$

*Check:*

$$2x + 3 = 7 - y \qquad\qquad 5x - 2y = 4x - 3$$
$$(2 \cdot 1) + 3 = 7 - 2 ? \qquad (5 \cdot 1) - (2 \cdot 2) = (4 \cdot 1) - 3 ?$$
$$2 + 3 = 7 - 2 \qquad\qquad 5 - 4 = 4 - 3$$
$$5 = 5 \checkmark \qquad\qquad\qquad 1 = 1 \checkmark$$

Each of the following sample problems are partially done.

**2.** Solve and check:  $4(x - 3) = 3(y + 3)$
$$8 - (2y + 1) = 17 - 2x$$

*Solution:*

Equation $4(x - 3) = 3(y + 3)$ is equivalent to $4x - 3y = 21$

Equation $8 - (2y + 1) = 17 - 2x$ is equivalent to $2x - 2y = 10$

*Answer:* $x = 6$, $y = 1$, or $(6, 1)$

**3.** Solve and check:  $.04x + .02y = 5$
$$.5(x - 2) - .4y = 29$$

*Solution:*

Equation: $.04x + .02y = 5$ is equivalent to $4x + 2y = 500$

Equation: $.5(x - 2) - .4y = 29$ is equivalent to $5x - 4y = 300$

*Answer:* $x = 100$, $y = 50$, or $(100, 50)$

**4.** Solve and check:  $\dfrac{x}{5} + \dfrac{y}{2} = 4$

$$\frac{x - 1}{3} = \frac{y + 2}{2}$$

*Solution:*

Equation: $\dfrac{x}{5} + \dfrac{y}{2} = 4$ is equivalent to $2x + 5y = 40$

Equation: $\dfrac{x - 1}{3} = \dfrac{y + 2}{2}$ is equivalent to $2x - 3y = 8$

*Answer:* $x = 10$, $y = 4$, or $(10, 4)$

**5.** Solve and check:  $\dfrac{2}{x} + \dfrac{3}{y} = 2$

$$\frac{10}{x} - \frac{6}{y} = 3$$

*Solution:*

Equation: $\dfrac{2}{x} + \dfrac{3}{y} = 2$ is equivalent to $\dfrac{4}{x} + \dfrac{6}{y} = 4$

Equation: $\dfrac{10}{x} - \dfrac{6}{y} = 3$ already has the term $\dfrac{6}{y}$. Eliminate the term $\dfrac{6}{y}$.

*Answer:* $x = 2$, $y = 3$, or $(2, 3)$

## DIAGNOSTIC TEST

Solve and check:

**1.** $6x + 3y = 19 - 2x$
   $9x - 13 = 5y$

**2.** $7x - 3y = 16 + (x - y)$
   $2(x - 2) = 4 - (y + 1)$

**3.** $x + y = 1,500$
   $.04x + .06y = 74$

**4.** $\dfrac{x}{3} + \dfrac{y}{4} = 4$

   $\dfrac{x}{y} = \dfrac{3}{4}$

**5.** $\dfrac{3}{x} + \dfrac{4}{y} = 5$

   $\dfrac{6}{x} - \dfrac{8}{y} = 2$

## *RELATED PRACTICE EXAMPLES*

Solve and check:

**1. a.** $x = 2y + 1$
   $y = 2x + 1$

**b.** $5y = 2x - 4$
   $3x = 8y + 5$

**c.** $4x + 8y = 5y$
   $3x + 1 = -2y$

**d.** $6x = 8 - 15y$
   $4x - 4 = -6y$

**e.** $3x + 5y = 2x$
   $x + 3y = y$

**f.** $2x + 3 = 7y$
   $4 + 3x + 5y = 15$

**g.** $x + 3 = 2 - y$
   $2x + 1 = 3y - 1$

**h.** $3x + 2y = 2x + 2$
   $2x - 6y = 15 + y$

**i.** $6x - 3y = 3x - y$
   $4x - 5y = 11 + 2x$

**j.** $2x + 6y = 13 - 2x$
   $9y + 6 = 8x + 15$

**2. a.** $2(x + 4) = 3(y + 2)$
   $4(x - 2) = 5(y - 2)$

**b.** $3 - (x - 5) = y + 2$
   $2(x - y) = 4 - 3y$

**c.** $8x - (2y + 5) = x$
   $y - (3x - 1) = 4$

**d.** $(x - 2) - (y - 3) = 8$
   $4(x + 5) = y + 3$

**e.** $3(x + 2) = 2(y + 4)$
   $2x - 3(y - 10) = 23$

**f.** $4(6 - x) - 6(5 - 2y) = -26$
   $9y - 5(-3x + 7) + 14 = 0$

3. **a.** $.6x + .2y = 2.2$
   $.5x - .2y = 1.1$

   **b.** $.3x + .4y = 2.4$
   $.5x - .3y = 1.1$

   **c.** $.12x + .02y = .78$
   $.03x + .04y = .3$

   **d.** $.6x + y = 7$
   $.2x + .5y = 3$

   **e.** $1.5x + .1y = 6.2$
   $3x - .4y = 11.2$

   **f.** $x + y = 1800$
   $.05x + .06y = 96$

   **g.** $x + y = 700$
   $.15x + .25y = 125$

   **h.** $x - y = 400$
   $.05x - .06y = 4$

   **i.** $.04(x - 10) = .08(y - 3)$
   $.5(x + 4) = 2y$

   **j.** $.04(x - 3) = .05(y + 2)$
   $.2(x + 2) = .1(6y + 8)$

4. **a.** $x = 2y$
   $$\frac{x}{3} + y = 10$$

   **b.** $\dfrac{x}{y} = \dfrac{2}{3}$
   $$\frac{x + 8}{3} = \frac{3y - 2}{4}$$

   **c.** $\dfrac{x}{4} + \dfrac{y}{2} = 5$
   $$\frac{x}{6} - \frac{y}{18} = 1$$

   **d.** $\dfrac{x}{2} - \dfrac{y}{3} = 1$
   $$\frac{x}{4} - \frac{y}{9} = \frac{2}{3}$$

   **e.** $\dfrac{x}{10} = \dfrac{y}{5}$
   $$\frac{x}{8} + \frac{y}{3} = \frac{7}{4}$$

   **f.** $\dfrac{3x - y}{5} = 2y - 1$
   $$\frac{3x}{8} - \frac{y}{4} = \frac{1}{2}$$

   **g.** $\dfrac{x}{3} = \dfrac{y}{4}$
   $$\frac{x - 4}{4} - \frac{y - 13}{3} = 1$$

   **h.** $\dfrac{x}{4} - 3 = \dfrac{y}{6}$
   $$\frac{x}{2} - y = -2$$

   **i.** $\dfrac{x + y}{2} = \dfrac{1}{2} + \dfrac{x - y}{3}$
   $$\frac{x + 2}{2} - \frac{y + 4}{3} = 4$$

5. **a.** $\dfrac{1}{x} + \dfrac{1}{y} = \dfrac{1}{2}$
   $$\frac{1}{x} - \frac{1}{y} = \frac{1}{6}$$

   **b.** $\dfrac{1}{x} + \dfrac{1}{y} = \dfrac{3}{8}$
   $$\frac{1}{x} - \frac{1}{y} = \frac{1}{8}$$

   **c.** $\dfrac{1}{x} + \dfrac{1}{y} = 5$
   $$\frac{1}{x} - \frac{1}{y} = 1$$

   **d.** $\dfrac{4}{x} + \dfrac{6}{y} = 4$
   $$\frac{6}{x} - \frac{6}{y} = 1$$

   **e.** $\dfrac{5}{x} + \dfrac{3}{y} = 26$
   $$\frac{4}{x} + \frac{3}{y} = 22$$

   **f.** $\dfrac{8}{x} - \dfrac{3}{y} = 5$
   $$\frac{4}{x} - \frac{1}{y} = 3$$

   **g.** $\dfrac{4}{x} + \dfrac{3}{y} = 7$
   $$\frac{5}{x} - \frac{5}{y} = 0$$

   **h.** $\dfrac{5}{x} - \dfrac{9}{y} = \dfrac{1}{6}$
   $$\frac{3}{x} - \frac{3}{y} = \frac{1}{2}$$

   **i.** $\dfrac{7}{x} + \dfrac{8}{y} = 2$
   $$\frac{2}{x} + \frac{12}{y} = 20$$

# Problem Solving | EXERCISE 13-6

**I. Aim**  To solve a system of two linear equations in two variables verbal problems involving two unknowns such as:
A. Number Problems,  B. Coin and Mixture Problems,
C. Investment Problems,  D. Motion Problems,  E. Digit Problems, and F. Problems Solved by Graphs.

**II. Procedure**

1. Read the problem carefully to determine what is required and any facts which are related to the unknown values.
2. Represent one of the unknown values by one variable and the second unknown by a second value.
3. Translate the related facts which involve the unknown values into equations in two variables.
4. Solve this system of equations.
5. Check answers directly with the facts of the problem, not the equation.
6. Note that several topics include additional instructional procedures.

## DIAGNOSTIC TEST

Solve and check each of the following problems:

1. The sum of two numbers is 25. The difference between five times the larger number and three times the smaller number is 61. What are the numbers?
2. Steve is 2 years younger than Scott. The sum of five times Scott's age and four times Steve's age is 136 years. Find their ages.
3. The school office used 250 stamps, some 15¢ stamps and the rest 18¢ stamps, costing $41.40 in all. How many stamps of each kind were used?
4. A grocer wishes to mix cookies worth 96¢ a pound with cookies worth 72¢ a pound to make a mixture to sell at 76¢ a pound. If she requires 60 pounds of mixed cookies, how many pounds of each kind should she use?
5. Tickets for a school game cost 85¢ for each adult and 35¢ for each student. If 612 tickets were sold for a total of $358.70, how many tickets of each kind were sold?

6. A man invested $10,500, part at 6% and the rest at 7% per year. The total annual income from these investments is $667. How much did he invest at each rate?

7. An airplane flew 1,680 miles with the wind in 6 hours. The return trip over the same distance against the wind took 7 hours. Find the rate of speed of the airplane in still air (air-speed) and the rate at which the wind was blowing (wind velocity).

8. The sum of the numbers named by the digits of a two-digit numeral is 7. If the digits are interchanged, the new number named by this reverse two-digit numeral is 9 more than the original number. Write the two-digit numeral naming the original number.

9. An airplane left Kennedy Airport at 2 P.M. flying due west at 360 mph. Two hours later another airplane left the same airport headed for the same destination but it flew at a speed of 600 mph. Determine by graph the number of hours it took the second airplane to overtake the first airplane. What time was it?

## A. Number Problems

*Sample Problem*

The sum of two numbers is 32 and their difference is 4. Find the numbers.

*Solution:*

Let     $x$ = one number

and    $y$ = other number

The equations are determined by the following facts:

The sum of the two numbers equals 32.

The difference of the two numbers equals 4.

$$x + y = 32$$
$$\underline{x - y = 4}$$
$$2x \quad\;\; = 36$$
$$x = 18, \text{ one number}$$

$$x + y = 32$$
$$18 + y = 32$$
$$18 - 18 + y = 32 - 18$$
$$y = 14, \text{ other number}$$

*Check:*

| 18 | 18 |
|----|----|
| +14 | −14 |
| 32✔ | 4✔ |

*Answer:* 18 and 14

## *RELATED PRACTICE PROBLEMS*

Solve and check each of the following problems:

**1. a.** The sum of two numbers is 45 and their difference is 7. Find the numbers.

   **b.** The difference between two numbers is 24 and their sum is 48. Find the numbers.

   **c.** The sum of two numbers is 22. Five times one number is equal to six times the second number. Find the numbers.

   **d.** The difference between two numbers is 4. Twice the larger number is equal to 3 times the smaller number increased by 2. Find the numbers.

   **e.** Separate 24 into two numbers so that 4 less than six times the smaller number equals 5 more than three times the larger number. Find the numbers.

**2. a.** John and Harry together weigh 210 pounds. The difference between three times John's weight and twice Harry's weight is 30 pounds. Find the weight of each boy.

   **b.** The difference in ages of two boys is 5 years. The sum of their ages is 19 years. Find their ages.

   **c.** There were 242 admission tickets sold. Three times the number of 50¢ tickets was 12 more than four times the number of 35¢ tickets. Find the number of 50¢ tickets sold.

   **d.** The difference between the length and width of a rectangle is 7 centimeters. The perimeter of the rectangle is 50 centimeters. Find the length and width.

   **e.** Maria has 28 coins. Some are nickels and some are dimes. The sum of the number of nickels and three times the number of dimes is 40. How many nickels and how many dimes does Maria have?

**B. Coin and Mixture Problems**

*Sample Problem*

Juan has 25 coins; some nickels and the rest dimes. The total value of all the coins is $1.80. Find the number of each kind of coin.

*Solution:* Let $\quad x = $ no. of nickels

and $\quad y = $ no. of dimes

then $\quad 5x = $ value of nickels in cents

$\quad 10y = $ value of dimes in cents

The equations are determined by the following facts:
Numbers of nickels plus the number of dimes $= 25$.
Value of nickels plus value of dimes $= $ total value.

$$x + y = 25$$
$$5x + 10y = 180$$
$$\overline{\phantom{xxxxxxxx}}$$
$$10x + 10y = 250$$
$$5x + 10y = 180$$
$$\overline{\phantom{xxxxxxxx}}$$
$$5x \phantom{+ 10y} = 70$$
$$x = 14 \text{ nickels}$$

$$x + y = 25$$
$$14 + y = 25$$
$$14 - 14 + y = 25 - 14$$
$$y = 11 \text{ dimes}$$

*Check:*

| 14 | 14 |
|----|-----|
| +11 | $.05 |
| 25 | $.70 |
| | |
| 11 | $1.10 |
| $.10 | + .70 |
| $1.10 | $1.80 ✔ |

*Answer:* 14 nickels and 11 dimes

## RELATED PRACTICE PROBLEMS

Solve and check each of the following problems:

3. **a.** Andrew has 18 coins, some quarters and the rest dimes. The total value of all the coins is $3.45. Find the number of each kind of coin.

   **b.** Angela has $91 in $5 bills and $2 bills. She has 23 bills in all. How many $5 bills does she have?

   **c.** Scott bought $15 worth of 15-cent and 20-cent stamps numbering 90 stamps in all. How many stamps of each kind did he buy?

   **d.** Robert was given $18 to buy 15-cent and 18-cent stamps. He returned from the post office with 113 stamps and 36 cents in change. How many stamps of each kind did he buy?

   **e.** When changing a $20 bill, Marilyn received 4 more $1 bills than three times the number of $5 bills. How many $1 bills did she receive?

4. **a.** A garden supply dealer mixes rye grass worth 40¢ a pound with blue grass worth $1.30 a pound. If she wishes to make a mixture of 50 pounds to sell at 67¢ a pound, how many pounds of each kind should she use?

   **b.** A grocer mixes walnuts selling at $1.65 a pound and almonds selling at $2.10 a pound, making a mixture of 30 pounds to sell at $1.83. How many pounds of walnuts should he use?

   **c.** How many pounds of candy at 80¢ a pound should be mixed with candy at $1.50 a pound to make 20 pounds of candy to sell at $1.01 a pound?

   **d.** A grocer mixes tea worth $1.60 a pound with tea worth $2.20 a pound, making a blend to sell for $1.80 a pound. How many pounds of each kind should he use if he plans to blend 75 pounds?

   **e.** How many pounds of dried prunes selling at $1.05 a pound should be mixed with dried apricots selling at $2.10 a pound to make a mixture of 100 pounds to sell at $1.26 per pound?

5. **a.** If 4 hamburgers and 3 bags of fries cost $5.33 but 5 hamburgers and 2 bags of fries cost $5.63, how much does a hamburger cost? A bag of fries?

**b.** There were 2,500 persons watching a baseball game. The adults paid 75¢ for their admission tickets but the children paid only 25¢. If the total receipts amounted to $1,503, how many adults and how many children saw the game?

**c.** Chocolates cost 80¢ a pound more than hard candy. If Doris paid $11.60 for 4 pounds of chocolates and 2 pounds of hard candy, what did she pay per pound for the chocolates?

**d.** Seats in the reserved section at the school play cost $1.50 each and in the regular section $1.00 each. How many tickets of each kind were sold if the total receipts for 980 tickets amounted to $1,265?

**e.** Perennial rye grass costs twice as much per pound as domestic rye grass. If Ronald paid $18.20 for 10 pounds of perennial rye grass and 15 pounds of domestic rye grass, how much was he charged per pound for each kind of grass?

**C.** Investment Problems

*Sample Problem*

A man invested $1,800; part at 4% and the rest at 6% per year. If he receives an annual income of $84 from these investments, how much did he invest at each rate?

*Solution:* Let $\quad x =$ amount in dollars invested at 4%

and $\quad y =$ amount in dollars invested at 6%

$.04x =$ annual income from 4% investment

$.06y =$ annual income from 6% investment

Equations are determined by the following facts:

Amount invested at 4% plus amount invested at 6% equal $1,800.

Income from 4% investment plus the income from the 6% investment equal $84.

$$x + y = 1,800$$
$$.04x + .06y = 84$$

$$x + y = 1,800$$
$$4x + 6y = 8,400$$

We clear second equation of decimals.

$$.04x + .06y = 84$$
$$4x + 6y = 8,400$$

We then solve the system:

$$x + y = 1,800$$
$$4x + 6y = 8,400$$

$$6x + 6y = 10,800$$
$$4x + 6y = 8,400$$
$$2x \qquad = 2,400$$
$$x = \$1,200 \text{ at } 4\%$$
$$x + y = 1,800$$
$$1,200 + y = 1,800$$
$$1,200 + (-1,200) + y = 1,800 + (-1,200)$$
$$y = \$600 \text{ at } 6\%$$

*Answer:* $1,200 at 4% and $600 at 6%

*Check:*

| $1,200 | $1,200 | $600 | $48 |
|--------|--------|------|-----|
| + 600 | × .04 | × .06 | + 36 |
| $1,800 ✔ | $48.00 | $36.00 | $84 ✔ |

## *RELATED PRACTICE PROBLEMS*

Solve and check each of the following problems:

**a.** A woman invested $4,000; part at 5% and the rest at 9% per year. If she receives $260 income for the year from these investments, how much did she invest at each rate?

**b.** Mr. Adams invested a part of his savings at 8% and the rest at 6% per year. If he receives an annual income of $240  from the total investment of $3,400, how much did he invest at 8%?

**c.** A 7% investment brings an annual return of $36 more than a 9% investment. The total amount invested is $1,200. Find the amount invested at each rate.

**d.** A man invested a certain amount of money at 8% per year and $2,000 more than that amount at 10% per year. If the total annual income is $524, how much did he invest at 10%?

**e.** Mr. Jones invested $500 more at 7% per year than he did at 12% per year. If the annual income he receives from the 12% investment is $90 more than the income from the 7% investment, how much did he invest at each rate?

**D.** Motion Problems

Observe in the following model solution that the rate of the boat downstream means the rate with the current, and the rate of the boat upstream is the rate against the current.

*Sample Problem*

A rower can row 24 miles downstream in 3 hours. However, the trip back over the same distance upstream takes her 6 hours. Find the rower's rate of rowing in still water and the rate of the current.

*Solution:*

Let $x$ = rate of rowing in still water in mph.

and $y$ = rate of current in mph.

|              | Rate of boat | Time | Distance |
|--------------|--------------|------|----------|
| Downstream   | $x + y$      | 3    | 24       |
| Upstream     | $x - y$      | 6    | 24       |

The equations are determined by the following facts:
Rate of boat downstream $(x + y)$ equals distance divided by time $(24 \div 3) = 8$.
Rate of boat upstream $(x - y)$ equals distance divided by time $(24 \div 6) = 4$.

$$x + y = 8$$
$$x - y = 4$$
$$2x = 12$$
$$x = 6 \text{ mph., rate of rowing in still water}$$
$$x + y = 8$$
$$6 + y = 8$$
$$6 + (-6) + y = 8 + (-6)$$
$$y = 2 \text{ mph., rate of current}$$

*Answer:* Rate of rowing in still water is 6 mph and rate of current is 2 mph.

*Check:*
Rate of boat downstream = 6 + 2 or 8 mph.
Rate of boat upstream = 6 − 2 or 4 mph.

| Downstream | Upstream |
|------------|----------|
| 3 hr. ✔ | 6 hr. ✔ |
| 8)24 miles | 4)24 miles |

## RELATED PRACTICE PROBLEMS

Solve and check each of the following problems:

7. a. A rower can row 12 miles downstream in 2 hours. The return trip over the same distance upstream takes him 4 hours. Find the rate at which the rower can row in still water and the rate of the current.

b. An airplane flies 360 miles with the wind in 2 hours. When flying against the wind, it takes 3 hours to fly the same distance. Find the rate of speed of the airplane in still air (air-speed) and the rate at which the wind is blowing (wind velocity).

**c.** A motorboat can go 120 miles downstream in 4 hours, but the trip back takes $1\frac{1}{2}$ times as long. Find the rate of speed of the motorboat in still water and the rate of the current.

**d.** An airplane traveled 600 miles in 3 hours, flying with the wind. On the return trip, however, flying against the wind, it took 4 hours to travel 640 miles. Find the rate at which the wind was blowing (wind velocity) and the rate of speed of the airplane in still air (air-speed).

**e.** A girl walks from her home in the city in 4 hours, but she travels the same distance on her bicycle in one hour. If she rides 6 miles per hour faster than she walks, how fast does she ride on her bicycle?

## E. Digit Problems

The number that has 3 as the tens digit and 6 as the units digit is *3 tens plus 6 units* or $(3 \times 10) + 6$ or 36. Thus observe in the following model solution that the number that has $x$ as the tens digit and $y$ as the units digit is *x tens plus y units* or $(x \cdot 10) + y$ or $10x + y$.

The sum of the numbers named by the digits of a two-digit numeral is 9. What is the number named by this two-digit numeral if it is six times the number named by the units digit?

*Solution:*                 *Check:*

Let $x$ = tens digit          3

and $y$ = units digit       $\underline{+6}$     $36 = 6 \times 6 \; ✔$

then $10x + y$ = the number    $9 \; ✔$

The equations are determined by the following facts:

The sum of the digits equals 9.

The number is 6 times the units digits.

$$x + y = 9$$
$$10x + y = 6y$$

Transform the second equation to standard form:

$$10x + y = 6y$$
$$10x + y - 6y = 6y + (-6y)$$
$$10x - 5y = 0$$

Then solve the system:

$$x + y = 9$$
$$10x - 5y = 0$$

*Answer:* 36

Check (right column):

$$x + y = 9$$
$$10x - 5y = 0$$

$$5x + 5y = 45$$
$$\underline{10x - 5y = 0}$$

$$15x \qquad = 45$$
$$x = 3, \text{ tens digit}$$
$$x + y = 9$$
$$3 + y = 9$$
$$3 - 3 + y = 9 - 3$$
$$y = 6, \text{ units digit}$$

Then $10x + y = 30 + 6 = 36$, the number

## *RELATED PRACTICE PROBLEMS*

**8. a.** The sum of the numbers named by the digits of a two-digit numeral is 8. What is the number named by this two-digit numeral if it is seven times the number named by the units digit?

**b.** The sum of the numbers named by the digits of a two-digit numeral is 13. What is the number named by this two-digit numeral if twice the number named by its tens digit increased by two equals five times the number named by its units digit?

**c.** The number named by the units digit of a certain two-digit numeral is 7 more than the number named by the tens digit. What is the number named by the two-digit numeral if it is 4 less than three times the sum of the numbers named by both digits?

**d.** The sum of the numbers named by the digits of a two-digit numeral is 12. If the digits are interchanged, the new number named by this reverse two-digit numeral is 18 less than the original number. Write the two-digit numeral for the original number.

**e.** The sum of the numbers named by the digits of a two-digit numeral is 7. If the digits are interchanged, the new number named by this reverse two-digit numeral increased by 3 equals four times the original number. Write the two-digit numeral for the original number.

**F. Solution of Problems by Graphs**

Graphs may be used to solve problems.

*Sample Problem*

How far from the depot and at what time will a truck, leaving at noon and traveling at an average speed of 40 mph, overtake another truck which left the same depot 2 hours earlier and was averaging 30 mph?

The graph of $d = 30t$ represents the path of the first truck and the graph of $d = 40t$ represents the path of the second truck.

*Answer:* The graph shows the first truck is overtaken 240 miles from the depot and at 6 P.M.

$d = vt$

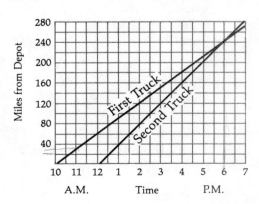

### RELATED PRACTICE PROBLEMS

Use graphs to solve each of the following problems:

**a.** A freight train leaves a station at 4 P.M. traveling at 15 mph. A passenger train leaves 3 hours later for the same destination but travels at 45 mph. At what time will the passenger train overtake the freight?

**b.** A cruiser left port at 4 A.M. Four hours later a destroyer followed. How many hours will it take the destroyer, averaging 30 knots, to overtake the cruiser, averaging 18 knots?

**c.** Show by graph when the rate of 12 cents for printing each negative with free developing is cheaper than the rate of 9 cents for each print with a charge of 30 cents for developing.

**d.** One laundry charges $2.50 for the first 12 pieces and 20¢ for each additional piece. Another company charges $3.00 for the first 12 pieces and 15¢ for each additional piece. Show by graph when one rate is cheaper than the other.

**e.** Show by graph when a $50 weekly salary and 10% commission on sales is better than an $80 weekly salary and 6% commission.

# REVIEW

Solve each of the following systems of equations, then check:

**1.** Solve graphically:
    **a.** $y = 3x$
        $5x - 2y = 0$
    **b.** $x = 6$
        $4y - x = 2$
    **c.** $2x - y = 14$
        $3x + 4y = -1$

**2.** Solve by the addition or subtraction method:
    **a.** $3x - 4y = 6$
        $5x - 4y = 2$
    **b.** $5x - 6y = 4$
        $4x + 7y = 15$
    **c.** $8x - 9y = -26$
        $6x + 12y = 18$

**3.** Solve by the substitution method:
    **a.** $y = x - 5$
        $4x - y = 17$
    **b.** $x = -2y$
        $7x - 3y = 51$
    **c.** $3x + y = -9$
        $x - 5y = -19$

**4.** Solve and check:
    **a.** $8x - 3(y + 2) = 15$
        $(4x - 5) - (10 - 4y) = 1$
    **b.** $.6x + .5y = 18$
        $.8x - .3y = 12.4$
    **c.** $\dfrac{2x + 7}{3} = \dfrac{7y - 1}{4}$
        $3x = 4y$
    **d.** $\dfrac{2x - y}{10} - \dfrac{x - 4y}{12} = \dfrac{7}{6}$
        $x = 10$

**5.** A woman invested $7,500 more at 12% than at 10% per year. If her annual interest from the two investments is $1,890, how much did she invest at each rate?

## Keyed Achievement Test

The colored numeral indicates where help may be found.
1. Draw in the number plane the graph of $2x - y = -12$. [12-3]
2. Draw the graph of $9x + 3y = -15$ by the slope-intercept method. [12-6]
3. Find the product of: **a.** $(7b - 2c)^2$ **b.** $(6a^2 - 5b)(6a^2 + 5b)$
   [9-4 and 9-3]
4. Factor: **a.** $m^2 - 5m - 24$ **b.** $100r^2 - 81s^2$ [9-10 and 9-8]
5. Write $\dfrac{4x^2 - 4}{8x^2 - 40x + 32}$ in simplest form. [10-2]

6. Multiply as indicated: [10-3]
$$24x^5y^2 \cdot \frac{3x - 5}{8x^2y}$$

7. Divide as indicated: [10-4]
$$\frac{a^2 + 2ax + x^2}{a^2 - x^2} \div \frac{a^2 + 4ax - 5x^2}{(a - x)^2}$$

8. Add as indicated: [10-9]
$$\frac{3n}{n - 2} + \frac{n + 2}{n - 4}$$

9. Solve and check: [11-1]
$$\frac{x + 4}{9} + \frac{3x - 1}{2} - \frac{2x + 5}{3} = 3$$

10. Find the value of $r$ when $S = 62$, $l = 32$, and $a = 2$, using the formula: $S = \dfrac{rl - a}{r - 1}$. [11-4]

Solve each of the following systems of equations, then check:
11. Solve graphically: [13-2]
    **a.** $y = x - 2$       **b.** $3x - y = 0$       **c.** $5x + 10 = 0$
       $x + 4y = 12$          $2x - 3y = 7$          $4x - 5y = -13$
12. Solve by the addition or subtraction method: [13-3]
    **a.** $8x - 3y = 31$    **b.** $9x + 10y = -27$   **c.** $5x - 7y = 17$
       $5x - 6y = 7$          $15x - 4y = 38$         $3x + 8y = -2$
13. Solve by the substitution method: [13-4]
    **a.** $y = -2x$          **b.** $y = 5 - x$        **c.** $x - 5y = 4$
       $6x - y = -16$         $4x - 2y = -28$         $2x + y = -14$
14. Solve and check: [13-5]
    **a.** $5(x - y) = 28 - x$                **b.** $\dfrac{5x + 4}{6} = \dfrac{3y - 1}{5}$
       $9x - 2(x + y) = 5(3 - y)$
    **c.** $x + y = 6{,}000$                  $\dfrac{x + 2y}{3} - \dfrac{4x - y}{9} = 5$
       $.05x + .06y = 280$
15. The sum of two numbers is 30. Four times the larger number decreased by 5 equals 7 more than eight times the smaller number. Find the numbers. [13-6]

# INTRODUCTION

Skill in determining powers and roots of monomials and in simplifying and operating radicals is essential in the study of radical and quadratic equations and their applications.

In this unit we deal with irrational numbers, usually expressed as a square root of positive numbers other than perfect squares (numbers having the exact square root). These irrational numbers are represented by radicals, each of which is an indicated root of a number such as $\sqrt{5}$, $3\sqrt{2}$, or $7\sqrt[3]{10}$.

We shall learn to simplify these radicals by making the radicand (number under the radical sign $\sqrt{\phantom{x}}$) the smallest possible whole number containing no factors for which the indicated root can be found exactly.

To simplify radicals we use the principle that the square root of a product of two numbers is equal to the product of the square roots of the numbers. For example, $\sqrt{9 \cdot 4} = \sqrt{9} \cdot \sqrt{4}$.

To multiply radicals we use the principle that the product of the square roots of two numbers is equal to the square root of the product of the two numbers. For example, $\sqrt{9} \cdot \sqrt{4} = \sqrt{9 \cdot 4}$.

To divide radicals we use the principle that the quotient of the square roots of the two numbers is equal to the square root of the quotient of the two numbers. For example, $\dfrac{\sqrt{36}}{\sqrt{9}} = \sqrt{\dfrac{36}{9}}$.

However, note that the square root of the sum of two numbers is not equal to the sum of the square roots of the numbers. For example, $\sqrt{9 + 4} \neq \sqrt{9} + \sqrt{4}$.

When rationalizing the denominator (page 558), an inverse operation is used. The square of the square root of a positive number is the number itself.

**Powers; Roots; Radicals**

14

# Power of a Power

**I. Aim**   To raise a monomial expressed in exponential form to a given power.

**II. Procedure**

1. Raise the numerical coefficient, if any, to the indicated power. Determine its value following the rules of signs for multiplication.
   a. When a positive number is raised to either an odd or even power, the answer is a positive number.
   b. When a negative number is raised to an even power, the answer is a positive number.
   c. When a negative number is raised to an odd power, the answer is a negative number.
2. Multiply the exponent of each of the variable factors by the exponent of the power to which it is to be raised.
   In general, $(a^m)^n = a^{mn}$ and $(ab)^m = a^m b^m$ where $a$ and $b$ are any real numbers and $m$ and $n$ are positive integers.
3. If the given expression is a fraction, raise both the numerator and the denominator of the fraction to the indicated power.
4. Note that the product of equal factors is called the *power* of the factor. Review Exercise 1-4 on page 20.

**III. Sample Solutions**

Raise to indicated power:

1. $(4a^2xy^3)^2$      *Answer:* $16a^4x^2y^6$

2. $(-3xy^4)^2$      *Answer:* $9x^2y^8$

3. $(2bx^2)^3$      *Answer:* $8b^3x^6$

4. $(-5m^4)^3$      *Answer:* $-125m^{12}$

5. $(-3c^3d)^4$      *Answer:* $81c^{12}d^4$

6. $(2a^2b)^5$      *Answer:* $32a^{10}b^5$

7. $(\frac{2}{3}m^2y^3)^2$      *Answer:* $\frac{4}{9}m^4y^6$

8. $(.3b^3x)^2$      *Answer:* $.09b^6x^2$

9. $(c^2[x-y]^4)^2$      *Answer:* $c^4[x-y]^8$

10. $\left(\dfrac{d^2m}{5c^2x^5}\right)^2$      *Answer:* $\dfrac{d^4m^2}{25c^4x^{10}}$

## DIAGNOSTIC TEST

Raise to indicated power:

1. $(7ab^4)^2$  2. $(-4x^2y^3)^2$  3. $(m^6x^4)^3$  4. $(-5b^2x^5)^3$

5. $(-2a^3x)^4$  6. $(-3b^2c^4)^5$  7. $(\frac{1}{2}m^4n^5)^2$  8. $(-.3b^2)^3$

9. $(2a[x+y]^3)^2$

10. $\left(\dfrac{4cd^2}{-5b^3}\right)^2$

## RELATED PRACTICE EXAMPLES

Raise to indicated power:

1. **a.** $(x^2)^2$  **b.** $(c^5y^3)^2$  **c.** $(5ab^2)^2$  **d.** $(8b^3)^2$
   **e.** $(3a^3b^6)^2$  **f.** $(2x^2y^3)^2$  **g.** $(12x^3y^4)^2$  **h.** $(6a^5x^4y^3)^2$

2. **a.** $(-b)^2$  **b.** $(-x^3y^7)^2$  **c.** $(-2a)^2$  **d.** $(-5x^5)^2$
   **e.** $(-3a^2b)^2$  **f.** $(-7b^2x^4)^2$  **g.** $(-6ab^2c^3)^2$  **h.** $(-10x^4yz^6)^2$

3. **a.** $(x^2)^3$  **b.** $(a^4y)^3$  **c.** $(3x^8)^3$  **d.** $(5cd^2)^3$
   **e.** $(4b^5x^4)^3$  **f.** $(6ax^2y^3)^3$  **g.** $(2m^5n^3x^2)^3$  **h.** $(8bc^4d^7)^3$

4. **a.** $(-b)^3$  **b.** $(-c^5d^3)^3$  **c.** $(-2a^2)^3$  **d.** $(-5m^4)^3$
   **e.** $(-3r^2s^5)^3$  **f.** $(-8d^4xy^2)^3$  **g.** $(-4m^5n^7y^8)^3$  **h.** $(-10c^2dx^3y)^3$

5. **a.** $(a^6)^4$  **b.** $(x^5y)^4$  **c.** $(2b^4)^4$  **d.** $(3b^7x^2)^4$
   **e.** $(-bc^3)^4$  **f.** $(-4axy^8)^4$  **g.** $(-3b^3c^2d)^4$  **h.** $(-5m^4xy^5z)^4$

6. **a.** $(a^2)^5$  **b.** $(xy^3)^5$  **c.** $(2c^4d^6)^5$  **d.** $(6b^4c^3d^7)^5$
   **e.** $(-x)^5$  **f.** $(-3b^3)^5$  **g.** $(-4c^3dx^2)^5$  **h.** $(-5m^2x^5y)^5$

7. **a.** $(\frac{3}{4}x^5y)^2$  **b.** $(\frac{1}{3}m^4x^3)^2$  **c.** $(\frac{4}{5}ac^3d^2)^2$  **d.** $(\frac{1}{2}ab^4)^3$
   **e.** $(\frac{2}{3}m^2n)^3$  **f.** $(\frac{1}{5}x^5y^3)^2$  **g.** $(\frac{3}{5}a^3b^6x^2)^3$  **h.** $(\frac{5}{6}c^7d^2x^5)^2$

8. **a.** $(.2c)^2$  **b.** $(.6b^5x^2)^2$  **c.** $(.3a^5y^6)^2$  **d.** $(-.1b^3c^4)^2$
   **e.** $(.4c^2d)^3$  **f.** $(-.9r^5s^4)^3$  **g.** $(.2mx^2)^4$  **h.** $(-2.4a^2xz^3)^2$

9. **a.** $(x[a+b])^2$  **b.** $(c^3[x-y]^2)^2$  **c.** $(3b[c+d]^4)^2$  **d.** $(-5x^2[x-y]^5)^2$
   **e.** $(2m^2[c+d]^4)^3$  **f.** $(3b^4[a-b]^2)^3$  **g.** $(-7a^2[c-4]^5)^2$  **h.** $(6y^2[b+c]^6)^2$

10. **a.** $\left(\dfrac{x}{b}\right)^2$  **b.** $\left(\dfrac{-4b^2}{5c}\right)^2$  **c.** $\left(-\dfrac{3a^2b^4}{2x^2}\right)^2$  **d.** $\left(-\dfrac{4r^2d^3}{3xy^4}\right)^3$

# Square Root

**I. Aim**    To find the square root of a number.

**II. Procedure**

**1.** Square Root by Estimation, Division, and Average

The approximate square root of a number may be found as follows:

    **a.** Estimate the square root of the given number.

    **b.** Divide the given number by the estimated square root.

    **c.** Find the average of the resulting quotient and estimate the square root.

    **d.** Divide the given number by this average (step c).

    **e.** Find the average of the divisor used and quotient found in step d.

    **f.** Continue this process to obtain a greater degree of approximation as the divisor and quotient will eventually approximate each other. See sample solution 1.

**2.** Square Root by Algorithm

    **a.** Write the numeral under the square root symbol ($\sqrt{\phantom{x}}$) and separate the numeral into groups of two figures each, starting at the decimal point and forming the groups by 2, first to the left and then to the right of the decimal point.

    If there is an odd number of figures to the left of the decimal point, there will be one group containing a single figure. However, if there is an odd number of figures to the right of the decimal point, annex a zero so that each group contains two figures.

    **b.** Find the largest square which can be subtracted from the first group at the left. Write it under the first group.

    **c.** Write the square root of this largest square above the first group as the first figure of the square root.

    **d.** Subtract the square number from the first group. Annex the next group to the remainder.

    **e.** Form the trial divisor by multiplying the root already found by ~~step b~~ *two* and annexing a zero which is not written but is used mentally.

    **f.** Divide the dividend (step d) by the trial divisor (step e). Annex the quotient to the root already found; also annex it to the trial divisor to form the complete divisor.

**g.** Multiply the complete divisor by the new figure of the root.

**h.** Subtract this product (step g) from the dividend (step d).

**i.** Continue this process until all groups have been used or the desired number of decimal places has been obtained.

**j.** Since each figure of the root is placed directly above its corresponding group, the decimal point in the root is placed directly above the decimal point in the numeral for the given number. See sample solution 2.

3. Square Root by Use of Table

Square roots of whole numbers 1 to 100 inclusive and of the perfect squares (squares of whole numbers) given in the table may be found directly from the table of squares and square roots on page 644.

**a.** To find the square root of any whole number from 1 to 100 inclusive, first locate the given number in the "No." column and then move to the right to the corresponding "Square Root" column to obtain the required square root.

**b.** To find the square root of a perfect square given in the table, first locate this number in the "Square" column and then move to the left to the corresponding "No." column to obtain the required square root.

## III. Sample Solutions

1. Find the square root of 30 by estimation, division, and average.
   *Solution:*
   Since 30 is between 25 and 36, its square root will be between 5 and 6.

   **a.** Use 5.4 as the estimate.

   **b.** Divide 30 by 5.4 to get the quotient 5.5.

   **c.** The average of 5.4 and 5.5 is 5.45.

$$\begin{array}{r} 5.5 \\ 5.4\overline{)30.00} \end{array}$$

$$\begin{array}{r} 5.4 \\ +5.5 \\ \hline 2\overline{)10.9}\,(5.45 \end{array}$$

   **d.** Divide 30 by 5.45 to get the quotient 5.50.

   **e.** The average of 5.45 and 5.50 is 5.475.

$$\begin{array}{r} 5.50 \\ 5.45\overline{)30.0000} \end{array}$$

$$\begin{array}{r} 5.45 \\ +5.50 \\ \hline 2\overline{)10.95}\,(5.475 \end{array}$$

   **f.** Divide 30 by 5.475 to get the quotient 5.479.

   **g.** The average of 5.475 and 5.479 is 5.477.

$$\begin{array}{r} 5.479 \\ 5.475\overline{)30.000000} \end{array}$$

$$\begin{array}{r} 5.475 \\ +5.479 \\ \hline 2\overline{)10.954}\,(5.477 \end{array}$$

*Answer:* 5.477

**2.** Find the square root of 328,329
*Solution:*

$$\begin{array}{r} 5\ \ 7\ \ 3 \\ \sqrt{32\ 83\ 29} \\ 25 \\ \hline 107\overline{)783} \\ 749 \\ \hline 1143\overline{)3429} \\ 3429 \\ \hline \cdot\ \cdot\ \cdot\ \cdot \end{array}$$

*Answer:* 573

---

## DIAGNOSTIC TEST

Find the square root of each of the following numbers by estimation, division, and average to 3 decimal places:

**1.** 39                                      **2.** 7.8

Find the square root of each of the following numbers by algorithm. If there is a remainder, find answer correct to nearest hundredth:

**3.** 6,400      **4.** 7,569      **5.** 39,204      **6.** 209,764      **7.** 8,934,121
**8.** 55,756,089   **9.** 256,036   **10.** .0961   **11.** 15          **12.** 768.3

Find the square root of each of the following numbers directly from the table on page 644.

**13.** 34                                     **14.** 7,056

---

## RELATED PRACTICE EXAMPLES

Find the square root of each of the following numbers by estimation, division, and average to 3 decimal places:

1. **a.** 7         **b.** 13        **c.** 34        **d.** 27        **e.** 43        **f.** 84
   **g.** 60        **h.** 110       **i.** 88        **j.** 125       **k.** 240       **l.** 300
2. **a.** 4.8       **b.** 2.7       **c.** 10.4      **d.** 18.1      **e.** 28.5      **f.** 40.2
   **g.** 6.75      **h.** 11.58     **i.** 18.25     **j.** 39.43     **k.** 108.3     **l.** 246.67

Find the square root of each of the following numbers by algorithm.
If there is a remainder, find answer correct to nearest hundredth:

3. **a.** 49     **b.** 8,100     **c.** .25     **d.** .0009     **e.** 400
4. **a.** 576     **b.** 841     **c.** 7,744     **d.** 5,184     **e.** 3,249
5. **a.** 21,609     **b.** 99,225     **c.** 62,001     **d.** 29,584     **e.** 80,656
6. **a.** 889,249     **b.** 136,161     **c.** 343,396     **d.** 589,824     **e.** 913,936
7. **a.** 5,692,996     **b.** 1,580,049     **c.** 7,606,564     **d.** 9,790,641     **e.** 2,241,009
8. **a.** 72,914,521     **b.** 58,491,904     **c.** 35,952,016     **d.** 71,791,729     **e.** 46,036,225
9. **a.** 94,864     **b.** 368,449     **c.** 592,900     **d.** 4,020,025     **e.** 81,036,004
10. **a.** 7.84     **b.** .0324     **c.** 88.5481     **d.** 3.0276     **e.** 1,070.5984
11. **a.** 8     **b.** 39     **c.** 983     **d.** 2,382     **e.** 5,000
12. **a.** .05     **b.** 147.6     **c.** 25.9     **d.** .924     **e.** 0.2

Find the square root of each of the following numbers directly from the table on page 644.

13. **a.** 16     **b.** 28     **c.** 7     **d.** 86     **e.** 63     **f.** 39     **g.** 95
    **h.** 74     **i.** 43     **j.** 80     **k.** 91     **l.** 59     **m.** 22     **n.** 67
14. **a.** 1,521     **b.** 676     **c.** 7,396     **d.** 4,624     **e.** 729
    **f.** 3,481     **g.** 8,836     **h.** 5,625     **i.** 6,889     **j.** 2,304

## MAINTENANCE PRACTICE IN ARITHMETIC

1. Add:

459,678
927,586
359,969

2. Subtract:

20,509,000
9,618,932

3. Multiply:

6985
4967

4. Divide:

725)506,050

5. Add:

$6\frac{5}{8} + 4\frac{1}{2} + 3\frac{5}{6}$

6. Subtract:

$10\frac{3}{4} - 9\frac{11}{20}$

7. Multiply:

$7\frac{1}{4} \times 2\frac{4}{5}$

8. Divide:

$9 \div \frac{2}{3}$

9. Add:

$4.59 + .38 + .7$

10. Subtract:

$41.56 - 3.842$

11. Multiply:

$6.37 \times .594$

12. Divide:

.1)6.72

13. Find 8% of $52.60
14. What percent of 20 is 25?
15. 12 is 15% of what number?
16. Find the square root of 93,025.

Use the equation method in each of the following:

17. $\frac{2}{5}$ of what number is 140?
18. .07 of what number is 42?
19. 75% of what amount is $255?
20. 3 is 12% of what number?
21. Find the regular price of a television set that sold for $294 at a 30% reduction sale.
22. How much money must be invested at the annual rate of 8% to earn $10,000 per year?

# Roots of Monomials

**I. Aim**    To find the root of a monomial, indicated by the index of the radical.

**II. Procedure**

1. Use the *radical* sign ($\sqrt{\phantom{x}}$) and a small numeral, called the *index*, written in the upper left of the radical sign to indicate the root of a given number.
   **a.** The index 3 in $\sqrt[3]{\phantom{x}}$ indicates the cube root.
   **b.** The index 4 in $\sqrt[4]{\phantom{x}}$ indicates the fourth root.
   **c.** The index 5 in $\sqrt[5]{\phantom{x}}$ indicates the fifth root.
   **d.** The index 2 is not written for the square root. The symbol without any index $\sqrt{\phantom{x}}$ means $\sqrt[2]{\phantom{x}}$.

2. Note that:
   **a.** The numeral or algebraic expression under the radical sign is called the *radicand*.
   **b.** The entire symbol, consisting of the radical sign and radicand, is called the *radical*. In the radical $\sqrt{36}$, the radicand is 36.

3. To find the root of a monomial:
   **a.** Determine the root of its numerical coefficient, if any.
   **b.** Divide the exponent of each variable by the index of the radical.
   **c.** Write as a product the root of the numerical coefficient and the roots of the variables prefixed by the proper sign.

4. Since a positive number has two square roots, one positive and the other negative:
   **a.** Use the radical symbol ($\sqrt{\phantom{x}}$) with no sign or preceded by a positive sign to indicate a positive square root as $\sqrt{36} = 6$ or $+\sqrt{36} = +6$. This positive square root is called the *principal square root*.
   **b.** Use the radical symbol ($\sqrt{\phantom{x}}$) preceded by a negative sign to indicate a negative square root as $-\sqrt{36} = -6$.
   **c.** Use the radical symbol ($\sqrt{\phantom{x}}$) preceded by the $\pm$ sign to indicate both roots, positive and negative, as $\pm\sqrt{36} = \pm 6$ meaning $+\sqrt{36} = +6$ and $-\sqrt{36} = -6$.
   **d.** There is no real number, positive or negative, to represent the square root of a negative number such as $\sqrt{-36}$. Therefore, the variables in this unit are restricted to non-negative numbers.

5. Use the above symbolism (steps 4.a. thru 4.c.) to indicate positive and negative roots for all even roots like fourth root, sixth root, etc.

6. When the required root is odd, like cube root, fifth root, etc., find the indicated root and prefix it with the sign of the given radicand.

7. To find the root of a fractional radicand, divide the root of the numerator by the root of the denominator.

8. To evaluate an expression consisting of a square root of a perfect square prefixed by a numerical coefficient, multiply the square root of the perfect square by this coefficient. See sample solution 12.

9. To evaluate an expression consisting of a square root of a positive number, not a perfect square, prefixed by a numerical coefficient, multiply the square root of this number found in the table of square roots on page 644 by the coefficient. See sample solution 13.

10. Note in this exercise only roots of perfect monomial squares (having a numerical coefficient, if any, which is a perfect square and variables with even exponents only), perfect monomial cubes, etc., are treated.

## III. Sample Solutions

Find the root indicated in each of the following:

1. $\sqrt{4a^2b^6}$      Solution: $\sqrt{4a^2b^6}=2ab^3$      *Answer:* $2ab^3$

2. $-\sqrt{.49x^{12}y^4}$      Solution: $-\sqrt{.49x^{12}y^4}=-.7x^6y^2$      *Answer:* $-.7x^6y^2$

3. $\pm\sqrt{25c^8d^2}$      Solution: $\pm\sqrt{25c^8d^2}=\pm5c^4d$      *Answer:* $\pm5c^4d$

4. $\sqrt{\frac{16}{81}a^4x^2y^{10}}$      Solution: $\sqrt{\frac{16}{81}a^4x^2y^{10}}=\frac{4}{9}a^2xy^5$      *Answer:* $\frac{4}{9}a^2xy^5$

5. $\sqrt{9b^2(r+s)^6}$      Solution: $\sqrt{9b^2(r+s)^6}=3b(r+s)^3$      *Answer:* $3b(r+s)^3$

6. $\sqrt{\dfrac{25m^4}{36c^2d^8}}$      Solution: $\sqrt{\dfrac{25m^4}{36c^2d^8}}=\dfrac{5m^2}{6cd^4}$      *Answer:* $\dfrac{5m^2}{6cd^4}$

7. $\sqrt[3]{64c^3d^9}$      Solution: $\sqrt[3]{64c^3d^9}=4cd^3$      *Answer:* $4cd^3$

8. $\sqrt[3]{-125m^6x^{12}}$      Solution: $\sqrt[3]{-125m^6x^{12}}=-5m^2x^4$      *Answer:* $-5m^2x^4$

9. $\sqrt[4]{81b^{16}}$      Solution: $\sqrt[4]{81b^{16}}=3b^4$      *Answer:* $3b^4$

10. $\sqrt[5]{243x^{10}}$      Solution: $\sqrt[5]{243x^{10}}=3x^2$      *Answer:* $3x^2$

11. $\sqrt[5]{-32c^{20}d^5}$      Solution: $\sqrt[5]{-32c^{20}d^5}=-2c^4d$      *Answer:* $-2c^4d$

12. Evaluate: $3\sqrt{25}$      Solution: $3\sqrt{25}=3\cdot5=15$      *Answer:* 15

13. Find the value of $5\sqrt{3}$, using the table of square roots on page 644.
Solution: $5\sqrt{3}=5\times1.732\approx8.660$
The symbol $\approx$ means "is approximately equal to."

*Answer:* 8.660

## DIAGNOSTIC TEST

Find the roots indicated in each of the following in the set of real numbers:

1. $\sqrt{64}$　　　2. $-\sqrt{81}$　　　3. $\pm\sqrt{144}$　　　4. $\sqrt{-49}$

5. $\sqrt{a^{12}b^8c^2}$　6. $\sqrt{100m^6n^{10}}$　7. $\sqrt{\frac{4}{25}}$　　　8. $\sqrt{\frac{49}{100}}$

9. Find the indicated square: $(\sqrt{4})^2$

10. Write the index, the root to be taken, and the radicand in $\sqrt[3]{-8c^6x^9}$.

Find the root indicated in each of the following:

11. $\sqrt[3]{27a^9b^{12}c^6}$　12. $\sqrt[3]{-125x^3y^{18}}$　13. $\sqrt[4]{256x^8y^{12}z^4}$　14. $\sqrt[5]{243m^{10}n^{15}}$

15. $\sqrt{.49r^2t^{10}}$　16. $\sqrt{\frac{9}{64}x^4y^{16}}$　17. $\sqrt{\frac{25b^4c^2}{81m^8n^6}}$　18. $\sqrt[3]{\frac{27a^{15}}{8b^3c^9}}$

19. Evaluate: $\dfrac{5\sqrt{64}+2\sqrt{100}}{6\sqrt{25}}$

20. Find the value of $3\sqrt{5}$, using the table of square roots on page 644.

## RELATED PRACTICE EXAMPLES

Find the roots indicated in each of the following:

1. a. $\sqrt{25}$　　　b. $\sqrt{4}$　　　c. $\sqrt{81}$　　　d. $\sqrt{144}$　　　e. $\sqrt{400}$

2. a. $-\sqrt{9}$　　b. $-\sqrt{49}$　　c. $-\sqrt{121}$　　d. $-\sqrt{169}$　　e. $-\sqrt{225}$

3. a. $\pm\sqrt{16}$　b. $\pm\sqrt{1}$　　c. $\pm\sqrt{100}$　　d. $\pm\sqrt{256}$　　e. $\pm\sqrt{196}$

4. Find the root indicated in each of the following, using real numbers:

a. $\sqrt{-25}$　　　b. $\sqrt{-100}$　　　c. $\sqrt{-81}$　　　d. $\sqrt{-144}$

Find the root indicated in each of the following:

5. a. $\sqrt{c^8d^4}$　　　b. $\sqrt{a^4x^2z^{12}}$　　　c. $-\sqrt{x^{10}y^6}$　　　d. $-\sqrt{r^{16}s^4t^8x^{12}}$

6. a. $\sqrt{9b^2}$　　　b. $\sqrt{16c^6}$　　　c. $\sqrt{49x^8y^2}$　　　d. $\sqrt{100m^4x^6z^{12}}$

e. $\sqrt{9b^2c^8d^2}$　f. $\sqrt{36d^4r^2t^{14}}$　g. $-\sqrt{81a^4x^2}$　h. $-\sqrt{25x^2y^{10}}$

i. $-\sqrt{4c^6d^{12}x^{18}}$　j. $\sqrt{36a^2b^4c^{10}}$　k. $-\sqrt{64x^{14}y^{16}z^6}$　l. $\sqrt{144m^8n^6x^{20}}$

7. a. $\sqrt{\frac{25}{49}}$　　　b. $\sqrt{\frac{1}{16}}$　　　c. $\sqrt{\frac{36}{169}}$　　　d. $\sqrt{\frac{81}{25}}$

8. a. $-\sqrt{\frac{9}{64}}$　　b. $\pm\sqrt{\frac{4}{121}}$　　c. $-\sqrt{\frac{1}{100}}$　　d. $\pm\sqrt{\frac{49}{169}}$

9. Find the indicated square of each of the following:
   a. $(\sqrt{49})^2$
   b. $(\sqrt{100})^2$
   c. $(\sqrt{5})^2$
   d. $(\sqrt{\frac{3}{4}})^2$

10. Write the index, the root to be taken, and the radicand in each of the following:
    a. $\sqrt{20xy}$
    b. $\sqrt[4]{3a^2b}$
    c. $\sqrt[3]{9cd^2}$
    d. $\sqrt[6]{40c^4}$
    e. $\sqrt[5]{m^4n^2x^3}$
    f. $\sqrt[8]{6b^5c^4}$
    g. $\sqrt{\frac{3a}{5b}}$
    h. $\sqrt[3]{-7r^2s^2}$

Find the root indicated in each of the following:

11. a. $\sqrt[3]{b^3c^{18}}$  b. $\sqrt[3]{27a^6x^{15}}$  c. $\sqrt[3]{125d^9m^{21}}$  d. $\sqrt[3]{8r^3s^6t^9}$

12. a. $\sqrt[3]{-a^6}$  b. $\sqrt[3]{-64c^3d^3}$  c. $\sqrt[3]{-8b^6c^{12}d^{18}}$  d. $\sqrt[3]{-27m^9x^{24}y^{27}}$

13. a. $\sqrt[4]{a^4b^{20}}$  b. $\sqrt[4]{625m^8}$  c. $\sqrt[4]{16x^{16}y^4}$  d. $\sqrt[4]{81a^8b^{12}c^{24}}$

14. a. $\sqrt[5]{b^5c^{20}}$  b. $\sqrt[5]{32m^{25}}$  c. $\sqrt[5]{-x^{10}}$  d. $\sqrt[5]{-243c^{15}m^{10}x^5}$

15. a. $\sqrt{.04a^{14}}$  b. $\sqrt{.64b^2m^8}$  c. $\sqrt{.0009a^2n^6}$  d. $\sqrt{1.44b^4c^{10}d^{12}}$

16. a. $\sqrt{\frac{1}{16}b^{16}}$  b. $\sqrt{\frac{4}{49}a^4c^8}$  c. $\sqrt{\frac{9}{64}n^6x^{18}}$  d. $\sqrt{\frac{25}{81}c^{12}d^2x^6}$

17. a. $\sqrt{\frac{x^4}{y^{20}}}$  b. $\sqrt{\frac{9b^2}{x^2y^4}}$  c. $\sqrt{\frac{25m^2n^6}{81a^{10}x^4}}$  d. $\sqrt{\frac{16r^4}{49d^8t^2}}$

18. a. $\sqrt[3]{\frac{m^6}{n^{12}}}$  b. $\sqrt[3]{-\frac{8x^9y^3}{27x^{12}}}$  c. $\sqrt[4]{\frac{81a^{12}}{16b^{16}}}$  d. $\sqrt[5]{\frac{c^{20}}{32d^{30}}}$

19. Evaluate each of the following expressions:
    a. $2\sqrt{36}$
    b. $5\sqrt{49}$
    c. $8\sqrt{9}$
    d. $4\sqrt{100}$
    e. $-3\sqrt{81}$
    f. $\sqrt{25}+\sqrt{4}$
    g. $\sqrt{49}-\sqrt{9}$
    h. $\sqrt{81}+\sqrt{64}$
    i. $\sqrt{144}-\sqrt{36}$
    j. $\sqrt{10+6}$
    k. $\sqrt{48-12}$
    l. $\sqrt{25\cdot4}$
    m. $\sqrt{9\cdot16}$
    n. $\sqrt{58+6}+\sqrt{16}$
    o. $\sqrt{30-5}-\sqrt{5\cdot5}$
    p. $4\sqrt{64}+2\sqrt{25}$
    q. $5\sqrt{9}+3\sqrt{16}$
    r. $6\sqrt{49}+\sqrt{1}$
    s. $3\sqrt{36}-2\sqrt{4}$
    t. $8\sqrt{100}-5\sqrt{9}$
    u. $\frac{2\sqrt{25}+3\sqrt{16}}{\sqrt{4}}$
    y. $\frac{6\sqrt{36}-4\sqrt{25}}{2\sqrt{16}}$
    w. $\frac{3\sqrt{49}-2\sqrt{9}}{\sqrt{25}}$
    x. $\frac{9\sqrt{64}-5\sqrt{81}}{\sqrt{9}}$
    y. $\frac{6\sqrt{100}+8\sqrt{144}}{2\sqrt{4}}$

20. Using the table of square roots on page 644, find the value of each of the following:
    a. $11\sqrt{2}$
    b. $3\sqrt{6}$
    c. $8\sqrt{7}$
    d. $2\sqrt{10}$
    e. $4\sqrt{11}$

# Fractional Exponents

## I. Aim

To find the root of a monomial, indicated by a fractional exponent.

## II. Procedure

1. Follow the procedure outlined in Exercise 14-3, using the denominator of the fractional exponent as the index to indicate the root.

   a. Since $x^{\frac{1}{2}} \cdot x^{\frac{1}{2}} = x^{\frac{1}{2}+\frac{1}{2}} = x^1 = x$, then $x^{\frac{1}{2}}$ is one of the two equal factors of $x$ and $x^{\frac{1}{2}} = \sqrt{x}$.

   b. Since $x^{\frac{1}{3}} \cdot x^{\frac{1}{3}} \cdot x^{\frac{1}{3}} = x^{\frac{1}{3}+\frac{1}{3}+\frac{1}{3}} = x^1 = x$, then $x^{\frac{1}{3}}$ is one of three equal factors of $x$ and $x^{\frac{1}{3}} = \sqrt[3]{x}$.

   c. In a similar manner $x^{\frac{1}{4}} = \sqrt[4]{x}$, $x^{\frac{1}{5}} = \sqrt[5]{x}$, etc.

2. Or multiply the exponent of each factor by the fractional exponent. See sample solution 2.b. Also see Exercise 14-1.

## III. Sample Solutions

Find the indicated root in each of the following:

1. $(49)^{\frac{1}{2}}$

   *Solution:* $(49)^{\frac{1}{2}} = \sqrt{49} = 7$

   *Answer:* 7

2. $(x^{12}y^8)^{\frac{1}{4}}$

   *Solution a.:* $(x^{12}y^8)^{\frac{1}{4}} = \sqrt[4]{x^{12}y^8} = x^3y^2$

   *Answer:* $x^3y^2$

   *Solution b.:* $(x^{12}y^8)^{\frac{1}{4}} = x^{\frac{1}{4}\cdot 12}y^{\frac{1}{4}\cdot 8} = x^3y^2$

   *Answer:* $x^3y^2$

3. $(-64b^6)^{\frac{1}{3}}$

   *Solution:* $(-64b^6)^{\frac{1}{3}} = \sqrt[3]{-64b^6} = -4b^2$

   *Answer:* $-4b^2$

4. $\left(\dfrac{4x^6}{81y^2}\right)^{\frac{1}{2}}$

   *Solution:* $\left(\dfrac{4x^6}{81y^2}\right)^{\frac{1}{2}} = \sqrt{\dfrac{4x^6}{81y^2}} = \dfrac{2x^3}{9y}$

   *Answer:* $\dfrac{2x^3}{9y}$

## *RELATED PRACTICE EXAMPLES*

**1.** Rename each of the following, using a radical sign:

   **a.** $(9a)^{\frac{1}{2}}$          **b.** $(8b^3c^6)^{\frac{1}{3}}$          **c.** $(x^{15}y^{10})^{\frac{1}{5}}$

   **d.** $(81r^8t^4)^{\frac{1}{4}}$          **e.** $(m^{24}n^{18})^{\frac{1}{6}}$

**2.** Rename each of the following, using a fractional exponent:

   **a.** $\sqrt[3]{27d^9}$          **b.** $\sqrt{36b^{12}c^2}$          **c.** $\sqrt[4]{16y^{16}}$

   **d.** $\sqrt[5]{32w^{15}x^5}$          **e.** $\sqrt[3]{-125b^{15}}$

Find the root indicated in each of the following:

**3. a.** $(4)^{\frac{1}{2}}$     **b.** $(25)^{\frac{1}{2}}$     **c.** $(100)^{\frac{1}{2}}$     **d.** $(1)^{\frac{1}{2}}$     **e.** $(144)^{\frac{1}{2}}$

**4. a.** $(x^2)^{\frac{1}{2}}$     **b.** $(x^4y^8)^{\frac{1}{2}}$     **c.** $(c^{10}d^8)^{\frac{1}{2}}$     **d.** $(b^2c^{10}d^{14})^{\frac{1}{2}}$     **e.** $(m^2n^4x^6)^{\frac{1}{2}}$

**5. a.** $(36a^2)^{\frac{1}{2}}$     **b.** $(81y^{10})^{\frac{1}{2}}$     **c.** $(4x^2y^4)^{\frac{1}{2}}$     **d.** $(25a^6b^2c^8)^{\frac{1}{2}}$     **e.** $(100d^6r^8t^4)^{\frac{1}{2}}$

**6. a.** $(1)^{\frac{1}{3}}$     **b.** $(27)^{\frac{1}{3}}$     **c.** $(-8)^{\frac{1}{3}}$     **d.** $(-125)^{\frac{1}{3}}$     **e.** $(64)^{\frac{1}{3}}$

**7. a.** $(x^6)^{\frac{1}{3}}$     **b.** $(-a^3)^{\frac{1}{3}}$     **c.** $(b^9m^3)^{\frac{1}{3}}$     **d.** $(x^{12}y^{21})^{\frac{1}{3}}$     **e.** $(c^6d^{18})^{\frac{1}{3}}$

**8. a.** $(8b^3)^{\frac{1}{3}}$                    **b.** $(64x^9)^{\frac{1}{3}}$                    **c.** $(125d^3t^6)^{\frac{1}{3}}$

   **d.** $(-27x^6y^{15})^{\frac{1}{3}}$          **e.** $(-64a^{24}b^{30}c^3)^{\frac{1}{3}}$

**9. a.** $(81)^{\frac{1}{4}}$                    **b.** $(b^{12})^{\frac{1}{4}}$                    **c.** $(x^4y^8)^{\frac{1}{4}}$

   **d.** $(256a^8)^{\frac{1}{4}}$          **e.** $(625m^{16}n^{20})^{\frac{1}{4}}$

**10. a.** $(32)^{\frac{1}{5}}$                    **b.** $(-x^{10}y^{20})^{\frac{1}{5}}$                    **c.** $(a^{15}b^5c^{25})^{\frac{1}{5}}$

   **d.** $(243m^{10})^{\frac{1}{5}}$          **e.** $(-32x^{30}y^{15})^{\frac{1}{5}}$

**11. a.** $\left(\dfrac{c^4}{36d^2}\right)^{\frac{1}{2}}$          **b.** $\left(\dfrac{9d^2x^8}{100b^4y^6}\right)^{\frac{1}{2}}$          **c.** $\left(\dfrac{64m^6}{n^{15}y^3}\right)^{\frac{1}{3}}$

   **d.** $\left(-\dfrac{8a^{12}}{125b^9}\right)^{\frac{1}{3}}$          **e.** $\left(\dfrac{81t^{16}}{625r^{20}}\right)^{\frac{1}{4}}$

# Simplification of Radicals— Square Roots of Whole Numbers

EXERCISE
14-5

**I. Aim**    To simplify radicals that are square roots of whole numbers containing a perfect square factor.

## II. Procedure

1. Express the radicand as two factors, one of which is the largest possible perfect square.
2. Replace this square root of the product of the factors by the product of the square roots of the factors. The square root of the product of two numbers is equal to the product of the square roots of the numbers. For example: $\sqrt{25 \cdot 3} = \sqrt{25} \cdot \sqrt{3}$.
3. Find the square root of the perfect square factor and write it as the coefficient of the radical whose radicand is the other factor. See sample solution 1.
4. If there is a coefficient in the given radical, multiply it by the square root found in above step 3. See sample solution 3.
5. Observe that a *radical* is in *simplest form* when its radicand has no factor raised to a power equal to or greater than the index of the radical and there is no fraction in the radicand.

## III. Sample Solutions

Simplify each of the following radicals:

1. $\sqrt{75}$
   *Solution:* $\sqrt{75} = \sqrt{25 \cdot 3} = \sqrt{25} \cdot \sqrt{3} = 5\sqrt{3}$
   *Answer:* $5\sqrt{3}$

2. $\sqrt{72}$
   *Solution:* $\sqrt{72} = \sqrt{36 \cdot 2} = \sqrt{36} \cdot \sqrt{2} = 6\sqrt{2}$
   *Answer:* $6\sqrt{2}$

3. $5\sqrt{18}$
   *Solution:* $5\sqrt{18} = 5\sqrt{9 \cdot 2} = 5\sqrt{9} \cdot \sqrt{2} = 5 \cdot 3\sqrt{2} = 15\sqrt{2}$
   *Answer:* $15\sqrt{2}$

4. $-\frac{1}{2}\sqrt{20}$
   *Solution:* $-\frac{1}{2}\sqrt{20} = -\frac{1}{2}\sqrt{4 \cdot 5} = -\frac{1}{2}\sqrt{4} \cdot \sqrt{5} = -\frac{1}{2} \cdot 2\sqrt{5} = -\sqrt{5}$
   *Answer:* $-\sqrt{5}$

## DIAGNOSTIC TEST

Simplify each of the following radicals:

1. $\sqrt{60}$    2. $\sqrt{48}$    3. $10\sqrt{96}$    4. $-5\sqrt{150}$    5. $\frac{1}{3}\sqrt{45}$

## *RELATED PRACTICE EXAMPLES*

Simplify each of the following radicals:

1. a. $\sqrt{12}$    b. $\sqrt{50}$    c. $\sqrt{54}$    d. $\sqrt{18}$    e. $\sqrt{40}$
   f. $\sqrt{125}$    g. $\sqrt{98}$    h. $\sqrt{147}$    i. $\sqrt{52}$    j. $\sqrt{117}$
2. a. $\sqrt{32}$    b. $\sqrt{200}$    c. $\sqrt{180}$    d. $\sqrt{112}$    e. $\sqrt{96}$
   f. $\sqrt{162}$    g. $\sqrt{108}$    h. $\sqrt{360}$    i. $\sqrt{405}$    j. $\sqrt{432}$
3. a. $3\sqrt{8}$    b. $2\sqrt{60}$    c. $2\sqrt{150}$    d. $5\sqrt{80}$    e. $3\sqrt{48}$
4. a. $-\sqrt{44}$    b. $-5\sqrt{28}$    c. $-6\sqrt{128}$    d. $-10\sqrt{90}$    e. $-12\sqrt{192}$
5. a. $\frac{1}{3}\sqrt{18}$    b. $\frac{1}{2}\sqrt{20}$    c. $\frac{2}{3}\sqrt{54}$    d. $-\frac{1}{4}\sqrt{48}$    e. $-\frac{3}{8}\sqrt{40}$

## *CUMULATIVE REVIEW*

1. Raise each of the following to the indicated power:

   a. $(7m^9n^6)^2$    b. $(-5a^7x^4y)^3$    c. $(4b^4c^8d^2)^4$    d. $\left(-\dfrac{9r^6x^5}{2t}\right)^2$

2. Find the square root of the following numbers correct to the nearest hundredth:

   a. 8,836    b. 43    c. 92,294,449    d. 54.76    e. 6.4

   Find the root indicated in each of the following:

3. a. $\sqrt{64x^8y^2}$    b. $\sqrt[4]{625a^{20}m^{12}}$    c. $\sqrt[5]{32b^{15}x^{10}y^5}$    d. $\sqrt{100c^6d^{12}z^4}$

4. a. $(49a^6)^{\frac{1}{2}}$    b. $(-27b^6x^{18})^{\frac{1}{3}}$    c. $(16r^8s^{20}t^4)^{\frac{1}{4}}$    d. $(25x^{10}y^2z^{14})^{\frac{1}{2}}$

5. Simplify:

   a. $\sqrt{27}$    b. $\sqrt{56}$    c. $\sqrt{72}$    d. $\sqrt{300}$
   e. $4\sqrt{32}$    f. $8\sqrt{18}$    g. $-6\sqrt{80}$    h. $-\frac{4}{5}\sqrt{75}$

6. Evaluate each of the following:

   a. $6\sqrt{81}$    b. $\sqrt{100}-\sqrt{49}$    c. $\sqrt{25\cdot36}$    d. $\dfrac{4\sqrt{16}+5\sqrt{64}}{7\sqrt{4}}$

7. Using the table of square roots on page 644, find the value of each of the following:

   a. $5\sqrt{7}$    b. $2\sqrt{6}$    c. $8\sqrt{2}$    d. $4\sqrt{5}$    e. $11\sqrt{3}$

# Simplification of Radicals– Square Roots of Monomials

### I. Aim

To simplify radicals that are square roots of monomials containing a perfect square factor. The variables in these monomials represent only non-negative numbers.

### II. Procedure

1. Follow the procedure outlined in Exercise 14-5.
2. Any even power of a variable is a perfect square.
3. If the radicand contains an odd power of a variable, use the next lower even power of this variable as the largest possible perfect square.

### III. Sample Solutions

Simplify each of the following radicals:

1. $\sqrt{x^4yz^2}$

   Solution: $\sqrt{x^4yz^2} = \sqrt{x^4z^2 \cdot y} = \sqrt{x^4z^2} \cdot \sqrt{y} = x^2z\sqrt{y}$

   Answer: $x^2z\sqrt{y}$

2. $\sqrt{5a^3b^7}$

   Solution: $\sqrt{5a^3b^7} = \sqrt{a^2b^6 \cdot 5ab} = \sqrt{a^2b^6} \cdot \sqrt{5ab} = ab^3\sqrt{5ab}$

   Answer: $ab^3\sqrt{5ab}$

3. $2d\sqrt{9c^5d}$

   Solution: $2d\sqrt{9c^5d} = 2d\sqrt{9c^4 \cdot cd} = 2d\sqrt{9c^4} \cdot \sqrt{cd} = 2d \cdot 3c^2\sqrt{cd}$

   Answer: $6c^2d\sqrt{cd}$

4. $\sqrt{28x^6y^9}$

   Solution: $\sqrt{28x^6y^9} = \sqrt{4x^6y^8 \cdot 7y} = \sqrt{4x^6y^8} \cdot \sqrt{7y} = 2x^3y^4\sqrt{7y}$

   Answer: $2x^3y^4\sqrt{7y}$

---

## DIAGNOSTIC TEST

Simplify each of the following radicals:

1. $\sqrt{a^2b}$
2. $\sqrt{x^3}$
3. $\sqrt{25x^2y^5}$
4. $\sqrt{18b^7}$
5. $3a\sqrt{12a^4b^3}$
6. $-x^2y\sqrt{72x^3y^9}$
7. $\dfrac{3}{4x}\sqrt{128ax^5}$

## *RELATED PRACTICE EXAMPLES*

Simplify each of the following radicals:

1. **a.** $\sqrt{a^2b}$    **b.** $\sqrt{mn^8}$    **c.** $\sqrt{5x^2}$    **d.** $\sqrt{3b^{10}c^4}$    **e.** $\sqrt{11x^2yz^6}$

2. **a.** $\sqrt{x^3}$    **b.** $\sqrt{d^9}$    **c.** $\sqrt{m^5y^{11}}$    **d.** $\sqrt{19r^7}$    **e.** $\sqrt{5b^9x}$

3. **a.** $\sqrt{9m}$    **b.** $\sqrt{36xy^2}$    **c.** $\sqrt{16m^9n^5}$    **d.** $\sqrt{81d^7rt^2}$    **e.** $\sqrt{64r^6s^8t^3}$

4. **a.** $\sqrt{27a^4}$    **b.** $\sqrt{32m^3}$    **c.** $\sqrt{40x^5y^7}$    **d.** $\sqrt{27mn^5}$    **e.** $\sqrt{128x^8y^9z^{10}}$

5. **a.** $c\sqrt{c^4d^3}$      **b.** $bx^2\sqrt{a^5b^2x^6}$      **c.** $2\sqrt{100a}$
    **d.** $4x\sqrt{50x^2}$      **e.** $2x^2y\sqrt{28x^2y^7}$

6. **a.** $-\sqrt{54x^2y^3}$      **b.** $-c\sqrt{b^5c^3}$      **c.** $-3a\sqrt{20a^7}$
    **d.** $-2c^2d\sqrt{45cd^4}$      **e.** $-7xy\sqrt{72x^3y^9}$

7. **a.** $\dfrac{1}{3x}\sqrt{27x^5}$    **b.** $\dfrac{3}{2m}\sqrt{112a^2m^5}$    **c.** $-\dfrac{3}{8b}\sqrt{96ab^7}$    **d.** $\dfrac{3a}{4b}\sqrt{32a^3b^4}$

## *COMPETENCY CHECK TEST*

In each of the following select the letter corresponding to your answer.

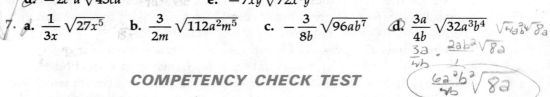

1. The number pair that satisfies both $\begin{array}{l}4x + y = 15\\ x + 3y = 23\end{array}$ is:

   **a.** (1, 11)      **b.** (4, −1)      **c.** (−4, 9)      **d.** (2, 7)

2. The system of equations which has as graphs a pair of intersecting lines is:

   **a.** $\begin{array}{l}3x + 2y = 12\\ 9x - 6y = 36\end{array}$   **b.** $\begin{array}{l}4x - 7y = 15\\ 8x - 14y = 20\end{array}$   **c.** $\begin{array}{l}2x + 5y = -4\\ 6x + 15y = -12\end{array}$   **d.** answer not given

3. $(3a^3mx^7)^4$ raised to the indicated power is:

   **a.** $12a^{12}mx^{28}$      **b.** $81a^{12}m^4x^{28}$      **c.** $12a^7m^4x^{11}$      **d.** $7a^7mx^{11}$

4. The square root of 9,216 is:

   **a.** 96      **b.** 34      **c.** 94      **d.** 36

5. The root of $\sqrt{\dfrac{16x^6}{81c^8d^2}}$ is:   **a.** $\dfrac{4x^4}{9c^6d}$   **b.** $\dfrac{2x}{3cd}$   **c.** $\dfrac{2x^2}{3c^3d}$   **d.** $\dfrac{4x^3}{9c^4d}$

6. The value of $6\sqrt{49} - 4\sqrt{25}$ is:   **a.** $2\sqrt{24}$   **b.** 190   **c.** 22   **d.** answer not given

7. The radical $\sqrt{108}$ in simplest form is:

   **a.** $3\sqrt{12}$      **b.** $6\sqrt{3}$      **c.** $2\sqrt{27}$      **d.** $3\sqrt{6}$

8. The radical $\sqrt{63a^7b^2}$ in simplest form is:

   **a.** $3a^3b\sqrt{7a}$      **b.** $9a^6b^2\sqrt{7a}$      **c.** $9a^3b\sqrt{7a}$      **d.** $9a^7b^2\sqrt{7}$

9. The root of $(64m^6n^{12})^{\frac{1}{3}}$ is:

   **a.** $8m^2n^4$      **b.** $4m^3n^9$      **c.** $8m^3n^9$      **d.** $4m^2n^4$

# Simplification of Radicals— Square Roots of Fractions

I. **Aim**    To simplify radicals that are square roots of fractions. The variables in these fractions represent only positive numbers.

II. **Procedure**

1. Multiply both the numerator and denominator of the fraction by the smallest possible number which will make the denominator a perfect square.
2. Express the new radicand as two factors, one of which is the greatest possible perfect square. This square is a fraction with the new denominator as its denominator.
3. Then follow the procedures outlined in Exercises 14-5 and 14-6.
4. If the radicand is a mixed number, first change it to a fraction.
5. If the radicand has an odd power of a variable in the denominator, first multiply both the numerator and denominator by the variable to make the denominator a perfect square. An even power of a variable is a perfect square. See sample solution 4.

III. **Sample Solutions**

Simplify each of the following radicals:

**1.** $\sqrt{\frac{2}{9}}$   *Solution:* $\sqrt{\frac{2}{9}} = \sqrt{\frac{1}{9} \cdot 2} = \sqrt{\frac{1}{9}}\sqrt{2} = \frac{1}{3}\sqrt{2}$

*Answer:* $\frac{1}{3}\sqrt{2}$

**2.** $\sqrt{\frac{4}{5}}$   *Solution:* $\sqrt{\frac{4}{5}} = \sqrt{\frac{4}{5} \cdot 1} = \sqrt{\frac{4}{5} \cdot \frac{5}{5}} = \sqrt{\frac{20}{25}} = \sqrt{\frac{4}{25} \cdot 5} = \sqrt{\frac{4}{25}} \cdot \sqrt{5} = \frac{2}{5}\sqrt{5}$

*Answer:* $\frac{2}{5}\sqrt{5}$

**3.** $\sqrt{\frac{7}{8}}$   *Solution:* $\sqrt{\frac{7}{8}} = \sqrt{\frac{7}{8} \cdot 1} = \sqrt{\frac{7}{8} \cdot \frac{2}{2}} = \sqrt{\frac{14}{16}} = \sqrt{\frac{1}{16} \cdot 14} = \sqrt{\frac{1}{16}} \cdot \sqrt{14} = \frac{1}{4}\sqrt{14}$

or: $\sqrt{\frac{7}{8}} = \sqrt{\frac{7}{8} \cdot 1} = \sqrt{\frac{7}{8} \cdot \frac{8}{8}} = \sqrt{\frac{56}{64}} = \sqrt{\frac{4}{64} \cdot 14} = \sqrt{\frac{4}{64}} \cdot \sqrt{14} = \frac{2}{8}\sqrt{14} = \frac{1}{4}\sqrt{14}$

*Answer:* $\frac{1}{4}\sqrt{14}$

**4.** $\sqrt{\frac{a}{x^3}}$   *Solution:* $\sqrt{\frac{a}{x^3}} = \sqrt{\frac{a}{x^3} \cdot 1} = \sqrt{\frac{a}{x^3} \cdot \frac{x}{x}} = \sqrt{\frac{ax}{x^4}} = \sqrt{\frac{1}{x^4} \cdot ax}$

$= \sqrt{\frac{1}{x^4}} \cdot \sqrt{ax} = \frac{1}{x^2}\sqrt{ax}$

*Answer:* $\frac{1}{x^2}\sqrt{ax}$

## DIAGNOSTIC TEST

Simplify each of the following radicals:

1. $\sqrt{\frac{5}{16}}$  2. $\sqrt{\frac{2}{3}}$  3. $\sqrt{\frac{5}{8}}$  4. $\sqrt{\frac{4}{15}}$  5. $\sqrt{\frac{12}{7}}$

6. $\sqrt{3\frac{1}{2}}$  7. $\sqrt{\frac{b}{a^2}}$  8. $\sqrt{\frac{x^2}{y}}$  9. $\sqrt{\frac{2ab^2}{c^2d}}$  10. $\sqrt{\frac{x}{2}}$

11. $\sqrt{\frac{2x}{3y}}$  12. $5d\sqrt{\frac{5c}{6d^3}}$  13. $-\sqrt{\frac{3b^7}{8a^3x^5}}$  14. $\frac{5bx}{c^2}\sqrt{\frac{4c^4}{5bx^2}}$

## RELATED PRACTICE EXAMPLES

Simplify each of the following radicals:

1. **a.** $\sqrt{\frac{3}{4}}$  **b.** $\sqrt{\frac{7}{9}}$  **c.** $\sqrt{\frac{15}{64}}$  **d.** $\sqrt{\frac{11}{36}}$

2. **a.** $\sqrt{\frac{1}{2}}$  **b.** $\sqrt{\frac{3}{10}}$  **c.** $\sqrt{\frac{2}{7}}$  **d.** $\sqrt{\frac{7}{15}}$

3. **a.** $\sqrt{\frac{3}{8}}$  **b.** $\sqrt{\frac{5}{12}}$  **c.** $\sqrt{\frac{7}{20}}$  **d.** $\sqrt{\frac{13}{24}}$

4. **a.** $\sqrt{\frac{4}{3}}$  **b.** $\sqrt{\frac{9}{10}}$  **c.** $\sqrt{\frac{25}{32}}$  **d.** $\sqrt{\frac{16}{7}}$

5. **a.** $\sqrt{\frac{24}{25}}$  **b.** $\sqrt{\frac{8}{5}}$  **c.** $\sqrt{\frac{18}{7}}$  **d.** $\sqrt{\frac{27}{2}}$

6. **a.** $\sqrt{2\frac{1}{4}}$  **b.** $\sqrt{1\frac{3}{4}}$  **c.** $\sqrt{2\frac{1}{2}}$  **d.** $\sqrt{1\frac{7}{8}}$

7. **a.** $\sqrt{\frac{c}{x^2}}$  **b.** $\sqrt{\frac{d}{m^2}}$  **c.** $\sqrt{\frac{b}{a^4}}$  **d.** $\sqrt{\frac{2b}{c^2d^4}}$

8. **a.** $\sqrt{\frac{a}{b}}$  **b.** $\sqrt{\frac{m^4}{n}}$  **c.** $\sqrt{\frac{x^2}{y^3}}$  **d.** $\sqrt{\frac{3c^6}{b^7}}$

9. **a.** $\sqrt{\frac{a}{xy^2}}$  **b.** $\sqrt{\frac{d^2}{r^4t}}$  **c.** $\sqrt{\frac{bc}{m^3n^4}}$  **d.** $\sqrt{\frac{3x}{b^5c^2}}$  **e.** $\sqrt{\frac{4a^2b}{x^8y^7}}$

10. **a.** $\sqrt{\frac{a}{4}}$  **b.** $\sqrt{\frac{b}{3}}$  **c.** $\sqrt{\frac{x}{5}}$  **d.** $\sqrt{\frac{m^2}{2}}$  **e.** $\sqrt{\frac{ab^2}{8}}$

11. **a.** $\sqrt{\frac{3a}{4b}}$  **b.** $\sqrt{\frac{13m}{16n}}$  **c.** $\sqrt{\frac{7a^2}{8cd}}$  **d.** $\sqrt{\frac{5a^2y}{12bx^2}}$  **e.** $\sqrt{\frac{2c^4d^3}{5x^3y}}$

12. **a.** $2\sqrt{\frac{4}{9y}}$  **b.** $x\sqrt{\frac{5d}{8x^2}}$  **c.** $4m\sqrt{\frac{2a^2}{3m^2n}}$  **d.** $6rt^2\sqrt{\frac{3c^3}{5r^7t^3}}$

13. **a.** $-\sqrt{\frac{b^4}{4c^3}}$  **b.** $-3\sqrt{\frac{7xy^2}{12b}}$  **c.** $-2a\sqrt{\frac{5c^5}{6a^2b}}$  **d.** $-mx\sqrt{\frac{3r^3x}{2m^4t^2}}$

14. **a.** $\frac{2}{3}\sqrt{\frac{x}{2y^2}}$  **b.** $\frac{1}{2}\sqrt{\frac{4m^2n}{5}}$  **c.** $-\frac{3}{4}\sqrt{\frac{8cd^3}{9xy}}$  **d.** $\frac{a}{b}\sqrt{\frac{2b^2c}{3ad^2}}$

# Addition and Subtraction of Radicals

**I. Aim**   To add and subtract radicals.

**II. Procedure**

**1.** Simplify all radicals.
**2. a.** Add or subtract like radicals just as you do with monomials.
  **b.** Or combine like radicals by adding their coefficients algebraically.
  **c.** The addition or subtraction of unlike radicals can only be indicated.
**3.** Observe that:
  **a.** Radicals which have the same radicand and the same index are called *like radicals.* $6\sqrt{5}$ and $11\sqrt{5}$ are like radicals.
  **b.** *Unlike radicals* are radicals which have either different radicands, when simplified, or different indices or both.
    $6\sqrt{5}$ and $11\sqrt{15}$ are unlike radicals but
    $6\sqrt{5}$ and $7\sqrt{20}$ are like radicals because $7\sqrt{20} = 14\sqrt{5}$.
    $4\sqrt[3]{2}$ and $5\sqrt[4]{2}$ are unlike radicals.
**4.** The variables in this exercise represent only non-negative numbers.

**III. Sample Solutions**

**1.** Add as indicated: $6\sqrt{5} + 11\sqrt{5}$
   *Solution:* $6\sqrt{5} + 11\sqrt{5} = (6 + 11)\sqrt{5} = 17\sqrt{5}$
*Answer:* $17\sqrt{5}$

**2.** Subtract as indicated: $7\sqrt{2} - 3\sqrt{2}$
   *Solution:* $7\sqrt{2} - 3\sqrt{2} = (7 - 3)\sqrt{2} = 4\sqrt{2}$
*Answer:* $4\sqrt{2}$

**3.** Add $3\sqrt{6}$ and $2\sqrt{7}$.
*Answer:* $3\sqrt{6} + 2\sqrt{7}$

**4.** Subtract $5\sqrt{10}$ from $8\sqrt{5}$.
*Answer:* $8\sqrt{5} - 5\sqrt{10}$

**5.** Combine $6\sqrt{11} + 8\sqrt{7} + 3\sqrt{11} - 2\sqrt{7}$.
   *Solution:*   $6\sqrt{11} + 8\sqrt{7} + 3\sqrt{11} - 2\sqrt{7}$
      $= 6\sqrt{11} + 3\sqrt{11} + 8\sqrt{7} - 2\sqrt{7}$
      $= (6 + 3)\sqrt{11} + (8 - 2)\sqrt{7}$
      $= 9\sqrt{11} + 6\sqrt{7}$
*Answer:* $9\sqrt{11} + 6\sqrt{7}$

**6.** Combine: $10\sqrt{\frac{1}{5}} + 4\sqrt{18} + 3\sqrt{45} - 8\sqrt{\frac{1}{2}}$

*Solution:* $10\sqrt{\frac{1}{5}} + 4\sqrt{18} + 3\sqrt{45} - 8\sqrt{\frac{1}{2}}$

$= 10\sqrt{\frac{5}{25}} + 4\sqrt{9\cdot2} + 3\sqrt{9\cdot5} - 8\sqrt{\frac{2}{4}}$

$= 10\sqrt{\frac{1}{25}\cdot5} + 4\cdot3\sqrt{2} + 3\cdot3\sqrt{5} - 8\sqrt{\frac{1}{4}\cdot2}$

$= 10\cdot\frac{1}{5}\sqrt{5} + 12\sqrt{2} + 9\sqrt{5} - 8\cdot\frac{1}{2}\sqrt{2}$

$= 2\sqrt{5} + 12\sqrt{2} + 9\sqrt{5} - 4\sqrt{2}$

$= 11\sqrt{5} + 8\sqrt{2}$

*Answer:* $11\sqrt{5} + 8\sqrt{2}$

---

### DIAGNOSTIC TEST

**1.** Add: $11\sqrt{19}+9\sqrt{19}$    **2.** Subtract: $7\sqrt{3}-6\sqrt{3}$

**3.** Combine: $9\sqrt{5}-7\sqrt{5}+2\sqrt{5}$    **4.** Add $4\sqrt{7}$ and $3\sqrt{17}$.

**5.** Subtract $2\sqrt{6}$ from $5\sqrt{5}$.

**6.** Combine: $6\sqrt{3}+2\sqrt{2}-7\sqrt{3}+5\sqrt{2}$

Simplify and combine:

**7.** $\sqrt{96}+\sqrt{54}-\sqrt{24}$    **8.** $3\sqrt{75}-2\sqrt{48}+\sqrt{24}$

**9.** $\sqrt{\frac{2}{3}}-\sqrt{\frac{1}{6}}+\sqrt{\frac{3}{2}}$    **10.** $4\sqrt{\frac{1}{2}}+2\sqrt{18}-6\sqrt{\frac{2}{9}}$

**11.** $8\sqrt{x}-6\sqrt{x}+a\sqrt{x}$    **12.** $4\sqrt{c^3d^3}+3d\sqrt{4c^3d}-2c\sqrt{9cd^3}$

**13.** $12\sqrt{\frac{1}{3}}-\sqrt{80}-\sqrt{27}+\sqrt{125}$

---

### RELATED PRACTICE EXAMPLES

**1.** Add as indicated:

  **a.** $8\sqrt{3}+5\sqrt{3}$    **b.** $\sqrt{7}+\sqrt{7}$    **c.** $9\sqrt{5}+\sqrt{5}$

  **d.** $16\sqrt{11}+21\sqrt{11}$    **e.** $6\sqrt{2}+7\sqrt{2}+14\sqrt{2}$    **f.** $8\sqrt{6}+3\sqrt{6}+7\sqrt{6}$

**2.** Subtract as indicated:

  **a.** $9\sqrt{6}-2\sqrt{6}$    **b.** $3\sqrt{10}-\sqrt{10}$    **c.** $2\sqrt{7}-\sqrt{7}$

  **d.** $12\sqrt{2}-12\sqrt{2}$    **e.** $6\sqrt{15}-5\sqrt{15}$    **f.** $3\sqrt{21}-7\sqrt{21}$

  **g.** $8\sqrt{5}-2\sqrt{5}$    **h.** $11\sqrt{13}-15\sqrt{13}$

3. Combine as indicated:

    **a.** $8\sqrt{6}+3\sqrt{6}+\sqrt{6}$         **b.** $3\sqrt{7}-4\sqrt{7}+2\sqrt{7}$

    **c.** $3\sqrt{3}-8\sqrt{3}-5\sqrt{3}$      **d.** $\frac{1}{2}\sqrt{2}+\frac{1}{3}\sqrt{2}-\frac{1}{6}\sqrt{2}$

4. Add each of the following:

    **a.** $\sqrt{2}$ and $\sqrt{3}$            **b.** $7\sqrt{6}$ and $3\sqrt{5}$

    **c.** $8\sqrt{7}$ and $2\sqrt{11}$       **d.** $-4\sqrt{3}$ and $8\sqrt{15}$

    **e.** $5\sqrt{17}$ and $\sqrt{6}$        **f.** $3\sqrt{10}$ and $4\sqrt{2}$

5. **a.** Subtract $2\sqrt{6}$ from $5\sqrt{2}$.     **b.** From $10\sqrt{3}$ take $6\sqrt{7}$.

    **c.** Take $3\sqrt{5}$ from $\sqrt{10}$.       **d.** From $\sqrt{17}$ subtract $\sqrt{2}$.

    **e.** Subtract $9\sqrt{11}$ from $8\sqrt{11}$.    **f.** From $8\sqrt{6}$ take $3\sqrt{6}$.

6. Combine:

    **a.** $3\sqrt{17}+5\sqrt{11}+9\sqrt{17}+6\sqrt{11}$     **b.** $9\sqrt{2}+3\sqrt{5}-8\sqrt{5}+5\sqrt{2}$

    **c.** $6\sqrt{a}+5\sqrt{a}-3\sqrt{b}+4\sqrt{b}$      **d.** $2b\sqrt{3c}+b\sqrt{5c}+b\sqrt{3c}-2b\sqrt{5c}$

Simplify and combine:

7. **a.** $\sqrt{48}+\sqrt{12}+\sqrt{27}$         **b.** $\sqrt{98}-\sqrt{8}-\sqrt{32}$

    **c.** $\sqrt{80}+\sqrt{45}-\sqrt{20}$        **d.** $\sqrt{50}-\sqrt{72}+\sqrt{18}$

8. **a.** $2\sqrt{40}+7\sqrt{90}+5\sqrt{160}$     **b.** $8\sqrt{150}-4\sqrt{96}-3\sqrt{600}$

    **c.** $2\sqrt{162}-\sqrt{32}+6\sqrt{128}$     **d.** $\frac{1}{3}\sqrt{147}+\frac{2}{3}\sqrt{27}-\sqrt{108}$

9. **a.** $\sqrt{\frac{1}{5}}+\sqrt{\frac{5}{4}}+\sqrt{\frac{4}{5}}$         **b.** $\sqrt{\frac{2}{25}}+\sqrt{\frac{1}{2}}-\sqrt{\frac{9}{2}}$

    **c.** $10\sqrt{\frac{2}{5}}-\sqrt{\frac{9}{10}}-\sqrt{\frac{1}{10}}$    **d.** $\sqrt{1\frac{1}{8}}+4\sqrt{\frac{1}{8}}+\sqrt{\frac{8}{9}}$

10. **a.** $\sqrt{18}+4\sqrt{\frac{1}{2}}+3\sqrt{32}$       **b.** $5\sqrt{\frac{1}{5}}+7\sqrt{5}-2\sqrt{20}$

    **c.** $6\sqrt{3}-2\sqrt{75}+4\sqrt{\frac{3}{16}}$     **d.** $\sqrt{24}-12\sqrt{\frac{1}{6}}+6\sqrt{\frac{2}{3}}$

11. **a.** $2\sqrt{a}+7\sqrt{a}-3\sqrt{a}$         **b.** $3x\sqrt{b}-2x\sqrt{b}+4x\sqrt{b}$

    **c.** $2a\sqrt{xy}+a\sqrt{xy}+4b\sqrt{xy}-2b\sqrt{xy}$    **d.** $5\sqrt{mn}-3\sqrt{mn}+b\sqrt{mn}$

12. **a.** $\sqrt{100x}-\sqrt{9x}+\sqrt{25x}$

    **b.** $\sqrt{x^3y}+\sqrt{4x^3y}+2x\sqrt{xy}$

    **c.** $a\sqrt{ab^3}+ab\sqrt{ab}+b\sqrt{a^3b}$

    **d.** $3\sqrt{27xy^4}-y\sqrt{48xy^2}+2y^2\sqrt{75x}$

13. **a.** $\sqrt{50}+\sqrt{98}-\sqrt{75}+\sqrt{27}$

    **b.** $2\sqrt{63}+5\sqrt{54}-\sqrt{28}-3\sqrt{24}$

    **c.** $8\sqrt{12}-10\sqrt{\frac{1}{5}}-\sqrt{108}+\sqrt{125}$

    **d.** $4\sqrt{\frac{3}{8}}+\frac{1}{4}\sqrt{48}+2\sqrt{96}-8\sqrt{\frac{3}{4}}$

# Multiplication of Radicals

I. **Aim**  To multiply: a. Two monomial radicals

b. A binomial radical by a monomial radical

c. Two binomial radicals

## II. Procedure

Only square root radicals are studied in this exercise. In this chapter only radicals with the same indices are multiplied.

**1.** To multiply two monomial radicals:

  **a.** Multiply the coefficients of the given radicals to obtain the coefficient of the product.

  **b.** Multiply the radicands of the given radicals to obtain the radicand of the product.

  **c.** If possible, simplify the resulting radical.

  **d.** Note that: (1) The product of the square roots of the two numbers is equal to the square root of the product of the numbers. (2) The square of the square root of a positive number is the number itself.

**2.** To multiply a binomial radical by a monomial radical, multiply each term of the binomial by the monomial, simplifying the resulting radical.

**3.** To multiply two binomial radicals:

  **a.** Follow the procedure for multiplying polynomials. See Exercise 5-9.

  **b.** Or use the procedures for special products. See Chapter 9.

  **c.** Simplify each resulting radical.

**4.** The variables in this exercise represent only non-negative numbers.

## III. Sample Solutions

Multiply and simplify in each of the following:

**1.** $\sqrt{5} \cdot \sqrt{6}$  *Solution:* $\sqrt{5} \cdot \sqrt{6} = \sqrt{30}$  *Answer:* $\sqrt{30}$

**2.** $\sqrt{6} \cdot \sqrt{8}$  *Solution:* $\sqrt{6} \cdot \sqrt{8} = \sqrt{48} = \sqrt{16 \cdot 3} = \sqrt{16} \cdot \sqrt{3} = 4\sqrt{3}$
*Answer:* $4\sqrt{3}$

**3.** $(-7\sqrt{5})(2\sqrt{15})$  *Solution:* $(-7\sqrt{5})(2\sqrt{15}) = (-7 \cdot 2)(\sqrt{5} \cdot \sqrt{15})$
$$= -14\sqrt{75} = -14\sqrt{25 \cdot 3} = 14\sqrt{25} \cdot \sqrt{3}$$
$$= -14 \cdot 5\sqrt{3} = -70\sqrt{3} \quad \textit{Answer: } -70\sqrt{3}$$

**4.** $\sqrt{ax^3} \cdot \sqrt{a^2x^5}$  *Solution:* $\sqrt{ax^3} \cdot \sqrt{a^2x^5} = \sqrt{a^3x^8} = \sqrt{a^2x^8 \cdot a}$
$$= \sqrt{a^2x^8} \cdot \sqrt{a} = ax^4\sqrt{a} \quad \textit{Answer: } ax^4\sqrt{a}$$

**5.** $(-5\sqrt{6})^2$  *Solution:* $(-5\sqrt{6})^2 = (-5\sqrt{6} \cdot -5\sqrt{6}) = 25\sqrt{36} = 25 \cdot 6 = 150$
*Answer:* 150

6. $2\sqrt{5}(3\sqrt{6}+\sqrt{10})$  *Solution:* $2\sqrt{5}(3\sqrt{6}+\sqrt{10})=(2\sqrt{5}\cdot3\sqrt{6})+$
$(2\sqrt{5}\cdot\sqrt{10})=6\sqrt{30}+2\sqrt{50}=6\sqrt{30}+2\sqrt{25\cdot2}=6\sqrt{30}+2\cdot5\sqrt{2}$
$=6\sqrt{30}+10\sqrt{2}$
*Answer:* $6\sqrt{30}+10\sqrt{2}$

7. $(5\sqrt{2}+\sqrt{3})(5\sqrt{2}-\sqrt{3})$  *Solution:* Using special products, $(5\sqrt{2}+\sqrt{3})$
$(5\sqrt{2}-\sqrt{3})=(5\sqrt{2})^2-(\sqrt{3})^2=(25\cdot2)-(3)=50-3=47$  *Answer:* 47

8. $(3\sqrt{5}+2\sqrt{6})^2$  *Solution:* Using special products, $(3\sqrt{5}+2\sqrt{6})^2$
$=(3\sqrt{5})^2+2(3\sqrt{5})(2\sqrt{6})+(2\sqrt{6})^2=(9\cdot5)+2(6\sqrt{30})+(4\cdot6)$
$=45+12\sqrt{30}+24=69+12\sqrt{30}$  *Answer:* $69+12\sqrt{30}$

## DIAGNOSTIC TEST

Multiply and simplify:

1. $\sqrt{2}\cdot\sqrt{3}$   2. $\sqrt{3}\cdot\sqrt{15}$   3. $\sqrt{7}\cdot\sqrt{7}$   4. $\sqrt{\frac{3}{5}}\cdot\sqrt{\frac{2}{3}}$
5. $\sqrt{a}\cdot\sqrt{x}$   6. $\sqrt{6a}\cdot\sqrt{12a}$   7. $\sqrt{8x^3}\cdot\sqrt{5y^2}$   8. $2\sqrt{3}\cdot5\sqrt{7}$
9. $2\sqrt{5}\cdot4\sqrt{15}$  10. $3\sqrt{5}\cdot3\sqrt{5}$  11. $4\cdot8\sqrt{6}$  12. $5\sqrt{2}\cdot\sqrt{8}$
13. $\frac{1}{2}\sqrt{3}\cdot6\sqrt{12}$   14. $(-3\sqrt{3})(-4\sqrt{6})$   15. $(2\sqrt{bx^2})(x\sqrt{10ab})$

Square each of the following as indicated, and simplify:

16. $(2\sqrt{2})^2$   17. $(-6\sqrt{3})^2$

Multiply and simplify:

18. $5(3\sqrt{5}-6)$   19. $3(2\sqrt{3}-3\sqrt{2})$
20. $\sqrt{6}(\sqrt{2}+4)$   21. $\sqrt{3}(\sqrt{18}-2\sqrt{12})$
22. $\frac{1}{2}\sqrt{2}(\sqrt{2}+\frac{2}{3}\sqrt{12})$   23. $-\sqrt{2}(\sqrt{8}-5\sqrt{6})$
24. $(2\sqrt{5}+3)(2\sqrt{5}-3)$   25. $(3\sqrt{2}+5)(3\sqrt{2}+5)$
26. $(4\sqrt{3}-6)(2\sqrt{3}+7)$   27. $(3\sqrt{2}-\sqrt{6})(3\sqrt{2}+\sqrt{6})$
28. $(2\sqrt{8}-3\sqrt{3})(2\sqrt{8}-3\sqrt{3})$   29. $(4\sqrt{6}-3\sqrt{5})(2\sqrt{6}-5\sqrt{5})$
30. $(3-2\sqrt{5})^2$

## *RELATED PRACTICE EXAMPLES*

Multiply and simplify:

1. **a.** $\sqrt{5}\cdot\sqrt{2}$   **b.** $\sqrt{3}\cdot\sqrt{7}$   **c.** $\sqrt{11}\cdot\sqrt{5}$   **d.** $\sqrt{7}\cdot\sqrt{6}$
2. **a.** $\sqrt{2}\cdot\sqrt{6}$   **b.** $\sqrt{8}\cdot\sqrt{3}$   **c.** $\sqrt{12}\cdot\sqrt{6}$   **d.** $\sqrt{5}\cdot\sqrt{15}$
3. **a.** $\sqrt{3}\cdot\sqrt{3}$   **b.** $\sqrt{5}\cdot\sqrt{5}$   **c.** $\sqrt{2}\cdot\sqrt{8}$   **d.** $\sqrt{27}\cdot\sqrt{3}$
4. **a.** $\sqrt{\frac{1}{2}}\cdot\sqrt{6}$   **b.** $\sqrt{\frac{4}{5}}\cdot\sqrt{\frac{1}{4}}$   **c.** $\sqrt{\frac{3}{4}}\cdot\sqrt{\frac{1}{2}}$   **d.** $\sqrt{\frac{2}{3}}\cdot\sqrt{\frac{2}{3}}$

5. **a.** $\sqrt{x} \cdot \sqrt{y}$ $\qquad$ **b.** $\sqrt{2x} \cdot \sqrt{3y}$

 **c.** $(-\sqrt{m})(-\sqrt{2n})$ $\qquad$ **d.** $(-\sqrt{5a})(\sqrt{2b})$

6. **a.** $\sqrt{5b} \cdot \sqrt{5b}$ $\qquad$ **b.** $\sqrt{2a} \cdot \sqrt{6a}$

 **c.** $\sqrt{8bc} \cdot \sqrt{4bc}$ $\qquad$ **d.** $(-3\sqrt{3mn})(-\sqrt{3mn})$

7. **a.** $\sqrt{ax} \cdot \sqrt{a}$ $\quad$ **b.** $\sqrt{a^3b} \cdot \sqrt{bc^3}$ $\quad$ **c.** $(\sqrt{2c^3})(-\sqrt{5cd})$ $\quad$ **d.** $\sqrt{6r^4} \cdot \sqrt{3rs^2}$

8. **a.** $3\sqrt{5} \cdot 4\sqrt{3}$ $\quad$ **b.** $2\sqrt{7} \cdot 2\sqrt{6}$ $\quad$ **c.** $5\sqrt{5} \cdot 3\sqrt{14}$ $\quad$ **d.** $6\sqrt{13} \cdot 2\sqrt{3}$

9. **a.** $2\sqrt{14} \cdot 4\sqrt{2}$ $\quad$ **b.** $5\sqrt{8} \cdot 3\sqrt{10}$ $\quad$ **c.** $4\sqrt{2} \cdot 2\sqrt{27}$ $\quad$ **d.** $7\sqrt{3} \cdot 9\sqrt{24}$

10. **a.** $2\sqrt{2} \cdot 2\sqrt{2}$ $\quad$ **b.** $5\sqrt{6} \cdot 5\sqrt{6}$ $\quad$ **c.** $2\sqrt{18} \cdot 6\sqrt{2}$ $\quad$ **d.** $3\sqrt{20} \cdot 6\sqrt{5}$

11. **a.** $5 \cdot 2\sqrt{6}$ $\quad$ **b.** $8 \cdot 3\sqrt{5}$ $\quad$ **c.** $2\sqrt{2} \cdot 6$ $\quad$ **d.** $7\sqrt{6} \cdot 3$

12. **a.** $4\sqrt{3} \cdot \sqrt{7}$ $\quad$ **b.** $5\sqrt{10} \cdot \sqrt{4}$ $\quad$ **c.** $\sqrt{5} \cdot 2\sqrt{18}$ $\quad$ **d.** $\sqrt{2} \cdot 3\sqrt{2}$

13. **a.** $\frac{1}{3}\sqrt{3} \cdot \sqrt{3}$ $\quad$ **b.** $\frac{1}{2}\sqrt{2} \cdot 4\sqrt{10}$ $\quad$ **c.** $\sqrt{15} \cdot \frac{1}{5}\sqrt{5}$ $\quad$ **d.** $\frac{1}{4}\sqrt{20} \cdot \frac{4}{5}\sqrt{5}$

14. **a.** $(-\sqrt{5})(-\sqrt{8})$ $\qquad$ **b.** $(-2\sqrt{12})(4\sqrt{5})$

 **c.** $(6\sqrt{3})(-\sqrt{18})$ $\qquad$ **d.** $(-\frac{1}{2}\sqrt{6})(4\sqrt{3})$

15. **a.** $2\sqrt{b} \cdot 3\sqrt{ab^3}$ $\qquad$ **b.** $a\sqrt{2x} \cdot x\sqrt{6x}$

 **c.** $2m\sqrt{7mn} \cdot 3\sqrt{7m}$ $\qquad$ **d.** $3y\sqrt{6x^3y} \cdot 2x\sqrt{8xy^4}$

Square each of the following as indicated, and simplify:

16. **a.** $(\sqrt{5})^2$ $\quad$ **b.** $(\sqrt{8})^2$ $\quad$ **c.** $(3\sqrt{3})^2$ $\quad$ **d.** $(4\sqrt{6})^2$ $\quad$ **e.** $(x\sqrt{2a})^2$

17. **a.** $(-\sqrt{3})^2$ $\quad$ **b.** $(-4\sqrt{5})^2$ $\quad$ **c.** $(-2\sqrt{2})^2$ $\quad$ **d.** $(-3\sqrt{x})^2$ $\quad$ **e.** $(-2a\sqrt{3b})^2$

Multiply and simplify:

18. **a.** $2(4\sqrt{2} + 1)$ $\qquad$ **b.** $3(3 + 5\sqrt{3})$ $\qquad$ **c.** $8(2 - 3\sqrt{5})$

19. **a.** $4(3\sqrt{7} + \sqrt{6})$ $\qquad$ **b.** $9(4\sqrt{3} - 2\sqrt{10})$ $\qquad$ **c.** $2(3\sqrt{12} - 5\sqrt{8})$

20. **a.** $\sqrt{2}(\sqrt{3} + 3)$ $\qquad$ **b.** $\sqrt{3}(\sqrt{27} + 4)$ $\qquad$ **c.** $\sqrt{8}(2\sqrt{3} - 5)$

21. **a.** $\sqrt{6}(\sqrt{2} - \sqrt{12})$ $\qquad$ **b.** $\sqrt{2}(3\sqrt{2} + \sqrt{18})$ $\qquad$ **c.** $\sqrt{12}(2\sqrt{5} - 4\sqrt{2})$

22. **a.** $3\sqrt{2}(\sqrt{3} - 5\sqrt{8})$ $\quad$ **b.** $5\sqrt{5}(3\sqrt{5} + 2\sqrt{6})$ $\quad$ **c.** $\frac{1}{3}\sqrt{3}(\sqrt{3} - \frac{1}{3}\sqrt{6})$

23. **a.** $-2(5\sqrt{3} - 4\sqrt{7})$ $\quad$ **b.** $-\sqrt{2}(-\sqrt{8} + 2\sqrt{2})$ $\quad$ **c.** $-3\sqrt{3}(3\sqrt{2} - 5\sqrt{6})$

24. **a.** $(\sqrt{2} + 4)(\sqrt{2} - 4)$ $\qquad$ 25. **a.** $(\sqrt{2} + 2)(\sqrt{2} + 2)$

 **b.** $(2\sqrt{3} - 3)(2\sqrt{3} + 3)$ $\qquad$ **b.** $(5 + 3\sqrt{3})(5 + 3\sqrt{3})$

 **c.** $(6 - 3\sqrt{8})(6 + 3\sqrt{8})$ $\qquad$ **c.** $(5\sqrt{18} - 3)(5\sqrt{18} - 3)$

26. **a.** $(\sqrt{3} + 4)(\sqrt{3} + 2)$ $\qquad$ 27. **a.** $(\sqrt{3} + \sqrt{2})(\sqrt{3} - \sqrt{2})$

 **b.** $(3\sqrt{2} - 4)(2\sqrt{2} + 7)$ $\qquad$ **b.** $(6\sqrt{5} + \sqrt{7})(6\sqrt{5} - \sqrt{7})$

 **c.** $(2\sqrt{12} - 5)(3\sqrt{12} - 4)$ $\qquad$ **c.** $(5\sqrt{6} - 2\sqrt{3})(5\sqrt{6} + 2\sqrt{3})$

28. **a.** $(\sqrt{5} + \sqrt{3})(\sqrt{5} + \sqrt{3})$ $\qquad$ 29. **a.** $(3\sqrt{6} - 2\sqrt{3})(2\sqrt{6} - 4\sqrt{3})$

 **b.** $(3\sqrt{8} - \sqrt{6})(3\sqrt{8} - \sqrt{6})$ $\qquad$ **b.** $(6\sqrt{5} - 3\sqrt{10})(4\sqrt{5} + \sqrt{10})$

 **c.** $(6\sqrt{10} + 7\sqrt{18})(6\sqrt{10} + 7\sqrt{18})$ $\quad$ **c.** $(2\sqrt{12} + 5\sqrt{8})(4\sqrt{12} - 2\sqrt{8})$

30. **a.** $(\sqrt{8} + 2)^2$ $\quad$ **b.** $(2\sqrt{3} - 1)^2$ $\quad$ **c.** $(\sqrt{2} + \sqrt{3})^2$ $\quad$ **d.** $(5\sqrt{6} - 6\sqrt{5})^2$

# Division of Radicals

### I. Aim
a. To divide a monomial radical or a polynomial radical by a monomial radical, when expressed in fraction form.
b. To rationalize the denominator.

### II. Procedure
Only square root radicals are studied in this exercise. Radicals must have the same indices to be divided.
1. To divide a monomial radical by a monomial radical:
   a. Divide the coefficient of the numerator (dividend) by the coefficient of the denominator (divisor) to obtain the coefficient of the answer (quotient).
   b. Divide the radicand of the radical in the numerator (dividend) by the radicand of the radical in the denominator (divisor) to obtain the radicand of the quotient.
   c. Simplify wherever possible.
   d. Note that the quotient of the square roots of the two numbers is equal to the square root of the quotient of the numbers.
2. To divide a polynomial radical by a monomial radical, divide each term of the polynomial radical by the monomial radical.
3. To rationalize the denominator, multiply the numerator and denominator by a radical having a number for its radicand that will make the radicand in its denominator a perfect square. Simplify if possible.

   To *rationalize the denominator* means to change a fraction which has a radical in its denominator to an equivalent fraction with only a rational number in the denominator.
4. First rationalize the denominator and then use the table of square roots found on page 644 to evaluate expressions of the type $\dfrac{3}{\sqrt{2}}$.
5. The variables in this exercise represent only positive numbers.

### III. Sample Solutions
Divide and simplify in each of the following:

1. $\dfrac{\sqrt{6}}{\sqrt{3}}$

   Solution: $\dfrac{\sqrt{6}}{\sqrt{3}} = \sqrt{\tfrac{6}{3}} = \sqrt{2}$

   Answer: $\sqrt{2}$

2. $\dfrac{8\sqrt{30}}{2\sqrt{5}}$

   Solution: $\dfrac{8\sqrt{30}}{2\sqrt{5}} = 4\sqrt{6}$

   Answer: $4\sqrt{6}$

3. $\dfrac{\sqrt{28}}{\sqrt{7}}$

   Solution: $\dfrac{\sqrt{28}}{\sqrt{7}} = \sqrt{4} = 2$

   Answer: $2$

**4.** $\dfrac{\sqrt{5}}{\sqrt{3}}$   *Solution:* $\dfrac{\sqrt{5}}{\sqrt{3}} = \sqrt{\tfrac{5}{3}} = \sqrt{\tfrac{5}{3}\cdot 1} = \sqrt{\tfrac{5}{3}\cdot\tfrac{3}{3}} = \sqrt{\tfrac{15}{9}} = \sqrt{\tfrac{1}{9}\cdot 15} = \sqrt{\tfrac{1}{9}}\cdot\sqrt{15}$

$\qquad\qquad\qquad = \tfrac{1}{3}\sqrt{15}$   *Answer:* $\tfrac{1}{3}\sqrt{15}$

Or solution by rationalizing the denominator:

$$\frac{\sqrt{5}}{\sqrt{3}} = \frac{\sqrt{5}}{\sqrt{3}}\cdot 1 = \frac{\sqrt{5}}{\sqrt{3}}\cdot\frac{\sqrt{3}}{\sqrt{3}} = \frac{\sqrt{5}\cdot\sqrt{3}}{\sqrt{3}\cdot\sqrt{3}} = \frac{\sqrt{15}}{\sqrt{9}} = \frac{\sqrt{15}}{3}\ \text{or}\ \tfrac{1}{3}\sqrt{15}$$

Shortened method: $\dfrac{\sqrt{5}}{\sqrt{3}} = \dfrac{\sqrt{5}}{\sqrt{3}}\cdot\dfrac{\sqrt{3}}{\sqrt{3}} = \dfrac{\sqrt{15}}{3}$   *Answer:* $\dfrac{\sqrt{15}}{3}$ or $\tfrac{1}{3}\sqrt{15}$

**5.** $\dfrac{32\sqrt{a^3b^2}}{4\sqrt{ab}}$   *Solution:* $\dfrac{32\sqrt{a^3b^2}}{4\sqrt{ab}} = 8\sqrt{a^2b} = 8\sqrt{a^2}\cdot\sqrt{b} = 8a\sqrt{b}$   *Answer:* $8a\sqrt{b}$

**6.** $\dfrac{\sqrt{24}+2\sqrt{60}-\sqrt{3}}{\sqrt{3}}$   *Solution:* $\dfrac{\sqrt{24}+2\sqrt{60}-\sqrt{3}}{\sqrt{3}} = \dfrac{\sqrt{24}}{\sqrt{3}} + \dfrac{2\sqrt{60}}{\sqrt{3}} - \dfrac{\sqrt{3}}{\sqrt{3}}$

*Answer:* $2\sqrt{2}+4\sqrt{5}-1$   $\qquad = \sqrt{8}+2\sqrt{20}-1 = 2\sqrt{2}+4\sqrt{5}-1$

**7.** Evaluate $\dfrac{3}{\sqrt{2}}$ by rationalizing the denominator and using the table of square roots.   *Solution:* $\dfrac{3}{\sqrt{2}} = \dfrac{3}{\sqrt{2}}\cdot\dfrac{\sqrt{2}}{\sqrt{2}} = \dfrac{3\sqrt{2}}{2} = \dfrac{3\times 1.414}{2} = 2.121$

*Answer:* 2.121

## DIAGNOSTIC TEST

Divide as indicated and simplify:   **1.** $\dfrac{\sqrt{6}}{\sqrt{2}}$   **2.** $\dfrac{\sqrt{27}}{\sqrt{3}}$

**3.** $\dfrac{\sqrt{60}}{\sqrt{5}}$   **4.** $\dfrac{12\sqrt{98}}{4\sqrt{2}}$   **5.** $\dfrac{\sqrt{32a^4}}{\sqrt{2a}}$   **6.** $\dfrac{10\sqrt{16x^3y^2}}{2\sqrt{2xy}}$

Rationalize the denominator of each of the following:

**7.** $\dfrac{\sqrt{5}}{\sqrt{2}}$   **8.** $\dfrac{\sqrt{3}}{\sqrt{24}}$   **9.** $\dfrac{9}{\sqrt{3}}$   **10.** $\dfrac{2}{5\sqrt{6}}$

Divide as indicated and simplify:

**11.** $\dfrac{4\sqrt{12}-2\sqrt{6}+6\sqrt{96}}{2\sqrt{3}}$   **12.** $\dfrac{\sqrt{8a^5b^5}+\sqrt{18a^3b^3}-\sqrt{24ab}}{\sqrt{2ab}}$

**13.** Evaluate $\dfrac{6}{\sqrt{2}}$ by first rationalizing the denominator and using the table of square roots found on page 644.

## RELATED PRACTICE EXAMPLES

Divide as indicated and simplify:

1. **a.** $\dfrac{\sqrt{12}}{\sqrt{6}}$    **b.** $\dfrac{\sqrt{15}}{\sqrt{5}}$    **c.** $\dfrac{\sqrt{30}}{\sqrt{6}}$    **d.** $\dfrac{\sqrt{39}}{\sqrt{3}}$    **e.** $\dfrac{\sqrt{56}}{\sqrt{8}}$

2. **a.** $\dfrac{\sqrt{75}}{\sqrt{3}}$    **b.** $\dfrac{\sqrt{63}}{\sqrt{7}}$    **c.** $\dfrac{\sqrt{96}}{\sqrt{6}}$    **d.** $\dfrac{\sqrt{128}}{\sqrt{2}}$    **e.** $\dfrac{\sqrt{147}}{\sqrt{3}}$

3. **a.** $\dfrac{\sqrt{24}}{\sqrt{3}}$    **b.** $\dfrac{\sqrt{90}}{\sqrt{2}}$    **c.** $\dfrac{\sqrt{108}}{\sqrt{6}}$    **d.** $\dfrac{\sqrt{140}}{\sqrt{7}}$    **e.** $\dfrac{\sqrt{120}}{\sqrt{5}}$

4. **a.** $\dfrac{6\sqrt{8}}{3\sqrt{8}}$    **b.** $\dfrac{8\sqrt{20}}{2\sqrt{5}}$    **c.** $\dfrac{12\sqrt{12}}{4\sqrt{4}}$    **d.** $\dfrac{15\sqrt{96}}{5\sqrt{2}}$    **e.** $\dfrac{9\sqrt{125}}{9\sqrt{5}}$

5. **a.** $\dfrac{\sqrt{ax}}{\sqrt{a}}$    **b.** $\dfrac{\sqrt{x^3y^4}}{\sqrt{xy}}$    **c.** $\dfrac{\sqrt{32b^3}}{\sqrt{8b}}$    **d.** $\dfrac{\sqrt{30m^3n}}{\sqrt{5m}}$    **e.** $\dfrac{\sqrt{x^2y}}{\sqrt{xy^2}}$

6. **a.** $\dfrac{4\sqrt{a^2b}}{2\sqrt{ab}}$    **b.** $\dfrac{12\sqrt{b^5c^2}}{3\sqrt{b^3c}}$    **c.** $\dfrac{16\sqrt{8m^8}}{4\sqrt{4m^3}}$    **d.** $\dfrac{30\sqrt{27x^5y^3}}{6\sqrt{3xy^3}}$    **e.** $\dfrac{42\sqrt{40r^3t^2}}{3\sqrt{5rt}}$

Rationalize the denominator of each of the following:

7. **a.** $\dfrac{\sqrt{3}}{\sqrt{2}}$    **b.** $\dfrac{\sqrt{4}}{\sqrt{3}}$    **c.** $\dfrac{\sqrt{7}}{\sqrt{2}}$    **d.** $\dfrac{\sqrt{8}}{\sqrt{5}}$    **e.** $\dfrac{\sqrt{10}}{\sqrt{6}}$

8. **a.** $\dfrac{\sqrt{1}}{\sqrt{2}}$    **b.** $\dfrac{\sqrt{5}}{\sqrt{8}}$    **c.** $\dfrac{\sqrt{8}}{\sqrt{12}}$    **d.** $\dfrac{\sqrt{10}}{\sqrt{18}}$    **e.** $\dfrac{\sqrt{8}}{\sqrt{20}}$

9. **a.** $\dfrac{1}{\sqrt{2}}$    **b.** $\dfrac{4}{\sqrt{6}}$    **c.** $\dfrac{6}{\sqrt{8}}$    **d.** $\dfrac{5}{\sqrt{5}}$    **e.** $\dfrac{2}{\sqrt{3}}$

10. **a.** $\dfrac{2}{3\sqrt{2}}$    **b.** $\dfrac{5}{2\sqrt{5}}$    **c.** $\dfrac{1}{4\sqrt{2}}$    **d.** $\dfrac{3}{5\sqrt{6}}$    **e.** $\dfrac{15}{4\sqrt{10}}$

Divide as indicated and simplify:

11. **a.** $\dfrac{\sqrt{18}+\sqrt{50}}{\sqrt{2}}$    **b.** $\dfrac{\sqrt{80}+\sqrt{90}-\sqrt{15}}{\sqrt{5}}$    **c.** $\dfrac{6\sqrt{12}-4\sqrt{27}+2\sqrt{108}}{2\sqrt{3}}$

12. **a.** $\dfrac{\sqrt{a^3}+\sqrt{a^2}}{\sqrt{a}}$      **b.** $\dfrac{\sqrt{x^3y^3}+\sqrt{x^2y^2}-\sqrt{xy}}{\sqrt{xy}}$

    **c.** $\dfrac{\sqrt{4a^4b}-\sqrt{2a^5b}+\sqrt{8a^2b^2}}{\sqrt{2a^2b}}$      **d.** $\dfrac{\sqrt{12m^2x^2}-\sqrt{24m^2x}-\sqrt{30mx^2}}{\sqrt{3mx}}$

13. Evaluate each of the following by rationalizing the denominator and using the table of square roots found on page 644.

    **a.** $\dfrac{2}{\sqrt{5}}$    **b.** $\dfrac{5}{\sqrt{3}}$    **c.** $\dfrac{4}{\sqrt{6}}$    **d.** $\dfrac{\sqrt{5}}{\sqrt{8}}$    **e.** $\dfrac{\sqrt{3}}{3\sqrt{2}}$    **f.** $\dfrac{1}{2\sqrt{7}}$

# Radical Equations

**I. Aim**    To solve radical equations which contain only square root radicals.

**II. Procedure**

1. Transform the given equation to an equivalent equation which has only the radical itself as one member of the equation and a constant or other terms as the second member.
2. Square both members to eliminate the square root radical.
3. Solve the resulting equation.
4. Check by substituting the root in the given equation. Sometimes extraneous roots are introduced when raising both sides of the equation to a power.
5. Observe that a *radical equation* is an equation containing one or more radicals with the variable in the radicand.

**III. Sample Solutions**

1. Solve and check: $\sqrt{5x} - 3 = 12$

   *Solution:*
   $$\sqrt{5x} - 3 = 12$$
   $$\sqrt{5x} - 3 + 3 = 12 + 3$$
   $$\sqrt{5x} = 15$$
   $$(\sqrt{5x})^2 = (15)^2$$
   $$5x = 225$$
   $$x = 45$$

   *Answer: 45*

   *Check:*
   $$\sqrt{5} - 3 = 12$$
   $$\sqrt{5 \cdot 45} - 3 = 12\,?$$
   $$\sqrt{225} - 3 = 12$$
   $$15 - 3 = 12$$
   $$12 = 12 \checkmark$$

2. Solve and check: $2\sqrt{4x - 3} = 10$

   $\left( 2\sqrt{4x} = 13 \right.$

   *Solution:*
   $$2\sqrt{4x - 3} = 10$$
   $$\sqrt{4x - 3} = 5$$
   $$(\sqrt{4x - 3})^2 = (5)^2$$
   $$4x - 3 = 25$$
   $$4x - 3 + 3 = 25 + 3$$
   $$4x = 28$$
   $$x = 7$$

   *Answer: 7*

   *Check:*
   $$2\sqrt{4x - 3} = 10$$
   $$2\sqrt{4 \cdot 7 - 3} = 10\,?$$
   $$2\sqrt{28 - 3} = 10$$
   $$2\sqrt{25} = 10$$
   $$2 \cdot 5 = 10$$
   $$10 = 10 \checkmark$$

## DIAGNOSTIC TEST

Solve and check:

1. $\sqrt{x}=3$     2. $\sqrt{2n}=4$     3. $3\sqrt{y}=6$

4. $\sqrt{x}+1=5$     5. $\sqrt{3x}-2=4$     6. $2\sqrt{2y+3}=11$

7. $\sqrt{x+2}=3$     8. $4\sqrt{2x-1}=12$     9. $\sqrt{x^2+8}=x+2$

10. $8-\sqrt{2x}=2$     11. $\sqrt{\dfrac{x}{2}}=5$     12. $\sqrt{\dfrac{2n}{3}}=6$

13. $2\sqrt{2y}=5$     14. $3\sqrt{3x-2}=4$

## RELATED PRACTICE EXAMPLES

Solve and check:

1. **a.** $\sqrt{x}=2$    **b.** $\sqrt{n}=5$    **c.** $\sqrt{y}=4$    **d.** $6=\sqrt{x}$

2. **a.** $\sqrt{3x}=6$    **b.** $\sqrt{5x}=5$    **c.** $\sqrt{4n}=8$    **d.** $10=\sqrt{2w}$

3. **a.** $4\sqrt{x}=8$    **b.** $2\sqrt{n}=10$    **c.** $3\sqrt{2x}=12$    **d.** $5\sqrt{3y}=60$

4. **a.** $\sqrt{x}+4=6$                        **b.** $\sqrt{2y}+11=15$
    **c.** $12+\sqrt{4x}=20$            **d.** $15=\sqrt{3x}+9$

5. **a.** $\sqrt{x}-2=3$                        **b.** $\sqrt{5y}-8=2$
    **c.** $6=\sqrt{3x}-3$              **d.** $\sqrt{4n}-7=1$

6. **a.** $3\sqrt{n}+4=10$                   **b.** $4\sqrt{3y}+9=21$
    **c.** $6\sqrt{2x}-3=45$          **d.** $25=2\sqrt{5x}-5$

7. **a.** $\sqrt{x+1}=4$                       **b.** $\sqrt{n-3}=8$
    **c.** $5=\sqrt{2x+3}$            **d.** $\sqrt{5y-4}=9$

8. **a.** $2\sqrt{n+3}=10$                 **b.** $18=6\sqrt{y-4}$
    **c.** $3\sqrt{2x+5}=9$          **d.** $4\sqrt{3x-2}=16$

9. **a.** $\sqrt{x^2+3}=x+1$           **b.** $\sqrt{x^2-35}=x-5$
    **c.** $x-4=\sqrt{x^2-32}$     **d.** $\sqrt{y^2+27}=y+3$

10. **a.** $9-\sqrt{x}=2$                   **b.** $12-\sqrt{2n}=4$
    **c.** $25-2\sqrt{5x}=5$        **d.** $50-3\sqrt{8x}=2$

11. **a.** $\sqrt{\dfrac{x}{3}}=1$    **b.** $\sqrt{\dfrac{y}{5}}=2$    **c.** $\sqrt{\dfrac{x}{3}}=5$    **d.** $2=\sqrt{\dfrac{x}{4}}$

12. **a.** $\sqrt{\dfrac{2x}{3}}=4$    **b.** $\sqrt{\dfrac{3x}{5}}=6$    **c.** $\sqrt{\dfrac{5n}{2}}=10$    **d.** $3=\sqrt{\dfrac{3x}{4}}$

13. **a.** $2\sqrt{n}=3$    **b.** $5\sqrt{y}=4$    **c.** $4\sqrt{5x}=6$    **d.** $3=6\sqrt{3y}$

14. **a.** $3\sqrt{3y-1}=1$    **b.** $4\sqrt{5x-4}=2$    **c.** $4=5\sqrt{2x-1}$    **d.** $2\sqrt{3n-5}=3$

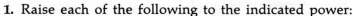

# REVIEW

1. Raise each of the following to the indicated power:

   **a.** $(9r^7t^{11})^2$    **b.** $(-10x^8y)^2$    **c.** $(4b^6x^4z^3)^3$    **d.** $(-5m^5n^6)^5$

   **e.** $(6ab^9c^6)^4$   **f.** $\left(\dfrac{7a^6x^2}{23b^3y^8}\right)^2$   **g.** $(-11r^{12}st^7)^3$   **h.** $(-\frac{3}{4}c^3d^8x)^4$

2. Find the square root of each of the following numbers correct to the nearest hundredth:

   **a.** 4,489    **b.** 74    **c.** 37,027,225    **d.** 9409    **e.** 8.1

3. Find the root indicated in each of the following:

   **a.** $\sqrt{49m^{10}n^6}$    **b.** $\sqrt[4]{256r^{12}y^{16}}$    **c.** $\sqrt[3]{-125b^{12}}$    **d.** $\sqrt{\frac{4}{81}c^{14}d^6y^2}$

   **e.** $\sqrt[5]{x^{25}y^{10}z^{20}}$   **f.** $\sqrt[3]{\dfrac{8c^{15}}{27x^9y^{18}}}$   **g.** $(64m^6n^{12})^{\frac{1}{2}}$   **h.** $(64m^6n^{12})^{\frac{1}{3}}$

4. Simplify each of the following radicals:

   **a.** $\sqrt{176}$    **b.** $6\sqrt{128}$    **c.** $\sqrt{72x^5y^2}$    **d.** $4xy^2\sqrt{25x^7y^2}$

   **e.** $\sqrt{27s^8t^2}$   **f.** $\sqrt{\frac{5}{6}}$   **g.** $\dfrac{m}{n^2}\sqrt{\dfrac{8n^5}{3m^2}}$   **h.** $\sqrt{\dfrac{16b^7}{9c^9}}$

5. Simplify and combine.

   **a.** $3\sqrt{48}+2\sqrt{108}-\sqrt{75}$

   **b.** $3\sqrt{192}-6\sqrt{\frac{4}{3}}+\sqrt{147}$

   **c.** $5\sqrt{96}+4\sqrt{\frac{5}{8}}-18\sqrt{\frac{2}{3}}+\sqrt{90}$

   **d.** $d\sqrt{98c^5d}-c\sqrt{128c^3d^5}+3cd\sqrt{50c^3d^3}$

6. Multiply and simplify:

   **a.** $3\sqrt{2}\cdot\sqrt{18}$    **b.** $(-2\sqrt{3})^2$    **c.** $11(3\sqrt{6}-7)$

   **d.** $5\sqrt{6}(\sqrt{2}-4\sqrt{12})$   **e.** $(7\sqrt{3}+8)(7\sqrt{3}-8)$   **f.** $(2\sqrt{5}-3\sqrt{6})^2$

7. Divide and simplify:

   **a.** $\dfrac{\sqrt{48}}{\sqrt{6}}$    **b.** $\dfrac{\sqrt{7}}{\sqrt{5}}$    **c.** $\dfrac{14\sqrt{75}}{7\sqrt{3}}$    **d.** $\dfrac{3\sqrt{18}+12\sqrt{80}-6\sqrt{2}}{3\sqrt{2}}$

8. First rationalize the denominator of each of the following, then evaluate by using the table of square roots on page 644:

   **a.** $\dfrac{-9}{\sqrt{3}}$    **b.** $\dfrac{1}{\sqrt{6}}$    **c.** $\dfrac{\sqrt{5}}{\sqrt{10}}$    **d.** $\dfrac{3}{2\sqrt{2}}$    **e.** $\dfrac{\sqrt{11}}{\sqrt{5}}$

9. Solve and check:

   **a.** $6\sqrt{x}=12$    **b.** $\sqrt{x^2+45}=x+3$    **c.** $\sqrt{2x}-4=6$    **d.** $\sqrt{\dfrac{x}{5}}=4$

# INTRODUCTION

Many scientific and mathematical formulas contain variables raised to the second power. Also, equations formulated from facts given in problems frequently contain the variable raised to the second power. To solve for these quantities it will be necessary to understand the principles used in the solution of the quadratic equation.

A *quadratic equation in one variable* is an equation where the variable is raised to the second power (but no higher power). It is also called an *equation of the second degree*. If the equation contains the second power but not the first power of the variable, it is called an *incomplete quadratic equation*.

$$9x^2 - 16 = 0 \text{ and } x^2 = 64$$
are incomplete quadratic equations.

The general quadratic equation in one variable is expressed as $ax^2 + bx + c = 0$ where $a$, $b$, and $c$ are all rational numbers and $a \neq 0$.

Incomplete quadratic equations may be solved by a special method based on the axiom "square roots of equals are equal." Some quadratic equations may be solved by factoring, using the principle that if the product of two factors is zero, at least one of the factors is zero. However, all quadratic equations may be solved by the general quadratic formula. This formula is derived by solving the general quadratic equation $ax^2 + bx + c = 0$ for $x$ in terms of $a$, $b$, and $c$, using the method of completing the square. In the quadratic formula $a$ represents the numerical coefficient of $x^2$, $b$ the numerical coefficient of $x$, and $c$ the constant term. Every quadratic equation has two roots.

In this chapter several methods of solving quadratic equations in one variable are studied.

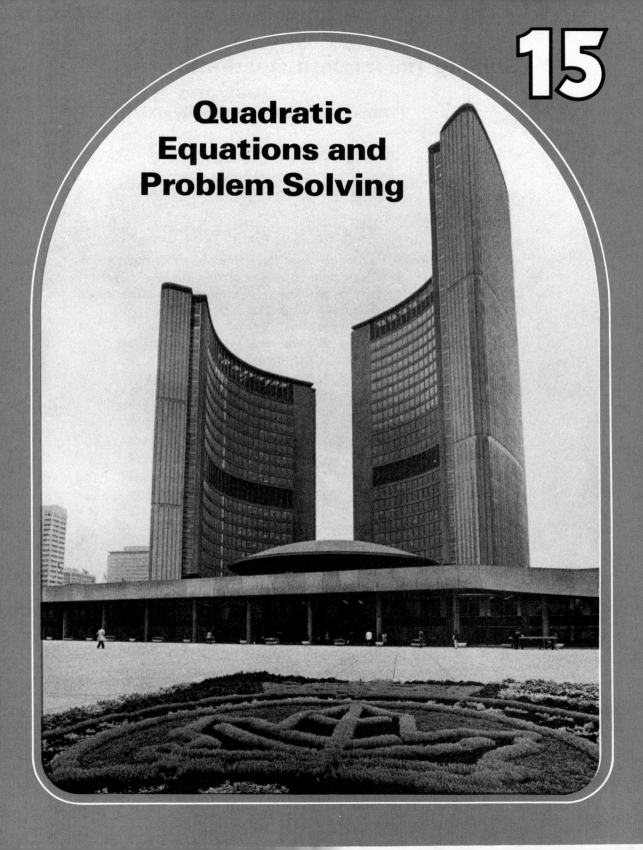

# Quadratic Equations and Problem Solving

**15**

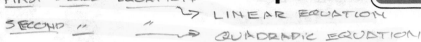

# Solving Incomplete Quadratic Equations

**I. Aim**     To solve an incomplete quadratic equation when the term containing the first degree of the unknown is missing.

## II. Procedure

1. Transform the given equation to an equivalent equation in which one member consists only of the square of the variable.
   a. If the equation is fractional, clear the equation of all fractions by multiplying by their least common denominator (LCD).
   b. If the equation contains parentheses, first remove the parentheses.
2. Divide both members by the coefficient of the variable.
3. Take the square root of both members, prefixing a $\pm$ sign to the root. Every quadratic equation has two answers.
4. If necessary, simplify the radicals in the roots.
5. Check each root by substituting it in the original equation.

## III. Sample Solutions:

1. Solve and check: $x^2 = 25$

   *Solution:*          *Check:*

   $x^2 = 25$         $x^2 = 25$                 $x^2 = 25$

   $x = \pm\sqrt{25}$      $(5)^2 = 25\,?$            $(-5)^2 = 25\,?$

   $x = \pm 5$         $25 = 25\ \checkmark$             $25 = 25\ \checkmark$

   *Answers:* $5,\ -5$

2. Solve and check: $x^2 = 24$

   *Solution:*          *Check:*

   $x^2 = 24$         $x^2 = 24$                 $x^2 = 24$

   $x = \pm\sqrt{24}$      $(2\sqrt{6})^2 = 24\,?$        $(-2\sqrt{6})^2 = 24\,?$

   $x = \pm\sqrt{4 \cdot 6}$     $24 = 24\ \checkmark$             $24 = 24\ \checkmark$

   $x = \pm 2\sqrt{6}$

   *Answers:* $2\sqrt{6},\ -2\sqrt{6}$

3. Solve and check: $x^2 = 23$

   *Solution:*          *Check:*

   $x^2 = 23$         $x^2 = 23$                 $x^2 = 23$

   $x = \pm\sqrt{23}$      $(\sqrt{23})^2 = 23\,?$        $(-\sqrt{23})^2 = 23\,?$

                             $23 = 23\ \checkmark$             $23 = 23\ \checkmark$

   *Answers:* $\sqrt{23},\ -\sqrt{23}$

**4.** Solve and check: $4x^2 - 27 = x^2$

*Solution:*

$$4x^2 - 27 = x^2$$
$$4x^2 - x^2 - 27 + 27 = x^2 - x^2 + 27$$
$$3x^2 = 27$$
$$\tfrac{1}{3} \cdot 3x^2 = \tfrac{1}{3} \cdot 27$$
$$x^2 = 9$$
$$x = \pm 3$$

*Answers:* $3, -3$

*Check:*

| | |
|---|---|
| $4x^2 - 27 = x^2$ | $4x^2 - 27 = x^2$ |
| $4(3)^2 - 27 = (3)^2\,?$ | $4(-3)^2 - 27 = (-3)^2\,?$ |
| $(4 \cdot 9) - 27 = 9$ | $4(9) - 27 = 9$ |
| $36 - 27 = 9$ | $36 - 27 = 9$ |
| $9 = 9 \; ✔$ | $9 = 9 \; ✔$ |

**5.** Solve and check: $6x^2 = 1$

*Solution:*       *Check:*

| | | |
|---|---|---|
| $6x^2 = 1$ | $6x^2 = 1$ | $6x^2 = 1$ |
| $x^2 = \tfrac{1}{6}$ | $6(\tfrac{1}{6}\sqrt{6})^2 = 1\,?$ | $6(-\tfrac{1}{6}\sqrt{6})^2 = 1\,?$ |
| $x = \pm\sqrt{\tfrac{1}{6}}$ | $6(\tfrac{1}{6}) = 1$ | $6(\tfrac{1}{6}) = 1$ |
| $x = \pm\tfrac{1}{6}\sqrt{6}$ | $1 = 1 \; ✔$ | $1 = 1 \; ✔$ |

*Answers:* $\tfrac{1}{6}\sqrt{6}, \; -\tfrac{1}{6}\sqrt{6}$

## DIAGNOSTIC TEST

Solve and check:

1. $x^2 = 9$      2. $x^2 = 48$      3. $x^2 = 15$

4. $6x^2 = 96$      5. $4x^2 = 100$      6. $x^2 = 40$

7. $5x^2 = 35$      8. $x^2 - 16 = 0$      9. $x^2 - 18 = 0$

10. $x^2 - 17 = 0$      11. $3x^2 + 2x^2 = 20$      12. $2x^2 + x^2 = 24$

13. $9x^2 - 5x^2 = 28$      14. $2x^2 + 9 = 81 - 4x^2$      15. $\dfrac{4 - x}{3} = \dfrac{4}{4 + x}$

16. $(x + 1)(x - 1) = 15$      17. $x^2 = \tfrac{4}{25}$      18. $9x^2 = 64$

19. $16x^2 - 49 = 0$      20. $x^2 = \tfrac{7}{8}$      21. $16x^2 = 3$

22. $5x^2 - 2 = 0$      23. $7x^2 + 3 = 3x^2 + 8$      24. $\dfrac{5x + 2}{8} = \dfrac{4}{5x - 2}$

## RELATED PRACTICE EXAMPLES

Solve and check:

1. **a.** $x^2 = 4$    **b.** $x^2 = 49$    **c.** $x^2 = 81$    **d.** $x^2 = 64$    **e.** $144 = x^2$

2. **a.** $x^2 = 32$    **b.** $x^2 = 75$    **c.** $x^2 = 20$    **d.** $x^2 = 54$    **e.** $80 = x^2$

3. **a.** $x^2 = 3$    **b.** $x^2 = 10$    **c.** $x^2 = 14$    **d.** $x^2 = 38$    **e.** $41 = x^2$

4. **a.** $3x^2 = 12$                           **b.** $7x^2 = 63$
   **c.** $6x^2 = 150$                   **d.** $5x^2 = 180$
   **e.** $405 = 5x^2$                  **f.** $2x^2 = 200$

5. **a.** $4x^2 = 36$                         **b.** $9x^2 = 144$
   **c.** $16x^2 = 64$                   **d.** $25x^2 = 400$
   **e.** $441 = 49x^2$                **f.** $36x^2 = 1$

6. **a.** $2x^2 = 24$                         **b.** $3x^2 = 135$
   **c.** $7x^2 = 189$                   **d.** $9x^2 = 360$
   **e.** $320 = 10x^2$              **f.** $5x^2 = 270$

7. **a.** $5x^2 = 10$                         **b.** $7x^2 = 42$
   **c.** $8x^2 = 88$                    **d.** $9x^2 = 90$
   **e.** $20 = 4x^2$                  **f.** $11x^2 = 77$

8. **a.** $x^2 - 1 = 0$                      **b.** $x^2 - 36 = 0$
   **c.** $x^2 - 81 = 0$                 **d.** $4x^2 - 100 = 0$
   **e.** $0 = x^2 - 9$                 **f.** $7x^2 - 175 = 0$

9. **a.** $x^2 - 27 = 0$                  **b.** $x^2 - 80 = 0$
   **c.** $x^2 - 72 = 0$                 **d.** $x^2 - 96 = 0$
   **e.** $0 = x^2 - 18$              **f.** $8x^2 - 104 = 0$

10. **a.** $x^2 - 2 = 0$                     **b.** $x^2 - 21 = 0$
   **c.** $x^2 - 34 = 0$                 **d.** $x^2 - 42 = 0$
   **e.** $0 = x^2 - 19$              **f.** $6x^2 - 18 = 0$

11. **a.** $x^2 + x^2 = 18$              **b.** $4x^2 + 3x^2 = 175$
   **c.** $8x^2 - 3x^2 = 320$         **d.** $288 = 2x^2 + 6x^2$

12. **a.** $5x^2 + x^2 = 48$            **b.** $2x^2 + 7x^2 = 180$
   **c.** $9x^2 - 5x^2 = 200$         **d.** $244 = 10x^2 - 2x^2$

13. **a.** $x^2 + 3x^2 = 20$            **b.** $4x^2 + 5x^2 = 63$
   **c.** $8x^2 - x^2 = 77$           **d.** $248 = x^2 + 7x^2$

14. **a.** $5x^2 - 128 = 3x^2$         **b.** $3x^2 = 220 - 2x^2$
   **c.** $4x^2 - 62 = 98 - 6x^2$    **d.** $3x^2 - 10 = x^2 + 214$

**15.** **a.** $\dfrac{x}{2} = \dfrac{2}{x}$                                    **b.** $\dfrac{5}{x} = \dfrac{x}{8}$

      **c.** $\dfrac{x+3}{2} = \dfrac{8}{x-3}$                     **d.** $\dfrac{6+x}{4} = \dfrac{5}{6-x}$

**16.** **a.** $(x+3)(x-3) = 18$               **b.** $(x-5)(x+5) = 39$

      **c.** $(x-4)(x+7) = 3x$                **d.** $(x-3)(x+2) + x = 0$

**17.** **a.** $x^2 = \frac{1}{4}$                                   **b.** $x^2 = \frac{16}{49}$

      **c.** $x^2 = \frac{25}{144}$                              **d.** $x^2 = \frac{81}{64}$

      **e.** $\frac{4}{9} = x^2$                                  **f.** $x^2 = \frac{121}{169}$

**18.** **a.** $16x^2 = 9$                            **b.** $25x^2 = 4$

      **c.** $9x^2 = 49$                             **d.** $100x^2 = 81$

      **e.** $144 = 121x^2$                          **f.** $36x^2 = 225$

**19.** **a.** $9x^2 - 4 = 0$                      **b.** $36x^2 - 25 = 0$

      **c.** $81x^2 - 64 = 0$                   **d.** $121x^2 = 144$

      **e.** $0 = 100x^2 - 49$                **f.** $49x^2 - 169 = 0$

**20.** **a.** $x^2 = \frac{1}{2}$                                  **b.** $x^2 = \frac{3}{4}$

      **c.** $x^2 = \frac{5}{6}$                                 **d.** $x^2 = \frac{2}{5}$

      **e.** $\frac{9}{10} = x^2$                                **f.** $x^2 = \frac{125}{18}$

**21.** **a.** $3x^2 = 1$                               **b.** $4x^2 = 5$

      **c.** $6x^2 = 3$                             **d.** $8x^2 = 9$

      **e.** $12 = 5x^2$                          **f.** $15x^2 = 32$

**22.** **a.** $2x^2 - 1 = 0$                      **b.** $9x^2 - 2 = 0$

      **c.** $8x^2 - 7 = 0$                    **d.** $12x^2 - 5 = 0$

      **e.** $0 = 4x^2 - 6$                    **f.** $20x^2 - 9 = 0$

**23.** **a.** $5x^2 + 3x^2 = 5$                **b.** $90x^2 - 49 = 9x^2$

      **c.** $4x^2 + 5 = x^2 + 6$            **d.** $6x^2 - 1 = 3 - 3x^2$

**24.** **a.** $\dfrac{3x+4}{3} = \dfrac{3}{3x-4}$           **b.** $\dfrac{4+x}{2x} = \dfrac{x}{4-x}$

      **c.** $\dfrac{x^2-3x}{3} = \dfrac{4x^2-5x-8}{5}$        **d.** $\dfrac{2-3x}{3x} - \dfrac{2x}{2+3x} = 0$

# Multiplicative Property of Zero

### I. Aim
To determine the values of the variable in two factors whose product is zero.

### II. Procedure
1. Note that the product of any number and zero is zero. The product is zero when:
   **a.** A non-zero number is multiplied by zero. $(0 \times 9 = 0)$
   **b.** Zero is multiplied by a non-zero number. $(5 \times 0 = 0)$
   **c.** Zero is multiplied by zero. $(0 \times 0 = 0)$
   Thus, if the product of two numbers is zero, then one of the factors is zero or both factors are zero.
2. To find the value of the variable that will make the value of the given expression zero, write the expression equal to zero and solve the resulting equation.
3. To determine the values of the variable in two factors whose product is zero, write each expression equal to zero and solve the resulting equation.

### III. Sample Solutions:
1. What value of $x$ will make the value of the expression $x + 7$ zero?
   *Solution:*
   $$x + 7 = 0$$
   $$x + 7 - 7 = 0 - 7$$
   $$x = -7$$

   *Answer:* $-7$

2. What values of the variable make the sentence $(x - 4)(x + 9) = 0$ true?

   **a.** *Solution:*
   $$x - 4 = 0$$
   $$x - 4 + 4 = 0 + 4$$
   $$x = 4$$

   *Check:*
   $$(x - 4)(x + 9) = 0$$
   $$(4 - 4)(4 + 9) = 0\,?$$
   $$(0)(13) = 0$$
   $$0 = 0 \checkmark$$

   **b.** *Solution:*
   $$x + 9 = 0$$
   $$x + 9 - 9 = 0 - 9$$
   $$x = -9$$

   *Check:*
   $$(x - 4)(x + 9) = 0$$
   $$(-9 - 4)(-9 + 9) = 0\,?$$
   $$(-13)(0) = 0$$
   $$0 = 0 \checkmark$$

   *Answers:* 4 and $-9$

## DIAGNOSTIC TEST

**1. a.** If $x \cdot y = 0$ and $y = 10$, what is the value of $x$?
   **b.** If $m \cdot n = 20$, can $m = 4$? Can $m = 0$? Can $n = 10$? Can $n = 0$?
**2.** What is the value of $(x + 12)(2x - 9)$ when $x = -12$? When $x = 4\frac{1}{2}$?
**3.** Find the value of $y$ that will make the expression $(3y - 5)$ zero.
**4.** Find the values of $n$ which make each of the following factors zero: $(n - 6)(3n + 5)$.
**5.** What values of $x$ make the sentence $(x + 10)(x - 2) = 0$ true?

## RELATED PRACTICE EXAMPLES

1. **a.** If $m \cdot n = 0$ and $n = 5$, what is the value of $m$?
   **b.** If $r \cdot s = 0$ and $r = 10$, what is the value of $s$?
   **c.** If $b \cdot c = 0$ and $b = 0$, what is the value of $c$? Can $c = 0$?
   **d.** If $x \cdot y = 0$ and $y = 0$, what is the value of $x$? Can $x = 0$?
   **e.** If $c \cdot d = 30$, can $c = 5$? Can $c = 0$? Can $d = 15$? Can $d = 0$?
   **f.** If $m \cdot x = 24$, can $m = 6$? Can $m = 0$? Can $x = 12$? Can $x = 0$?

2. What is the value of:
   **a.** $n - 2$ when $n = 2$?
   **b.** $x + 6$ when $x = -6$?
   **c.** $5(b - 4)$ when $b = 4$?
   **d.** $(a - 7)(a + 3)$ when $a = 7$? when $a = -3$?
   **e.** $(2x - 5)(3x + 2)$ when $x = 2\frac{1}{2}$? when $x = -\frac{2}{3}$?
   **f.** $y(y - 6)$ when $y = 0$?

3. In each of the following, find the value of the variable which makes the expression zero:
   **a.** $(n - 2)$  **b.** $(c + 8)$  **c.** $(4t - 2)$  **d.** $(6t + 1)$  **e.** $(5m - 12)$

4. In each of the following, find the values of the variable which make each of the factors zero:
   **a.** $(r - 7)(r - 11)$   **b.** $(b + 1)(b + 10)$   **c.** $(n - 5)(n + 8)$
   **d.** $(2x - 3)(x + 4)$   **e.** $(3n - 9)(5n + 2)$   **f.** $(8w + 6)(9w - 15)$

5. What values of the variable make each of the following sentences true?
   **a.** $(c - 2)(c - 3) = 0$        **b.** $(n + 18)(n + 4) = 0$
   **c.** $(x - 13)(x + 5) = 0$       **d.** $(10y - 7)(2y + 7) = 0$
   **e.** $(3b - 12)(8b - 4) = 0$     **f.** $(9x + 8)(12x - 15) = 0$

# Solving Quadratic Equations by Factoring

## I. Aim

To solve quadratic equations by factoring when one of the members is a binomial or trinomial which can be factored and the other member is zero.

## II. Procedure

1. If necessary, transform the given equation to an equivalent equation with all the terms on one side and zero on the other side.
   **a.** If the equation is fractional, first clear the equation of all fractions.
   **b.** If the equation contains parentheses, first remove the parentheses.
2. Combine like terms.
3. Factor the polynomial to obtain two factors.
4. Write each factor containing the variable equal to zero.
5. Solve each of the resulting equations.
6. Check each root by substituting it in the given equation.
7. Note that when the polynomial is a perfect trinomial square, its two factors are identical, producing the same root twice. This *double root* is written only once in the answer.
8. When solving a quadratic equation that has the constant term missing, like $x^2 = 7x$, do not divide by the variable. See sample solution 3.

## III. Sample Solutions

1. Solve $x^2 - 3x - 10 = 0$ by factoring and check.
   *Solution:*
$$x^2 - 3x - 10 = 0$$
$$(x - 5)(x + 2) = 0$$
   Thus, $(x - 5) = 0$ or $x + 2 = 0$
   If $\quad x - 5 = 0 \qquad$ If $\quad x + 2 = 0$
   then $\quad x = 5 \qquad\quad$ then $\quad x = -2$

*Answers:* 5 and $-2$

   *Check:*

| $x^2 - 3x - 10 = 0$ | $x^2 - 3x - 10 = 0$ |
|---|---|
| $(5)^2 - (3 \cdot 5) - 10 = 0\,?$ | $(-2)^2 - (3 \cdot [-2]) - 10 = 0\,?$ |
| $25 - 15 - 10 = 0$ | $4 + 6 - 10 = 0$ |
| | $0 = 0$ ✔ |

572

VOCAB P. 143
FACTORING CHAPTER 9

**2.** Solve $x^2 - 10x + 25 = 0$ by factoring and check.

*Solution:*
$$x^2 - 10x + 25 = 0$$
$$(x - 5)(x - 5) = 0$$

Thus, $(x - 5) = 0$   or   $(x - 5) = 0$

If $\qquad x - 5 = 0$

then $\qquad x = 5$

*Check:*
$$x^2 - 10x + 25 = 0$$
$$(5)^2 - (10 \cdot 5) + 25 = 0?$$
$$25 - 50 + 25 = 0$$
$$0 = 0 ✔$$

*Answer:* Double root: 5

**3.** Solve: $x^2 = 7x$

*Solution:*
$$x^2 = 7x$$
$$x^2 - 7x = 7x - 7x$$
$$x^2 - 7x = 0$$
$$x(x - 7) = 0$$

Not this:
$$x^2 = 7x$$
$$\frac{x^2}{x} = \frac{7x}{x}$$
$$x = 7$$

Thus, $x = 0$   or   $x - 7 = 0$

If $\quad x - 7 = 0$

then $\qquad x = 7$

*Answers:* 0 and 7

## DIAGNOSTIC TEST

Solve by factoring and check:

**1.** $x^2 - 5x + 6 = 0$      **2.** $x^2 + 12x + 35 = 0$      **3.** $x^2 + x - 12 = 0$

**4.** $x^2 - 1 = 0$      **5.** $4x^2 - 25 = 0$      **6.** $x^2 - 7x = 0$

**7.** $x^2 + 2x = 0$      **8.** $3x^2 - 2x = 0$      **9.** $4x^2 + x = 0$

**10.** $x^2 - 8x + 16 = 0$      **11.** $x^2 + 24x + 144 = 0$      **12.** $x^2 + 7x = 18$

**13.** $3x^2 - 3x = x^2 + 4x$      **14.** $81x^2 = 49$      **15.** $18x^2 - 5 = 2x^2 + 20$

**16.** $4x^2 + 6x = 3x^2 - 8$      **17.** $x(x + 5) = -4$      **18.** $2x^2 + 7x + 6 = 0$

**19.** $\dfrac{x}{3} - 1 = \dfrac{6}{x}$      **20.** $x^2 = 8x$

## *RELATED PRACTICE EXAMPLES*

Solve by factoring and check:

1. **a.** $x^2 - 7x + 12 = 0$   **b.** $x^2 - 12x + 32 = 0$   **c.** $0 = x^2 - 10x + 16$
2. **a.** $x^2 + 14x + 45 = 0$   **b.** $x^2 + 6x + 8 = 0$   **c.** $0 = x^2 + 16x + 63$
3. **a.** $x^2 + 3x - 18 = 0$   **b.** $x^2 - 4x - 12 = 0$   **c.** $x^2 + 15x - 54 = 0$
4. **a.** $x^2 - 9 = 0$   **b.** $x^2 - 64 = 0$   **c.** $x^2 - 144 = 0$
5. **a.** $9x^2 - 16 = 0$   **b.** $36x^2 - 25 = 0$   **c.** $0 = 81x^2 - 100$
6. **a.** $x^2 - 6x = 0$   **b.** $x^2 - 3x = 0$   **c.** $0 = x^2 - x$
7. **a.** $x^2 + 3x = 0$   **b.** $x^2 + 5x = 0$   **c.** $x^2 + 10x = 0$
8. **a.** $2x^2 - 5x = 0$   **b.** $3x^2 - 7x = 0$   **c.** $8x^2 - 4x = 0$
9. **a.** $5x^2 + 3x = 0$   **b.** $4x^2 + 6x = 0$   **c.** $0 = 7x^2 + 10x$
10. **a.** $x^2 - 16x + 64 = 0$   **b.** $x^2 - 4x + 4 = 0$   **c.** $x^2 - 20x + 100 = 0$
11. **a.** $x^2 + 6x + 9 = 0$   **b.** $x^2 + 14x + 49 = 0$
    **c.** $0 = x^2 + 2x + 1$
12. **a.** $x^2 - 3x = 40$   **b.** $x^2 + 15 = 8x$
    **c.** $10x = 24 - x^2$
13. **a.** $2x^2 - 3x = x^2 - 6x$   **b.** $5x^2 - 8x = 4x^2 + x$
    **c.** $8x^2 + 7x = 6x - 4x^2$
14. **a.** $4x^2 = 1$   **b.** $25x^2 = 49$
    **c.** $16 = 49x^2$
15. **a.** $11x^2 - 3 = 2x^2 + 1$   **b.** $38x^2 - 2 = 7 - 11x^2$
    **c.** $98x^2 + 4 = 5 - 46x^2$
16. **a.** $4x^2 - x = 3x^2 + 2$   **b.** $x^2 - 3x = 4x - 12$
    **c.** $x^2 + 16x + 10 = 46$
17. **a.** $x(x + 4) = 12$   **b.** $x(x - 3) = 2(10 - x)$
    **c.** $x(x - 5) = 5(x - 5)$
18. **a.** $3x^2 + 5x + 2 = 0$   **b.** $8x^2 - 22x - 21 = 0$
    **c.** $15x^2 - 20x = 11x - 10$   **d.** $6x^2 - 19x + 8 = 0$
    **e.** $8x^2 - 6x - 5 = 0$   **f.** $15x^2 + 14x = 8$
    **g.** $18x^2 - 69x + 56 = 0$   **h.** $4x^2 + 35x + 24 = 0$
    **i.** $12x^2 = 63 - x$
19. **a.** $x = \dfrac{10}{x - 3}$   **b.** $\dfrac{x}{2} - 2 = \dfrac{6}{x}$

    **c.** $\dfrac{2}{x} + 4 = x - \dfrac{3}{x}$   **d.** $\dfrac{4}{x + 5} = \dfrac{x}{x + 3}$

    **e.** $\dfrac{x - 5}{x + 1} = \dfrac{2}{x - 4}$   **f.** $\dfrac{1}{x + 3} = \dfrac{3 - x}{7 + x}$

20. **a.** $x^2 = 2x$   **b.** $x^2 = x$
    **c.** $x^2 = -3x$   **d.** $x^2 = -9x$

# Solving Quadratic Equations by Completing the Square

## I. Aim    To solve quadratic equations by completing the square.

## II. Procedure

1. Transform the given equation, if necessary, to an equivalent equation in which the terms containing the variable form one member and the constant term is the other member.

2. If the coefficient of the variable squared term ($x^2$) is a number other than 1 or its equivalent, divide each term of the given quadratic by this coefficient.

3. Add to both members the square of one-half of the coefficient of the variable ($x$). This is a fractional number when the coefficient of $x$ is an odd number.

4. Factor the perfect square trinomial in the left member. Simplify the right member. See Exercise 9-9 on page 377.

5. Find the square root of both members. Prefix a $\pm$ sign to the square root of the right member.

6. Solve the two resulting equations.

7. Check each root by substituting it in the given equation.

## III. Sample Solutions

1. Solve $x^2 + 6x = 16$ by completing the square and check.

   *Solution:*    $x^2 + 6x = 16$

   $$x^2 + 6x + 9 = 16 + 9$$
   $$(x + 3)^2 = 25$$
   $$x + 3 = \pm 5$$

   Thus, $x + 3 = 5$      or      $x + 3 = -5$

   | When $x + 3 = 5$ | When $x + 3 = -5$ |
   |---|---|
   | $x + 3 - 3 = 5 - 3$ | $x + 3 - 3 = -5 - 3$ |
   | $x = 2$ | $x = -8$ |

   *Check:*

   | $x^2 + 6x = 16$ | $x^2 + 6x = 16$ |
   |---|---|
   | $(2)^2 + (6 \cdot 2) = 16\,?$ | $(-8)^2 + (6 \cdot [-8]) = 16\,?$ |
   | $4 + 12 = 16$ | $64 - 48 = 16$ |
   | $16 = 16\ \checkmark$ | $16 = 16\ \checkmark$ |

*Answers:* 2 and $-8$

**2.** Solve $3x^2 - 7x - 6 = 0$ by completing the square and check.

*Solution:*

$$3x^2 - 7x - 6 = 0$$
$$3x^2 - 7x = 6$$
$$x^2 - \tfrac{7}{3}x = 2$$
$$x^2 - \tfrac{7}{3}x + (\tfrac{7}{6})^2 = 2 + (\tfrac{7}{6})^2$$
$$x^2 - \tfrac{7}{3}x + \tfrac{49}{36} = 2 + \tfrac{49}{36}$$
$$(x - \tfrac{7}{6})^2 = \tfrac{121}{36}$$
$$x - \tfrac{7}{6} = \pm\tfrac{11}{6}$$

Thus, $x - \tfrac{7}{6} = \tfrac{11}{6}$  or  $x - \tfrac{7}{6} = -\tfrac{11}{6}$

When $x - \tfrac{7}{6} = \tfrac{11}{6}$

$$x - \tfrac{7}{6} + \tfrac{7}{6} = \tfrac{11}{6} + \tfrac{7}{6}$$
$$x = \tfrac{18}{6}$$
$$x = 3$$

When $x - \tfrac{7}{6} = -\tfrac{11}{6}$

$$x - \tfrac{7}{6} + \tfrac{7}{6} = -\tfrac{11}{6} + \tfrac{7}{6}$$
$$x = -\tfrac{4}{6}$$
$$x = -\tfrac{2}{3}$$

*Check:*

$$3x^2 - 7x - 6 = 0$$
$$3(3)^2 - (7 \cdot 3) - 6 = 0\,?$$
$$3 \cdot 9 - 21 - 6 = 0$$
$$27 - 21 - 6 = 0$$
$$0 = 0 \;\checkmark$$

$$3x^2 - 7x - 6 = 0$$
$$3(-\tfrac{2}{3})^2 - (7 \cdot [-\tfrac{2}{3}]) - 6 = 0\,?$$
$$3(\tfrac{4}{9}) + \tfrac{14}{3} - 6 = 0$$
$$\tfrac{4}{3} + \tfrac{14}{3} - 6 = 0$$
$$0 = 0 \;\checkmark$$

*Answers:* 3 and $-\tfrac{2}{3}$

## DIAGNOSTIC TEST

Solve by factoring the left member of each of the following equations and check:

**1.** $x^2 - 16x + 64 = 121$      **2.** $x^2 + 10x + 25 = 9$

**3.** $x^2 - 20x + 100 = 20$      **4.** $x^2 - 5x + \tfrac{25}{4} = \tfrac{81}{100}$

Add a term to each of the following expressions to make a perfect square trinomial:

**5.** $x^2 + 12x + ?$      **6.** $x^2 - 50x + ?$      **7.** $x^2 + 7x + ?$

**8.** $x^2 - 19x + ?$      **9.** $x^2 + \tfrac{1}{3}x + ?$      **10.** $x^2 - \tfrac{4}{5}x + ?$

Solve by completing the square and check:

**11.** $x^2 + 4x = 12$      **12.** $x^2 + 8x = -12$      **13.** $x^2 - 10x = 39$

**14.** $x^2 - 16x = -48$      **15.** $x^2 + 3x = 10$      **16.** $x^2 - 5x = 24$

**17.** $x^2 + 2x - 24 = 0$      **18.** $x^2 - 8x - 20 = 0$      **19.** $x^2 - x = \tfrac{3}{4}$

**20.** $x^2 + \tfrac{3}{2}x = 1$      **21.** $3x^2 + 3x - 168 = 0$      **22.** $4x^2 - x = 6$

## *RELATED PRACTICE EXAMPLES*

Solve by factoring the left member of each of the following equations and check:

1. **a.** $x^2 - 14x + 49 = 100$
   **b.** $x^2 - 18x + 81 = 64$
   **c.** $x^2 - 24x + 144 = 400$

2. **a.** $x^2 + 4x + 4 = 25$
   **b.** $x^2 + 20x + 100 = 81$
   **c.** $x^2 + 2x + 1 = 49$

3. **a.** $x^2 - 16x + 64 = 20$
   **b.** $x^2 + 8x + 16 = 32$
   **c.** $x^2 - 10x + 25 = 15$

4. **a.** $x^2 + x + \frac{1}{4} = \frac{9}{16}$
   **b.** $x^2 - 3x + \frac{9}{4} = \frac{49}{64}$
   **c.** $x^2 - \frac{2}{3}x + \frac{1}{9} = \frac{5}{6}$

Add a term to each of the following expressions to make a perfect square trinomial:

5. **a.** $x^2 + 8x + ?$   **b.** $x^2 + 14x + ?$
   **c.** $x^2 + 30x + ?$   **d.** $x^2 + 48x + ?$

6. **a.** $x^2 - 10x + ?$   **b.** $x^2 - 22x + ?$
   **c.** $x^2 - 28x + ?$   **d.** $x^2 - 40x + ?$

7. **a.** $x^2 + 5x + ?$   **b.** $x^2 + 11x + ?$
   **c.** $x^2 + x + ?$   **d.** $x^2 + 17x + ?$

8. **a.** $x^2 - 9x + ?$   **b.** $x^2 - 3x + ?$
   **c.** $x^2 - 15x + ?$   **d.** $x^2 - x + ?$

9. **a.** $x^2 + \frac{1}{2}x + ?$   **b.** $x^2 + \frac{2}{5}x + ?$
   **c.** $x^2 + \frac{5}{8}x + ?$   **d.** $x^2 + \frac{4}{9}x + ?$

10. **a.** $x^2 - \frac{2}{3}x + ?$   **b.** $x^2 - \frac{3}{4}x + ?$
    **c.** $x^2 - \frac{7}{10}x + ?$   **d.** $x^2 - \frac{5}{6}x + ?$

Solve by completing the square and check:

11. **a.** $x^2 + 2x = 8$    **b.** $x^2 + 8x = 20$    **c.** $x^2 + 10x = 24$
12. **a.** $x^2 + 8x = -15$    **b.** $x^2 + 6x = -5$    **c.** $x^2 + 18x = -56$
13. **a.** $x^2 - 10x = 11$    **b.** $x^2 - 16x = 80$    **c.** $x^2 - 8x = 84$
14. **a.** $x^2 - 6x = -5$    **b.** $x^2 - 12x = -20$    **c.** $x^2 - 22x = -72$
15. **a.** $x^2 + 3x = 18$    **b.** $x^2 + 7x = 8$    **c.** $x^2 + x = 20$
16. **a.** $x^2 - 5x = 36$    **b.** $x^2 - 11x = 42$    **c.** $x^2 - 3x = 70$
17. **a.** $x^2 + 10x + 9 = 0$    **b.** $x^2 + 4x - 32 = 0$    **c.** $x^2 + 20x + 36 = 0$
18. **a.** $x^2 - 12x + 27 = 0$    **b.** $x^2 + 26x + 48 = 0$    **c.** $x^2 - 22x - 75 = 0$
19. **a.** $x^2 + 2x = -\frac{3}{4}$    **b.** $x^2 - x = -\frac{2}{9}$    **c.** $x^2 + 3x + \frac{20}{9} = 0$
20. **a.** $x^2 + \frac{x}{2} = 3$    **b.** $x^2 - \frac{7}{2}x = -\frac{3}{2}$    **c.** $x^2 + \frac{11}{3}x + 2 = 0$
21. **a.** $7x^2 - 7x = 140$    **b.** $2x^2 + 48x + 126 = 0$    **c.** $5x^2 - 80x + 240 = 0$
22. **a.** $2x^2 - 7x = 4$    **b.** $3x^2 - 4x = -1$    **c.** $4x^2 - 13x - 12 = 0$

# Solving Quadratic Equations by Formula

## I. Aim  To solve quadratic equations by formula.

## II. Procedure

**1.** Transform, if necessary, the given equation into an equivalent equation in which the terms are so arranged that they correspond to the general quadratic equation $ax^2 + bx + c = 0$.

**a.** If the equation is fractional, first clear the equation of all fractions.

**b.** If the equation contains parentheses, first remove the parentheses.

**2.** Compare the equation to the general quadratic equation $ax^2 + bx + c = 0$ and find the numerical values of $a$, $b$, and $c$.

Note: Since $5x^2 - 4 = 0$ may be thought of as $5x^2 + 0x - 4 = 0$ and $2x^2 - 3x = 0$ as $2x^2 - 3x + 0 = 0$, use zero (0) as the $b$ value or $c$ value in the corresponding missing terms.

**3.** Substitute these numerical values for $a$, $b$, and $c$ in the formula $x = \dfrac{-b \pm \sqrt{b^2 - 4ac}}{2a}$ which indicates the values of the variable in terms of the coefficients and constant term of the given quadratic equation.

**4.** Perform all the necessary operations to determine the two roots. Simplify when possible.

**5.** Sometimes the two are identical, a double root.

**6.** When the roots are irrational, express them in simplest form.

**7.** To find irrational roots correct to the nearest hundredth, replace the radical by its decimal equivalent and perform the necessary operations.

**8.** Check each root by substituting it in the given equation.

**9.** By solving the general quadratic equation $ax^2 + bx + c = 0$ for the variable $x$ in terms of $a$, $b$, and $c$, using the method of completing the square, develop the formula (above step 3) that is used to solve quadratic equations in one variable. See top of page 579 for this development.

$$ax^2 + bx + c = 0$$
$$ax^2 + bx = -c$$
$$x^2 + \frac{b}{a}x = -\frac{c}{a}$$
$$x^2 + \frac{b}{a}x + \left(\frac{b}{2a}\right)^2 = \left(\frac{b}{2a}\right)^2 - \frac{c}{a}$$

$$\frac{b \pm \sqrt{b^2 - 4ac}}{2a}$$

$$x^2 + \frac{b}{a}x + \frac{b^2}{4a^2} = \frac{b^2}{4a^2} - \frac{c}{a}$$

$$\left(x + \frac{b}{2a}\right)^2 = \frac{b^2 - 4ac}{4a^2}$$

$$x + \frac{b}{2a} = \pm\frac{\sqrt{b^2 - 4ac}}{2a}$$

$$x = -\frac{b}{2a} \pm \frac{\sqrt{b^2 - 4ac}}{2a}$$

$$x = \frac{-b \pm \sqrt{b^2 - 4ac}}{2a}$$

The formula $x = \dfrac{-b \pm \sqrt{b^2 - 4ac}}{2a}$ represents $x = \dfrac{-b + \sqrt{b^2 - 4ac}}{2a}$

and $x = \dfrac{-b - \sqrt{b^2 - 4ac}}{2a}$. The $\pm$ symbol is read "plus or minus."

## III. Sample Solutions

1. Solve $x^2 + 4x + 3 = 0$ by formula and check.
   *Solution:*

   $a = 1$      $x = \dfrac{-b \pm \sqrt{b^2 - 4ac}}{2a}$

   $b = 4$

   $c = 3$      $x = \dfrac{-4 \pm \sqrt{16 - 12}}{2}$

   $\qquad\qquad x = \dfrac{-4 \pm \sqrt{4}}{2}$

   $\qquad\qquad x = \dfrac{-4 \pm 2}{2}$

   $x = \dfrac{-4 + 2}{2}$  or  $x = \dfrac{-4 - 2}{2}$

   $x = \dfrac{-2}{2}$        $x = \dfrac{-6}{2}$

   $x = -1$         $x = -3$

   *Check:*

   | | |
   |---|---|
   | $x^2 + 4x + 3 = 0$ | $x^2 + 4x + 3 = 0$ |
   | $(-1)^2 + (4 \cdot [-1]) + 3 = 0\,?$ | $(-3)^2 + (4 \cdot [-3]) + 3 = 0\,?$ |
   | $1 - 4 + 3 = 0$ | $9 - 12 + 3 = 0$ |
   | $0 = 0\ ✔$ | $0 = 0\ ✔$ |

*Answers:* $-1$ and $-3$

**2.** Solve $2x^2 - 4x = 1$ by formula and check:

*Solution:*                                    *Check:*

$a = 2 \qquad x = \dfrac{-b \pm \sqrt{b^2 - 4ac}}{2a}$ $\qquad 2\left(\dfrac{2 + \sqrt{6}}{2}\right)^2 - 4\left(\dfrac{2 + \sqrt{6}}{2}\right) = 1$

$b = -4$

$c = -1 \qquad x = \dfrac{4 \pm \sqrt{16 + 8}}{4}$ $\qquad 5 + 2\sqrt{6} - 4 - 2\sqrt{6} = 1\,?$

$\qquad\qquad\qquad\qquad\qquad\qquad\qquad\qquad 1 = 1\;\checkmark$

$\qquad\qquad x = \dfrac{4 \pm \sqrt{24}}{4}$ $\qquad 2\left(\dfrac{2 - \sqrt{6}}{2}\right)^2 - 4\left(\dfrac{2 - \sqrt{6}}{2}\right) = 1$

$\qquad\qquad x = \dfrac{4 \pm 2\sqrt{6}}{4}$ $\qquad 5 - 2\sqrt{6} - 4 + 2\sqrt{6} = 1\,?$

$\qquad\qquad\qquad\qquad\qquad\qquad\qquad\qquad 1 = 1\;\checkmark$

$\qquad\qquad x = \dfrac{2(2 \pm \sqrt{6})}{4}$

$\qquad\qquad x = \dfrac{2 \pm \sqrt{6}}{2}$ $\qquad$ *Answer:* $x = \dfrac{2 \pm \sqrt{6}}{2}$

Note: To find irrational roots correct to the nearest hundredth, replace radical by decimal equivalent and perform the necessary operations.

$x = \dfrac{2 + \sqrt{6}}{2} = \dfrac{2 + 2.449}{2} = \dfrac{4.449}{2} = 2.22$

$x = \dfrac{2 - \sqrt{6}}{2} = \dfrac{2 - 2.449}{2} = \dfrac{-.449}{2} = -.22$

The roots of sample solution 2 are to the right.

## DIAGNOSTIC TEST

1. What numerical values in $5x^2 - 4x - 7 = 0$ correspond to $a$, $b$, and $c$ of the general quadratic equation $ax^2 + bx + c = 0$?

Solve by formula and check:

| | | |
|---|---|---|
| **2.** $x^2 + 7x + 12 = 0$ | **3.** $x^2 - 8x + 12 = 0$ | **4.** $x^2 + 2x - 15 = 0$ |
| **5.** $x^2 - 4x - 32 = 0$ | **6.** $3x^2 - 8x + 4 = 0$ | **7.** $x^2 + 2x = 8$ |
| **8.** $x^2 - 4x + 4 = 0$ | **9.** $x^2 + 3x = 0$ | **10.** $9x^2 - 4 = 0$ |
| **11.** $x(x + 2) + 2(x - 2) = 1$ | **12.** $x^2 - 2 = 0$ | **13.** $2x^2 - 16 = 0$ |
| **14.** $4x^2 - 3 = 0$ | **15.** $x^2 - 3x - 2 = 0$ | **16.** $x^2 + x - 11 = 0$ |
| **17.** $3x^2 + 4x - 1 = 0$ | **18.** $\dfrac{x}{6} - \dfrac{1}{3} = \dfrac{20}{x}$ | |

Find roots correct to nearest hundredth:

**19.** $3x^2 - 6x + 2 = 0$ $\qquad$ **20.** $x^2 + .2x - .08 = 0$

## RELATED PRACTICE EXAMPLES

1. Find the numerical values corresponding to $a$, $b$, and $c$ of the general quadratic equation $ax^2 + bx + c = 0$ for each of the following quadratic equations:

   **a.** $6x^2 + 9x + 5 = 0$     **b.** $x^2 - 3x + 7 = 0$     **c.** $10x^2 - x - 6 = 0$

   **d.** $9x^2 + 11x - 2 = 0$     **e.** $x^2 - 4x - 9 = 0$     **f.** $4x^2 + 6x - 5 = 0$

Solve by formula and check:

2. **a.** $x^2 + 6x + 5 = 0$     **b.** $x^2 + 9x + 20 = 0$     **c.** $x^2 + 12x + 35 = 0$
3. **a.** $x^2 - 9x + 8 = 0$     **b.** $x^2 - 8x + 15 = 0$     **c.** $x^2 - 12x + 27 = 0$
4. **a.** $x^2 + 2x - 3 = 0$     **b.** $x^2 + 4x - 12 = 0$     **c.** $x^2 + 5x - 14 = 0$
5. **a.** $x^2 - 3x - 4 = 0$     **b.** $x^2 - 9x - 36 = 0$     **c.** $x^2 - 4x - 12 = 0$
6. **a.** $2x^2 + 3x - 20 = 0$     **b.** $3x^2 + 5x + 2 = 0$     **c.** $5x^2 - 13x + 6 = 0$
   **d.** $2x^2 - x - 6 = 0$     **e.** $6x^2 - 17x + 12 = 0$     **f.** $4x^2 + 4x - 8 = 0$
7. **a.** $x^2 + 8x = -12$     **b.** $6x^2 - 13x = 5$     **c.** $3x^2 + 6x = 24$
8. **a.** $x^2 + 10x + 25 = 0$     **b.** $x^2 - 14x + 49 = 0$     **c.** $x^2 - 8x + 16 = 0$
9. **a.** $x^2 - 4x = 0$     **b.** $2x^2 - 7x = 0$     **c.** $6x^2 + 5x = 0$
10. **a.** $x^2 - 49 = 0$     **b.** $4x^2 - 1 = 0$     **c.** $16x^2 - 9 = 0$
11. **a.** $2x^2 - 3x + 6 = x^2 + 2x$     **b.** $x^2 - 3(x + 7) = x$     **c.** $(x + 2)^2 = 2(5x - 2)$
12. **a.** $x^2 - 6 = 0$     **b.** $2x^2 - 14 = 0$
    **c.** $4x^2 - 40 = 0$
13. **a.** $x^2 - 12 = 0$     **b.** $3x^2 - 60 = 0$
    **c.** $2x^2 - 36 = 0$
14. **a.** $4x^2 - 5 = 0$     **b.** $5x^2 - 9 = 0$
    **c.** $8x^2 - 3 = 0$
15. **a.** $x^2 + 3x - 1 = 0$     **b.** $2x^2 + 9x + 3 = 0$
    **c.** $4x^2 + x - 1 = 0$
16. **a.** $x^2 + 7x + 1 = 0$     **b.** $5x^2 + 5x - 1 = 0$
    **c.** $2x^2 + 7x - 13 = 0$
17. **a.** $x^2 + 3x - 9 = 0$     **b.** $3x^2 - 6x + 1 = 0$
    **c.** $4x^2 + 2x - 3 = 0$
18. **a.** $\dfrac{x^2}{2} + \dfrac{3x}{4} = 11$     **b.** $\dfrac{x}{4} + \dfrac{1}{2} = \dfrac{2}{x}$

    **c.** $\dfrac{x}{2x - 1} = \dfrac{2x + 3}{15}$

*Howdy Greg, howz your day going? My is not going at all. Have a fun time in Alb. i Like Mom says " get your homework donbe. "*

Solve by formula and check. Find roots correct to nearest hundredth:

19. **a.** $x^2 - 5x + 3 = 0$       20. **a.** $x^2 - 1.7x - .6 = 0$
    **b.** $2x^2 + 5x + 1 = 0$          **b.** $.16x^2 + 1.6x - 12 = 0$
    **c.** $3x^2 + 4x - 3 = 0$          **c.** $.2x^2 - 1.75x + 1.2 = 0$

I. Aim    To determine the value of the variable that is indicated as squared in the given formula when the values of all the other variables are known.

II. Procedure
1. Copy the formula.
2. Substitute the given values for the variables.
3. Perform the necessary operations.
4. Solve the resulting equation for the value or values of the required variable.
5. Since a quadratic equation has two roots, discard any value that is meaningless.

III. Sample Solution

Find the value of $I$ in the formula $W = I^2R$ when $W = 100$ and $R = 25$.

$$W = I^2R$$
$$100 = I^2 \cdot 25$$
$$100 = 25I^2$$
$$4 = I^2$$
$$2 = I^* \qquad \text{* The negative root is meaningless.}$$
$$I = 2$$

*Answer: $I = 2$*

---

**DIAGNOSTIC TEST**

Find the value of:
1. $V$ when $K = 2{,}700$ and $m = 6$, using the formula $K = \frac{1}{2}mV^2$
2. $t$ when $s = 64$ and $v = 80$, using the formula $s = vt - 16t^2$
3. The formula $s = 16t^2$ may be used to find the distance ($s$) in feet that an object falls in a given number of seconds ($t$) or the time the object requires to fall a given distance.
   a. How many feet does an object fall in 3 seconds? 5 seconds? 11 seconds?
   b. How many seconds does it take an object to fall 256 feet? 1,296 feet? 3,600 feet?

### *RELATED PRACTICE EXAMPLES*

Find the value of:

**1. a.** $s$ when $A = 36$, using the formula $A = s^2$
   **b.** $s$ when $A = 150$, using the formula $A = 6s^2$
   **c.** $t$ when $s = 784$, using the formula $s = 16t^2$
   **d.** $d$ when $A = 1,963.5$, using the formula $A = .7854d^2$
   **e.** $I$ when $W = 72$ and $R = 8$, using the formula $W = I^2R$
   **f.** $r$ when $\pi = 3.14$ and $A = 50.24$, using the formula $A = \pi r^2$
   **g.** $r$ when $\pi = \frac{22}{7}$ and $A = 1,232$, using the formula $A = 4\pi r^2$
   **h.** $r$ when $V = 62.8$, $\pi = 3.14$, and $h = 5$, using the formula $V = \pi r^2 h$
   **i.** $t$ when $s = 400$ and $g = 32$, using the formula $s = \frac{1}{2}gt^2$
   **j.** $V$ when $K = 1,024$ and $m = 8$, using the formula $K = \frac{1}{2}mV^2$

**2. a.** $v$ when $F = 3$, $m = 8$, $g = 32$, and $r = 3$, using the formula $F = \dfrac{mv^2}{gr}$

   **b.** $b$ when $h = 25$ and $a = 20$, using the formula $h^2 = a^2 + b^2$

   **c.** $r$ when $A = 1,496$, $\pi = \frac{22}{7}$, and $l = 20$, using the formula $A = \pi r^2 + \pi rl$

   **d.** $t$ when $s = 24$ and $v = 100$, using the formula $s = vt - 16t^2$

**3. a.** The formula $V = \frac{1}{4}\pi d^2 h$ expresses the relation between the volume, diameter, and height of a cylinder.

   (1) Find the volume of a cylinder if the diameter is 21 inches and the height is 15 inches.

   (2) What must the diameter of a tank be to hold 1,155 gallons if the height is 4 feet? A cubic foot holds about $7\frac{1}{2}$ gallons.

   **b.** The height ($s$) of an object in feet at the end of $t$ seconds, when thrown upward with a velocity ($v$) in feet per second, may be determined by the formula $s = vt - 16t^2$.

   (1) Find the height of an object at the end of 5 seconds when thrown upward at the rate of 120 feet per second.

   (2) How many seconds will it take an object to reach a height of 126 feet if it is thrown upward at the rate of 90 feet per second? As the object falls, when is it again at 126 feet?

   **c.** The dimensions of a well-proportioned rectangle may be determined by the formula $\dfrac{w}{l} = \dfrac{l}{w + l}$, with $l$ representing the length and $w$ the width. Using this formula, find to the nearest hundredth:

   (1) The length of a rectangle when the width is 6 inches.
   (2) The width of a rectangle when the length is 15 centimeters.
   (3) The length and width of a rectangle when the perimeter is 8 meters.

# Problem Solving

I. **Aim**    To solve verbal problems which involve the use of quadratic equations and their solutions.

II. **Procedure**

1. Use the directions outlined in the introduction to Chapter seven on page 244 to obtain a quadratic equation.

2. Solve this quadratic equation.

3. Check both answers directly with the facts of the given problem, discarding any value that is meaningless in the problem.

III. **Sample Solution**

The length of a rectangle is 3 centimeters more than the width. Its area is 40 square centimeters. Find the length and width of the rectangle.

*Solution:* Let $x$ = width of rectangle in centimeters

Then $x + 3$ = length of rectangle in centimeters

The equation is determined by the fact: The area of a rectangle equals the length times the width.

$$x(x + 3) = 40$$
$$x^2 + 3x = 40$$
$$x^2 + 3x - 40 = 0$$
$$(x - 5)(x + 8) = 0$$

Thus $x - 5 = 0$   or   $x + 8 = 0$

If    $x - 5 = 0$            If    $x + 8 = 0$
then        $x = 5$ cm wide       then        $x = -8$, meaningless root
      $x + 3 = 8$ cm long

*Check:*

| 8 cm long | 8 cm long |
| $-5$ cm wide | $\times 5$ cm wide |
| 3 cm more ✔ | 40 sq. cm, area ✔ |

*Answer:* 8 cm long, 5 cm wide.

## DIAGNOSTIC TEST

Solve each of the following problems:
1. The square of a certain number added to five times this number is 104. What is the number?   8, or −13
2. Find the length and width of a rectangle when its perimeter is 48 centimeters and its area is 135 square centimeters.

## RELATED PRACTICE PROBLEMS

Solve each of the following problems:
1. Number Problems
   a. One number 9 more than a second number. If the product of the two numbers is 90, what are the numbers?

   b. Find two consecutive positive integers whose product is 992.

   c. If the square of a number is increased by three times the number, the result is 28. Find the number.

   d. The sum of the two numbers is 8 and the sum of the squares of these numbers is 40. Find the numbers.

   e. Seven times a certain number is equal to twice the square of this number decreased by 15. Find the number.

2. Geometry Problems
   a. The length of a rectangle is 4 meters more than the width. The area is 45 square meters. Find the length and width of the rectangle.

   b. The width of a rectangle is 3 inches less than the length. The area is 70 square inches. Find the length and width of the rectangle.

   c. The perimeter of a rectangle is 22 meters and its area is 24 square meters. Find the length and width of the rectangle.

   d. If two parallel sides of a square are each increased by 6 inches and the other two parallel sides are each decreased by 1 inch, a rectangle is formed having twice the area of the square. How long is each side of the square?

   e. The perimeter of a rectangle is 26 meters. If the length is increased by 4 meters and the width is increased by 3 meters, the area of the new rectangle is 96 square meters. Find the length and width of the original rectangle.

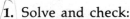

# REVIEW

1. Solve and check:
   a. $8x^2 = 96$

   b. $5x^2 - 4 = 0$

   c. $(x + 7)(x - 7) = 32$

   d. $\dfrac{2x - 3}{3} = \dfrac{5}{2x + 3}$

2. Solve by factoring and check:
   a. $x^2 - x - 72 = 0$    b. $4x^2 - 3x = 0$    c. $x(x - 9) = -8$

3. Solve and check:
   a. $x^2 + 2x + 1 = 81$   b. $x^2 - 5x + \frac{25}{4} = \frac{9}{4}$   c. $x^2 + \frac{2}{3}x + \frac{1}{9} = 8$

4. Solve by completing the square and check:

   a. $x^2 + 4x = 32$    b. $2x^2 - x - 10 = 0$    c. $\dfrac{x - 6}{3} = \dfrac{7}{x - 2}$

5. Solve by formula and check:

   a. $x^2 - 7x - 5 = 0$   b. $8x^2 - 6x - 9 = 0$   c. $\dfrac{x}{2} - \dfrac{3}{4} = \dfrac{x + 1}{x}$

6. Find the value of $w$ when $l = 10$, using the formula $\dfrac{w}{l} = \dfrac{l}{w + l}$.

7. Find two consecutive positive integers whose product is 7,832.

## COMPETENCY CHECK TEST

In each of the following, select the letter corresponding to your answer:

1. When $\sqrt{125} - 2\sqrt{27} + \sqrt{48} - 10\sqrt{\frac{1}{5}}$ is simplified and combined, the result is:

   a. $2\sqrt{6} - 3\sqrt{3}$   b. $6\sqrt{5} - 7\sqrt{2}$   c. $\sqrt{2}$      d. $3\sqrt{5} - 2\sqrt{3}$

2. The solution of $\sqrt{x + 3} = 7$ is:

   a. 46          b. 10          c. 4          d. 16

3. The solutions of $x^2 - 3x - 10 = 0$ are:

   a. 5 and 2      b. $-5$ and 2      c. 5 and $-2$   d. $-5$ and $-2$

4. Using the formula $s = 16t^2$ and $s = 784$, the value of $t$ is:

   a. 800         b. 7          c. 49         d. 14

5. The length of a rectangle is five times the width. If the area is 245 square meters, the length measures:

   a. 49 meters      b. 1,225 meters   c. 35 meters   d. 25 meters

## Keyed Achievement Test

The numeral at the end of each problem indicates the section where help may be found.

1. Divide as indicated: $\dfrac{a^2 - 25a - 54}{6a^2} \div (3a - 81)$  [10-4]

2. Solve and check: $\dfrac{3x + 1}{10} - \dfrac{x - 6}{4} = 2$  [11-1]

3. Find the value of $T'$ when $V = 400$, $V' = 500$, and $T = 280$, using the formula $\dfrac{V}{V'} = \dfrac{T}{T'}$.  [11-4]

Solve each of the following systems of equations:

4. Graphically: [13-2]  
$2x - 3y = 2$  
$3x + y = 14$

5. By addition or substitution method: [13-3] [13-4]  
$5x - 9y = 24$  
$8x + 4y = 20$

6. Raise to the indicated power: **a.** $(9c^6d^3x)^2$  **b.** $(-4m^5n^4y^8)^3$  [14-1]

7. Find the root indicated: **a.** $\sqrt{49c^8d^4x^2}$  **b.** $(-8r^{12}t^3y^9)^{\frac{1}{3}}$  
[14-3 and 14-4]

8. Simplify: **a.** $\sqrt{45}$  **b.** $\sqrt{216x^7y}$  **c.** $\sqrt{\dfrac{2b^4}{3c^2}}$  [14-5 thru 14-7]

9. Simplify and combine: $10\sqrt{\tfrac{4}{5}} + 3\sqrt{72} + 2\sqrt{20} - 5\sqrt{\tfrac{1}{2}}$  [14-8]

10. Multiply and simplify: **a.** $6\sqrt{12} \cdot 2\sqrt{5}$  **b.** $(4\sqrt{7} + 3\sqrt{6})^2$  [14-9]

11. Divide and simplify: $\dfrac{15\sqrt{60} - 25\sqrt{75} + 10\sqrt{3}}{5\sqrt{3}}$  [14-10]

12. Solve and check: $\sqrt{x^2 + 21} = x + 3$  [14-11]

13. Solve and check: $5x^2 - 90 = 0$  [15-1]

14. Solve by factoring and check: $x^2 + 16x - 36 = 0$  [15-3]

15. Solve and check: $x^2 - 26x + 169 = 144$  [15-4]

16. Solve $x^2 - 9x = 22$ by completing the square and check.  [15-4]

17. Solve $6x^2 - 7x - 20 = 0$ by formula and check.  [15-5]

18. Find the value of $V$ when $K = 2{,}700$ and $m = 6$, using the formula $K = \tfrac{1}{2}mV^2$.  [15-6]

19. The area of a rectangle is 168 square centimeters and its perimeter is 52 centimeters. Find its length and width.  [15-7]

# INTRODUCTION

The formulas $d = rt$, $t = \dfrac{d}{r}$, and $r = \dfrac{d}{t}$ express the interdependence between distance ($d$), rate of speed ($r$), and the time of travel ($t$). If a student knows only the formula $d = rt$, by employing the principles of solving equations, the other forms could easily be derived. Proficiency in the skill of transforming a formula is time-saving. Instead of memorizing each form of a given formula, the student is required to remember only one form. This rearrangement of formulas is also known as "changing the subject of the formula."

Also in this chapter ratio and proportions are studied and used with the principles of variation to evaluate formulas.

In formulas and equations having two variables, any change in one variable will produce a change in the other. Some quantities do not change in value. They are called *constants*. Any arithmetic number is a constant.

When one variable depends upon another for its value, the first variable is said to be a *function* of the second variable.

In the formula $c = 3.14d$, $c$ is a function of $d$.

A formula shows the relationship between quantities by means of numerals, variables, and equality and operational symbols. In this chapter the principles of direct and inverse variation are developed and used to determine what effect a change in one quantity of a formula will have upon another and to evaluate formulas.

# 16

# Ratio
# and Proportion;
# Formulas; Variation

# Ratio and Proportion

**I. Aim**   To use proportions to solve problems.

**II. Procedure**

**1.** Ratio

When two quantities are compared by division, the answer obtained is called the *ratio* of the two quantities.

**a.** Express the ratio of the two quantities that are being compared as a fraction in lowest terms.

The ratio of 4 compared to 10 is $\frac{4}{10}$ or $\frac{2}{5}$.

**b.** A colon may be used instead of a fraction bar.

The ratio $\frac{2}{5}$ is sometimes written as 2:5 and both are read "2 to 5" (two to five).

**c.** When the quantities compared are denominate numbers, first express them in the same units.

**d.** A ratio may be used to express a rate.

**e.** Ratios such as $\frac{2}{4}, \frac{3}{6}, \frac{4}{8}, \frac{5}{10}$, etc. express the same comparison or rate and are called *equivalent ratios*. They all equal the ratio $\frac{1}{2}$.

**2.** Proportion

*A proportion is a mathematical sentence that states that two ratios are equivalent.*

**a.** Express the proportion by using either two fractions or colon arrangement.

The proportion $\frac{4}{6} = \frac{10}{15}$ or 4:6 = 10:15 are both read
"4 is to 6 as 10 is to 15" or
"4 compared to 6 is the same as 10 compared to 15."

**b.** Observe that there are four terms in a proportion. For example,

$$\begin{array}{l} \text{first} \rightarrow \frac{4}{6} = \frac{10}{15} \leftarrow \text{third} \\ \text{second} \rightarrow \phantom{\frac{4}{6}} \phantom{=} \phantom{\frac{10}{15}} \leftarrow \text{fourth} \end{array}$$

$$\begin{array}{cccc} 1^{st} & 2^{nd} & 3^{rd} & 4^{th} \\ \downarrow & \downarrow & \downarrow & \downarrow \\ 4 & : 6 & = 10 & : 15 \end{array}$$

The first term and the fourth term of a proportion are called the *extremes* and the second term and the third term are called the *means*.

**c.** Also note that the *product of the means is equal to the product of the extremes.* These products are equal only when the ratios are equivalent.

In the proportion $4:6 = 10:15$, the product $6 \times 10$ is equal to the product of $4 \times 15$.

In the form $\frac{4}{6} = \frac{10}{15}$, these products are the cross products $\frac{4}{6} \bowtie \frac{10}{15}$.

**d.** To check whether two ratios are equivalent:
   (1) Express each ratio in simplest form.
   (2) Or determine whether the product of the means is equal to the product of the extremes.
   (3) Or determine whether the cross products are equal.

**e.** If any three of the four terms of a proportion are known quantities, determine the fourth term by:
   (1) Considering the proportion as an equation with fraction coefficient, or as a fractional equation and solving it. See Exercises 11-1 and 11-2.
   (2) Or transforming the proportion to a simple equivalent equation by writing the product of the means equal to the product of the extremes and then solving it.

## III. Sample Solutions

Find the ratio of:

**1.** 18 to 6   *Solution:* $\frac{18}{6} = \frac{3}{1}$
*Answer:* $\frac{3}{1}$ or 3 to 1 or $3:1$

**2.** 6 to 18   *Solution:* $\frac{6}{18} = \frac{1}{3}$
*Answer:* $\frac{1}{3}$ or 1 to 3 or $1:3$

**3.** 9 in. to 2 ft.   *Solution:* $\frac{9}{24} = \frac{3}{8}$
*Answer:* $\frac{3}{8}$ or $3:8$

**4.** Write as a proportion: Some number $x$ compared to 36 is the same as 7 compared to 9.   *Answer:* $\dfrac{x}{36} = \dfrac{7}{9}$ or $7:9$

**5.** Solve and check: $\dfrac{n}{45} = \dfrac{7}{15}$

| *Solution (1):* | *Solution (2):* | *Check:* |
|---|---|---|
| $\dfrac{n}{45} = \dfrac{7}{15}$ | $\dfrac{n}{45} = \dfrac{7}{15}$ | $\dfrac{n}{45} = \dfrac{7}{15}$ |
| $45 \cdot \dfrac{n}{45} = 45 \cdot \dfrac{7}{15}$   or | $15 \times n = 7 \times 45$ | $\dfrac{21}{45} = \dfrac{7}{15}$ ? |
|  | $15n = 315$ | $\dfrac{7}{15} = \dfrac{7}{15}$ ✔ |
| $n = 21$ | $n = 21$ |  |

Short method: Since 45 is 3 times 15, then $n$ must be 3 times 7 or 21.

*Answer:* 21

## DIAGNOSTIC TEST

**1.** Use the fraction bar to express the ratio of nine to fourteen.

**2.** Use the colon to express the ratio of ten to three.

Find in simplest form the ratio of:

   **3.** 7 to 12    **4.** 15 to 35    **5.** 11 to 4    **6.** 25 to 5    **7.** 24 to 18

Find the ratio of:

   **8.** $x$ to $y$          **9.** 24 sec. to 40 sec.     **10.** 1 yd. to 20 in.

**11.** a nickel to a dime   **12.** 8 things to a dozen

**13.** Express the rate 240 miles on 15 gallons of gasoline as a ratio.

**14.** The compression ratio is the ratio of the total volume of a cylinder to its clearance volume. Find the compression ratio of a cylinder of an engine if its total volume is 77 cu. in. and its clearance volume is 14 cu. in.

**15.** Which of the following are pairs of equivalent ratios?

   **a.** $\frac{6}{9}$ and $\frac{9}{12}$    **b.** $\frac{18}{27}$ and $\frac{28}{42}$    **c.** $\frac{30}{36}$ and $\frac{65}{78}$    **d.** $\frac{48}{54}$ and $\frac{72}{84}$

Write each of the following as a proportion:

**16.** 12 compared to 15 is the same as 36 compared to 45.

**17.** Some number $x$ is to 16 as 65 is to 40.

Which of the following are true?

**18. a.** $\frac{63}{84}=\frac{45}{60}$   **b.** $\frac{27}{24}=\frac{56}{48}$   **19. a.** $8:20=30:100$   **b.** $25:6=125:30$

Solve and check:

**20. a.** $\dfrac{n}{54}=\dfrac{7}{9}$   **b.** $\dfrac{80}{48}=\dfrac{15}{x}$   **c.** $\dfrac{x}{60-x}=\dfrac{2}{3}$   **d.** $\dfrac{m+3}{m-3}=\dfrac{m+15}{m}$

Using proportions, solve each of the following:

**21.** 32 compared to what number is the same as 56 compared to 14?

**22.** A motorist travels 192 miles in 4 hours. How long will it take at that rate to travel 324 miles?

## RELATED PRACTICE EXAMPLES

**1.** Use the fraction bar to express each of the following ratios:

   **a.** Ratio of seven to three        **b.** Ratio of six to eleven

   **c.** Ratio of twenty to one         **d.** Ratio of two to fifteen

**2.** Use the colon to express each of the following ratios:

   **a.** Ratio of eight to nineteen      **b.** Ratio of twenty to nine

   **c.** Ratio of one to seven           **d.** Ratio of five to four

Find in simplest form the ratio of:

3. **a.** 4 to 7 **b.** 5 to 13 **c.** 1 to 9 **d.** 3 to 8 **e.** 10 to 21
4. **a.** 2 to 8 **b.** 6 to 10 **c.** 12 to 36 **d.** 30 to 75 **e.** 64 to 96
5. **a.** 9 to 2 **b.** 5 to 3 **c.** 20 to 11 **d.** 7 to 6 **e.** 36 to 25
6. **a.** 10 to 5 **b.** 28 to 7 **c.** 50 to 10 **d.** 72 to 24 **e.** 120 to 12
7. **a.** 12 to 8 **b.** 16 to 6 **c.** 20 to 15 **d.** 45 to 25 **e.** 90 to 40
8. **a.** $x$ to 6 **b.** 5 to $m$ **c.** $r$ to $s$ **d.** $4n$ to $n$ **e.** $5 - x$ to $x$

Find the ratio of:

9. **a.** 4 in. to 9 in. **b.** 45 min. to 18 min. **c.** 6 oz. to 14 oz.
   **d.** 15 qt. to 6 qt. **e.** 100 ft. to 1,000 ft. **f.** 48 hr. to 72 hr.
10. **a.** 2 ft. to 1 yd. **b.** 1 hr. to 48 min. **c.** 40 sec. to 2 min.
    **d.** 1 pt. to 1 gal. **e.** 2 lb. to 6 oz. **f.** 18 mo. to 4 yr.
11. **a.** A quarter to a dollar **b.** A dime to a nickel
    **c.** A half-dollar to a dime **d.** A dollar to 3 quarters
    **e.** 7 nickels to 6 dimes **f.** 64 pennies to 8 nickels
12. **a.** 4 things to a dozen **b.** 1 dozen to 9 things
    **c.** 10 things to a dozen **d.** 6 things to 2 dozen
    **e.** 3 dozen to 20 things **f.** 15 things to 6 dozen
13. Express each of the following rates as a ratio:
    **a.** 400 L in 5 min. **b.** 700 ft. in 10 sec. **c.** 180 mi. in 3 hr.
    **d.** 156 mi. on 12 gal. **e.** 8,000 revolutions in 5 min. **f.** 280 km in 4 hr.
    **g.** 1,000 liters in 8 min. **h.** 4,500 revolutions in 3 sec.
14. **a.** The pitch of a roof is the ratio of the rise to the span. What is the pitch of a roof with a rise of 8 feet and a span of 24 feet?
    **b.** The aspect ratio is the ratio of the length of an airplane wing to its width. What is the aspect ratio of a wing 91 feet long and 13 feet wide?
    **c.** Find the ratio of a 36-tooth gear to a 48-tooth gear.
    **d.** There are 20 girls and 12 boys in a mathematics class. What is the ratio of: boys to girls? girls to boys? boys to the entire class? girls to the entire class?
    **e.** A mixture used to make concrete contains 1 part cement, 2 parts sand, 4 parts gravel, and water. If 6 bags of sand are used, how many bags of cement and bags of gravel are required?
15. Which of the following are pairs of equivalent ratios?
    **a.** $\frac{8}{12}$ and $\frac{6}{8}$ **b.** $\frac{6}{15}$ and $\frac{4}{10}$ **c.** $\frac{9}{24}$ and $\frac{21}{56}$
    **d.** $\frac{27}{36}$ and $\frac{20}{40}$ **e.** $\frac{40}{48}$ and $\frac{15}{18}$ **f.** $\frac{35}{100}$ and $\frac{30}{75}$
    **g.** $\frac{36}{54}$ and $\frac{60}{90}$ **h.** $\frac{21}{84}$ and $\frac{48}{144}$ **i.** $\frac{27}{63}$ and $\frac{39}{91}$

Write each of the following as a proportion:

16. **a.** 9 compared to 12 is the same as 6 compared to 8.
    **b.** 20 is to 16 as 30 is to 24.
    **c.** 17 compared to 68 is the same as 18 compared to 72.

**d.** 54 is to 108 as 400 is to 800.

**e.** 96 compared to 64 is the same as 9 compared to 6.

**17. a.** Some number $n$ compared to 10 is the same as 8 compared to 15.

    **b.** Some number $x$ is to 45 as 80 is to 54.     **c.** 18 is to $n$ as 6 is to 5.

    **d.** 75 compared to 200 is the same as some number $y$ compared to 40.

    **e.** 84 is to 144 as $300 - x$ is to $x$.

Which of the following are true?

**18. a.** $\frac{12}{15} = \frac{3}{5}$            **b.** $\frac{9}{16} = \frac{27}{48}$            **c.** $\frac{24}{36} = \frac{18}{27}$

    **d.** $\frac{28}{63} = \frac{20}{45}$            **e.** $\frac{48}{51} = \frac{60}{65}$            **f.** $\frac{75}{50} = \frac{120}{80}$

**19. a.** $6:12 = 8:15$          **b.** $16:36 = 12:20$        **c.** $18:14 = 27:21$

    **d.** $54:45 = 12:10$        **e.** $9:7 = 81:72$           **f.** $96:144 = 120:168$

**20.** Solve and check:

    **a.** $\dfrac{n}{8} = \dfrac{5}{8}$      **b.** $\dfrac{a}{36} = \dfrac{2}{3}$      **c.** $\dfrac{3}{7} = \dfrac{d}{28}$      **d.** $\dfrac{4}{25} = \dfrac{c}{5}$

    **e.** $\dfrac{x}{4} = \dfrac{6\frac{1}{2}}{3}$      **f.** $\dfrac{x}{10} = \dfrac{4}{15}$      **g.** $\dfrac{n}{24} = \dfrac{9}{16}$      **h.** $\dfrac{15}{8} = \dfrac{x}{10}$

    **i.** $\dfrac{32}{n} = \dfrac{4}{2\frac{3}{4}}$      **j.** $\dfrac{1}{15} = \dfrac{3}{y}$      **k.** $\dfrac{6}{b} = \dfrac{90}{16}$      **l.** $\dfrac{108}{144} = \dfrac{18}{n}$

    **m.** $\dfrac{30}{d} = \dfrac{5}{14}$      **n.** $\dfrac{2x}{15} = \dfrac{2}{5}$      **o.** $\dfrac{700 - x}{x} = \dfrac{9}{5}$      **p.** $\dfrac{x}{x+2} = \dfrac{5}{7}$

    **q.** $\dfrac{8n}{100} = \dfrac{n+15}{120}$    **r.** $\dfrac{x+5}{x-5} = \dfrac{x-1}{x-7}$    **s.** $\dfrac{x}{4-x} = \dfrac{9-2x}{2x-5}$    **t.** $\dfrac{x+3}{2} = \dfrac{8}{x-3}$

**21.** Using proportions, solve each of the following:

    **a.** What number compared to 60 is the same as 9 compared to 20?

    **b.** 24 compared to 56 is the same as what number compared to 35?

    **c.** 45 compared to 60 is the same as 93 compared to what number?

    **d.** 84 compared to what number is the same as 63 compared to 18?

    **e.** What number is to 72 as 42 is to 216?

**22. a.** A motorist uses 25 gallons of gasoline to travel 350 miles. At this rate of gasoline consumption, how many gallons of gasoline will be consumed when the motorist travels 462 miles?

    **b.** At the rate of 6 items for 14¢, how many items can you buy for 70¢?

    **c.** Andrew's father pays $950 taxes on a house assessed at $25,000. Using the same rate, find the taxes on a house assessed at $30,400.

    **d.** Marilyn saved $75 in 6 weeks. At that rate how long will it take her to save $650?

    **e.** Mr. A and Mr. B, partners in a business, agree to divide their profits in the ratio $3:4$, respectively. How much did each receive if the profits for a year were $22,400?

# Transforming Formulas

**I. Aim** To apply the principles of solving equations to a given formula to derive other forms of the formula.

## II. Procedure

**1.** To transform a formula:

**a.** Copy the given formula.

**b.** Clear the formula of all fractions when necessary.

**c.** Solve for the required variable in terms of the other variables of the formula.

**d.** When the required variable appears in more than one term, factor when necessary.

**2.** To derive a new formula from two or more given formulas having a common variable by applying the principles of solving a system of equations:

**a.** Eliminate a common variable by replacing this variable in one formula by an equivalent expression found in the second formula.

**b.** Then transform and simplify as necessary.

## III. Sample Solutions

**1.** Transform $V = Bh$, solving for $h$.

*Solution:*

$$V = Bh$$

$$\frac{V}{B} = \frac{Bh}{B}$$

$$\frac{V}{B} = h$$

$$h = \frac{V}{B}$$

*Answer:* $h = \dfrac{V}{B}$

**2.** Transform $A = p + prt$, solving for $r$.

*Solution:*

$$A = p + prt$$

$$A - p = p - p + prt$$

$$A - p = prt$$

$$\frac{A - p}{pt} = \frac{prt}{pt}$$

$$\frac{A - p}{pt} = r$$

$$r = \frac{A - p}{pt}$$

*Answer:* $r = \dfrac{A - p}{pt}$

595

**3.** Transform $I = \dfrac{E}{R}$, solving for $R$.

*Solution:*

$$I = \dfrac{E}{R}$$

$$I \cdot R = \dfrac{E}{R} \cdot R$$

$$IR = E$$

$$\dfrac{IR}{I} = \dfrac{E}{I}$$

$$R = \dfrac{E}{I} \quad \textit{Answer: } R = \dfrac{E}{I}$$

**4.** Transform $W = I^2R$ solving for $I$.

*Solution:*

$$W = I^2R$$

$$\dfrac{W}{R} = \dfrac{I^2R}{R}$$

$$\dfrac{W}{R} = I^2$$

$$\sqrt{\dfrac{W}{R}} = I$$

$$I = \sqrt{\dfrac{W}{R}} \quad \textit{Answer: } I = \sqrt{\dfrac{W}{R}}$$

**5.** Derive a formula for $W$ in terms of $I$ and $R$, using the formulas $W = IE$ and $E = IR$.

*Solution:* Substitute $IR$ for $E$ in the formula $W = IE$.

$$E = IR \qquad W = IE = I \cdot IR = I^2R \quad \textit{Answer: } W = I^2R$$

## DIAGNOSTIC TEST

Transform each of the following formulas:

**1.** $W = Fd$; solve for $d$

**2.** $A = bh$; solve for $b$

**3.** $V = lwh$; solve for $w$

**4.** $A = L + C$; solve for $L$

**5.** $s = c + m$; solve for $m$

**6.** $i = A - p$; solve for $A$

**7.** $M = 180 - N$; solve for $N$

**8.** $E = Ir + IR$; solve for $r$

**9.** $v = V + gt$; solve for $t$

**10.** $d = \dfrac{C}{3.14}$; solve for $C$

**11.** $L = \dfrac{W}{A}$; solve for $W$

**12.** $R = \dfrac{s}{c}$; solve for $c$

**13.** $\dfrac{V}{V'} = \dfrac{T}{T'}$; solve for $V'$

**14.** $\dfrac{1}{R} = \dfrac{1}{R_1} + \dfrac{1}{R_2}$; solve for $R_1$

**15.** $A = p + prt$; solve for $p$

**16.** $l = a + (n - 1)d$; solve for $n$

**17.** $d = 16t^2$; solve for $t$

**18.** $V = \sqrt{\dfrac{2K}{m}}$; solve for $K$

**19.** Derive a formula for $s$ in terms of $g$ and $t$, using the formulas $s = vt$ and $v = \frac{1}{2}gt$.

## *RELATED PRACTICE EXAMPLES*

Transform each of the following formulas:

1. **a.** $p = 4s$; solve for $s$  **b.** $d = 2r$; solve for $r$  **c.** $A = lw$; solve for $w$
   **d.** $C = \pi d$; solve for $d$  **e.** $W = IE$; solve for $E$  **f.** $v = at$; solve for $t$

2. **a.** $W = Fd$; solve for $F$  **b.** $P = HD$; solve for $H$  **c.** $A = ab$; solve for $a$
   **d.** $p = ns$; solve for $n$  **e.** $A = lw$; solve for $l$  **f.** $p = br$; solve for $b$

3. **a.** $C = 2\pi r$; solve for $r$  **b.** $A = \pi ab$; solve for $b$
   **c.** $i = prt$; solve for $p$  **d.** $V = lwh$; solve for $l$
   **e.** $F = AHD$; solve for $H$  **f.** $T.S. = \pi dN$; solve for $d$

4. **a.** $A = p + i$; solve for $p$  **b.** $s = c + g$; solve for $c$
   **c.** $A = C + 273$; solve for $C$  **d.** $A + B = 90$; solve for $A$
   **e.** $m = p + e$; solve for $p$  **f.** $l = n + d$; solve for $n$

5. **a.** $s = c + m$; solve for $m$  **b.** $A = L + C$; solve for $C$
   **c.** $M + N = 180$; solve for $N$  **d.** $D_t = D_w - D_p$; solve for $D_p$
   **e.** $s = p + c$; solve for $c$  **f.** $A = p + i$; solve for $i$

6. **a.** $g = s - c$; solve for $s$  **b.** $d = l - n$; solve for $l$
   **c.** $p = A - i$; solve for $A$  **d.** $l = c - s$; solve for $c$
   **e.** $C = A - 273$; solve for $A$  **f.** $L = A - C$; solve for $A$

7. **a.** $l = c - s$; solve for $s$  **b.** $g = s - c$; solve for $c$
   **c.** $L = A - C$; solve for $C$  **d.** $p = A - i$; solve for $i$
   **e.** $s = m - d$; solve for $d$  **f.** $B = 90 - A$; solve for $A$

8. **a.** $p = b + 2e$; solve for $b$  **b.** $v = V + gt$; solve for $V$
   **c.** $l = a + (n-1)d$; solve for $a$  **d.** $p = 2l + 2w$; solve for $l$
   **e.** $F = 1.8C + 32$; solve for $C$  **f.** $S = 180n - 360$; solve for $n$

9. **a.** $p = 2l + 2w$; solve for $w$  **b.** $E = Ir + IR$; solve for $R$
   **c.** $a = S - Sr$; solve for $r$  **d.** $A = p + prt$; solve for $t$
   **e.** $A = \pi r^2 + \pi rl$; solve for $l$  **f.** $A = 2\pi r^2 + 2\pi rh$; solve for $h$

10. **a.** $r = \dfrac{d}{2}$; solve for $d$  **b.** $m = \dfrac{s}{60}$; solve for $s$  **c.** $V = \dfrac{Bh}{3}$; solve for $h$

   **d.** $S = \dfrac{\pi dN}{12}$; solve for $d$  **e.** $H.P. = \dfrac{TV}{550}$; solve for $T$

11. **a.** $d = \dfrac{m}{v}$; solve for $m$  **b.** $P = \dfrac{F}{A}$; solve for $F$  **c.** $a = \dfrac{v}{t}$; solve for $v$

   **d.** $R = \dfrac{s}{c}$; solve for $s$  **e.** $P = \dfrac{Fd}{t}$; solve for $F$  **f.** $W = \dfrac{wl}{L}$; solve for $l$

12. **a.** $a = \dfrac{360}{n}$; solve for $n$  **b.** $r = \dfrac{p}{b}$; solve for $b$  **c.** $L = \dfrac{W}{A}$; solve for $A$

   **d.** $v = \dfrac{s}{t}$; solve for $t$  **e.** $P = \dfrac{Fd}{t}$; solve for $t$  **f.** $F = \dfrac{Wv^2}{gr}$; solve for $r$

13. **a.** $\dfrac{V}{V'} = \dfrac{P'}{P}$; solve for $V$   **b.** $\dfrac{W_1}{W_2} = \dfrac{L_2}{L_1}$; solve for $L_2$

**c.** $\dfrac{F}{W} = \dfrac{h}{d}$; solve for $W$   **d.** $\dfrac{R_1}{R_2} = \dfrac{R_3}{R_4}$; solve for $R_4$

**e.** $\dfrac{D}{d} = \dfrac{r}{R}$; solve for $d$   **f.** $\dfrac{PV}{T} = \dfrac{P'V'}{T'}$; solve for $T'$

14. **a.** $F = \frac{9}{5}C + 32$; solve for $C$   **b.** $I = \dfrac{E}{R_1 + R_2}$; solve for $E$

**c.** $S = \dfrac{a}{1-r}$; solve for $r$   **d.** $\dfrac{1}{f} = \dfrac{1}{D} + \dfrac{1}{d}$; solve for $d$

**e.** $I = \dfrac{nE}{R+nr}$; solve for $R$   **f.** $S = \dfrac{rl-a}{r-1}$; solve for $a$

15. **a.** $E = Ir + IR$; solve for $I$   **b.** $a = S - Sr$; solve for $S$

**c.** $I = \dfrac{nE}{R+nr}$; solve for $n$   **d.** $S = \dfrac{rl-a}{r-1}$; solve for $r$

16. **a.** $p = 2(l+w)$; solve for $l$   **b.** $A = p(1+rt)$; solve for $t$

**c.** $A = 2\pi r(r+h)$; solve for $h$   **d.** $S = \dfrac{n}{2}(a+l)$; solve for $n$

**e.** $l = a + (n-1)d$; solve for $d$   **f.** $C = 180 - (A+B)$; solve for $A$

17. **a.** $A = s^2$; solve for $s$   **b.** $A = 6s^2$; solve for $s$   **c.** $A = 4\pi r^2$; solve for $r$

**d.** $s = \frac{1}{2}gt^2$; solve for $t$   **e.** $V = \pi r^2 h$; solve for $r$   **f.** $F = \dfrac{mv^2}{gr}$; solve for $v$

18. **a.** $r = \sqrt{\dfrac{A}{\pi}}$; solve for $A$   **b.** $t = \pi\sqrt{\dfrac{l}{g}}$; solve for $l$

**c.** $I = \sqrt{\dfrac{P}{R}}$; solve for $R$   **d.** $h = \sqrt{a^2 + b^2}$; solve for $b$

19. Derive a formula:

**a.** For $A$ in terms of $p$, $r$, and $t$, using the formulas $A = p + i$ and $i = prt$.

**b.** For $C$ in terms of $\pi$ and $d$, using the formulas $C = 2\pi r$ and $r = \dfrac{d}{2}$.

**c.** For $A$ in terms of $p$, using the formulas $p = 4s$ and $A = s^2$.

**d.** For $P$ in terms of $F$ and $v$, using the formulas $P = \dfrac{W}{t}$, $W = Fd$, and $v = \dfrac{d}{t}$.

**e.** For $S$ in terms of $n$, $a$, and $d$, using the formulas $l = a + (n-1)d$ and $S = \dfrac{n}{2}(a+l)$.

# Direct Variation

**I. Aim**   To use the principles of direct variation to evaluate formulas.

**II. Procedure**

**1. Observe that:**

**a.** Two variables, which are so related that their ratio is constant are said to *vary directly.*

Thus, if the variables $x$ and $y$ are so related that the ratio $\dfrac{y}{x}$ equals a constant $k$, then the variable $y$ varies directly as $x$. This constant $k$ is called the *constant of variation* or the *constant of proportionality.*

**b.** When a change in one variable produces the same change in a second variable, the second variable *varies directly* as the first variable.

For example, assume values for $s$ in formula $p = 4s$ to obtain corresponding values for $p$ as follows:

| $s$ | 2 | 4 | 5 | 6 | 10 |
|---|---|---|---|---|---|
| $p$ | 8 | 16 | 20 | 24 | 40 |

When each value of $p$ is compared to its corresponding value of $s$, the ratio is always 4. Therefore $k = 4$.

The above table reveals that:

Doubling the value of $s$ (from 2 to 4) doubles the value of $p$ (from 8 to 16).

Tripling the value of $s$ (from 2 to 6) triples the value of $p$ (from 8 to 24).

Halving the value of $s$ (from 10 to 5) halves the value of $p$ (from 40 to 20).

A change in the value of $s$ produces the same change in the value of $p$. Or $p$ *varies directly* as $s$, or $p$ is *directly proportional* to $s$.

**c.** Since $\dfrac{y}{x} = k$, then $y = kx$.

Thus in formulas of the type $d = 2r$, $c = 3.14d$, $p = 3s$, etc., where one variable is equal to a constant $(k)$ times a second variable, the two variables vary directly.

2. To determine the value of the constant $k$ from a pair of corresponding values of two variables that vary directly, substitute the values in $\dfrac{y}{x} = k$ or $y = kx$ and solve the resulting sentence.

3. To determine the corresponding changed value of the second variable when two variables vary directly and the value of one variable is changed, substitute the known values in the proportion $\dfrac{y_1}{x_1} = \dfrac{y_2}{x_2}$ and solve. See sample solution 2.

4. Assume that all variables in a formula given in this exercise other than the two specified variables remain unchanged (or are considered constants).

Thus, in the formula $A = lw$,

$A$ varies directly as $w$ when $l$ remains constant and

$A$ varies directly as $l$ when $w$ remains constant.

## III. Sample Solutions

1. When $y$ varies directly as $x$, what is the value of $k$ if $x = 3$ and $y = 18$?

Given: $x = 3$
$\quad\quad\; y = 18$
$\quad\quad\; k = ?$

*Solution:* $\dfrac{y}{x} = k$   or   *Solution:*  $y = kx$
$\quad\quad\quad \dfrac{18}{3} = k$ $\quad\quad\quad\quad\quad 18 = k \cdot 3$
$\quad\quad\quad\; 6 = k$ $\quad\quad\quad\quad\quad\quad 18 = 3k$
$\quad\quad\quad\; k = 6$ $\quad\quad\quad\quad\quad\quad\; 6 = k$
$\quad\quad\quad\quad\quad\quad\quad\quad\quad\quad\quad\quad k = 6$

*Answer:* 6

2. Suppose $y$ varies directly as $x$, and $y = 30$ when $x = 6$. What is the value of $y$ when $x = 10$?

*Solution:*

Given: $x_1 = 6$
$\quad\quad\; y_1 = 30$
$\quad\quad\; x_2 = 10$
$\quad\quad\; y_2 = ?$

$\dfrac{y_1}{x_1} = \dfrac{y_2}{x_2}$

$\dfrac{30}{6} = \dfrac{y_2}{10}$

$6y_2 = 300$
$\;\; y_2 = 50$   *Answer:* 50

*Alternate Solution:*

Observe that $x_2$ is $\frac{10}{6}$ or $\frac{5}{3}$ times $x_1$.

Therefore $y_2$ is $\frac{5}{3}$ times $y_1$.

$\quad\quad \frac{5}{3} \times 30 = 50$

## DIAGNOSTIC TEST

1. Determine in each of the following tables whether variable $y$ varies directly as $x$. If it does, find the constant of variation.

   **a.**

   | $x$ | 5 | 6 | 10 | 30 |
   |---|---|---|---|---|
   | $y$ | 18 | 5 | 9 | 3 |

   **b.**

   | $x$ | 5 | 6 | 10 | 18 |
   |---|---|---|---|---|
   | $y$ | 60 | 72 | 120 | 216 |

2. In which of the following do the variables vary directly?

   **a.** $dr = 2$   **b.** $\dfrac{m}{n} = 18$   **c.** $\dfrac{18}{m} = n$   **d.** $d = 2r$

3. Write as an equation, using $k$ as the constant of variation:
   The perimeter $(p)$ of a regular hexagon varies directly as the length of its side $(s)$.

4. Find the missing word:
   In the formula $W = 5R$, if $R$ is doubled, then $W$ is _____.

5. What effect does halving the value of $v$ have on the value of $P$ in the formula $P = Fv$ when $F$ is constant?

6. When $y$ varies directly as $x$, what is the value of $k$ if $x = 7$ and $y = 42$?

7. Suppose $y$ varies directly as $x$, and $y = 56$ when $x = 8$. What is the value of $y$ when $x = 32$?

8. In the formula $A = lw$, $A = 42$ when $l = 6$. Find the value of $A$ when $l = 18$ and $w$ is constant.

## RELATED PRACTICE EXAMPLES

1. Determine in each of the following tables whether variable $y$ varies directly as $x$. If it does, find the constant of variation.

   **a.**

   | $x$ | 3 | 4 | 5 | 9 |
   |---|---|---|---|---|
   | $y$ | 15 | 20 | 25 | 45 |

   **b.**

   | $x$ | 2 | 3 | 4 | 8 |
   |---|---|---|---|---|
   | $y$ | 3 | 6 | 1 | 5 |

**c.**

| $x$ | 5 | 6 | 8 | 12 |
|---|---|---|---|---|
| $y$ | 35 | 42 | 56 | 84 |

**d.**

| $x$ | 2 | 4 | 5 | 8 |
|---|---|---|---|---|
| $y$ | 20 | 10 | 8 | 5 |

**e.**

| $x$ | 2 | 4 | 6 | 18 |
|---|---|---|---|---|
| $y$ | 1 | 2 | 3 | 9 |

**f.**

| $x$ | 5 | 10 | 25 | 50 |
|---|---|---|---|---|
| $y$ | 20 | 10 | 4 | 2 |

2. In which formulas and equations do the variables vary directly?

    **a.** $p = 10s$      **b.** $y = 7x$      **c.** $m = \dfrac{s}{60}$      **d.** $xy = 30$

    **e.** $\dfrac{x}{y} = 12$      **f.** $y = \dfrac{12}{x}$      **g.** $H.P. = \dfrac{W}{746}$      **h.** $C = \dfrac{44}{7}r$

    **i.** $mn = 1$      **j.** $\dfrac{x}{5} = y$      **k.** $m = \dfrac{i}{39.37}$      **l.** $d = \dfrac{200}{t}$

3. Write each relationship as an equation. Use $k$ as the constant of variation:
    **a.** The circumference ($c$) of a circle varies directly as the length of the diameter ($d$).
    **b.** The total cost ($c$) of a given number of a particular item varies directly as the price per item ($p$).
    **c.** The area ($A$) of rectangles having the same length varies directly as the width ($w$).
    **d.** The distance ($d$) traveled by a vehicle at a uniform rate of speed varies directly as the time ($t$) it travels.
    **e.** The annual interest ($i$) at 5% varies directly as the principal ($p$).

4. Find the missing word (increased, decreased, doubled, tripled, halved):
    **a.** In the formula $p = 8s$, if $s$ is increased, then $p$ is _____.
    **b.** In the formula $c = 3.14d$, if $d$ is halved, then $c$ is _____.
    **c.** In the formula $h = \dfrac{m}{60}$, if $m$ is tripled, then $h$ is _____.

5. What effect does:
    **a.** Doubling the value of $d$ have on the value of $r$ in the formula $r = \dfrac{d}{2}$?
    **b.** Halving the value of $l$ have on the value of $A$ in the formula $A = lw$ when $w$ is constant?
    **c.** Tripling the value of $I$ have on the value of $E$ in the formula $E = IR$ when $R$ is constant?
    **d.** Doubling the radius of a circle have on the value of the circumference of the circle. Formula: $C = 2\pi r$.

6. When $y$ varies directly as $x$, find the value of the constant $k$:
   **a.** If $x=2$ and $y=20$   **b.** If $x=3$ and $y=48$   **c.** If $x=9$ and $y=72$
   **d.** If $x=4$ and $y=2$   **e.** If $x=16$ and $y=4$   **f.** If $x=30$ and $y=150$
   **g.** If $x=8$ and $y=12$   **h.** If $x=10$ and $y=7$   **i.** If $x=6$ and $y=54$

7. Suppose $y$ varies directly as $x$, and:
   **a.** $y=15$ when $x=3$.   Find $y$ when $x=12$.
   **b.** $y=36$ when $x=9$.   Find $y$ when $x=45$.
   **c.** $y=63$ when $x=7$.   Find $y$ when $x=28$.
   **d.** $y=72$ when $x=6$.   Find $y$ when $x=9$.
   **e.** $y=48$ when $x=16$.   Find $y$ when $x=4$.

8. In the formula:
   **a.** $A=ab$, $A=35$ when $a=5$. Find $A$ when $a=15$ and $b$ is constant.
   **b.** $d=rt$, $d=240$ when $r=40$. Find $d$ when $r=60$ and $t$ is constant.
   **c.** $W=Fd$, $W=800$ when $d=8$. Find $W$ when $d=11$ and $F$ is constant.
   **d.** $V=lwh$, $V=96$ when $w=6$. Find $V$ when $w=30$ and $l$ and $h$ are constant.
   **e.** $C=2\pi r$, $C=44$ when $r=7$. Find $C$ when $r=42$.

## CUMULATIVE REVIEW

1. Find in simplest form the ratio of:
   **a.** 20 to 5   **b.** 40 min. to 1 hour   **c.** 1 km to 200 m

2. Solve and check:   **a.** $\dfrac{n}{60} = \dfrac{11}{15}$   **b.** $\dfrac{x + 6}{18} = \dfrac{3x - 10}{12}$

3. Using proportions, find how many centimeters each part measures when a line segment measuring 35 centimeters is divided into two parts having the ratio 2 to 3.

4. Transform each of the following formulas:

   **a.** $W = IE$; solve for $E$   **b.** $P = \dfrac{F}{A}$; solve for $F$

   **c.** $\dfrac{1}{f} = \dfrac{1}{D} + \dfrac{1}{d}$; solve for $f$   **d.** $I = \dfrac{E}{R_1 + R_2}$; solve for $R_1$

5. Derive a formula for $A$ in terms of $C$, using the formulas $A = \pi r^2$ and $C = 2\pi r$.

6. Suppose $y$ varies directly as $x$, and $y = 29$ when $x = 7$. Find $y$ when $x = 35$.

7. In the formula $C = \frac{22}{7}d$, if $d$ is tripled, what effect does this have on the value of $C$?

8. In the formula $d = rt$, $d = 630$ when $r = 200$. Find $d$ when $r = 300$ and $t$ is constant.

# Inverse Variation

**I. Aim**    To use the principles of inverse variation to evaluate formulas.

**II. Procedure**

**1.** Observe that:

**a.** Two variables, which are so related that their product is a constant are said to *vary inversely.*

Thus, if the variables $x$ and $y$ are so related that the product $xy$ equals a constant $k$, then the variable $y$ varies inversely as $x$.

**b.** When a change in one variable produces an inverse change in a second variable, the second variable *varies inversely* as the first variable.

For example, assume values for $r$ in the formula $t = \dfrac{150}{r}$ to obtain corresponding values for $t$ as follows:

| $r$ | 15 | 25 | 30 | 50 | 75 |
|-----|----|----|----|----|----|
| $t$ | 10 | 6 | 5 | 3 | 2 |

The product of the two variables in each case is 150. Therefore, $k = 150$.

The above table reveals that:

Doubling the value of $r$ (from 15 to 30) halves the value of $t$ (from 10 to 5).

Tripling the value of $r$ (from 25 to 75) makes $t$ equal one-third its original value (from 6 to 2).

Halving the value of $r$ (from 50 to 25) doubles the value of $t$ (from 3 to 6).

A change in the value of $r$ produces an inverse change in the value of $t$. Or, $t$ *varies inversely* as $r$. Or $t$ is *inversely proportional* to $r$.

**c.** Since $xy = k$, then $y = \dfrac{k}{x}$.

Thus in formulas of the type $a = \dfrac{360}{n}$ where one variable is equal to the quotient of a constant divided by a second variable, the two variables vary inversely.

2. To determine the value of the constant $k$ from a pair of corresponding values of the two variables that vary inversely, substitute the values in $xy = k$ and multiply.

3. To determine the corresponding changed value of the second variable when two variables vary inversely and the value of one variable is changed, substitute the known values in the proportion $\dfrac{y_1}{y_2} = \dfrac{x_2}{x_1}$ and solve. See sample solution.

4. Assume that all variables in a formula given in this exercise other than the two specified variables remain unchanged (or are considered constants).

Thus, in the formula $V = Bh$ when transformed to $B = \dfrac{V}{h}$, $B$ varies inversely as $h$ when $V$ remains constant.

## III. Sample Solutions

1. When $y$ varies inversely as $x$, what is the value of $k$ if $x = 3$ and $y = 18$?

Given: $x = 3$      *Solution:*    $xy = k.$
        $y = 18$             $3 \cdot 18 = k$
        $k = ?$              $54 = k$
                         $k = 54$    *Answer: 54*

2. Suppose $y$ varies inversely as $x$, and $y = 30$ when $x = 6$. What is the value of $y$ when $x = 10$?

*Solution:*

Given: $x_1 = 6$     $\dfrac{y_1}{y_2} = \dfrac{x_2}{x_1}$
        $y_1 = 30$
        $x_2 = 10$      $\dfrac{30}{y_2} = \dfrac{10}{6}$
        $y_2 = ?$
                 $10y_2 = 180$
                   $y_2 = 18$

*Alternate Solution:*

Observe that $x_2$ is $\frac{10}{6}$ or $\frac{5}{3}$ times $x_1$. Therefore $y_2$ is $\frac{3}{5}$ times $y_1$.

       $\frac{3}{5} \times 30 = 18$    *Answer: 18*

## DIAGNOSTIC TEST

1. Determine in each of the following tables whether variable $y$ varies inversely as $x$. If it does, find the constant of variation.

| $x$ | 3 | 6 | 15 | 18 |
|---|---|---|---|---|
| $y$ | 27 | 54 | 135 | 162 |

| $x$ | 2 | 8 | 16 | 24 |
|---|---|---|---|---|
| $y$ | 72 | 18 | 9 | 6 |

2. In which of the following do the variables vary inversely?

    **a.** $\dfrac{r}{s} = 20$      **b.** $\dfrac{20}{s} = r$      **c.** $12 = ab$      **d.** $b = 12a$

3. Write as an equation, using $k$ as the constant of variation:
   The pitch ($p$) of a roof varies inversely as the span ($s$).

4. Find the missing word:

   In the formula $d = \dfrac{120}{v}$, if $v$ is doubled, then $d$ is _____.

5. What effect does halving the value of $I$ have on the value of $E$ in

   the formula $E = \dfrac{W}{I}$ when $W$ is constant?

6. When $y$ varies inversely as $x$, what is the value of $k$ if $x = 4$ and
   $y = 6$?

7. Suppose $y$ varies inversely as $x$, and $y = 15$ when $x = 10$. What is
   the value of $y$ when $x = 2$?

8. In the formula $B = \dfrac{V}{h}$, $B = 48$ when $h = 6$. Find $B$ when $h = 36$

   and $V$ is constant.

## *RELATED PRACTICE EXAMPLES*

1. Determine in each of the following tables whether variable $y$ varies inversely as $x$. If it does, find the constant of variation.

**a.**

| $x$ | 1 | 7 | 14 | 35 |
|---|---|---|---|---|
| $y$ | 70 | 10 | 5 | 2 |

**b.**

| $x$ | 3 | 6 | 9 | 18 |
|---|---|---|---|---|
| $y$ | 24 | 12 | 8 | 4 |

**c.**

| $x$ | 3 | 6 | 9 | 18 |
|---|---|---|---|---|
| $y$ | 6 | 12 | 18 | 36 |

**d.**

| $x$ | 1 | 4 | 8 | 10 |
|---|---|---|---|---|
| $y$ | 2 | 32 | 128 | 200 |

**e.**

| $x$ | 2 | 4 | 10 | 40 |
|---|---|---|---|---|
| $y$ | 60 | 30 | 12 | 3 |

**f.**

| $x$ | 25 | 50 | 75 | 100 |
|---|---|---|---|---|
| $y$ | 12 | 6 | 4 | 3 |

2. In which of the following formulas and equations do the variables vary inversely?

    **a.** $mn = 10$      **b.** $a = \dfrac{25}{b}$      **c.** $m = 9n$      **d.** $E = \dfrac{200}{I}$

    **e.** $R = \dfrac{s}{30}$      **f.** $LA = 1{,}000$      **g.** $d = \dfrac{C}{3.14}$      **h.** $96 = Bh$

    **i.** $\dfrac{v}{t} = 60$      **j.** $Fd = 24$      **k.** $C = \pi d$      **l.** $\dfrac{160}{x} = y$

3. Write each of the following relationships as an equation, using $k$ as the constant.

    **a.** When the area of a rectangle is constant, its width ($w$) varies inversely as its length ($l$).

    **b.** The angular measure ($a$) of a central angle of a polygon varies inversely as the number ($n$) of sides.

    **c.** The yield index ($y$) of a bond varies inversely as its price ($p$).

    **d.** Electric current ($I$) varies inversely as the resistance ($R$).

4. Find the missing word (increased, decreased, doubled, halved, etc.) in each of the following:

    **a.** In the formula $a = \dfrac{360}{n}$, if $n$ is increased, then $a$ is _____.

    **b.** In the formula $r = \dfrac{25}{b}$, if $b$ is halved, then $r$ is _____.

5. What effect does:

    **a.** Doubling the value of $A$ have on the value of $L$ in the formula $L = \dfrac{W}{A}$ when $W$ is constant?

    **b.** Halving the value of $t$ have on the value of $a$ in the formula $a = \dfrac{v}{t}$ when $v$ is constant?

    **c.** Tripling the base of a parallelogram have on the value of its height if the area remains the same? Formula: $A = bh$

6. When $y$ varies inversely as $x$, find the value of the constant $k$:

    **a.** If $x = 2$ and $y = 7$      **b.** If $x = 8$ and $y = 16$      **c.** If $x = 9$ and $y = 3$

    **d.** If $x = 10$ and $y = 2$      **e.** If $x = 5$ and $y = 15$      **f.** If $x = 4$ and $y = 4$

    **g.** If $x = 6$ and $y = 20$      **h.** If $x = 14$ and $y = 11$      **i.** If $x = 3$ and $y = 45$

7. Suppose $y$ varies inversely as $x$, and:

    **a.** $y = 40$ when $x = 2$.      Find $y$ when $x = 10$.

    **b.** $y = 54$ when $x = 6$.      Find $y$ when $x = 18$.

    **c.** $y = 10$ when $x = 12$.      Find $y$ when $x = 3$.

    **d.** $y = 24$ when $x = 8$.      Find $y$ when $x = 4$.

    **e.** $y = 60$ when $x = 5$.      Find $y$ when $x = 12$.

8. In the formula:

    **a.** $r = \dfrac{400}{t}$, $r = 50$ when $t = 8$. Find $r$ when $t = 80$.

    **b.** $n = \dfrac{360}{a}$, $n = 72$ when $a = 5$. Find $n$ when $a = 15$.

    **c.** $I = \dfrac{E}{R}$, $I = 55$ when $R = 4$. Find $I$ when $R = 10$ and $E$ is constant.

1. Find in simplest form the ratio of:
   a. 21 to 14
   b. 6 to 15
   c. 8 in. to 1 ft.
   d. 2 hr. to 20 min.
   e. 5 dimes to 3 quarters
   f. 18 things to 1 dozen

2. In a school where there are 1,170 students and 45 teachers, what is the ratio of the number of students to the number of teachers?

3. Solve and check:

   a. $\dfrac{x}{48} = \dfrac{5}{12}$
   b. $\dfrac{11}{b} = \dfrac{7}{20}$
   c. $\dfrac{15}{9} = \dfrac{400 - y}{y}$
   d. $\dfrac{x - 5}{4} = \dfrac{6}{x + 5}$

4. Todd earned \$73.50 in 3 days. At that rate how many working days will it take him to earn \$245?

5. Transform each of the following formulas:

   a. $p = b + 2e$; solve for $e$
   b. $\dfrac{F}{W} = \dfrac{h}{d}$; solve for $d$

   c. $E = Ir + IR$; solve for $I$

   d. $\dfrac{1}{R} = \dfrac{1}{R_1} + \dfrac{1}{R_2} + \dfrac{1}{R_3}$; solve for $R_2$

   e. $S = \dfrac{n}{2}(a + l)$; solve for $a$
   f. $S = \dfrac{rl - a}{r - 1}$; solve for $r$

6. Derive a formula for $P$ in terms of $E$ and $R$, using the formulas $P = IE$ and $E = IR$.

7. a. Suppose $y$ varies directly as $x$, and $y = 108$ when $x = 9$. Find $y$ when $x = 27$.
   b. Suppose $y$ varies inversely as $x$, and $y = 108$ when $x = 9$. Find $y$ when $x = 27$.

8. What effect does doubling the side of a square have on its perimeter?

9. What effect does halving the radius of a circle have on its circumference?

10. In the formula:

    a. $H.P. = \dfrac{W}{746}$, does $H.P.$ vary directly or inversely as $W$?

    b. $a = \dfrac{360}{n}$, does $a$ vary directly or inversely as $n$?

## Keyed Achievement Test

The colored numeral indicates where help may be found.

1. Multiply as indicated: $\dfrac{b^2 - 3b + 2}{b^2 - 4b + 4} \cdot \dfrac{b^2 - 3b - 4}{b^2 - 1}$  [10-3]

Solve each of the following systems of equations:

2. Graphically: [13-2]
   $2x - 3y = 11$
   $x + 5y = -14$

3. By substitution method: [13-4]
   $y = 8 - 5x$
   $6x - 5y = 22$

4. Raise to the indicated power: **a.** $(-6a^8b^4c^5)^2$ **b.** $(-5x^4yz^7)^3$ [14-1]

5. Find each root: **a.** $\sqrt{81a^{10}x^6y^2}$ **b.** $(64m^{12}n^{21}x^9)^{\frac{1}{3}}$  [14-5 thru 14-7]

6. Simplify and combine: $4\sqrt{63} - 7\sqrt{40} + 10\sqrt{\frac{2}{5}} - 12\sqrt{175}$ [14-8]

7. Solve and check: $6\sqrt{3x + 4} = 42$  [14-11]

8. Solve and check: $\dfrac{n - 8}{9} = \dfrac{4}{n + 8}$  [15-1]

9. Solve by factoring and check: $x^2 - x - 72 = 0$  [15-3]

10. Solve $x^2 - 3x = 10$ by completing the square and check. [15-4]

11. Solve $8x^2 + 22x - 21 = 0$ by formula and check.  [15-5]

12. Find the value of $s$ when $A = 726$, using the formula $A = 6s^2$. [15-6]

13. There are 660 boys and 720 girls enrolled in a school. What is the ratio of: boys to girls? girls to boys? boys to entire enrollment? girls to the entire enrollment?  [16-1]

14. Solve and check:     **a.** $\dfrac{n}{72} = \dfrac{19}{24}$     **b.** $\dfrac{n-8}{n+6} = \dfrac{n+2}{n-4}$  [16-1]

15. Transform each of the following formulas:

    **a.** $d = 16t^2$; solve for $t$     **b.** $I = \dfrac{E}{R_1 + R_2}$; solve for $R_2$  [16-2]

16. Derive a formula for $A$ in terms of $\pi$, $r$, and $h$, using the formulas $A = \pi d(h + \frac{1}{2}d)$ and $d = 2r$.  [16-2]

17. **a.** Suppose $y$ varies inversely as $x$, and $y = 4$ when $x = 15$. Find $y$ when $x = 5$.  [16-4]

    **b.** Suppose $y$ varies directly as $x$, and $y = 144$ when $x = 9$. Find $y$ when $x = 72$.  [16-3]

18. In the formula $r = \dfrac{300}{t}$, does $r$ vary directly or inversely as $t$? If $t$ is doubled, what happens to $r$?  [16-3]

# INTRODUCTION

Since it is impossible to measure all distances and heights directly, engineers, surveyors, astronomers, and navigators find it necessary to employ indirect measurement. In this chapter we shall study several methods of measuring indirectly. The Pythagorean Theorem and the essentials of numerical trigonometry will be treated as applications of the algebraic processes and principles studied in the preceding chapters.

*Trigonometry* means triangle measure. We shall see that the sides and angles of a triangle are so related that when certain parts of a triangle are known, the other parts can be determined indirectly.

In the trigonometry of a right triangle we use standard labels as follows:

$\angle C$ is the right angle (90°).

$\angle A$ is the base acute angle.

$\angle B$ is the second acute angle.

$c$ (or side $AB$) is the hypotenuse and is opposite to $\angle C$.

$a$ (or side $BC$) is the altitude and is opposite to $\angle A$.

$b$ (or side $AC$) is the base and is opposite $\angle B$.

Observe that while $a$ is the side opposite to $\angle A$, it is an adjacent side of $\angle B$. Also while $b$ is the side opposite to $\angle B$, it is an adjacent side of $\angle A$.

In trigonometry, ratios of the sides of the right triangle and their relation to the acute angles are used to determine the unknown values of certain parts of the triangle when the values of the other parts are known.

The ratios of the sides of a right triangle are related to the acute angles as follows:

a. The ratio of the side opposite an acute angle in a right triangle to the adjacent side is called the *tangent* of the angle (abbreviated tan). Thus in right triangle *ABC*, $\tan A = \dfrac{a}{b}$ and $\tan B = \dfrac{b}{a}$.

b. The ratio of the side opposite and acute angle in a right triangle to the hypotenuse is called the *sine* of the angle (abbreviated sin).

Thus, $\sin A = \dfrac{a}{c}$ and $\sin B = \dfrac{b}{c}$.

c. The ratio of the adjacent side of an acute angle in a right triangle to the hypotenuse is called the *cosine* of the angle (abbreviated cos).

Thus, $\cos A = \dfrac{b}{c}$ and $\cos B = \dfrac{a}{c}$.

   The tangent ratio for a given angle is the same for right triangles of all sizes. This is also true of the sine and cosine ratios.

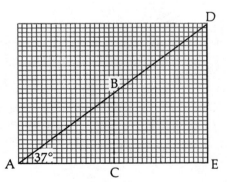

In triangle *ABC*, $\tan 37° = \dfrac{\text{side } BC}{\text{side } AC} = \dfrac{15}{20}$ or $.75$,

and in triangle *ADE*, $\tan 37° = \dfrac{\text{side } DE}{\text{side } AE} = \dfrac{30}{40}$ or $.75$.

In triangle *ABC*, $\sin 37° = \dfrac{\text{side } BC}{\text{side } AB} = \dfrac{15}{25}$ or .6,

and in triangle *ADE*, $\sin 37° = \dfrac{\text{side } DE}{\text{side } AD} = \dfrac{30}{50} = .6.$

In triangle *ABC*, $\cos 37° = \dfrac{\text{side } AC}{\text{side } AB} = \dfrac{20}{25}$ or .8,

and in triangle *ADE*, $\cos 37° = \dfrac{\text{side } AE}{\text{side } AD} = \dfrac{40}{50} = .8.$

These ratios depend upon the size of the acute angles and not upon the size of the triangle. The table on page 633 gives the trigonometric values of these ratios for angles ranging from 0° to 90°.

Angles are measured both vertically and horizontally. An angle measured vertically between the horizontal line and the observer's line of sight to an object is called the *angle of elevation* when the object is above the observer and the *angle of depression* when the object is below the observer.

# Trigonometry

# Complementary Angles in a Right Triangle

I. **Aim**  To determine the measure of an acute angle of a right triangle when the measure of the other acute angle is known.

II. **Procedure**
1. Observe that:
   a. The sum of measures of the three angles of any triangle is $180°$.
   b. *Complementary angles* are two angles whose sum of measures is $90°$. Each angle is said to be the complement of the other.
   c. The symbol $m \angle B$ is read "the measure of angle B."
   d. Since the right angle in the right triangle measures $90°$, the sum of measures of the remaining two angles is $90°$ ($180° - 90° = 90°$). Each of these remaining two angles is an acute angle and each is the complement of the other.

   That is, in right triangle $ABC$, if $m \angle C = 90°$,
   $m \angle A = 90° - m \angle B$ or simply $A = 90° - B$,
   and $m \angle B = 90° - m \angle A$ or simply $B = 90° - A$.

2. To determine the measure of one of the acute angles of a right triangle when the measure of the other acute angle is known:
   a. Select the proper formula.
   b. Substitute the given value.
   c. Then subtract as indicated.

3. Since $90° = 89°60'$ (eighty-nine degrees, 60 minutes), to find the angular measure of the complement of an angle whose measure is in degrees and minutes, subtract this measure from $89°60'$.

III. **Sample Solutions**
In right triangle $ABC$ if $m \angle C = 90°$,

1. Find $m \angle A$ if $m \angle B = 40°$
   Solution:
   $m \angle A = 90° - m \angle B$
   $m \angle A = 90° - 40°$
   $m \angle A = 50°$

   *Answer:* $m \angle A = 50°$

2. Find $m \angle B$ if $m \angle A = 28°40'$
   Solution:
   $m \angle B = 90° - m \angle A$
   $m \angle B = 90° - 28°40'$
   $m \angle B = 61°20'$

   *Answer:* $61°20'$

## DIAGNOSTIC TEST

1. What is the measure of the angle that is the complement of an angle whose measure is 57°?
2. What is the measure of the angle that is the complement of an angle whose measure is 18°25′?

In right triangle $ABC$ if $m\angle C = 90°$, find $m\angle A$:

3. When $m\angle B = 68°$.  4. When $m\angle B = 21°40′$.

In right triangle $ABC$ if $m\angle C = 90°$, find $m\angle B$:

5. When $m\angle A = 11°$.  6. When $m\angle A = 77°8′$.

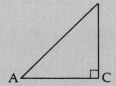

## *RELATED PRACTICE EXAMPLES*

1. Find the measure of the angle that is the complement of an angle whose measure is:

   **a.** 30°  **b.** 76°  **c.** 17°  **d.** 2°  **e.** 45°  **f.** 84°  **g.** 27°  **h.** 63°

2. Find the measure of the angle that is the complement of an angle whose measure is:

   **a.** 43°15′  **b.** 11°48′  **c.** 72°50′  **d.** 37°23′  **e.** 22°5′
   **f.** 89°12′  **g.** 66°35′  **h.** 50°45′  **i.** 4°8′  **j.** 78°29′

In right triangle $ABC$ if $m\angle C = 90°$, find $m\angle A$ when:

3. **a.** $m\angle B = 59°$  **b.** $m\angle B = 28°$   4. **a.** $m\angle B = 18°30′$  **b.** $m\angle B = 61°45′$
   **c.** $m\angle B = 82°$  **d.** $m\angle B = 9°$     **c.** $m\angle B = 73°6′$  **d.** $m\angle B = 5°27′$

In right triangle $ABC$ if $m\angle C = 90°$, find $m\angle B$ when:

5. **a.** $m\angle A = 33°$  **b.** $m\angle A = 62°$   6. **a.** $m\angle A = 19°15′$  **b.** $m\angle A = 70°48′$
   **c.** $m\angle A = 7°$  **d.** $m\angle A = 55°$     **c.** $m\angle A = 3°24′$  **d.** $m\angle A = 45°45′$

## *MAINTENANCE PRACTICE IN ARITHMETIC*

1. Add:

   829,674
   978,597

2. Subtract:

   28,708,006
   9,619,096

3. Multiply:

   8,409
   9,567

4. Divide:

   $785\overline{)712,780}$

5. Add:

   $8\frac{3}{4} + 1\frac{5}{6} + 3\frac{1}{3}$

6. Subtract:

   $6 - 5\frac{11}{12}$

7. Multiply:

   $16 \times 1\frac{11}{24}$

8. Divide:

   $2\frac{2}{3} \div 1\frac{3}{5}$

9. Add:

   $.056 + .2 + 3.7$

10. Subtract:

    $.004 - .0003$

11. Multiply:

    $.025 \times .08$

12. Divide:

    $1.5\overline{)6}$

13. Find 100% of 67.

14. 25 is what percent of 40?

15. 25% of what number is 40?

# Pythagorean Theorem

**I. Aim**  To solve problems finding the length of a side of a right triangle using the Pythagorean theorem.

**II. Procedure**

**1.** Observe that:

   **a.** The Pythagorean theorem expresses the relationship of the sides of a right triangle as follows:

   The square of the length of the hypotenuse (side opposite the right angle) is equal to the sum of the squares of the lengths of the other two sides.

   **b.** This relationship is expressed by the formula $c^2 = a^2 + b^2$ where $c$ represents the hypotenuse; $a$, the altitude; and $b$, the base.

   **c.** The diagram at the right illustrates this theorem by showing that the area of the square drawn on the hypotenuse (25 square units) is equal to the sum of the areas of the squares drawn on the altitude and base (16 square units and 9 square units respectively).

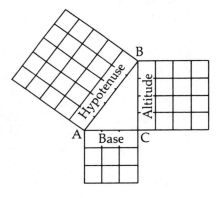

**2.** If any two sides of a right triangle are known, find the third side by the Pythagorean relation expressed in one of the following simplified forms:

$$c = \sqrt{a^2 + b^2} \qquad a = \sqrt{c^2 - b^2} \qquad b = \sqrt{c^2 - a^2}$$

**3.** To solve problems:

   **a.** Draw a right triangle and indicate the standard labels and given measurements.

   **b.** Select the proper formula.

   **c.** Substitute the given numbers in the formula.

   **d.** Perform the necessary operations.

## III. Sample Solutions

**1.** In the right triangle *ABC,* find the hypotenuse if the altitude is 36 meters and the base is 15 meters.

*Solution:*

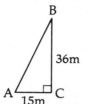

$a = 36\text{m}$
$b = 15\text{m}$
$c = ?$

*Answer:* 39m

$c = \sqrt{a^2 + b^2}$
$c = \sqrt{(36)^2 + (15)^2}$
$c = \sqrt{1296 + 225}$
$c = \sqrt{1521}$
$c = 39$ meters

$$\begin{array}{r} 1296 \\ +\ 225 \\ \hline 1521 \end{array}$$

$\sqrt{1521} = 39$

**2.** In the right triangle *ABC,* find the altitude if the hypotenuse is 40 feet and the base is 24 feet.

*Solution:*

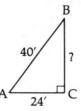

$c = 40$ ft.
$b = 24$ ft.
$a = ?$

*Answer:* 32 ft.

$a = \sqrt{c^2 - b^2}$
$a = \sqrt{(40)^2 - (24)^2}$
$a = \sqrt{1600 - 576}$
$a = \sqrt{1024}$
$a = 32$ ft.

$$\begin{array}{r} 1600 \\ -\ 576 \\ \hline 1024 \end{array}$$

$\sqrt{1024} = 32$

## DIAGNOSTIC TEST

**1.** Find the hypotenuse of the following right triangle:

**2.** Find the hypotenuse of a right triangle if its altitude is 135 in. and its base is 84 in.

**3.** Find the altitude of the following right triangle:

**4.** Find the altitude of a right triangle if its hypotenuse is 200 ft. and its base is 56 ft.

**5.** Find the base of the following right triangle:

**6.** Find the base of a right triangle if its hypotenuse is 122 m and its altitude is 120 m.

**7.** How high up on a wall does a 25-foot ladder reach if the foot of the ladder is 7 feet from the wall?

## *RELATED PRACTICE EXAMPLES*

1. Find the hypotenuse **a.**   **b.**   **c.**   **d.**
   of each:

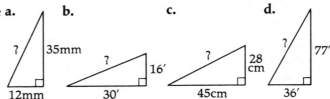

2. Find to the nearest hundredth the hypotenuse of a right triangle if:
   **a.** altitude = 21 cm    **b.** altitude = 70 yd.    **c.** altitude = 100 ft.
        base = 72 cm.        base = 24 yd.        base = 125 ft.

3. Find the altitude **a.**   **b.**   **c.**   **d.**
   of each:

 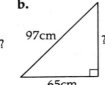  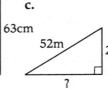

4. Find to the nearest hundredth the altitude of a right triangle if:
   **a.** hypotenuse = 65 in. **b.** hypotenuse = 81 m **c.** hypotenuse = 255 ft.
        base = 63 in.          base = 59 m.          base = 108 ft.

5. Find the base **a.**   **b.**   **c.**   **d.**
   of each:

 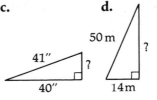 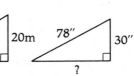

6. Find to the nearest hundredth the base of a right triangle if:

   **a.** hypotenuse = 146 yd. **b.** hypotenuse = 78 ft. **c.** hypotenuse = 325 m
        altitude = 96 yd.        altitude = 60 ft.        altitude = 165 m.

7. **a.** An empty lot is 216 meters long and 195 meters wide. Find the distance
   a boy walks if he crosses the lot diagonally from one corner to another.
   **b.** What is the distance from home plate to second base if the distance
   between bases is 90 feet?
   **c.** Two telephone poles, 32 feet and 39 feet high respectively, are 80 feet
   apart. How long is the wire between the tops of the poles?
   **d.** A girl lets out 100 feet of string on a kite. The distance from a point on
   the ground directly under the kite to where she stands is 60 feet. If the
   girl holds the string 4 feet from the ground, how high is the kite?

# Tangent Ratio

## I. Aim

To solve problems by the tangent ratio.

## II. Procedure

1. Draw a right triangle and indicate the standard labels and given measurement.
2. Select the proper formula.
3. Substitute the given values, using the table of trigonometric values to find the trigonometric value of the tangent of the given angle.
4. Solve the resulting equation.
5. When the measure of the given angle is expressed in degrees and minutes, add the calculated tangent value of the minutes to the tangent value of the degrees. See sample solution 2.
6. To find the angle when the trigonometric value of the tangent of the angle is known:
   a. Find this value in the table in the tangent column and take the corresponding degree measure from the degree column.
   b. If the given value is not in the table, calculate the number of minutes as illustrated in sample solution 4.

## III. Sample Solutions

1. Find the value of $\tan 53°$.     $\tan 53° = 1.3270$    *Answer:* 1.3270

2. Find the value of $\tan 31°45'$.
   *Solution:*

   | | | |
   |---|---|---|
   | $\tan 32° = .6249$ | $\frac{45}{60} = \frac{3}{4}$ | $\tan 31° = .6009$ |
   | $\tan 31° = \underline{.6009}$ | $\frac{3}{4} \times .0240 = .0180$ | $\frac{3}{4} \times .0240 = \underline{.0180}$ |
   | difference    .0240 | | $\tan 31°45' = .6189$ |

   *Answer:* .6189

3. Find $m \angle B$ if $\tan B = .8693$.
   $\tan 41° = .8693$     Thus $m \angle B = 41°$    *Answer:* 41°

4. Find $m \angle A$ if $\tan A = 4.8146$.
   *Solution:* The value 4.8146 is between 4.7046 and 5.1446 in the tangent column.

   | | | |
   |---|---|---|
   | $\tan 79° = 5.1446$ | $\tan A = 4.8146$ | $\frac{.1100}{.4400} = \frac{1}{4}$ |
   | $\tan 78° = \underline{4.7046}$ | $\tan 48° = \underline{4.7046}$ | |
   | difference 0.4400 | 0.1100 | $\frac{1}{4} \times 60' = 15'$ |
   | | Thus $m \angle A = 78°15'$ | |

   *Answer:* 78°15'

**5.** Find side $a$ if $m\angle A = 64°$ and $b = 125$ ft. Find answer to the nearest tenth.

*Solution:*

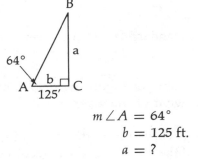

$$m\angle A = 64°$$
$$b = 125 \text{ ft.}$$
$$a = ?$$

*Answer:* 256.3 ft.

$$\tan A = \frac{a}{b}$$

$$\tan 64° = \frac{a}{125}$$

$$2.0503 \approx \frac{a}{125}$$

$$2.0503 \times 125 \approx a$$
$$256.2875 \approx a$$
$$a \approx 256.3 \text{ ft.}$$

**6.** Find side $a$ if $m\angle B = 22°$ and $b = 80.8$ meters.

*Solution 1*      or      *Solution 2*

$$\tan B = \frac{b}{a}$$

$$\tan 22° = \frac{80.8}{a}$$

$$.404 \approx \frac{80.8}{a}$$

$$.404a \approx 80.8$$

$$a \approx 200 \text{ meters}$$

$$m\angle A = 90° - m\angle B$$
$$m\angle A = 90° - 22°$$
$$m\angle A = 68°$$

$$\tan A = \frac{a}{b}$$

$$\tan 68° = \frac{a}{80.8}$$

$$2.4751 \approx \frac{a}{80.8}$$

$$2.4751 \times 80.8 \approx a$$
$$199.98808 \approx a$$
$$a \approx 200 \text{ meters}$$

$$m\angle B = 22°$$
$$b = 80.8 \text{ m}$$
$$a = ?$$

*Answer:* 200 meters

**7.** Find $m\angle A$ if $a = 182$ m and $b = 500$ m.

*Solution:*

$$a = 182 \text{ meters}$$
$$b = 500 \text{ meters}$$
$$m\angle A = ?$$

$$\tan A = \frac{a}{b}$$

$$\tan A = \frac{182}{500}$$

$$\tan A = .3640$$
$$m\angle A = 20°$$

*Answer:* 20°

## DIAGNOSTIC TEST

Use the table of trigonometric values on page 633 to determine the following:

Find the value of:

**1.** tan 32°                  **2.** tan 64°30′

**3.** Find $m \angle A$ if tan $A = .4245$     **4.** Find $m \angle B$ if tan $B = 1.5852$

Before solving any of the following problems, draw for each a right triangle, if one is not given, and indicate the standard labels and given measurements.

**5.** Find the indicated part of the following right triangle:

**6.** Find side $a$ to nearest tenth if $m \angle A = 20°$ and $b = 50$ meters.

**7.** Find side $b$ to nearest tenth if $m \angle B = 53°$ and $a = 200$ feet.

**8.** Find side $b$ to nearest tenth if $m \angle A = 29°$ and $a = 25$ millimeters.

**9.** Find the indicated part of the following right triangle:

**10.** Find side $a$ to nearest tenth if $m \angle B = 68°$ and $b = 150$ yards.

**11.** $m \angle A$ in the following right triangle:

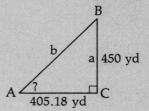

**12.** Find $m \angle A$ if $a = 132.7$ meters and $b = 100$ meters.

**13.** Find $m \angle B$ if $a = 500$ feet and $b = 404.9$ feet.

**14.** A lighthouse is 200 feet high. The angle of elevation of its top as viewed from a ship at sea is 15°. How far from the foot of the lighthouse is the ship?

## RELATED PRACTICE EXAMPLES

Use the table of trigonometric values on page 633 to determine the following:

1. Find the value of:
   a. tan 42°      b. tan 30°      c. tan 17°      d. tan 55°      e. tan 84°
   f. tan 8°       g. tan 45°      h. tan 27°      i. tan 64°      j. tan 73°

2. Find the value of:
   a. tan 27°30′       b. tan 67°30′       c. tan 9°15′        d. tan 82°45′
   e. tan 42°20′       f. tan 12°10′       g. tan 85°50′       h. tan 61°25′

3. Find $m \angle A$ in each of the following if:
   a. tan A = 1.3764        b. tan A = .9004        c. tan A = .2867
   d. tan A = 1.2799        e. tan A = 4.0108       f. tan A = .1405
   g. tan A = 8.1443        h. tan A = .7265        i. tan A = 1.0000

4. Find $m \angle B$ in each of the following if:
   a. tan B = .8542        b. tan B = .2962        c. tan B = 1.6323
   d. tan B = .2555        e. tan B = 3.1260       f. tan B = .9408
   g. tan B = 2.5726       h. tan B = .6049        i. tan B = 1.0540

Before solving any of the following problems, draw for each a right triangle, if one is not given, and indicate the standard labels and given measurements.

5. Find the indicated parts of the following right triangles:

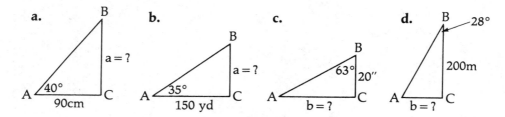

6. Find side *a* to nearest tenth if:
   a. $m \angle A = 45°$ and $b = 25$ mm.          b. $m \angle a = 61°$ and $b = 50$ ft.
   c. $m \angle A = 49°$ and $b = 100$ ft.          d. $m \angle A = 20°$ and $b = 75$ m.
   e. $m \angle A = 74°$ and $b = 18$ in.           f. $m \angle A = 58°45′$ and $b = 500$ yd.

7. Find side *b* to nearest tenth if:
   a. $m \angle B = 25°$ and $a = 10$ mi.           b. $m \angle B = 60°$ and $a = 30$ in.
   c. $m \angle B = 31°$ and $a = 25$ km.           d. $m \angle B = 75°$ and $a = 150$ ft.
   e. $m \angle B = 16°$ and $a = 32$ m.            f. $m \angle B = 23°30′$ and $a = 80$ yd.

8. Find side *b* to nearest tenth if:
   a. $m \angle A = 35°$ and $a = 50$ mm.           b. $m \angle A = 18°$ and $a = 200$ yd.
   c. $m \angle A = 29°$ and $a = 100$ ft.          d. $m \angle A = 74°$ and $a = 75$ cm.
   e. $m \angle A = 82°$ and $a = 270$ ft.          f. $m \angle A = 41°20′$ and $a = 60$ in.

9. Find the indicated parts of the following right triangles:

**a.** 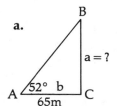   **b.** 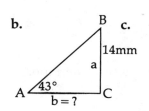   **c.** 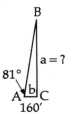   **d.**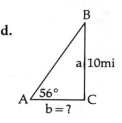

10. Find side $a$ to nearest tenth if:
   **a.** $m\angle B=39°$ and $b=30$ ft.        **b.** $m\angle B=48°$ and $b=100$ yd.
   **c.** $m\angle B=14°$ and $b=40$ in.        **d.** $m\angle B=22°$ and $b=75$ ft.
   **e.** $m\angle B=67°$ and $b=20$ mi.        **f.** $m\angle B=56°10'$ and $b=750$ yd.
11. **a.** Find $m\angle A$        **b.** Find $m\angle B$        **c.** Find $m\angle A$        **d.** Find $m\angle B$

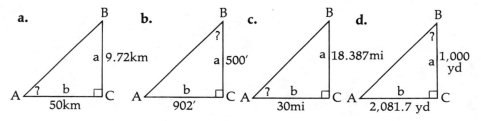

12. Find $m\angle A$ if:
   **a.** $a=3.64$ ft. and $b=10$ ft.          **b.** $a=373.21$ yd. and $b=100$ yd.
   **c.** $a=9.72$ km. and $b=50$ km.          **d.** $a=1,327$ ft. and $b=1,000$ ft.
   **e.** $a=144.97$ mi and $b=140$ mi.        **f.** $a=1,231.15$ ft. and $b=250$ ft.
13. Find $m\angle B$ if:
   **a.** $a=30$ m. and $b=7.479$ m.           **b.** $a=350$ ft. and $b=141.4$ ft.
   **c.** $a=750$ yd. and $b=3,008.1$ yd.      **d.** $a=340$ mm. and $b=340$ mm.
   **e.** $a=1,500$ ft. and $b=974.1$ ft.      **f.** $a=200$ in. and $b=189.82$ in.
14. In each of the following problems, first draw the figure, then label it:
   **a.** At a point 175 feet from the foot of a building the angle of elevation of
   the top is 60°. How high is the building?
   **b.** A chimney casts a shadow 62 feet long when the angle of the sun is
   41°45'. How high is the chimney?
   **c.** The angle of depression of a boat at sea is 29°. If a cliff is 180 feet
   above the river, how far is the boat from the foot of the cliff?
   **d.** A road rises 24 m in a distance of 240 m. Find its angle of elevation.
   **e.** An airplane has an altitude of 1,200 meters when it is directly over
   point $C$. If the pilot at that point finds that the angle of depression of
   a distant airport is 4°15', how far is point $C$ from the airport?

I. **Aim**    To solve problems by the sine ratio.

II. **Procedure**

Follow the procedure outlined in Exercise 17-3. However, use the trigono-metric values of the sine column instead of the tangent column of the table of trigonometric values.

III. **Sample Solutions**

**1.** Find side $a$ if $m \angle A = 74°$ and $c = 50$ ft. Find answer to nearest tenth.

*Solution:*

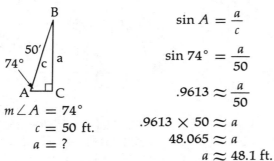

$m \angle A = 74°$
$c = 50$ ft.
$a = ?$

$$\sin A = \frac{a}{c}$$

$$\sin 74° = \frac{a}{50}$$

$$.9613 \approx \frac{a}{50}$$

$$.9613 \times 50 \approx a$$
$$48.065 \approx a$$
$$a \approx 48.1 \text{ ft.}$$

*Answer:* 48.1 ft.

**2.** Find $c$ if $m \angle A = 50°$ and $a = 383$ millimeters.

*Solution:*

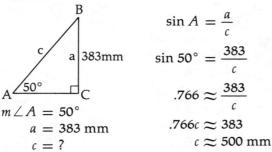

$m \angle A = 50°$
$a = 383$ mm
$c = ?$

$$\sin A = \frac{a}{c}$$

$$\sin 50° = \frac{383}{c}$$

$$.766 \approx \frac{383}{c}$$

$$.766c \approx 383$$
$$c \approx 500 \text{ mm}$$

*Answer:* 500 millimeters

**3.** Find $m \angle A$ if $a = 136.4$ ft. and $c = 200$ ft.

*Solution:*

$$\sin A = \frac{a}{c}$$

$$\sin A = \frac{136.4}{200}$$

$$\sin A = .6820$$

*Answer:* $43°$

$$m \angle A = 43°$$

## DIAGNOSTIC TEST

Use the table of trigonometric values on page 633 to:
Find the value of:

**1.** $\sin 43°$          **2.** $\sin 74°15'$

**3.** Find $m \angle B$ if $\sin B = .8572$     **4.** Find $m \angle A$ if $\sin A = .3665$

Before solving any of the following problems, draw for each a right triangle, if one is not given, and indicate the standard labels and given measurements.

**5.** Find the indicated part of the following right triangle:

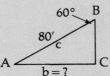

**6.** Find side $a$ to nearest tenth if $m \angle A = 32°$ and $c = 20$ kilometers.

**7.** Find side $b$ to nearest tenth if $m \angle B = 69°$ and $c = 75$ feet.

**8.** Find side $b$ to nearest tenth if $m \angle A = 47°$ and $c = 130$ yards.

**9.** Find side $a$ to nearest tenth if $m \angle B = 16°$ and $c = 10$ miles.

**10.** Find the hypotenuse of the following right triangle:

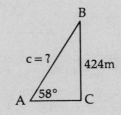

**11.** Find side $c$ if $m \angle A = 20°$ and $a = 342$ meters.

**12.** Find side $c$ if $m \angle B = 66°$ and $b = 1,827$ feet.

**13.** Find $m \angle B$ in the following right triangle:

**14.** Find $m \angle A$ if $a = 424$ meters and $c = 500$ meters.

**15.** Find $m \angle B$ if $b = 45$ yards and $c = 200$ yards.

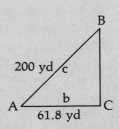

**16.** How high is a kite if the string makes an angle of $48°$ with the ground and 200 feet of string is let out?

## *RELATED PRACTICE EXAMPLES*

Use the table of trigonometric values on page 633 to determine the following:

1. Find the value of:
    **a.** sin 60°      **b.** sin 36°      **c.** sin 74°      **d.** sin 5°      **e.** sin 52°
    **f.** sin 41°      **g.** sin 30°      **h.** sin 24°      **i.** sin 57°      **j.** sin 86°

2. Find the value of:
    **a.** sin 42°30′      **b.** sin 22°45′      **c.** sin 65°15′      **d.** sin 37°40′
    **e.** sin 71°10′      **f.** sin 32°50′      **g.** sin 14°12′      **h.** sin 47°36′

3. Find $m \angle B$ in each of the following if:
    **a.** sin $B = .9511$          **b.** sin $B = .6018$          **c.** sin $B = .8480$
    **d.** sin $B = .4067$          **e.** sin $B = .0698$          **f.** sin $B = .7660$
    **g.** sin $B = .1908$          **h.** sin $B = .4848$          **i.** sin $B = .9986$

4. Find $m \angle A$ in each of the following if:
    **a.** sin $A = .1822$          **b.** sin $A = .9170$          **c.** sin $A = .7518$
    **d.** sin $A = .3638$          **e.** sin $A = .9824$          **f.** sin $A = .2868$
    **g.** sin $A = .0756$          **h.** sin $A = .9979$          **i.** sin $A = .5892$

Before solving any of the following problems, draw for each a right triangle if one is not given, and indicate the standard labels and given measurements.

5. Find the indicated parts of the following right triangles:

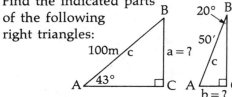
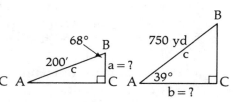

6. Find side $a$ to nearest tenth if:
    **a.** $m \angle A = 30°$ and $c = 20$ mm.          **b.** $m \angle A = 18°$ and $c = 50$ ft.
    **c.** $m \angle A = 5°$ and $c = 25$ cm.          **d.** $m \angle A = 66°$ and $c = 500$ yd.
    **e.** $m \angle A = 83°$ and $c = 32$ mi.          **f.** $m \angle A = 33°30′$ and $c = 100$ yd.

7. Find side $b$ to nearest tenth if:
    **a.** $m \angle B = 56°$ and $c = 30$ ft.          **b.** $m \angle B = 13°$ and $c = 20$ mi.
    **c.** $m \angle B = 35°$ and $c = 75$ mm.          **d.** $m \angle B = 50°$ and $c = 340$ yd.
    **e.** $m \angle B = 74°$ and $c = 60$ mi.          **f.** $m \angle B = 86°$ 24′ and $c = 200$ m.

8. Find side $b$ to nearest tenth if:
    **a.** $m \angle A = 27°$ and $c = 50$ yd.          **b.** $m \angle A = 45°$ and $c = 75$ km.
    **c.** $m \angle A = 66°$ and $c = 100$ mi.          **d.** $m \angle A = 54°$ and $c = 83$ m.
    **e.** $m \angle A = 15°$ and $c = 300$ ft.          **f.** $m \angle A = 41°15′$ and $c = 250$ yd.

9. Find side $a$ to nearest tenth if:
    **a.** $m \angle B = 77°$ and $c = 10$ in.          **b.** $m \angle B = 21°$ and $c = 1,000$ ft.
    **c.** $m \angle B = 16°$ and $c = 500$ m.          **d.** $m \angle B = 47°$ and $c = 55$ km.
    **e.** $m \angle B = 54°$ and $c = 150$ mi.          **f.** $m \angle B = 23°12′$ and $c = 2,000$ yd.

**10.** Find the hypotenuse of each of the following right triangles:

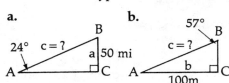

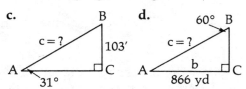

**11.** Find side $c$ to nearest tenth if:
    **a.** $m\angle A = 13°$ and $a = 450$ ft.    **b.** $m\angle A = 30°$ and $a = 100$ in.
    **c.** $m\angle A = 56°$ and $a = 829$ yd.    **d.** $m\angle A = 67°$ and $a = 150$ ft.
    **e.** $m\angle A = 86°$ and $a = 200$ yd.    **f.** $m\angle A = 41°30'$ and $a = 500$ yd.

**12.** Find side $c$ to nearest tenth if:
    **a.** $m\angle B = 43°$ and $b = 341$ ft.    **b.** $m\angle B = 27°$ and $b = 227$ yd.
    **c.** $m\angle B = 52°$ and $b = 394$ yd.    **d.** $m\angle B = 18°$ and $b = 618$ ft.
    **e.** $m\angle B = 84°$ and $b = 400$ yd.    **f.** $m\angle B = 66°18'$ and $b = 1{,}000$ ft.

**13. a.** Find $m\angle A$    **b.** Find $m\angle B$    **c.** Find $m\angle A$    **d.** Find $m\angle B$

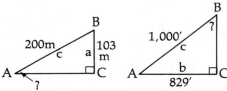

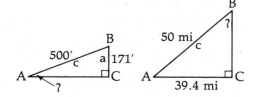

**14.** Find $m\angle A$ if:
    **a.** $a = 20$ km. and $c = 40$ km.    **b.** $a = 90$ yd. and $c = 400$ yd.
    **c.** $a = 103$ m. and $c = 200$ m.    **d.** $a = 1{,}272$ ft. and $c = 1{,}500$ ft.
    **e.** $a = 89.1$ ft. and $c = 1{,}000$ ft.    **f.** $a = 25$ in. and $c = 30$ in.

**15.** Find $m\angle B$ if:
    **a.** $b = 32$ cm. and $c = 64$ cm.    **b.** $b = 433$ yd. and $c = 500$ yd.
    **c.** $b = 1{,}023$ yd. and $c = 1{,}500$ yd.    **d.** $b = 41.2$ m. and $c = 80$ m.
    **e.** $b = 53.46$ ft. and $c = 60$ ft.    **f.** $b = 11$ mi. and $c = 16$ mi.

**16.** In each of the following problems, first draw the figure, then label it:
    **a.** A 40-foot ladder makes an angle of 78° with the ground. How far up the side of the building does it reach?
    **b.** How high is a kite if 180 feet of string is let out and the string makes an angle of 47° with the ground?
    **c.** From the top of a cliff 200 feet high, the angle of depression of a boat is 54°. Find the distance from the boat to the top of the cliff.
    **d.** The angle of elevation of a bridge approach is 5°. How high above the street is an auto after it is driven 0.3 km up the approach?
    **e.** A 400-meter ramp rises a vertical distance of 60 meters at its peak. Find the angle of elevation of the ramp.

# Cosine Ratio

## I. Aim    To solve problems by the cosine ratio.

## II. Procedure

Follow the procedure outlined in 17-3 but with the following exceptions:
1. Use the trigonometric value of the cosine column instead of the tangent column of the table of trigonometric values.
2. To find the trigonometric value of the cosine of an angle given in degrees and minutes, subtract the calculated cosine value of the minutes from the cosine value of the degrees. See sample solution 1.
3. To find the angle when the trigonometric value of the cosine of the angle is not on the table, calculate the number of minutes as illustrated in sample solution 2.

## III. Sample Solutions

1. Find the value of 4°30′.

$$\cos 4° = .9976$$
$$\cos 5° = .9962$$
difference    .0014

$$\tfrac{30}{60} = \tfrac{1}{2}$$
$$\tfrac{1}{2} \times .0014 = .0007$$

$$\cos 4° = .9976$$
$$\tfrac{1}{2} \times .0014 = .0007$$
$$.9969$$

*Answer: 9969*

2. Find $m \angle A$ if $\cos A = .6665$.
   *Solution:* The value .6665 is between .6691 and .6665 in the cosine column.

$$\cos 48° = .6691$$
$$\cos 49° = .6561$$
difference    .0130

$$\cos 48° = .6691$$
$$\cos A = .6665$$
difference    .0026

$$\frac{.0026}{.0130} = \frac{1}{5}$$
$$\frac{1}{5} \times 60' = 12'$$
Thus $m \angle A = 48°12'$

*Answer: 48°12′*

3. Find side $b$ if $m \angle A = 57°$ and $c = 20$ in. Find answer to nearest tenth.
   *Solution:*

$$m \angle A = 57°$$
$$c = 20 \text{ in.}$$
$$b = ?$$

$$\cos A = \frac{b}{c}$$

$$\cos 57° = \frac{b}{20}$$

$$.5446 \approx \frac{b}{20}$$

$$.5446 \times 20 \approx b$$
$$10.892 \approx b$$
$$b \approx 10.9 \text{ in.} \quad \textit{Answer: 10.9 in.}$$

628

## DIAGNOSTIC TEST

Use the table of trigonometric values on page 633 to determine the following:

Find the value of:

**1.** cos 26°                                          **2.** cos 34°15′

**3.** Find $m \angle A$ if cos $A$ = .5736          **4.** Find $m \angle B$ if cos $B$ = .7518

Before solving any of the following problems, draw for each a right triangle, if one is not given, and indicate the standard labels and given measurements.

**5.** Find the indicated part of the following right triangle:

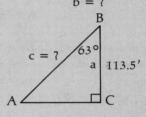

**6.** Find side $b$ to nearest tenth if $m \angle A$ = 55° and $c$ = 60 km.

**7.** Find side $a$ to nearest tenth if $m \angle B$ = 17° and $c$ = 25 m.

**8.** Find the hypotenuse of the following right triangle:

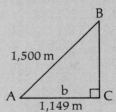

**9.** Find side $c$ if $m \angle A$ = 36° and $b$ = 161.8 ft.

**10.** Find side $c$ if $m \angle B$ = 77° and $a$ = 450 km.

**11.** Find $m \angle A$ in the following right triangle:

**12.** Find $m \angle A$ if $b$ = 227 meters and $c$ = 500 meters.

**13.** Find $m \angle B$ if $a$ = 70 yd. and $c$ = 4,000 yd.

**14.** The angle of elevation of a ramp is 6°. How long is the ramp if its horizontal distance is 300 feet?

## RELATED PRACTICE EXAMPLES

Use the table of trigonometric values on page 633 to determine the following:

1. Find the value of:

   **a.** cos 72°     **b.** cos 14°     **c.** cos 86°     **d.** cos 31°     **e.** cos 5°

   **f.** cos 47°     **g.** cos 22°     **h.** cos 64°     **i.** cos 50°     **j.** cos 29°

2. Find the value of:

   **a.** cos 24°30′     **b.** cos 34°15′     **c.** cos 75°45′     **d.** cos 6°20′

   **e.** cos 42°40′     **f.** cos 56°12′     **g.** cos 17°48′     **h.** cos 69°50′

3. Find $m \angle A$ in each of the following if:
   **a.** $\cos A = .7771$      **b.** $\cos A = .2419$      **c.** $\cos A = .5878$
   **d.** $\cos A = .9976$      **e.** $\cos A = .3256$      **f.** $\cos A = .0872$
   **g.** $\cos A = .7314$      **h.** $\cos A = .6691$      **i.** $\cos A = .4384$
4. Find $m \angle B$ in each of the following if:
   **a.** $\cos B = .7132$      **b.** $\cos B = .3543$      **c.** $\cos B = .1779$
   **d.** $\cos B = .8526$      **e.** $\cos B = .1103$      **f.** $\cos B = .9977$
   **g.** $\cos B = .4514$      **h.** $\cos B = .9603$      **i.** $\cos B = .5962$

Before solving any of the following problems, draw for each a right triangle, if one is not given, and indicate the standard labels and given measurements.

5. Find the indicated parts of the following right triangles:

**a.**      **b.**      **c.**      **d.**

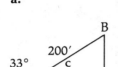   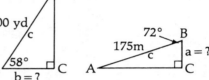

6. Find side $b$ to nearest tenth if:
   **a.** $m \angle A = 65°$ and $c = 30$ mm.      **b.** $m \angle A = 28°$ and $c = 80$ yd.
   **c.** $m \angle A = 49°$ and $c = 16$ in.      **d.** $m \angle A = 16°$ and $c = 35$ mi.
   **e.** $m \angle A = 72°$ and $c = 100$ km.      **f.** $m \angle A = 38°30'$ and $c = 300$ ft.

7. Find side $a$ to nearest tenth if:
   **a.** $m \angle B = 40°$ and $c = 50$ ft.      **b.** $m \angle B = 35°$ and $c = 25$ km.
   **c.** $m \angle B = 81°$ and $c = 100$ yd.      **d.** $m \angle B = 4°$ and $c = 250$ m.
   **e.** $m \angle B = 23°$ and $c = 425$ mi.      **f.** $m \angle B = 46°15'$ and $c = 200$ ft.

8. Find the hypotenuse of each of the following right triangles:

**a.**      **b.**      **c.**      **d.**

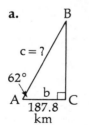 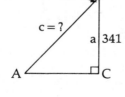 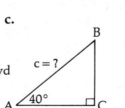 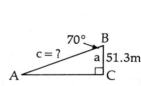

9. Find side $c$ to the nearest tenth if:
   **a.** $m \angle A = 60°$ and $b = 16$ km.      **b.** $m \angle A = 77°$ and $b = 45$ ft.
   **c.** $m \angle A = 32°$ and $b = 424$ mi.      **d.** $m \angle A = 70°$ and $b = 17.1$ m.
   **e.** $m \angle A = 47°$ and $b = 102.3$ yd.
   **f.** $m \angle A = 23°12'$ and $b = 36.764$ mi.

10. Find side $c$ to the nearest tenth if:
    a. $m \angle B = 7°$ and $a = 397$ ft.       b. $m \angle B = 59°$ and $a = 515$ m.
    c. $m \angle B = 84°$ and $a = 2.09$ mi.     d. $m \angle B = 36°$ and $a = 24.27$ yd.
    e. $m \angle B = 24°$ and $a = 182.7$ km.
    f. $m \angle B = 56°30'$ and $a = 27.595$ mi.

11. a. Find $m \angle A$       b. Find $m \angle B$       c. Find $m \angle A$       d. Find $m \angle B$

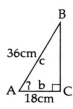 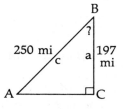 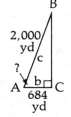 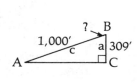

12. Find $m \angle A$ if:
    a. $b = 90$ m. and $c = 400$ m.          b. $b = 227$ yd. and $c = 500$ yd.
    c. $b = 829$ ft. and $c = 1,000$ ft.     d. $b = 412$ mi. and $c = 800$ mi.
    e. $b = 43.3$ km. and $c = 50$ km.       f. $b = 5$ ft. and $c = 13$ ft.

13. Find $m \angle B$ if:
    a. $a = 341$ ft. and $c = 500$ ft.        b. $a = 248$ yd. and $c = 496$ yd.
    c. $a = 189.1$ mi. and $c = 200$ mi.      d. $a = 40.45$ km. and $c = 50$ km.
    e. $a = 47.541$ ft. and $c = 65$ ft.      f. $a = 36$ cm. and $c = 39$ cm.

14. In each of the following problems, first draw the figure, then label it:
    a. The base of a right triangle is 20 centimeters and the angle opposite the altitude is 43°. Find the hypotenuse.

    b. Find the distance from a point on the ground directly under a kite to where the boy flying it stands if 250 feet of string has been let out and the string makes an angle of 53° with the ground.

    c. A boat sails 75 kilometers on a course $E\ 32°\ N$. How far east did the boat sail?

    d. The angle of elevation of a ramp is 7°. How long is the ramp if its horizontal distance is 220 feet.

    e. A 30-foot ladder leans against a wall and its foot is 8 feet from the wall. First determine the measure of the angle formed by the ladder and the ground. Then determine the measure of the angle formed by the top of the ladder and the wall.

# REVIEW

1. In each of the following, find the measure of the third angle of a triangle when the other two angles measure:
   **a.** 52° and 17°    **b.** 41°32′ and 74°26′.
2. In right triangle $ABC$ if $m\angle C = 90°$, find:
   **a.** $m\angle B$ when $m\angle A = 34°$.    **b.** $m\angle A$ when $m\angle B = 81°43′$.
3. Find to the nearest hundredth:
   **a.** The base of a right triangle when its hypotenuse is 65 cm and its altitude is 33 cm.
   **b.** The hypotenuse of a right triangle when its altitude is 48 yd. and its base is 55 yd.
4. An airplane, flying 180 miles from airport $M$ due east to airport $N$, drifts off its course in a straight line and is 75 miles due north of airport $N$. What distance did the airplane actually fly?
5. Use the table of trigonometric values:
   **a.** To determine the value of: **(1)** tan 57° **(2)** sin 15° **(3)** cos 81° **(4)** tan 9°45′ **(5)** cos 26°30′ **(6)** sin 70°20′
   **b.** To find: **(1)** $m\angle A$ if tan $A = .5543$ **(2)** $m\angle B$ if sin $B = .7986$ **(3)** $m\angle A$ if cos $A = .7193$
6. **a.** Find to nearest tenth:
   **(1)** Side $a$ if $m\angle A = 40°$ and $b = 100$ ft.
   **(2)** Side $b$ if $m\angle A = 62°$ and $a = 50$ yd.
   **(3)** Side $b$ if $m\angle B = 19°$ and $c = 200$ ft.
   **(4)** Side $c$ if $m\angle A = 54°$ and $a = 30$ km.
   **(5)** Side $b$ if $m\angle A = 73°$ and $c = 400$ yd.
   **b.** **(1)** Find $m\angle B$ if $a = 91$ mi. and $b = 250$ mi.
   **(2)** Find $m\angle A$ if $a = 63$ ft. and $b = 126$ ft.
7. Find all the missing measures (sides to nearest tenth and angles to nearest degree) in each of the following right triangles:

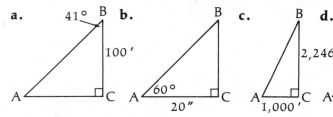

8. Find the height of a flagpole if, at a point 200 meters from the foot of the pole, the angle of elevation of the top is 12°.

# TABLE OF TRIGONOMETRIC VALUES

| Angle | Sine | Cosine | Tangent | Angle | Sine | Cosine | Tangent |
|-------|------|--------|---------|-------|------|--------|---------|
| 0° | .0000 | 1.0000 | .0000 | 46° | .7193 | .6947 | 1.0355 |
| 1° | .0175 | .9998 | .0175 | 47° | .7314 | .6820 | 1.0724 |
| 2° | .0349 | .9994 | .0349 | 48° | .7431 | .6691 | 1.1106 |
| 3° | .0523 | .9986 | .0524 | 49° | .7547 | .6561 | 1.1504 |
| 4° | .0698 | .9976 | .0699 | 50° | .7660 | .6428 | 1.1918 |
| 5° | .0872 | .9962 | .0875 | 51° | .7771 | .6293 | 1.2349 |
| 6° | .1045 | .9945 | .1051 | 52° | .7880 | .6157 | 1.2799 |
| 7° | .1219 | .9925 | .1228 | 53° | .7986 | .6018 | 1.3270 |
| 8° | .1392 | .9903 | .1405 | 54° | .8090 | .5878 | 1.3764 |
| 9° | .1564 | .9877 | .1584 | 55° | .8192 | .5736 | 1.4281 |
| 10° | .1736 | .9848 | .1763 | 56° | .8290 | .5592 | 1.4826 |
| 11° | .1908 | .9816 | .1944 | 57° | .8387 | .5446 | 1.5399 |
| 12° | .2079 | .9781 | .2126 | 58° | .8480 | .5299 | 1.6003 |
| 13° | .2250 | .9744 | .2309 | 59° | .8572 | .5150 | 1.6643 |
| 14° | .2419 | .9703 | .2493 | 60° | .8660 | .5000 | 1.7321 |
| 15° | .2588 | .9659 | .2679 | 61° | .8746 | .4848 | 1.8040 |
| 16° | .2756 | .9613 | .2867 | 62° | .8829 | .4695 | 1.8807 |
| 17° | .2924 | .9563 | .3057 | 63° | .8910 | .4540 | 1.9626 |
| 18° | .3090 | .9511 | .3249 | 64° | .8988 | .4384 | 2.0503 |
| 19° | .3256 | .9455 | .3443 | 65° | .9063 | .4226 | 2.1445 |
| 20° | .3420 | .9397 | .3640 | 66° | .9135 | .4067 | 2.2460 |
| 21° | .3584 | .9336 | .3839 | 67° | .9205 | .3907 | 2.3559 |
| 22° | .3746 | .9272 | .4040 | 68° | .9272 | .3746 | 2.4751 |
| 23° | .3907 | .9205 | .4245 | 69° | .9336 | .3584 | 2.6051 |
| 24° | .4067 | .9135 | .4452 | 70° | .9397 | .3420 | 2.7475 |
| 25° | .4226 | .9063 | .4663 | 71° | .9455 | .3256 | 2.9042 |
| 26° | .4384 | .8988 | .4877 | 72° | .9511 | .3090 | 3.0777 |
| 27° | .4540 | .8910 | .5095 | 73° | .9563 | .2924 | 3.2709 |
| 28° | .4695 | .8829 | .5317 | 74° | .9613 | .2756 | 3.4874 |
| 29° | .4848 | .8746 | .5543 | 75° | .9659 | .2588 | 3.7321 |
| 30° | .5000 | .8660 | .5774 | 76° | .9703 | .2419 | 4.0108 |
| 31° | .5150 | .8572 | .6009 | 77° | .9744 | .2250 | 4.3315 |
| 32° | .5299 | .8480 | .6249 | 78° | .9781 | .2079 | 4.7046 |
| 33° | .5446 | .8387 | .6494 | 79° | .9816 | .1908 | 5.1446 |
| 34° | .5592 | .8290 | .6745 | 80° | .9848 | .1736 | 5.6713 |
| 35° | .5736 | .8192 | .7002 | 81° | .9877 | .1564 | 6.3138 |
| 36° | .5878 | .8090 | .7265 | 82° | .9903 | .1392 | 7.1154 |
| 37° | .6018 | .7986 | .7536 | 83° | .9925 | .1219 | 8.1443 |
| 38° | .6157 | .7880 | .7813 | 84° | .9945 | .1045 | 9.5144 |
| 39° | .6293 | .7771 | .8098 | 85° | .9962 | .0872 | 11.4301 |
| 40° | .6428 | .7660 | .8391 | 86° | .9976 | .0698 | 14.3007 |
| 41° | .6561 | .7547 | .8693 | 87° | .9986 | .0523 | 19.0811 |
| 42° | .6691 | .7431 | .9004 | 88° | .9994 | .0349 | 28.6363 |
| 43° | .6820 | .7314 | .9325 | 89° | .9998 | .0175 | 57.2900 |
| 44° | .6947 | .7193 | .9657 | 90° | 1.0000 | .0000 | |
| 45° | .7071 | .7071 | 1.0000 | | | | |

# Keyed Inventory Test

The numeral at the end of each problem indicates the exercise where help may be found.

1. Express as a formula: The area of a circle $(A)$ is equal to one-fourth the product of Pi $(\pi)$ and the square of the diameter $(d)$. $\boxed{1\text{-}8}$

2. **a.** Add: $\boxed{2\text{-}11}$   **b.** Subtract: $\boxed{2\text{-}12}$ **c.** Multiply: $\boxed{2\text{-}13}$ **d.** Divide: $\boxed{2\text{-}14}$
   $(-6) + (-7)$    $(-3) - (+5)$    $(-9)(-2)$     $(+20) \div (-4)$

3. **a.** Find the opposite of $-10$. $\boxed{2\text{-}4}$     **b.** Find the additive inverse of $+.8$. $\boxed{2\text{-}4}$

   **c.** Find the absolute value:     **d.** Find the multiplicative
   $|-9| = ?$ $\boxed{2\text{-}3}$       inverse of $-\frac{3}{4}$. $\boxed{2\text{-}13}$

4. On a number line draw the graph of: $-6, -2, 0, 1, 3, 5$ $\boxed{2\text{-}7}$

5. Find the value of $3a^2 - 5(a - 4)$ when $a = 2$. $\boxed{3\text{-}4 \text{ and } 3\text{-}6}$

6. What is the value of $v$ when $V = 45$, $g = 32$, and $t = 10$, using the formula $v = V + gt$? $\boxed{3\text{-}7}$

7. Write a formula expressing the number of grams $(g)$ in $m$ milligrams. $\boxed{4\text{-}1}$

8. Find the sum of $12x^4 + 3x^2y^2 - 2y^4$, $9y^4 - 15x^4$, and $2x^4 - 8x^2y^2$. $\boxed{5\text{-}3}$

9. From $4r^2 - 9rs - s^2$ subtract $4r^2 - 3s^2$. $\boxed{5\text{-}5}$

10. Multiply $5m - 9n$ by $6m - 7n$. $\boxed{5\text{-}5}$

11. Divide $12c^2 + 4cd - 5d^2$ by $2c - d$. $\boxed{5\text{-}13}$

12. Remove parentheses and add like terms:
    $9b - (4b - 7) + (-8b + 3) - 10$ $\boxed{5\text{-}14}$

13. Solve and check: $\boxed{6\text{-}4 \text{ thru } 6\text{-}8}$

    **a.** $x + 6.5 = 9$    **b.** $r - \frac{2}{3} = 0$    **c.** $-24 = -6y$    **d.** $\frac{n}{9} = -1$

    **e.** $5s - 3 = 29 - 3s$        **f.** $(x - 5)(x - 2) = (x - 7)(x - 1)$

14. Draw on a number line the graph of $6x + 1 = -11$. $\boxed{6\text{-}11}$

15. Find the value of $I$ when $E = 126$, $r = 9$, and $R = 12$, using the formula $E = Ir + IR$. $\boxed{6\text{-}10}$

16. The length of a rectangle is 15 meters more than the width. Its perimeter is 86 meters. What do the length and width each measure? $\boxed{7\text{-}5}$

When the replacements for the variable are all the: $\boxed{8\text{-}4 \text{ thru } 8\text{-}8}$

17. Whole numbers, solve $3n - 5 < 10$.

18. Real numbers, draw the graph of $2x + 1 > 9$ on a number line. $\boxed{8\text{-}9}$

19. Find the product of each of the following at sight: $\boxed{9\text{-}1 \text{ thru } 9\text{-}5}$
    **a.** $(4x - 3y)(4x + 3y)$       **b.** $(6c^2 - 5d)^2$
    **c.** $(a + 7)(a - 9)$          **d.** $(3m - 4n)(7m + 8n)$

20. Factor completely: $\boxed{9\text{-}6 \text{ thru } 9\text{-}12}$
    **a.** $100b^2 - 49c^6d^8$       **b.** $16x^6y^2 + 48x^4y^4 + 64x^2y^8$
    **c.** $20a^2 + 41a - 9$        **d.** $36x^2 + 60xy + 25y^2$

**21.** Express in simplest form:

$$\frac{a^2 - 4a - 5}{a^2 - 10a + 25} \quad \boxed{10\text{-}2}$$

**22.** Divide: $\boxed{10\text{-}4}$

$$\frac{x^2 + 4x - 12}{3x + 12} \div \frac{x^2 + 9x + 18}{6x + 24}$$

**23.** Add or subtract as indicated: $\boxed{10\text{-}5 \text{ thru } 10\text{-}11}$

a. $\dfrac{1}{r} + \dfrac{1}{s}$   b. $\dfrac{d - t}{dt} - \dfrac{r - t}{rt}$   c. $\dfrac{3b}{b - c} + \dfrac{2c^2}{b^2 - c^2} - \dfrac{4c}{b + c}$

**24.** Solve and check: $\boxed{11\text{-}1 \text{ thru } 11\text{-}3}$

a. $\dfrac{7n}{12} - \dfrac{3n}{2} = 22$   b. $\dfrac{3x + 2}{4} - \dfrac{2x - 5}{5} = \dfrac{x}{2}$

c. $\dfrac{y - 4}{y - 2} = \dfrac{y - 5}{y - 4}$   d. $\dfrac{n}{n + 3} + \dfrac{n - 2}{n - 3} = \dfrac{2n^2 + 2}{n^2 - 9}$

e. $C + .35C = 405$   f. $.06a + .11(500 - a) = 45$

**25.** Find the value of $R_2$ when $I = 5$, $E = 110$, and $R_1 = 9$, using the formula

$$I = \frac{E}{R_1 + R_2}. \quad \boxed{11\text{-}4}$$

**26.** Without drawing the graph, determine the slope and the coordinates of the point where the graph of $3x + y = 5$ crosses the $Y$-axis. $\boxed{12\text{-}4 \text{ and } 12\text{-}5}$

**27.** Draw in the number plane the graph of $x - 2y = 6$. $\boxed{12\text{-}3}$

**28.** Solve graphically: $\begin{array}{l} 4x - y = 14 \\ x + 2y = -1 \end{array}$  **29.** Solve and check: $\begin{array}{l} 2x - 3y = 11 \\ 5x + 2y = 18 \end{array}$

$\boxed{13\text{-}2}$                    $\boxed{13\text{-}3 \text{ and } 13\text{-}4}$

**30.** Mrs. Smith invests $10,000, part at 8% and the rest at 10% annual interest. How much does she invest at each rate if her total annual interest from both investments is $870? $\boxed{13\text{-}6}$

**31.** Simplify: $\sqrt{\dfrac{x^8 y^4}{9z^{10}}}$ $\boxed{14\text{-}3}$

**32.** Simplify and combine:

$$2\sqrt{48} + 4\sqrt{50} - \sqrt{98} + 5\sqrt{27} \quad \boxed{14\text{-}8}$$

Solve and check:

**33.** $6\sqrt{2y} - 25 = 23$ $\boxed{14\text{-}11}$   **34.** $3x^2 - 48 = 0$ $\boxed{15\text{-}1}$

**35.** $2x^2 + x - 15 = 0$   **36.** Find the value of $s$ when $A = 486$, using the formula $A = 6s^2$. $\boxed{15\text{-}6}$

$\boxed{15\text{-}3 \text{ thru } 15\text{-}5}$

**37.** Transform formula $\dfrac{F}{W} = \dfrac{h}{d}$; solve for $d$. $\boxed{16\text{-}2}$

**38.** Derive a formula for $P$ in terms of $E$ and $R$, using the formula $P = IE$ and $E = IR$. $\boxed{16\text{-}2}$

**39.** Suppose $y$ varies directly as $x$, and $y = 18$ when $x = 6$. Find $y$ when $x = 24$. $\boxed{16\text{-}3}$

**40.** The angle of depression of a boat at sea is 5° as viewed from a cliff 80 meters above the sea. How far is the boat from the foot of the cliff?

$\boxed{17\text{-}3 \text{ thru } 17\text{-}5}$

## COMPETENCY CHECK TEST

In each of the following, select the letter corresponding to your answer.

1. The sentence "The perimeter of a square is equal to four times the length of its side" is expressed as a formula as:

    a. $s = 4p$      b. $p = \dfrac{s}{4}$      c. $s = \dfrac{p}{4}$      d. $p = 4s$

2. The sum of $-5b^3$ and $3b^3$ is:
    a. $-8b^6$      b. $-2b^6$      c. $-2b^3$      d. $+2b^3$

3. When $6r^2$ is subtracted from $-4r^2$, the answer is:
    a. $2r^2$      b. $-2r^2$      c. $-10r^2$      d. 2

4. The product of $-2a^3b^4$ and $-7a^4b^2$ is:
    a. $14a^7b^6$      b. $14a^{12}b^8$      c. $-14a^{12}b^8$      d. $-9a^7b^6$

5. When $-16m^9n^6$ is divided by $8m^3n^2$, the quotient is:
    a. $-2m^3n^3$      b. $-2m^6n^4$      c. $-8m^3n^3$      d. $-24m^{12}n^8$

6. The value of $2a^2 - 5a + 6$ when $a = 4$ is:
    a. 50      b. 6      c. 38      d. 18

7. The value of $V$ in the formula $V = Bh$ when $B = 12$ and $h = 3$ is:
    a. 15      b. 9      c. 36      d. 4

8. The formula expressing the number of minutes ($m$) in $s$ seconds is:

    a. $s = 60m$      b. $s = \dfrac{60}{m}$      c. $m = \dfrac{s}{60}$      d. $m = 60s$

9. The solution of $c + .4c = 98$ is:
    a. 94      b. 97.6      c. 7      d. 70

10. The solution of $2n - 5(n - 3) = 27$ is:
    a. 6      b. $-4$      c. 9      d. answer not given

11. The regular price of a typewriter that sells for $189 at a 30% reduction sale is: a. $219      b. $270      c. $300      d. $250

12. When the replacements for the variable are all the whole numbers, the solutions of $3x + 2 < 8$ are:
    a. 1 and 2      b. 0 and 1      c. 1      d. 0, 1, and 2

13. When the replacements for the variable are all the real numbers, the graph of $4y > 12$ is:

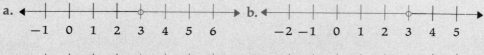

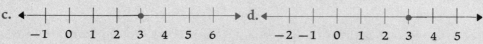

14. When the formula $E = IR$ is used and $E = 220$ and $I = 10$, the value of $R$ is: a. 2200      b. 230      c. 210      d. 22

15. The factors of $x^2 - 11x - 12$ are:
    a. $(x - 3)(x + 4)$          b. $(x - 12)(x + 1)$
    c. $(x + 6)(x - 2)$          d. $(x - 7)(x - 4)$

16. $\dfrac{2a - 4}{a^2 - 4}$ expressed in simplest form is:

    a. $\dfrac{2}{a + 2}$     b. $\dfrac{2}{a}$     c. $\dfrac{1}{a + 1}$     d. answer not given

17. Multiply: $45r^6 \cdot \dfrac{6r - 7}{9r^2}$   The product is:

    a. $30r^4 - 35r^3$   b. $36r^4 - 42$   c. $30r^5 - 35r^4$   d. $30r - 7$

18. Add: $\dfrac{2x}{x + 3y} + \dfrac{6y}{x + 3y}$   The sum is:

    a. $2x + 6y$     b. $\dfrac{8xy}{x + 3y}$     c. $2$     d. $4$

19. The solution of $n - \dfrac{n}{3} + \dfrac{n}{5} = 52$ is:

    a. $15$          b. $30$          c. $60$          d. $75$

20. The coordinates of the point that is on the line which forms the graph of
    the equation $3x - y = 10$ are:
    a. $(-2, 4)$       b. $(1, 7)$       c. $(-3, 1)$       d. $(2, -4)$

21. The solutions of: $\begin{array}{l} 2x + y = 8 \\ x - 2y = 9 \end{array}$ are:
    a. $x = 6$        b. $x = 3$        c. $x = 5$        d. $x = 4$
       $y = -4$          $y = 2$           $y = -2$          $y = 1$

22. An airplane carrier left port at 6 A.M. Three hours later a cruiser
    followed. The cruiser, averaging 28 knots, will overtake the carrier,
    averaging 21 knots at:   a. 9 A.M.   b. 12 noon   c. 3 P.M.   d. 6 P.M.

23. When $4\sqrt{75} - 10\sqrt{\frac{4}{5}}$ is simplified and combined, the result is:
    a. $-2\sqrt{5}$       b. $-6\sqrt{60}$       c. $-14\sqrt{15}$       d. answer not given

24. The solutions of $x^2 - 5x - 36 = 0$ are:
    a. 6 and $-6$     b. 12 and 3     c. $-4$ and $+9$   d. $-18$ and $-2$

25. When the formula $d = \dfrac{m}{v}$ is solved for $v$, the new formula is:

    a. $v = \dfrac{m}{d}$       b. $v = dm$       c. $v = \dfrac{d}{m}$       d. answer not given

26. Suppose $y$ varies inversely as $x$, and $y = 100$ when $x = 12$. When $x = 48$,
    $y$ is:          a. 400          b. 25          c. 200          d. 50

## THE EQUATION

1. Write an equation for the sentence:
   **a.** Some number ($n$) increased by nine is equal to fifteen.
   **b.** Twice the length $x + 8$ subtracted from ten times the width $x$ equals three times the sum of the length and width.
2. Which of the following equations have $-3$ as the solution?
   **a.** $5x - 2 = 13$      **b.** $-8x = 24$      **c.** $x = -3$      **d.** $2x + 9 = 3$
   Which of the above equations are equivalent?
   Which equations of the equivalent equations are in the simplest form?

Solve and check:

3. **a.** $n + 17 = 32$      **b.** $a - 23 = 8$      **c.** $18y = 54$      **d.** $\dfrac{c}{6} = 9$

4. **a.** $p + .08p =$                       **b.** $13 - 3n = 16 - 12n$
   **c.** $8b + 23 - 4b = 31 + 6b$      **d.** $9 - 2x + 7 + 5x = x - 6 + 22$
5. **a.** $7x - (15 - x) = 17$            **b.** $3y - 5(y - 9) = 11$
   **c.** $50a + 25(95 - a) = 4{,}250$   **d.** $(x + 5)(x - 3) = (x - 2)(x - 3)$

6. **a.** $r - \dfrac{r}{2} + \dfrac{r}{3} - \dfrac{r}{4} = 28$      **b.** $\dfrac{x}{x - 7} = \dfrac{x - 6}{x + 8}$

   **c.** $\dfrac{4n - 1}{3} - \dfrac{n - 2}{5} = 8$      **d.** $\dfrac{n + 6}{n + 1} - \dfrac{3n + 8}{n - 4} = \dfrac{4 - 2n^2}{h^2 - 3n - 4}$

7. **a.** $y = -2x + 3$            **b.** $8m - 3n = -25$
   $\phantom{a.}\ 3x - 4y = 21$            $\phantom{b.}\ 6m + 5n = 3$

   **c.** $x + y = 7{,}000$            **d.** $\dfrac{x}{6} + \dfrac{y}{4} = 6;\quad \dfrac{x - y}{2} = \dfrac{x + y}{10}$
   $\phantom{c.}\ .02x + .03y = 165$

8. **a.** $\sqrt{6n} = 12$            **b.** $\sqrt{2x - 1} = 7$
   **c.** $\sqrt{x^2 + 40} = x + 2$      **d.** $4\sqrt{6x + 7} = 28$
9. **a.** $\dfrac{x + 4}{5} = \dfrac{4}{x - 4}$      **b.** $4x^2 - 9x = 0$

   **c.** $x^2 - 3x - 10 = 0$      **d.** $12x^2 - 28x + 15 = 0$
10. Draw on a number line the graph of the equation $5x + 11 = 1$
11. Draw in the number plane the graph of the equation $4x - y = 3$
12. Solve graphically: $\begin{array}{l} 2x - 3y = 11 \\ x + 5y = -14 \end{array}$

## THE FORMULA

1. Express as a formula: The altitude ($a$) of a right triangle is equal to the square root of the difference between the squares of the hypotenuse ($h$) and the base ($b$).

2. The amount is equal to the principal plus the interest. Write a formula expressing the interest ($i$) in terms of principal ($p$) and amount ($A$).

3. **a.** Find $l$ when $a = 15$, $n = 12$, and $d = 4$, using $l = a + (n - 1)d$.
   **b.** Find $p$ when $A = 696$, $r = .04$, and $t = 5$, using $A = p + prt$.
   **c.** Find the value of $C$ when $F = -13$, using the formula $F = 1.8C + 32$.

4. Transform each of the following formulas:
   **a.** By removing parentheses: $A = 180 - (B + C)$
   **b.** By multiplying the right member: $A = \pi d(h + \frac{1}{2}d)$
   **c.** By factoring the right member: $A = \pi R^2 - \pi r^2$
   **d.** By simplifying the right member: $R = \dfrac{1}{\dfrac{1}{R_1} + \dfrac{1}{R_2}}$

5. Find the value of $V'$ when $P = 24$, $V = 100$, $T = 300$, $P' = 30$, and $T' = 360$, using the formula $\dfrac{PV}{T} = \dfrac{P'V'}{T'}$.

6. Rearrange formula $s = \dfrac{rl - a}{r - a}$ to solve for $a$. Solve for $l$. Solve for $r$.

7. In the formula $A = bh$, $A = 204$ when $b = 12$. Find $A$ when $b = 3$ and $h$ is constant.

8. In the formula $a = \dfrac{y}{t}$, $a = 6$ when $t = 10$. Find $a$ when $t = 30$ and $v$ is constant.

9. Draw the graphs of:    **a.** $p = 5s$    **b.** $d = 40t$    **c.** $i = .03p$

10. Using the formulas $A = \pi d(h + \frac{1}{2}d)$ and $d = 2r$, derive a formula for $A$ in terms of $\pi$, $r$, and $h$.

11. Find the value of $F$ when $m = 768$, $V = 100$, $g = 32$, and $r = 600$, using the formula $F = \dfrac{mV^2}{gr}$.

12. Find the values of $t$ when $s = 240$ and $v = 128$, using the formula $s = vt - 16t^2$.

13. Rearrange formula $s = 16t^2$, solving for $t$.

14. Using the right triangle $ABC$ (see page 610) find:
   **a.** Side $a$ if $c = 140$ meters and $m\angle A = 13°$.
   **b.** Side $c$ if side $a = 84$ centimeters and side $b = 135$ centimeters.
   **c.** Side $b$ if side $a = 500$ yards and $m\angle B = 61°$.

15. Using the formula $i = prt$, determine how much money must be invested for 10 years at 6% interest to bring the same interest as $1,000 invested for 12 years at 4%.

16. Using the formula $A = s^2$ and $p = 4s$, find the perimeter of a square having an area of 529 square meters.

## *PROBLEM SOLVING*

Use the equation method to solve each problem. Also check.

1. How many girls are in the freshman class of 247 students if there are 29 more girls than boys?

2. A board 4.2 meters long is to be cut into two pieces, one 1.5 meters longer than twice the other. Find the length of each piece.

3. A man is four times as old as his daughter. In 5 years he will be only three times as old. Find their ages now.

4. At what price must a dealer sell a mirror which costs $73.50 to realize a profit of 30% on the selling price?

5. Jim can wash the family's car in 50 minutes and his father can do it in 40 minutes. How long will it take both together to wash the car?

6. What is the radius of action of an airplane (how far can it fly from the base and yet return) in 8 hours if its outgoing speed is 490 mph and its return speed is 294 mph?

7. When changing a $1 bill, Mary received 6 nickels more than 5 times the number of dimes. How many nickels did she receive?

8. How many ounces of alcohol must be added to 10 ounces of a 20% iodine solution to make a 15% solution?

9. Angle R is 10 degrees more than angle S. Angle T is 15 degrees less than 3 times angle R. Find the measure of each angle in this triangle.

10. A woman invests $900 more at 8% than she does at 10%. The income from the 8% investment is $10 less than the income at 10%. How much was invested at 8%?

11. An airplane flew 1,440 miles with the wind in 4 hours. The return trip against the wind took $4\frac{1}{2}$ hours. Find the air speed (speed in still air) of the airplane and the wind velocity.

12. How many pounds of grass seed worth $1.30 a pound should be mixed with 20 pounds of grass seed worth $1.60 a pound to make a mixture to sell at $1.50 a pound?

13. Find the dimensions of a rectangle having a perimeter of 130 meters and an area of 1,000 square meters.

14. Find the list price of a pin that sold for $81.60 at a 15% reduction sale?

15. Find the selling price of a house if the owner received $59,055 after paying a 7% commission to the real estate agent.

16. How high up on a wall does a 13-foot ladder reach if the foot of the ladder is 5 feet from the wall?

17. What is the elevation of a road if the road rises 8.75 meters in a horizontal distance of 100 meters?

## *ALGEBRAIC OPERATIONS*

**1.** Add:
$(-8) + (+5)$

**2.** Subtract:
$(-3) - (-10)$

**3.** Multiply:
$(-5)(-6)$

**4.** Divide:
$(+32) \div (-4)$

**5.** Add:

**a.** $-5x$
$\underline{-6x}$

**b.** $2n^2 - 3n + 8$
$\underline{5n^2 + 3n - 9}$

**6.** Subtract:

**a.** $8t$
$\underline{10t}$

**b.** $3x^2 - 5xy - y^2$
$\underline{x^2 - 7xy + 2y^2}$

**7.** Multiply:

**a.** $-9x^3y^2$
$\underline{-3xy^8}$

**b.** $-2m^2(3m^2 - m + 6)$

**c.** $8bx^3(-b^3 - 2bx^2 + 3x^5)$

**8.** Divide:  **a.** $\dfrac{-16r}{+4r}$

**b.** $\dfrac{12a^4c^5 - 9a^2c^4 - 3ac^3}{-3ac^3}$

**9.** Combine: $3bc - 2b^2 - c^2 - 2bc - 2c^2 - 3b^2 - c^2$

**10.** Subtract: $4a^2 - 3a - 6$ from $3a^2 - 6$

**11.** Multiply: $7m^2 - 2mx - x^2$ by $5m - 2x$

**12.** Divide: $24b^3 - 41b^2c - 3bc^2 + 20c^3$ by $3b - 4c$

**13.** Remove parentheses and add like terms:
$8x - x(2x - 5) + (x - 1)(x - 4)$

**14.** Inclose the last three terms within parentheses preceded by a minus sign: $5x^4 - 2x^3 - x^2 + 6x - 1$

**15.** Multiply at sight:

**a.** $(\tfrac{3}{4}x - 5y^2)(\tfrac{3}{4}x + 5y^2)$

**b.** $(6s^2 + 5xy)^2$

**c.** $(a - 9)(a + 8)$

**d.** $(5m - 8n)(8m - 6n)$

**16.** Factor:

**a.** $25a^2 - 100b^2$

**b.** $t^2 - t - 56$

**c.** $81x^2 + 144xy^2 + 64y^4$

**d.** $6s^2x - 18sx + 12x$

**17.** Express in simplest form:
$$\frac{24x - 12y}{144x^2 - 36y^2}$$

**18.** Multiply:
$$\frac{9b^3c^2}{25x^2y} \cdot \frac{10y}{3b^4c}$$

**19.** Divide:
$$\frac{8ax - 6bx}{7ab} \div 4ax^3$$

**20.** Add:
$$\frac{1}{m} + \frac{1}{n}$$

**21.** Subtract:
$$\frac{9}{x^2 - 25} - \frac{4}{x^2 - 8x + 15}$$

**22.** Combine:
$$\frac{3}{a + 7} - \frac{6a}{a^2 - 49} - \frac{5}{7 - a}$$

**23.** Combine: $\dfrac{4x + y}{x - 3y} - 1$

**24.** Simplify: $\sqrt{117a^4b^9}$

**25.** Simplify: $\sqrt{\dfrac{27c}{4a}}$

**26.** Simplify and combine: $3\sqrt{180} + \sqrt{147} - 10\sqrt{\dfrac{1}{5}} + 4\sqrt{432}$

**27.** Multiply: $(5\sqrt{6} - 7\sqrt{2})^2$

**28.** Divide: $\dfrac{16\sqrt{8} - 20\sqrt{10} + 4\sqrt{16}}{4\sqrt{2}}$

# Keyed Achievement Test

The colored numeral indicates the exercise where help may be found.

**1.** Express as a formula: The sum of the measures of angles $A$, $B$, $C$, and $D$ of the quadrilateral $ABCD$ is equal to $360°$. [1-8]

**2. a.** Add: [2-11]   **b.** Subtract: [2-12] **c.** Multiply: [2-13] **d.** Divide: [2-14]
[2-4] $(-9) + (-4)$   $(+2) - (+8)$   $(-5)(+9)$        $(-28) \div (-7)$

**3. a.** Find the opposite of $+31$.   **b.** Find the additive inverse of $-17$.
   **c.** Find the absolute value: $|-.8| = ?$ [2-3]                           [2-4]

   **d.** Find the multiplicative inverse of $-\dfrac{7}{6}$. [2-13]

**4.** On a number line draw the graph of: $-4$, $-2$, $-1$, $0$, $3$, $4$ [2-7]

**5.** Find the value of $5x^2 + 3xy - 4y^2$ when $x = 6$ and $y = -2$. [3-6]

**6.** What is the value of $l$ when $a = 9$, $n = 7$, and $d = 3$ using the formula $l = a + (n - 1)d$? [3-7]

**7.** Write a formula expressing the number of centiliters ($c$) in $l$ liter. [4-1]

**8.** Find the sum of $2y^2 - x^2$, $3xy - 5y^2$, and $4x^2 - 3xy$. [5-3]

**9.** Subtract $m^2 - 2mn + n^2$ from $m^2 - n^2$. [5-5]

**10.** Multiply $6a^2 - 3ab + b^2$ by $2a - 3b$. [5-9]

**11.** Divide $c^5 + x^5$ by $c + x$. [5-13]                              [5-14]

**12.** Remove parentheses and add like terms: $4n - 2n(n - 5) + (n - 2)(n - 3)$

**13.** Solve and check: [6-4 thru 6-8]

   **a.** $12x - 6 = 0$   **b.** $5 = \dfrac{n}{-5}$   **c.** $s - .3s = ?$   **d.** $16 - r = r$

   **e.** $6(3x + 8) - (-7x + 5) = -7$   **f.** $(n + 7)(n - 1) = (n + 1)(n + 3)$

**14.** Draw on a number line the graph of $2x - 5 = 3$. [6-11]

**15.** Find the value of $t$ when $A = 980$, $p = 700$, and $r = .08$, using the formula $A = p + prt$. [6-10]

**16.** How long will it take a seaplane, flying at a speed of 180 knots, to overtake a destroyer if the destroyer left the naval base 8 hours before the seaplane and was averaging 30 knots? [7-5]

When the replacements for the variable are all the: [8-4 thru 8-8]

**17.** One-digit prime numbers, solve $-7x < -27$.

**18.** Real numbers, draw the graph of $3x + 16 \leq 4$ on a number line. [8-9]

**19.** Find the product of each of the following at sight: [9-1 thru 9-5]
   **a.** $(8x - 7y)(8x - 7y)$                     **b.** $(3c - 5d)(3c + 5d)$
   **c.** $(6b + 1)(6b - 5)$                        **d.** $(4r - 9s)(2r + 3s)$

**20.** Factor completely: **a.** $25a^2 + 400b^2$        **b.** $y^2 - 10y - 24$
   [9-6 thru 9-12]          **c.** $36m^2 - 132m + 121$   **d.** $ar^5 - ar$

**21.** Express in simplest form:

$$\frac{8c^2 - 2c}{4c + 2}$$  `10-2`

**22.** Divide: `10-4`

$$\frac{3x - 12y}{x^2 - xy - 12y^2} \div \frac{x^2 - 6xy + 9y^2}{x^2 - 9y^2}$$

**23.** Add or subtract as indicated: `10-5 thru 10-11`

**a.** $\dfrac{x - 5y}{6} - \dfrac{3x - 2y}{4} + \dfrac{6x + y}{8}$   **b.** $\dfrac{2c}{c + d} - \dfrac{3d}{c - d} + \dfrac{3d^2}{c^2 - d^2}$

**24.** Solve and check: `11-1 thru 11-3`

**a.** $\dfrac{5}{4x} - \dfrac{3}{16} = \dfrac{9}{8x} - \dfrac{1}{8}$   **b.** $\dfrac{4n - 1}{3} - \dfrac{n - 2}{5} = 8$

**c.** $\dfrac{x - 2}{x - 4} + \dfrac{8}{x + 1} = 1$   **d.** $\dfrac{2}{a - 5} - \dfrac{1}{a - 2} = \dfrac{9}{a^2 - 7a + 10}$

**e.** $x - .07x = 837$   **f.** $.08n + .12(2{,}500 - n) = 228$

**25.** Find the value of $r$ when $S = 186$, $a = 6$, and $l = 96$, using the formula $S = \dfrac{rl - a}{r - 1}$.  `11-4`

**26.** Without drawing the graph, determine the slope and the coordinates of the point where the graph of $4x - y = 8$ crosses the $Y$-axis. `12-4 thru 12-5`

**27.** Draw in the number plane the graph of $3x - 2y = 6$. `12-3`

**28.** Solve graphically: $2x - y = 6$   **29.** Solve and check: $7x + 2y = -13$
`13-2`           $x + 2y = 8$   `13-3 and 13-4`    $3x - 9y = -45$

**30.** How many kg of grass seed selling for \$4.77/kg should be mixed with grass seed selling at \$1.47/kg to make 200 kg to sell at \$1.80/kg? `13-6`

**31.** Simplify: $\sqrt{\dfrac{25r^4}{4s^8t^2}}$  `14-3`   **32.** Simplify and combine: `14-8`
$$\sqrt{294} + 12\sqrt{\tfrac{2}{3}} - 3\sqrt{150}$$

Solve and check:

**33.** $x^2 - 32 = x - 2$   **34.** $8x^2 - 6x - 9 = 0$   **35.** $\dfrac{x}{3} - \dfrac{6}{x} = 1$ `11-1 thru 11-3`
`14-3`            `15-3 thru 15-5`

**36.** Find the value of $w$ to the nearest tenth when $l = 4$, using the formula $\dfrac{w}{l} = \dfrac{l}{l + w}$. `15-6`

**37.** Transform formula $I = \dfrac{nE}{R + nr}$, solving for $r$. `16-2`

**38.** Derive a formula for $d$ in terms of $A$ and $\pi$, using the formulas $A = \pi r^2$ and $d = 2r$. `16-2`

**39.** Using the formula $v = \dfrac{s}{t}$, $v = 32$ when $t = 2$, find the value of $v$ when $t = 8$ and $s$ remains constant. `16-4`

**40.** What is the angle of elevation of the sun, to the nearest degree, when a 30-foot flagpole casts a shadow of 18 feet? `17-3 thru 17-5`

## TABLE OF SQUARES AND SQUARE ROOTS

| No. | Square | Square Root | No. | Square | Square Root | No. | Square | Square Root |
|---|---|---|---|---|---|---|---|---|
| 1 | 1 | 1.000 | 34 | 1,156 | 5.831 | 67 | 4,489 | 8.185 |
| 2 | 4 | 1.414 | 35 | 1,225 | 5.916 | 68 | 4,624 | 8.246 |
| 3 | 9 | 1.732 | 36 | 1,296 | 6.000 | 69 | 4,761 | 8.307 |
| 4 | 16 | 2.000 | 37 | 1,369 | 6.083 | 70 | 4,900 | 8.367 |
| 5 | 25 | 2.236 | 38 | 1,444 | 6.164 | 71 | 5,041 | 8.426 |
| 6 | 36 | 2.449 | 39 | 1,521 | 6.245 | 72 | 5,184 | 8.485 |
| 7 | 49 | 2.646 | 40 | 1,600 | 6.325 | 73 | 5,329 | 8.544 |
| 8 | 64 | 2.828 | 41 | 1,681 | 6.403 | 74 | 5,476 | 8.602 |
| 9 | 81 | 3.000 | 42 | 1,764 | 6.481 | 75 | 5,625 | 8.660 |
| 10 | 100 | 3.162 | 43 | 1,849 | 6.557 | 76 | 5,776 | 8.718 |
| 11 | 121 | 3.317 | 44 | 1,936 | 6.633 | 77 | 5,929 | 8.775 |
| 12 | 144 | 3.464 | 45 | 2,025 | 6.708 | 78 | 6,084 | 8.832 |
| 13 | 169 | 3.606 | 46 | 2,116 | 6.782 | 79 | 6,241 | 8.888 |
| 14 | 196 | 3.742 | 47 | 2,209 | 6.856 | 80 | 6,400 | 8.944 |
| 15 | 225 | 3.873 | 48 | 2,304 | 6.928 | 81 | 6,561 | 9.000 |
| 16 | 256 | 4.000 | 49 | 2,401 | 7.000 | 82 | 6,724 | 9.055 |
| 17 | 289 | 4.123 | 50 | 2,500 | 7.071 | 83 | 6,889 | 9.110 |
| 18 | 324 | 4.243 | 51 | 2,601 | 7.141 | 84 | 7,056 | 9.165 |
| 19 | 361 | 4.359 | 52 | 2,704 | 7.211 | 85 | 7,225 | 9.220 |
| 20 | 400 | 4.472 | 53 | 2,809 | 7.280 | 86 | 7,396 | 9.274 |
| 21 | 441 | 4.583 | 54 | 2,916 | 7.348 | 87 | 7,569 | 9.327 |
| 22 | 484 | 4.690 | 55 | 3,025 | 7.416 | 88 | 7,744 | 9.381 |
| 23 | 529 | 4.796 | 56 | 3,136 | 7.483 | 89 | 7,921 | 9.434 |
| 24 | 576 | 4.899 | 57 | 3,249 | 7.550 | 90 | 8,100 | 9.487 |
| 25 | 625 | 5.000 | 58 | 3,364 | 7.616 | 91 | 8,281 | 9.539 |
| 26 | 676 | 5.099 | 59 | 3,481 | 7.681 | 92 | 8,464 | 9.592 |
| 27 | 729 | 5.196 | 60 | 3,600 | 7.746 | 93 | 8,649 | 9.644 |
| 28 | 784 | 5.292 | 61 | 3,721 | 7.810 | 94 | 8,836 | 9.695 |
| 29 | 841 | 5.385 | 62 | 3,844 | 7.874 | 95 | 9,025 | 9.747 |
| 30 | 900 | 5.477 | 63 | 3,969 | 7.937 | 96 | 9,216 | 9.798 |
| 31 | 961 | 5.568 | 64 | 4,096 | 8.000 | 97 | 9,409 | 9.849 |
| 32 | 1,024 | 5.657 | 65 | 4,225 | 8.062 | 98 | 9,604 | 9.899 |
| 33 | 1,089 | 5.745 | 66 | 4,356 | 8.124 | 99 | 9,801 | 9.950 |

# INDEX

# SELECTED ANSWERS

## Chapter 1 *Language of Algebra*

### DIAGNOSTIC TESTS

**1–1** (p. 13) **1.** 45 **3.** none **5.** Zero, two, four, and so on up to and including twenty-eight. **7.** 1, 2, 3, . . . **9.** 15, 17, 19, 21, 23, 25, 27, 29, 31, 33, 35

**1–2** (p. 15) **1.** $7 + 2$ **3.** $5 \times 11$

**1–3** (p. 18) **1.** 5364 **3.** 765,700 **5.** $8\frac{1}{2}$ **7.** 20 **9.** 29.12 **11.** .009 **13.** \$1.26 **15.** 250 **17.** $+; 3$ **19.** $x; 6$

**1–4** (p. 21) **1. a.** eleven to the sixth power **b.** seven squared **c.** six cubed **3.** 7 **5.** $6^7$ **7.** $\sqrt{91}$ **9.** 54

**1–5** (p. 24) **1.** $x \div y$ **3.** $l - d$ **5.** $lw - lh$ **7.** $a^2 + b^2$ **9.** $\sqrt{h^2 - a^2}$

**1–6** (p. 28) **1.** is less than **3.** Six tenths plus seven tenths is greater than one. **5.** Four plus eleven is not greater than six times three. **7.** Twelve divided by three is not less than nine minus two. **9.** $45 = 9 \times 5$ **11.** $72 \neq 4 \times 15$ **13.** $12 + 11 \not> 90 \div 3$ **15.** none

**1–7** (p. 31) **1.** $25 > 16 + n$ **3.** Six times some number $x$ minus one is equal to twenty-three. **5.** Twenty-five is greater than each number $n$ plus two. **7.** Three times each number $b$ divided by two is not greater than twelve. **9.** $c - 1 = 12$ **11.** $t \div 4 < 13$ **13.** $4x \not< 51$

**1–8** (p. 33) **1. a.** $H.P. = \dfrac{DV}{550}$ **b.** $p = 2(l + w)$ **c.** $b = \sqrt{h^2 - a^2}$

### CHAPTER REVIEW (p. 36)

**1. a.** $10 + 4$ **b.** $3 \times 9$ **c.** $20 - 12$ **d.** $\frac{60}{5}$ **e.** $7^6$ **f.** $\sqrt{57}$ **g.** $6(11 - 2)$ **h.** $2^9$ **2. a.** $\dfrac{d}{c}$ **b.** $2n - 8$ **c.** $R + S + T$ **d.** $16t^2$ **e.** $\sqrt{xy}$ **f.** $(c + d)(a - b)$ **3. a.** Six minus two is less than seventeen. **b.** One and five tenths divided by three is not equal to five divided by two and five tenths. **c.** Eight and three fourths is greater than eleven minus three and one fourth. **4. a.** $\frac{40}{5} \not> 6 + 4$ **b.** $7 \times 12 = 100 - 16$ **c.** $63 \not< 9 \times 7$ **5.** b., c., e., g., h., k., and l. are true **6. a.** not open **b.** open; inequality **c.** open **d.** not open; inequality **e.** open; inequality **f.** open **7. a.** Eight times each number $x$ plus six is greater than eighty. **b.** Ten times each number $c$ minus nine is not less than nineteen. **c.** Three times each number $n$ minus seven is not equal to the same number $n$ plus four. **8. a.** $n + 4 < 12$ **b.** $2x - 10 \not> 3$ **c.** $21 = 3y + 9$ **9. a.** $D = AS$ **b.** $C + D = 90°$ **c.** $p = \dfrac{r}{s}$ **10. a.** The perimeter of a regular octagon ($p$) is equal to three times the length of the side ($s$). **b.** The power in watts ($W$) is equal to the current in amperes ($I$) squared times the resistence in ohms ($R$). **c.** The area of a trapezoid ($A$) is equal to one half the height ($h$) times the sum of the length of the lower base ($b$) and the length of the upper base ($b'$)

### KEYED ACHIEVEMENT TEST (p. 37)

**1. a.** $11 \times 9$ **b.** $8 + 4$ **2. a.** $d - r$ **b.** $10 \div t$ **3. a.** Seventy five divided by five is less than eight times four. **b.** Six times each number $x$ plus seven is greater than forty three. **c.** Each number $r$ minus three is not equal to four times the same number $r$.
**4. a.** $25 - 11 < 15$ **b.** $n + 5 = 30$ **5.** $n = \dfrac{1}{p}$

**A1**

## Chapter 2 *Positive and Negative Numbers*

### DIAGNOSTIC TESTS

**2–1** (p. 42) **1.** negative nine **3.** positive forty-eight **5.** negative ninety-six hundredths **7.** positive two and three-fourths **9.** $+\frac{9}{10}$; $+87$; $+6\frac{2}{3}$ **11.** $+21$ **13.** $-\frac{5}{9}$

**2–2** (p. 44) **1.** A downward force of 70 pounds. **3.** $-\$3$

**2–3** (p. 46) **1.** The absolute value of negative twenty-nine. **3.** $+15$ **5.** $+\frac{3}{5}$

**2–4** (p. 48) **1. a.** $-23$ **b.** $+\frac{3}{8}$ **3. a.** $-(-7)$ or $+7$ **b.** $-18$ **c.** $-(-12)$ or $+12$ **5. a.** $+58$ **b.** $-8.3$

**2–5** (p. 51) **1.** $-4$; $-\frac{30}{5}$; $+\sqrt{25}$ **3.** $-\sqrt{95}$ **5.** $-2, -1, 0, 1, 2, 3, 4$

**2–6** (p. 53) **1.** $+30$; $-24$; $+102$ **3.** $-1, 0, +1$ **5.** $-12, -10$

**2–7** (p. 56) **1. a.** $-1$ **b.** $+5$ **c.** $-5$ **d.** $+3$ **e.** $-7$

**3.**

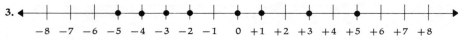

```
◄———●———|———●———|———●———|———●———|———●———►
   -4  -3  -2  -1   0  +1  +2  +3  +4
```

**2–8** (p. 59) **1. a.** R **b.** P **3. a.** J **b.** F **5. a.** $+8$ **b.** $-3$ **c.** $+9$ **d.** 0 **e.** $+1$ **7. a.** yes **b.** yes **c.** yes **d.** yes **9.** $+17, +8, +6, +2, +1, 0, -2, -4, -9, -11$ **11.** $-8$

**2–9** (p. 63) **1. a.** 14 units to the left. **b.** 23 units to the right **3. a.** $-4$ **b.** $+1$ **5. a.** $+4$ **b.** $-3$

**2–10** (p. 67) **1.** $+7$ **3.** $+2$ **5.** $-5$ **7.** 0

**2–11** (p. 70) **1.** $+41$ **3.** $-54$ **5.** $+3$ **7.** $-5$ **9.** $-6$ **11. a.** $-6$ **b.** $+2$ **13. a.** $+\frac{7}{8}$ **b.** $-1\frac{19}{30}$ **15.** $-1$ **17.** $+2$ **19.** 4

**2–12** (p. 76) **1. a.** $+9$ **b.** $-8$ **c.** $-4$ **d.** $-5$ **e.** $+4$ **f.** 0 **3.** $-3$ **5.** $+13$ **7.** $+51$ **9.** $-37$ **11.** $+6$ **13.** $-64$ **15.** $-16$ **17.** $+1$ **19. a.** $-4\frac{1}{8}$ **b.** $+\frac{1}{30}$ **21. a.** $-43$ **b.** 58 **23.** $-9$ **25. a.** $-1$ **b.** $-3$ **27. a.** $-17$ **b.** 2 **29. a.** $-12$ **b.** $-21$ **31.** no

**2–13** (p. 84) **1. a.** nine times six **b.** negative eleven times positive fifteen **c.** eight times negative five **3. a.** $+63$ **b.** $-54$ **5. a.** $-60$ **b.** $-180$ **7. a.** 0 **b.** 0 **9. a.** $-45$ **b.** $+9$ **11. a.** $-4$ **b.** $+144$ **13. a.** $+8$ **b.** $-60$ **c.** $-10$ **15. a.** $-72$ **b.** $+7$ **17.** $-\frac{4}{3}$; $+\frac{1}{10}$; $-\frac{3}{5}$

**2–14** (p. 89) **1. a.** $+8$ **b.** $+6$ **3. a.** $-7$ **b.** $-5$ **5. a.** $+1$ **b.** $-1$ **7. a.** $-4$ **b.** $-7$ **9. a.** $+9$ **b.** $-.3$ **11. a.** $+1$ **b.** $-6$ **13.** $-\frac{1}{2}$

### CHAPTER REVIEW (p. 92)

**1.** $-2.75$, $-43$, $-\frac{7}{9}$, $-1\frac{5}{6}$, $-500$ **2.** Integers; $-13$, $+\frac{25}{5}$ Rationals; $-.1$, $+2\frac{1}{8}$, $-13$, $+\frac{25}{5}$, $-\frac{11}{16}$ Reals; all    Irrationals; $-\sqrt{13}$

**3.**

```
◄——|——|——|——●——●——●——●——|——●——●——|——●——|——|——|——►
  -8 -7 -6 -5 -4 -3 -2 -1  0 +1 +2 +3 +4 +5 +6 +7 +8
```

**4.** $-5, -4, -3, -2, 0, +1, +3, +5$   **5.** 225 feet below sea level   **6.** a., b., and c. are true   **7.** $+14, +12, +9, +6, 0, -1, -5, -8, -10, -11$   **8. a.** $+70$ **b.** $-.2$ **c.** (1) 57 (2) $\frac{7}{12}$ **d.** (1) $-1\frac{2}{3}$ (2) $+64$ **e.** (1) $-5$ (2) $-\frac{1}{20}$ (3) $+\frac{9}{8}$

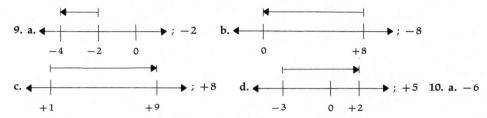

**9. a.** ; $-2$    **b.** ; $-8$

**c.** ; $+8$    **d.** ; $+5$   **10. a.** $-6$

**b.** $-7$ **c.** 0 **d.** $+7$   **11. a.** $+13$ **b.** $-20$ **c.** $-74$ **d.** $+39$   **12. a.** $-60$ **b.** $+81$ **c.** $-40$ **d.** $-350$   **13. a.** $+25$ **b.** $-4$ **c.** $-18$ **d.** $-5$   **14. a.** $-10$ **b.** $-5$   **15.** $+7$

## KEYED ACHIEVEMENT TEST   (p. 93)

**1.** $5x - 7 > 48$   **2.** $s = \pi dR$

**3.**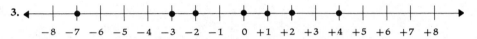

$$-8 \quad -7 \quad -6 \quad -5 \quad -4 \quad -3 \quad -2 \quad -1 \quad 0 \quad +1 \quad +2 \quad +3 \quad +4 \quad +5 \quad +6 \quad +7 \quad +8$$

**4. a.** (1) 36 (2) $\frac{9}{25}$ **b.** (1) $-57$ (2) $+8$ (3) $+\frac{4}{7}$ **c.** (1) $-2\frac{1}{2}$ (2) $+61$ **d.** (1) $+5\frac{3}{4}$ (2) $-.04$ (3) $+101$ **e.** (1) $-\frac{1}{15}$ (2) $+\frac{13}{11}$ (3) $-\frac{2}{3}$   **5. a.** $-8; -20; 0$ **b.** $-21; +6; -21$ **c.** $-48; +35; -24$ **d.** $-4; -1; +9$

## Chapter 3 *Evaluation*

## DIAGNOSTIC TESTS

**3-1**   (p. 97)   **1.** 28   **3.** 76   **5.** 168   **7.** 3   **9.** 111   **11.** 141   **13.** 132   **15.** 45   **17.** 6

**3-2**   (p. 100)   **1.** 6   **3.** $-16$   **5.** $-3$   **7.** 2   **9.** $-6$   **11.** $-24$   **13.** $-14$   **15.** 19

**3-3**   (p. 102)   **1.** 30   **3.** 18   **5.** 32   **7.** 380   **9.** 204   **11.** 18   **13.** 18   **15.** 80   **17.** 36

**3-4**   (p. 104)   **1.** $-2$   **3.** 28   **5.** $-4$   **7.** 340   **9.** $-4$   **11.** 3

**3-5**   (p. 106)   **1.** 9   **3.** 12   **5.** 144   **7.** 64   **9.** 144   **11.** 243   **13.** 30   **15.** 22   **17.** 46   **19.** $6\frac{3}{4}$   **21.** 5   **23.** $\frac{1}{5}$

**3-6**   (p. 109)   **1.** 36   **3.** $-96$   **5.** 52,488   **7.** $-108$   **9.** 9   **11.** $-180$   **13.** 2

**3-7**   (p. 111)   **1.** 208   **3.** 27   **5.** 113.04   **7.** $-25$

## CHAPTER REVIEW   (p. 116)

**1. a.** 57 **b.** $-32$ **c.** 90 **d.** $-8$   **2. a.** $-5, -2, 1, 4, 7, 10, 13, 16, 19$ **b.** 170, 124, 78, 32, $-14, -60, -106, -152, -198, -242$ **c.** 16, 0, $-5, -9, -8, 0, 7$   **3. a.** 162; 1296 **b.** $-45; -24$ **c.** $-18; -12$   **4. a.** 180 **b.** 23 **c.** 144 **d.** 132 **e.** $-4$

## KEYED ACHIEVEMENT TEST   (p. 117)

**1.** $(bh - lw)^2$   **2.** $l = \dfrac{v}{n}$

**3.**

**4. a.** $-22$; $0$; $3$; $+14$ **b.** $+10$; $-2$; $-19$; $-18$ **c.** $+27$; $-60$; $-8$; $-35$ **d.** $-60$; $+225$; $-81$; $+256$ **5. a.** $3.2$ **b.** $-5\frac{3}{4}$; $+93$ **c.** $+49$; $-\frac{5}{12}$ **d.** $+\frac{1}{20}$; $-\frac{3}{25}$ **6. a.** $58$ **b.** $+39$ **c.** $-\frac{2}{3}$ **d.** $299$ **e.** $2112$

## Chapter 4 *Making Formulas*

### DIAGNOSTIC TESTS

4-1 (p. 121) **1.** $120$; $360$; The air pressure in pounds ($p$) is 15 times the area in square inches ($A$); $P = 15(A)$ **3.** \$4.18; \$9.40; The interest ($i$) is one hundredth of the principal ($p$); $i = \dfrac{p}{100}$ **5.** \$3.22; \$4.38; After the first print each additional print ($n$) is \$.29 more than the cost ($c$) of the first print.; $c = \$.29n + \$.90$ **7.** \$.52; \$.64; The cost ($c$) equals \$.06 times the weight in ounces ($w$) minus two plus \$.10; $c = \$.06(w - 2) + \$.10$

4-2 (p. 128) **1. a.** Face of note is equal to proceeds plus bank discount; $f = p + d$ **b.** Bank discount is equal to face of note minus proceeds; $d = f - p$ **c.** Proceeds is equal to face of note minus bank discount; $p = f - d$ **3. a.** The measure of angle $a$ is equal to 180° minus the measure of angle $b$; $m\angle a = 180° - m\angle b$ **b.** The measure of angle $a$ is equal to the measure of angle $c$; $m\angle a = m\angle c$ **c.** The measure of angle $b$ is equal to the measure of angle $d$; $m\angle b = m\angle d$ **d.** The measure of angle $c$ is equal to 180° minus the measure of angle $d$; $m\angle c = 180° - m\angle d$ **e.** The sum of the measures of angles $b$ and $c$ are 180°; $m\angle b + m\angle c = 180°$

### CHAPTER REVIEW (p. 136)

**1. a.** $30$; $54$; $y = 6x$ **b.** $18$; $26$; $n = m + 5$ **c.** $27$; $47$; $S = 4R + 3$ **2.** 3,000 m; 15,000 m; $m = 1,000 k$ **3.** $A = bh + \frac{1}{2}c(a + h)$

### KEYED ACHIEVEMENT TEST (p. 137)

**1.** Rationals; $-\frac{5}{6}$, $\frac{27}{9}$, 16%, $-53$, $+7.2$, $+8\frac{2}{3}$, $-.009$ Integers; $\frac{27}{9}$, $-53$

**2.**       **3. a.** $-17$ **b.** $-24$

**c.** $-12$ **d.** $-8$ **e.** $-5$ **f.** $6$ **g.** $+5$ **h.** $-8$ **4. a.** $-4$ **b.** $103$
**5.** $A = 744$ **6.** $5$; $10$; Each number in row $B$ is 5 less than each corresponding number in row $A$; $B = A - 5$ **7.** $p = 26 + 2\pi r$ **8.** 300 minutes; 720 minutes; $m = 60h$

## Chapter 5 *Operations with Polynomials*

### DIAGNOSTIC TESTS

5-1 (p. 143) **1. a.** $2$ **b.** $4$ **c.** $1$ **3. a.** none or 0 **b.** $25$ **c.** $1$ **5.** $3a^4$; $-5a^3b$; $2a^2b^2$; $-ab^3$; $10b^4$ **7.** $-11$ **9.** The opposite of $by^2$ or the additive inverse of $by^2$

**11.** $-5c^4 + 4c^3 + 3c^2 - c + 7$  **13.** $5m^4 + 10m^3x - 10m^2x^2 - 3mx^3 + 12x^4$
or $12x^4 - 3mx^3 - 10m^2x^2 + 10m^3x + 5m$

**5-2** (p. 147)  **1.** $+10m$  **3.** $-4b$  **5.** $2b^2$  **7.** $-2x$  **9.** $-s^2$  **11.** $+y$  **13.** $-3a^2b$
**15.** $7(a + b)$  **17.** $-.2b$  **19.** $b^3 + b^3 - 2b$  **21.** $-13y$

**5-3** (p. 151)  **1.** $5a - 2$  **3.** $b + 1$  **5.** $0$  **7.** $16r - 6t$  **9.** $-3m + 5r^2$  **11.** $6x^2 + 4x + 2$
**13.** $x^2 - 9x + 3$

**5-4** (p. 155)  **1.** $+2x$  **3.** $12abc$  **5.** $-d$  **7.** $+5a$  **9.** $-8a^2$  **11.** $-5.1x^2$  **13.** $-8x$
**15.** $.94p$  **17.** $-7x$  **19.** $+7c$  **21.** $10x + 9y$

**5-5** (p. 160)  **1.** $5a - 13$  **3.** $3x^2 - x + 8$  **5.** $0$  **7.** $-x^3 + x^2 - 4x$  **9.** $x^2 - 2x + 1$
**11.** $-x^2 + 3xy$

**5-6** (p. 165)  **1. a.** $c^8$  **b.** $x^4y^5z$  **3.** $a^9$  **5.** $m^{11}n^3$  **7.** $m^2x^4$  **9.** $bcx^2$

**5-7** (p. 167)  **1.** $8c^6$  **3.** $-20m^{10}$  **5.** $-6a^3b^3c^5$  **7.** $-5ax$  **9.** $-3ac^2$  **11.** $-28a^3x^3y$
**13.** $\frac{1}{2}c^7d^5$  **15.** $15ab$  **17.** $-8b^6$

**5-8** (p. 169)  **1.** $3a + 12$  **3.** $-8m - 6$  **5.** $18x - 24y$  **7.** $x^2 + x$
**9.** $3b^6 - b^5 - b^4 + 2b^3$  **11.** $2a^2b - 4ab^2$  **13.** $4x^4 - 3x^3y^6 + 5x^2y^5$
**15.** $-16c^5d^7 + 32c^4d^8 + 24c^3d^9$  **17.** $a^6b^4 + .3a^5b^3 - .05a^4b^2$

**5-9** (p. 173)  **1.** $x^2 + 7x + 12$  **3.** $y^2 - 4y - 21$  **5.** $x^2 + 2xy + y^2$  **7.** $c^2 - 4$
**9.** $m^4 - 6m^2 - 27$  **11.** $30 - 7b - 2b^2$  **13.** $10s^2 - 7s - 12$
**15.** $5a^4 - 17a^2b^2 + 12b^4$  **17.** $x^3 + 2x^2 + x$  **19.** $a^2 + ab + ac + bc$
**21.** $28x^3 - 3x^2y + 18xy^2 + 27y^3$

**5-10** (p. 177)  **1.** $a^4$  **3.** $b^3d^3n^4$  **5.** $mr^2$

**5-11** (p. 179)  **1.** $4b$  **3.** $-3m^3n^2$  **5.** $4c^6$  **7.** $-7b^2x$  **9.** $-7s^3t^2$  **11.** $-7$  **13.** $-4a^6b$
**15.** $-5xy$

**5-12** (p. 181)  **1.** $3x + 2$  **3.** $a + 2b$  **5.** $2b^2 - 3bc + 4c^2$  **7.** $m^2 - m$  **9.** $5d^2 + 3$
**11.** $-4c + 1$  **13.** $4a^2 - 3b$  **15.** $m^3nx - 2m^2x^3 + 3mx^3 - 4$

**5-13** (p. 186)  **1.** $c + 7$  **3.** $m + 2$  **5.** $s - 3$  **7.** $x - 2$  **9.** $4b^3 + 5$  **11.** $7m - 2n$
**13.** $2a^2b + 5x$  **15.** $n^2 + 7n - 6$  **17.** $a - c$  **19.** $b^3 - 2b^2 + 4b - 8$

**5-14** (p. 190)  **1.** $11b + 7$  **3.** $-6a + 13$  **5.** $4c^2 - 7cd + d^2$  **7.** $-7a + 4$  **9.** $-x^2 - 4x$
**11.** $-2c$  **13.** $5x - 5y$  **15.** $2a^3 - 7a^2 - 3a$  **17.** $x - 14$
**19.** $-4x^2 - 2x + 4$  **21.** $6m - 2n$  **23.** $4a$  **25.** $-14b + 11$  **27.** $6y - 14$
**29.** $-3b^2 + 9b$  **31.** $5x^2 + 16x$  **33.** $b^2 - 21b + 48$  **35.** $7m - 32$

**5-15** (p. 195)  **1.** $2a + 3b + (4c + 5d)$  **3.** $4b^3 + 5b^2 + (-5b - 8)$
**5.** $c^5 - 5c^4 - 10c^3 + (-10c^2 + 5c - 1)$  **7.** $9a^2 + 4b^2 - (3c^2 - 5d^2)$
**9.** $m^4 - 8m^3 + 12m^2 - (-16m + 18)$

## CHAPTER REVIEW  (p. 197)

**1. a.** $-16bx^3$  **b.** $-13a^2$  **c.** $1.14p$  **2. a.** $5x^2 - 12xy + 3y^2$  **b.** $8b^2 + 3bc + 4c^2$  **c.** $x^2 - x - 9$
**3. a.** $3a^4b^3$  **b.** $-8x + 5y$  **c.** $-28c^2$  **4. a.** $-7x^2 + 2x + 6$  **b.** $-2m^2 + 2mn - 9n^2$
**c.** $6a^2 - 9ab + 5b^2$  **5.** $a^{11}b^8$  **6. a.** $24x^9y^7z^7$  **b.** $-108cx$  **c.** $-8b^6$  **7. a.** $24ab - 6bc$
**b.** $42x^9y^5 - 35x^6y^8 + 14x^2y^{15}$  **c.** $-60a^{12}n^{13} + 54a^{10}n^{11} - 6a^8n^{15}$  **8. a.** $90x^2 - 219x + 132$
**b.** $20c^4 - 32c^3d - 5c^2d^2 + 14cd^3 - 3d^4$  **c.** $56a^2 + 28ab - 40ac - 20bc$  **9.** $a^3$  **10. a.** $8x^7y^7$
**b.** $9b^6n^3$  **c.** $-6$  **11. a.** $-2 + b$  **b.** $3x^4 - 9x^2y + 6y^2$  **c.** $-6x^6 + 4x^5 - 7x^3 + 5x^2 + 1$
**12. a.** $c - 8$  **b.** $7x - 5y$  **c.** $2a^2 - 9a + 7$ R $3a + 2$  **13. a.** $8m - 4n$  **b.** $-42x + 69$

**c.** $10c^2 - 12c + 4$ **d.** $-2a^2 + 22a + 21$   **14. a.** $6b^3 + 4b^2 + (-7b + 8)$
**b.** $12x^4 - 5x^3y + 7x^2y^2 - (+xy^3 - 2y^4)$

## KEYED ACHIEVEMENT TEST  (p. 199)

**1.** Integers; $-108$, $\frac{56}{8}$ Rationals; $-3.7$, $50\%$, $-108$, $\frac{56}{8}$, $-4\frac{7}{12}$, $-\frac{13}{25}$   **2.** a., b., c., e. are true

**3.**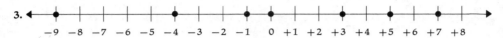
$$-9 \quad -8 \quad -7 \quad -6 \quad -5 \quad -4 \quad -3 \quad -2 \quad -1 \quad 0 \quad +1 \quad +2 \quad +3 \quad +4 \quad +5 \quad +6 \quad +7 \quad +8$$

**4.** 26, 16, 8, 2, $-2$, $-4$, $-4$   **5.** 270   **6.** 15, 35; $s$ is equal to four times $r$ minus 1;
$s = 4r - 1$   **7.** $-3x^2 - 18x + 7$   **8.** $-4c^2 - 3d^2$   **9.** $24n^3 - 66n^2 + 63n - 27$
**10.** $7a - 5b$   **11.** $-2m + 11$

### Chapter 6  *Equations in One Variable*

### DIAGNOSTIC TESTS

6–1   (p. 203)   **1.** 15   **3.** $-4$   **5.** $x - 1 = -3$, $x = -2$; $x - 1 = -3$ and $x = -2$ are
equivalent equations; $x = -2$   **7.** $8t - t = 0$ and $11 - x = x$ are
conditional equations; $2x + 5 = 5 + 2x$ is satisfied by all replacements;
$n + 7 = n$ has no solution

6–2   (p. 206)   **1. a.** $-2$ **b.** 9   **3. a.** $\div$; 7 **b.** $\div$; 2   **5. a.** $x$; 6 **b.** $x$; 12′   **7.** $+9.7$   **9.** $-\frac{8}{3}$

6–3   (p. 209)   **1. a.** 18 **b.** 3   **3. a.** 8 **b.** 12   **5.** yes, illustrations will vary; see bottom of
page 209   **7.** yes, illustrations will vary; see page 209

6–4   (p. 212)   **1.** 29   **3.** 18   **5.** $1\frac{1}{4}$   **7.** 0   **9.** $-32$

6–5   (p. 215)   **1.** 20   **3.** 10   **5.** $2\frac{1}{2}$   **7.** $-18$

6–6   (p. 219)   **1.** 7   **3.** 1   **5.** $2\frac{3}{4}$   **7.** $\frac{5}{9}$   **9.** 300   **11.** 36   **13.** $-12$   **15.** 7

6–7   (p. 222)   **1.** 21   **3.** 80   **5.** 64   **7.** 0   **9.** 18   **11.** $-24$

6–8   (p. 226)   **1.** 8   **3.** 8   **5.** 8   **7.** 2   **9.** 5   **11.** $1\frac{1}{4}$   **13.** 4   **15.** 70   **17.** 5   **19.** 7   **21.** 9
**23.** 13   **25.** 1   **27.** 11   **29.** $\frac{1}{3}$   **31.** $-3$   **33.** 8

6–9   (p. 230)   **1. a.** 65 and $-65$ **b.** $-126$ and 126

6–10  (p. 232)   **1.** 5   **3.** $-34$   **5.** .05   **7.** 9   **9.** 4   **11.** 5   **13.** 4   **15.** 225   **17.** $\frac{1}{4}$

6–11  (p. 239)   **1.** $-1$   **3.**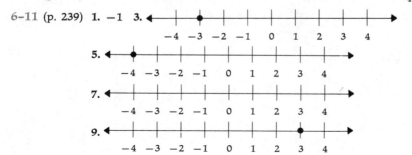

## CHAPTER REVIEW  (p. 242)

**1.** a., c., and d.; a, c, and d; $x = -1$   **2.** a. 13 b. $-7$ c. $-8$   **3.** a. 10 b. 28 c. $-3$
**4.** a. $-1\frac{2}{3}$ b. 125 c. 13   **5.** a. 48 b. $-40$ c. $-20$   **6.** a. $-3$ b. $-13$ c. 3   **7.** a. 15 or $-15$
b. 13 or $-13$   **8.** 56   **10.** 9   **11.** $-100$   **12.** 5   **13.** 750

**14. a.**

**b.**

**c.**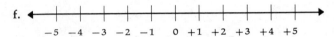

**d.**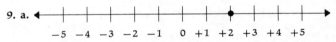

**e.** no graph

**f.**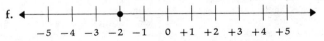

## KEYED ACHIEVEMENT TEST  (p. 243)

**1.** $-3a^2 - 6ab - 13b^2$   **2.** $n^2 - n - 4$   **3.** $-10x^2 + 21xy - y^2$
**4.** $18b^4 - 42b^3c + 29b^2c^2 - 3bc^3 - 2c^4$
**5.** $m^5 + 10m^4n + 20m^3n^2 + 40m^2n^3 + 80mn^4 + 160n^5$ R $384n^6$   **6.** $10b^2 - 25b - 15$
**7.** a. 66 b. 7 c. $-3$ d. $\frac{1}{2}$ e. 343 f. $-3$ g. 11 h. 10 i. 2 j. 34   **8.** a. 7 b. 48

**9. a.**

**b.**

**c.**

**d.** no graph **e.**

**f.**

## Chapter 7  *Problem Solving*

## DIAGNOSTIC TESTS

7-1  (p. 247)   **1.** $x$ = no. of dimes, $4x$ = no. of nickels   **3.** $n$ = Scott's money in
dollars, $100 - n$ = Peter's money in dollars or vise-versa

7-2  (p. 251)   **1.** $85° - d$   **3.** $s + t$; $3st$   **5.** $b + 2$

7–3 (p. 255) **1. a.** $x + x = 46$ **b.** $6x - 10 = 81$ **c.** $\dfrac{x}{12} = -7$

7–4 (p. 260) **1.** 52 **3. a.** 22.4 **b.** 1,960 **5. a.** $\frac{1}{15}$ **b.** $\frac{2}{3}$ **7. a.** 75% **b.** 18% **9. a.** 600
**b.** 640 **11.** 731 students **13.** 130, 132, and 134

7–5 (p. 263) **1.** $125,000 **3.** $125 **5.** $79 **7.** Marilyn, 2 yr.; Charlotte, 11 yr.
**9.** $1\frac{2}{3}$ hr.

7–6 (p. 280) **1. a.** 59 meters **b.** 1,386 sq. meters **c.** 25 meters **3.** 57 teeth **5.** 200 lbs.
**7.** 8 ft.; 8 to 1; 16 lb. per sq. ft.

## CHAPTER REVIEW  (p. 298)

**1.** $5a = 3a + 26$ **2.** $2x - 20 = x + 15 + 10$ **3. a.** 48 **b.** 245 **c.** 10 and 7 **d.** 86, 88, 90
**4.** $48.75 **5.** 91 inches **6.** 21 $5 bills **7.** $1\frac{3}{4}$ hr. **8.** Scott, $14\frac{2}{3}$ yr., Peter, $5\frac{1}{3}$ yr. **9.** 1,386 sq.
in. **10.** $22\frac{2}{9}°$ **11.** 6 amperes; $18\frac{1}{3}$ ohms

## KEYED ACHIEVEMENT TEST  (p. 299)

**1.** $m^2 - 7mn + 3n^2$ **2.** $-4a^2 + 6ab - 5b^2$ **3.** $30x^3 - 16x^2y + 37xy^2 - 7y^3$
**4.** $7n^2 + 2n - 21$ R 106 **5. a.** 6 **b.** $-\frac{1}{2}$ **c.** $-6$ **d.** 19 **e.** 2 **f.** $-7$ **6. a.** 11 **b.** 6 **c.** 900
**7.** $4,500 **8.** Kim, 15 yr., Brother, 12 yr. **9.** 21,177 sq. ft. **10.** $900

## Chapter 8 *Inequalities in One Variable*

## DIAGNOSTIC TESTS

8–1 (p. 304) **1. a.** Six times some number $x$ minus eight is equal to fifteen. **b.** Each
number $t$ plus nine is less than seven. **c.** Twenty-five is greater than each
number $n$ plus two. **d.** Eleven times each number $c$ is not less than fifty.
**e.** Nine times each number $a$ minus five is not equal to four times this
number $a$ plus thirteen. **f.** Three times each number $b$ divided by two is
not greater than ten. **3. a.** Six times each number $y$ is greater than or
equal to forty-eight. **b.** Three times each number $x$ plus seven is less
than or equal to negative nine. **c.** Forty-six is not less than or not equal
to seven times each number $n$ minus eleven. **d.** Each number $m$ divided
by eight is not greater than or not equal to fifteen. **5. a.** yes **b.** no **c.** yes
**d.** yes **7. a.** Each number $b$ is greater than six and less than sixteen.
**b.** Each number $n$ is less than two and greater than negative three.
**c.** Each number $x$ is less than or equal to ten and greater than or equal
to one. **d.** Five times each number $d$ is greater than or equal to negative
eleven and less than or equal to zero. **9. a.** Each number $x$ is greater
than negative two and less than or equal to seven. **b.** Each number $h$ is
less than or equal to fifteen and greater than four. **11. a.** $10 < c$
**b.** $-5 > r$ **c.** $16 \not> x$ **d.** $2 \leq y$

8–2 (p. 310) **1.** $6 > 3$ **3.** $12 < 30$ **5.** reverse **7.** 6 **9.** 12

8–3 (p. 315) **1. a.** 0, 1, 2, . . ., 11 **b.** none **c.** 0 **3. a.** 1 **b.** 1, 2, 3, 4, 5, 6, 7 **c.** none
**5. a.** 10, All real numbers greater than 10 **b.** 1, All real numbers greater
than 1 **c.** $-2$, All real numbers greater than $-2$ **7. a.** All integers
except $-8$ **b.** All integers except 11 **c.** All integers except 1 **9a.** 2,
3, 5, 7, 11, 13 **b.** 2, 3, 5, 7 **c.** 2, 3, 5, 7, 11, 13, 17, 19, 23

8-4   (p. 321)   **1. a.** All real numbers greater than 15 **b.** All real numbers greater than −4 **c.** All real numbers greater than 0   **3. a.** All real numbers except 7 **b.** All real numbers except −7 **c.** All real numbers except −8   **5. a.** 3, All real numbers greater than 3 **b.** −7, All real numbers greater than −7 **c.** .8, All real numbers greater than .8   **7. a.** 13, All real numbers less than 13 **b.** 0, All real numbers less than 0 **c.** −10, All real numbers less than −10   **9. a. (1)** All real numbers less than −5 **(2)** −12, All real numbers less than −12 **(3)** 0, All real numbers greater than 0 **b. (1)** −16, −14, −12, . . . **(2)** . . . , 2, 4, 6 **(3)** All even integers

8-5   (p. 325)   **1. a.** All real numbers greater than 28 **b.** All real numbers greater than $6\frac{2}{3}$ **c.** All real numbers greater than 15   **3. a.** All real numbers except 31 **b.** All real numbers except $1\frac{3}{4}$ **c.** All real numbers except 0   **5. a.** 54, All real numbers greater than 54 **b.** −6, All real numbers greater than −6 **c.** 98, All real numbers greater than 98   **7. a.** 8, All real numbers less than 8 **b.** 8.25, All real numbers less than 8.25 **c.** 21, All real numbers less than 21   **9. a. (1)** All real numbers greater than 22 **(2)** −3, All real numbers greater than −3 **(3)** 10, All real numbers less than 10 **b. (1)** . . . , −1, 1, 3 **(2)** 3, 5, 7, . . . **(3)** . . . , −5, −3, −1

8-6   (p. 331)   **1. a.** All real numbers greater than 19 **b.** All real numbers greater than 400 **c.** All real numbers greater than 18 **d.** All real numbers greater than −3   **3. a.** All real numbers except 4 **b.** All real numbers except 9 **c.** All real numbers except 70 **d.** All real numbers except $-\frac{3}{5}$   **5. a.** 12, All real numbers greater than 12 **b.** 700, All real numbers greater than 700 **c.** 72, All real numbers greater than 72 **d.** $-6\frac{1}{3}$, All the real numbers greater than $-6\frac{1}{3}$   **7. a.** $8\frac{1}{2}$, All real numbers less than $8\frac{1}{2}$ **b.** .7, All real numbers less than .7 **c.** 98, All real numbers less than 98 **d.** $-1\frac{1}{3}$, All the real numbers less than $-1\frac{1}{3}$   **9. a.** All real numbers greater than $-\frac{1}{2}$ **b.** All real numbers greater than 5   **11. a.** 0, All real numbers greater than 0 **b.** 3, All real numbers greater than 3   **13. a.** $-1\frac{1}{4}$, All real numbers less than $-1\frac{1}{4}$ **b.** 3, All real numbers less than 3   **15. a. (1)** 1, 2, 3, . . . , 8 **(2)** none **(3)** 1, 2, 3, . . . **b. (1)** . . . , −10, −8, −6 **(2)** 1, 2, 3, . . . **(3)** 81, 82, 83, . . . **c. (1)** 2, 3, 5, 7, 11, 13, 17 **(2)** 2, 3, 5 **(3)** 2, 3, 5, 7

8-7   (p. 336)   **1.** All real numbers greater than 55 **3.** All real numbers except 3 **5.** 0, All real numbers greater than 0 **7.** 180, All real numbers less than 180   **9.** All real numbers greater than 8   **11.** 64, All real numbers greater than 64   **13.** 0, All real numbers less than 0   **15. a. (1)** 1, 2, 3, . . . , 13 **(2)** 31, 32, 33, . . . **(3)** 1, 2, 3, . . . , 40 **b. (1)** . . . , −15, −14, −13 **(2)** 1, 2, 3, . . . **(3)** All integers except 21 **c. (1)** 2, 3, 5, 7, 11, 13 **(2)** 2, 3, 5, 7, 11, 13, 17, 19, 23 **(3)** 2, 3, 5, 7, 11

8-8   (p. 341)   **1. a.** All real numbers greater than 7 **b.** All real numbers less than 5 **c.** All real numbers except 3 **d.** $3\frac{1}{6}$, All real numbers less than $3\frac{1}{6}$ **e.** $5\frac{1}{4}$, All real numbers greater than $5\frac{1}{4}$ **f.** −7, All real numbers greater than −7 **g.** −6, All real numbers less than −6   **3. a.** All real numbers greater than 28 **b.** All real numbers greater than −7 **c.** All real numbers except 5 **d.** 9, All real numbers less than 9 **e.** 7, All real numbers less than 7 **f.** −7, All real numbers greater than −7 **g.** −5, All real numbers greater than −5   **5. a.** All real numbers greater than −7 **b.** All real numbers greater than 2 **c.** All real numbers except 3 **d.** 4, All real numbers less than 4 **e.** −9, All real numbers greater than −9 **f.** 2, All real numbers greater than 2 **g.** 3, All real numbers less than 3   **7. a.** 1, 2, 3, . . . , 7 **b.** . . . , −16, −14, −12 **c.** none

8-9   (p. 348)   The graph is:
**1.** the half-line to the right of the point whose coordinate is: **a.** 1 **b.** −4
**3.** the entire number line excluding the point whose coordinate is: **a.** 3

**b. 4  5.** the ray to the left of but including the point whose coordinate is: **a.** 6 **b.** −2  **7.** the ray to the right of but including the point whose coordinate is: **a.** −6 **b.** 5  **9.** the interval between the points whose coordinates are 0 and −3  **11.** the line segment whose endpoints have the coordinate 9 and 1  **13.** the part of the number line between the points whose coordinates are −4 and 4 but includes the endpoint whose coordinate is −4  **15.** the ray to the left of but including the point whose coordinate is −2  **17.** the half-line to the left of the point whose coordinate is −4  **19.** the half-line to the left of the point whose coordinate is 5  **21.** the ray to the left of but including the point whose coordinate is 5  **23.** the half-line to the right of the point whose coordinate is 6  **25.** the half-line to the right of the point whose coordinate is −4

## CHAPTER REVIEW  (p. 352)

**1. a.** $4x \not> 6$ **b.** $9n - 2 \leq 49$  **2. a. (1)** 0, 1, 2, 3, 4, 5, 6, 7 **(2)** 16, 17, 18, . . . **b. (1)** . . . , 2, 4, 6 **(2)** . . . , −6, −4, −2, 0 **c. (1)** 2, 3, 5 **(2)** 2, 3 **d. (1)** All real numbers greater than or equal to 4.5. **(2)** All real numbers less than 14.  **3. a. (1)** 0, 1, 2, 3, 4, 5, 6, 7, 8, 9, 10 **(2)** All whole numbers except 2. **b. (1)** All real numbers greater than $\frac{2}{3}$. **(2)** All real numbers greater than 9. **c. (1)** 8, 9, 10 **(2)** 0, 1, 2, 3, 4, 5 **d. (1)** none **(2)** −4, −3, −2, −1

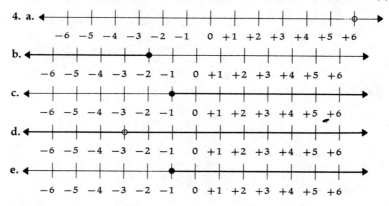

**f.** the entire number line

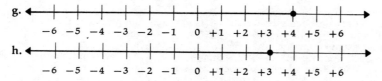

**5. a.** $x < 3$ **b.** $x \geq 0$ **c.** $x \neq -4$ **d.** $-7 \leq x \leq -1$

## KEYED ACHIEVEMENT TEST  (p. 353)

**1. a.** $y = 3$ **b.** $x = \frac{7}{6}$ **c.** $n = 11$ **d.** $y = 7$

**2. a.**

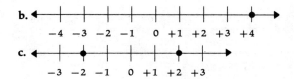

**3. a.** $-52.78$ **b.** $172.06$ **4. a.** 11 quarters, 19 dimes **b.** \$77.35 **c.** width 59 m, length 91 m
**5. a.** All real numbers less than 24 **b.** 5, 6, 7, . . . **c.** All real numbers less than or equal to 5
**d.** $-9, -7, -5$ **e.** All real numbers greater than $-9$.

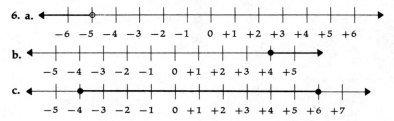

## Chapter 9 *Special Products and Factoring*

## DIAGNOSTIC TESTS

**9–1** (p. 357)    **1.** $54n^2 + 93n + 35$    **3.** $14a^2 + 13a - 10$    **5.** $24w^2 + 10w - 99$
         **7.** $15c^2 - cd - 40d^2$    **9.** $50b^2c^2 - 40abcx + 6a^2x^2$
         **11.** $42a^4 - 25a^2x^3y + 3x^3y$    **13.** $9a^2 + 66ax + 121x^2$
         **15.** $20bc + 35by - 8cx - 14xy$

**9–2** (p. 360)    **1.** $81$    **3.** $\frac{49}{64}$    **5.** $625$    **7.** $5.76$    **9.** $a^8b^6c^{18}$    **11.** $100d^{16}r^{12}s^{14}$

**9–3** (p. 361)    **1.** $r^2 - 81$    **3.** $w^2 - v^2$    **5.** $4 - n$    **7.** $a^2b^4 - c^2$    **9.** $49a^2 - 36b^2$
         **11.** $.81x^2 - .25y^2$    **13.** $b^4 - 16$

**9–4** (p. 364)    **1.** $(7x - 3y)^2$    **3.** $t^2 + 12t + 36$    **5.** $d^2 + 2dr + r^2$    **7.** $a^2 + 2amy + m^2y^2$
         **9.** $w^2 + 24w + 144$    **11.** $x^2 + 2xz + z^2$    **13.** $64a^2 + 48ab + 9b^2$
         **15.** $t^2 - 2tw + w^2$    **17.** $25a^6 - 80a^3cx + 64c^2x^2$    **19.** $m^2 - 2mx + x^2$
         **21.** $36n^4 - 24n^2rx + 4r^2x^2$    **23.** $.81a^2 - 2.16ab + 1.44b^2$
         **25.** $81c^4 - 18c^2d^2 + d^4$

**9–5** (p. 369)    **1.** $x^2 + 10x + 24$    **3.** $t^2 + 8t - 20$    **5.** $y^2 - 6y - 27$    **7.** $n^2 - n - 90$
         **9.** $9x^2 + 9x - 40$    **11.** $a^2b^4 + 4ab^2 - 21$    **13.** $49a^2 + 16.1a + 7.6$
         **15.** $25x^2 - 75xy + 44y^2$

**9–6** (p. 371)    **1.** $24$    **3.** $25m^5n^3$    **5.** $2b^2 - 3b + 4$    **7.** $c(x - y + z)$    **9.** $x(a - b + 1)$
         **11.** $a^3(a + 2b)$    **13.** $4x^4y^3(3x^4y^3 - 2x^2 + 6y^2)$

**9–7** (p. 373)    **1.** $6, -6$    **3.** $\frac{4}{5}, -\frac{4}{5}$    **5.** $r^3st^5, -r^3st^5$    **7.** $.2w^2x^6y^3, -.2w^2x^6y^3$    **9.** $11$

**9–8** (p. 375)    **1.** $(n + 2)(n - 2)$    **3.** $(9x + 4)(9x - 4)$    **5.** $(7c + 12d)(7c - 12d)$
         **7.** $(c + .6x)(c - .6x)$

**9–9** (p. 378)    **1.** the trinomials in b. and d.    **3.** $(x + 4)^2$    **5.** $(3 + w)^2$    **7.** $(a^2 + b^2)^2$
         **9.** $(9 - 2x)^2$    **11.** $(7x + 6y)^2$    **13.** $(n + \frac{9}{2})^2$

**9–10** (p. 381)    **1.** 9 and 2    **3.** 12 and $-8$    **5.** 4 and $-14$    **7.** $(x - 4)(x - 2)$
         **9.** $(y + 2)(y - 9)$

**9–11** (p. 383)  **1.** $(4x + 3)(2x + 5)$  **3.** $(3a + 5)(4a - 1)$  **5.** $(3 - 4a)(2 - 5a)$
**7.** $(4a + 1)(4a - 5)$  **9.** $(4bd + 3c)(6bd - 5c)$  **11.** $(3n - 8)^2$

**9–12** (p. 385)  **1.** $a(x + 1)(x - 1)$  **3.** $a(b^2 + c^2)(b + c)(b - c)$  **5.** $2b(b + 1)(b - 5)$
**7.** $(b - c + d)(b - c - d)$

**9–13** (p. 387)  **1.** 2,464  **3.** 4,450  **5.** 74.4

## CHAPTER REVIEW  (p. 388)

**1. a.** $24b^2 + 46b + 21$ **b.** $72x^2 - 31x - 5$ **c.** $10c^2 - 53cd + 45d^2$ **d.** $48a^2b^2 + 42abc^2 - 33c^4$
**2. a.** $\frac{49}{64}$ **b.** $r^4s^6t^{10}$ **c.** $81x^8y^2$ **d.** $.09a^{12}t^4$ **3. a.** $b^2 - 144$ **b.** $100 - 81m^2$ **c.** $16a^4b^2 - 225x^2y^4$
**d.** $16m^4 - n^4$ **4. a.** $r^2 - 18r + 81$ **b.** $49x^2 + 140x + 100$ **c.** $16b^2 - 88bc + 121c^2$
**d.** $36m^4 + 60m^2n^3 + 25n^6$ **5. a.** $c^2 + 7c - 30$ **b.** $d^2 - d - 20$ **c.** $49m^2 + 98mn + 48n^2$
**d.** $60 - 51a + 9a^2$ **6.** $6a^3c^2$ **7. a.** $7b^3(3 - 2b^4)$ **b.** $4r^4s^3(4r^2s - 3rs^5 + 1)$
**c.** $-a^2(8a^2 + 5ab + 4b^2)$ **d.** $(x + 6)(x - 3)$ **8. a.** 10 **b.** $9b^6c^3d$ **c.** $.2a^5x^2$ **d.** $\frac{4}{5}r^4s^2$
**9. a.** $(7c + y)(7c - 9)$ **b.** $(3 + 11b^2c^3)(3 - 11b^2c^3)$ **c.** $(m^2 + x^2)(m + x)(m - x)$ **10. a.** 64
**b.** $168cd$ **11. a.** $(116 + 4c)^2$ **b.** $(9x^2y^3 - 5)^2$ **12. a.** $(w + 2)(w - 12)$ **b.** $(t + 8)(t - 7)$
**c.** $(8x + 9y)(8y + 3y)$ **d.** $(3r - 5s)(7r - 10s)$ **13. a.** $50(c^2 + 2d^2)(c^2 - 2d^2)$
**b.** $7a(a + 1)(a - 4)$ **c.** $(m - n + y)(m - n - y)$ **d.** $(x + 5)(x - 5)(x + 4)(x - 4)$
**14. a.** 1,596 **b.** 9,216 **15.** 5000

## KEYED ACHIEVEMENT TEST  (p. 389)

**1. a.** $n = -4$ **b.** $c = 6$ **c.** $x = 17$ **d.** $x = -\frac{1}{2}$

**2.**

**3. a.** 0, 1 **b.** 6, 8, 10, . . .

**4.**

**5. a.** $108m^2 - 69m - 24$

**b.** $50c^2 - 130cd + 72d^2$ **6. a.** $64a^2 - 25$ **b.** $m^4 - 16n^4$ **7. a.** $16r^2 - 72rs + 81s^2$
**b.** $9a^2b^2 - 6abc^2 + c^4$ **8. a.** $t^2 + 3t - 88$ **b.** $9w^2 - 51w + 70$ **9. a.** $9b^4x^2(4b^2 - 9bx^5 + 7x^8)$
**b.** $(n + 7)(n - 2)$ **10. a.** $(8c - 9)(8c + 9)$ **b.** $(x^2 + 4)(x - 2)(x + 2)$ **11. a.** $(3w + 5)^2$
**b.** $(11x^2y - 7z)^2$ **12. a.** $(m + 7)(m - 8)$ **b.** $(r + 9)(r - 2)$ **13. a.** $(4a + 3)(9a + 5)$
**b.** $(7x - 9)(6x + 1)$ **14. a.** $4(1 - a)(1 + a)$ **b.** $9(n - 6)(n - 1)$

## Chapter 10  *Operations with Algebraic Fractions*

## DIAGNOSTIC TESTS

**10–1**  (p. 393)  **1.** $\dfrac{w}{w + d}$  **3.** $\dfrac{c + d}{m}$  **5. a.** 0, 5 **b.** 6, −3 **7.** 1

**10–2**  (p. 398)  **1.** $\frac{19}{34}$  **3.** the expressions in b. and c.  **5.** $\dfrac{ab}{2}$  **7.** $\dfrac{a^2x^2}{y}$  **9.** $\dfrac{14c}{d}$  **11.** $\dfrac{1}{3a}$

**13.** $\dfrac{a}{3}$  **15.** $\dfrac{x(a + 7)}{2y(a + 4)}$  **17.** $\dfrac{a}{b}$  **19.** $\dfrac{2x}{x - y}$  **21.** $\dfrac{x - 1}{x + 3}$

**23.** $\dfrac{x - 6}{x + 8}$, $x \neq -5$, $x \neq -8$

**10-3**   (p. 402) **1.** $\frac{27}{48}$   **3.** $\frac{b^6 d^2}{a^4 c}$   **5.** $\frac{3by^3}{4a^2}$   **7.** 1   **9.** $\frac{(a-2)(a-1)}{(a+4)(a+1)}$   **11.** $\frac{2}{x+3}$

   **13.** $\frac{3n(n+3)}{2}$

**10-4**   (p. 405) **1. a.** $\frac{10}{7}$  **b.** $-\frac{1}{5}$   **3.** $\frac{3}{4}$   **5.** $\frac{3c}{2ax}$   **7.** $\frac{3(a+b)}{a-b}$

**10-5**   (p. 408) **1.** $\frac{7}{8}$   **3.** $\frac{8b}{4}$   **5.** $\frac{x+y}{3}$   **7.** $\frac{1}{a}$   **9.** $\frac{c+d}{c-d}$   **11.** 3   **13.** $\frac{x}{5}$   **15.** $\frac{a}{2}$   **17.** $\frac{2}{c}$

   **19.** $\frac{n^2}{2x^2}$   **21.** $\frac{2x}{x+y}$   **23.** $\frac{7x+y}{5}$   **25.** $\frac{c-6d}{4}$   **27.** 2   **29.** $\frac{1}{a+b}$

   **31.** $\frac{b^2+25}{b^2+2b-15}$

**10-6**   (p. 414) **1.** 9, 18, 27, 36, 45, 54, 63, 72   **3.** 60   **5.** $x^2+2x-24$
   **7.** $(n+3)(n+4)(n-9)$   **9.** 80   **11.** $c^4 d^5$   **13.** $4ab(4a-b)(a-4b)$

**10-7**   (p. 418) **1.** $\frac{42}{72}$   **3.** $\frac{3(c-6)(c+4)}{24c(c-5)(c+4)}$

**10-8**   (p. 420) **1.** $\frac{43}{48}$   **3.** $\frac{a}{2}$   **5.** $\frac{37c}{30}$   **7.** $\frac{9b+10c}{12}$   **9.** $\frac{11}{5a}$   **11.** $\frac{20a+21b}{24c}$   **13.** $\frac{3ab+2c}{6a^3}$

   **15.** $\frac{6cy+5cx}{8xy}$   **17.** $\frac{x+7}{6}$   **19.** $\frac{2x+2y}{3}$   **21.** $\frac{8d-17}{10}$

   **23.** $\frac{3y+4xy+2x}{xy}$   **25.** $\frac{3b^2 c-2a^2 c+6ab^2-abc}{6abc}$

**10-9**   (p. 424) **1.** $\frac{5a+12}{(a+2)(a+3)}$   **3.** $\frac{2(x+12)}{(x-3)(x+3)}$   **5.** $\frac{m(7m+16)}{(m-2)(m+4)}$

   **7.** $\frac{c^2+4cd+d^2}{cd(c+d)}$   **9.** $\frac{-4bx}{(b+x)(b-x)}$   **11.** $\frac{13}{2(b+2)}$   **13.** $\frac{9}{(a-3)^2}$

   **15.** $\frac{4a^2-ab+2b^2}{4(a+2b)(2a-b)}$

**10-10**   (p. 426) **1.** $\frac{3b-3}{2b+4}$   **3.** $\frac{4m^2-24m+8}{m^2-10m+25}$   **5.** $\frac{-3(a+x)(a-x)}{2(2a+x)(a+2x)}$

**10-11**   (p. 429) **1.** $a, b, c$   **3.** $\frac{-1}{b+a}$   **5.** $-1$

**10-12**   (p. 432) **1.** $\frac{10x^2-7}{3x}$   **3.** $\frac{b-4x}{2}$   **5.** $\frac{x^2}{x+y}$   **7.** $x^2-xy$   **9.** $c-d$

## CHAPTER REVIEW   (p. 435)

**1. a.** 0 **b.** 5, $-\frac{2}{3}$ **c.** 7, $-6$   **2. a.** $\frac{4b^3 x^5}{9y^2}$   **b.** $\frac{5}{3a+8}$   **c.** $\frac{8x-3y}{2x+5}$   **3. a.** $\frac{48s^2}{25rxy}$

**b.** $\frac{(n+6)(n-3)}{(n+2)(n+5)}$   **4. a.** $\frac{a-3b}{18a^2 b^3}$   **b.** $\frac{c+2d}{5cd}$   **5. a.** $\frac{3x}{y}$ **b.** 4 **c.** $\frac{n-5}{n+5}$   **6. a.** $72r^3 x^9$

**b.** $mn(m+3n)^2(m^3-9mn)$   **7. a.** $xy(x-y)$ **b.** $(c-8)^2(c+8)(c-2)$   **8. a.** $6a^6 c$

**b.** $2x(x-4)(x-8)$ **9. a.** $\dfrac{-15s^2 - 6r^2 + 16rs}{36rs}$ **b.** $\dfrac{45c + 33c^2d - 12c^3d^2}{126d^4}$

**10.** $\dfrac{(3n+2)(n+4)}{3n}$ **11.** $\dfrac{x-y}{xy}$ **12.** $\dfrac{1}{d(c-d)}$ **13. a.** $\dfrac{15m}{m+4n}$ **b.** $\dfrac{2a^2 - 6ax - 5x^2}{a(a+2x)(2a+x)}$

**c.** $\dfrac{(2x+y)(x-5y)}{x(x+y)(x-y)}$ **d.** $\dfrac{-n^2 + 27n - 42}{(n-1)(n+1)^2(n-4)}$ **14.** $\dfrac{(-3x-2)(x-1)}{(x-2)(x+1)}$ **15.** $\dfrac{c^2 - cd + d^2}{c^2 + cd - d^2}$

## KEYED ACHIEVEMENT TEST  (p. 437)

**1. a.** $54w^2 - 87w + 35$ **b.** $72x^2 + 37xy - 24y^2$ **2. a.** $n^2 - 14n + 49$ **b.** $100t^2 - 81x^2$
**3. a.** $r^2 - 18r + 81$ **b.** $49x^2 - 42xy^2 + 9y^4$ **4. a.** $a^2 - 5a - 150$ **b.** $16c^2 + 64c + 55$
**5. a.** $25(n^2 + 4y^2)$ **b.** $9a^3c^3(6a^4 - 9ac^2 + 8c^6)$ **6. a.** $(4x - 3y)(4x + 3y)$
**b.** $(5a^2c - 7d^3)(5a^2c + 7d^3)$ **7. a.** $(9s - 4)(9s - 4)$ **b.** $(7m + 11n)(7m + 11n)$
**8. a.** $(r + 3)(r - 21)$ **b.** $(w - 2)(w + 1)$ **9. a.** $(9b - 10)(2b + 9)$ **b.** $(6r - 7)(8r - 5)$

**10. a.** $16a^2(2y - 1)(2y + 1)$ **b.** $x(x - 3)(x + 3)(x^2 + 9)$ **11. a.** $\dfrac{3yz^2}{8a^2x}$ **b.** $-1$ **c.** $\dfrac{2}{3n-2}$

**12. a.** $\dfrac{5s^7}{12r^2x^4}$ **b.** $\dfrac{4x(x-6)^2}{(x+3)(x+6)^2}$ **13. a.** $\dfrac{20b^4d^2}{a^2c}$ **b.** $\dfrac{(n-2)(n-5)}{4(n+4)(n+2)}$ **14. a.** $3c + d$

**b.** $\dfrac{7rs - 16s^2 - 12r^2}{24r^2s^2}$ **c.** $\dfrac{9b + 68}{b^2 - 7b - 18}$ **d.** $\dfrac{7y^2 + 45y - 200}{y(y-5)^2(y+5)}$ **15.** $\dfrac{1}{a-3b}$

## Chapter 11  *Fractional Equations and Problem Solving*

## DIAGNOSTIC TESTS

**11–1** (p. 442) **1.** $\dfrac{4}{5} \times y;\ \dfrac{4y}{5}$ **3.** 20 **5.** 72 **7.** 20 **9.** $4\frac{1}{2}$ **11.** 15 **13.** 1 **15.** 1 **17.** 10
**19.** 5

**11–2** (p. 447) **1.** 3 **3.** 5 **5.** 8 **7.** 3 **9.** 2 **11.** 6

**11–3** (p. 450) **1.** .9 **3.** 3 **5.** 60

**11–4** (p. 452) **1.** $t = 27$ **3.** $R_4 = 63$ **5.** $F = 50$

**11–5** (p. 454) **1.** 1 **3.** $3\frac{3}{5}$ hr. **5.** 3 L

## CHAPTER REVIEW  (p. 460)

**1. a.** 96 **b.** 16 **c.** 45 **d.** 12 **2. a.** $2\frac{1}{15}$ **b.** 9 **c.** $3\frac{1}{2}$ **d.** 8 **3. a.** .19 **b.** 80 **c.** 2 **d.** 900 **4. a.** $d = 27$
**b.** $b' = 16$ **c.** $f = 21$ **d.** $R = -28$ **e.** $r = \frac{1}{2}$ **5. a.** 72 **b.** 80 min. **c.** 13 oz.

## KEYED ACHIEVEMENT TEST  (p. 461)

**1. a.** $36t^2 - 21t - 30$ **b.** $m^2 - 27m + 176$ **c.** $144b^2c^4 - 24bc^2x^3 + x^6$ **d.** $9x^4y^2 - 16z^2$
**2. a.** $7s^3t^4(5r^9t^4 + 9r^6t - 2s^2)$ **b.** $(10c^2 + 9d)(10c^2 - 9d)$ **c.** $(8n - 5)^2$ **d.** $(m + 14)(m - 4)$

**e.** $(7x + 6)(5x + 3)$ **f.** $ab(1 + b^2)(1 + b)(1 - b)$ **3.** $\dfrac{6b - 5x}{8b^2}$ **4.** $\dfrac{x}{6(x-5)}$ **5.** $\dfrac{7y^2}{10a^2x^3}$

**6. a.** $\dfrac{-7c - 38}{24}$ **b.** $\dfrac{x^2 - 16x + 18}{x(x-2)(x-3)}$ **7. a.** $-2$ **b.** 7 **c.** 65 **d.** 200 **8.** $n = 5$ **9.** 4.8 hours

## Chapter 12 *Graphing Linear Equations*

### DIAGNOSTIC TESTS

**12-1** (p. 466) **1.** $(-5, -1)$ **3.** $A(2, -8)$, $B(-5, 8)$, $C(7, 6)$, $D(0, 0)$, $E(-4, -4)$, $F(-6, 0)$, $G(0, -3)$ **5. a.** III **b.** II **c.** I **d.** IV **e.** II **f.** III **7.** $y = 7 + x$

**12-2** (p. 470) **1.** $(4, 3)$, $(0, -3)$, $(-6, -12)$ **3.** $(-2, -14)$, $(0, -9)$, $(4, 1)$

**12-3** (p. 474) **1-5.**

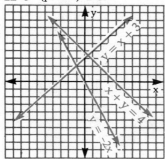

**7-11.**

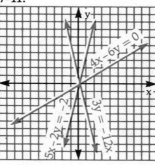

**13.** $(-4, -4)$, $(4, 2)$, and $(0, -1)$ **15.** $(0, 2)$

**12-4** (p. 480) **1.** positive number: d; negative number: b; Zero: c **3. a.** fall **b.** rise **5.** the graph is a line parallel to or equal to the line containing: **a.** $(0, 0)$ and $(1, 5)$ **b.** $(0, 0)$ and $(1, -4)$ **c.** $(0, 0)$ and $(5, 2)$ **d.** $(0, 0)$ and $(2, -3)$ **7.** 6

**12-5** (p. 484) **1. a.** $-7$ **b.** $-6$ **c.** $-5$ **3.** $(0, -4)$

**12-6** (p. 487) **1.** A line with $y$-intercept $-4$, and slope $\frac{5}{2}$.

**12-7** (p. 489) **1.** The graph is the line containing $(0, 180)$ and $(30, 150)$; horizontal scale: values for $N$; vertical scale: values for $M$

**12-8** (p. 492) **1. a.** $y = 3x + 2$ **b.** $y = -\frac{2}{5}x - 7$ **3. a.** $x + 2y = 19$ **b.** $3x - y = -13$ **5.** $y + 2x = -6$

### CHAPTER REVIEW (p. 496)

**1.**

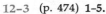

**2.** $y = 9 + x$

**3.** $(1, -1)$

**4.** $y = \frac{5}{3}x + \frac{2}{3}$

**5.** $(-1, -7)$, $(-1, -4)$, $(4, 8)$

**6.**

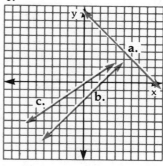

**7. a.** (0, 3) **b.** (−5, 4) and (2, 0)   **8. a.** (−5, 0)
   **b.** (3, 0) **c.** ($\frac{1}{2}$, 0)   **9. a.** 3 **b.** $\frac{9}{4}$   **10. a.** 2 **b.** 11
**11. a.** slope, −4; $y$-intercept, 7; (0, 7)
   **b.** slope, $\frac{5}{8}$; $y$-intercept, 2; (0, 2) **c.** slope, $-\frac{3}{2}$;
   $y$-intercept, $\frac{3}{4}$; (0, $\frac{3}{4}$)

**12.**

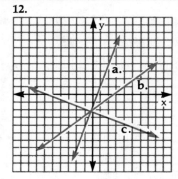

**13.**

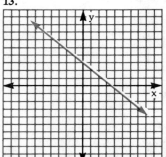

**14.**

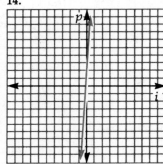

**15. a.** $y = -5x - 4$

   **b.** $y = 2x - 7$

   **c.** $y = x + 7$

## KEYED ACHIEVEMENT TEST   (p. 497)

**1. a.**

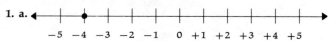

   **b.**                                                               **c.** no graph

   **d.**

   **e.**                                                               **f.** all numbers

**2.**

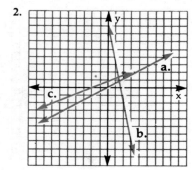

**3. a.** (3, −7) **b.** (−3, −1)   **4. a.** 3 **b.** $-\frac{5}{2}$   **5. a.** slope,
6; $y$-intercept, −2; (0, −2) **b.** slope, 2; $y$-intercept,
−5; (0, −5) **c.** slope, $\frac{3}{2}$; $y$-intercept, $\frac{5}{6}$; (0, $\frac{5}{6}$)

**6.** 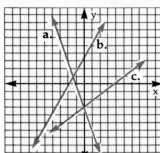   **7.** 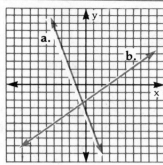   **8.** $5y = -11x + 3$

## Chapter 13   *Systems of Linear Equations in Two Variables and Problem Solving*

### DIAGNOSTIC TESTS

**13–1** (p. 501)  **1.** $7x - 2y - 4 = 0$  **3.** $(-5, -3)$  **5.** the system given in b.  **7.** the system given in b.

**13–2** (p. 506)  **1.** $(2, 3)$  **3.** $(-2, 5)$  **5.** $(0, 4)$  **7.** $(5, -6)$

**13–3** (p. 510)  **1.** $x$  **3.** $(7, 2)$  **5.** $(6, 1)$  **7.** $(-3, -2)$  **9.** $(3, 2)$  **11.** $(2, -2)$

**13–4** (p. 515)  **1.** $(2, 2)$  **3.** $(-3, 1)$  **5.** $(3, 6)$  **7.** $(7, 4)$  **9.** $(2, 3)$

**13–5** (p. 519)  **1.** $(2, 1)$  **3.** $(800, 700)$  **5.** $(1, 2)$

**13–6** (p. 521)  **1.** 17 and 8  **3.** 120 15-cent and 130 18-cent stamps  **5.** 289 adult and 323 student tickets  **7.** 260 mph air speed and 20 mph wind velocity  **9.** 7 P.M.

### CHAPTER REVIEW   (p. 530)

**1. a.** $(0, 0)$ **b.** $(6, 2)$ **c.** $(5, -4)$  **2. a.** $(-2, -3)$ **b.** $(2, 1)$ **c.** $(-1, 2)$  **3. a.** $(4, -1)$ **b.** $(6, -3)$ **c.** $(-4, 3)$  **4. a.** $(3, -1)$ **b.** $(20, 12)$ **c.** $(4, 3)$ **d.** $(10, 0)$  **5.** \$4,500 at 10%, \$12,000 at 12%

### KEYED ACHIEVEMENT TEST   (p. 531)

**1–2.**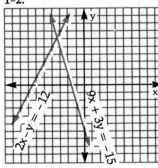

**3. a.** $49b^2 - 28bc + 4c^2$ **b.** $36a^4 - 25b^2$
**4. a.** $(m - 8)(m + 3)$ **b.** $(10r - 9s)(10r + 9s)$
**5.** $\dfrac{x + 1}{2(x - 4)}$  **6.** $9x^4y - 15x^3y$
**7.** $\dfrac{(a + x)}{(a + 5x)}$ **8.** $\dfrac{4n^2 - 12n - 4}{(n - 2)(n - 4)}$ **9.** $x = 5$
**10.** $r = 2$  **11. a.** $(4, 2)$ **b.** $(-1, -3)$ **c.** $(-2, 1)$ **12. a.** $(5, 3)$ **b.** $\left(\dfrac{272}{186}, \dfrac{747}{186}\right)$ **c.** $(2, -1)$
**13. a.** $(-2, 4)$ **b.** $(-3, 8)$ **c.** $(-6, -2)$ **14. a.** $x = 3$, $y = -2$ **b.** $x = 4$, $y = 7$ **c.** $x = 8,000$, $y = -7,000$
**15.** 21 and 9

## Chapter 14   *Powers; Roots; Radicals*

### DIAGNOSTIC TESTS

14–1   (p. 535) **1.** $49a^2b^8$   **3.** $m^{18}x^{12}$   **5.** $16a^{12}x^4$   **7.** $\frac{1}{4}m^8n^{10}$   **9.** $4a^2(x+3)^6$

14–2   (p. 538) **1.** 6.245   **3.** 80   **5.** 198   **7.** 2,989   **9.** 506   **11.** 3.87   **13.** 5.831

14–3   (p. 542) **1.** 8   **3.** $\pm 12$   **5.** $a^6b^4c$   **7.** $\frac{2}{5}$   **9.** 4   **11.** $3a^3b^4c^2$   **13.** $4x^2y^3z$   **15.** $.7rt^5$

            **17.** $\dfrac{5b^2c}{9m^4n^3}$   **19.** 2

14–4   (p. 545) **1.** $\sqrt[3]{64r^6s^9}$   **3.** 9   **5.** $7a^3bc^5$   **7.** $r^3s^2t^5$   **9.** $3a^5$   **11.** $\dfrac{2b^6}{3c^{12}}$

14–5   (p. 547) **1.** $2\sqrt{15}$   **3.** $40\sqrt{6}$   **5.** $\sqrt{5}$

14–6   (p. 548) **1.** $a\sqrt{ab}$   **3.** $5xy^2\sqrt{y}$   **5.** $6a^3b\sqrt{3b}$   **7.** $3x\sqrt{2ax}$

14–7   (p. 551) **1.** $\frac{1}{4}\sqrt{5}$   **3.** $\frac{1}{4}\sqrt{10}$   **5.** $\frac{2}{7}\sqrt{21}$   **7.** $\dfrac{1}{a}\sqrt{b}$   **9.** $\dfrac{b}{cd}\sqrt{2ad}$   **11.** $\dfrac{1}{3y}\sqrt{6xy}$

            **13.** $-\dfrac{b^3}{4a^2x^3}\sqrt{6abx}$

14–8   (p. 553) **1.** $20\sqrt{19}$   **3.** $4\sqrt{5}$   **5.** $5\sqrt{5}-2\sqrt{6}$   **7.** $5\sqrt{6}$   **9.** $\frac{2}{3}\sqrt{6}$   **11.** $(2+a)\sqrt{x}$

            **13.** $\sqrt{3}+\sqrt{5}$

14–9   (p. 556) **1.** $\sqrt{6}$   **3.** 7   **5.** $\sqrt{ax}$   **7.** $2xy\sqrt{10x}$   **9.** $40\sqrt{3}$   **11.** $32\sqrt{6}$   **13.** 18

            **15.** $2bx^2\sqrt{10a}$   **17.** 108   **19.** $6\sqrt{3}-9\sqrt{2}$   **21.** $3\sqrt{6}-12$

            **23.** $-4+10\sqrt{3}$   **25.** $43+30\sqrt{2}$   **27.** 12   **29.** $123-26\sqrt{30}$

14–10   (p. 559) **1.** $\sqrt{3}$   **3.** $2\sqrt{3}$   **5.** $4a\sqrt{a}$   **7.** $\dfrac{\sqrt{10}}{2}$   **9.** $3\sqrt{3}$   **11.** $4+11\sqrt{2}$   **13.** 4.242

14–11   (p. 562) **1.** 9   **3.** 4   **5.** 12   **7.** 7   **9.** 1   **11.** 50   **13.** $\frac{25}{8}$

## CHAPTER REVIEW   (p. 563)

**1. a.** $81r^{14}t^{22}$ **b.** $100x^{16}y^2$ **c.** $64b^{18}x^{12}z^9$ **d.** $-3125m^{25}n^{30}$ **e.** $129a^4b^{36}c^{24}$ **f.** $\dfrac{49a^{12}x^4}{529b^6y^{16}}$

**g.** $-1331r^{36}s^3t^{21}$ **h.** $\dfrac{81c^8d^{32}x^4}{256}$   **2. a.** 67 **b.** 8.6 **c.** 6,085 **d.** 97 **e.** 2.85 **3. a.** $7m^5n^3$ **b.** $4r^3y^4$

**c.** $-5b^4$ **d.** $\frac{2}{9}c^7d^3y$ **e.** $x^5y^2z^4$ **f.** $\dfrac{2c^5}{3x^3y^6}$ **g.** $8m^3n^6$ **h.** $4m^2n^4$ **4. a.** $4\sqrt{11}$ **b.** $48\sqrt{2}$ **c.** $6x^2y\sqrt{2x}$

**d.** $20x^4y^3\sqrt{x}$ **e.** $3s^4t\sqrt{3}$ **f.** $\dfrac{\sqrt{30}}{6}$ **g.** $2\sqrt{\dfrac{2n}{3}}$ **h.** $\dfrac{4b^3}{3c^4}\sqrt{\dfrac{b}{c}}$ **5. a.** $19\sqrt{3}$ **b.** $27\sqrt{3}$

**c.** $14\sqrt{6}+4\sqrt{10}$ **d.** $(7c^2d+7c^2d^2)\sqrt{2cd}$ **6. a.** 18 **b.** 12 **c.** $33\sqrt{6}-\sqrt{77}$ **d.** $10\sqrt{3}-120\sqrt{2}$

**e.** 83 **f.** $74-12\sqrt{30}$ **7. a.** $2\sqrt{2}$ **b.** $\dfrac{\sqrt{35}}{5}$ **c.** 10 **d.** $1+8\sqrt{10}$ **8. a.** 5.196 **b.** .408 **c.** .707

**d.** 1.061 **e.** 1.483 **9. a.** 4 **b.** 6 **c.** 50 **d.** 80

## Chapter 15   *Quadratic Equations and Problem Solving*

### DIAGNOSTIC TESTS

15–1   (p. 567) **1.** $3, -3$   **3.** $\sqrt{15}, -\sqrt{15}$   **5.** $5, -5$   **7.** $\sqrt{7}, -\sqrt{7}$   **9.** $3\sqrt{2}, -3\sqrt{2}$

**11.** 17, $-17$ **13.** $\sqrt{7}, -\sqrt{7}$ **15.** 2, $-2$ **17.** $\frac{2}{5}, -\frac{2}{5}$ **19.** $\frac{7}{4}, -\frac{7}{4}$
**21.** $\frac{\sqrt{3}}{4}, \frac{-\sqrt{3}}{4}$ **23.** $\frac{\sqrt{5}}{2}, \frac{-\sqrt{5}}{2}$

**15-2** (p. 571) **1. a.** 0 **b.** yes; no; yes; no **3.** $1\frac{2}{3}$ **5.** $-10, 2$

**15-3** (p. 573) **1.** 6, $-1$ **3.** 3, $-4$ **5.** $2\frac{1}{2}, -2\frac{1}{2}$ **7.** 0, $-2$ **9.** 0, $-\frac{1}{4}$ **11.** $-12$
**13.** 0, $3\frac{1}{2}$ **15.** $1\frac{1}{4}, -1\frac{1}{4}$ **17.** $-4, -1$ **19.** 6, $-3$

**15-4** (p. 576) **1.** 19, $-3$ **3.** $10 + 2\sqrt{5}, 10 - 2\sqrt{5}$ **5.** 36 **7.** $\frac{49}{4}$ **9.** $\frac{1}{36}$ **11.** 2, $-6$
**13.** 13, $-3$ **15.** 2, $-5$ **17.** 4, $-6$ **19.** $1\frac{1}{2}, -\frac{1}{2}$ **21.** 7, $-8$

**15-5** (p. 580) **1.** $a = 5, b = -4, c = -7$ **3.** 6, 2 **5.** 8, $-4$ **7.** 2, $-4$ **9.** 0, $-3$
**11.** 1, $-5$ **13.** $2\sqrt{2}, -2\sqrt{2}$ **15.** $\frac{3 \pm \sqrt{17}}{2}$ **17.** $\frac{-2 \pm \sqrt{7}}{3}$
**19.** 1.58, .42

**15-6** (p. 582) **1.** $V = 30$ **3. a.** 144 ft.; 400 ft.; 1,936 ft. **b.** 4 sec.; 9 sec.; 15 sec.

**15-7** (p. 585) **1.** 8 or $-13$

## CHAPTER REVIEW (p. 586)

**1. a.** $2\sqrt{3}, -2\sqrt{3}$ **b.** $\frac{2\sqrt{5}}{5}, -\frac{2\sqrt{5}}{5}$ **c.** 9, $-9$ **d.** $\sqrt{6}, -\sqrt{6}$ **2. a.** $-8, 9$ **b.** 0, $\frac{3}{4}$ **c.** 1, 8

**3. a.** 8, $-10$ **b.** 1, 4 **c.** $\frac{-1 \pm 6\sqrt{2}}{3}$ **4. a.** 4, $-8$ **b.** $2\frac{1}{2}, -2$ **c.** 9, $-1$ **5. a.** $\frac{7 \pm \sqrt{69}}{2}$
**b.** $1\frac{1}{2}, -\frac{3}{4}$ **c.** 4, $-\frac{1}{2}$ **6.** $w = -5 \pm 5\sqrt{5}$ **7.** 88 and 89

## KEYED ACHIEVEMENT TEST (p. 587)

**1.** $\frac{a + 2}{18a^2}$ **2.** 8 **3.** $T' = 350$ **4.** $(4, 2)$ **5.** $(3, -1)$ **6. a.** $81c^{12}d^6x^2$ **b.** $-64m^{15}n^{12}y^{24}$

**7. a.** $7c^4d^2x$ **b.** $-2r^4ty^3$ **8. a.** $3\sqrt{5}$ **b.** $6x^3\sqrt{6xy}$ **c.** $\frac{b^2}{3c}\sqrt{6}$ **9.** $8\sqrt{5} + \frac{13}{2}\sqrt{2}$ **10. a.** $24\sqrt{15}$
**b.** $166 + 24\sqrt{42}$ **11.** $6\sqrt{5} - 23$ **12.** 2 **13.** $3\sqrt{2}, -3\sqrt{2}$ **14.** $-18, 2$ **15.** 1, 25
**16.** 11, $-2$ **17.** $2\frac{1}{2}, -1\frac{1}{3}$ **18.** $V = 30$ **19.** $l = 14$ cm, $w = 12$ cm

## Chapter 16  *Ratio and Proportion; Formulas; Variation*

## DIAGNOSTIC TESTS

**16-1** (p. 592) **1.** $\frac{9}{14}$ **3.** $\frac{7}{12}$ **5.** $\frac{11}{4}$ **7.** $\frac{4}{3}$ **9.** $\frac{3}{5}$ **11.** $\frac{1}{2}$ **13.** 16 mi./gal. **15.** the pairs
given in b. and c. **17.** $\frac{x}{16} = \frac{65}{40}$ **19.** the statement in b. **21.** 8

**16-2** (p. 596) **1.** $d = \frac{W}{F}$ **3.** $w = \frac{V}{hl}$ **5.** $m = s - c$ **7.** $N = 180 - M$ **9.** $t = \frac{v - V}{g}$
**11.** $W = LA$ **13.** $V' = \frac{VT'}{T}$ **15.** $p = \frac{A}{l + rt}$ **17.** $t = \frac{d}{4}$ **19.** $s = \frac{1}{2}gt^2$

**16-3** (p. 601) **1.** no; yes, 12 **3.** $p = ks$ **5.** $P$ is halved **7.** 224

**16-4** (p. 605) **1.** no; yes, 144 **3.** $ps = k$ **5.** $E$ is doubled **7.** 75

## CHAPTER REVIEW  (p. 608)

**1. a.** $\frac{3}{2}$ **b.** $\frac{2}{5}$ **c.** $\frac{2}{3}$ **d.** $\frac{6}{1}$ **e.** $\frac{2}{3}$ **f.** $\frac{3}{2}$  **2.** $\frac{26}{1}$  **3. a.** 20  **b.** $31\frac{3}{7}$ **c.** 150 **d.** 7 or $-7$  **4.** 10 days

**5. a.** $e = \dfrac{p - b}{2}$ **b.** $d = \dfrac{Wh}{F}$ **c.** $I = \dfrac{E}{r + R}$ **d.** $R_2 = \dfrac{RR_1R_3}{R_1R_3 - RR_3 - RR_1}$ **e.** $a = \dfrac{2S - nl}{n}$

**f.** $r = \dfrac{a - S}{l - S}$  **6.** $P = \dfrac{E^2}{R}$  **7. a.** 324 **b.** 36  **8.** The perimeter is doubled.  **9.** The circumference is halved.  **10. a.** directly **b.** inversely

## KEYED ACHIEVEMENT TEST  (p. 609)

**1.** $\dfrac{b - 4}{b - 2}$  **2.** $(1, -3)$  **3.** $(2, -2)$  **4. a.** $36a^{16}b^8c^{10}$ **b.** $-125x^{12}y^3z^{21}$  **5. a.** $9a^5x^3y$ **b.** $4m^3n^7x^3$
**6.** $-48\sqrt{7} - 12\sqrt{10}$  **7.** 15  **8.** 10, $-10$  **9.** $-8, 9$  **10.** 5, $-2$  **11.** $\frac{3}{4}, -3\frac{1}{2}$
**12.** $s = 11$  **13.** $\frac{11}{12}; \frac{12}{11}; \frac{11}{23}; \frac{12}{23}$  **14. a.** 57 **b.** 1  **15.** 112 mi  **16. a.** $t = \dfrac{\sqrt{d}}{4}$ **b.** $R_2 = \dfrac{E - IR_1}{I}$
**17.** $A = 2\pi r(h + r)$  **18. a.** 12 **b.** 1,152

## Chapter 17  *Trigonometry*

## DIAGNOSTIC TESTS

**17-1**  (p. 615)  **1.** 33°  **3.** 22°  **5.** 79°

**17-2**  (p. 617)  **1.** 17 ft.  **3.** 21 cm  **5.** 135 ft.  **7.** 24 ft.

**17-3**  (p. 621)  **1.** .6249  **3.** 23°  **5.** 140.04 ft.  **7.** 265.4 ft.  **9.** 753.9 m  **11.** 48°  **13.** 39°

**17-4**  (p. 625)  **1.** .6820  **3.** 59°  **5.** 69.3 ft.  **7.** 70.0 ft.  **9.** 9.6 mi.  **11.** 1,000 m  **13.** 18°
**15.** 13°

**17-5**  (p. 629)  **1.** .8988  **3.** 55°  **5.** 251.6 ft.  **7.** 24.0 m  **9.** 200 ft.  **11.** 40°  **13.** 89°

## CHAPTER REVIEW  (p. 632)

**1. a.** 111° **b.** 64°2′  **2. a.** 56° **b.** 8°17′  **3. a.** 56 ft. **b.** 73 yd.  **4.** 195 mi.  **5. a. (1)** 1.5399
**(2)** .2588 **(3)** .1564 **(4)** .1718 **(5)** .8949 **(6)** .9416 **b. (1)** 29° **(2)** 53° **(3)** 44°  **6. a. (1)** 83.9 ft.
**(2)** 26.6 yd. **(3)** 65.1 ft. **(4)** 37.1 mi. **(5)** 177.0 yd. **b. (1)** 20° **(2)** 26°34′  **7. a.** $m\angle A = 49°$;
$b = 86.9$ ft.; $c = 132.5$ ft. **b.** $m\angle B = 30°$; $b = 34.6$ in.; $c = 40$ in. **c.** $m\angle A = 66°$;
$m\angle B = 24°$; $c = 2458.6$ ft. **d.** $m\angle A = 50°$; $m\angle B = 40°$; $b = 321.4$ ft.  **8.** 42.5 meters

## KEYED INVENTORY TEST  (p. 634)

**1.** $A = \frac{1}{4}\pi d^2$  **2. a.** $-13$ **b.** $-8$ **c.** 18 **d.** $-5$  **3. a.** $+10$ **b.** $-.8$ **c.** 9 **d.** $-\frac{4}{3}$

**4.**    **5.** 22  **6.** 365

**7.** $g = \dfrac{m}{1,000}$  **8.** $-x^4 - 5x^2y^2 + 7y^4$  **9.** $-9rs + 2s^2$  **10.** $30m^2 - 89mn + 63n^2$
**11.** $6c + 5d$
**12.** $-3b$  **13. a.** 2.5 **b.** $\frac{2}{3}$ **c.** 4 **d.** $-9$  **14.**
**e.** 4 **f.** $-3$

**15.** $l = 6$  **16.** $w = 14$ meters, $l = 29$ meters  **17.** 0, 1, 2, 3, 4

**18.**

$$-2 \quad -1 \quad \ 0 \quad +1 \quad +2 \quad +3 \quad +4$$

**19. a.** $16x^2 - 9y^2$

**b.** $36c^4 - 60c^2d^2 + 25d^2$ **c.** $a^2 - 2a - 63$ **d.** $21m^2 - 4mn - 32n^2$

**20. a.** $(10b - 7c^3d^4)(10b + 7c^3d^4)$ **b.** $16x^2y^2(x^4 + 3x^2y^2 + 4y^6)$ **c.** $(5a - 1)(4a + 9)$

**d.** $(6x + 5y)(6x + 5y)$  **21.** $\dfrac{a+1}{a-5}$  **22.** $\dfrac{2(x-2)}{x+3}$  **23. a.** $\dfrac{s+r}{rs}$ **b.** $\dfrac{rd - rt - dr + dt}{drt}$

**c.** $\dfrac{3b^2 - bc + 6c^2}{b^2 - c^2}$  **24. a.** $n = -24$ **b.** $x = 10$ **c.** $y = 6$ **d.** $n = -4$ **e.** $C = 300$ **f.** $a = 200$

**25.** $R_2 = 13$  **26.** Slope $= -3$; $y$-intercept $= 5$

**27.** A line with $y$-intercept $-3$, slope $\frac{1}{2}$.  **28.** Solution, $(3, -2)$.

**29.** $x = 4$ and $y = -1$  **30.** \$6,500 at 8%, \$3,500 at 10%  **31.** $\dfrac{x^4y^2}{3z^5}$  **32.** $23\sqrt{3} + 13\sqrt{2}$

**33.** $y = 32$  **34.** $x = \pm 4$  **35.** $x = \frac{5}{2}$ and $-3$  **36.** $s = \pm 9$  **37.** $d = \dfrac{hW}{F}$  **38.** $P = \dfrac{E^2}{R}$

**39.** 72  **40.** 914.408 meters

## COMPETENCY CHECK TEST  (p. 636)

**1.** d  **2.** c  **3.** c  **4.** a  **5.** b  **6.** d  **7.** c  **8.** c  **9.** d  **10.** b  **11.** b  **12.** b  **13.** a  **14.** d
**15.** b  **16.** a  **17.** c  **18.** c  **19.** c  **20.** d  **21.** c  **22.** d  **23.** d  **24.** c  **25.** a  **26.** b

## THE EQUATION  (p. 638)

**1. a.** $n + 9 = 15$ **b.** $10x - 2(x + 8) = 3(2x + 8)$  **2.** b, c, and d; b, c, and d; c
**3. a.** $n = 15$ **b.** $a = 31$ **c.** $y = 3$ **d.** $c = 54$  **4. a.** $1.08p$ **b.** $n = \frac{1}{3}$ **c.** $b = -4$ **d.** $x = 0$
**5. a.** $x = 4$ **b.** $y = 17$ **c.** $a = 75$ **d.** $x = 3$  **6. a.** $r = 48$ **b.** $x = 2$ **c.** $x = 7$ **d.** $n = -4$
**7. a.** $x = 3$ and $y = -3$ **b.** $m = -2$ and $n = 3$ **c.** $y = 4500$ and $x = 2500$ **d.** $x = 18$ and
$y = 12$  **8. a.** $n = 24$ **b.** $x = 32$ **c.** $x = 9$ **d.** $x = 7$  **9. a.** $x = \pm 6$ **b.** $x = 0$ or $\frac{9}{4}$ **c.** $x = 5$ or
$-2$ **d.** $x = \frac{3}{2}$ or $\frac{5}{6}$

**10.**

$$-3 \quad -2 \quad -1 \quad \ 0 \quad +1 \quad +2 \quad +3$$

**11.** A line with $y$-intercept $-3$, slope 4.  **12.** Solution, $(1, -3)$.

## THE FORMULA  (p. 638)

**1.** $a = h^2 - b^2$  **2.** $i = A - p$  **3. a.** $l = 59$ **b.** $p = 580$ **c.** $C = -25$

**4. a.** $A = 180 - B - C$ **b.** $A = \pi dh + \pi\dfrac{d^2}{2}$ **c.** $A = \pi(R^2 - r^2)$ **d.** $R = \dfrac{R_1R_2}{R_2 + R_1}$

**5.** $V' = 96$  **6.** $a = \dfrac{r(s-l)}{s-1}$; $l = \dfrac{sr - sa + a}{r}$; $r = \dfrac{a(s-1)}{s-l}$  **7.** $A = 51$  **8.** $a = 2$

**9. a.** $(1, 5)$ and $(2, 10)$; horizontal scale: length of side ($s$);
    vertical scale: perimeter ($p$).

  **b.** $(1, 40)$ and $(2, 80)$; horizontal scale: time ($t$); vertical
    scale: distance ($d$)

  **c.** $(100, 3)$ and $(200, 6)$; horizontal scale: principal ($p$);
    vertical scale: interest ($i$)

**10.** $A = 2\pi r(h + r)$  **11.** $F = 400$  **12.** $t = 3$ and $5$  **13.** $t = \sqrt{\dfrac{s}{4}}$  **14. a.** 31.5 meters
**b.** 159 cm  **c.** 902 yds.  **15.** \$800  **16.** 92 meters

## PROBLEM SOLVING  (p. 640)

**1.** 138 girls  **2.** .9 meters, 3.3 meters  **3.** daughter = 10 yr.; father = 40 yr.  **4.** \$95.55
**5.** 22.2 minutes  **6.** 1,470 miles  **7.** 16 nickels  **8.** $3\frac{1}{3}$ ounces  **9.** $S = 31°$; $R = 41°$;
$T = 108°$  **10.** \$5,000  **11.** 340 m.p.h.  **12.** 10 pounds  **13.** width = 25 m; length = 40 m
**14.** \$96  **15.** \$63,500  **16.** 12 ft.  **17.** 5°

## ALGEBRAIC OPERATIONS  (p. 641)

**1.** $-3$  **2.** 7  **3.** 30  **4.** $-8$  **5. a.** $-11x$  **b.** $7n^2 - 1$  **6. a.** $-2t$  **b.** $2x^2 + 2xy - 3y^2$
**7. a.** $27x^4y^{10}$  **b.** $-6m^4 + 2m^3 - 12m^2$  **c.** $8b^4x^3 - 16b^2x^5 + 24bx^8$  **8. a.** $-4$
**b.** $-4a^3c^2 + 3ac + 1$  **9.** $-4c^2 - 5b^2 + bc$  **10.** $-a^2 + 3a$
**11.** $35m^3 - 24m^2x - mx^2 + 2x^3$  **12.** $8b^2 - 3bc - 5c^2$  **13.** $-x^2 + 8x + 4$
**14.** $5x^4 - 2x^3 - (x^2 - 6x + 1)$  **15. a.** $\frac{9}{16}x^2 - 25y^4$  **b.** $36s^4 + 60s^2xy + 25x^2y^2$
**c.** $a^2 - a - 72$  **d.** $40m^2 - 94mn + 48n^2$  **16. a.** $25(a - 2b)(a + 2b)$  **b.** $(t + 7)(t - 8)$

**c.** $(9x + 8y^2)(9x + 8y^2)$  **d.** $6x(s - 2)(s - 1)$  **17.** $\dfrac{1}{3(2x + y)}$  **18.** $\dfrac{6c}{5x^2b}$  **19.** $\dfrac{4a - 3b}{14abx^2}$

**20.** $\dfrac{n + m}{mn}$  **21.** $\dfrac{5x - 47}{(x - 5)(x + 5)(x - 3)}$  **22.** $\dfrac{2}{a - 7}$  **23.** $\dfrac{3x + 4y}{x - 3y}$  **24.** $3a^2b^4\sqrt{13b}$  **25.** $\dfrac{3}{2}\sqrt{\dfrac{3c}{a}}$

**26.** $16\sqrt{5} + 55\sqrt{3}$  **27.** $248 - 140\sqrt{3}$  **28.** $8 - 5\sqrt{5} + 2\sqrt{2}$

## KEYED ACHIEVEMENT TEST  (p. 642)

**1.** $A + B + C + D = 360°$  **2. a.** $-13$  **b.** $-6$  **c.** $-45$  **d.** 4  **3. a.** $-31$  **b.** $+17$  **c.** .8
$100c$
**d.** $-\frac{6}{7}$  **4.**  **5.** 128  **6.** 27  **7.** $c = 100l$

**8.** $-3y^2 + 3x^2$  **9.** $2mn - 2n^2$  **10.** $12a^3 - 24a^2b + 11ab^2 - 3b^3$
**11.** $c^4 - c^3x + c^2x^2 - cx^3 + x^4$  **12.** $-n^2 + 9n + 6$
**13. a.** $x = \frac{1}{2}$  **b.** $n = -25$  **c.** $.7s$  **d.** $r = 8$  **e.** $x = -2$  **f.** $n = 5$

**14.**  **15.** $t = 5$  **16.** 1 hr. 36 min.  **17.** 5, 7

**18.**  **19. a.** $64x^2 - 112xy + 49y^2$

**b.** $9c^2 - 25d^2$  **c.** $36b^2 - 24b - 5$  **d.** $8r^2 - 6rs - 27s^2$  **20. a.** $25(a^2 + 16b^2)$

**b.** $(y + 2)(y - 12)$  **c.** $(6m - 11)^2$  **d.** $ar(r - 1)(r + 1)(r^2 + 1)$  **21.** $\dfrac{c(4c - 1)}{2c + 1}$  **22.** $\dfrac{3}{x - 3y}$

**23. a.** $\dfrac{4x - 5y}{24}$  **b.** $\dfrac{c(2c - 5d)}{c^2 - d^2}$  **24. a.** $x = 2$  **b.** $n = 7$  **c.** $x = 3$  **d.** $a = 8$  **e.** $x = 900$
**f.** $n = 1800$  **25.** $r = 2$  **26.** Slope $= 4$; $y$-intercept $= -8$
**27.** A line with $y$-intercept $-3$, slope $\frac{3}{2}$.  **28.** Solution $(4, 2)$.  **29.** $x = -3$ and $y = +4$
**30.** 20 kg  **31.** $\dfrac{5r^2}{25s^4t}$  **32.** $-4\sqrt{6}$  **33.** $x = 6$ and $-5$  **34.** $x = \frac{3}{2}$ and $-\frac{3}{4}$

**35.** $x = 6$ and $-3$  **36.** $w = 2.5$ or $-6.5$  **37.** $r = \dfrac{nE - IR}{In}$  **38.** $d = 2\sqrt{\dfrac{A}{\pi}}$  **39.** 8  **40.** 59°